Technology of Paints, Varnishes and Lacquers

Edited by

CHARLES R. MARTENS

Director of Associated Products
The Sherwin-Williams Company
Cleveland, Ohio

Instructor Organic Coating Technology
Cuyahoga Community College

ROBERT E. KRIEGER PUBLISHING COMPANY
MALABAR, FLORIDA

Original Edition 1968
Reprint Edition 1974, w/corrections

Printed and Published by
ROBERT E. KRIEGER PUBLISHING COMPANY, INC.
KRIEGER DRIVE
MALABAR, FLORIDA 32950

Library of Congress Cataloging-in-Publication Data

Martens, Charles R
     Technology of paints, varnishes, and laquers.

     Includes bibliographies.
     1. Paint. 2. Varnish and varnishing. 3. Lacquer
and lacquering. I. Title.
[TP935.M28 1974]   667'.6   73-92865
ISBN 0-88275-154-9

10   9   8   7   6

Contributors

F. C. Bertsch, The Sherwin-Williams Company, Chicago, Illinois

F. A. Bonzagni, Monsanto Company, Springfield, Massachusetts

H. L. Cahn, General Electric Company, Waterford, New York

D. L. Davies, Imperial Chemical Industries, Ltd., Runcorn, Cheshire, England

R. H. Duzy, Union Carbide Corporation, New York, New York

W. M. Edwards, E. I. du Pont de Nemours & Company, Wilmington, Delaware

D. M. Gans, Coatings Research Group, Inc., Cleveland, Ohio

N. Grier, Merck & Company, Rahway, New Jersey

F. J. Hahn, Monsanto Company, Springfield, Massachusetts

H. R. Hicks, The Sherwin-Williams Company, Gibbsboro, New Jersey

D. F. Householder, The Sherwin-Williams Company, Cleveland, Ohio

S. J. Huey, The Sherwin-Williams Company, Cleveland, Ohio

R. G. Hummeldorf, The Sherwin-Williams Company, Cleveland, Ohio

J. M. Klarquist, Shell Chemical Company, Union, New Jersey

A. F. Knoll, Alcan Metal Powders, Inc., Elizabeth, New Jersey

E. C. Larson, Shell Development, Emeryville, California

M. L. Levy, Crobaugh Laboratories, Cleveland, Ohio

M. Malaga, The Glidden Company, Cleveland, Ohio

J. Marchbank, Sprayon Products Company, Cleveland, Ohio

F. J. Martinek, The Sherwin-Williams Company, Cleveland, Ohio

S. Meinstein, Witco Chemical Company, Inc., Chicago, Ill.

E. P. Miller, Ransburgh Electro-Coating Corporation, Indianapolis, Indiana

G. Mutersbaugh, The Glidden Company, Cleveland, Ohio

K. V. McCullough, Union Carbide Corporation, Bound Brook, New Jersey

W. H. McKnight, Union Carbide Corporation, Bound Brook, New Jersey

G. S. PEACOCK, Union Carbide Corporation, Bound Brook, New Jersey

R. S. RADCLIFFE, The Dimlich-Radcliffe Company, Cleveland, Ohio

S. H. RICHARDSON, Plastics Division, Union Carbide Corporation, Bound Brook, New Jersey

K. SCHREIBER, Pittsburgh Plate Glass Company, Cleveland, Ohio

H. R. SPIELMAN, Witco Chemical Company, Chicago, Illinois

F. STESLOW, The Sherwin-Williams Company, Chicago, Illinois

F. B. STIEG, Titanium Pigment Corporation, New York, New York

K. TATOR, Kenneth Tator Associates, Corapolis, Pennsylvania

I. B. THOMAS, The DeVilbiss Company, Toledo, Ohio

W. H. TOMC, The Glidden Company, Cleveland, Ohio

K. C. WALDO, The Sherwin-Williams Company, Cleveland, Ohio

E. R. WELLS, Mobay Chemical Company, Pittsburgh, Pennsylvania

Preface

The technology of paints, varnishes and lacquers is changing rapidly and becoming more complex each day. Whereas fifty years ago the protective coating industry field was principally an art, today it has become a science. The paint industry is an important segment of the chemical industry. Paint technology utilizes the sciences of chemistry, physics and engineering. Paint technology overlaps into such fields as inks, plastics, rubber, adhesives, etc. This book is intended to present the latest technical information in this field.

During the past few years emphasis has been on the development of non-polluting coatings. The lead and mercury content of coatings has been reduced. In order to reduce air pollution, coatings have been developed using non-photo active solvents, water borne systems, high solid systems and powder coatings. Recently, because of the energy crisis, reduction in the amount of energy used in curing coatings has become very important.

The technology of raw materials, formulation, production, testing and application of protective coatings is covered. The performance of specific coatings for trade sales, industrial usage and maintenance is detailed. Topics such as color science, aerosols, paint and varnish removers, and safety are included in the book.

There are many excellent publications available in this field such as the *Journal of Paint Technology, Paint and Varnish Production, American Paint Journal, Industrial Finishing* and the *Journal of Oil and Colour Chemists' Association* to supplement this book.

I would like to express my deep appreciation to the many technical experts and their companies who have contributed the excellent chapters to this book.

CHARLES R. MARTENS

November 6, 1967

v

Contents

Preface ... v

1. Introduction to the Coating Industry, *C. R. Martens* ... 1
2. Film Formation, *David M. Gans* ... 12
3. Raw and Processed Oils, *C. R. Martens* ... 23
4. Alkyd Resins, *C. R. Martens* ... 33
5. Aminoplast Resins, *F. A. Bonzagni and F. J. Hahn* ... 49
6. Phenolic Resins, *S. H. Richardson and K. V. McCullough* ... 67
7. Epoxy Resins, *John M. Klarquist* ... 82
8. Hydrocarbon Resins, *C. R. Martens* ... 107
9. Acrylic Resins, *W. M. Edwards* ... 111
10. Cellulosic Film Formers, *F. Steslow* ... 132
11. Vinyl Resins for Coatings, *W. H. McKnight and G. S. Peacock* ... 147
12. Chlorinated Rubber, *D. L. Davies* ... 188
13. Urethane Coatings, *E. R. Wells* ... 205
14. Silicones, *Harold L. Cahn* ... 223
15. Miscellaneous Resins, *F. J. Martinek* ... 258
16. Plasticizers, *C. R. Martens* ... 273
17. Hydrocarbon Solvents, *E. C. Larson* ... 281
18. Oxygenated Solvents, *R. H. Duzy* ... 308
19. Pigments—General Classification and Description, *Robert S. Radcliffe* ... 327
    Part A: Pigment Classification ... 327
    Part B: Supplemental Pigments ... 331
20. White Pigments, *F. B. Stieg* ... 347
21. Colored Pigments, *M. Malaga* ... 363
22. Metallic Soaps: Driers, Suspending Agents, Flow Modifiers, Flatting Agents and Sanding Aids, *Siegfried Meinstein and H. Russell Spielman* ... 390

23. PRESERVATIVES AND FUNGICIDES, *Nathaniel Grier*   422

24. TESTING OF RAW MATERIALS FOR COATINGS, *Morton L. Levy*   442

25. COLOR IN PAINT, *S. J. Huey*   456

26. PAINT FORMULATION, *W. H. Tomc*   495

27. PIGMENT DISPERSION, *R. G. Hummeldorf*   505

28. EMULSION PAINTS, *C. R. Martens*   513

29. TRADE SALES PAINTS, *H. Roy Hicks and D. F. Householder*   531

30. INDUSTRIAL FINISHES, *Floyd C. Bertsch*   554

31. MAINTENANCE PAINTS, *Kenneth Tator*   575

32. ALUMINUM PIGMENTS AND PAINTS, *Alex F. Knoll*   589

33. AEROSOL COATINGS, *Joseph Marchbank*   607

34. PAINT AND VARNISH REMOVERS, *C. R. Martens*   613

35. SURFACE PREPARATION, *Kenneth C. Waldo*   619

36. APPLICATION
    PART A: BRUSH, ROLLER, DIP AND FLOW COAT,
    Kenneth C. Waldo   644
    PART B: SPRAY PAINTING, *Irvin B. Thomas*   656
    PART C: ELECTROSTATIC SPRAYING AND ELECTRODEPOSITION,
    Emery P. Miller   666

37. PRODUCTION
    PART A: RESIN AND VEHICLE MANUFACTURE, *C. R. Martens*   690
    PART B: PAINT MANUFACTURING, *Gordon Mutersbaugh*   699

38. PAINT TESTING, *K. Schreiber*   710

39. FIRE PROTECTION, SAFETY AND HEALTH, *C. R. Martens*   724

40. GLOSSARY   730

    INDEX   737

1

*Introduction to the Coating Industry**

The paint industry is an important segment of the chemical industry. Paint technology utilizes the sciences of chemistry, physics and engineering. A plant manufacturing a broad line of trade sales, maintenance and industrial paints uses raw materials numbered in the thousands. Oils, resins, solvents, plasticizers, pigments, extenders, dyes, driers, etc., are various types of materials used in paint manufacturing. The paint industry has changed from an art to a science in the last hundred years. At the beginning of this century, paint manufacturers began to hire chemists. Paint technology overlaps such fields as inks, plastics, rubber, adhesives, etc.

FUNCTION OF PAINT

The two primary functions of paint are decoration and protection. Decorative effects may be produced by color, gloss or texture, or a combination of these. The buyer who purchases paint for living room walls is primarily interested in decoration. The paint should be durable and washable, but a room is usually repainted to change color and not because the paint has failed. A secondary decorative function of paint is lighting. The color of the surface affects the reflectance. The proportion of light that is reflected by a surface is expressed as a percentage of complete reflectance. The following are approximate reflectance values for various colors:

White	90-80%	Medium to dark tints	60-20%
Very light tints	80-70%	Deep color	20-3%
Light tints	70-60%	Black	2-1%

*By C. R. Martens, The Sherwin-Williams Co., Cleveland, Ohio.

Because there is a multiple reflection of light from one surface to another, the effect of color on room lighting is much more pronounced than indicated in the above figures.

The protective function of paints includes resistance to air, water, organic liquids and chemicals (i.e., acids, alkalies and atmospheric fumes) as well as improved mechanical properties on some materials through better hardness, abrasion resistance, etc.

The protective coating may be the paint on a wooden boat for protection against rotting, interior lining of metal cans or drums to prevent corrosion for foods and chemicals, coatings on electrical parts for excluding moisture, fire-retardant paints to protect combustible surfaces, coatings on porous surfaces such as concrete and plaster for ease of cleaning, and many others.

SIZE AND GROWTH

The paint industry has continued to expand steadily. However, industry growth lags behind that of the national economy. Since 1955, sales volume has increased on the average of 3% a year, compared to a 5% annual rate increase for gross national product (see Table 1.1).

TABLE 1.1. Total U.S. Sales—Paints, Varnishes and Lacquers

Year	Million Dollars	Million Gallons
1909	110	—
1919	300	—
1929	501	—
1939	443	—
1949	1107	—
1959	1727	—
1960	1764	663
1961	1750	623
1962	1833	643
1964	1939	700
1965	2100	760
1966	2200	795
1966	2364	837
1967	2348	781
1969	2776	880
1970	2731	826
1971	2813	880
1972	3010	927
1973(est.)	3152	902

The lag in the industry's growth rate reflects in part the success of the producers and users of coatings in developing more durable and efficient products. Automobile coatings today may last twice as long as those used ten years ago. Because of improved performance, film thicknesses of coatings used on appliances have been reduced 25 to 50% during this period. Precoated products such as aluminum siding have eliminated the need for many house paint repaint jobs. Many new products made of plastics and other materials do not require paint. In the building field, products such as floor tile, vinyl wall coverings, stainless steel panels, reinforced plastic panels, etc., have eliminated both the original and the maintenance markets for paint.

The outlook for the next few years is for continued sales growth, and it is estimated that the industry's sales volume will reach 6 billion dollars by 1980.

No one really knows how many companies in the United States make paint, but it is estimated that there are about 1800 in the field. The size of the companies varies from one-man establishments to large companies that have several thousand employees. However, it is estimated that the 20 largest companies make over 50% of all the paint. Two-thirds of all paint companies have less than 20 employees. Because of low capital investments, return on investment can be considered excellent compared to other industries. Net profit margins amount to about 6% of sales.

PRODUCTION

Paint making does not lend itself to automation since it is still largely a batch process. Many different raw materials are used, often in relatively small amounts, to produce a multitude of products. Some materials like solvents, oils, glycols, molten phthalic, etc., are handled in large quantities with tank car or tank wagon deliveries and tank storage. Labor costs in this industry are around 10% of the cost of the product.

Large paint makers have done much in recent years to improve the output per man-hour. Costs have been reduced by turning paint out in batches as large as 4000 to 6000 gallons. Resins and vehicles are also produced in larger batches of over 5000 gallons. High-speed dispersion equipment reduces the dispersion time. Handling of the finished product has become automated; cans are filled, sealed, labeled, placed in cartons and shipped without being touched by hand.

Color matching, which is time consuming in production, is now being facilitated by the use of electronic instruments and computers. Color matching can be reduced to two or three operations by the use of such equipment.

Paint exports have been rising steadily at an annual average rate of 8% since 1967 and reached $82 million in 1973. U.S. imports of paint are of little consequence, amounting to less than one-tenth of one percent of the domestic shipments.

U.S. Department of Commerce figures show that about 75% of all paint is shipped for distances under 500 miles and 25% for distances under 100 miles. Paint is generally produced locally because of the high shipping costs associated with a heavy product.

Trucks are used primarily for the shipment of paint since 87% of the product is shipped by motor carrier or private trucks.

CLASSIFICATION

The protective coatings industry is divided into two broad categories. Trade sales or shelf goods include products sold to consumers, contractors and professional painters for use on new construction or maintenance. Industrial finishes or product finishes include a variety of products for application to manufactured products such as automobiles, appliances, furniture, etc.

The requirements for each of these are quite different. Shelf goods are usually air dry finishes. Industrial finishes are most likely to be baking-type finishes.

Each of these fields uses somewhat different vehicle systems and has different application methods so that the formulation of products differs considerably in shelf goods and industrial finishes.

HISTORY

Paint technology goes back to the beginning of history. The caveman first expressed his artistic abilities by daubing mud and colored clay on the walls of his cave. The Egyptians decorated their walls with paintings executed in distemper. The Romans combined pigments with plaster to decorate interior surfaces. The ancient Hebrews used milk or curd paints to brighten their walls. Whitewash was used extensively during colonial days as an all-purpose finish.

The first U.S. paint patent was issued in 1865 to D. P. Flinn; it covered a water-base paint of zinc oxide, potassium hydroxide, resin, milk and linseed oil.

Oil paints originated when the artists of the early centuries (A.D.) added drying oils to their colors. The literature is not defined in regard to the exact date, but it is generally associated with the great painters of the fifteenth century.

The date of the first incorporation of resin with oil is likewise indefinite. There is reference to the use of resin with drying oil, as well as the addition of driers, in these early centuries. In the eleventh century, a monk, Theophilus, described the solution of molten resin in a hot oil. In the fifteenth century, painters seem to have been familiar with both oleoresinous and spirit-type varnishes.

The use of solvents with varnishes to reduce viscosity seems to have been well known at this time. Japanese lacquer (400 A.D.) was made from the sap of the varnish tree (*Rhus verniciflua*).

The art of making paint and varnish remained in the hands of the painter until the Civil War. In fact, the first paint mill in this country was owned by a Boston painter named Thomas Child around 1700. It consisted of a granite trough in which was placed an 18-inch granite ball. The first paint mills used commercially were of the buhrstone type. As the industry progressed, other mills such as the roller, pebble and steel ball mills came into use.

Transition from Art to Science

In the early 1900's, it took a week to ten days to finish an automobile with primer, surfacer, glazing putty, ground coat, finish coat, color coat and finishing varnishes. The installation of baking equipment reduced the time to two or three days. However, the introduction of lacquer revolutionized the industry. As soon as the solvent evaporated from the lacquer, a matter of not more than 30 minutes, the car could be recoated. Not only did lacquer make mass production of automobiles possible, but it entered the furniture and other industrial fields as well.

With the introduction of phenolic and maleic resins about 1926, it was possible to make wood-oil varnishes which dried in four hours, and varnishes regained some of the market lost to lacquers.

It was not until the alkyd-type enamel was marketed about 1930 that the luster retention and durability of the nitrocellulose-type finish was surpassed. The alkyd-type finish had the additional advantage that the solids at spraying viscosity were higher than in a lacquer. Two coats of alkyd enamel have the same build as four coats of lacquer.

The use of alkyd vehicles in lacquers as plasticizers also improved their luster retention and helped increase the solids so that the two finishes nearly approached each other in durability and luster retention.

The introduction of titanium dioxide in 1924 contributed to the production of high-hiding white finishes. This pigment had four to six times the hiding power of the principal types of white pigments such as lithophone, zinc oxide, white lead, etc., which were in use at the time.

In 1925, the best white finishes were produced with either pigmented white lacquer or pigmented white dammar varnish. With the introduction of alkyd vehicles, it was possible to make white synthetic enamels which would bake to a hard, tough, flexible finish in one hour at 300°F. A few years later the introduction of urea-formaldehyde resin with alkyds made

TABLE 1.2 History of the Coatings Industry in the United States

1804	First white lead production in United States
1815	First varnish production in United States
1865	First U.S. water paint patent—U.S. Patent 50,068 (to D. P. Flinn)
1867	Production of ready-mixed paints in United States
1910	Casein powder paints on market
1923	Nitrocellulose lacquer
1924	Titanium dioxide introduced
1924	Modified phenolic resin commercially available
1927	Alkyd resins commercially available
1928	Oil-soluble phenolics
1929	Urea-formaldehyde resin for modification of alkyd resin
1930	Chlorinated rubber first produced
1930	Casein paste paints put on market
1930	Molybdate orange introduced
1933	Vinyl copolymer resins introduced
1934	Oil-base emulsion paints put on market
1937	Phthalocyanine blue introduced
1939	Commercial production of dehydrated castor oil
1939	Melamine-formaldehyde resins available
1939	Ethylcellulose introduced
1939	Polyurethane resins introduced
1944	Silicone resins commercially available
1948	Latex wall finishes based on styrene-butadiene put on market
1950	Epoxy resins commercially available
1950	Polyvinyl acetate and acrylic copolymer latices commercially available
1957	Latex house paints introduced
1962	Electrodeposition of water-base paints
1966	Solvent regulations Rule 66
1968	Semi-gloss latex paints introduced

it possible to produce a harder finish in a shorter time, 30 minutes at 300°F. This finish was used on appliances such as refrigerators, washing machines, etc.

In the 1930's, casein paints were the principal water-type paint in use. Casein paints contained high refractive index pigments and were low in cost. The addition of a drying oil to a casein paint produced an improved paint. As the amount of oil increased, the evolution from a casein paint to an emulsion took place. Emulsion oil paints enjoyed great popularity during World War II when the limited amounts of oils available for decorative painting were more effectively utilized in emulsion paints.

At the end of World War II, the era of synthetic latex paints began. Wall finishes based on styrene-butadiene latex were introduced and they became very popular because of ease of application and easy cleanup. About ten years later, latex house paints containing acrylic latices were introduced.

TABLE 1.3.   Properties of Coating Resins[a,b]

	Alkyd	Alkyd Amine	Styrenated Alkyd	Acrylic	Nitrocellulose	Catalyzed Epoxy	Epoxy Ester	Phenolic	Silicone	Urethane	Vinyl	Chlorinated Rubber
Exterior durability	E	E	G	E	E	G	G	E	E	E	E	E
Salt spray	E	G	G	E	E	E	E	E	E	E	E	E
Alkali resistance	P	F	F	F	P	E	G	G	P	F	E	G
Solvent aliphatic hydrocarbon	G	E	F	F	F	E	G	E	G	F	E	G
Solvent ester ketones	P	P	P	P	P	G	F	G	P	F	P	P
Flexibility	E	G	G	E	E	F	G	G	F	E	E	G
Impact	G	E	G	E	E	G	E	G	G	E	E	G
Heat resistance	G	G	G	G	P	G	G	E	E	E	E	E
Color retention	G	G	G	E	E	F	G	P	G	G	G	G
Gloss retention	E	G	G	E	G	P	F	F	E	P	F	F

[a] Code: E, excellent; G, good; F, fair; P, poor.
[b] Ratings are approximate. Properties vary in each group depending on specific formulation and application and whether air dried or baked.

The coatings industry uses many different vehicle systems for various applications and environmental conditions. A wide range of properties can be obtained (see Table 1.3). The most important vehicles used in coatings are alkyds. They account for almost 50% of the total. Usage had dropped recently due to inroads of latex systems. (See Table 1.4.)

TABLE 1.4   Alkyd Production in the United States

Year	Million Pounds
1940	100
1945	200
1950	400
1955	530
1960	520
1965	560
1970	600

A summary of the resins used in the coatings industry is shown in Table 1.5.

TABLE 1.5   Consumption of Resins in Coatings 1970

	Million Pounds
Alkyds	600
Acrylics — Solvent	158
— Latex	62
Vinyls — Solvent	75
— Latex	135
Cellulosics	75
Epoxies	68
Rosin Esters	70
Amino	60
Phenolics	33
Styrene	37
Polyurethanes	45
Hydrocarbon	30
Polyesters	7
Natural	50
Others	50
Resin Total (dry)	1555

Government Regulations

In recent years the paint industry has been subjected to new laws and regulations at every government level—federal, state and local. Most of these regulations are concerned with the environment, involving the use of toxic materials such as lead and mercury, and the pollution of water and the atmosphere. Other regulations are concerned with the use and shipment of paints.

Lead, like most heavy metals, is toxic when inhaled or digested in sufficient quantities. Recent interest in environmental pollution and reported cases of lead poisoning among children in city slums have spurred the federal government to enact regulations limiting its use in paints. Some major cities and a few states previously had laws regulating the sale and use of lead paints.

On January 1, 1973, the Food & Drug Administration (FDA) ordered the reduction in the lead content in paints used in or around the household to 0.5%.

Most victims of lead poisoning or plumbism are children between the ages of 1 and 6 years. Either as an aspect of pica or of the oral exploration usual to children of this age, they may pick up and eat paint chips found on floors, or chew on cribs or window sills that were long ago covered with high lead content paint.

Pica, a very common precondition, is the medical term for an unnatural craving for dirt or other non-food items. Some studies found that from 70 to 90 percent of children suffering from plumbism had a history of pica.

The preponderence of evidence shows that lead poisoning to children is caused by paints applied at least 30 years ago when the lead content was as high as 50%.

Mercury, another heavy metal used in paints, is also toxic to humans at some levels of exposure. Organic compounds of mercury, usually phenyl-mercuric salts of acetic, oleic or propionic acid are used in almost all water-based paints as can preservatives to prevent bacterial action which can lead to the formation of odoriferous compounds, loss of viscosity and formation of gas, resulting in pressure build-up in cans. In addition, mercury fungicides are used in some exterior paints to prevent fungi from attacking the paint film. General practice in the paint industry has been to use up to 0.2% level of mercury in paints for outdoor use and up to 0.02% level for indoor paints. The actual amount of mercury used in paint is generally much less than the above levels.

At present time, government regulations on mercury have not been finalized.

The Environmental Protection Agency (EPA) has established national standards on air quality with the enforcement of these standards by the states. The Federal guidelines of most concern to the paint industry and its customers, involve the reduction in the emission of solvents and other organic compounds. Most of the guidelines are similar to those established in Los Angeles County in 1966 when its Rule 66 went into effect.

Rule 66 controls the emissions of photochemically reactive hydrocarbons and oxidents, which react with nitric oxide in the presence of ultraviolet radiation to form smog causing eye irritation and other deleterious effects. The regulations as amended in 1971, states that no facility is allowed to discharge more than 8 lbs. per hour or 40 day of organic materials containing photochemically reactive solvents unless the total volume of organic material discharged is reduced by 85%. In addition, no facility is allowed to discharge more than 3 lbs. per hour of 15 lbs. per day of organic material containing any organic solvent, photochemically reactive or not, if the solvent comes into contact with heat in the presence of oxygen, unless the total of organic material is again reduced by 85%. Thus, all heat-curing techniques using any organic solvents are affected.

Los Angeles defines a photochemically reactive solvent as any solvent with an aggregate of more than 20% of its total volume composed of the chemical compounds classified below or which exceeds the individual limitations listed.

(1)   A combination of hydrocarbons, alcohols, aldehydes, esters, ethers or ketones having an olefinic or cyclo-olefinic type of unsaturation :5%.

(2)   A combination of aromatic compounds with eight or more carbon atoms to the molecule except ethyl benzene :8%.

(3)   A combination of ethyl benzene, ketones having branched hydrocarbon structures, trichlorocthylene, or toluene :20%.

Rule 66 prohibits the sale or use of all architectural coatings containing photochemical reactive solvents in one quart containers or larger. Exempt are paints whose volatile components consist of 80% water as long as the remaining 20% is not photochemically reactive.

Regulations similar to Rule 66 have been adopted by many other cities. Water pollution will be regulated with the aim to eliminate water pollutants in all discharges by 1985.

Paints are often considered hazardous materials and are thus subject to a number of government regulations concerning their manufacture, shipping and use. The office of Occupational Safety & Health Administration (OSHA) of the Department of Labor requires all paint manufacturers to furnish Material Safety Data Sheets on each shipment with information on solvents, etc. used in manufacture.

Shipping regulations are administered by the National Transportation Safety Board of the Department of Transportation. Regulations are primarily concerned with the type and construction of containers used in shipping with specifications for flash points, pressures, viscosities and weights.

FUTURE TRENDS

At the present time, about 48% of the total gallons and about 55% of the dollar volume of the paint business are trade sales. However, industrial finishes, which account for 45% of all paint business, will probably grow faster than trade sales. One reason is the trend toward factory-finished building products. Many factors are accelerating the trend to factory finishing to replace on-site painting. Factory painting under carefully controlled conditions insures the production of the best possible coating of uniform thickness. Coating materials have been so improved in recent years that warranties up to 15 years or more are given on many factory-finished exterior building products. Higher labor costs have accelerated the trend to production line finishing. It is estimated that labor costs account for 80% of on-site finishing costs.

Wood siding, paneling, flooring, trim, kitchen cabinets and interior doors are all finished on fully automated production lines without hand labor. Curtain coaters, roll coaters, reverse roll coaters and automatic hot spray and airless spray units are used. Production lines of this type travel at speeds up to 150 ft/min.

With the growing scarcity of woods such as pecan, walnut, cherry, maple, etc., the use of printed hardwood, particle board and plywood will increase. The graining is accomplished by a multicolor printing in which the grain is reproduced by a photographic process. The overall system consists of a base coat, first feature print, second accent print, third detail print and a clear top coat. The traditional hand application, hand rubbing and hand sanding operations in the furniture industry are being replaced by automated operations.

Coil coating or strip coating is a process in which long rolls of metal strip, usually aluminum or steel are continually coated on one or both sides by rubber rollers. Strips in widths up to 66 inches are coated at speeds up to 300 ft/min. The metal may be fabricated without further painting into many types of finished products such as venetian blinds, siding, awning, roof decking, signs, etc. Advantages include lower cost, better control of film thickness and elimination of nearly all waste of paint. It is predicted that the overall market for coil coatings will grow at the rate of 15 to 20% annually in the next few years (see Table 1.6).

Table 1.6   Paints for Prefinished Stock

Year	Coil Coating	Wood
1962	16	12
1964	19	14
1966	29	24
1968	39	34
1970	58	46
1972	71	64

Electrodeposition of organic coatings is similar to electroplating of metallic films. The metallic object to be coated acts as the anode and the tank containing the coating material serves as the cathode. Anionic resins and pigments are dispersed or dissolved in water. When the current is applied, the negatively charged paint and pigment particles are deposited on the anode or metal surface, and it is essentially dry as it leaves the tank. This process has the advantage of applying a uniform coating on irregular surfaces and inaccessible areas.

At the present time there are over 150 electrodeposition tanks in operation in the United States. These used more than 3 million gallons of paint last year valued at more than 20 million dollars. Motor vehicles and parts were the greatest volume, followed by appliances, electrical equipment, metal furniture and steel products.

The use of water-base paints will continue to grow at a rate faster than overall paint sales. For trade sales, water-base paints offer the advantages of ease of application and easy cleanup. The annual growth rate is 10%. In industrial coatings, water-base paints offer the additional benefits of reduction in fire hazard and a nontoxic solvent. Adoption of water-type paints to industrial applications has been slower than predicted, but within five years they should have about a 20% share of the industrial market (see Figure 1.3).

Powdered coatings have been used in Europe for some time but have only recently become an important method in this country for metal finishes. Powder coatings contain 100% solids, including resins, pigments, extenders, flow control agents and catalysts but no solvents. Two methods of applying powders are used: (1) the fluidized bed, and (2) electrostatic spraying.

The fluidized bed process involves blowing fine powder with air to form a uniform solids-in-gas suspension in a coating vessel. The object to be coated is submerged in this fluid bed under conditions which will cause the plastic powder to adhere to the surface. This is commonly accomplished by preheating the parts to be coated to a temperature above the fusion point of the plastic. The thickness of the coating can be controlled by the duration of immersion. Film thicknesses up to 60 mils can be obtained by this method.

Electrostatic spraying of powder coatings is a more versatile system which allows film thicknesses as thin as 3-4 mils. Particles of the powder are given an electric charge at the nozzle of the spray gun, generating an electrostatic field between the gun and the object to be coated, which is grounded. The particles are thus attracted to the substrate where they are cured by heat to a homogeneous film.

Vinyls, nylons, cellulosics, epoxies, polyesters, polyolefins and other coatings are applied by this manner. The current market for powdered coatings is estimated to be over 40 million pounds a year, and is expected to be about 6% of the industrial coating used by 1980.

Another trend in the coating industry is high solids coatings. This reduces or eliminates the use of solvents along with the problem of air pollution. Although these coatings may be as high as 100% solids, they are actually applied as liquids.

Radiation curing has become commercial. The coating is formulated with monomers or prepolymers and is cured in seconds or less on exposure to an electron discharge or ultraviolet light. As the curing occurs at room temperature, it can be used on substrates such as plastics or wood which cannot normally be subjected to heat curing techniques.

Fire-retardant coatings swell when subjected to fire, creating an insulation surface of charred foam which prevents further burning of the combustible surface such as wood to which it is applied. Fire-retardant coatings, while available for many years, have not been used to any great extent because of shortcomings such as high cost and lack of durability. However, new, improved products are being developed which are not excessive in cost so that sales of products should increase.

The use of spray paints in aerosol packages will continue to grow. In terms of the amount of paint in the can, aerosol paints are expensive but the consumer likes the convenience of such a package.

The coatings industry has changed from an art to a science in the past hundred years. The progress of the coatings industry has paralleled that of the chemical industry. As new improved materials are developed, these will be utilized to produce improved coatings.

REFERENCES

Marketing Guide to the Paint Industry—Fairfield, N.J., Charles H. Kline & Co. (1972)
Current Industrial Reports M28F Paint, Varnish & Lacquer, Bureau of Census, Washington
   D.C. (Annual)
U.S. Production & Sales of Synthetic Organic Chemicals, U.S. Tariff Commission, U.S.
   Government Printing Office, Washington, D.C. (Annual)

2

*Film Formation**

A mixture of pigment and vehicle is no more a paint before it is spread on a surface than an egg is a chicken before it is hatched. The word paint implies the ultimate existence of a coating with a considerable measure of physical integrity as a film. The conversion of the material in a container into an adherent, durable coating is the process of film formation. This process may be divided into three steps, which we shall call application, fixation and cure.

Application is the spreading of the composition into a thin layer several thousandths of an inch thick over the surface to be painted. This may be accomplished with a brush, a roller, a spray gun, a curtain coater, etc. It may also be achieved by more sophisticated methods, such as the process of fluid bed coating, in which the preheated object is exposed to a cloud of the coating-to-be as a finely divided, fusible powder which attaches itself adhesively to the hot surface, on which it is then flowed out by reheat into a continuous film; or the process of electrodeposition in which the coating, present as microscopic particles that carry electric charges by reason of their dispersion in an aqueous medium, is plated out on the surface to be painted by means of an electric potential applied between the paint and the object to be coated.

After application, fixation is the step of stabilizing the coating on the painted work against any tendency to run off or to form an undesirably uneven layer. The most common means for achieving fixation is by evaporation of volatile solvent from an organic solvent system or of water from a latex system. The residue becomes a nonflowing film because of the growth within it of rheological resistance to flow as the nonvolatiles concentrate. Fixation may also result from thixotropy built into the paint by the formulator.

Cure is the conversion of a fixed film into a more durable coating, to

*By David M. Gans, Coatings Research Group Inc., Cleveland, Ohio.

enable it to withstand better the rigors of exposure for which it was designed. Thus the oxidation of the oleoresinous ingredients in a house paint, thereby changing them from a liquid to a nonfluid state, is one type of cure. Another is the baking of a thermosetting enamel.

Not all coatings systems demand all three steps for film formation. For example, ordinary lacquers combine fixation and cure in the evaporation of the solvent during application and shortly thereafter. They may slowly undergo oxidation subsequently, but this is often an undesirable degradation of the film rather than a planned extension of cure. As another example, fixation and cure overlap in the chemical conversion after application of a fluid solution of an unsaturated polyester in styrene to give the nonflowing reaction product of the two.

Let us take a more detailed view of some of the aspects of these phenomena as a qualitative survey of coatings technology. Much of the subject matter is treated in depth in the following chapters of this volume. Allied reviews are available in the literature as well as other approaches to the general subject of film formation[7,9,22,29,30]

APPLICATION[16,21,24,27]

Application of coatings is not entirely a matter of mechanical engineering. A new method of application often requires the paint technologist to modify the coating as packaged to lend itself to the new method, even when the final cured film is the same in both cases.

Thus one method may call for less solvent or for more slowly evaporating solvent than another, or for no solvent at all. A paint to be applied to a vertical surface, as by brush or spray, may call for built-in resistance to sagging, perhaps at the partial sacrifice of another desirable characteristic, whereas the same paint, applied to a horizontal surface, as by spray or curtain coater, would not need such anti-sag insurance and could therefore be adjusted for greater flow and better leveling.

It is not within the present scope to examine the mechanical aspects of application equipment. Let it suffice to list a number of approaches.

(A) Hand brushing,[26] an ancient and time-honored method.
(B) Hand roller-coating, popularized in recent years by do-it-yourself amateurs.
(C) Spraying
    (1) Conventional, in which the paint is atomized with compressed air.
    (2) Steam spraying, in which the compressed carrier is steam.
    (3) Electrostatic spraying, which uses a potential difference be-

tween the work piece and the spray gun to attract most of the spray to the object to be coated.

    (4) Hot spray, employing hot paint under pressure in which the solvent explodes the paint into a spray when it is expelled into the atmosphere at normal pressure.

    (5) Airless spray, where the paint is broken into a spray by mechanical means rather than by a compressed carrier.

    (6) Flame spray, in which the coating as a powder is sprayed through the action of a flame, thereby melting the suspended particles which resolidify on striking the surface being coated.

(D) Roller coating

    (1) Direct roller coating, in which the paint as a thin layer on the roller is transferred to the work by rolling over it.

    (2) Reverse roller coating, in which the roller rotates in the direction opposite to the travel of the work, thereby allowing more intimate contact and generally a smoother surface.

    (3) Coil coating, a high-speed roller coating process for long strips of metal or other substrate supplied in coils·rather than in individual flat sheets.

(E) Knife coating, using a blade as a spreader for the paint instead of a roller.

(F) Curtain coating, in which a thin film of paint in the form of a dropping curtain, such as would escape from a slot in the bottom of a paint reservoir, is laid down on the work passing on a conveyor through the curtain.

(G) Flow coating, wherein paint is squirted through judiciously placed orifices at the work piece and then drains away to leave a uniform coating even on a comparatively irregular object.

(H) Dipping,[11] which entails the immersion of the work in a tank of paint, followed by controlled withdrawal to give a uniform film.

(I) Tumbling, in which a metered amount of paint is distributed over a considerable number of work pieces by tumbling in a closed container.

(J) Electrodeposition, described above.[8,18,28,31]

(K) Fluid bed coating, also already mentioned.[6,17]

FIXATION

For almost all methods of application, the ideal paint is one which is free flowing during application at the fluidity best suited for the particular method of application but which, once it encounters the work surface and spreads out into a uniform film, immediately loses its ability to flow and thus remains in place on the work.

In real paints, however, there is a time interval between the flowing and the nonflowing states. The real paint may never actually be completely free flowing. Undesirable as it may seem, some resistance to flow may in most cases have to be built into the paint in the can as a head start if it is to achieve a nonflowing state soon enough on the object to be coated. Once on the surface, the length of the time period during which the film loses its ability to flow depends on a number of factors. One is the film thickness necessary for the desired protective or decorative effect. Another is the existence of solvents of sufficient dissolving power for the resin in the paint, coupled with the proper degree of solvent evaporation to permit rapid buildup of film viscosity, but not so rapid as to leave an undesirably uneven surface.

All this points to the fact that there are two major influences that contribute to the fixation of the applied paint film: the evaporative and the rheological.

The Evaporation Factor[1,34]

The nonvolatile binder in some types of paint is in itself a liquid of low enough viscosity not to require the use of a solvent to make a sufficiently free-flowing paint. House paints based on raw and modified vegetable-type oils are of this kind. But most other coating compositions need solvent to fluidize the solid resins in the binder. These solvents are not wanted in the final film, and they satisfy this fugitive requirement by being volatile.

In the simplest model, evaporation takes place through at least two mechanisms:[5,14,25] that of solvent evaporation from the very surface of the film, in a manner akin to, and controlled by the same factors as, evaporation from the surface of the pure solvent by itself; that of solvent diffusion, in which the limiting factor on solvent loss is the rate at which it diffuses within the body of the film to the air surface from the lower depths, thus replenishing the solvent at the surface of release. Both mechanisms appear to be operative, with diffusion gaining in importance as solvent leaves. Generally, however, the evaporation mechanism predominates during the initial time period that is important for fixation. For water-base systems, the initial loss of water is at a rate close to that for pure water and does not fall below this rate appreciably, except when large amounts of evaporation depressants such as glycols are present.

It should not be supposed that the vapor tension of a pure solvent is necessarily a measure of the degree to which it leaves a paint film in the final stages of solvent evaporation.[5,6,15,20] The nonvolatile vehicular portions of various coatings vary widely in their ability to hang on to substantial amounts of solvent, for days and even months. But this takes place after the film has practically established its identity and is therefore outside the scope of the present discussion.

Initial solvent loss can be a turbulent process, albeit on a scale invisible to the naked eye. As the film thickens and becomes immobile, a microscopic geologic record of this turbulence is left behind by swirling pigment particles. A common pattern is a honeycomb of vortexes[19,32] which were active as cellular disturbances during solvent evaporation.

The Rheological Factor[4,12,26]

A paint must have more than mere reluctance to flow in order to keep from sagging on a vertical surface. It must not flow at all. Even if it is extremely viscous, a liquid will still move downward, no matter how slowly, under such conditions. To eradicate flow entirely, some physical structure within the paint must prevent it. Advantage may be taken of two types of such structure—that associated with flocculation and that described as thixotropic. The building blocks in these structures are the finely divided powders in a paint, i.e., the pigments and the extenders.

The pigment and extender particles can be made repulsively independent of each other by the addition of traces of suitable surface-active agents. Such a paint is free flowing. However, by alteration of the surfaces of the pigment and extender, particle by particle, through a different choice of surface-active agents, a measure of attraction between the particles may instead be created, generally adjustable from weak to strong. The interlinked chains of particles developed by flocculation form a network that resists distortion and flow. The term "yield value" is used in the simplest description of the phenomenon. Force must be applied in excess of the yield value of the system to get the paint to move at all. The yield value is thus a threshold value. If a great enough force is brought to bear by the brush, the roller or the knife blade, the paint can be spread. The moment this force is discontinued, the paint sets because the attractive forces between the particles are such that the flocculation network immediately reconstitutes itself. How great a yield value is acceptable in a paint depends on how much force is available in the method of application selected. What may do for a reverse roller coating may be excessively high for a paint to be applied by a brush.

Unlike yield value structure, thixotropic structure involves a time factor. When mechanically disturbed, as by the commotion of application, a thixotropic composition thins out to become comparatively free flowing. Left to itself, it then builds up structure, not instantaneously as with yield value, but in time. The time, depending on the actual composition, may be a matter of a fraction of a second to minutes or even hours, and the ultimate level of structure may be low or high. The phenomenon of thixotropy may therefore be harnessed to give an approximation of an ideal coating, i.e., one that is as free flowing as desired during application but rigid immediately thereafter.   Regrettably, however, thixotropy is not a simple property to control easily in practice from batch to batch of the same formulation.   It is a more temperamental variable in many cases than yield value.

Typical yield value figures and thixotropic ratings for several types of commercial paints follow:[12]

	Yield Value (dynes/cm^2)	Thixotropy Rating
Enamels, gloss	0–30	Nil to slight
Enamels, semigloss	50–120	Slight
Flat paints	20–100	Slight to marked
Wall coatings, water-base	10–100	Slight to marked
Primers, metal	0–100	Nil to marked
Varnishes	0	Nil

As solvent evaporates from a coating during fixation, the yield value may change, as may also the top level of thixotropy.   The possibility of such behavior may therefore be exploited in formulating a paint with the most desirable flow properties attainable for each stage of the fixation period.

CURE[2]

Once the fixation period is passed, the curing stage is reached, although the dividing line between the two is not always sharp.   Curing is often very complex and is different in kind from one type of coating to the next or, more specifically, from one type of binder to the next.

Many coatings are already in or nearly in their final state of cure when they complete their fixation period.   Lacquers, broadly defined, are a case in point.   In fact, lacquers are those coatings which are cured simply by solvent evaporation.   Vestigial solvent often takes a long time to leave some films, and to the extent that such residual solvent acts as a plasticizer (in addition to any plasticizer intentionally incorporated into what would otherwise be too brittle a coating), such films are not fully cured.

Curing of the kind that enhances the film properties after fixation results from chemical reaction within the coating, generally catalyzed, or from fusion or coalescence into a continuous film of finely divided binder particulars such as those in plastisols and in latex-based compositions.

Chemical Reaction

Traditional curing methods involving chemical reaction are those of oxidation and those that are triggered by a rise in temperature, i.e, baking or stoving.   Reaction between two major binder components at room temperature is a more recent variety. Most modern are those in which radiation from radioactive sources or from high-voltage electron generators initiates the selected binder reaction.

Details concerning the curing requirements of each class of binders are associated in the following chapters with the discussion of each type of vehicles. Here we shall touch on a few examples.

Conventional oleoresinous house paints contain little volatile solvent. The total vehicle, largely vegetable oil, is liquid. Body built into the paint keeps it from sagging at practical film thicknesses. However, such paints would be quite useless except as insect traps unless they converted in time into the dry leathery films with which we are familiar. With such oil-base house paints, this takes hours to days for an acceptable dry coat. The formulator sets the period by the use of the proper blend of drier catalysts, commonly the lead, manganese and cobalt salts of organic acids such as naphthenic, abietic and octoic, or the acid mixtures constituting tall oil. As a rule, the oxidation process docs not start as soon as the paint is spread and exposed to the air. An induction period is involved, a matter of an hour or two, after which the unsaturated vegetable-type acid esters in the paint start to absorb oxygen to create chemically reactive sites in the constituent molecules; these combine to form larger molecules which result in the familiar house paint films. Actually, the drying process never stops. It slows down greatly and must be controlled to avoid the ultimate formation of an excessively brittle film. Because the access of oxygen is from one side of the film, the drying paint is not uniform from top to underside. This can be a boon if rapid freedom from surface tack is sought, or a defect if a tight dry skin is created over a soft interior.

Increase of temperature increases the rate of chemical reaction. Curing processes that take place at normal temperatures do so much faster under the influence of heat. Some coatings, most notably conventional appliance enamels, do not cure at room temperatures and fortunately remain stable in the paint container for extended periods of time, but convert rapidly, in a fraction of an hour, when exposed to heat. For appliance enamels, for example, the temperatures are in the range of 280 to 350°F, with baking schedules of 10 to 30 minutes. Convertible coatings of this type[2] are numerous. Amine resins are one of the classes. They divide, broadly speaking, into urea-formaldehyde derivatives and melamine-formaldehyde derivatives. When heated, these tend to yield such brittle reaction products that they are formulated with plasticizers, usually alkyd resins, to give the most commonly employed baking enamels.

Among the most durable of coatings are several types in which two components, each a major constituent, react on intimate mixing, even at room temperature, to effect cure in minutes to hours. Obviously, such coatings must be supplied in two packages which are blended when used, possibly with a catalyst if it cannot be included initially in one of the components.

One example of such a coating is the epoxy-polyamide system. The epoxy resins are characterized by their chemically reactive oxygen atoms, each bound in the molecule to two adjacent carbon atoms to form a three-membered ring. Such rings are sensitive to a number of other reactive groups. The polyamide resins are low molecular weight relatives of the nylons. Their amide groups can tie up rapidly at room temperature with epoxy groups to form large, bulky molecules that are excellent binders in coatings. The cure can be accelerated by raising the temperature.

Another type of two-component system employs two unsaturated materials, each of which is capable of combining with the other at the unsaturated sites. Thus, an unsaturated polyester can be one component and styrene the other. Styrene is present in a double role—as a partner in the reaction and also as a solvent for the system. Since styrene is a volatile liquid, a slight excess of it is used to allow for the loss by evaporation after the coating is applied. Catalysts regulate the progress of the conversion. This is an interesting instance where a solvent never gets full opportunity to evaporate, because it is chemically trapped as cure proceeds.

Most recent is the use of high-voltage electrons to effect cure. The high-energy electron stream for this purpose is pulled out of an incandescent metallic filament in high vacuum by an electrical potential of 8.1 million to 8.5 million volts. The electrons escape through a thin window of a metal such as titanium and impinge on the applied coating. Cure takes place in a matter of seconds, on a variety of substrates. The method is limited to such binder compositions as can properly be energized and transformed into molecularly reactive groups by such electrons. The coatings are generally solvent free or contain a solvent like styrene, again acting in a double function. Since the electrical equipment is bulky, the method is best suited for conveyorized or production line painting.

Fusion[13] and Coalescence[3,23,30]

High molecular weight variations of certain resins are tougher and more durable, but unfortunately much less soluble than their chemical counterparts of lower molecular weight. As a consequence, the high molecular weight materials cannot be dissolved in solvents to give vehicles for what would on pigmentation become very desirable coating compositions. However, if such a resin in fine powder form is mixed with pigment and with a nonvolatile liquid which subsequently serves as plasticizer for the resin, the resulting product may be spread on a surface and then heated to soften the resin and induce its amalgamation with the plasticizer and the pigment, to yield a tough, durable coating. Such compositions are called plastisols and are commonly based on high molecular weight polyvinyl chloride and related polymers. These are instances of cure by fusion.

If the plastisol mixture is too high in viscosity to be applied properly by the selected method, it may be thinned with a minor amount of solvent for the plasticizer. These variations are called organosols. After the solvent evaporates from the film, the cure is also by fusion.[10]

Remotely related to the organosols in structure are the coating compositions based on latices; their popularity has revolutionized the paint industry in recent years. Here the volatile liquid, present to a much greater extent than the solvent in an organosol, is water. The water phase carries the pigments, the plasticizers, and a variety of thickeners and other additives. It also perforce carries the binder in the form of high molecular weight, microscopically finely divided globules of acrylic ester, vinyl ester, styrene-butadiene or other type, polymerized in an aqueous medium as the latex component. Emulsions of oleoresinous binders fall in a related category, but a distinction may be made between them and latices, since the binder first exists in bulk and is then emulsified in the case of oil emulsions, whereas the simple molecular components are dispersed in an aqueous medium and then polymerized in the case of latices.

A coat of latex paint dries in two stages. First, the water evaporates; as it does so, all the ingredients other than the latex globules are concentrated in the interstices between the globules. These become more closely packed in a geometric array and ultimately touch each other. The film at this juncture has no particular strength and may be wiped away as a powder. As with an organosol, if such a film were warmed sufficiently, the globules would sinter at their points of contact. This would lend integrity to the coating. Such heating is, of course, impractical for paint on a wall. Fortunately, it is possible to design latex compositions that sinter or coalesce at ordinary ambient temperatures, interior and exterior. In other words, it is possible to create latex-based systems in which coalescence does not take place in the bulk package in storage but proceeds smoothly during the curing period of the film and progresses sufficiently in a short time for favorable comparison with non-latex paints for the same use.

The forces responsible for coalescence are well understood[3] and will be discussed in detail in a later chapter. It should be mentioned, however, that the process may never reach completion in real paints. Visual vestiges of the original individual latex globules may be found months and years after the coating is laid down. The nonvolatile ingredients that are not soluble in the globule substance are trapped in the intersticial cells between the globules when the volatiles escape, and migration of resinous material into these pockets is very slow at best.

Cure is a relative term. What may be an adequately cured film for gentle handling may still be raw in respect to solvent attack, for example. Methods[33] for assaying the degree of cure are numerous. Each end use dictates its own pertinent method.

REFERENCES

1. Bews, I. C. R. (editor), "Non-convertible Coatings," in (Taylor, C. J. A., and Marks, S., editors-in-chief) "Paint Technology Manuals," Part 1, New York, Reinhold Publishing Corp., 1961.
2. Bews, I. C. R. (editor), "Convertible Coatings," in (Taylor, C. J. A. and Marks, S., editors-in-chief) "Paint Technology Manuals," Part 3, New York, Reinhold Publishing Corp., 1962.
3. Brown, G. L., *J. Polymer Sci.*, **22**, 423(1956).
4. Doherty, D. J., "Rheological Characteristics," in (Chatfield, H. W., editor) "The Science of Surface Coatings," Ch. 13, Princeton, N.J., D. Van Nostrand Co., 1962.
5. Doolittle, A. K., "The Technology of Solvent and Plasticizers," New York, John Wiley & Sons, 1954.
6. Elbling, I. N., *Offic. Dig. Federation Paint Varnish Prod. Clubs*, **31**, 1625 (1959).
7. Filson, A. C., and Jones, P. H., *J. Oil Colour Chemists' Assoc.*, **46**, 809 (1963).
8. Finn, S. R., and Mell, C. C., *J. Oil Colour Chemists' Assoc.*, **47**, 219 (1964).
9. Fisk, P. M., "The Principles of Film Formation," in (Chatfield, H. W., editor) "The Science of Surface Coatings," Ch. 2, Princeton, N.J., D. Van Nostrand Co., 1962.
10. Fitch, R. M., *Offic. Dig. Federation Soc. Paint Technol.*, **37**, 243 (1965).
11. Foley, E. W., *J. Oil Colour Chemists' Assoc.*, **47**, 427 (1964).
12. Gans, D. M., "Some Theoretical Considerations Regarding the Significance of the Pigment Dispersion Problem in Paint Formulation," in (Von Fischer, W., and Bobalek, E. G., editors) "Organic Protective Coatings," Ch. 3, New York, Reinhold Publishing Corp., 1953.
13. Glaser, M. A., and Rosenberg, P., *J. Paint Technol.*, **39**, 605 (1964).
14. Griebel, R. D., and Reynolds, W. W., *Offic. Dig. Federation Soc. Paint Technol.*, **33**, 921 (1961).
15. Hays, D. R., *Offic. Dig. Federation Soc. Paint Technol.*, **36**, 605 (1964).
16. Hislop, D. W., "Application Techniques," in (Chattfield, H. W., editor) "The Science of Surface Coatings," Ch. 18, Princeton, N. J., D. Van Nostrand Co., 1962.
17. Kutik, L., *J. Paint Technol.*, **38**, 493 (1966).
18. LeBras, L. R., *J. Paint Technol.*, **38**, 85 (1966).
19. LeBras, L. R., Bobalek, E. G. Von Fischer, W., and Powell, A. S., *Offic. Dig. Federation Paint Varnish Prod. Clubs*, **27**, 607 (1955).
20. Murdock, R. E., and Carney, J. A., Jr., *Offic. Dig. Federation Soc. Paint Technol.*, **33**, 181 (1961).
21. Newton, D. S. (editor), "The Application of Surface Coatings," in (Taylor, C. J. A., and Marks, S., editors-in-chief) "Paint Technology Manuals," Part 4, New York, Reinhold Publishing Corp., 1965.
22. Payne, H. F., "Organic Coating Technology," Vol. 1, Ch. 1, New York, John Wiley & Sons, 1954.
23. Redknap, E. F., *J. Oil Colour Chemists' Assoc.*, **49**, 1023 (1966).

24. Sawyer, G. B., Lammiman, L. W., and Miller, E. P., "Methods of Applying Surface Coatings," (Von Fischer, W., editor) "Paint and Varnish Technology," Ch. 26, New York, Reinhold Publishing Corp., 1948.
25. Sletmoe, G. M., *J. Paint Technol.*, **38,** 642 (1966).
26. Smith, N. D. P., Orchard, S. E., and Rhind-Tutt, A. J., *J. Oil Colour Chemists' Assoc.*, **44,** 618 (1961).
27. Stordy, J. J., and Appleton, W. G. J., *J. Oil Colour Chemists' Assoc.*, **39,** 565 (1956).
28. "Symposium on Electrodeposition of Coatings," *J. Paint Technol.*, **38,** 421 (1966).
29. "Symposium on the Physical Processes of Drying and Aging," *Offic. Dig. Federation Soc. Paint Technol.*, **33,** 915 (1961).
30. Talen, H. W., *J. Oil Colour Chemists' Assoc.*, **45,** 387 (1962).
31. Tawn, A. R. H., and Berry, J. R., *J. Oil Colour Chemists' Assoc.*, **48,** 790 (1965).
32. Van Loo, M., *Offic. Dig. Federation Paint Varnish Prod. Clubs*, **28,** 1126 (1956).
33. Werner, H. M., *Offic. Dig. Federation Soc. Paint Technol.*, **33,** 948 (1961).
34. Wright, P. D., *J. Oil Colour Chemists' Assoc.*, **39,** 129 (1956).

3

*Raw and Processed Oils**

Oils used in coatings are classified broadly as vegetable oils and marine oils. Chemically they are triglycerides, i.e., compounds of one molecule of glycerin and three molecules of long chain fatty acids. Fats are also triglycerides. They differ from oils in that they are solids rather than liquids at normal temperatures. Oils vary in drying properties depending on the degree of unsaturation of the fatty acid. Drying oils are converted by the oxygen of air to dry, hard, insoluble resinous materials. Oils are used by themselves as vehicles for coatings or in varnishes, alkyds, epoxies, urethanes and other polymers to impart drying properties as well as flexibility to the coating. Nondrying oils serve as plasticizers for many polymers used in coatings.

Not only the physical properties but the economics determine which oils are used in coatings at any particular time. Drying oils, being natural products, have been subject to wide price fluctuations over the years, and if the price of one particular oil increases out of proportion to the others, this oil may be replaced with another one or with a combination of oils.

COMPOSITION

Triglycerides have the following composition:

Glycerol   *Fatty Acid*

*By C. R. Martens, The Sherwin-Williams Co., Cleveland, Ohio.

Fatty acid chain length can vary from C_9 to C_{22}.

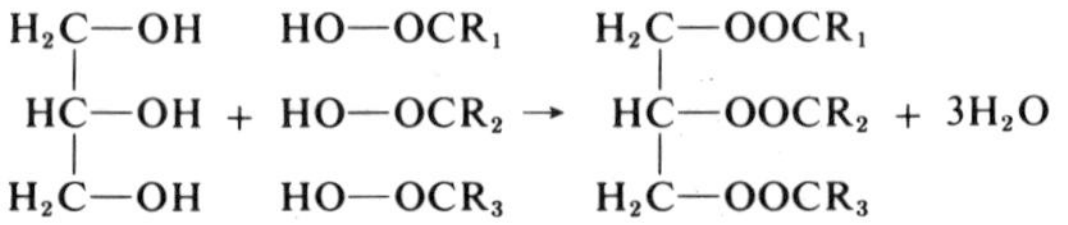

$$
\begin{array}{lll}
H_2C\text{—}OH & HO\text{—}OCR_1 & H_2C\text{—}OOCR_1 \\
HC\text{—}OH \;+\; & HO\text{—}OCR_2 \;\rightarrow\; & HC\text{—}OOCR_2 \;+\; 3H_2O \\
H_2C\text{—}OH & HO\text{—}OCR_3 & H_2C\text{—}OOCR_3
\end{array}
$$

1 Mole of Glycerol   *3 Moles of Fatty Acids*   *Triglyceride*   *3 Moles of Water*

In the formulas shown above, R_1, R_2 and R_3 stand for fatty acid chains. If R_1, R_2 and R_3 are the same, a simple triglyceride results. If they are not the same, the triglyceride is mixed. Triglycerides occurring in nature are usually of this type. This is a reversible reaction so that if we hydrolyze an oil we obtain glycerine and fatty oils.

Fatty Acids

The nature of the fatty acid present in an oil determines its characteristics. The fatty acid consists of a carboxyl group attached to a hydrocarbon chain. Saturated fatty acids have hydrocarbon chains containing no double bonds, each carbon having a least two hydrogen atoms. Fatty acids with chains containing double bonds are termed *unsaturated*. They may have one, two, three or more double bonds, whose position in the chain may vary. Two double bonds separated by a single bond are called *conjugated*.

The double bonds in unsaturated fatty acids are chemically reactive sites. The reaction of oxygen with the oil molecule at the double bond results in drying; usually the greater the unsaturation, the better is the drying. Saturated fatty acids are nondrying. Fatty acids with a single double bond are essentially nondrying. Fatty acids with three double bonds dry the most rapidly. However, in addition to the number of double bonds, the position of the double bond is important. Conjugated double bonds polymerize and dry more rapidly than isolated double bonds.

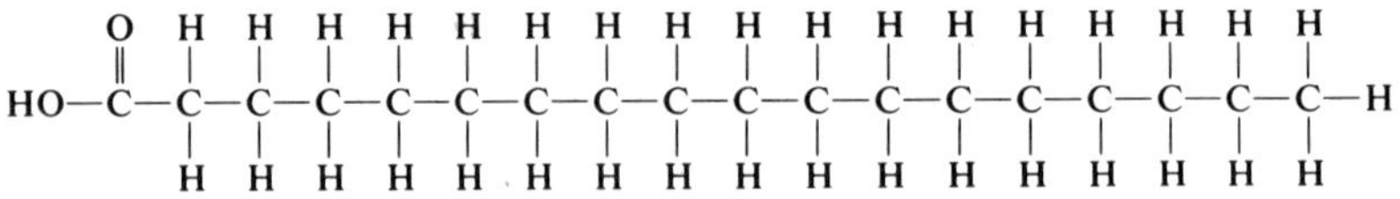

Stearic Acid

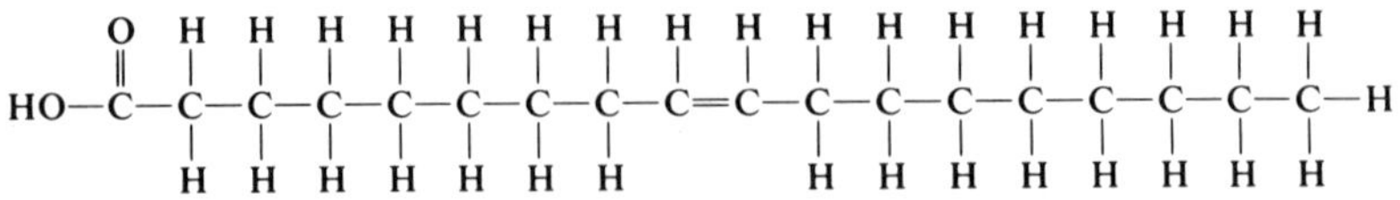

(9)

Oleic Acid

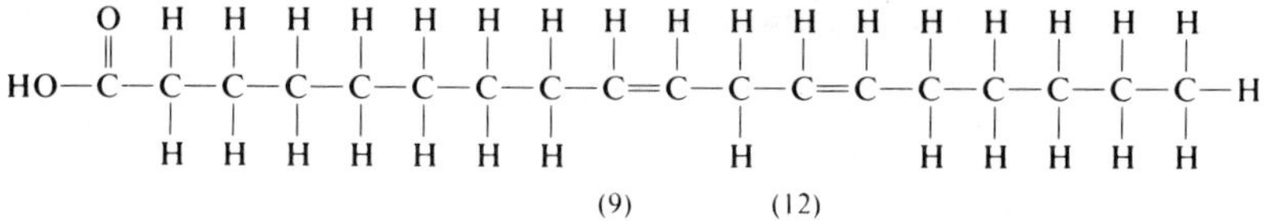

$$\text{HO—C(=O)—CH}_2\text{—CH}_2\text{—CH}_2\text{—CH}_2\text{—CH}_2\text{—CH}_2\text{—CH}_2\text{—CH=CH—CH}_2\text{—CH=CH—CH}_2\text{—CH}_2\text{—CH}_2\text{—CH}_2\text{—CH}_3$$

(9)   (12)

Linoleic Acid

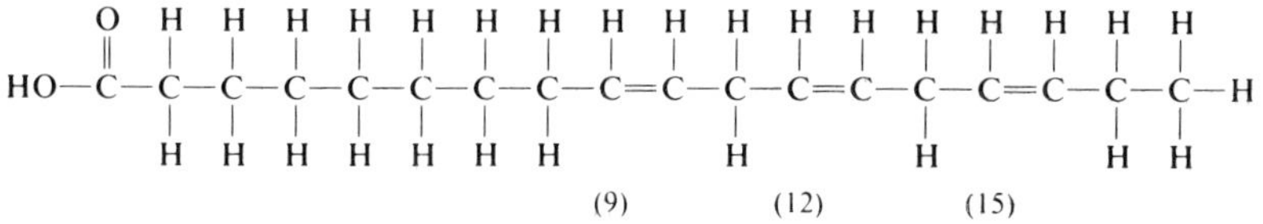

$$\text{HO—C(=O)—CH}_2\text{—CH}_2\text{—CH}_2\text{—CH}_2\text{—CH}_2\text{—CH}_2\text{—CH}_2\text{—CH=CH—CH}_2\text{—CH=CH—CH}_2\text{—CH=CH—CH}_2\text{—CH}_3$$

(9)   (12)   (15)

Linolenic Acid

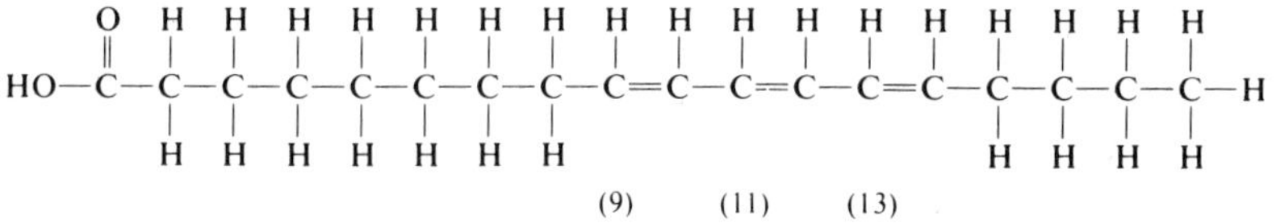

$$\text{HO—C(=O)—CH}_2\text{—CH}_2\text{—CH}_2\text{—CH}_2\text{—CH}_2\text{—CH}_2\text{—CH}_2\text{—CH=CH—CH=CH—CH=CH—CH}_2\text{—CH}_2\text{—CH}_2\text{—CH}_3$$

(9)   (11)   (13)

Eleostearic Acid

$$\text{HO—C(=O)—CH}_2\text{—CH}_2\text{—C(=O)—CH}_2\text{—CH}_2\text{—CH}_2\text{—CH}_2\text{—CH=CH—CH=CH—CH=CH—CH}_2\text{—CH}_2\text{—CH}_2\text{—CH}_3$$

(4)   (9)   (11)   (13)

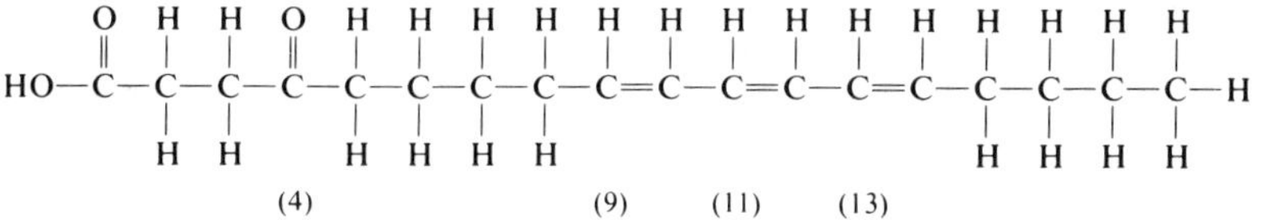

Licanic Acid

RAW OILS

Oils, being natural products, vary in composition. The geographical area and the weather during the growing season affect the composition of an oil. Composition can also vary among particular varieties of the plant family.

Variety of Oils

Linseed Oil.   Linseed oil is obtained from the seed of the common flax plant, *Linum usitatissimum*, L. Flaxseed has an average oil content of 35%. Linseed oil, historically, has been the most important oil in the coatings industry. Although its use has decreased because of other materials, linseed oil still leads in volume in the coatings industry. Linseed oil has a moderate drying rate. Because of its high linolenic acid content, linseed oil yellows on aging. It is used as the principal vehicle in house paints and also as the drying oil in some alkyds.

Safflower Oil.   Safflower oil was introduced in this country about 1955. The safflower plant is grown in the western part of the United States. The seed contains about 30% oil. Since it contains only minor amounts of linolenic acid, it is nonyellowing. Safflower oil is somewhat between linseed oil and soya oil in its properties. Its principal use is in alkyds.

Soya Oil. This oil is obtained from the seeds of various legumes known as *Soja max*. The oil content of soybeans averages 18%. It cannot be used in paints alone as it is semidrying. However, it can be processed in alkyds to give good color-retentive, air-drying and baking vehicles.

Tall Oil Fatty Acids. Tall oil is a by-product of the manufacture of sulfate process or Kraft paper. Crude tall oil contains 6 to 13% of unsaponifiable matter, the remainder being about equal parts of fatty acids and rosin acids. Tall oil can be fractionated by distillation down to a rosin content of 1%. Tall oil alkyds approach soya alkyds in performance.

Tung Oil. Tung oil or chinawood oil is obtained from the nut of the tung oil tree now grown in this country. The oil content is about 50%. Tung oil contains about 80% eleostearic acid which is a conjugated oil. Tung oil dries rapidly to a hard film which has good water and alkali resistance and good durability.

Oiticica Oil. Oiticica oil is obtained from the seed of the *Licania rigida* tree which is native to Brazil. Oiticica oil is unique in that it contains keto acid groups. Properties are somewhat similar to tung oil, but the oil heat-polymerizes less rapidly and films are more brittle and less water resistant.

Dehydrated Castor Oil. Castor oil is obtained from the castor bean. This is a nondrying oil containing an hydroxy acid group. By chemical dehydration the hydroxyl group is removed leaving a double bond. Dehydrated castor oil is a reactive drying oil with good color retention properties.

Fish Oil. Fish oils are obtained from sardines or menhaden. They contain a large amount of highly unsaturated fatty acids plus nondrying acids with no intermediate materials. For this reason they are fast drying but have a tendency to have after-tack.

Coconut Oil. Coconut oil is obtained from copra, the dried up kernel of coconuts which contain about 65% oil. This is nondrying with excellent flexibility and color retention.

Cottonseed Oil. Cottonseed, a by-product of cotton production, gives a nondrying oil lower in cost than coconut oil but inferior in properties.

Extraction of Oils

Oils are obtained from seeds, nuts or other material by pressure extraction, solvent extraction or a combination of both these methods. In pressure extraction, the main steps are cleaning, crushing, steam cooking and pressing on a plate press. In solvent extraction, the steps are the same but the oil is extracted with a solvent such as hexane. Yields are

TABLE 3.1. Fatty Acid Composition of Vegetable and Marine Oils

	Number of Carbons	Double Bonds	Tung	Oiticica	Dehy-drated Castor	Fish	Linseed	Safflower	Soya	Tall[a] Oil Acids	Cotton-seed	Coconut
Oleic	18	1	8	6	9	10	22	13	25	46	24	
Linoleic	18	2	4		82	15	16	75	51	41	40	7
Linolenic	18	3	3				52	1	9	3		2
Eleostearic	18	3	80									
Licanic	18	3		78								
Ricinoleic	18	1			9							
Palmitoleic	16	1										
Arachidonic	20	4				30						
Clupanodonic	22	5				25						
Stearic	18	0	1	5	2	2	4	4	4	3	4	6
Palmitic	16	0	8	7		12	6	6	11	5	29	11
Myristic	14	0				6					1	18
Lauric	12	0										44
Capric	10	0										6
Caprylic	8	0										6
Iodine value			160–175	140–160	125–140	165–195	170–190	140–150	120–140	128–138	99–113	7–10
Viscosity, Gardner-Holdt			I	X	G	A	A	A	A	A	A	A
Lb/gal			7.85	8.10	7.81	7.69	7.76	7.70	7.70	7.53	7.65	7.68
Color, Gardner			10	9	5	12	11	10	10	4	8	5
Acid value			8	8	4	6	4	4	3	194	1	2
Saponification value			190	190	190	190	190	190	190	196	190	250

[a] Distilled 1% rosin acids.

usually higher with a solvent extraction. The oil obtained by either method is known as raw oil.

Refining

Various methods of refining are used to remove impurities from raw oils. There are three general types of refining processes: mechanical, acid and alkali.

In mechanical refining, the oil is first treated with 2% water at 180°F and centrifuged. This coagulates and removes the foots. The resulting oil is a non-break oil which means there is no separation of solid material when heated to 600°F. For a paler color, the non-break raw oil is heat treated with an organic peroxide or a color absorbent such as fullers earth. This is called stainless raw oil.

Acid refining of oils is done by controlled treatment with sulfuric acid which chars and precipitates the break. When a lighter color is desired, bleaching clay is added prior to centrifuging or filtering.

Alkali refined oil is the type most widely used. A solution of sodium hydroxide is added to crude oil in sufficient amounts to reduce the acid value below 0.3. Low acidity favors fast bodying and good color in bodied oils and varnishes. The alkali also precipitates the foots, phosphatides and some color bodies.

PROCESSED AND TREATED OILS

Raw oils can be used in coating formulations. However, in most cases they are processed further for use.

Boiled oil is a term applied to linseed oil which is made by incorporating metallic driers and then heating.

Heat-polymerized oils are made by heating a break-free oil until the desired viscosity is obtained. The optimum temperature for heat bodying is in the range of 550 to 600°F. Viscosities as high as Z_9 may be obtained. This is a polymerization process. In heat polymerization the oil loses weight during the process as decomposition products are volatilized.

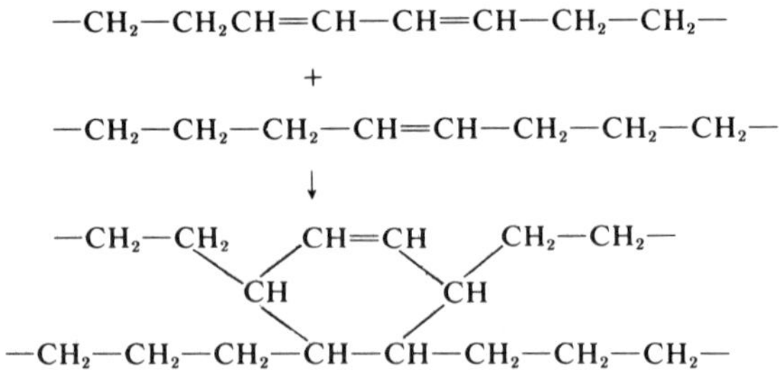

$$-CH_2-CH_2CH=CH-CH=CH-CH_2-CH_2-$$

$$+$$

$$-CH_2-CH_2-CH_2-CH=CH-CH_2-CH_2-CH_2-$$

$$\downarrow$$

$$-CH_2-CH_2\underset{\diagdown CH\diagup}{\quad}CH=CH\underset{\diagdown CH\diagup}{\quad}CH_2-CH_2-$$

$$-CH_2-CH_2-CH_2-CH-CH-CH_2-CH_2-CH_2-$$

Figure 3.1.   Polymerization of fatty acid chains.

Blown oils are made by bubbling air through oil at temperatures of 180 to 230°F. This produces a partial oxidation of the oil, a reduction in unsaturation and polymerization. During the blowing process, oils gain weight by the addition of oxygen.

Table 3.2 shows the changes in properties upon the bodying of an alkali refined linseed oil. During the early stages of heating, the color of the linseed oil lightens and then darkens. Formation of free acid and color are due mostly to oxidation decomposition during the heating in an open kettle. When bodying takes place under an inert atmosphere or reduced pressure, these changes are minimized. The decrease in iodine value is the result of saturation of carbon-carbon double bonds and the resulting growth in molecular weight as reflected in increases of specific gravity. Bodied linseed oil drys in about one-sixth the time required by raw oil. Generally speaking, there is good correlation between the drying properties of an oil and its bodying rate.

TABLE 3.2. Heat Bodying of Linseed Oil in Air at 580°F

Viscosity	Iodine No.	Acid No.	d 15/15
A_2	186	0.3	0.9327
B	180	0.8	0.9346
N	151	4.0	0.9508
V	146	5.1	0.9546
W	142	6.2	0.9591
X	140	6.6	0.9609
Z_1	137	8.0	0.9650
Z_3	136	9.0	0.9672
Z_4	135	10.0	0.9701
Z_5	133	11.3	0.9713
Z_6	133	12.0	0.9722
Z_7	132	12.5	0.9733

MODIFIED OILS

Maleic-treated Oils

These are made by heat treating an unsaturated oil with 2 to 10% maleic anhydride. The maleic anhydride reacts with the oil at the double bonds. Maleic anhydride reacts most rapidly with a conjugated double bond.

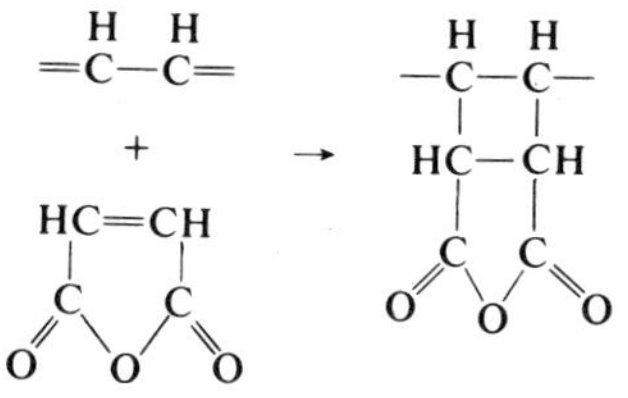

The maleic-treated oil can be made water soluble by reacting with ammonia or amines. For solvent systems, since the oil is acidic it is reacted and neutralized with a polyol. Maleic anhydride will also react with non-conjugated acids. The increased functionallity of maleic-treated oils explains their improved properties, particularly faster bodying rate, speed of dry and water resistance of treated soya and tall oil.

Copolymer Oils

Copolymer oils are made by heat processing an unsaturated oil with a reactive monomer such as styrene or vinyltoluene. With conjugated oils such tung oil copolymers are formed. However with oils such as linseed oil, a large amount of polymer is formed so that the oil is largely a mixture of homopolymers and oil. The major problem in oil styrenation is to obtain a homogenous product that dries to a clear film. The nature of the oil and the reaction conditions are important. Blown oils are sometimes used to give homogeneous products. Styrenated oils are fast drying and have good color and water resistance, but they have poor solvent resistance.

Cyclopentadiene will react with drying oils to give a fast-drying, hard oil. Polar catalysts such as boron fluoride are effective catalysts, and the reaction is carried out under pressure.

Epoxidized Oils

Epoxidized oils are produced by the reaction of unsaturated oils with peracetic acid. The double bonds are converted to epoxy or oxirane groups.

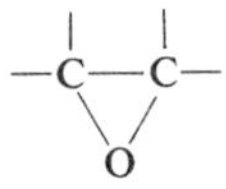

$$-\underset{\diagdown}{\overset{|}{C}}-\underset{\diagup}{\overset{|}{C}}-$$
$$O$$

These materials are used as plasticizers and are covered in more detail in Chapter 16.

DRYING PHENOMENA

When drying oil films are exposed to air, oxygen enters the wet film. The oxygen forms peroxides and hydroperoxides at the double bonds and adjacent $-CH_2-$ groups which are quite reactive, as are the double bonds themselves. Peroxide formation will remove double bonds which can be detected by a reduction in iodine value.

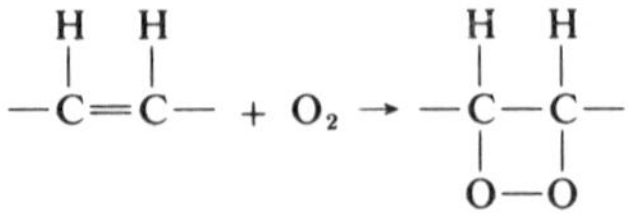

$$-\overset{\overset{\displaystyle H}{|}}{C}=\overset{\overset{\displaystyle H}{|}}{C}- \; + \; O_2 \; \rightarrow \; -\overset{\overset{\displaystyle H}{|}}{\underset{\underset{\displaystyle O}{|}}{C}}-\overset{\overset{\displaystyle H}{|}}{\underset{\underset{\displaystyle O}{|}}{C}}-$$

Hydroperoxide formation would take place as follows:

$$-\overset{\displaystyle H}{\underset{}{C}}=\overset{\displaystyle H}{\underset{}{C}}-\overset{\displaystyle H}{\underset{\displaystyle H}{C}}-\ +\ O_2\ \rightarrow\ -\overset{\displaystyle H}{\underset{}{C}}=\overset{\displaystyle H}{\underset{}{C}}-\overset{\displaystyle H}{\underset{\displaystyle O-O-H}{C}}-$$

It is now generally believed that oxidation polymerization of drying oils is produced by the formation of free radicals with the decomposition of peroxides and hydroperoxides. Addition polymerization initiated by free radicals then proceeds by cross-linking from chain to chain through the double bonds.

Certain metal compounds containing cobalt, manganese, lead, etc., accelerate this oxidation polymerization phenomena.

A broad classification of oils is as follows:

Drying	*Semi-drying*	*Nondrying*
Linseed	Safflower	Cottonseed
Tung	Soya	Castor
Oiticica	Tall oil acids	Coconut
Dehydrated castor		
Fish		

Drying oils are frequently classified as hard oils and soft oils. A hard oil is one that dries fast, gives a hard film and heat-polymerizes rapidly. Hard oils are tung, oiticica and dehydrated castor. Hard oils contain conjugated fatty acids which promote fast drying. In normal fatty acids, the double bonds are isolated as shown previously. On the other hand, in conjugated fatty acids the double bonds are in sequence and are thus more reactive.

OIL TESTS

Viscosity is usually measured by the Gardner-Holdt tube and given in letters. Color is given in the Gardner 1933 tube color standards which are number 1 to 18.

Specific gravity or pounds per gallon is most conveniently determined with a hydrometer.

Acid value indicates the amount of free acids in an oil. It is the number of milligrams of potassium hydroxide required to neutralize the acids in 1 gram of oil.

Iodine value indicates the degree of unsaturation of an oil. It is expressed as the centigrams of iodine absorbed by 1 gram of oil under controlled conditions.

Saponification value indicates the purity of the oil. Short chain length fatty acids give higher values. Saponification value is expressed as the number of milligrams of potassium hydroxide that reacts with 1 gram of oil.

Foots are the solid impurities that precipitate from an oil during storage.

Break means the impurities that separate from an oil when it is heated to 600°F.

Gel time applies mainly to reactive oils such as tung and oiticica. It is the time required to form a solid gel under specified conditions of temperature.

REFERENCES

1. Barley, A. E., "Industrial Oil and Fat Products," Third ed. New York, Interscience Publishers, 1964.
2. Eckey, E. W., "Vegetable Fats and Oils," New York, Reinhold Publishing Corp., 1954.
3. Kirschenbauer, H. G., "Fats and Oils," New York, Reinhold Publishing Corp., 1960.
4. Markley, K. S., "Fatty Acids," Vols. 1, 2, 3, New York, John Wiley & Sons, 1960, 1961, 1964.

4

*Alkyd Resins**

PROPERTIES

Alkyds possess most of the desirable properties required of a vehicle for protective coatings. Low-cost solvents can be used which give ease of application as well as a minimum of odor. Alkyds are ideal vehicles for pigmented coatings because they have good wetting and dispersing properties. Alkyd coatings are comparatively low in cost, have excellent durability, flexibility, gloss retention and good solvent resistance, toughness, heat resistance and color retention.

COMBINATIONS

Alkyds can be combined with the following:

Nitrocellulose
Urea-formaldehyde resins
Melamine-formaldehyde resins
Phenolic resins
Ethyl cellulose
Chlorinated rubber
Chlorinated paraffin
Epoxy resins
Polyisocyanates
Silicone resins
Polyamides
Natural resins
Cellulose acetobutyrate
Monomers
    (styrene, vinyl toluene, methyl methacrylate)
Synthetic latices
    (styrene-butadiene, polyvinyl acetate, acrylic)

DEFINITION

The term "alkyd" is applied to a group of synthetic resins which can be best described as oil-modified polyester resins. They are reaction products derived from a polyhydric alcohol, polybasic acid and a fatty monobasic acid. The word was coined from the "al" of alcohol and "kyd" which represents the last syllable of acid. Alkyds are therefore members of a large class of materials known as polymeric esters.

*By C. R. Martens, The Sherwin-Williams Co., Cleveland, Ohio.

ESTERIFICATION-POLYMERIZATION

Let us look at some of the fundamental chemistry involved. The most elementary esterification reaction is between a monofunctional acid, such as acetic acid, and a monofunctional alcohol, such as methyl alcohol, to produce methyl acetate and water [Figure 4.1(A)]. As this reaction is reversible, it is necessary to remove water to drive the reaction to completion. If the methyl alcohol is replaced with ethylene glycol, a larger molecule, ethylene diacetate, is produced [Figure 4.1(B)]. Going one step further, by replacing the acetic acid with a bifunctional acid, such as succinic, a complex molecule results. First the primary ester is formed which contains hydroxyl and carboxyl end groups. As the reaction progresses, the second molecule of succinic acid can esterify the hydroxyl group. Then the second molecule of ethylene glycol can esterify the carboxyl group. This series of reactions can continue with alternate glycol and acid groups adding to the chain until a long, linear molecule is formed [Figure 4.1(C)].

If the functionality of one of the reactants is greater than 2, the course of the reaction is changed further. If glycerin, which has a functionality of 3, is reacted with succinic acid, the reaction is at first similar to the previous one. The succinic acid will react first with the primary hydroxyl groups of the glycerin to form short, linear chains. As the reaction proceeds, the secondary alcohols react, forming a branched structure [Figure 4.1(D)]. Finally, branching or cross-linking proceeds to such an extent that the molecule is no longer soluble or fusible and reaches the gel state. Including a fatty acid we get an alkyd as shown in Figure 4.1(E).

RAW MATERIALS

Before discussing any particular class of alkyd resin, it will be of interest to consider the types of alcohols, acids and oils available for use in alkyds. The following materials listed are the more common ones, but there are many others available.

Polyols

The polyhydric alcohols shown in Table 4.1 have functionalities from 2 to 6. Glycerin was first obtained as a by-product of the splitting of fats and oils in the manufacture of soap. In the early 1940's, synthetic glycerin was produced commercially from petroleum sources. Glycerin has a functionality of 3 and contains both primary and secondary hydroxyl groups. The principal use of glycerin is in short and medium oil alkyds.

Pentaerythritol is a polyol second only to glycerin for alkyd usage and is produced by the condensation of acetaldehyde with formaldehyde in an aqueous alkaline medium. Pentaerythritol contains four primary hydroxyl groups and is outstanding for use in long oil alkyds.

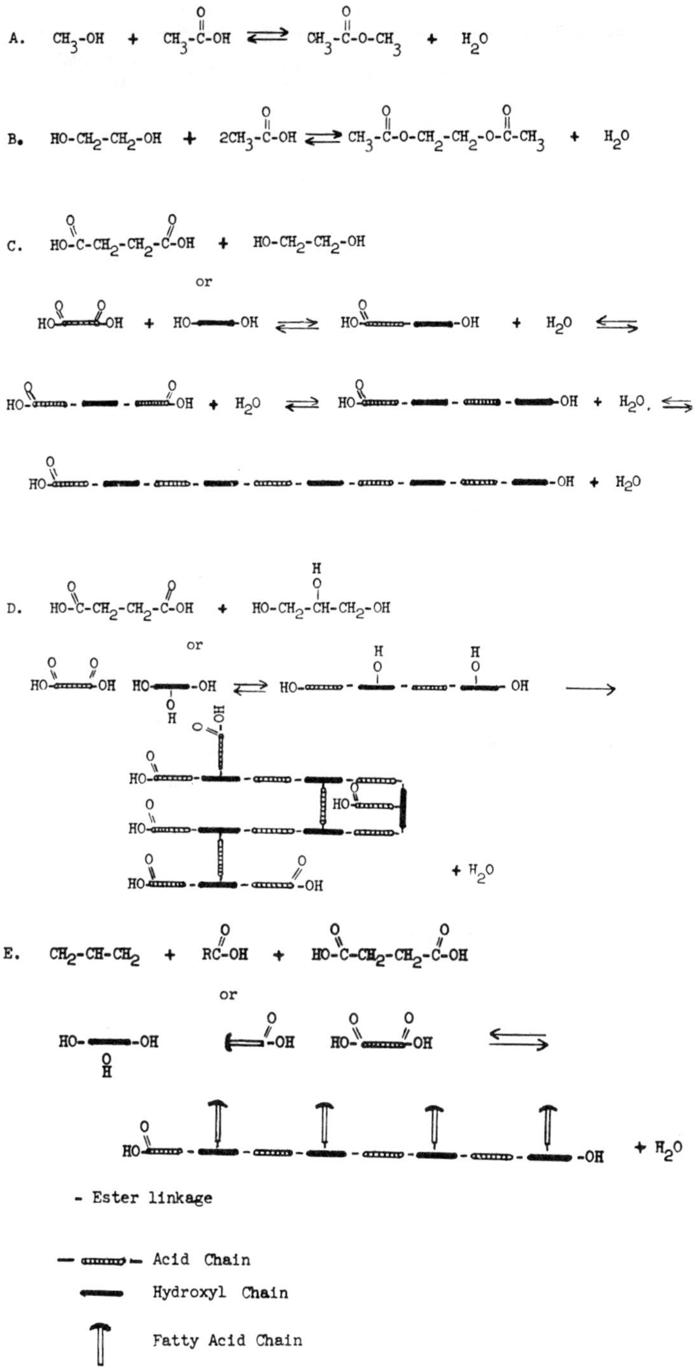

A. CH_3-OH + $CH_3-\overset{O}{\overset{\|}{C}}-OH$ ⇌ $CH_3-\overset{O}{\overset{\|}{C}}-O-CH_3$ + H_2O

B. $HO-CH_2-CH_2-OH$ + $2CH_3-\overset{O}{\overset{\|}{C}}-OH$ ⇌ $CH_3-\overset{O}{\overset{\|}{C}}-O-CH_2-CH_2-O-\overset{O}{\overset{\|}{C}}-CH_3$ + H_2O

C. $HO-\overset{O}{\overset{\|}{C}}-CH_2-CH_2-\overset{O}{\overset{\|}{C}}-OH$ + $HO-CH_2-CH_2-OH$

or

$HO-OH$ + $HO-OH$ ⇌ $HO-OH$ + H_2O ⇌

$HO-OH$ + H_2O ⇌ $HO-OH$ + H_2O, ⇌

$HO-OH$ + H_2O

D. $HO-\overset{O}{\overset{\|}{C}}-CH_2-CH_2-\overset{O}{\overset{\|}{C}}-OH$ + $HO-CH_2-\overset{\overset{H}{\overset{\|}{O}}}{CH}-CH_2-OH$

or

$HO-OH$ $HO-OH$ ⇌ $HO-OH$ ⟶

+ H_2O

E. $CH_2-CH-CH_2$ + $R\overset{O}{\overset{\|}{C}}-OH$ + $HO-\overset{O}{\overset{\|}{C}}-CH_2-CH_2-\overset{O}{\overset{\|}{C}}-OH$

or

$HO-OH$ $-OH$ $HO-OH$ ⇌

$HO-OH$ + H_2O

Figure 4.1. Esterification reaction.

TABLE 4.1. Polyols

Name	Formula	Form	Molecular Weight	Boiling Point (°F)
Ethylene glycol	H HC—OH | HC—OH H	Liquid	62.07	386
Diethylene glycol	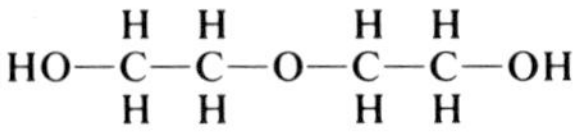 $HO-CH_2-CH_2-O-CH_2-CH_2-OH$	Liquid	106.12	473
Propylene glycol	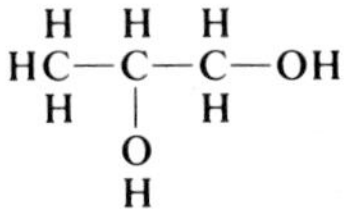 $HC-CH_2-CH_2-OH$ with OH	Liquid	76.09	374
Glycerine CP-95% glycerine Super-98% "	H HC—OH | HC—OH | HC—OH H	Liquid	92.09	554

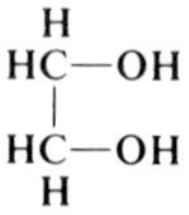

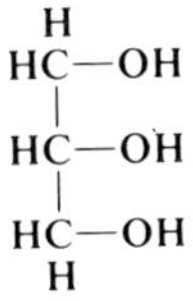

Name	Formula	Form	Molecular Weight	Melting Point (°F)
Trimethylolethane	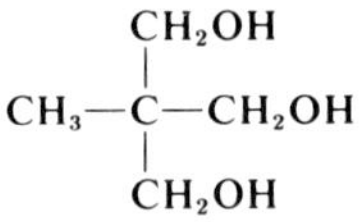 CH_2OH | CH_3-C-CH_2OH | CH_2OH	White solid	120.15	395
Trimethylol-propane	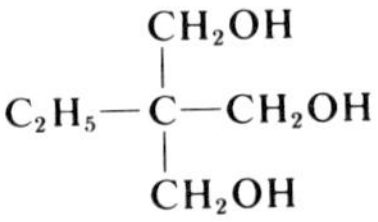 CH_2OH | $C_2H_5-C-CH_2OH$ | CH_2OH	White solid	134.18	136
Pentaerythritol	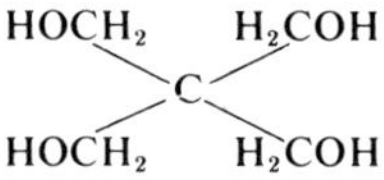 $HOCH_2$ H_2COH C $HOCH_2$ H_2COH	White solid	136.15	504
Sorbitol	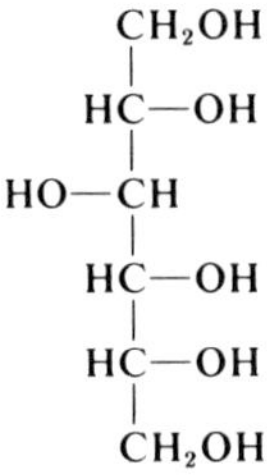 CH_2OH | HC—OH | HO—CH | HC—OH | HC—OH | CH_2OH	White solid	182.17	195–230

Polypentaerythritol, dipentaerythritol and tripentaerythritol are by-products of pentaerythritol manufacture. Because of their high functionality, (6 and 8, respectively), they are used in long oil alkyds.

Ethylene glycol is the most important glycol used in alkyd resins. At the present time, glycols are the lowest-cost polyols available. Their volatility is a disadvantage in alkyd manufacture. In many cases, glycols are combined with polyols of higher functionality such as pentaerythritol.

Trimethylolethane is made by the condensation of formaldehyde with propionaldehyde. It contains three primary hydroxyl groups.

Trimethylolpropane is produced from the condensation of formaldehyde with butyraldehyde.

Sorbitol is produced by the catalytic hydrogenation of glucose. Sorbitol contains six hydroxyl groups but its functionality is calculated as 4, since not all the hydroxyls will esterify under alkyd processing conditions.

Acids

Acidic materials may be in the form of acids or anhydrides. The anhydride is formed from two equivalents of acid minus a mole of water. Reaction rates are more rapid when the anhydride is used, and there is less water evolved from the reaction (Table 4.2).

Phthalic anhydride (*ortho*) is the principal dibasic acid used in alkyds. Unless specified, phthalic refers to the *ortho* form. It is produced from the catalytic oxidation of naphthalene or orthoxylene.

Isophthalic acid (*meta*) has become available commercially the past few years from the catalytic oxidation of petroleum xylenes. Tereperthalic acid (*para*) is difficult to use in alkyds because of its high melting point. Tetrahydrophthalic anhydride is produced commercially from the Diels-Alder reaction of maleic anhydride with unsaturated petroleum materials.

Maleic anhydride is obtained from the catalytic oxidation of naphthalene and is the *cis* form of butendioic anhydride. Maleic is unsaturated and will, therefore, cross-link with the double bonds of fatty acids. Since this increases the functionality of the system, it is sometimes added to alkyds to increase viscosity.

Adipic acid is obtained from the oxidation of cyclohexane. In an alkyd this material produces a softer, more flexible resin.

Benzoic acid cannot be used as the sole organic acid, since it is monofunctional. However, if a small amount of phthalic anhydride is replaced with benzoic this acts as a chain stopper and the alkyd can be cooked to lower acid values without gelation.

Oils and Fatty Acids

Oils or fatty acids impart flexibility and drying to an alkyd. The solvent first evaporates and then the unsaturated double bonds polymerize further by atmospheric oxidation. In general, the greater the unsaturation as

TABLE 4.2.   Acids and Anhydrides

Name	Formula	Form	Molecular Weight	Melting Point (°F)	Boiling Point (°F)
Phthalic anhydride (*ortho*)		White solid	148.11	270	544
Phthalic acid (*ortho*)		White solid	166.13	375–410 Decomposes	Decomposes
Isophthalic acid (*meta*)		White needles	166.13	650	Sublimes
Terephthalic acid (*para*)		White crystals	166.13	Sublimes >570	—

Name	Structure	Appearance	Mol. wt.		
Tetrahydro- phthalic acid	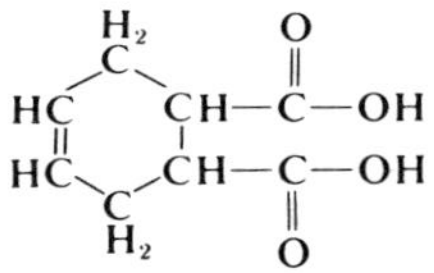	White solid	170.16	248	—
Hexahydro- phthalic acid	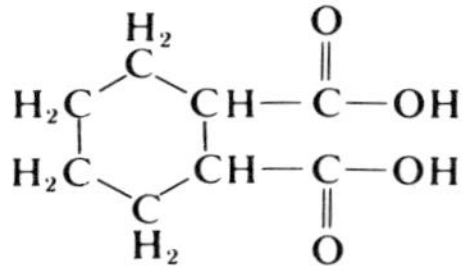	White solid	172.18	377	—
Benzoic acid	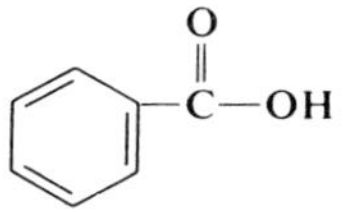	White solid	122.12	251	482
Maleic anhydride	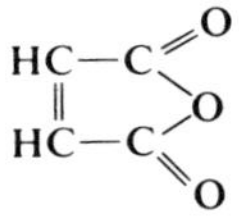	White solid	98.06	126	395
Maleic acid	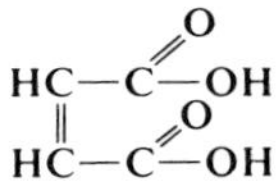	White solid	116.07	266	275

(Continued)

TABLE 4.2. Acids and Anhydrides (Continued)

Name	Formula	Form	Molecular Weight	Melting Point (°F)	Boiling Point (°F)
Fumaric acid	(structure)	White solid	116.07	548	554
Succinic anhydride	(structure)	White solid	100.07	248	502
Adipic acid	(structure)	White solid	146.14	306	509
Sebacic acid	$HO-\overset{O}{\overset{\|}{C}}-(CH_2)_8-\overset{O}{\overset{\|}{C}}-OH$	White solid	202.25	271	563 @ 100 min.

measured by the iodine value, the greater is the drying phenomena and (usually) the darker the color of the alkyd. Fatty oils are triglycerides of aliphatic acids having 18 carbons in their chain. The number of unsaturated double bonds varies from stearic acid which is completely saturated and therefore has no double bonds, to eleostearic which has three double bonds. These materials are described in detail in Chapter 3.

PREPARATION

Alkyds can be prepared directly from a fatty acid, polyol and acid or from the oil (triglyceride), polyol and acid. However, if an attempt is made to prepare an alkyd by reacting an oil, glycerin and phthalic anhydride together as the glyceryl phthalate forms, it would precipitate and be insoluble in the oil. Nevertheless, an alkyd can be prepared from these ingredients if the reaction is carried out in a slightly different manner. The oil is first converted to a monoglyceride by heating with a

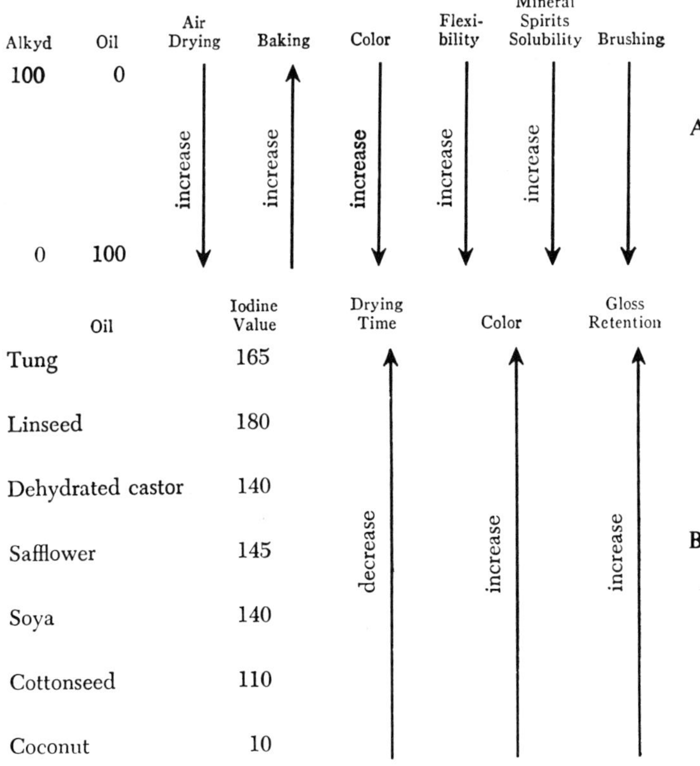

Figure 4.2.   A. Effect of amount of oil on properties.
B. Effect of type of oil on properties.

polyol in the presence of a catalyst. The monoglyceride is then reacted with phthalic anhydride to form an alkyd.

Variations of the type and amount of oil allows a wide variation of film properties from a soft, colorless plasticizing film to a hard-tough tack-free film (see Fig. 4.2). The percentage of oil contained in an alkyd classifies the alkyd as to end use since this affects such properties as speed of dry, flexibility, durability, etc.

	% Oil	*% P.A.*
Short oil	33–45	>35
Medium oil	46–55	30–35
Long oil	56–70	20–30
Very long oil	71 up	<20

STRUCTURE OF ALKYDS

Figure 4.3 shows the composition and simplified structure of a short, medium and long oil alkyd. Actually, the molecules are so complex that they cannot be shown. They are three dimensional, and the figure shows only a small portion of the molecule. Branching is not indicated. The illustration shows a glyceryl phthalate alkyd produced from phthalic

	I	II Short	III Medium	IV Long
Type		Short	Medium	Long
% Oil	0	32	55	74
% Phthalic Anhydride	66	40	33	15
Excess Hydroxy	50	29	6	0
Moles				
Phthalic Anhydride	6	6	6	6
Glycerine	6	6	6	6
Fatty Acids	--	2	5	6
Oil	--	--	--	3

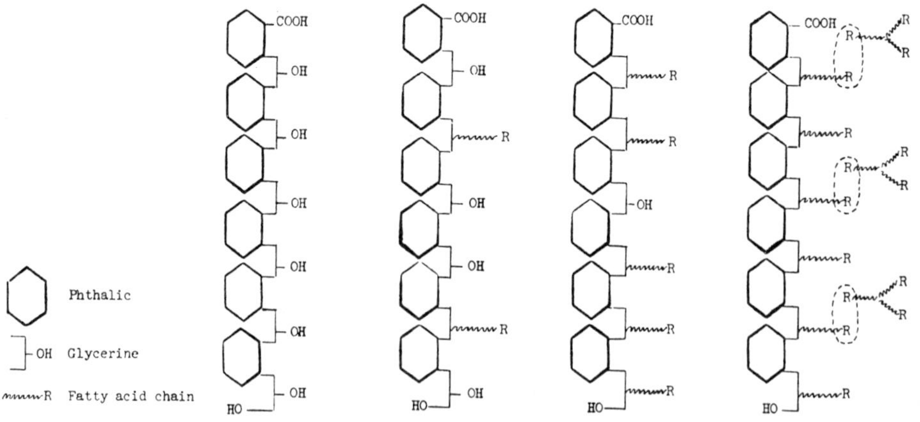

Phthalic

OH Glycerine

R Fatty acid chain

Figure 4.3. Composition of glyceryl phthalate alkyds containing various percentages of oil.

anhydride (bifunctional), glycerin (trifunctional) and fatty acids (mono-functional).

Type I is an oil-free alkyd or polyester produced with a large excess of hydroxyl groups. This resin is alcohol soluble in the early stages of condensation.

Type II alkyd is a short oil alkyd in which fatty acid chains have reacted with a portion of the hydroxyl groups, reducing the amount of excess hydroxyl. This resin is soluble in aromatic solvents.

Type III is a medium oil alkyd in which essentially all the hydroxyl groups are reacted with the fatty acid chain. Such an alkyd is soluble in aliphatic-type solvents.

Type IV is a long oil alkyd in which there is an excess of oil beyond that needed for esterification. The excess oil is either present but un-reacted with the alkyd or bound by heat polymerization with the fatty acid in the alkyd.

Calculations

Shown below are examples of calculations of alkyds prepared from a fatty acid and by alcoholysis:

Fatty Acid Cook

Composition	*Equivalent*		*Pounds*	*% in Final*
	COOH	*OH*		*Alkyd*
Soya fatty acids	1.20	—	336	58.0
Pentaerythritol	—	3.84	131	23.0
Phthalic anhydride	2.00	—	148	26.0
	3.20	3.84	615	107.0
Water loss			−40	−7.0
			575	100.0

(1) Excess hydroxyl: $\dfrac{100(3.84 - 3.20)}{3.20} = 20\%$

(2) Theoretical phthalic anhydride: $\dfrac{100(148)}{575} = 25\%$

(3) Fatty acids: $\dfrac{100(336)}{575} = 58\%$

(4) Oil: $(1.04)(58\%) = 60\%$

Alcoholysis Cook

Composition	*Equivalent*		*Pounds*	*% in Final*
	COOH	*OH*		*Alkyd*
Linseed oil	.50	.50	148	42.0
Glycerin	—	2.40	74	21.0
Phthalic anhydride	2.00	—	148	42.0
	2.50	2.90	370	105.0
Water loss			−18	−5.0
			352	100.0

(1) Excess hydroxyl: $\dfrac{100(2.90 - 2.50)}{2.50} = 16\%$

(2) Theoretical phthalic anhydride: $\dfrac{100(42)}{105} = 40\%$

(3) Glyceryl phthalate: $\dfrac{100(148 + 74 - 18)}{352} = 57\%$

(4) Oil: $\dfrac{100(148)}{352} = 42\%$

MANUFACTURE

Alkyd resins are produced by a two-step batch process. The first step is the esterification reaction in which the materials are reacted to a specified end point. Then the resin is partially cooled and dropped into solvent in the thinning tank. From this point, the alkyd is pumped to filter presses for clarification and then to storage tanks.

The esterification reaction can be carried by either the fusion or the solvent process. In the fusion method, the reactants are charged into the kettle and heated under an inert gas atmosphere. Near the end of the cook, inert gas is blown into the resin mass to carry off water and unreacted materials.

The solvent process uses a small amount of solvent (3 to 10%) in the reaction mixture to act as a reflux medium. The water on reaction is carried off by the solvent, separated out and the solvent is returned to the batch.

Short Oil Alkyds

These are drying and nondrying types. The latter are essentially plasticizers and are non-film-forming. Short oil alkyds containing soya or dehydrated castor oil are used in conjunction with amino resin for appliance finishes. Short oil alkyds containing nondrying oils such as coconut are used in conjunction with nitrocellulose for automotive lacquers. Linseed short oil alkyds by themselves can be used as fast air-drying or baking finishes.

The following is a short oil dehydrated castor alkyd made by the solvent process: Dehydrated castor oil is used and converted to the monoglyceride by the alcoholysis method. This alkyd is suitable for use as a baking vehicle either by itself or in combination with an amino resin.

	Moles	*Pounds*
Dehydrated castor oil	.183	160
Glycerin	1.210	110
Phthalic anhydride	1.210	180

	Moles	*Pounds*	
Litharge			4 oz.
		450	
Water loss		−22	
Base yield		428	
Xylene		428	
Fatty acids	35%	Excess hydroxyl	37.5%
Phthalic anhydride	42%	Functionality	1.95

Heat dehydrated castor oil, glycerin and litharge to 450°F. Hold for clear by methanol test. Cool to 280°F, add phthalic anhydride and 4% xylene (18 pounds) and heat to reflux temperature. Cook for approximately 6 hours until characteristics are obtained. Thin with equal parts by weight of xylene.

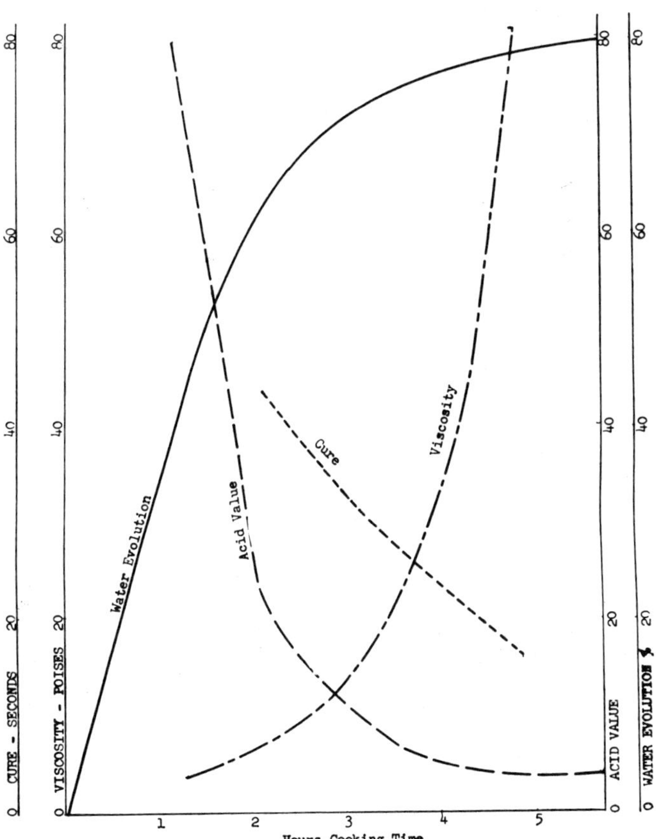

Figure 4.4.   Cooking of short oil alkyd.

Weight per gallon	8.25 lb
Viscosity	Z1
Color	6
Acid value	4.0
NVM	50%

Figure 4.4 shows progress of cooking of this alkyd.

Drying

Air dry set	.
Air dry hard	4–5 hr
Bake	25–50 min at 250–275°F
with 10–30% urea-formaldehyde resin	5–45 min at 300–325°F
with 10–30% melamine-formaldehyde resin	15–45 min at 250–275°F

Drier Recommendation

	Air Dry	*Bake*
Lead	None	None
Manganese	None	None
Cobalt	0.03–0.07%	0.01–0.02%
Calcium	0.2–0.4%	None
Zinc	None	0.1%

Medium Oil Alkyds

These are the most versatile of alkyds and are used both for air-drying and baking applications. The modifying oil is usually linseed, soya or tall oil. The following is a medium-length tall oil fatty acids cook prepared by the fusion method:

	Moles	*Pounds*
Tall oil fatty acids	.69	200
Pentaerythritol	2.92	100
Phthalic anhydride	1.61	120
		420
Water loss		−40
Base yield		380
Mineral spirits		380

Fatty acids	50%	Excess hydroxyl	27%
Phthalic anhydride	33%	Functionality	2.08

Heat tall oil fatty acids, pentaerythritol and phthalic anhydride to a temperature of 550°F in a kettle fitted with a steam condenser. After 1 hour of heating, sparge with inert gas until an acid value of 10 is reached. Total time will be about 6 hours. Reduce with equal parts of mineral spirits.

Weight per gallon	7.60 lb
Color	8
Acid value/solids	10
Viscosity	4
NVM	50%

Drying

Air dry set	30 min
Air dry hard	6–8 hr
Bake	1 hr at 275–300°F

Drier Recommendation

	Air Dry	*Bake*
Lead	0.1–0.3%	None
Manganese	0.01–0.03%	None
Cobalt	0.02–0.03%	0.01–0.02%
Calcium	0.2–0.4%	None
Zinc	None	0.11–0.2%

Long Oil Alkyds

Long oil alkyds are used for exterior applications where a flexible film is desired. The following is a long oil linseed glyceryl phthalate alkyd made by the fusion process. Alcoholysis step is used on linseed oil.

	Moles	*Pounds*	
Linseed oil	1.22	360	
Glycerin	2.57	80	
Phthalic anhydride	2.17	160	
Litharge			4 oz.
		600	
Water loss		−30	
Base yield		570	
Mineral spirits		244	

Heat linseed oil and glycerin to 500°F. Add litharge and hold for clear with methanol. Add phthalic anhydride. Cook until acid value of 10 is reached. Reduce to 70% solids with mineral spirits.

Oil	63%
Weight per gallon	8.10 lb
Color	10
Acid value	7
Viscosity	Z
NVM	70%

Drying and Drier Recommendation

Air dry set	$2\frac{1}{2}$ hr
Air dry hard	6–8 hr

	Air Dry
Lead	0.3–0.6%
Manganese	None
Cobalt	0.03–0.06%
Calcium	
(to replace lead)	0.3–0.6%
Zinc	None

REFERENCES

1. Martens, C. R., "Alkyd Resins," Reinhold Publishing Corp., New York, 1961.
2. Patton, T. C., "Alkyd Resin Technology," Interscience Publishers, New York, 1962.
3. "The Chemistry and Processing of Alkyd Resins," Monsanto Chemical Co., St. Louis, Mo., 1952.
4. "Alkyd Reports," Hercules Powder Co., Wilmington, Del., 1960.
5. "Alkyd Resins," *Paint Varnish Prod.*, **56** (1961): February, pp. 47–51; March, pp. 35–38; May, pp. 59–62; June, pp. 42–44; August, pp. 43–48; September, pp. 75–78; October, pp. 49–52; November, pp. 87–95; December, pp. 47–50.

5

*Aminoplast Resins**

This class of resins is employed in a variety of industrial bonding, molding and coatings uses. These uses include plywood adhesives, molding compositions, impregnated webs for laminating, textile treatment, wet strength paper treatment and protective coatings. The total aminoplast production and sales in the United States during 1964 was reported[1] as 526 million pounds apportioned between the various uses as shown in Table 5.1.

**TABLE 5.1. Aminoplast Thermosetting Resins
(thousands of pounds, dry basis, 1964)**

	Production	Sales
Urea and melamine resins—total	526,491	452,818
Textile treating and coating resins	46,664	41,806
Paper treating and coating resins	46,841	32,751
Bonding and adhesive resins for		
Laminating	54,512	37,445
Plywood	109,943	98,502
Fibrous and granulated wood	88,609	76,114
All other	15,681	15,946
Protective coatings	51,827	31,632
All other uses (including molding)	112,414	101,191
Sales for export		13,428

This chapter concerns itself with the approximately 52 million pounds that go into protective coatings. It is estimated that this quantity is divided approximately equally between urea and melamine types, and that in combination with greater amounts of alkyd or thermosettable acrylic prepolymers they represent approximately 250 million pounds of

*By F. A. Bonzagni and F. J. Hahn, Monsanto Company, Plastic Products & Resins Division, Springfield, Mass.

thermosettable binder employed in the highest-quality organic baking finishes.

Urea, melamine, substituted guanamine and related triazines have each been the subject of much effort in the search for improved aminoplasts. To date, economics and performance dictate a condition whereby urea- and melamine-based resins dominate all uses of aminoplasts, although benzoguanamine and melamine–toluenesulfonamide–formaldehyde resins have reached significant commercial acceptance in appliance and automotive coatings, respectively. In addition, ethylene urea- and thiourea-based resins have had some success in special textile applications. Without exception, however, the amine or amide is reacted with formaldehyde to form methylol groups which may or may not be temporarily blocked against excessive condensation by etherification with one of the lower alcohols.

CHEMISTRY AND COMPOSITION

Urea readily reacts with formaldehyde in neutral or mildly alkaline aqueous solution to form the mixed mono and dimethylol derivatives of urea shown in Figure 5.1. The kinetics of these reactions have been studied extensively.[4,6,20]

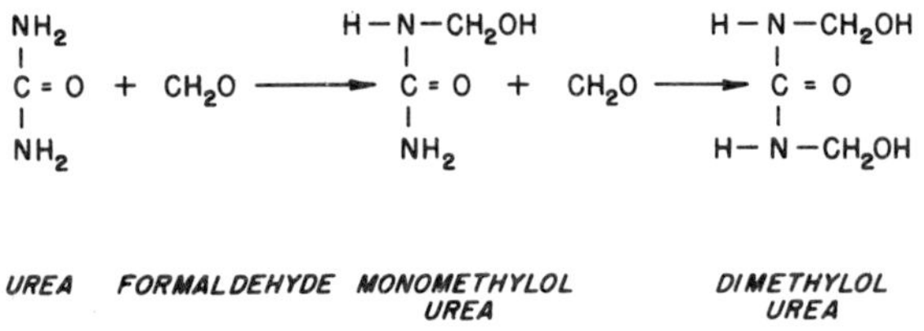

$$
\begin{array}{ccccc}
NH_2 & & H-N-CH_2OH & & H-N-CH_2OH \\
| & & | & & | \\
C=O \; + \; CH_2O & \longrightarrow & C=O \; + \; CH_2O & \longrightarrow & C=O \\
| & & | & & | \\
NH_2 & & NH_2 & & H-N-CH_2OH
\end{array}
$$

UREA   FORMALDEHYDE   MONOMETHYLOL   DIMETHYLOL
UREA   UREA

Figure 5.1.   Mono and dimethylol urea.

Both monomethylol urea and dimethylol urea have been isolated as crystalline solids, soluble in water, but insoluble in ether and aromatic hydrocarbons.[8] Both tri- and tetramethylol derivatives of urea have been detected but never isolated. Analysis of certain commercial urea resins revealed the presence of formaldehyde in amounts beyond that associated with dimethylol urea. However, Iliceto,[12] in confirming the presence of trimethylol urea in a condensation involving more than 2 moles of formaldehyde per mole of urea, found that even with 20 moles of formaldehyde per mole of urea the average ratio of formaldehyde reached only 2.81 moles per mole of urea.

Melamine and formaldehyde can react in neutral or mildly alkaline aqueous solution to yield the mono-, di-, tri-, tetra-, penta- or hexamethylol derivatives.[9,11] The reaction leading to hexamethylol melamine is given in Figure 5.2.

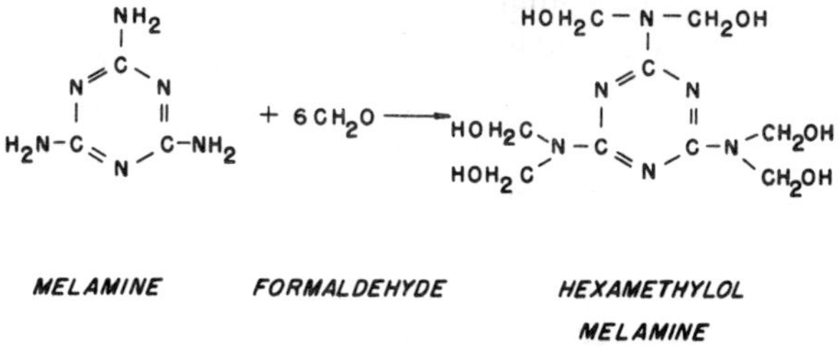

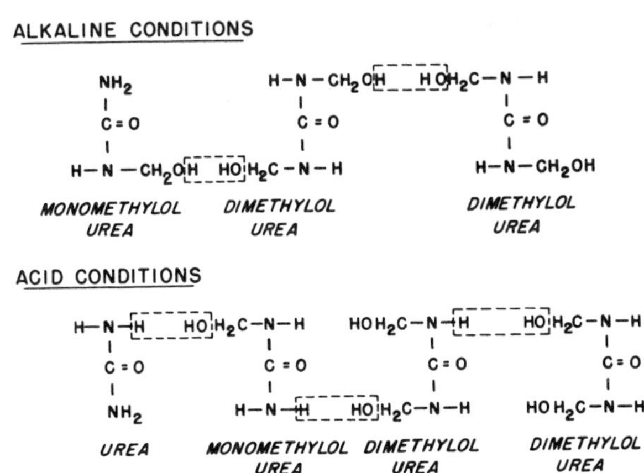

Figure 5.2. Hexamethylol melamine.

Separation of the various methylol melamines is extremely difficult because of their similar solubility characteristics, but isolation of hexamethylol melamine attests to the ability of all active hydrogens to be condensed with formaldehyde. These monomeric methylol melamines, like methylol ureas, tend to separate from the aqueous solution upon cooling. On being heated, they exhibit a strong tendency to condense.[2] In methylol ureas, under alkaline conditions, condensation proceeds between methylol groups to yield ether linkages, or under acid conditions via interaction between methylol and amide groups to form methylene linkages.[7,30] These condensation reactions are exemplified in Figure 5.3.

Figure 5.3. Condensation reactions of urea, monomethylol and dimethylol urea.

The formation of six-membered rings during condensation as postulated by Thurston,[23] Marvel *et al.*[14] and Wohnsiedler[29] has been challenged by Vale and Taylor.[24]

The mechanism of the methylol melamine condensation reactions has received far less attention than the methylol urea condensation, and less detailed information is available concerning the various reactions involved. It is assumed by analogy that acid promotes methylene linkages and base promotes ether linkages. Figure 5.4 shows the possible condensation reactions of trimethylol melamine. It is evident that the methylol melamine condensation products are more complex than the methylol ureas due to the increased functionality. Linear, branched and ring structures progressively develop.

Condensation which occurs immediately following the formation of the methylol derivatives gradually overcomes the tendency toward separation from solution associated with the initial reaction compounds. Thus a water-soluble, clear urea-formaldehyde or melamine-formaldehyde polymer solution is produced. Upon continued condensation, promoted by heating, the polymer exhibits a loss of solubility in water accompanied by a rapid increase in viscosity, ultimately leading to gelation.

To take advantage of the potential of aminoplasts, it is necessary to arrest condensation by cooling and rendering the solution slightly alkaline before water solubility is lost. The polymerization may be subsequently resumed at will by the application of heat and/or acid catalyst. Continued heating leads to hard, brittle, light-colored, chemical- and heat-resistant polymers. The partially self-condensed melamine polymers serve in laminating applications, i.e., overlays for counter and table tops. The use of aminoplasts as adhesives to provide wet strength to paper or to render textiles shrink proof are further examples where self-condensation is the primary reaction. A case of intermolecular cross-linking of cellulose by aminoplast can, however, be supported by experimental data.[26]

In surface coatings, when prolonged package stability is required, addi-

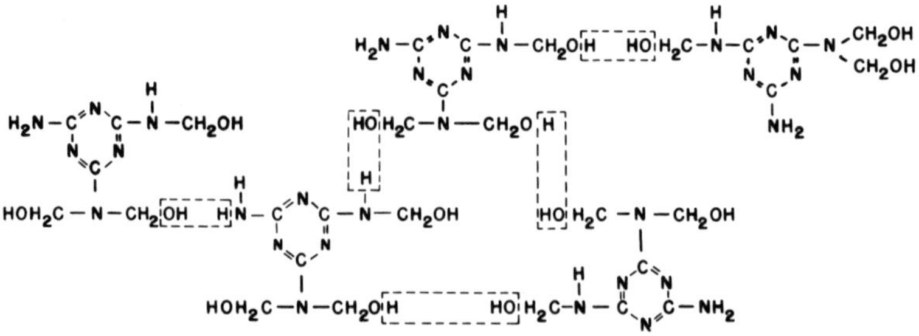

Figure 5.4. Condensation reactions of trimethylol melamine.

tional protection against the strong tendency of these polymers toward premature polymerization is necessary and may be accomplished by blocking the reactivity of the methylol groups through etherification with reactive alcohols. Such etherification promoted by the presence of acid is depicted in Figure 5.5 for *n*-butanol.

$$\underset{\substack{\text{METHYLOL} \\ \text{DERIVATIVE}}}{R-\overset{\overset{\displaystyle H}{|}}{N}-CH_2OH} \;+\; \underset{\text{n-BUTANOL}}{C_4H_9OH} \longrightarrow \underset{\text{n-BUTYL ETHER}}{R-\overset{\overset{\displaystyle H}{|}}{N}-CH_2OC_4H_9} \;+\; \underset{\text{WATER}}{H_2O}$$

Figure 5.5.   Etherification reaction.

In practice, some condensation takes place concurrently with etherification. For the solvent-soluble butylated resins, this condensation is fostered and controlled by adjusting processing conditions to develop the desired resin viscosity. Methyl ethers, on the other hand, are usually prepared under milder conditions to minimize condensation and loss of water solubility.[21]

As the number of methylol groups blocked by alkyl ether increases, the stability toward condensation also increases. The choice of alcohol further influences the resistance to condensation of the final resin. Ethers of the higher alcohols, being more stable, increase the threshold temperature at which the methylol group may be unblocked to exercise its potential reactivity. The type and amount of alcohol employed in etherification also determines the compatibility and solubility characteristics of the polymer. Thus a family of products ranging from stable, water-soluble methanol ethers to stable organo-soluble aminoplasts representing higher alkyl ethers, i.e., *n*-propyl, *n*-butyl, isobutyl, or higher, is possible. Secondary and tertiary butyl alcohols have been found insufficiently reactive to be of practical importance. To date, the butyl and isobutyl ethers have been employed most extensively in aminoplasts designed for use in conventional organic solvents.

More recently, the methyl ethers of methylol melamine, particularly those employing very high ratios of formaldehyde to melamine, have received particular attention in connection with the desire to eliminate organic solvents from industrial baking finishes. The hexamethyl ether of hexamethylol melamine, a compound first described in 1941,[9] being monomeric, offers advantages in improved compatibility, lower viscosity and greater potential for cross-linking with other polymers having complementary functionality. It is soluble in organic solvents and in water.

Such a product, being completely blocked, requires the presence of a strong acid to initiate unblocking at practical conversion temperatures.

The hexamethylol ethers, which employ somewhat less combined methanol, retain the dual solubility in organic solvents and water, and the advantages in compatibility and viscosity; yet they need no strong acid catalyst when the combined methanol approximates 3.5 moles. In addition, such less etherified products can undergo sufficient condensation to prevent the crystallization associated with the monomeric hexamethyl ether of hexamethylol melamine.

A great number of new aminoplast resins have been reported[5,10,27,28] during the past 20 years based on substituted guanamine and melamine compounds. Compounds such as benzoguanamine, N-*t*-octylmelamine and N-cyclohexylmelamine react readily with formaldehyde and can be further condensed and etherified as described above. Resins capable of imparting specific property improvements have been claimed. Except for the case of benzoguanamine, development of such resins has been retarded because of availability and price. Aminoplast ethers based upon melamine and toluenesulfonamide coreacted with formaldehyde[18] have enjoyed outstanding success in automotive topcoats due to superior gloss, leveling, flexibility and adhesion. Regardless of the amine or amide employed, the surface coatings industry confines its interest to etherified products only, because of its requirements of solubility, compatibility and stability.

COMMERCIAL PRACTICE AND COMPOSITION

An example of a butylated ether of urea-formaldehyde typical of a product suitable for use in enamels employing organic solvents might be prepared as follows: Heat 1 mole of urea with 2 to 3 moles of slightly alkaline aqueous formalin (pH 8 to 9) to form the methylol derivatives. Add 2 to 3 moles of butanol and adjust with acid to a pH of 3 to 6. Remove the bulk of the water by continuous azeotropic distillation and proceed to an atmospheric distillation for removal of the final traces of water. Solids are adjusted to 50 to 60% with butanol and xylene. High-assay formalin solution, paraform or butyl "Formcel" (40% formaldehyde, 51.5% *n*-butanol, 8.5% water) may be employed to shorten processing time since less water removal would be required.

The amount of combined formaldehyde and the relative rates of the competing etherification and polymerization reactions determine the structure and composition of the product until it is arrested from further reaction. The methods of preparation as well as the ratio of reactants offer means of control over the competing reactions which accompany water removal. In regard to the material charge, increased formaldehyde and/or butanol increases the tolerance for hydrocarbon solvent, increases the stability to gelation and decreases the molecular weight. Increased acid catalyst provides increased tolerance for hydrocarbon solvent and increased viscosity, but may favor gelation and insolubility if used to excess. Higher proportions of alcohol in the solvent portion provide greater tolerance of hydrocarbons and impart additional stability to gelation. The finished resinous solutions contain from 0.5 to 1.0 mole of combined butanol and about 2 moles of formaldehyde per mole of urea. The free formaldehyde and water content are generally less than 0.5% each.

The chemical composition and molecular weight of the solids portion of three typical butylated urea-formaldehyde resins are tabulated in Table 5.2. Resin No. 1 contains nine to ten urea units per polymer chain. Resins No. 2 and No. 3 contain approximately seven and six urea units per chain, respectively.

TABLE. 5.2.   Molar Composition—Urea Aminoplasts

| | *Mole Ratio* | | |
	Urea	*Formaldehyde*	*Butanol*	*Average Molecular Weight (mn)*
Resin No. 1	1	1.98	0.76	1415
Resin No. 2	1	2.08	0.91	1145
Resin No. 3	1	2.35	1.08	1065

An example of a melamine-formaldehyde ether typical of those employed in organic-soluble enamels follows: 1 mole of melamine, 4 to 8 moles formaldehyde (37%) and 5 to 10 moles of butanol are adjusted to a pH of 8.5 to 10.5 in a suitable reactor. After refluxing for 30 minutes to form the methylol melamine derivatives, the pH is adjusted to 4.5 to 6.5 with acid (oxalic, phosphoric, phthalic), and the solution is dehydrated by azeotropic separation. After the last traces of water are removed by distillation, the nonvolatile content is adjusted to 50 to 60% with additional butanol or xylol. Solvent solubility, resin compatibility and viscosity characteristics which reflect composition and structure are controlled by both process technology and the ratios of reactants. Increased formaldehyde and/or butanol increases tolerance for hydrocarbon solvent, increases stability to gelation, and lowers viscosity. Increased acid charge has a similar effect upon hydrocarbon tolerance to a point, but excessive amounts favor advancement in viscosity and reduced solubility. As with urea aminoplasts, more concentrated sources of formaldehyde may be employed in order to avoid an unduly long process cycle. Increased alcohol in the volatile content provides additional stability against gelation and higher tolerance for hydrocarbon solvents.

The chemical composition and average molecular weight of the solids portion of typically used butylated melamine-formaldehyde resins determined by analysis are given in Table 5.3.

TABLE 5.3. Molar Composition—Melamine Aminoplasts

	Mole Ratio			
	Melamine	*Formaldehyde*	*Butanol*	*Average Molecular Weight (mn)*
Resin No. 1	1	4.42	1.83	1650
Resin No. 2	1	4.48	2.06	2300
Resin No. 3	1	4.63	2.48	1630
Resin No. 4	1	4.77	3.36	1680

The above data indicate the chain length of the resins to be on the order of four to seven melamine units.

The commercial amino resins include a wide variety of products representing variations in the amount of combined formaldehyde, type and amount of combined and uncombined alcohol, and degree of condensation. These relationships, which are governed by the ratio of reactants, catalyst type and concentration, and process conditions (i.e., time, temperature, material balance, etc.), are reflected in the physical characteristics. The important physical characteristics thus controlled include solids, viscosity, tolerance for hydrocarbons and compatibility with adjunct film-forming components.

FUNCTIONAL USE AND MECHANISM

The value of the amino resins in surface coatings resides chiefly in their ability to react with other polymers having complimentary functional reactivity and to simultaneously undergo further self-condensation. Catalysis with small amounts of strong acid is often employed to speed these two reactions. Both reactions are essential to chemically link together the various polymeric species into a tough, hard, solvent-resistant, high molecular weight structure and to consume any excess reactive functionality from the aminoplast. Use of a stoichiometric excess of aminoplast coupled with its ability to self-condense removes the need for concern regarding excess functionality in the film. In practice, these reactions are caused to occur after the compounded finish is applied to a surface.

Industrial baking finishes must possess flexibility well beyond the capability of the self-condensed amino resins. Consequently, without exception, aminoplasts are employed along with other polymers having both complementary functionality and the capability of introducing flexibility into the cured polymer structure. The functionality found most useful in complementary adjunct prepolymers in the order of decreasing reactivity includes primary hydroxyl, secondary hydroxyl, carboxyl and amide. All have found use, but the hydroxyl has enjoyed the greatest success. The use of complementary adjunct prepolymers demands that such polymers, together with the amino resins, exhibit mutual compatibility and solubility in practical solvents. By compatibility is meant the ability to coexist in mutual solution without apparent phase separation either in the solution or in the film. This implies that both polymers be relatively low in molecular weight and possess similar polarity.

The advent of the aminoplast ethers and their use with alkyds as the complementary adjunct polymer (possessing reactive hydroxyl and carboxylic sites) exemplifies the most successful application of the aminoplasts in surface coating. Such usage initially permitted a threefold reduction in the 1-hour baking cycle associated with the drying oil alkyds of that time. Subsequent developments both in alkyds and in aminoplasts have continually shortened this time to a point where baking cycles of 30 seconds at 500° F are now employed in continuous coil coating operations with complete success. The contribution of the aminoplast resins toward meeting the demands for shorter baking cycles associated with the increased production rates essential to our high-level economy cannot be overestimated. The ability to provide rapid cure along with excellent color, gloss, hardness, durability, and chemical and solvent resistance is a unique combination whereby the aminoplasts have become established as the predominant cross-linking polymer type for thermosetting finishes.

Prior to the appearance of the aminoplast resins, thermosetting coating systems were highly restricted with regard to choice of materials. Such finishes were based either on the highly colored, brittle but chemical-resistant, alkali-catalyzed phenol-formaldehyde resins or on the more widely used drying oil alkyds or varnishes; the latter cured by the relatively slow mechanism involving polymerization via unsaturation in the fatty acid component.

It is quite possible that had there been no alkyds at the time urea and melamine resins appeared on the market, aminoplasts might never have found a place in the surface coatings industry. From their first use in 1936 and until about 1963, the role of flexibilizing adjunct co-condensate for aminoplasts in surface coatings uses was fulfilled almost exclusively by the oil-modified alkyd resins.

Alkyds designed for use with amino resins are relatively low in oil content (i.e., 35 to 45%) and deliberately possess unesterified hydroxyl groups as reactive sites to permit molecular weight growth through cross-linking to the required mechanical toughness. Such alkyds are available in a wide variety offering a choice in oil type and content, free hydroxyl, source of free hydroxyl, acidity and source of acidity, and are usually supplied at 50 to 60% solids in xylene solution. The minor amount of unesterified carboxylic acid groups present in alkyd resins serves to catalyze the co-condensation reactions occurring with aminoplasts. In addition, the carboxylic site provides additional complementary functionality.

More recently, alkyds have been displaced to a significant extent from the role of co-condensate polymer by acrylic copolymers in areas capable of tolerating added cost in order to achieve improved resistance to weathering or to chemicals. The newer acrylic copolymers represent alternate complementary adjunct co-condensates of increasing importance. They are of relatively low molecular weight and structurally are comprised of a carbon-to-carbon backbone with pendant groups attached. Variation in the pendant group is accomplished by means of vinyl monomer choice, and the selection depends upon the specific properties sought. For example, chemical resistance and hardness are contributed by phenyl groups, solubility and flexibility by carboxyl esters, and reactivity is introduced via hydroxyalkyl-, amide-, epoxide- or carboxylic-containing monomers (the latter type often uses diepoxide cross-linkers to better advantage than aminoplasts).

Other co-condensate copolymers which find use with aminoplasts include polyvinyl butyral, hydrolyzed vinyl chloride–vinyl acetate copolymer, epoxy resins and their fatty esters, nitrocellulose, etc. Useful members in this group, like the alkyds and thermosettable acrylic prepolymers, must share with aminoplasts the requisites of mutual compatibility, solubility and complementary reactivity, and they must have the ability to contribute flexibility to the final cross-linked structure.

It should be recognized that co-condensation between the complementary polymer species is only one of a number of competing reactions which occur simultaneously. The various competing reactions postulated to occur between aminoplasts and/or a hydroxyl-containing adjunct prepolymer are depicted in Figure 5.6. It is the relative rate of each of these reactions which determine the final structure in any given system and hence the performance properties.

No direct structural evidence of the presence or nature of intermolecular cross-links is available, but circumstantial evidence of its occurrence is very strong. Along this line, Wohnsiedler[29] and Seidler and Gratz[19] provide evidence supporting the case for etherification between hydroxyls

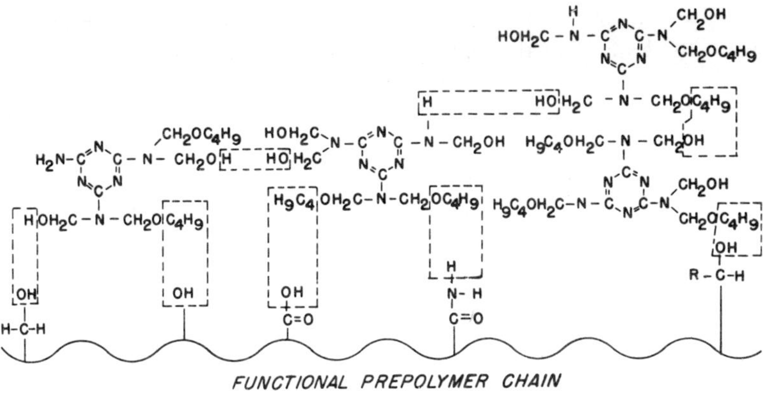

Figure 5.6. Competing condensation and co-condensation reactions.

of solvent-soluble alkyd and the methylol or butoxymethylol groups of solvent-soluble amino resins. Recent work[26] with water-soluble alkyd provides proof of co-condensation between alkyd and aminoplasts as follows:

Dimethylol cyclic ethylene urea (DMEU) and water-soluble alkyds (trimellitic anhydride, adipic acid, neopentylglycol), are each incapable of self-condensation to provide MEK-insoluble material when heated alone. However, they yield a tough film substantially insoluble in MEK after heating together 30 minutes at 250°F. As little as 10% DMEU provides 36% nonextractables. An increase in the ratio of DMEU increases the MEK nonextractable through a maximum which then decreases, presumably due to the depletion of alkyd carboxylic or hydroxyl groups essential for co-condensation. A comparison employing a conventional water-soluble methyl ether of melamine-formaldehyde showed that the latter more generously contributes to the nonextractable fraction, presumably through its capacity for self-condensation as well as cocondensation.

Similarly, hexamethoxymethylol melamine, also incapable of yielding an MEK-insoluble fraction upon heating alone, yielded 85% nonextractable material when employed at a level of 30% with a trimellitic water-soluble alkyd. Nitrogen analysis revealed that at a level of 20%, approximately 90% of it became incorporated into the MEK nonextractable fraction.

One might expect that since the hexamethyl ethers of hexamethylol melamine are blocked against self-condensation, they should possess greater potential reactivity for a subsequent cross-linking action than the less etherified amino resins in which a greater amount of activity is ordinarily consumed through prior condensation during resin manufacture. Evidence that this is the case is presented in the data and argument of Brett.[3]

Radiochemical studies[3] using C^{14}-labeled stearic acid provided strong evidence of direct reaction between carboxyl and amino resins during heating at 120°C. More recent studies[17] of the reaction between an acrylic acid copolymer and a melamine-formaldehyde condensate have confirmed the possibility of the esterification reaction. The presence of methylene ester bridges was established after heating at 150°C with or without the presence of acid catalyst.

FORMULATION

Organo-soluble thermosettable finishes employing aminoplasts conform generally to the pigment-vehicle-volatile relationships established for industrial alkyd or related finishes. Such finishes usually acquire their measure of flexibility by employing alkyds, or less frequently thermosettable acrylic prepolymers or epoxy esters, as the dominant binder component. Aromatic hydrocarbons, usually xylene, serve as the predominant volatile component but some minimum amount of active solvent, usually butanol, is ordinarily present to impart the needed package viscosity stability. The amount of butanol supplied with the melamine resin is often sufficient to impart the needed package stability, but where necessary this is supplemented with an additional amount of active solvent.

While urea and melamine ether resins are each employed to cross-link alkyds, thermosettable acrylics or other functional prepolymers, they differ in their relative influence upon individual film properties while performing this function. Table 5.4 presents two groups of film properties,

TABLE 5.4.   Relative Performance Advantages of Melamine versus Urea Aminoplasts in Enamels

Property Advantages of Melamine	*Property Advantages of Urea*
Gloss	
Gloss retention (UV or heat exposure)	
Color retention (heat exposure)	
Water resistance	
Cure response (uncatalyzed)	Low-temperature catalyzed cure response
Detergent and chemical resistance	
Solvent and stain resistance	Strippability, intercoat adhesion
Exterior durability	Cost
Hardness	Adhesion and flexibility

one favored by the choice of melamine and one by urea-based aminoplasts.

Although benzoguanamine aminoplasts surpass melamine aminoplasts in imparting flexibility and chemical resistance and have a slight advantage in heat resistance, they are severely deficient in exterior durability. This deficiency, together with higher cost, has restricted their commercial use to a minor role in total aminoplast usage.

Beyond any general classification, the members within each class represent alternate compromises of the relationship between the properties listed. These more subtle differences justify the large number of urea and melamine resins found on pp. 370A to 374 of the Resin Section of the *Raw Materials Index* issued by the National Paint, Varnish, and Lacquer Association. Each member product might be regarded as a tool designed for the variety of specific formulation needs[15,16] discussed below.

Industrial finishes may contain as little as 5% or as much as 30% melamine or even 50% urea resin. The lower amounts of aminoplast would be encountered in conjunction with drying oil alkyds or epoxy esters as a means of providing supplementary cure or chemical resistance. Primer surfacers currently serving under baked enamel automotive top coats exemplify the use of small amounts of aminoplast in conjunction with drying oil epoxy esters as in Table 5.5.

TABLE 5.5.   Automotive Primer Surfacer

10.00	Iron oxide
5.50	Lithopone
11.00	Barytes
9.10	Calcium silicate
16.15	40% linseed or tall oil epoxy ester
1.25	Butylated urea-formaldehyde (solids)
47.00	Volatile solvent[a] (xylene, butanol, aliphatic hydrocarbon)
100.00	

[a] Includes solvent from resin solutions.

More frequently, intermediate levels (approximately 20%) of aminoplast have been employed in conjunction with "short" drying oil alkyds to yield films capable of excellent cure, adhesion and toughness. During the period from 1947 to 1955 automotive topcoat baking enamels underwent an increase in melamine content from 7 to 20%, while the companion drying oil alkyd simultaneously underwent a shift toward the less conjugated and better color-retentive drying oil alkyds. Such a topcoat is shown in Table 5.6. Replacement of the melamine resin with urea resin in this same formula would yield an enamel having performance and economics well suited for a number of general industrial nonexterior metal finishing applications.

TABLE 5.6.   Grey Gloss Enamel

20.00	Rutile titanium dioxide
0.10	Lampblack
24.00	41% soya alkyd
6.00	Butylated toluenesulfonamid-modified melamine-formaldehyde (solids)
49.90	Volatile[a] (xylene and butanol)
100.00	

[a] Includes solvent from resin solids.

Washing machine primers require high corrosion and detergent resistance combined with moderate flexibility and good adhesion and intercoat adhesion. Since color is not of prime importance, this use has been served well by dehydrated castor esters of epoxy resins in conjunction with benzoguanamine-based aminoplasts at a level of 30 to 40%. More recently, styrene–allyl alcohol copolymer modified alkyds have begun to replace the epoxy ester in such use.

Applications requiring excellent color and gloss as well as chemical resistance must necessarily avoid drying oil alkyds. Top coats for the major appliances fall into this category. These have been served well by high levels of melamine or benzoguanamine aminoplast in conjunction with nondrying alkyds or, more recently, in conjunction with styrene–acrylic copolymers having the chemical functionality essential for cross-linking. Such systems in which the aminoplast cross-linking is the sole mechanism of molecular growth demand high levels of aminoplast for effective cure. A typical formula is shown in Table 5.7.

TABLE 5.7.   White Appliance Enamel

25.00	Rutile titanium dioxide
20.00	33% coconut alkyd (solids)
10.00	Butylated melamine or benzoguanamine formaldehyde (solids)
45.00	Volatile[a] (xylene, butanol)
100.00	

[a] Includes volatile from resin solutions.

After 1955, automotive top coat baking enamels underwent a major change. Drying oil alkyds were replaced with nondrying alkyds, and the melamine level was increased to approximately 30%. This resulted in improved color and gloss retention made necessary by the emphasis on lighter colors and highly metallic finishes. The nondrying alkyd has been replaced by thermosettable acrylic prepolymers which has resulted in still further gains in color and gloss retention upon exterior exposure. Unfortunately, departure from drying oil alkyds is accompanied by a loss in adhesion and flexibility, and a good deal of attention is being devoted to improvement of these properties. A typical metallic automotive top coat is shown in Table 5.8.

TABLE 5.8. Grey Metallic Automotive Tint Base

26.00	Hydroxy functional acrylic prepolymer
11.80	Butylated melamine-formaldehyde aminoplast (solids)
1.20	Nonleafing aluminum[a]
61.00	Volatile[b] (butanol, xylene, etc.)
100.00	

[a] Introduced as paste.
[b] Includes solvents from resin solutions.

Urea aminoplasts have enjoyed singular success over the others in the field of force-dried catalyzed wood finishes. Bakes varying from 5 minutes to 1 hour at temperatures as low as 120 to 160°F are frequently encountered; for this purpose, acid-catalyzed urea in conjunction with drying oil alkyd has had particular success. Such clear wood finishes involve the addition of a catalyst just prior to use in order to circumvent inadequate package stability which would be associated with earlier inclusion of acid catalyst. A formulation illustrating a clear catalyzed wood finish is shown in Table 5.9.

TABLE 5.9. Catalyzed Wood Finish

20.00	37% soya or dehydrated castor alkyd
20.00	Butylated urea-formaldehyde
59.00	Volatile[a] (xylene, butanol)
1.00	*para*-Toluenesulfonic acid[b]
100.00	

[a] Includes solvent from resin solutions.
[b] Introduced as 10% solution in xylene just before use.

Although the desirability of applying high-quality baking enamels from aqueous dispersions has been widely recognized, until recently the accomplishment of this feat remained beyond reach. Continued progress in both the water-soluble flexibilizing components and water-soluble aminoplast has made this desire a reality, at least in limited uses.[25] Rule 66 describes pending legislation in the County of Los Angeles aimed at alleviation of the smog problem, and this has motivated a much more diligent effort by the industry to develop practical aqueous enamel systems. Along this line, the recent introduction of methyl ethers of highly methylolated melamine has provided excellent tools for formulation of aqueous dispersed baking enamels. A typical formulation of such a system is shown in Table 5.10.

TABLE 5.10.   Aqueous Soluble Baking Enamel

25.00	Rutile titanium dioxide
20.00	Water-soluble nondrying alkyd
10.00	Partially methylated ether of highly methylolated melamine
45.00	Volatile (water and any co-solvents associated with resin solutions)
100.00	

In working with aminoplast ethers, the chemist should recognize that in any commercial product the combined and uncombined alcohol are in equilibrium.[22] Since all the practical solids tests involve heating, they shift the equilibrium in the direction of de-etherification, and hence report lower solids than those which actually exist. Consequently, the reported solids are arbitrary low values associated with a particular method of test specified by the supplier. ASTM method D-1259-61 B has been adopted in an attempt to standardize the method of determining solids. It should be noted that the analytical techniques associated with the data presented in Tables 5.2 and 5.3 were designed to be especially mild to avoid a shift in equilibrium and hence present a more realistic picture of the composition of these aminoplasts.

One problem occasionally encountered in the use of aminoplast enamels is that of surface distortion. This may vary from the extreme of large wrinkles down to a surface distortion so fine that it is often not recognized as a wrinkle but merely as low gloss. Such surface distortion results from the formation of a surface skin which becomes swollen from contact with the trapped underlying solvent. This swollen skin ultimately folds upon itself to conform to the area boundaries of the substrate during subsequent contraction of the film during curing. Other systems conducive to reaching the critical skin-solvent relationship, e.g., raw china wood oil, have the same problems.

The correction for this situation involves avoidance of the critical combination of skin and underlying solvent, usually by delaying the skin formation until more solvent has evaporated. In the case of aminoplast enamels, longer air dry times, lower oven entry temperatures, thinner films, avoidance of direct air impingment and avoidance of any acid vapor contact of the film either before or during the early baking stage are helpful. Another particularly effective technique to overcome the finer type of surface distortion is to include a very small amount of additive to reduce surface tension.

Aminoplasts, despite their relatively high cost, have become firmly established as the most important cross-linking polymers in the industry. Well over 50% of all industrial baking finishes employ aminoplasts to effect the cross-linking essential to achieve the required mechanical properties. Diepoxides and polyisocyanates, although important in specific areas, do not constitute a serious threat to the position enjoyed by aminoplasts.

REFERENCES

1. Anony., U. S. Tariff Commission, S.O.C. Series P-64-12 (1965).
2. Auten, R. W., in (Von Fischer, W., editor) "Paint and Varnish Technology," p. 147, New York, Reinhold Publishing Corp., 1948.
3. Brett, R. A., *J. Oil Colour Chemists' Assoc.*, **47**, 779, 787, 981 (1964).
4. Crowe, G. A., and Lynch, C. L., *J. Am. Chem. Soc.*, **70**, 3796 (1948); **71**, 3791 (1949); **72**, 3622 (1950).
5. Culbertson, H. M., and Hahn, F. J. (to Monsanto Company), U.S. Patent 2,959,558 (1960).
6. DeJong, and Dejonge, J., *Rec. Trav. Chim.*, **71**, 643,661,890 (1952).
7. DeJong, and Dejonge, J., *Rec. Trav. Chim.*, **72**, 139,202,213 (1953).
8. Einhorn, A., and Hamburger, A., *Ber. Dtsch. Chem. Ges.*, **41**, 24 (1908).
9. Gams, A., Widmer, G., and Fisch, W., *Brit. Plastics*, **14**, 508 (1943).
10. Gussman, S., and Rytina, A. W. (to Rohm & Haas Co.), U.S. Patent 2,951,048 (1960).
11. Hodgins, T. S., and Hovey, A. G., *Ind. Eng. Chem.*, **33**, 769 (1941).
12. Iliceto, A., *Ann. Chim. Rome*, **43**, 625 (1953).
13. Marvel, C. S., Elliot, J. R., Boettner, F. E., Yuska, H. J., *J. Am. Chem. Soc.*, **68**, 1681 (1946).
14. Marvel, C. S., and Horning, E. C., in (Gilman, H., editor) "Organic Chemistry," Vol. 1, pp. 701–778, 1943.
15. Parker, C. H., and Hahn, F. J., in (Von Fischer and Bobalek, editors) "Organic Protective Coatings," p. 215, New York, Reinhold Publishing Corp., 1953.
16. Payne, H. F., *Organic Coating Technology*, **1**, 342 (1954).
17. Saxon, R., and Daniel, J. H., *American Chemical Society Washington Meeting*, **1**, 277 (1962).
18. Scott, M. J., and Hahn, F. J. (to Monsanto Company), U.S. Patent 2,508,876 (1950).
19. Seidler, R., and Gratz, H. J., *FATIPEC*, **VI**, 282 (1962).
20. Smythe, L. E., *J. Phy. Sol. Chem.*, **51**, 369 (1947).
21. Swen, T. J. (to American Cyanamid Co.), U.S. Patent 2,529,856 (1950).
22. Touchin, H. R., *J. Oil Colour Chemists' Assoc.*, **39**, 653 (1956).
23. Thurston, J. T., Gibson Island Conference (1941).
24. Vale, C. P., and Taylor, W. G. K., "Aminoplasts," London, ILIFFE Books, Ltd., 1934.

25. Vale, C. P., *Brit. Plastics*, **38**, 492 (1965).
26. Valentine, L., *J. Oil Colour Chemists' Assoc.*, **47**, 981–989 (1964).
27. Widmer, G., and Fisch, W. (to Ciba Products Corp.) U.S. Patent 2,197,357 (1940).
28. Williams, B. L., and Culbertson, H. M. (to Monsanto Company), U.S. Patent 2,914,508 (1959); 2,957,835 (1960); 2,957,836 (1960); 2,980,636 (1961).
29. Wohnsiedler, H. P., *American Chemical Society, New York Meeting*, **20**, 53 (1960).
30. Ziguener, G., *Monatsh*, **82**, 175 (1951); **83**, 1099 (1952).

6

*Phenolic Resins**

Phenolic resins have been used in the coatings industry since about 1910. With the exception of "Celluloid" (nitrocellulose plasticized with camphor) and ester gum, they are the oldest synthetic resins used in this industry. Basically they are obtained by the reaction of phenol or a substituted phenol with an aldehyde, in almost all instances formaldehyde. In their final form, they may be viscous liquids or brittle solids. Depending upon the ingredients used, their proportions, the catalysts employed and the reaction conditions, they may or may not be heat reactive.** As the terms imply, non-heat-reactive resins do not change appreciably when heated above their melting points and held at elevated temperatures. On the other hand, heat-reactive resins increase in viscosity and molecular weight under the influence of heat and will crosslink, if sufficient functionality is present, to infusible, insoluble gels.

The fact that phenol could be reacted with formaldehyde was known before 1880. The reaction had been studied by Baeyer as early as 1872. However, work in this area was discontinued because products of known composition could not be obtained. At that time, the emphasis in organic chemistry was on the preparation of compounds which could be purified by crystallization. In the early 1900's, Baekeland decided to reinvestigate the reaction of phenol with formaldehyde with the intention of developing products of industrial importance. His initial work was directed to the preparation of molding materials. It was not long, however, before he started to use solutions of the resins made for molding purposes as coatings. Modified formulas made specifically for use as baked coatings on metal quickly followed.

Since then, phenolic resins have been in continuous use in the coatings industry. Coatings represent one specialized application of phenolic resin

*By S. H. Richardson and K. V. McCullough, Plastics Division, Union Carbide Corp.

**Non-heat-reactive resins are also called novolacs or two-step resins. Heat-reactive resins are known as resoles and/or one-step resins.

technology. These resins are currently used in molding materials, industrial and decorative laminates, for rubber modification (tackifiers, vulcanizing, etc.) as binders for thermal insulation, grinding wheels and brake lining, and in adhesives for making plywood, bonding metals, and other materials. A voluminous literature has been built up in all these fields, and reference should be made to it for a fuller understanding of the technology and potential for these resins. The following data from the 1964 U.S. Tariff Commission report, "Synthetic Organic Chemicals" shows the relation between total phenolic resin production and the amount consumed by the coatings industry.

	Pounds
Total production of phenolic resins	832,500,000
Production of unmodified and modified (except by rosin) coating resins	32,300,000
Rosin esters modified by phenolic and other tar acids (hard resins)[a]	26,378,000

[a] From the 1961 U.S. Tariff Commission report. In following years, this information was not presented separately.

The annual production of phenolic coating resins for the past ten years has remained essentially level between 27 and 32 million pounds annually.

Figure 6.1.   Phenolic coated ducts. These exhaust fume ducts are being coated with "Bakelite" phenolic resin baking finishes. Resistance to most solvents and chemicals at elevated processing temperatures is an inherent property of this type resin. Other uses include linings for lard pails, tanks, pails, drums and coatings for pipe.

Figure 6.2. Phenolic maintenance coatings. This bridge has been painted only four times in the past 20 years. Phenolic finishes resist weather extremes and salt atmosphere.

During this period, there has been a considerable change in the end use pattern of these resins. Old uses have been decreasing, while new uses have developed. Competition has also been intense from new types of polymers entering the coatings market. Despite all these pressures, there is current expectation that the volume of phenolic resins will grow moderately as new products come along.

The uses of phenolic resins in the coatings industry are quite diverse. As a consequence, it is difficult to present an easily understandable picture of the field. Table 6.1 has been prepared in an attempt to accomplish this.

The chemistry of phenolic resins is complex and not fully understood. However a great deal has been learned about it.[1] Phenol will react with formaldehyde under acid or alkaline conditions. In the presence of an alkaline catalyst, the reaction may be portrayed as follows:

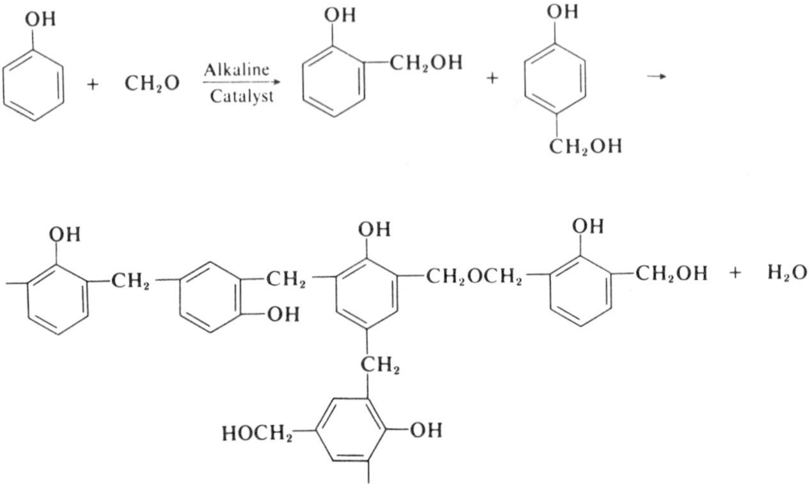

$$\text{phenol} + CH_2O \xrightarrow[\text{Catalyst}]{\text{Alkaline}} o\text{-methylolphenol} + p\text{-methylolphenol} \rightarrow$$

(1)

$$\text{(cross-linked phenol-formaldehyde structure)} + H_2O$$

Insofar as reaction with formaldehyde is concerned phenol has a functionality of 3. Formaldehyde has a functionality of 2. The reactive sites on phenol are the two *ortho* positions and the *para* position on the ring. The above sketch is intended to portray a three-dimensional structure with the phenol molecules linked together at both the *ortho* and *para* positions. The —CH$_2$OH groups at points along the chain are capable of reaction with the available *ortho* and *para* hydrogens on adjacent phenolic rings, with the elimination of water. Reaction can also occur with another methylol group, with the formation of a methylene ether. Upon further heating, this may give off formaldehyde and leave a —CH$_2$— bond between adjacent rings. It is believed that phenol alcohols as shown in Equation (1) are formed under both acid and alkaline conditions. However, under acid conditions, the phenol alcohol reacts so quickly with another phenol that it cannot be isolated.

The reaction of phenol and formaldehyde under acid conditions may be illustrated as follows.

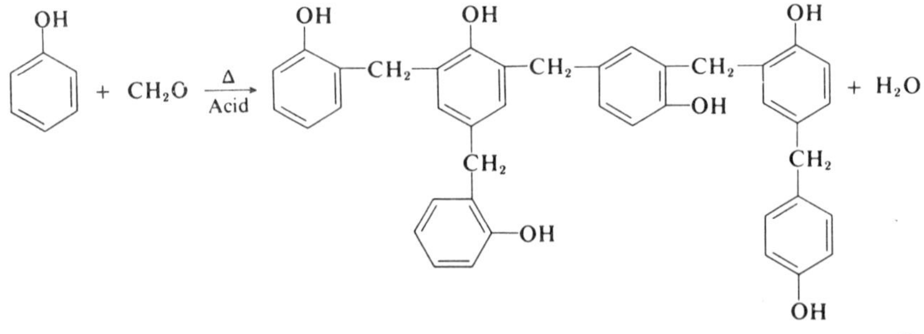

$$\text{phenol} + CH_2O \xrightarrow[\text{Acid}]{\Delta} \text{(novolak structure)} + H_2O$$

(2)

TABLE 6.1.   Phenolic Resins

Type	*Non-oil-soluble Resins*		*Oil-soluble Resins*		
	Heat-reactive	*Non-heat-reactive*	*Modified Phenolic*	*Heat-reactive*	*Non-heat-reactive*
Type	Phenol Cresol Xylenols	Phenol Cresol Xylenols	Phenol Substituted phenols Bisphenol A [2,2-bis(4-hydroxy phenyl)propane]	Substituted phenols Bisphenol A	Substituted phenols Bisphenol A
Soluble in	Alcohols Ketones Esters Glycol ethers	Alcohols Ketones Esters Glycol ethers	Vegetable oils Aromatic hydrocarbons	Vegetable oils Aromatic hydrocarbons Ketones	Vegetable oils Aromatic hydrocarbons Ketones
Modifier	—	—	Ester gum rosin	—	—
Film former as is	Yes	No	No	No	No
Used in	Can coatings, pail and drum liners, chemically resistant coatings for pipes and tanks, military hardware. Can be combined with urea, melamine and epoxy resins.	In combination with hexamethylenetetramine as adhesion promoters for vinyl plastisols.	Low-cost trade sales varnishes and printing inks.	In combination with blown oils and alkyds as coil impregnating varnishes. To modify rosin and ester gum. To make modified phenolic resin varnishes. In combination with neoprene to make chemically resistant coatings and adhesives.	In combination with vegetable oils to make spar varnishes, primers, aluminum paints and food can coatings. To modify alkyds.
Notes	Must be baked at 275–400°F to obtain optimum properties. Cured films are glass hard and low in distensibility. Have excellent resistance to solvents, water and heat. Are poor in alkali resistance. May be reacted with epoxies and polyamides to increase flexibility.	Must be baked at least 10 min at 350°F in a plastisol formulation.	Varnishes have fair to poor durability on exterior exposure.	Not recommended for use on exterior exposure.	Can be formulated to obtain excellent durability on exterior exposure.

Again a three-dimensional structure can be formed but there are no methylol groups and no methylene ethers present.

If a substituted phenol is reacted with formaldehyde, a simpler structure will usually result. Most synthetic substituted phenols have the added groups *ortho* or *para* to the hydroxyl. In effect, this blocks one of the positions available for reaction with formaldehyde. As a result, only linear chain growth is possible. Thus for an alkaline catalysed resin:

$$(3)$$

For an acid-catalyzed phenolic resin,

$$(4)$$

There are many substituted phenols available and many more could be made if the need were established.

Forty-four different substituted phenols were tested for resin making properties and performance in oleoresinous varnishes by Turkington and Allen.[2] This work clearly showed that practically all the *ortho* compounds gave resins of poor color stability. The *para*-substituted phenols were much better in this respect and the *meta* compounds were intermediate.

Despite the large number of substituted phenols that might be used in making oil-soluble phenolic resins, relatively few are of commercial importance. These are:

Compound	*Formula*	*Molecular Weight*	*Freezing Point (°C)*
p-Cresol		108	33

Compound	Formula	Molecular Weight	Freezing Point (°C)
p-Phenylphenol		170	166
p-tert-Butylphenol		150	97
p-tert-Amylphenol		164	90
p-Octylphenol		206	75–80
Bisphenol A [2,2-bis(4-hydroxy-phenyl)propane]		228	152

As will be noted, all are *para* substituted.

The heat-reactive resins made from phenol and cresol have been in use for the protection of metal surfaces for more than 50 years. During this time, numerous minor modifications have been made, but the performance expected from this class of material is well defined. In general, these resins are sold as solutions in alcohol and are used as baking coatings. However, some lump solids are also available. Typical cures vary from 1 hour at 275°F to 15 minutes at 400°F. The cured resin is almost as hard as glass and is unaffected by prolonged immersion in most organic solvents, water or dilute mineral acids. Acetone is a very strong solvent for many coatings. A properly baked phenolic coating will be completely unaffected by this solvent and can be used to line containers in which it is stored. The alkali resistance of these resin films is generally poor, and signs of failure in a 5% caustic solution at room temperature can be expected in 5 to 15 minutes. The baked film is very tightly cross-linked and has a distensibility generally of less than 1%. Because of the cross-linking that occurs during baking, these resins have very limited compatibility with plasticizing materials.

An example of a typical phenolic resin solution is the following quoted from U.S. Patent No. 2,253,235.[3]

"A pure phenol-formaldehyde resin is prepared by incorporating with 100 parts of phenol, 150 parts of 40% by volume formaldehyde and 1 part of ammonium hydroxide (26° Bé). The ammonium hydroxide acts as a catalyst and any other suitable catalyst may be employed in its place. These ingredients are boiled in a reflux condenser until resin precipitates and the condensation is then continued for $\frac{1}{2}$ hour more. The water present is then removed by vacuum distillation until the temperature of the resin reaches 108 to 115°C and a sample remains clear and is without stickiness under cold water. The resultant resin is then dissolved in ethyl alcohol to form an approximately 85% solution of the resin."

While it was earlier stated that there are not many plasticizers for the baking phenolic solution coatings, there are a few. Among these are polyvinyl butyral, polyvinyl formal, a few carefully formulated alkyds, epoxy resins, phenoxy resins (bisphenol A, epichlorohydrin polymer) and polyamides. These are not universally compatible, but are all compatible in specific instances. The best compatibility is obtained when a reaction occurs between the phenolic resin and the plasticizer. In general, it may be said that flexibility is obtained at the expense of solvent resistance. However, adequate solvent resistance for all but the most difficult exposures usually remains.

Three types of reaction on baking can be postulated for these plasticizers. These are:

(1) Etherification of those materials containing aliphatic hydroxyl groups by reaction with the methylol groups on the phenolic resin. This

is facilitated by the presence of a catalyst such as phosphoric acid. Thus

$$R_1OH + R_2CH_2OH \rightarrow R_1OCH_2R_2 + H_2O$$

Examples of R_1 are polyvinyl butyral, polyvinyl formal and phenoxy resins and R_2 is a heat-reactive phenolic resin. If an occasional ether link can be formed between these materials, compatibility is helped.

(2) Reaction between phenolic hydroxyls and epoxy groups. This reaction is rapid and more certainly established by experimental evidence than the preceding one. Trace amounts of alkaline catalysts such as lithium naphthenate and α-methylbenzyldimethylamine are beneficial. Again an ether is formed, but at a different site on the phenolic resin:

Transferring the phenolic hydroxyl to an ether greatly helps the alkali resistance of the baked coating, improves its color and helps its impact resistance.

(3) Reactive polyamides, such as those made from C_{36} dimer acids derived from 2 moles of unsaturated vegetable acids and reacted with polyamines, can be chemically combined with phenolic resins to obtain films having excellent properties. The amine is its own catalyst in this reaction as follows:

The proportions of polyamide and phenolic resin to combine are best determined by test rather than by stoichiometric considerations. U.S. Patent 2,891,023[4] describes such coatings. Again improvement is obtained in flexibility and alkali resistance.

Useful coatings can also be obtained by combining the baking-type phenol-formaldehyde resins with urea-formaldehyde and/or melamine-formaldehyde resins. While many different proportions of the two types of resin may be blended for specific purposes, the most common is a 1:1 mixture on a solids basis. Such combinations are frequently used to protect polished metal household objects and costume jewelry from tarnishing. The phenolic provides water and perspiration resistance, and the urea or melamine resin reduces the color of the phenolic to a scarcely noticeable level. Again, judicious selection must be made to obtain compatibility. Phosphoric acid is frequently used in small amounts to catalyze copolymerization, and minimal bakes are used to obtain the lightest color possible. A typical baking schedule is 30 minutes at 250°F.

Two other materials should be considered under the category of baking phenolics—trimethylol phenol and its allyl ether. These are essentially monomers and are usually combined with other materials to obtain useful films. Their formulas are:

Trimethylol Phenol *Allyl Ether of Trimethylol Phenol*

The preparation of these materials is described in several U.S. Patents.[5-7] Briefly they are made by the reaction of phenol at a relatively low temperature (45°C) with a large excess of formaldehyde in the presence of a stoichiometric or greater amount of sodium, barium or calcium hydroxide. The alkaline salt may be acidified to give trimethylol phenol, or it may be treated with an alkyl chloride such as allyl chloride to yield the ether of trimethylol phenol. Trimethylol phenol is available as a 70% nonvolatile aqueous solution and finds use principally in cross-linking hydroxyl-containing polymers such as starch, dextrin and polyvinyl alcohol.

The allyl ether is available as the allyl ethers of mono-, di- and trimethylol phenols. The cured material has excellent resistance to alkalies, acids, detergents and solvents. It is most widely used in combination with epoxy resins and polyvinyl acetals. The use of a small amount of phosphoric acid speeds the cure of such mixtures.

Many of the uses listed for the baking phenolics in Table 6.1 require pigmentation. Most commonly used are iron oxide and chromium oxide pigments. These reinforce the film, improve its resistance to many corrosive aqueous solutions, and make possible the application of somewhat thicker coatings than can be used with the unpigmented resins. Hard settling of these pigments is a frequent problem which can be overcome by the use of suspending agents. Among the most effective are some of the chemically modified bentonite clays. Carbon black and titanium dioxide are also used with these resins and generally present no problems with respect to pigment settling. A baked film containing titanium dioxide will vary in color from medium yellow to dark brown. Reactive pigments such as zinc oxide should be used with care. A few resins can be found which will not gel with this pigment, but the acidic nature of this class of resins in general is such that reaction can be expected in most cases.

Reference was made above to coating thickness. The problem here is basically removal of the volatile solvents present and the volatile products formed during the baking operation (formaldehyde and water) which result from the further condensation of the resin. Since these latter products are liberated only after most of the solvent has been removed and the film has started to harden, there is a tendency for the gaseous water and formaldehyde to cause blistering. In clear films, a dry thickness of 0.5 to 0.7 mil is frequently the maximum possible without blistering. There are many applications which do not require this thickness, e.g., coatings for tin cans and hardware. Anything which reduces the volatiles liberated during baking helps to minimize blistering. Among these may be listed: (1) reducing the methylol groups present in the resin so that the volatile products formed during condensation will be lessened, (2) using epoxy resins with the phenolic which react without generating gaseous products and (3) pigmentation. The latter is helpful in reducing blistering by facilitating the escape of volatiles or by absorbing them on their surface. By these means satisfactory films of $1\frac{1}{2}$ to 2 mils may be obtained.

Table 6-1 shows the use of novolacs, (non-heat-reactive, non-oil-soluble phenolic resins) as a coating ingredient. This is essentially a new use and is based on the fact that these resins when mixed with 6 to 10% of hexamethylenetetramine aid the adhesion of plastisols to metal when added in amounts up to 25% based on the weight of vinyl resin present. The resulting product shows excellent adhesion to steel, but is markedly yellow in color and likely to be somewhat porous due to decomposition of some of the hexamethylenetetramine to formaldehyde and ammonia, and consequent bubble formation. The principal use of plastisols formulated in this manner is for gasketing and caulking compounds which are cured by baking for about 10 minutes at 350°F.

At the time of the first commercial use of phenolformaldehyde resins in coatings there were few resins available that were soluble in vegetable oils. The only products of this class were rosin, ester gum and natural resins which had been made oil-soluble by heat treatment (i.e., kauri, Congo, etc.). Naturally with the advent of a new resin it was tested for solubility in vegetable oils. Resins made from phenol were found to be insoluble, but a resin made from *o*-cresol by Aylsworth described in a patent[8] filed June 12, 1912 was oil soluble. Actually this resin was of little practical value because it gave products which dried slowly and were quite dark in color, but it is of historical interest because of its early date.

Other methods were sought to achieve solubility. One successful way was the combination of a relatively small amount of phenolic resin with a large amount of rosin and subsequent esterification of the rosin acids with glycerin. This esterification gave a stable resin with reactive pigments such as zinc oxide and one of better water resistance than previously possible.

In 1912 a patent was also filed by Beatty[9] covering the preparation of bisphenol A [2, 2-bis(4-hydroxyphenyl)propane]. This was followed by a patent[10] granted in the United States in 1927, after filing in the United States in 1924 and in Germany in 1917, on the use of bisphenol A formaldehyde condensates modified with rosin and esterified with glycerin. This represented a large step forward because this bisphenol, with its *para* positions blocked, gives resins with good color and color retention. Essentially this is the way in which most modified phenolic resins are made to this day.

The phenolic resin content of modified resins usually lies between 8 and 12% by weight. The ball and ring softening points are much higher than ester gum and in some cases reach 320°F. Compared with ester gum, the modified phenolics give varnishes which dry faster, are more water and alkali resistant, and are more durable on exterior exposure.

Another approach to oil solubility is the reaction of phenol with tung oil and then formaldehyde. Products of this type were patented by Byck[11] in 1926. However their usefulness was limited since the only satisfactory oil was tung oil and the resulting product had to be baked.

More generally useful resins were obtained by reacting phenol with dipentene or α-pinene.

Resins of this type find use primarily in adhesives but to some extent are used in printing inks and coatings. Several of the terpenes may be reacted with phenol in the presence of BF_3 or zinc chloride, making in effect substituted phenols. It is possible to react more than one terpene with a single phenol molecule and also to polymerize the terpene with itself. Small amounts of formaldehyde may also be used to combine two or more of these highly substituted phenols to yield resins. A number of patents exist in this area, the most generally useful being one by Powers.[12] Terpene phenolic resins have good color, and many of them dry well in linseed and dehydrated castor oil varnishes. However, they are not very durable on exterior exposure and are not very effective in "gas-proofing"* tung oil varnishes.

The advent of the substituted phenolic resins, beginning about 1929,

---

*Raw tung oil dries with a flat matte finish crisscrossed with lines and cracks. Heating the oil minimizes this tendency. However, the products of combustion from an ordinary gas flame causes this flatting and wrinkling to occur in many tung oil varnishes. Acid-catalyzed substituted phenolic resins are quite effective in preventing this phenomenon, but phenol terpene resins are not. Neither are ester gums and most modified phenolic resins.

was another step forward in coatings technology. The first synthetic substituted phenol resin to achieve commercial importance was based on *p*-phenylphenol.[13] This was followed by *p-tert*-amylphenol and *p-tert*-butylphenol in the early 1930's.[14,15] Both acid- and alkali-catalyzed resins were described. A typical formula for the preparation of an acid-catalyzed butylphenol resin is as follows:

p-tert-Butylphenol	100 parts
37% formaldehyde solution	81 parts
5% HCl solution	1 part

These ingredients are refluxed under atmospheric conditions for 1 hour and then dehydrated by heating to 220°C and holding at this temperature for 30 minutes. The resin is then discharged into pans and allowed to cool. The finished product has a softening point of about 115°C (ball and ring method) and a weight average molecular weight of about 600.

A heat-reactive resin (alkali catalyzed) may be prepared as follows:

p-tert-Butylphenol	100 parts
37% formaldehyde solution	85 parts
10% caustic soda solution	193 parts
50% sulfuric acid	60 parts

The butylphenol is dissolved in the caustic soda solution by heating to 90°C. The solution is cooled to about 60°C and the formaldehyde slowly added. The mixture is heated to 65°C and held for 3 hours. It is then neutralized with the sulfuric acid solution and allowed to settle for 3 hours. The aqueous layer is removed, and the resin is washed three times with 150 parts of water. The resin is then vacuum dehydrated to 140°C and held at this temperature until a ball and ring softening point of 50°C is obtained. The resin is then discharged into pans and cooled.

Both types of resin are soluble in vegetable oils in all proportions and may be used to make varnishes of many types. Experience has shown, however, that the acid-catalyzed resins are best suited for use in spar varnishes, maintenance primers and aluminum paints. The heat-reactive resins find use in coil impregnating varnishes, in short oil baking varnishes and in combination with various synthetic rubbers.

The acid-catalyzed phenolic resins, when combined with vegetable oils, are among the most durable coatings for use clear (unpigmented) over wood on exterior exposure. These resins are much more satisfactory for use on exterior exposure than are alkali-catalyzed resins. The best varnishes can be expected to show little or no failure after 18 months when exposed on maple or mahogany panels inclined 45° to the horizontal and facing south at Miami, Florida. The ideal oil length is 25 to 33 gallons. The preferred oil is tung oil, and the preferred resin is based on *p*-phenylphenol. Excellent results can also be obtained with some butylphenol resins. The addition of a UV absorber equivalent to 4 to 5% of the total solids will definitely improve the life of phenolic varnishes on outdoor exposure.

In general, the best 100% phenolic varnishes are superior to alkyds, oil-extended urethanes and most two-package or moisture-cured urethanes on exterior exposure. A few reactive urethane systems have been tested which show better durability, but they present problems with adhesion when overcoating becomes necessary. Phenolic coatings have suffered from competition from newer materials in recent years. However, they have survived these attacks and are regaining some of their former position.

One of the limitations on the oil-soluble resins has been their dependence upon tung oil. This oil is ideally suited to varnish use. However, over the past 20 years its price has fluctuated widely, and there have been times when it was in short supply. Every adverse turn in the history of tung oil has resulted in substitution of alternate materials, with the result that phenolic varnishes have also been replaced. Satisfactory varnishes can also be made from linseed, dehydrated castor and oiticica oils. However, phenolic resins cannot be used successfully with soya, safflower and tall oil. This has been an added handicap.

Recently a technique has been developed which permits the combination of a phenolic resin with slow-drying oils in such a manner that excellent color is maintained, satisfactory viscosity is developed and good drying speed is obtained. Briefly, a diglyceride of a vegetable oil is made. This is reacted with 1 mole of tolylenediisocyanate (TDI), leaving one isocyanate group available for further reaction. This is combined with the hydroxyl group on an oil-soluble phenolic resin. Every phenolic hydroxyl in the resin is thus combined with a vegetable oil diglyceride by means of TDI. Obviously other polyols and different fatty acids can be used. This method permits the use of the slow-drying oils heretofore unsatisfactory for varnish use. The resulting varnish has better exterior durability than most alkyds and almost as good as the best phenolic–tung oil coatings. A varnish of this type does not show the brown discoloration characteristic of TDI systems when exposed out of doors.

As can be seen from the preceding material, technical progress is still being made in the use of phenolic resins in coatings. Much more progress can be made in the application of modern instrumental analytical methods. Surprisingly little is known with certainty as to the amount of cure and the type of cross-linking that occurs when a phenolic is baked, particularly if a second compound is present. A number of reactions can be postulated for what happens when a phenolic resin is "cooked" with a vegetable oil. High temperatures and relatively long periods of time are involved in this operation (450 to 585°F, 1 to 4 hours). Limited studies have been made in this field, and most of these have been devoted to investigation of physical properties rather than chemical changes. If progress is made in these areas, faster-baking resins should be forthcoming, hopefully with greater flexibility and better color, without sacrificing solvent resistance; in addition, resins better suited for use in combination with slow-drying vegetable oils and alkyds should be developed.

REFERENCES

1. See for instance: Martin, R. W., "The Chemistry of Phenolic Resins," New York, John Wiley & Sons, 1956; Megson, N. J. L., "Phenolic Resin Chemistry," New York, Academic Press, 1938.
2. Turkington, V. H., and Allen, Ivy, Jr., "Oil Soluble Phenolic Resins," *Ind. Eng. Chem.*, **33**, 966 (1941).
3. Hempel, C. H. (to Heresite and Chemical Company), U. S. Patent 2,253,235 (Aug. 19, 1941).
4. Purman, D. E., and Floyd, D. E. (to General Mills), U. S. Patent 2,891,023 (June 16, 1959).
5. Martin, R. W. (to General Electric), U.S. Patent 2,579,329 (Dec. 18, 1951).
6. Meyers, C. Y. (to Union Carbide), U.S. Patent 2,889,374 (June 2, 1959).
7. Martin, R. W. (to General Electric), U.S. Patent 2,579,330 (Dec. 18, 1951).
8. Aylsworth, J. W. (to Condensite Company), U.S. Patent 1,111,289 (Sept. 22, 1914).
9. Beatty, W. A. (to George W. Beadle of New York), U.S. Patent 1,225,748 (May 15, 1917).
10. Anann, A., and Fonrabert, E. (to Chemische Fabriken die Kurt Albert), U.S. Patent 1,623,901 (April 5, 1957).
11. Byck, L. C. (to Bakelite Corporation), U.S. Patent 1,590,079 (June 22, 1926).
12. Powers, Paul O. (to Armstrong Cork Company), U.S. Patent 2,343,845 (Mar. 1944).
13. Turkington, V. H., and Butler, W. H. (to Bakelite Corporation), U.S. Patent 2,017,877 (Oct. 22, 1935).
14. Honel, H. (to Beck Koller and Company), U.S. Patent 2,052,093 (Aug. 25, 1936).
15. Turkington, V. H., and Butler, W. H. (to Bakelite Company), U.S. Patent 2,173,346 (Sept. 19, 1939).

7

*Epoxy Resins**

Epoxy resins have been available in substantial commercial quantities for about 15 years. Over that span of time, consumption has risen steadily, and in 1965, their annual sales exceeded 100 million pounds. The surface coatings industry accounted for approximately half of this volume and thus continues to be the largest single consumer of these interesting resins. Table 7.1 shows one recent estimate[3] of epoxy resin sales to the surface coatings industry. It will be noted that these figures are broken down into major end use areas.

TABLE 7.1.   Estimated 1965 Epoxy Coating Market

End Use	Million Pounds
Plant maintenance	14.5
Automotive primers	9.5
Can and drum coatings	6.5
Pipe coatings	6.0
Appliance coatings	5.0
Others[a]	6.5
Total	48.0

[a]Includes trade sales, adhesives for home and industry, and military and aerospace usage.

Epoxy resins are no longer a novelty to the coatings industry. The prime attributes of these resins are broadly recognized and utilized by the paint formulator. Coatings based on epoxy resins are generally characterized by their excellent adhesion and overall chemical resistance. They also exhibit a high degree of resistance to impact, abrasion and other types of physical abuse. It is therefore not at all surprising that epoxies have found utility in the end use areas shown in Table 7.1.

*By John M. Klarquist, Shell Chemical Co., Union, N. J.

Epoxy resins also find application in industries other than surface coatings. The electrical and electronic industries, for example, consume a wide variety of epoxy formulations in potting and encapsulating of components and in electrical laminates. In the field of tooling, epoxies are used to manufacture molds, dies, jigs and checking fixtures. The high strength-to-weight ratio of epoxy-glass composites is utilized in the manufacture of structural laminates for the aircraft and other industries. Filament-wound composites represent a rapidly expanding field for epoxy resins. Much of the development work in filament winding has centered around rocket cases and related items of military and aerospace hardware. In the civilian field, filament winding of corrosion-resistant epoxy pipe and storage tanks is growing in importance.

PHYSICAL AND CHEMICAL CHARACTERISTICS OF EPOXY RESINS

The term "epoxy resin", when used in the broad sense, implies a polymeric material containing epoxide or oxirane groups:

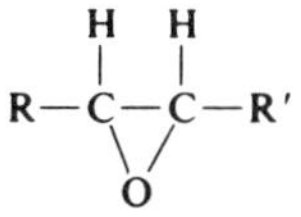

$$R-\overset{\overset{\displaystyle H}{|}}{C}-\overset{\overset{\displaystyle H}{|}}{\underset{\diagdown\diagup}{C}}-R'$$
$$O$$

In many cases, R in the above structure will simply be a hydrogen atom, such as is the case where the epoxide group is in a terminal position on the polymer:

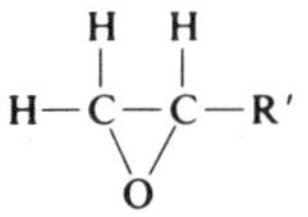

$$H-\overset{\overset{\displaystyle H}{|}}{C}-\overset{\overset{\displaystyle H}{|}}{\underset{\diagdown\diagup}{C}}-R'$$
$$O$$

Manufacture

In the surface coatings industry, by far the most commonly used epoxy resins are of the glycidyl ether type, particularly those derived from bisphenol A and epichlorohydrin. The manufacture of this resin type can be represented in simplified form as follows:

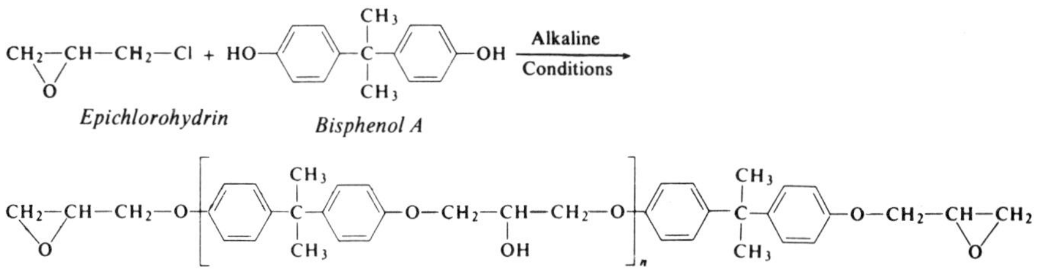

$$CH_2\underset{\diagdown\diagup}{-CH-}CH_2-Cl \;+\; HO-\!\!\langle\ \rangle\!\!-\overset{CH_3}{\underset{CH_3}{C}}-\!\!\langle\ \rangle\!\!-OH \quad\xrightarrow[\text{Conditions}]{\text{Alkaline}}$$
$$O$$

Epichlorohydrin   *Bisphenol A*

$$CH_2\underset{\diagdown\diagup}{-CH-}CH_2-O-\!\!\left[\langle\ \rangle\!\!-\overset{CH_3}{\underset{CH_3}{C}}-\!\!\langle\ \rangle\!\!-O-CH_2-\overset{}{\underset{OH}{CH}}-CH_2-O\right]_n\!\!\langle\ \rangle\!\!-\overset{CH_3}{\underset{CH_3}{C}}-\!\!\langle\ \rangle\!\!-O-CH_2\underset{\diagdown\diagup}{-CH-}CH_2$$
$$O \qquad\qquad\qquad\qquad\qquad\qquad\qquad\qquad\qquad\qquad\qquad\qquad\qquad\qquad O$$

A brief perusal of the epoxy resin molecule itself will give insight into some of the characteristics displayed by epoxy resin-based coatings:

(1) *Adhesion:*   The polar hydroxyl groups evenly spaced along the molecule contribute to hydrogen bonding, thus providing a high order of adhesion.

(2) *Chemical resistance:*   Along the backbone of the polymer only carbon-to-carbon and ether linkages are present.   Since both are stable bonds, epoxy systems possess a high order of inertness to most corrosive media.

(3) *Thermal stability and rigidity:*   The presence of aromatic ring structures in the polymer chain imparts these characteristics to the resin.

(4) *Reactivity:*   The presence of both epoxy and hydroxyl groups suggests that a variety of mechanisms may be used to build molecular weight during the curing process.

Where n is equal to zero in the foregoing molecular formula, we find the simplest version of the ECH/BPA polymer, i.e., the diglycidyl ether of bisphenol A. As the value of n increases, there is a corresponding increase in molecular weight, viscosity and melting point of the resin. A rather wide spectrum of these resins is commercially available with n values ranging from essentially zero to over 15. In Table 7.2 is a listing of the physical characteristics of several commercially available epoxy resins used by the coatings industry.

TABLE 7.2.   Physical Characteristics of Several Commercial Epoxy Resins

WPE^a	Molecular Weight (Average)	Esterification Equivalent Weight[b]	Melting Point ($°C^c$)	Viscosity at 25°C[d]
175–195	330	85	Liquid	500–700 cps
175–190	370	85	Liquid	5,000–10,000 cps
180–195	380	85	Liquid	10,000–16,000 cps
230–280	470	105	Liquid	>90,000 cps
425–550	900	145	65–75	D-G
875–1025	1400	175	95–105	Q-U
2000–2500	2900	200	125–135	Y-Z_1
2500–4000	3750	220	145–155	Z_2-Z_5

[a] WPE = weight per epoxide, i.e., grams of resin containing 1 gram-equivalent of epoxide. (WPE is synonymous with epoxide equivalent weight.)

[b] Grams of resin required to esterify completely 1 gram-mole of monobasic acid, e.g., 280 g C_{18} fatty acid or 60 g acetic acid.

[c] By Durran's mercury method.

[d] Values reported in cps are viscosities on resins as supplied; letter values are Gardner-Holdt viscosities of resins at 40% by wt in butyl "Carbitol."

Curing Mechanisms

Only rarely are epoxy resins used "as is" in paint technology.* The molecular weight of this family of resins is too low for them to be useful film formers without building them to higher molecular weight and complexity via a curing process. As noted earlier, epoxy resins of the glycidyl ether type possess two types of reactive sites which make possible this curing process: terminal epoxide groups and secondary hydroxyl groups.

Among the commonly used curing agents (also referred to as cross-linkers or hardeners) are:

(1) Aliphatic polyamines and their adducts,
(2) Aromatic polyamines,
(3) Tertiary amines,
(4) Polyamide resins and polyamide adducts,
(5) Ketimines,
(6) Acids and acid anhydrides,
(7) Lewis acids (particularly boron trifluoride complexes),
(8) Urea- and melamine-formaldehyde resins,
(9) Phenol-formaldehyde resins,
(10) Isocyanates,
(11) Mercaptans.

The applicability of the commercially important cross-linking agents and their curing mechanisms will be discussed under the appropriate sections following.

Other Epoxy Resins

While the great bulk of epoxy resins consumed in surface coatings are of the epichlorohydrin/bisphenol A type already described, glycidated novolacs (another member of the glycidyl ether family) are also commercially available. To date, however, their consumption in surface coatings has been small.

A newer and entirely different group of resins, the cycloaliphatic epoxies, are derived from the epoxidation of olefins. Like the glycidated novolacs, the role of these resins in surface coatings is still relatively minor.

Epoxidized oils are well known as stabilizers and plasticizers to the vinyl formulator, but a discussion of this product class is beyond the scope of this chapter.

---

*Two notable exceptions are (1) the use of relatively low molecular weight epoxy resins as stabilizers in vinyl coatings, and (2) the use of the new thermoplastic epoxies, which possess sufficient molecular weight to be film formers in their own right.

TWO-PACKAGE OR AMINE-CURED EPOXY COATINGS

In the last ten years, two-package or amine-cured coatings have become well established in areas where both physical and corrosion resistance are demanded of a paint film. For example, this type of epoxy coating is used in the industrial and marine maintenance fields, interior coatings for pipelines, aircraft primers and top coats, tank linings, railroad hopper car linings, high-glaze wall coatings, and heavy-duty floor coatings. The two-package system is also finding acceptance in the trade sale field for a variety of uses around the home.

As the name implies, coatings of this broad type consist of two separate components which are mixed just prior to use. Once mixed, the pot life of these materials will be limited, ranging from minutes up to several days, depending on the formulation. These materials cure readily at ambient temperatures down to about 50°F to yield films having chemical and physical resistance properties of a baked coating. If a faster cure is desired, most two-package systems can also be force-dried or even baked at elevated temperatures. Actually, this type of coating is used in fair volume in product finishing, being baked or cured at temperatures ranging from 140 to 400°F.

The term "amine-cured epoxy" is frequently used to describe this entire class of coatings, but strictly it is incorrect. In the early days of two-package or "cold-cured" epoxies, virtually every such coating on the market was indeed cured with diethylenetriamine (DTA), triethylenetetramine (TETA) or a similar aliphatic polyamine. Over the years, however, research and development in the field of curing agents has led to more sophisticated materials such as amine adducts, polyamide resins, imidazolines, ketimines and polyamide adducts. Nonetheless, all of these newer curing agents still depend on the reaction between the terminal epoxide groups of the epoxy resin and the amine hydrogen atoms of primary or secondary amines.

Basic Curing Mechanism and Curing Agent Requirements[10]

The manner in which an amine-type curing agent reacts with an epoxy resin to produce a cured three-dimensional network is illustrated below. To accomplish this type of cure, however, the curing agent must necessarily have an average functionality in excess of 2, i.e., more than two amine hydrogens per molecule.

It should be recognized that the above reaction involves a stoichiometric relationship between the curing agent and the epoxy resin. Therefore, the two ingredients must be combined in proper quantities to achieve

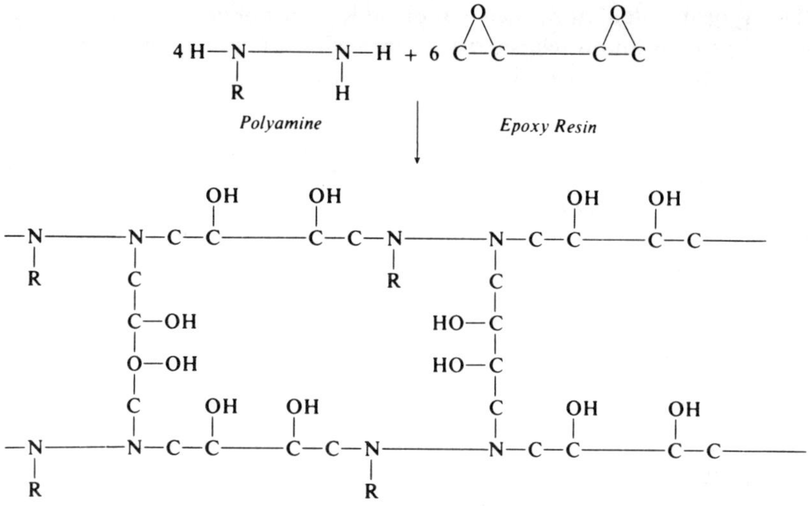

Polyamine *Epoxy Resin*

Cured or Cross-linked Resin

optimum cure. The stoichiometric amount of any given amine-type curing agent to be used with a specific epoxy resin is readily calculated as follows:

$$\text{Stoichiometric Amount of Amine (phr)*} = \frac{\text{Equivalent Weight of Amine}}{\text{Epoxide Equivalent of Resin}} \times 100$$

where

$$\text{Equivalent Weight of Amine} = \frac{\text{Molecular Weight of Amine}}{\text{Number of Amine Hydrogens}}$$

and

$$\text{Epoxide Equivalent of Resin (or WPE)} = \text{Grams of Resin Containing 1 gram-equivalent of Epoxide}$$

Let us assume we wish to use DTA to cure an epoxy resin with a WPE or epoxide equivalent of 490. Since DTA has a molecular weight of 103 and contains five amine hydrogens, the equivalent weight is 103/5 or 20.6.

Thus, the stoichiometric amount of DTA to use with this particular resin is:

$$\text{phr of DTA} = \frac{20.6}{490} \times 100 = 4.2$$

*phr = Parts (by weight) of amine per 100 parts (by weight) of epoxy resin.

This type of calculation is also applicable to a polymeric or proprietary amine-type compound whose exact composition is not known. The supplier of such a curing agent will be able to provide the formulator with the equivalent weight or combining weight of his product.

In practice, useful concentrations of curing agent usually fall within 25% of the calculated stoichiometric amount, i.e., that amount required to provide one amine hydrogen for each epoxide group in the resin. The formulator usually uses this calculated amount only as a starting point in his work. He will then determine the optimum curing agent level for a specific coating end use by running an experimental "ladder" of curing agent levels between 75 and 125% of the calculated stoichiometric value.

Curing Agent Selection

There are several types of curing agents commercially available. Since each type has both advantages and weaknesses, the formulator must take these factors into account when developing an epoxy coating to do a specific job.

As noted earlier, aliphatic polyamines such as DTA and TETA were the first curing agents to be used with two-package epoxy coatings. These amines are relatively inexpensive and they yield cured films with excellent overall chemical and solvent resistance. On the other hand, aliphatic amines exhibit curing agent blush or "sweat" when the coating is applied under conditions of high humidity. This phenomenon in a coating is esthetically undesirable, being typified by low gloss and sometimes marked discoloration and/or a surface tack on the cured paint film. By allowing the amine to age or "sweat in" the epoxy component for about 1 hour before application, curing agent blush will be reduced. Aliphatic amines can also cause dermatitic or respiratory problems where the coating user does not exercise proper precautions during application.

Polyamine adducts provide cured films with physical and chemical resistance properties at least equivalent to those provided by unmodified aliphatic amines. An adduct, however, exhibits a reduced tendency toward curing agent blush under high humidity. Most adducts are reaction products between an epoxy resin and a sufficiently large excess of aliphatic polyamine to completely consume all epoxy groups and leave residual active amine groups at each end of the adduct molecule. A curing agent of this sort has a much lower vapor pressure and lower dermatitic potential than the parent aliphatic polyamine used in its preparation.

Amine-terminated polyamide resins represent another major group of curing agents used with two-package epoxy coatings. These materials are reaction products of a polyamine and dimerized vegetable oil acids. Polyamide-cured epoxy films are more flexible than amine-cured films and are only slightly lower in chemical resistance. The polyamides tend to wet the substrate better than either amines or amine adducts and are therefore better suited for use on surfaces that may be slightly damp or contain small amounts of contamination. Another advantage of polyamide resins is their relatively low order of irritation potential.

Polyamide resin adducts[7] have recently appeared on the market. These are reaction products of a polyamide resin and an epoxy resin. It is claimed that these curing agents will provide epoxy coatings with improved gloss and flow, better water and chemical resistance, and will offer greater freedom from surface exudation problems if the coating is applied under adverse weather conditions.

Tertiary amines find occasional usage in epoxy coating formulations, but primarily as accelerators for polyamide-cured systems. It is interesting to note that when a tertiary amine is used as the sole curing agent for an epoxy resin, the curing mechanism involved is distinctly different from that of primary and secondary aliphatic amines. A tertiary amine causes homopolymerization of the epoxy resin through the epoxide groups, resulting in a chain of ether bridges. Thus, a tertiary amine behaves as a true catalyst, while aliphatic polyamines and the other curing agents discussed above enter the reaction stoichiometrically.

Selection of Epoxy Resin Type

Epoxy resins having a molecular weight of around 900 are the most commonly used resins in the formulation of solvent-based two-package coatings. While this molecular weight range provides the best overall balance of film properties, it is often necessary to use other epoxy grades to meet specific end use requirements. As molecular weight of the chosen epoxy resin is increased, the following trends will be observed in the cured coating:

(1) Flexibility, flexibility retention, and flow properties improve.
(2) Pot life at application viscosity increases.
(3) Nonvolatile content at application viscosity decreases.
(4) Cross-link density of the film decreases. This will be reflected in lower solvent resistance and lower resistance to some chemicals.

In solventless or "super-high solids" coatings, the liquid epoxy resins (molecular weight = 330 to 380) are customarily used. Coatings based on these lower molecular weight resins exhibit a very high order of chemical and solvent resistance by virtue of their high cross-link density, but as might be expected, they are also less flexible than coatings based on the higher molecular weight resins.

Solvent Selection

The most desirable solvent combination for two-package epoxy coatings is a blend of oxygenated solvents (usually ketones and alcohols) and aromatic hydrocarbons. To obtain a good balance between flow properties and solvent release, the oxygenated solvent portion should consist partly of moderately high boiling solvents (e.g., diacetone alcohol and methyl isobutyl carbinol) and the remainder should be lower boilers (e.g., methyl isobutyl ketone and isopropyl alcohol). Acetone, methyl ethyl ketone and other very low boilers are attractive because of their extreme solvent power, but they should be used in small quantities where good gloss and flow are important.

Ester solvents are usually not recommended for use with two-package coatings, since these solvents may react with the amine-type curing agent, thus lessening the ultimate degree of cure. Certain very high boiling oxygenated solvents such as cyclohexanol and the glycol ethers contribute to good gloss and flow, but they tend to be retained for long periods in the deposited film. Such solvents should therefore be used sparingly to avoid a marked reduction in film resistance to fresh water, solvents and chemicals.

Later in this section suggested formulations will be found which illustrate solvent blends suitable for various types of two-package coatings.

Two-package Maintenance Coatings

As noted earlier, epoxies are finding increasing use as maintenance coatings in moderate to severe corrosive environments. As is the case with any high-quality protective coating, proper surface preparation is a most important part of an epoxy coating job. For general maintenance painting on steel, a commercial gray sandblast is adequate. Where the corrosive environment is severe or where immersion service is involved, the surface should be sandblasted to white metal.

One common maintenance coating system consists of two coats of a two-package epoxy primer, followed by two enamel top coats. Since most formulations allow a build of $1\frac{1}{2}$ to 2 mils per coat without sagging, a coating scheme of this sort results in a total film thickness of 6 to 8 mils.

Most primers of this type are formulated at pigment volume concentrations in the 35 to 45% range. Corrosion-inhibiting pigments such as red lead, lead silicochromate, and zinc chromate are commonly used. Enamel top coats normally use nonchalking rutile titanium dioxide as the prime white hiding pigment at conventional pigment loadings.

Epoxy coatings will begin to chalk after 6 to 12 months of exterior exposure, and this is accompanied by a loss of gloss. While this characteristic makes epoxies unsuitable for some end uses, this is usually not the case with maintenance finishes. Chalking actually assists in keeping the painted surface clean and free of surface soiling. It should also be recognized that the chalking exhibited by epoxy coatings is not of a rapid, erosive sort. On the contrary, this appears to be a relatively slow process.

For example, several oil storage tanks coated with epoxy systems over eight years ago have shown no significant loss in film thickness over that time. In certain areas, however, some film thickness loss will be seen where water runoff or other abrasion of the painted surface occurs. A good example of this situation is found on cone roof storage tanks. For example, over a six-year span, the cone roofs on another group of storage tanks had lost about 2 mils from an original thickness of 6 mils. Film thickness measurements on the vertical shells of these same tanks showed no measurable loss.

The following is a formulation for a typical maintenance primer which has exhibited very good corrosion resistance in moderate to severe atmospheres. This primer would normally be top-coated with an enamel based upon the same epoxy resin/amine adduct binder system as is used in the primer. The four-coat scheme mentioned earlier would be an excellent coating system.

The primer can be prepared by first making a 60% solution of the epoxy resin in part of the indicated solvent blend. A portion of this vehicle is

FORMULATION 1. Maintenance Primer (Amine Adduct Type)

Base Component	*Pounds*	*Gallons*
Red lead (97%)	729.6	9.78
"Celite" 266 (Johns-Manville Products Corp.)	68.8	4.13
"Asbestine" 3X (International Talc Co.)	56.6	2.37
Aluminum stearate	3.4	0.41
"Epon" 1001 (Shell Chemical Co.)	170.5	16.97
"Beetle" 216-8 (American Cyanamid Co.)	10.4	1.22
MIBK	80.4	12.04
Ethylene glycol monobutyl ether	9.0	1.20
Toluene	89.5	12.34
Total base component	1218.2	60.46
Curing Agent Component		
"Epon Curing Agent" C-111 (Shell Chemical Co.)	88.9	10.70
MIBK	80.3	12.03
Ethylene glycol monobutyl ether	9.0	1.20
Toluene	90.4	12.48
Ethyl alcohol (proprietary solvent)	21.2	3.13
Total curing agent component	289.8	39.54
Total formulation	1508.0	100.00
Properties of Mixed Formulation		
Total nonvolatile, % by wt	71.1	
Pigment volume concentration, %	43.9	
Weight per gallon, lb	15.1	
Usable pot life, hr	8	

used to disperse the pigments and extenders using a three-roll or ball mill. The grind is let down with the remaining epoxy resin solutions, flow control agent ("Beetle" 216-8), and solvent blend. The curing agent component is prepared by a simple mixing operation.

Obviously, the two components are packaged separately and are mixed just prior to use. Because of the limited pot life of the mixed formulation, do not mix any more material than can be used during the working day.

This formulation should be at suitable viscosity for spray application with conventional spray equipment. Should additional thinning be required, the following thinner may be used in small amounts:

	Per Cent Weight
MIBK	40
Ethylene glycol monobutyl ether	5
Toluene	45
Ethyl alcohol (proprietary solvent)	10
Total	100

Heavy-duty Coatings[16]

A logical outgrowth of the two-package epoxy coating is the high-build heavy-duty coating. Coatings of this sort can be formulated to give dry film thicknesses of 6 to 40 mils in a single coat without sagging, depending on the identity and level of the extenders and thixotroping agents employed. Since these coatings are normally formulated at high pigment loadings, gloss of the cured system will range from semigloss to dead flat, again depending on the formulation. A solid grade epoxy resin (or blends of two solid resins) cut in a ketone/alcohol/aromatic hydrocarbon solvent system would be a typical binder. If solvent or fuel resistance is of paramount importance, an amine or amine adduct is used as the curing agent. For other uses, a polyamide resin may be used.

This type of coating has been used successfully in tank and tank car linings, hopper car linings, and as a heavy-duty industrial maintenance coating. One typical tank lining system involves two coats of this heavy-duty material resulting in a final dry film thickness of 12 to 14 mils. Similar schemes can also be used in such heavy-duty areas as offshore drilling rigs and platforms.

Trade Sales Formulations

In the last few years, several paint companies have begun to supply epoxy two-package coatings as a trade sales item. These coatings are frequently designed to allow a one-to-one mix of the two components and they normally contain a high percentage of slow-evaporating solvents to facilitate brushing. Despite the limited pot life involved, this type of coating has apparently captured the fancy of the homeowner. These coatings are used on floors, walls and household equipment. Specially formulated versions are used on pleasure boats and as swimming pool coatings.

The most popular epoxy resin for such a coating is in the 900 molecular weight range, and polyamide resins are by far the most common type of curing agent.

Formulation 2 is a suggested starting point for a brushable trade sales-type enamel.

FORMULATION 2. Epoxy/Polyamide Brushing Enamel (Gray)

Base Component	Pounds	Gallons
"Epon" 1001-CX-75 (Shell Chemical Co.)	474.8	56.54
"Beetle" 216-8 (American Cyanamid Co.)	16.6	1.88
Titanium dioxide, Rutile NC	471.1	13.50
Talc No. 399 (Whittaker, Clark, and Daniels Co.)	47.1	2.10
"Bentone" 27/ethyl alcohol (1/1)	5.7	0.62
Lampblack	2.8	0.18
Diacetone alcohol	63.1	8.06
Heavy aromatic naphtha (KB = 90)	125.3	17.18
Total base component	1206.5	100.06
Curing Agent Component		
"Epon Curing Agent" V1-ET-60 (Shell Chemical Co.)	594.0	77.41
Heavy aromatic naphtha (KB = 90)	111.0	15.38
Ethylene glycol monoethyl ether	56.0	7.19
Total curing agent component	761.0	99.98
Properties of Mixed Formulation		
Total nonvolatile, % by wt	63.3	
Mixing ratio, by volume	1.0/1.0	
Weight per gallon, lb	9.8	
Usable pot life, hr	8	

Note: This enamel is designed for brush application, but it may be thinned for spraying by using a 1/1 blend of MIBK and xylene.

Epoxy/Coal Tar Coatings

Coatings based on the combination of epoxy resins and coal tar pitch are the subject of U. S. Patent 2,765,288 (Whittier) assigned to Pittsburgh Coke and Chemical Company. Coatings of this type are usually quite high in solids (80 to 85% by weight) and can be formulated to permit application of films up to 10 mils without sagging. A typical coating system would be two coats of an epoxy/coal tar system, yielding a total dry film thickness of 12 to 14 mils. These coatings are usually applied directly to clean, sandblasted steel without a primer.

Epoxy/coal tar coatings exhibit excellent resistance to fresh and salt water and to a wide variety of chemicals, including many inorganic acids. These coatings have fairly good resistance to attack by gasoline and other hydrocarbon fuels, but they are not suitable as a tank lining material for this service, since they tend to discolor clean fuel by a leaching process.

Epoxy/coal tar systems enjoy wide use in heavy-duty maintenance and in the marine field where they have been used extensively on ships and barges and on fixed marine structures such as pilings.

Solventless and Super-high Solids Coatings

Since there are epoxy resins available in liquid form at relatively low viscosity (see Table 7.2), solventless or high solids two-package coatings are readily formulated. One of the earliest forms of this coating type was the highly filled trowellable system which can be applied at thicknesses of $\frac{1}{8}$ to $\frac{1}{4}$ inch. This type of coating has been used for industrial flooring, tank repair compounds, and even as a lining for concrete electrolytic cells in a nickel refining process.[2]

Sprayable solventless coatings[9] based on low-viscosity epoxy resins and curing agents are used to some extent in industry today. Most formulations of this sort, however, have a very short pot life after the curing agent is added. The usable life of a solventless system is often only a few minutes. Since this situation leads to formidable application problems in standard spray equipment, special two-component metering and mixing equipment has been developed which circumvents the pot life and premature gelation problem. The two-component application equipment consists of two separate supply pots (one for epoxy resin and one for curing agent), a means of accurately metering the precise ratios of each material, a mixing chamber wherein the two components are rapidly and thoroughly blended, and finally a spray head for application. Since the mixed material has a very short residence time in the equipment, plugging by cured resin is avoided. Naturally, this type of special equipment is both complicated and rather costly.

Ketimines are a new class of curing agent which make possible the application of solventless or super-high solids epoxy coatings with standard spray equipment.[13] A ketimine under dry conditions would react with an epoxy resin at an extremely slow rate, but in the presence of water, such as atmospheric moisture, the ketimine is broken down into a polyamine and ketone as follows:

$$\begin{array}{c} R_1 \\ \diagdown \\ \diagup \\ R_2 \end{array} C{=}N{-}R{-}N{=}C \begin{array}{c} \diagup R_1 \\ \\ \diagdown R_2 \end{array} + 2H_2O \rightleftharpoons H_2N{-}R{-}NH_2 + 2R_1{-}\overset{\displaystyle O}{\overset{\|}{C}}{-}R_2$$

$$\text{Ketimine} \qquad + \text{ Water} \rightleftharpoons \text{ Polyamine} \quad + \quad \text{Ketone}$$

The polyamine thus released is free to cure the epoxy resin in the normal manner.

Most ketimine-cured epoxy coatings should not be applied at film thicknesses greater than 10 mils in a single coat in order to assure complete through-cure.

Coatings of this type are used as tank linings, hopper car linings and as high-glaze wall coatings.[1]

Zinc-rich Epoxy Primers

Zinc-rich primers employing epoxy/polyamide binder systems have been available for several years and have enjoyed commercial acceptance as preconstruction primers for steel, particularly in European shipyards.[20] These primers are usually applied to shotblasted or sandblasted steel at film thicknesses of 0.5 to 1 mil The primed steel can subsequently be cut, welded and formed without serious damage to the primer. Since zinc dust accounts for over 90% of the total primer solids, the coating cathodically protects the steel substrate and thus provides excellent corrosion resistance. This type of primer will accept a wide variety of top coats, but if the zinc-rich primer has been allowed to weather outdoors, the soluble white zinc salts which have formed should first be removed by scrubbing with fresh water before top coats are applied.

Government Specifications

Listed below are several typical government specifications covering two-package epoxy coatings.

Specification	Coating Type and Use
TT-C-535a	Epoxy/polyamide high gloss enamel for metal, wood, concrete, and masonry
MIL-C-22750B	Epoxy/polyamide aircraft top coat
MIL-P-22808A	Epoxy/polyamide coating resistant to hydraulic fluid
MIL-P-23236	Epoxy and epoxy/coal tar coatings for ship tank linings
MIL-P-23377	Epoxy/polyamide aircraft primer
MIL-P-46056	Epoxy/polyamide primer for glass fiber reinforced plastic
MIL-P-52192	Epoxy/amine adduct corrosion-resistant primer.

EPOXY ESTERS[11]

An epoxy ester is simply the esterification product of an epoxy resin and an acid, such as vegetable oil acids or rosin. Although these esters are inherently superior to alkyds in certain respects, they resemble alkyds in handling and processing characteristics. For example, both drying and nondrying epoxy esters can be prepared in conventional resin processing kettles. Standard paint formulation techniques are used to transform the esters into durable primers, enamels or varnishes.

Since they offer excellent corrosion resistance and will accept a variety of top coats, epoxy ester primers and primer surfacers have been used in large volume in the automotive industry in the United States and abroad. Other volume uses for epoxy esters include overprint varnishes, metal product finishes, floor varnishes and industrial maintenance coatings.

Selection of Epoxy Resins and Acids for Esterification

Virtually any organic acid will react with epoxy resins to form an ester. In this reaction, the epoxy resin acts as a resinous polyol containing epoxide and hydroxyl groups, both of which can be esterified. The epoxide groups have a hydroxyl functionality of 2, since two ester groups can be formed per epoxide.

Table 7.2 (p. 84) lists the major types of resins along with their molecular weights and esterification equivalent weights. Industrial experience has shown that an epoxy resin in the 1400 molecular weight range is the most useful for ester preparation since the resultant esters have the best balance of properties. The higher molecular weight resins (molecular weight 2900), however, are useful in cases where higher vehicle viscosity is desired or can be tolerated, since these resins provide improved film flexibility and toughness. Lower molecular weight resins (molecular weight 900) can be used to reduce vehicle viscosity and to give vehicles of greater compatibility with other film formers.

By proper selection of fatty acids, it is possible to formulate epoxy esters having drying, semi-drying, or nondrying characteristics. The type of fatty acid selected is governed by the same principles that apply to alkyd manufacture. For example, excellent epoxy ester baking enamels can be based on lauric or coconut fatty acids, while air-drying vehicles can be based on linseed acids. Occasionally, modest quantities of dimerized fatty acids are used to upgrade the toughness and flexibility of the epoxy ester. Certain monobasic organic acids, such as tertiary butylbenzoic acid, are useful modifiers in baking-type esters. Modifications of epoxy esters with styrene or vinyltoluene are also possible.

Formulation Principles and Calculations

Since the preparation of an epoxy ester involves mainly the reaction of a polyol with acids, the calculations involved in formulating are rather simple. A generalized version of the esterification reaction would be:

$$\text{Epoxy Resin} + \text{Fatty Acid} \xrightarrow{\text{Heat}} \text{Epoxy Ester} + \text{Water}$$

Under normal esterification conditions, the carboxyl groups present in the acid react preferentially with the epoxy groups, then with hydroxyl groups. It should be noted that no water is formed when a carboxyl group reacts with an epoxy group. This reaction results in the opening of an epoxide ring with subsequent formation of one ester group and one hydroxyl group. The newly formed hydroxyl group may later be esterified in conventional fashion to yield a molecule of water.

It is simplest to formulate epoxy esters on the basis of equivalents of reactants. Theoretically, 1 esterification equivalent of an epoxy resin will react with exactly 1 equivalent of a fatty acid to yield a completely esterified product. In actual practice, however, rarely more than 0.9 equivalent of acid is used per equivalent of resin. This excess hydroxyl content in the cook allows low acid numbers to be reached within a reasonable cooking time.

Laboratory and commercial experience indicates that virtually all useful epoxy esters fall within the range of 0.3 to 0.9 equivalent of acid per equivalent of epoxy resin. As with alkyd resins, a system or terminology for describing differences in "oil length" has been applied to epoxy esters:

Equivalents of Acid	Oil Length
0.3–0.4	Short
0.5–0.6	Medium
0.7–0.9	Long

Given below is a sample calculation for the formulation of a long oil ester based on an epoxy resin (molecular weight 1400), 0.7 equivalent of linseed fatty acid and 0.1 equivalent of rosin.

Material	Equivalents Used	Equivalent Weight	Weight Used	Per Cent Weight
Epoxy resin (molecular weight 1400)	1.0	175	175	43.0
Linseed fatty acid	0.7	282	197	48.4
Rosin	0.1	348	35	8.6

Manufacturing Procedure

Epoxy esters can be prepared in standard cooking equipment via either fusion or azeotropic methods. In fusion cooking, the use of an inert gas sparge is highly recommended, since it aids water removal and improves the color of the final ester.

Top processing temperatures of epoxy esters vary, depending on the oil length and type of acids employed. It is felt, however, that 500 to 525°F represents the best working range for most vehicles.

After an epoxy ester has been processed to the desired vehicle constants, it is cooled to about 300 to 350°F and thinned with the proper solvent to yield a finished vehicle. It is usually desirable to filter the vehicle while it is still hot, using conventional filter aids.

Drier Selection

The driers used in epoxy ester finishes are those conventionally used in alkyds. For air-drying esters, 0.04 to 0.05% cobalt (metal) based on vehicle solids is adequate. With certain esters, the incorporation of 0.02% calcium together with the normal amount of cobalt drier improves through-dry. Lead driers should be used only after careful testing for stability, since lead may precipitate from certain epoxy ester vehicles on aging.

When epoxy esters are baked without amino resins, the addition of 0.005 to 0.01% cobalt is recommended. In dehydrated castor and soya esters, 0.05% cerium will give the same film hardness as cobalt but with less effect on film color.

It is also desirable to include 0.25% antiskinning agent in air-drying esters to prevent skinning and to minimize "after-body" on storage. This level of antiskinning agent does not appreciably retard drying time.

Epoxy Esters from Liquid Epoxy Resins

In the last few years, another route to epoxy resin esters has been described.[8] The preparation of an epoxy ester is accomplished in two distinct phases. As a first step, a liquid epoxy resin (molecular weight about 380), bisphenol A, a catalyst such as lithium naphthenate, and a precalculated quantity of fatty acids are all charged to a kettle and heated rapidly to about 425°F. At this point, an exotherm will be observed, in some cases rising to 550°F. The vehicle is allowed to cool to 445°F, where it is held for about 1 hour or until constant vehicle viscosity is reached.

In this first stage, the carboxyl groups in the presence of the alkaline catalyst will first react with the epoxide groups on the resin, yielding a monoester. When the fatty acid carboxyls have been consumed, the monoester, bisphenol A and the remaining resin react, resulting in an acid-terminated resin of higher molecular weight.

In the second stage, remaining fatty acid is added, the temperature is increased to about 500°F and the vehicle is processed in a fashion similar to epoxy esters based on solid resins. The amount of acid added during this stage is obviously dependent on the oil length desired in the finished vehicle.

It is claimed that the liquid resin route to epoxy esters results in vehicles with film properties at least equivalent to the esters prepared from solid epoxy resins.

HIGH-PERFORMANCE BAKING FINISHES

The reactivity of epoxy resins with other thermosetting resins provides an intriguing route to the formulation of high-quality baking finishes.[14]

Epoxy resins may be coreacted with other thermosetting resins such as phenolics, urea-, melamine- and triazine-formaldehyde resins, and thermosetting acrylic resins. All these will react with epoxy resins (usually under the influence of heat) via the hydroxyl and/or epoxide groups present on the epoxy resin structure.

Epoxy-phenolic Coatings

Epoxy-phenolic systems are one of the most chemically resistant epoxy finishes available. Since they show excellent resistance to a host of alkalies, acids and solvents, epoxy-phenolics are widely used as linings for cans, drums, tanks and certain process equipment. These coatings, despite their chemical inertness, are quite flexible. This flexibility, together with good adhesion, allows an epoxy-phenolic coating to withstand a great deal of physical abuse. A coating of this type may be used in both clear and pigmented form, but the relatively dark vehicle color precludes it from those areas where light colors are required for esthetic reasons.

It is postulated that the major reaction during the cure of an epoxy-phenolic system occurs between the hydroxyl groups along the epoxy chain and the methylol groups present in the phenolic resin. Using a generalized notation for a phenol-formaldehyde resin, this reaction mechanism would be as follows:

$$R-\underset{OH}{C_6H_3}-CH_2OH \;+\; \left[O-CH_2-\underset{OH}{CH}-CH_2-O-C_6H_4-\underset{CH_3}{\overset{CH_3}{C}}-C_6H_4\right]_n$$

Phenolic   Epoxy Resin
Resin

$$\Big\downarrow H^+$$

$$\left[O-CH_2-\underset{\underset{OH}{\underset{|}{\underset{R-C_6H_3-CH_2}{|}}}}{CH}-CH_2-O-C_6H_4-\underset{CH_3}{\overset{CH_3}{C}}-C_6H_4\right]_n \;+\; H_2O$$

It should be noted that this reaction is catalyzed by acidic materials such as phosphoric acid and alkyl derivatives of phosphoric acid, e.g., butyl dihydrogen phosphate. *p*-Toluenesulfonic acid is an extremely effective catalyst.

It is also possible that the terminal epoxide groups of the epoxy resin may react with the hydroxyl groups of the phenolic resin to form ethers. A third competing curing mechanism is that of the phenolic resin reacting with itself.

Solid epoxy resins in the higher molecular weight ranges (2900 to 4000) are commonly used in this type formulation, since they are higher in hydroxyl content than the lower molecular weight species. As the molecular weight of the epoxy increases, the following trends in film properties will be noted:

(1) Flexibility and abrasion resistance increase.
(2) Overall solvent and chemical resistance increase.
(3) Flow and leveling of the film improve.
(4) Total solids (at application viscosity) decrease.

In the formulation of an epoxy-phenolic, care must be taken in the selection of flow control agents. Without such an additive, virtually any formulation will show a tendency to crawl, pinhole or crater. The most common flow control agents are silicone resins such as Silicone Resin SR-82 (General Electric Co.) or DC-840 Resin (Dow Corning Corporation) at levels of about 1% based on total resin solids. In some formulations, 0.2 to 0.3% of a medium-viscosity ethylcellulose is beneficial. To further minimize flow problems, it is desirable to filter the finished clear vehicle before packaging or proceeding with pigmentation.

Formulation 3 is illustrative of a pigmented epoxy-phenolic enamel suitable for spray application. A typical bake schedule for a single coat system would be 20 minutes at 385°F, but if multiple coats are used,

FORMULATION 3. Epoxy-phenolic Baking Enamel (Green)

Material	*Pounds*	*Gallons*
Chrome oxide	84.8	1.96
"Epon" 1007 (Shell Chemical Co.)	207.6	21.73
"Methylon" 75108 (General Electric Co.)	69.3	6.93
Silicone Resin SR-82 (General Electric Co.)	4.7	0.53
Phosphoric acid (85%)	5.0	0.35
n-Butanol	37.6	5.56
"Cellosolve" acetate	240.8	29.69
Xylene	240.8	33.25
Total	890.6	100.00

Formulation Constants

Total nonvolatiles (% by wt)	41.4
Pigment/binder (weight ratio)	23/77
Epoxy resin/phenolic resin (weight ratio)	75/25
Weight per gallon (lb)	8.9

only a light intermediate bake between coats is recommended (e.g., 10 minutes at 300°F). If a more rigorous bake is used between coats, intercoat adhesion will be adversely affected.

Preparation

Prepare a 40% solution of the epoxy resin in "Cellosolve" acetate/xylene (1/1) and disperse the pigment in a suitable portion of this vehicle via three-roll mill or other milling equipment. To prepare a composite letdown vehicle, mix the remaining epoxy solution and phenolic resin. Dilute the phosphoric acid catalyst with the butanol before adding to the letdown vehicle.*

For spray application, the enamel may be reduced to the desired viscosity with a thinner composed of equal parts of diacetone alcohol and xylene.

Amino Resin-converted Epoxy Coatings

Amino resin-converted epoxy finishes are similar to the phenolic-converted systems in both formulation and application. While the amino-cured films are distinctly lighter in color than their phenolic counterparts, they are slightly inferior in chemical resistance. Epoxy-amino coatings are used as clear coatings for costume jewelry, brass hardware, wire coatings, and certain container coatings applications. In pigmented form, they find use as primers and top coats for steel laboratory furniture, appliances, and scientific equipment where chemical resistance and corrosion resistance are demanded.

As with the epoxy-phenolic systems, epoxy resins in the 2900 to 4000 molecular weight range are preferred for conversion with amino resins. Urea-formaldehyde resins are the most commonly used, but certain melamine- and triazine-formaldehyde resins are also of value. With the latter two resin types, however, care must be exercised in resin selection, since they do not enjoy as wide a range of compatibility with epoxy resins as do the urea types.

An epoxy/amino resin ratio of 70/30 is usually near optimum for most uses, but this ratio can be varied between 85/15 to 60/40. As epoxy resin level is increased, the resultant film shows increasing flexibility and impact resistance, but at the expense of hardness and solvent resistance.

Since amino resin-converted systems respond to acid catalysis, the choice of converting resin has a distinct influence on cure rate. With the faster-curing (acid-containing) amino resins, a clear coating with the

Note: The clear letdown vehicle should be aged at least two days (preferably one week) prior to letting down the pigment dispersion. This aging process minimizes the danger of pigment flocculation during the letdown process.

classical 70/30 epoxy/amino resin ratio will cure in 20 minutes at 300°F. When pigmented, this same system will require 20 minutes at 385°F to achieve full cure. If one of the slower-curing, more flexible urea resins is used, bake schedules of 20 minutes at 385°F are required in both the clear and pigmented versions. It is also important to note that as the relative ratio of epoxy resin to amino resin is increased, curing schedules must also be increased.

Use of Epoxy Resins with Thermosetting Acrylics[19]

Although they are relative newcomers to the surface coatings industry, thermosetting acrylic resins have already found broad industrial acceptance. When formulated as baking finishes, they offer a combination of film hardness, mar-proofness, gloss and color retention, and chemical resistance. They are already being used as finishes for major household appliances and automobiles and in the metal decorating field.

One type of commercially available thermosetting acrylic resin contains pendant hydroxyl groups which may serve as reaction sites with cross-linking agents. The most common cross-linkers used with this resin type are melamine-formaldehyde resins, but modest amounts of epoxy resin may also be used in conjunction with melamine resins to upgrade detergent resistance and adhesion.

A group of rather flexible thermosetting acrylic resins is also available. These resins contain carboxyl groups which serve as the reactive site for cross-linking with epoxy resins. The epoxy resins best suited for this purpose have a molecular weight of about 900. Resins of higher molecular weight lead to compatibility problems, while lower molecular weight epoxies are inferior in film resistance and color.

Although there is some latitude in the amount of epoxy resin which may be used with thermosetting acrylic resins, it should be recognized that overbake resistance and exterior durability will decrease with increasing amounts of epoxy resin.

OTHER TYPES OF EPOXY COATINGS

Epoxy Powder Coatings[5,12]

The entire field of powder coatings has been receiving increasing attention in technical and trade publications in the last few years. Powder coatings can be based on either thermoplastic or thermosetting resins, but regardless of resin type, this new concept offers several inherent advantages over solvent-containing coatings:

(1) Powder coatings can be applied in very high film thicknesses in a single application without fear of film porosity caused by entrapped solvent.

(2) Irregular features of the substrate, such as sharp corners and edges, are readily coated in a single application.

(3) Atmospheric pollution and fire hazards are minimized since no solvent is present; however, a dust explosion hazard exists with many powders and must be taken into account when designing the application process.

Powder coatings based on epoxy resins possess most of the properties found in solvent-applied systems. They have excellent adhesion, impact resistance and corrosion resistance. They also enjoy a high order of electrical insulation resistance and thermal stability, making them attractive as coatings for electrical equipment.

Epoxy powders are usually based on one of the solid epoxy resins (molecular weight 900 to 2900). Several practical curing agents exist, including dicyandiamide, boron trifluoride complexes such as $BF_3 \cdot$ mono ethylamine, acid anhydrides, and aromatic amines. The dihydrazides[18] are of particular value in the formulation of epoxy powder coatings where very fast cure cycles are desired.

In addition to the epoxy resin and curing agent, a properly formulated powder coating will usually contain pigments, one or more flow control agents, and a thixotropic agent to prevent sagging when applied in thick films. Particle size distribution of the finished powder is controlled within rather narrow limits to insure proper application characteristics.

One of the best-known methods of applying powder coatings is the fluidized bed process which is the subject of a U. S. patent.[6] Using this process, the object to be coated is first preheated above the sintering point of the epoxy powder formulation and is then dipped into the fluidizing chamber. The powder in the chamber is maintained in a state of fluidity by an ascending flow of gas applied through a porous membrane at the bottom of the chamber. The powder coming in contact with the preheated object melts and fuses to the surface. The object is withdrawn from the chamber and transported to an oven where the coating is cured.

Powdered coatings may also be applied with flocking guns, electrostatic spray equipment and fluidizing chambers employing electrostatic principles.

Epoxy/Polyamide Splash Zone Coatings[4,15]

Marine structures such as offshore drilling platforms and pilings are particularly susceptible to corrosion in the splash zone, i.e., that area of the structure bounded by the line of low tide and a line a few feet above high tide. One of the newer means of arresting corrosion in this area is by the use of epoxy/polyamide compounds formulated to allow underwater cure. These materials are formulated to a stiff, putty-like consistency so as to resist deformation when exposed to the slapping of wave action.

These coatings are applied by hand or by trowel at film thicknesses of 1/8 to 1/4 inch. The polyamide curing agent displaces the water from the surface of the wet substrate, allowing the coating to adhere and cure to a dense, impermeable barrier.

Thermoplastic Epoxy Resin Coatings [17]

Earlier it was noted that most epoxy resins lack sufficient molecular weight to be true film formers until one of several curing mechanisms is employed to increase molecular weight. The thermoplastic epoxy resins are an exception to that rule. Like the epoxy resins already discussed, the thermoplastic epoxy resins are also based on bisphenol A and epichlorohydrin. Their basic structure may be represented as follows:

$$\left[-O-\underset{}{\bigcirc}-\underset{\underset{CH_3}{|}}{\overset{\overset{CH_3}{|}}{C}}-\bigcirc-O-CH_2-\underset{\underset{OH}{|}}{CH}-CH_2- \right]_n$$

Terminal epoxide groups have been omitted from the structural formula since the resin molecule is so large that the epoxide groups are relatively unimportant as reactive sites. In addition, some of the thermoplastic epoxy resins are deliberately phenol terminated.

One typical commercial resin has a number average ($\overline{M}_n$) molecular weight of about 7000, corresponding to some 25 of the repeating units (n) shown above. Measurements of weight average molecular weight ($\overline{M}_w$) on this same resin give values of 300,000 which corresponds to an n of roughly 100. Molecular weights in this range are clearly of an entirely different order of magnitude than those shown in Table 7.2, and as a result, the thermoplastic epoxy resins are true film formers in their own right. Solvent-deposited films of these polymers will dry by solvent evaporation without need for a curing agent.

Several types of coatings based on thermoplastic epoxy resins are commercially available. For example, quick drying one-package general-purpose primers of this type are used as maintenance and preconstruction primers. They may be coated with a wide variety of top coats after an overnight dry, or if used as a shop primer, they will also accept top coats even after prolonged exposure to weathering. Since this type of primer dries by solvent evaporation alone, application under cold or humid conditions is feasible. Primers based on thermoplastic epoxy resins are approximately equivalent to two-package epoxy formulations in overall corrosion resistance and are superior to the two-package systems in resistance to mechanical damage.

It is also possible to formulate single-package, zinc-rich primers based on thermoplastic epoxy resins with film performance approximating that of the epoxy/polyamide type.

These resins are suitable for baking finishes in clears, enamels or primers. These coatings have a high degree of flexibility and abrasion resistance allowing them to be used in post-forming applications. A preponderance of hydroxyl groups occurring along the polymer chain allows these resins to be cross-linked with amino or phenolic resins if desired. For example, the steam processing resistance of a clear thermoplastic epoxy baked coating can be upgraded considerably by the addition of only 3 to 5% of urea-formaldehyde resin. Since thermoplastic epoxy resins have a rather high hydroxyl functionality, seemingly insignificant amounts of amino resin have a pronounced effect on final film properties. Regardless of whether a cross-linking resin is used or not, these coatings can be cured at schedules ranging from 30 minutes at 300°F to 2 minutes at 450°F.

The hydroxyl functionality of thermoplastic epoxy resins may also be utilized as a reactive site for isocyanate cures. One example of such a formulating approach would be a two-package system employing a solution of thermoplastic epoxy resin (clear or pigmented) as the base component and "Mondur" CB-60* as the curing agent component. This system exhibits a usable pot life of about 8 hours after the two components are mixed. The NCO/OH ratio can be varied within the usual 1.1/1.0 to 0.25/1.0 limits depending on the specific film properties desired. This type of coating exhibits relatively quick cure and excellent resistance to water and acids.

REFERENCES

1. Anon., "New Civic Center Uses Epoxy in Place of Tile," *American Painting Contractor* (April 1964).
2. Anon., "Epoxy Resins Protect Concrete Tanks in Direct-Reduction Nickel Process," *Eng. Mining J.* (July 1960).
3. Anon., *Mod. Plastics*, **43**, No. 5, 116 (1966).
4. Drisko, R. W., Alumbaugh, R. L., and Cobb, J. W., *Materials Protection*, 24 (September 1965).
5. Elbling, I. N., *Offic. Dig. Federation Paint Varnish Prod. Clubs*, **31**, No. 419, 1625 (1959).
6. Gemmer, E. (to Knapsack-Grieshein A.G.), U.S. Patent 2,844,489 (1958).
7. General Mills Company preliminary bulletin "Versamid 230—A Polyamide/Epoxy Adduct," 1965.

*Mobay Chemical Co.

8. Richardson, S. H., and Wertz, W. I., *Offic. Dig. Federation Soc. Paint Technol.*, **33,** No. 441, 1310 (1961).
9. Richardson, S. H., *Offic. Dig. Federation Paint Varnish Prod. Clubs*, **31,** No. 417, 1300 (1959).
10. Shell Chemical Company Technical Bulletin SC: 59-300, 1959.
11. Shell Chemical Company Technical Bulletin SC: 54-46, 1954.
12. Shell Chemical Company Preliminary Information Sheet "Fluidized Bed Coatings Based on EPON Resins," 1960.
13. Shell Chemical Company Technical Bulletin SC: 62-36.
14. Shell Chemical Company Technical Bulletin SC: 60-303.
15. Shell Chemical Company Technical Bulletin SC: 62-42.
16. Somerville, G. R., "High Build Epoxy Maintenance Finishes," *Paint Ind.* (August 1959).
17. Somerville, G. R., "Properties and Applications of a Linear Epoxy Resin," *Paint Varnish Prod.* (November 1962).
18. Wear, R. L. (to Minnesota Mining and Manufacturing Co.), U.S. Patent 2,847,395 (1958).
19. Wetzel, L. A., and Mercurio, A., *Resin Review* (*Rohm and Haas*), **XV,** No. 3, 24 (1965).
20. Wilson, R. W., and Zonsveld, J. J., *Institution Transactions* (*North East Coast Institution of Engineers and Shipbuilders*), **78,** 277 (1962).

8

*Hydrocarbon Resins**

This chapter covers a group of resins designated loosely as "hydrocarbon resins." Chemically, these are made up of carbon and hydrogen, but in a few cases they do contain small amounts of oxygen.

Historical

Hydrocarbon resins were first produced commercially in the United States during the 1920's. These resins were made by polymerizing the fraction of coal tar distillates containing coumarone and indene, and thereby became known as coumarone-indene resins.

Many unsaturated hydrocarbons became available from the cracking of petroleum, and these were converted to various types of hydrocarbon resins. Styrene monomer became available during the 1930's and was combined with various hydrocarbons to make coating resins. During the late 1930's, polyterpene resins made by polymerizing β-pinene became available. Polydiene resins made by polymerizing olefins became commercially available in 1950.

Description

This classification covers low molecular weight (under 20,000) thermoplastic resins such as coumarone-indenes, polyterpenes, polydiene, polystyrene, miscellaneous petroleum resins, polyethylene, polypropylene and many rubbers.

The widespread use of hydrocarbon resins is primarily due to their neutral hydrocarbon nature, acid and alkali resistance, ease of processing, wide range of compatibility, solubility and melting points, and their low cost. Some hydrocarbon resins are available in emulsion form and are used to modify synthetic latices.

*By C. R. Martens, The Sherwin-Williams Co., Cleveland, Ohio.

Coumarone-Indene Resins

Coumarone-indene resins are produced from these two materials which are obtained from coal tar naphthas:

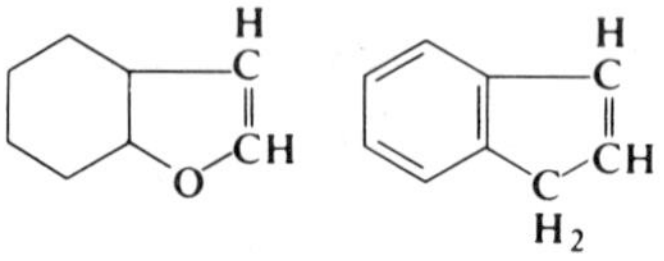

Coumarone   *Indene*

They are polymerized with sulfuric acid as the polymerization catalyst or, in some cases, aluminum, stannic or antimony chloride.

Coumarone-indene resins have melting points up to 320°F and are soluble in aromatic solvents and moderately soluble in petroleum solvents. They have low acid values and excellent alcohol, acid and alkali resistance, but they are somewhat yellow in color.

These resins are used in aluminum paints, wall sealers, concrete paints, insulating varnishes, container coatings and floor coatings.

Commercial resins of this type are "Picco,"* "Cumar,"** etc.

Petroleum Resins

Modern methods of cracking petroleum oils to increase the yields of gasoline have produced many unsaturated hydrocarbons in large quantities. These are undesirable in gasoline and are removed because they form gummy polymers on storage. Some of these are:

$$CH_2{=}CH_2 \qquad CH{\equiv}CH \qquad CH_2{=}CH{-}CH_2{-}CH_3 \qquad CH_2{=}C{\raise1ex\hbox{$\scriptstyle CH_3$}}_{CH_3}$$

Ethylene   *Acetylene*   *Butylene*   *Isobutylene*

$$CH_2{=}CH{-}CH{=}CH_2 \qquad\qquad CH_2{=}\underset{\underset{CH_3}{|}}{C}{-}CH{=}CH_2$$

Butadiene   *Isoprene*

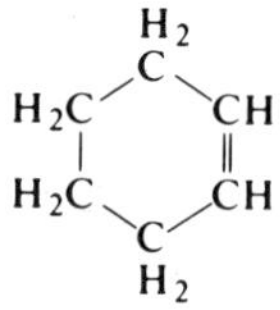

Cyclohexene 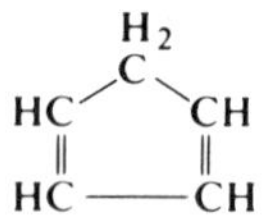   *Cyclopentadiene*

*Registered trademark, Pennsylvania Industrial Chemical Corporation.
**Registered trademark, Neville Chemical Co.

Polymers are formed very easily from these materials. One method is to use catalysts such as aluminum, antimony and stannic chloride at elevated temperatures. Polymers are obtained ranging in molecular weight from 250 to 1000, and in softening point up to 220°F. These products vary from rubbers to hard resins. Some polymers contain a considerable amount of unsaturation as indicated by iodine values of 190 to 300. They have colors from light yellow to almost black. Consequently, some are suitable for light colors but not for pure whites.

These resins may be dissolved in solvents and blended with bodied oils for aluminum paints. They can be combined with phenolics and alkyd resins to lower costs.

Examples of commercial resins of this type are "Velsicols,"* "Panarez,"** "Picodiene"† and "LX."‡

Terpene Resins

Turpentine which is obtained from the resin of the pine tree contains a number of compounds known as terpenes. These are cyclic hydrocarbons with double bonds. Some of these are:

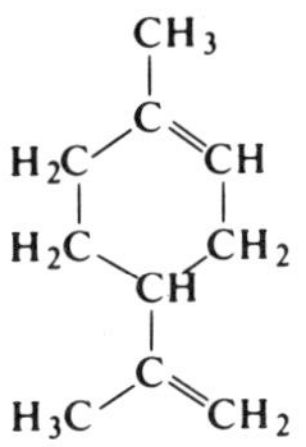

Dipentene

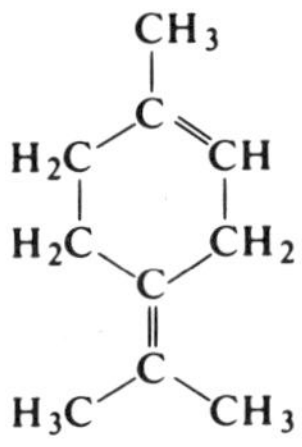

Terpinolene

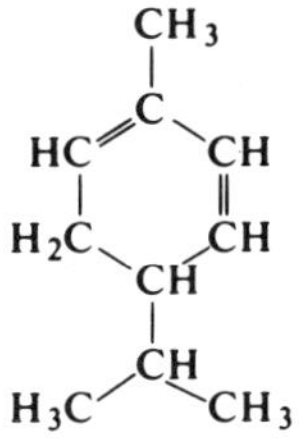

Phillandrene

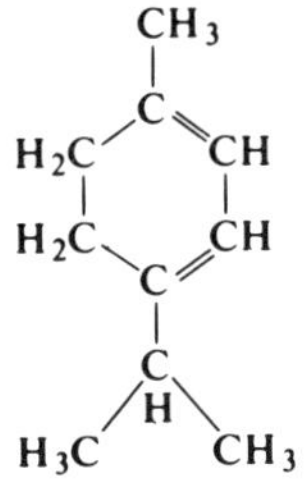

Terpinene

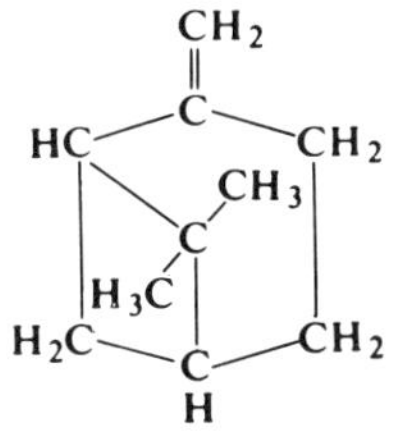

β-Pinene

---

*Registered trademark, Velsicol Chemical Corporation.
**Registered trademark, Amoco Chemicals Corp.
†Registered trademark, Pennsylvania Industrial Chemical Corporation.
‡Registered trademark, Neville Chemical Co.

The principal polyterpene resins are made by polymerizing β-pinene. Resins with melting points of 50 to 260°F, low acid values (< 4) and pale colors are obtained. These are soluble in aliphatic hydrocarbons and are used in architectural and concrete paints.

Terpenes can be combined with maleics, rosin and phenolics to make many other types of resins.

"Piccolyte"* resins are terpene resins which are commercially available. Also of hydrocarbon nature are certain rubbers which are used in protective coatings.

Styrene-Butadiene Resins

The high styrene-butadiene resins are copolymers which contain 85% or more styrene. "Pliolite" S-5** is a material of this type. The resins are soluble in aromatic hydrocarbons, ketones and esters. A mixture of aliphatic and aromatic solvents with a kauri-butanol value of 60 is recommended for brushing enamels. These resins are not generally compatible with alkyd resins and have limited compatibility with drying oils. These are compatible with rosin materials and coumarone-indene resins. Rubber resins are used in fast air-drying finishes which have good water and chemical resistance and good abrasion resistance.

REFERENCES

1. Jackson, K. E., *Offic. Dig. Federation Paint Varnish Prod. Clubs* (May 1956).
2. Koeneche, D. F., *Offic. Dig. Federation Soc. Paint Technol.* (June 1960).
3. Preuss, H. P., *Metal Finishing*, **61,** No. 7, 59 (1963).

*Registered trademark, Pennsylvania Industrial Chemical Corporation.
**Registered trademark, Goodyear Tire & Rubber Co.

9

*Acrylic Resins**

In the coatings field as well as in the plastics field, the term "acrylic" resins applies generally to the polymers and copolymers of the esters of methacrylic and acrylic acid. Frequently the copolymers of one or more of these esters with non-acrylic monomers such as styrene, butadiene or vinyl acetate are referred to as acrylic resins, but for best usage the term should be reserved for those resins which are predominantly of the characteristic acrylic or methacrylic structure given below:

$$\left[-CH_2-\underset{\underset{COOCH_3}{|}}{\overset{\overset{CH_3}{|}}{C}}-CH_2-\underset{\underset{COOCH_3}{|}}{\overset{\overset{CH_3}{|}}{C}}-\right]_n \qquad \left[-CH_2-\underset{\underset{COOC_2H_5}{|}}{\overset{\overset{H}{|}}{C}}-CH_2-\underset{\underset{COOC_2H_5}{|}}{\overset{\overset{H}{|}}{C}}-\right]_n$$

Poly(methyl methacrylate)   *Poly(ethyl acrylate)*

These resins, often modified by minor amounts of functional monomers of the same unique structure, have very characteristic properties which, for the most part, are not enhanced by the inclusion of non-acrylic segments in the polymer chain.

The acrylic resins are some of the most versatile used in the coatings field. They can be manufactured conveniently in several physical forms which can then be modified in the formulation of the final product. Solid and solution polymers are used in lacquer finishes for metallic, wood, leather, ceramic and plastic surfaces. Acrylic emulsions, manufactured directly in the latex form, are used in both indoor and outdoor paints. Organosols promise utility in high solids coatings for a variety of substrates. Thermosetting acrylic resins of various compositions are commercially available.

The properties imparted to coatings by the acrylic resins include stability (i.e., resistance to degradation by UV light, hydrolysis and corrosive chemicals), resistance to mechanical damage, and an eye-appealing, high-

*By W. M. Edwards, E. I. du Pont de Nemours & Co., Wilmington, Del.

quality appearance. Hardness and slip properties can be varied over a wide range to suit the demands of the application, and such properties as adhesion and solvent resistance are built into the resins by copolymerization with functional monomers, both for their effect *per se* and because they furnish a means by which the structures can subsequently be cross-linked if required.

Acrylic acid and its esters were prepared in Germany during the last part of the nineteenth century. However, it was not until 1901 that the first significant study of the acrylic resins was undertaken, and it was more than twenty years later before the first feasible process for manufacturing acrylic monomers was devised. Applications which have stimulated broad industrial usage of acrylic monomers and polymers include emulsions for leather finishing and paints, clear rigid poly(methyl methacrylate) sheet, molding and extrusion powders for plastics applications, and automotive lacquers. Cross-linking to improve high-temperature performance and solvent resistance (Strain, 1939) further widened the scope of acrylic resins in coatings. As increased production and improved technology have brought down the cost of acrylics, advantage of their unique properties has been taken in many new applications. The full potential of these versatile materials has certainly not yet been realized.

TYPES OF ACRYLIC RESINS

Acrylic resins which are available to the coatings industry include homopolymers and copolymers in solid, solution and emulsion form. Moreover, organosols are becoming increasingly important.

The solid acrylics used in coatings are thermoplastic and in the form of either small beads or irregularly shaped particles crushed from larger blocks. They are homopolymers of one of the methacrylate esters listed above, or copolymers of a methacrylate with an acrylate or a second methacrylate ester, with or without a lesser quantity of one of the functional monomers listed (see Table 9.1). In general, they are readily soluble in a variety of solvents or mixtures thereof, the ease of solution depending on particle size and molecular weight.

Solutions of many of these same resin compositions are also available, usually manufactured by solution polymerization processes. Films cast from them either are thermoplastic in nature or, if the resin contains enough of an appropriate functional monomer, can be cross-linked to thermoset coatings.

Two types of acrylic emulsions are generally available. Thermoplastic emulsion resins for use in latex paints are copolymers of methyl methacrylate with varying amounts of acrylate esters (and sometimes non-acrylic monomers) to lower the minimum film-forming temperature. Minor amounts of functional monomers are also used in the manufacture of these copolymers to improve pigment binding, emulsion stability and other properties of the latex and polymer.

Acrylic latices are also available with functional groups through which the final film can be cross-linked. This is accomplished by adding appropriate quantities of cross-linking reagents which react with the functional groups of the polymer when the film is baked or "cured." This transforms it into a thermoset polymer with attendant improvements in solvent and chemical resistance, toughness and, of course, greatly increased hardness at elevated temperatures.

PROPERTIES OF ACRYLIC RESINS

Stability

The term "stability" is used here to describe the relative inertness of acrylic resins, that is, their resistance to degradation by UV light, hydrolysis, acids, bases, oxidizing agents and other corrosive industrial chemicals. The methacrylate resins do not contain tertiary hydrogens attached directly to the polymer chain. The absence of such sites for photo-oxidative and thermo-oxidative attack renders them exceptionally stable to UV light and oxygen; indeed, the only resins of comparable light stability are the fluorocarbons—polymers and copolymers of tetrafluoroethylene. Since polymers of the acrylate esters do contain a tertiary hydrogen atom, subject to photo-oxidation, they are somewhat less resistant to both UV light and thermal oxidation. Copolymers which are primarily from methacrylates (with only minor amounts of acrylates) are nonetheless among the most light stable of the polymeric binders used in the coatings industry.

TABLE 9.1   Commercially Important Acrylic Monomers

Basic Monomers	*Functional Monomers*
Methyl methacrylate	Methacrylic acid
Ethyl methacrylate	Acrylic acid
n-Butyl methacrylate	Acrylamide
Isobutyl methacrylate	2-Hydroxyethyl methacrylate
Lauryl methacrylate	2-Hydroxypropyl methacrylate
Stearyl methacrylate	Glycidyl methacrylate
Methyl acrylate	Dimethylaminoethyl methacrylate
Ethyl acrylate	*tert*-Butylaminoethyl methacrylate
n-Butyl acrylate	Ethylene dimethacrylate
2-Ethylhexyl acrylate	Trimethylolpropane trimethacrylate
Cyclohexyl methacrylate	Butylene dimethacrylate
2-Ethylhexyl methacrylate	Diethylaminoethyl acrylate

The other major factor inducing the degradation of polymers on their exposure to weather is hydrolysis. Since acrylics as well as other polymers from "vinyl" monomers contain only carbon atoms in the polymer chain, they are quite resistant to hydrolysis. While hydrolysis of side-chain ester groups of some resins causes a more or less rapid change in properties (e.g., polyvinyl acetate), ester groups of the methacrylate polymers are quite resistant to hydrolysis, and property retention is consequently excellent. The polyacrylates, although somewhat less resistant to hydrolytic attack, are still considered good in this respect.

The acrylic resins, particularly the methacrylates, are also resistant to acids, bases, weak and moderately strong oxidizing agents, and many corrosive industrial gases and fumes. Stability toward these materials is largely responsible for their growing use in industrial protective finishes.

Hardness, Flexibility and Elongation

The interrelated properties of hardness, flexibility and elongation vary over wide ranges and depend upon the nature of the alcohol portion of the monomer. Polyacrylates are softer and more flexible than the corresponding polymethacrylates. Moreover, as the number of carbon atoms in the ester function increases, the hardness decreases, while flexibility and elongation increase in both resin families. This holds true with straight-chain alcohols up to about C_8 with the acrylates, and about C_{12} with the methacrylates. The hardness of resins from higher alcohols is increased by crystallization of the carbon chains in the alcohol. Acrylic resins from straight-chain or normal alcohol esters are softer and have higher elongation than resins from the isomeric branched chain esters.

These properties can be controlled readily since they are essentially linear functions of copolymer composition. Increasing amounts of ethyl acrylate thus reduce the hardness and increase the elongation of methyl methacrylate polymers; however, to achieve the same softening effect without loss of light stability, it is frequently better to use *n*-butyl methacrylate as the comonomer. The softening effect of *n*-butyl methacrylate when copolymerized with methyl methacrylate is shown in Figure 9.1.

Two factors make it very difficult to determine the hardness and strength of an acrylic resin by direct measurement on a cast film. One is the effect of the substrate on the hardness of a thin film, the other the difficulty in completely removing the solvent. Hence, it is common practice to determine this property, as well as several others, on a compression- or injection-molded sample of the solid resin. Values as determined represent those properties, then, of completely solvent-free resins.

The frictional properties of resin coatings depend on several phenomena, but when the chemical constitution is similar, as within the

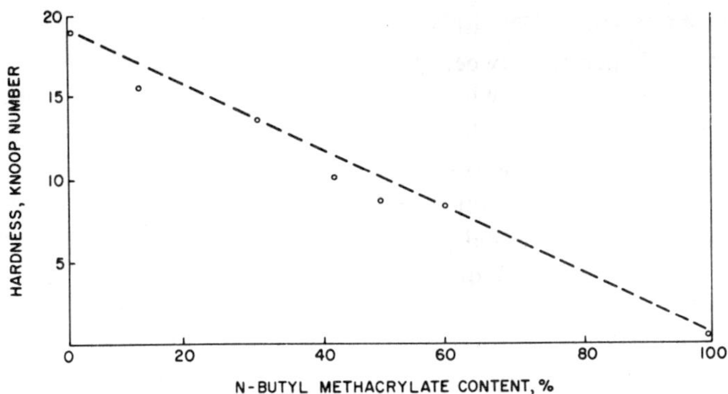

Figure 9.1.   Hardness of MMA/*n*-BMA copolymers.

family of acrylic resins, slip (static and dynamic coefficient of friction) is closely related to hardness. Poly(methyl methacrylate), being the hardest of the conventional acrylic homopolymers, also has the best slip properties. In general, the methacrylate polymers have better slip properties than the acrylates and are often used in lacquer formulations to impart this property to the coating.

Strength and Toughness

The actual tensile strength of the polymeric binders used in coating applications is not of paramount importance as long as it is at least several hundred pounds per square inch. Tensile strength, however, is of importance because of its influence on toughness. Depending on both tensile strength and elongation as well as other physical properties of the resin, toughness is of importance because it plays an important part in the abrasion resistance, scratch resistance, film integrity and impact resistance of the final coating. The area under the stress-strain curve, often used as a measure of toughness, does not seem to define this property adequately in connection with coatings, by and large, because it does not take into consideration notch sensitivity, crack propagation or the rate of impact loading.

Perhaps the most important manifestations of the toughness of coatings are abrasion resistance and impact resistance, particularly the latter in thermosetting appliance enamels. Abrasion resistance within any series of similar polymer structures depends primarily on molecular weight— the higher the molecular weight, the greater are the toughness and resistance to abrasion. There are many instances where the performance of an acrylic resin could be improved by raising its molecular weight, if the application were not limited by those properties adversely affected, i.e., solution viscosity and solvent release.

Second-order or Glass Transition

The physical properties of resinous materials are governed by the fundamental attraction between, and the mobility of segments within, the long polymer molecules. This in turn depends upon various polar or van der Waals forces, upon the size and spatial arrangements of segments of molecules, and upon the order or crystallinity within the plastic mass.

With the structurally similar family of acrylic resins, there is an obvious relationship between physical properties as the pendant ester groups are changed. The second-order or glass transition temperature, T_g, is a characteristic property of polymers and convenient for purposes of comparison. This is the temperature at which an amorphous material changes from a brittle glassy state to a rubbery state and is measured as the abrupt change in slope of the plot of some physical property (e.g., refractive index, specific volume, flexural modulus) against temperature. Since the acrylic resins are noncrystalline and have no definite melting point, the glass transition is useful in characterizing them.

Poly(methyl methacrylate) has the highest T_g (105°C) and poly(2-ethylhexyl acrylate) the lowest (−55°C), of the more common acrylic resins. Other predominantly acrylic polymers and copolymers lie between the two.[3] Like hardness, T_g of the acrylic resins depends upon the number of carbon atoms in the ester group, decreasing with the higher alcohols. Again, above eight carbons in the acrylates and about twelve carbons in the methacrylate series, T_g increases. This effect is shown in Figure 9.2.

While T_g may not be recognized as being critical in coating applications, it bears a direct relationship to such very important properties as soften-

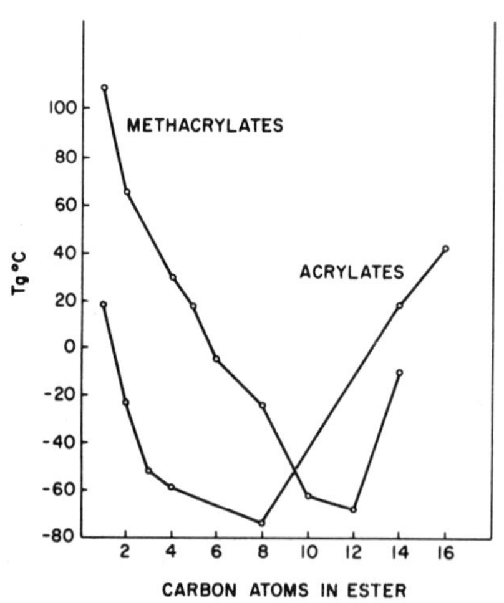

Figure 9.2. Transition temperatures of acrylic homopolymers.

ing point, brittle point, "tack temperature" and minimum film-forming temperature. These important, though less well-defined, thermal properties all increase in value as the transition temperatures of specific acrylic homopolymers and copolymers increase.

It may be helpful to bear in mind the following relationship between T_g and other physical properties when selecting the optimum acrylic resin coating for a particular application. As T_g increases:

Brittle point increases	Elongation decreases
Minimum film temperature increases	Flexibility decreases
Softening point increases	Tackiness decreases
Hardness increases	Blocking tendency
Tensile strength increases	decreases
Slip increases	Adhesion decreases

Adhesion

Within the family of acrylic resins, at least, the adhesive qualities of individual resins vary both with flexibility and with functionality. Adhesion to any substrate is improved by making the resin more flexible, preferably by copolymerization rather than plasticization. The practical limit of this approach, of course, is the maximum softness which can be tolerated in the application. On the other hand, adhesion is improved by copolymerization with functional monomers, e.g., acrylic and methacrylic acid. A high degree of functionality, however, is limited by the lowering of the water resistance of the final film.

Formulation can often be used to improve adhesion in two ways—compounding the acrylic with a compatible resin known for its adhesion to the particular substrate (e.g., vinyl acrylic top coats for vinyl upholstery fabric), and solvent formulation with a more active or more slowly evaporating solvent.

Solubility

The acrylic resins, by virtue of their structure, are characteristically soluble in solvents that are moderately hydrogen bonded—ketones, esters, tetrahydrofurane and unsymmetrical chlorinated hydrocarbons—and in aromatic hydrocarbons. Their solubility in alcohols, aliphatic hydrocarbons and naphtha depends upon the alcohol portion of the acrylate or methacrylate ester. As the length of the alcohol chain is increased, the solubility of both acrylate and methacrylate homopolymers in these weak solvents increases. Poly(methyl methacrylate) is completely insoluble in ethanol, whereas poly(ethyl methacrylate) is appreciably soluble. Polymers of butyl methacrylate and butyl acrylate are soluble in naphtha, alcohols and many aliphatic hydrocarbons. A convenient starting point in formulating a mixed solvent system for a specific acrylic resin is based on the solubility parameter concept.[2,4]

The molecular weight of an acrylic resin has more effect on solution viscosity than on the limits of solubility in specific solvents. However, since solution viscosity is of fundamental importance in coatings technology, the molecular weight of a polymer to be used in a given application is limited by the requirements of a practical solution viscosity at a practical solids content. Another polymer characteristic which influences solution viscosity is that of molecular weight distribution. Solutions of resins having a wide molecular weight distribution, i.e., significant fractions of exceedingly long molecules, are considerably more viscous at a given solids content than solutions of resins having the same average molecular weight but a narrow distribution. Since the physical properties important in coatings are not enhanced by a wide molecular weight distribution, resins manufactured by polymerization techniques leading to a narrow distribution are advantageous in affording a lower solution viscosity at a given solids content.

Plasticization

High molecular weight esters such as butyl benzyl phthalate, butyl phthalate and dioctyl phthalate are sometimes used as plasticizers for acrylic resins to decrease hardness and to increase flexibility and elongation. Since these materials may eventually migrate to the surface and escape from the resin, it is preferable, if possible, to achieve the property requirements by copolymerization rather than plasticization. Acrylates and higher alkyl methacrylates are used as comonomers to soften methyl methacrylate polymers. This usage is often referred to as internal plasticization.

Compatibility with Other Resins

Acrylic resins are often used in formulations containing other, cheaper resins to improve the properties of the latter. Occasionally by the addition of other resins to a predominantly acrylic formulation, one can improve flexibility, adhesion and block resistance.

Two resins are said to be compatible if clear films of the mixture are obtained from clear, or sometimes even cloudy, solutions. Acrylics are compatible with many polar resins, being relatively polar themselves. Moreover, within any family of resins, compatibility may vary, particularly with the molecular weight of the modifying resins, which should be approximately the same or somewhat lower.

Generally the improvements sought in alkyd coatings by the inclusion of acrylics include resistance to discoloration on aging and heating and resistance to staining. Compatible cellulosics are used with acrylics to improve flow properties, gloss and block resistance. The addition of vinyl chloride polymers and copolymers improves the flexibility and adhesion of acrylics, whereas the vinyl polymers are upgraded by acrylics in gloss and slip properties. Acrylics are also used with chlorinated rubber to improve gloss and slip. Epoxies and amino resins are blended with acrylics, as described under Thermosets, to bring about subsequent cross-linking. In Table 9.2 is listed the compatibility of several general classes of resins with hard, predominantly methyl methacrylate resins and with softer copolymers containing acrylates or higher methacrylates. This information has been generalized to indicate trends rather than the exact compatibility between specific resins.

TABLE 9.2. Compatibility[a] of Acrylics with other Resins

Non-acrylic Resin	Hard Acrylics	Soft Acrylics
Cellulose acetate butyrate, half sec	PC	C
Nitrocellulose, half sec	C	C
Ethyl cellulose	I	I
Polyvinyl chloride	C	C
Polyvinyl acetate	PC	I
Polyvinyl chloride–polyvinyl acetate copolymers	PC	C
Acrylic-modified alkyd	I	C
Rosin	C	C
Chlorinated rubber	I	C

[a]C = compatible; I = incompatible; PC = partially compatible.

Property Compromises

Most applications of acrylic resins, as well as other polymeric binder resins, involve property compromises. It is often not possible to achieve the ultimate in one property without detracting too much from another. A classic example is softness. A resin made harder to improve slip properties will naturally lose softness and flexibility. An excessively high solution viscosity may limit the molecular weight desired for maximum toughness. The practical limit on flexibility may be the solvent sensitivity of the softer polymers, and as mentioned previously, the limit of adhesion which can be achieved by functionality is usually sensitivity to water (see Figure 9.1).

Needless to say, the ideal compromise in resin properties is often not achieved because of lack of availability of the series of resins needed to define the optimum. Much can be done to improve the performance of acrylic resins in coatings as more polymers become available. To achieve the ideal property compromise, two important variables are at the disposal of the resin manufacturer: molecular weight and comonomer ratio. Molecular weight has its greatest effect on toughness and only to a minor extent on strength and extensibility. Comonomer ratio affects hardness, extensibility and some solubility characteristics. Molecular weight distribution, generally speaking, should always be as narrow as practical. Some important physical properties of homopolymers are listed in Table 9.3.

TABLE 9.3. Properties of Acrylic Homopolymers

Homopolymer	T_g	Hardness (Knoop)	Tensile Strength (psi)	Elongation (%)
Methyl methacrylate	105	19	10,000	2
Ethyl methacrylate	65	11	5,400	25
Butyl methacrylate	22	1	500	300
Isobutyl methacrylate	50	8	—	—
Methyl acrylate	8	—	1,000	750
Ethyl acrylate	−22	—	30	1800
Butyl acrylate	−56	—	5	2000

POLYMERIZATION OF ACRYLIC MONOMERS

The polymerization of acrylic monomers has much in common with other vinyl or free radical polymerizations conducted at atmospheric or slightly elevated pressure. The effects of initiators, inhibitors and chain stoppers are similar, and each monomer has its characteristic reactivity with other monomers which affects the homogeneity of the resulting polymer. The acrylics can be polymerized and copolymerized in bulk or suspension to obtain the unmodified resins, in solutions of appropriate solvents such as aromatic hydrocarbons or ketones, and in emulsion form with the addition of appropriate emulsifying aids. Since it is beyond the scope of this chapter to treat polymerization theory in detail, these topics will be discussed only as pertinent to acrylic polymerization. For a thorough theoretical treatment, see references 1, 10 and 11.

Reactivity of Monomers

Relatively speaking, the acrylic monomers are among the more reactive of those used to prepare polymers for coating applications. The acrylates are characteristically more reactive than the corresponding methacrylates, while in both series reactivity generally diminishes with an increase in the size of the alcohol portion of the ester. Both acrylic and methacrylic acid are highly reactive.

Inhibitors added to the monomers to prevent polymerization during storage, e.g., hydroquinone and the methyl ether of hydroquinone, are generally removed before polymerizing. However, with the availability of monomers containing very low levels, 0 to 15 ppm, it is becoming increasingly common to override the inhibitor with a slight excess of initiator.

Initiators

Free radical initiators are materials which, when decomposed thermally, chemically or sometimes photolytically, yield free radicals, i.e., fragments of molecules which are unstable due to their having an unpaired electron. These radicals attack vinyl monomers, such as acrylics, which are thereby converted into free radicals themselves and perpetuate the growth of the polymer molecule. The most common free radical initiator for acrylic polymerizations is probably benzoyl peroxide, although other peroxides are often used. Azobisisobutyronitrile, a faster and safer free radical source, is becoming increasingly important. Free radical fragments from the initiator are found as end groups in the polymer chain. Since both benzene rings and benzoyl groups are present in polymers polymerized with the aid of benzoyl peroxide and since both are attacked by UV light, to attain the ultimate in stability toward sunlight the azo initiator is better than benzoyl peroxide.

Initiating systems for emulsion polymerizations must generally be water soluble; hence inorganic peroxy compounds, usually potassium or ammonium persulfate, are used, often with a reducing agent (sodium bisulfite) and an activator in the form of a multivalent heavy-metal salt such as ferrous sulfate.

Copolymerization

The utility of acrylic resins in the coating field is due in great measure to the convenience in manufacturing copolymers that cover a wide range of the important properties—hardness, flexibility, adhesion and solubility. The majority of the resins available commercially are copolymers whose compositions have been arrived at to fill one or more specific coating needs. Moreover, the capabilities of acrylic copolymers are far from being fully exploited, as new monomers, particularly new functional monomers, continue to reach the market.[12]

Molecular Weight Control

The regulation of the molecular weight during polymerization depends entirely on controlling the relative frequency with which individual polymer chains are initiated and terminated, including transfer reactions.[7]

The rate of initiation depends upon the concentration of free radicals at any given time, which in turn is governed by the concentration of the initiator and the temperature (rate at which the initiator is decomposing into free radicals).

The termination rate is controlled by the relative concentration of chain transfer agents, added telogens or solvent molecules, as well as their relative efficiency. Many solvents act as moderately effective chain stoppers, a fact which makes the concentration of monomer in the solvent somewhat critical in solution polymerization. Chain transfer agents, notably mercaptans, are far more effective in regulating molecular weight and are often added for this purpose. Transfer reactions can, of course, occur between free radicals, from initiator or growing chains, and dead polymer molecules. In all transfer reactions, as the term implies, the radical becomes inactive and a new radical is formed which is capable of adding monomer and continuing the propagation step.

Apparently, disproportionation between two growing chains is the dominant termination reaction, but termination by combination has also been shown to occur.[8]

APPLICATIONS OF ACRYLIC POLYMERS

Acrylic resins are added to protective coatings to enhance both their decorative and protective function. From the decorative standpoint, their high gloss, pigment-binding characteristics and clarity are the important considerations. Their adhesion, hardness and durability are reasons for their use as a protective film.

Each of the three physical forms in which these resins are available has its characteristic uses and limitations. The solid types offer the formulator complete freedom in the choice of solvent. On the other hand, he must carry out the solution step which, in the case of high molecular weight resins or weak solvents, may be tedious. Use of solution-grade resins imposes the limitation of a partially fixed solvent system and perhaps more wastage in handling. Moreover, savings by avoiding the solution step may be more than offset by freight on a cheap solvent, if the solution must be shipped great distances. Emulsions are manufactured as such most economically and of course impose the aqueous system upon the formulator. While their wide utility is sufficient indication of their advantages, there are some limitations to their use, not the least of which is the residual emulsifying agent and other water-soluble impurities that remain in the coalesced film, decreasing its durability. Moreover, unless a final bake step can be used to coalesce the film, the resins must be limited to those with minimum film-forming temperatures below normal ambient conditions; these are generally the softer of the resins available.

Lacquers

Lacquers are solutions of resins in organic solvents which harden only as a result of evaporation of the solvent. A bake step is often used to speed evaporation.

Acrylics are used as the major or in some cases a minor component of lacquers for the protection and decoration of both ferrous and non-ferrous metals. Factory finishing of automobiles is one of the largest uses of acrylic lacquers. The advantages of the thermoplastic lacquers over thermosetting enamels in this use include thermal reflow and better dispersability of metallic colors. Applied over a primer and subsequently baked, a relatively hard thermoplastic acrylic copolymer has proved its utility by resisting many years of outdoor exposure very well. Even harder acrylics, copolymers consisting almost entirely of poly(methyl methacrylate), are used for durable and attractive finishes for toys. Automotive touch-up lacquers, some in the form of aerosol sprays, represent an area of large usage. In this case, lower molecular weight can be used for better sprayability and more rapid solvent release.

For coating nonferrous metals such as copper, brass and aluminum, softer acrylic copolymers rich in *n*-butyl methacrylate or other "softening" monomers are used. These resins usually contain a functional monomer (methacrylic or acrylic acid or amide) to further enhance the adhesion inherent in the flexible resins. Coil coating of aluminum strip and sheet, for protection of the metal during fabrication and construction, as well as afterwards, is consuming more and more of the acrylic resins, both soft thermoplastics and thermosetting compositions. Aerosol sprays to prevent tarnishing of copper, brass, silver and gold also make use of the clarity and general durability of the acrylics.

"Touch-up" colored and clear lacquers for use on wood, paper or metal substrates were among the earlier uses for acrylic lacquers. Clear acrylic lacquers, formulated for brush or spray application to paper, cloth, wood or metal are used almost exclusively for the protection of art work.

Protective and sealant coatings for a variety of masonry substrates, such as concrete blocks, ceramic tile, brick and the like, is a fast-growing application. In this use, rather soft but very tough (high molecular weight) copolymers containing polar functionality for adhesion are used. Tough pigmented lacquers for concrete floors have been formulated from these same resins.

Coatings for plastics substrates represent a growing market for acrylic resins, especially with the availability of more grades which are soluble in solvents which will not attack the plastic. These include ethanol, isopropanol, naphtha and straight-chain hydrocarbons. The resins used are low to medium molecular weight homopolymers and copolymers of ethyl and *n*-butyl methacrylate. Low molecular weight is desirable in this use for better solubility, sprayability and solvent release, whereas the ultimate in abrasion resistance is usually not necessary.

Typical uses in this area are base coats and top coats for vacuum metallized plastic parts (both first and second surfaces), back painting of thermoformed signs, and lacquers for imparting one or more colors for molded objects. Substrates include polystyrene, acrylonitrile-butadiene-styrene copolymers, molded, extruded and cast acrylics, and polyvinyl chloride. Acrylics are used in this area for their qualities of rapid drying, high gloss, hardness and durability as well as for their uniqueness in adhering to polystyrene and other plastic substrates. When abrasion resistance is of great importance, higher molecular weight copolymers can be used, with some penalty in solvent release. Lacquers to protect and improve the appearance of fiber glass reinforced polyester panels and molded parts, particularly after weathering, are being formulated from moderately hard acrylics of medium molecular weight.

Other substrates, the durability and appearance of which can be improved with coating of acrylic lacquers, are leather, asbestos and asphalt shingles, rubber and even glass.

The top coating of reinforced polyvinyl chloride upholstery fabric to improve slip and other properties, is a rather specialized but very important segment of both acrylic and PVC markets. Lacquers for this use are usually formulated from the harder methacrylates (methyl or ethyl) of very high molecular weight, especially where abrasion resistance is important, while a relatively high level of PVC homopolymer or copolymer is used with the acrylic to promote adhesion and improve its flexibility. If a somewhat softer acrylic, e.g., poly(ethyl methacrylate) is used in greater quantity it has an increased effect as a plasticizer barrier, preventing migration of the plasticizer to the surface with subsequent loss from the substrate.

Thermosetting Finishes

Thermoset acrylic finishes are generally harder, tougher and more resistant to heat and solvents than the thermoplastic finishes. On the other hand they are also generally less resistant to UV light and, of course, must be subjected to a thermal "cure" after application to attain the improvements in physical properties.

An important advantage to be gained in using thermosets, other than property advantages of the final finish, is in the application. Relatively low molecular weight copolymers of methyl methacrylate with other acrylic or non-acrylic monomers constitute the uncured resin. Functional monomers provide sites for subsequent cross-linking, usually by reaction with non-acrylic additives. Because the molecular weight is low (it is later to be converted into a three-dimensional network of "infinite" molecular weight) a solution of relatively high solids content is sprayable, permitting the deposition of a coating of the desired thickness with fewer spray applications. Moreover, the solution will level much better before cross-linking and result in a smooth coating. On the other hand, after curing, thermal reflow is not effective in improving the smoothness.

The functional groups incorporated into the base resin for subsequent cross-linking are of three general types:

(1) Carboxyl groups from acrylic or methacrylic acid for cross-linking with epoxy resins;

(2) Hydroxyl groups from hydroxyalkyl methacrylates for cross-linking with urea-formaldehyde or melamine-formaldehyde resins;

(3) Methylolated acrylamide groups which cross-link themselves when catalyzed or which can be cross-linked with epoxies. These copolymers often contain vinyl resins in the final formulation.

A number of other cross-linking systems have been studied, but the three mentioned are the only ones currently of commercial importance.

A detailed discussion of cross-linking reactions is to be found in references 6 and 7. In Figure 9.3 are outlined several representative reactions which take place when typical thermosetting resins are cured or cross-linked thermally at 250 to 350°F.

Most thermosetting acrylic resins available for commercial coating applications contain relatively high percentages of styrene, vinyltoluene, epoxides or amine resins to enhance their in-use properties, to lower their cost, or to effect the cross-linking. In fact, the term "acrylic thermoset" has sometimes been applied to resins which are acrylic only by virtue of the functional acrylic monomer establishing the sites for curing the resins. Consequently the large non-acrylic portion of these resins is likely to detract from the characteristic properties usually sought in an acrylic finish, notably resistance to UV light and to thermal discoloration. A general comparison of important properties of thermoset acrylic coatings is made in Table 9.4.

Coatings applications in which advantage can be taken of the thermosetting acrylics are limited to those substrates which can resist the rela-

TABLE 9.4. Characteristics of Thermosetting Acrylic Resins

Type	Hardness	Water Resistance	Toughness	Outdoor Durability	Adhesion
Carboxyl epoxy	Variable	Good	Excellent	Poor	Excellent
Hydroxylamine	High	Fair	Fair	Excellent	Good
Amide	Variable	Excellent	Variable	Poor to good	Excellent

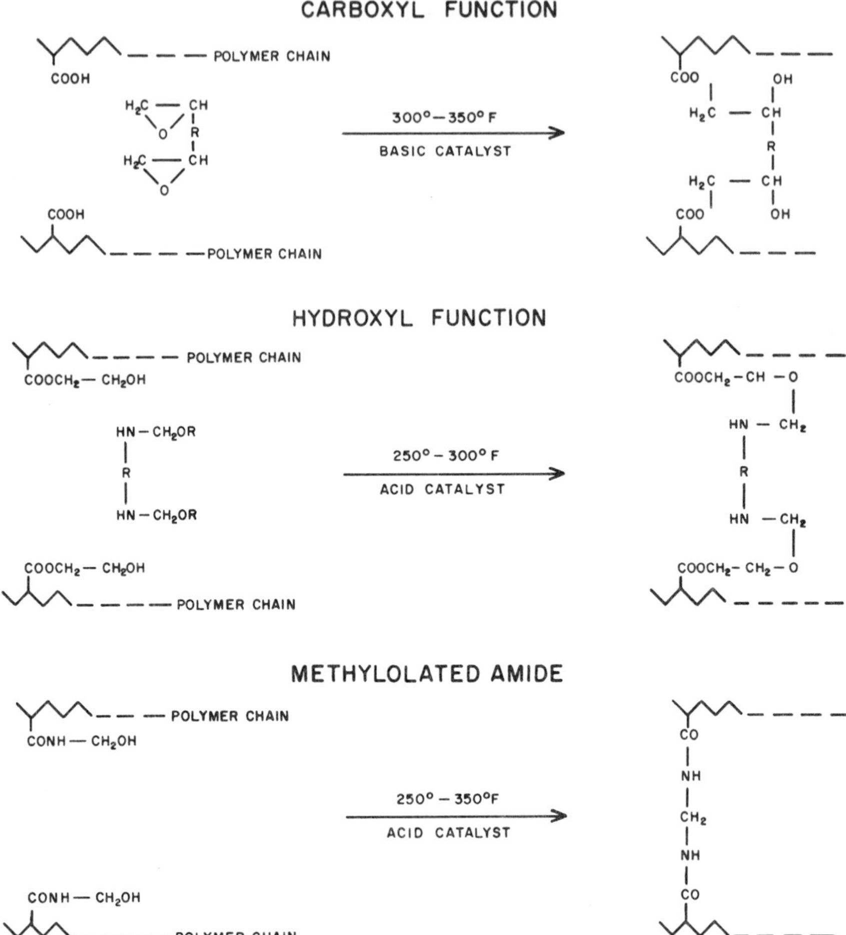

Figure 9.3.   Typical reactions used to cure acrylic thermosets.

tively high curing temperature required by currently available compositions. While this includes, for the most part, only metals and ceramics,
it nevertheless represents a large and growing market. Automotive finishes and appliance enamels are the two major use areas. Others include
can coatings, strip coatings, and many specialty coatings for small metal
parts, laboratory and hospital equipment and the like.

Emulsion Paints

The use of acrylic latices in paints, both trade sales and, more recently,
industrial finishes, has become tremendously important with the continu-

ing acceptance of water-based paint vehicles. Convenience of application and cleanup, and nonflammability, of course, represent two of the main factors contributing to the acceptance of this medium for which acrylic resins are well adapted. Because of their ease of manufacture and performance in use, acrylics are continuing to overcome their one disadvantage in this field—higher initial cost.[9]

The copolymer resins used for the purpose differ in composition according to whether they are designed for indoor or outdoor application. For both uses, the resins must have good pigment binding powers and relatively low film-forming temperatures. The ever-present cost factor dictates the use of the cheapest monomers available in the acrylic copolymer for indoor use where UV light resistance is not of paramount importance. In this area, vinyl acetate and styrene can be used acceptably and often account for a greater percentage of the copolymer than acrylics. For outdoor use, however, the photostability of the acrylic structure is the basis for its considerable usage. Poly(methyl methacrylate) itself is much too hard and has too high a minimum film-forming temperature for coalescing, so the resin must be softened by copolymerization, normally with such esters as ethyl- or 2-ethylhexyl acrylate. While small amounts of functional monomers are used to enhance pigment-binding capabilities, a high proportion must be avoided to maintain water-spot resistance.

Emulsions have two properties which inherently limit their utility in some applications—the retention of the emulsifying agent in the final film, which tends to lessen water resistance, and the fixed rate of evaporation of the water.

The acrylic resins are manufactured directly in emulsion form and are of relatively high molecular weight—a phenomenon inherent in the emulsion process. To improve leveling properties of the resin and to aid in coalescing, high-boiling fugitive plasticizers are used in considerable quantities.

Acrylics are also available in the emulsion form as thermosetting resins. In this form, the thermosets afford advantages similar to those of thermoplastic emulsions, i.e., higher solids at workable viscosities and the absence of a flammable solvent. When resins are cross-linked by baking, hardness and chemical resistance are improved in a manner analogous to the thermosetting acrylic solutions. Typical uses include primers and top coats for appliances and metal furniture.

The resin matrix of paints designed for application to dimensionally unstable porous substrates, such as many woods, must be flexible enough to expand and contract without cracking. This, too, is a function of copolymer composition and plasticization. On the other hand, emulsion paints for ceramic, metal and other dimensionally stable substrates are not limited by flexibility but nevertheless must have a relatively low coalescing temperature, unless a baking step is practical.

Miscellaneous

Several newer methods of applying decorative and protective coatings are well adapted to the use of acrylics. These include electrodeposition, electrostatic spray of solids and curable "100% solids" systems (polymer in monomer). Hot solvent dip processes are uniquely adaptable to acrylics because of their thermal and oxidative stability. Organosol and hydrosol forms have recently made their appearance, offered for use in automotive finishes, trade sales paints and the like. As the coatings industry becomes more aware of their versatile properties, acrylic resins will inevitably be used more and more in these and other applications.

STARTING FORMULATIONS

Several recipes are given below which represent starting points in formulating coatings based on acrylics.* Modification may be necessary to obtain the optimum blend for each use, taking into account differences in substrates, environments and methods of application. Specific resins and suppliers of ingredients have been listed in Table 9.5. Since many

TABLE 9.5.   Resins Used in Formulations

Soft solid acrylic resin #1-"Elvacite" 2044 acrylic resin (a)
Moderately hard acrylic solution resin #2-"Elvacite" 6014 acrylic resin (a)
Low molecular weight acrylic resin #3-"Elvacite" 6013 acrylic resin (a)
Alcohol-soluble acrylic resin #4-"Elvacite" 2045 acrylic resin (a)
Solid acrylic #5-"Elvacite" 2013 acrylic resin (a)
Methacrylate solution resin #6-"Elvacite" 6016 acrylic resin (a)
Hard methacrylate solution polymer #7-"Elvacite" 6011 acrylic resin (a)
Carboxyl-functional acrylic #8-"Acryloid" AT-70 (i)
Exterior acrylic emulsion #9-"Rhoplex" AC-34 (i)

Suppliers

(a) E. I. du Pont de Nemours & Co., Wilmington, Del.
(b) Shell Chemical Co., New York, N. Y.
(c) Union Carbide Corp., New York, N. Y.
(d) American Mineral Spirits Co., Murray Hill, N. J.
(e) Monsanto Co., St. Louis, Mo.
(f) Eastman Chemical Products, Inc., Kingsport, Tenn.
(g) Dow Corning Corporation, Midland, Mich.
(h) Raybo Chemical Co., Huntington, W. Va.
(i) Rohm & Haas Co., Philadelphia, Pa.
(j) Nopco Chemical Co., Harrison, N. J.
(k) Metalsalts Corporation, Hawthorne, N. J.

---

*Letters in parentheses refer to list of suppliers given in Table 9.5.

patents are involved in these formulations, applications and specific ingredients, this publication is not to be taken as a license to operate under or a recommendation to infringe them.

Soft Lacquer for Nonferrous Metals

	Parts by Weight
Soft solid acrylic resin (#1)	89
Thinner: 59% toluene, 25% MIBK, 10% iPA, 6% "Pentoxone" (b)	356
$\frac{1}{2}$-sec nitrocellulose (HB-14-P) (a)	87
MIBK	213

To spray, cut with more thinner to viscosity.

White Lacquer for Aluminum

	Parts by Weight
Grind Portion	
Moderately hard acrylic solution resin (#2)	105
Methyl ethyl ketone	17
"Cellosolve" (c)	15
"Ti-Pure" R-900 titanium dioxide (a)	40
Letdown Portion	
Methyl ethyl ketone	18
"Cellosolve" (c)	15
Acrylic resin (#2)	744
Toluene	197.2
Ethyl alcohol, 2B	50
Benzotriazole	4.4
Epoxidized soybean oil	4.4

Clear Aerosol Lacquer

	Parts by Weight
Acrylic resin (#3)	17.1
Toluene	14.9
Methylene chloride (or acetone)	13.8
MIBK	4.6
"Poly-Solv" EE Acetate (d) (or high flash naphtha)	3.6
"Santicizer" 160 (e)	1.0
"Freon-12" Propellant (a)	45.0

Alcohol-based Spray Lacquer (Useful under Adverse Humidity Conditions)

	Parts by Weight
Alcohol-soluble acrylic resin (#4)	10
Isopropyl alcohol	25
n-Propyl alcohol	40
"Pentoxone" (b)	25

Acrylic Concrete Sealer

	Parts by Weight
Acrylic resin (#5)	28
"Santicizer" 160 (e)	3
Xylene	27
Toluene	42

Acrylic-Butyrate Wood Lacquer (Nonyellowing)

	Parts by Weight
Acrylic resin (#6)	21.3
Cellulose acetate butyrate $\frac{1}{2}$ sec (f)	8.5
"Santicizer" 160 plasticizer (e)	3.0
DC-510 (1000 centistokes) fluid (g)	0.01
Eastman Inhibitor DOBP (f)	0.09
Toluene	32.1
"Tecsol," 95% (f)	10.0
Ethyl acetate	5.0
Isobutyl acetate	10.0
MIAK (methyl isoamyl ketone)	10.0

Steel Coating Lacquer[5]

	Parts by Weight
Grind Portion	
"Ti-Pure" R-610 titanium dioxide (a)	6.17
Carbon black	0.07
Hard methacrylate solution polymer (#7)	4.03
"Cellosolve" acetate (c)	2.53
Letdown Portion	
Hard methacrylate solution polymer (#7)	11.60
"Santicizer" 160 butyl benzyl phthalate (e)	3.76
Cellulose acetate butyrate, $\frac{1}{2}$ sec (25% solids)	10.03
"Cellosolve" acetate (c)	17.87
MEK	21.97
Toluene	21.97

Thermosetting Appliance Enamel

	Parts by Weight
Grind Portion	
"Ti-Pure" R-900 titanium dioxide (a)	27.4
Carboxyl functional acrylic (#8)	18.3
Letdown Portion	
Carboxyl functional acrylic (#8)	21.8
"Epon" 1001 (50% solids) (b)	26.7
Xylene	7.8
"Cellosolve" acetate (c)	2.6
Raybo 3 (anti-silk agent for smoothness) (h)	0.06
Reduce with xylene-"Cellosolve" acetate to desired viscosity	

Bake 30 minutes at 350°F

White Exterior House Paint

	Pounds per Hundred Gallons
Grind Portion	
Water	53.6
"Tamol" 731 (25%) (i)	10.7
"Triton" CF-10 (i)	2.5
"Nopco" NXZ (j)	1.0
Ethylene glycol	25.0
Pine oil	3.0
"Metasol" 57 (100%) (k)	1.8
Hydroxyethylcellulose (2.5% solution) (c)	115.0
"Ti-Pure" R-610 titanium dioxide (a)	240.0
"Ti-Pure" FF titanium dioxide (a)	10.0
Talc	100.0
Calcium carbonate	110.0
Letdown Portion	
Exterior acrylic emulsion (#9)	512.0
Water	7.7
"Nopco" NXZ (j)	1.0
Ammonium hydroxide (28%)	2.0

REFERENCES

1. Brown, W. T., and Miranda, T. J., *Offic. Dig. Federation Soc. Paint Technol.*, **36**, 92 (1964).
2. Burrell, H., *Offic. Dig. Federation Paint Varnish Prod. Clubs*, **27**, 726 (1955).
3. Burrell, H., *Offic. Dig. Federation Soc. Paint Technol.*, **34**, 131 (1962).
4. E. I. du Pont de Nemours & Co., "A New Dimension in Solvent Formulation-"Elvacite" Resins," Brochure A-44369, 1965.
5. Evans, J. L. (to E. I. du Pont de Nemours & Co.), U.S. Patent 2,849,409 (August 26, 1958).
6. Gerhart, H. L., *et al.*, *Offic. Dig. Federation Soc. Paint Technol.*, **33**, 679 (1961).
7. Hamm, S. P., "Copolymerization," p. 653, New York, Interscience Publishers, 1964.
8. Luskin, L. S., and Myers, R. J., "Acrylic Ester Polymers," in "Encyclopedia of Polymer Science and Technology," Volume I, p. 246, New York, Interscience Publishers, 1964.
9. Payne, H. F., "Organic Coating Technology," Vol. 1, p. 553, New York, John Wiley & Sons, 1965.
10. Riddle, E. H., "Monomeric Acrylic Esters," p. 29, New York, Reinhold Publishing Corp., 1954.
11. Schildknecht, C. E., "Vinyl and Related Polymers," p. 297, New York, John Wiley & Sons, 1952.
12. Trommsdorff, E., *et al.*, in "Methoden der Organische Chemie," Vol. 14, Part 1, p. 1010, Stuttgart, Georg Thieme Verlag, 1961.

10

*Cellulosic Film Formers**

Cellulose Structure

Cellulose chemistry has made great strides since its beginning in the early 1800's, but it still presents great difficulties. The present theory states that the cellulose polymer is a condensation derivative of the carbohydrate β-glucose, a six-carbon polyhydroxy alcohol whose long chain is formed by the loss of H and OH at the 1- and 4-positions, respectively.

The repeating linear chain of the cellulose molecule can be illustrated as:

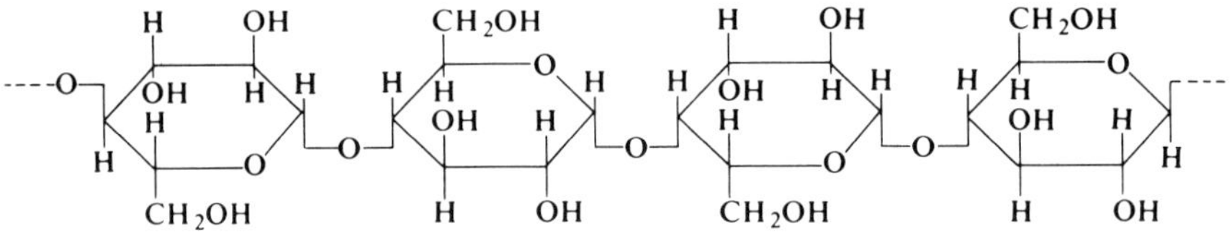

Cellulose

Cellulosic Derivatives

Cellulose, since it is essentially polyhydric, can form esters of organic acids, reaction products of inorganic acids, as well as ether compounds. These cellulose derivatives may be classified according to reaction product and include:

(1). Esters of Inorganic Acids
   Cellulose nitrate
(2). Esters of Organic Acids
   Cellulose acetate
   Cellulose acetate butyrate
(3). Ethers
   Methylcellulose
   Ethylcellulose
   Hydroxyethylcellulose
   Ethylhydroxyethylcellulose

*By F. Steslow, The Sherwin-Williams Co., Chicago, Ill.

These reaction products are the most important ones to the coatings industry. Other products have been formed but, because of physical and chemical properties as well as economics, have not been as successful. Thus the coatings industry has a wide variety of materials from which to select and create an end product with varying film properties.

Cellulosic Lacquers

The protective or decorative coatings formed by cellulosic products have been known as "lacquers." The term as it is used today does not confine the role only to cellulosics but includes other film formers. A more general definition of a lacquer is a coating whose primary film properties are formed by evaporation of the solvent. The paint chemist or formulator utilizes these cellulosic derivatives to make a clear or pigmented coating made up of the following components: (1) cellulosic, (2) volatiles, (3) modifying agents, (4) pigment. In a clear coating, the pigment is excluded.

The cellulosic used is dependent upon the ultimate properties and end use visualized by the paint chemist. Table 10.1 covers these properties both physical and chemical.

The volatile or solvent portion of the lacquer can be divided into three distinct constituents:

(1) Active solvent: a true solvent for the cellulosic.

(2) Latent solvent or co-solvent: by itself, it is not a true solvent, but in the presence of an active solvent it will improve the solvent power of the active solvent and will itself contribute considerable solvency.

(3) Diluent: a nonsolvent for the cellulosic.

The volatiles may be further classified as:

(1) Low boilers: boiling range below 100°C.

(2) Medium boilers: boiling range between 100 and 145°C.

(3) High boilers: boiling range between 145 and 170°C.

(4) Extra-high boilers: boiling range over 170°C.

The selection of the volatiles is very important not only because they do a particular job, but also because of their economics. The main point to keep in mind is to have a proper solvent balance of the three volatiles, namely active solvent, latent solvent and diluent, in such a ratio as to produce a cellulosic film with the optimum properties.

Cellulosics by themselves produce films or coatings with inferior properties. Therefore, they are usually used in conjunction with modifying agents, either plasticizers or resins. The plasticizers may be classed in two groups, nonsolvent and solvent. The main purpose of adding plasticizers to cellulosics is to increase the flexibility of the coatings. At the same time, they soften the coating and decrease its tensile strength. The

TABLE 10.1. Physical and Chemical Properties of Cellulosics

	Nitro-cellulose	Cellulose Acetate	Cellulose Acetate Butyrate High Acetyl	Cellulose Acetate Butyrate High Butyrl	Ethyl-cellulose	Methyl-cellulose	Ethylhydroxy-ethylcellulose
Bulking value, gal/lb	.0704	.0925	.104	.0965	.110	.108	.107
Specific gravity	1.70	1.30	1.25	1.17	1.09	1.29	1.12
Tensile strength, psi (1-mil film)	9000–16,000	8500–11,000	2500–6500		6800–10,500[a]	8500–11,400	3000–6000[a]
Elongation, psi (1-mil film)	10–50	4–55	40–60		7–30[a]	10–15	5–10[a]
Softening point, °F	155–220	235–255	240	165	160	190	175
Solubility	Esters, ketones, ether alcohols, glycol ethers	Ketones	Ketones, esters	Alcohols, aromatic hydrocarbons	Alcohols, aromatic hydrocarbons	Water	Aliphatic and aromatic hydrocarbons, esters, ketones, chlorinated solvents
Resin compatibility	Good	Limited	Limited	Wide	Wide	Limited	Wide
Plasticizer compatibility	Excellent	Limited	Limited	Wide	Wide	Limited	Wide
Resistance to:							
Weak acids	Fair	Good	Good	Poor	Good	Good	Good
Strong acids	Poor	Poor	Poor	Poor	Poor	Poor	Poor
Weak alkalies	Poor	Good	Poor	Poor	Good	Good	Good
Strong alkalies	Poor	Poor	Poor	Poor	Very poor	Poor	Good
Water	Good	Good	Good	Good	Good	Poor	Good
Discoloration							
Sunlight	Fair	Good	Excellent	Excellent	Good	Excellent	Good
Heat	Fair	Very good	Very good	Fair	Fair	Fair	Fair
Viscosity range	Wide	Limited	Fair	Fair	Wide	Wide	Limited
Flammability	High	Very low	Low	Low	Low	Low	Low
Color of film	Water white	Water white	Water white	Water white	Water white	Water white	Water white
Use in coatings	Wide	Limited	Fair	Wide	Wide	Limited	Limited

[a] 3 mil.

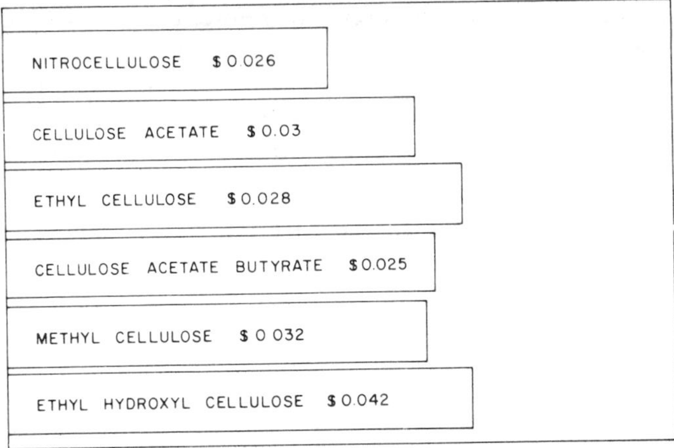

Calculations based on Union Carbide Plastics Div. Nomograph.

Figure 10.1.   Cost comparison of cellulosic resin (1966).   Cost ¢/1000 sq in. of 1 mil film.

resins may be natural or synthetic types and are used to increase adhesion of the film, improve light and weather resistance, increase gloss, increase moisture and water resistance, and reduce costs.

Pigments are added to the cellulosics to create a decorative effect in a wide range of hues. Pigments are also functional since they extend the service life of the cellulosics upon exterior exposure. The reaction of pig-

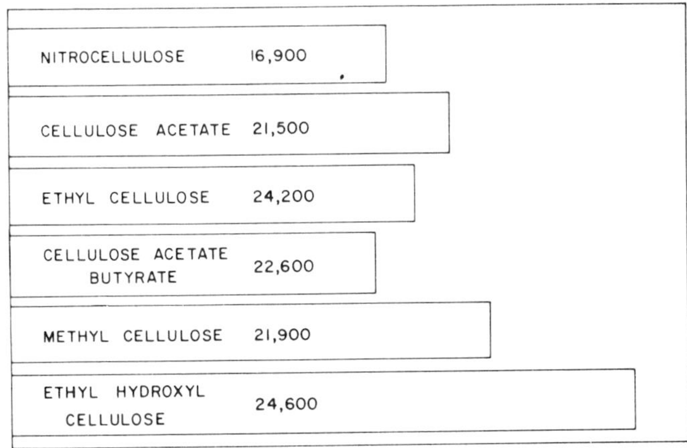

Calculations based on Union Carbide Plastics Div. Nomograph.

Figure 10.2.   Yield comparison of cellulosic resin (sq in. of 1 mil film/pound resin).

ments with these cellulosics can occur; therefore, care must be exercised in their selection.

The comparative properties of cellulosics are given in Table 10.1. Also, Figure 10.1 compares costs per 1000 square inches per mil of film and Figure 10.2 compares yields in square inches per pound of resin.

CELLULOSE NITRATE

History

Cellulose nitrate, commonly called nitrocellulose, is the only inorganic ester of cellulose which has gained commercial importance and large-volume usage as compared with other cellulosics. It is also the oldest of the cellulosic derivatives. Initially, the interest in nitrocellulose was in the field of explosives, and the early pioneers included Braconnet in France (1833), Schoenbein in Switzerland (1845), and Parkes in England (1855). It was not until 1882 with the production of amyl acetate by J. H. Stevens in the United States that the modern lacquer coatings industry was initiated. The greatest growth rate of nitrocellulose for the coatings industry, especially for finishing automobiles, occurred immediately after World War I; today, it is a multimillion dollar business (Figure 10.3).

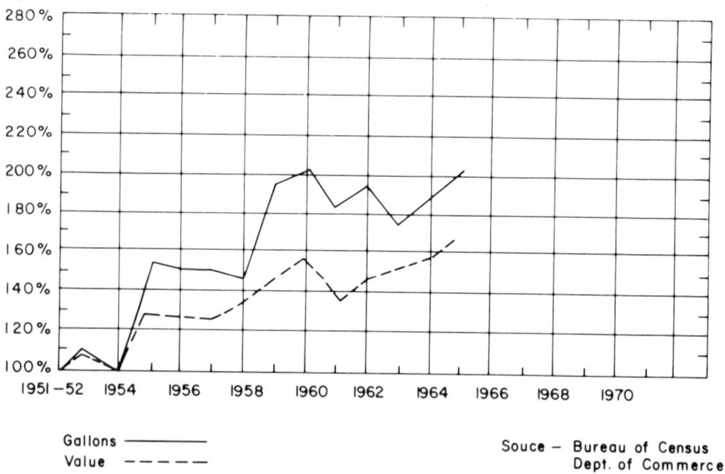

Figure 10.3. Industrial lacquers, percentage change

Chemistry

Cellulose nitrate is a reaction product of cellulose, nitric acid and sulfuric acid.

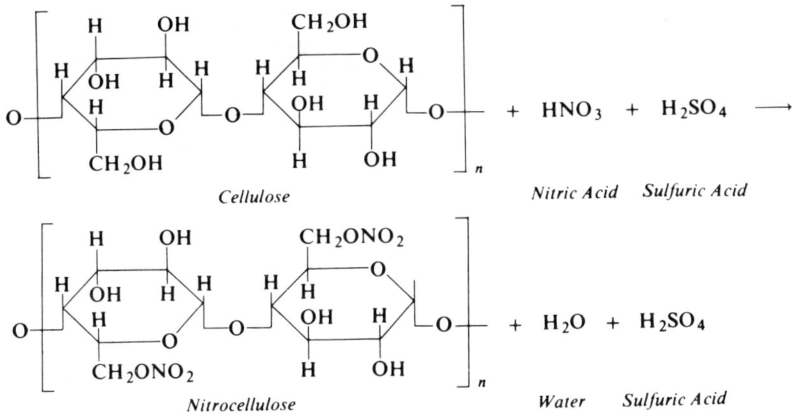

$$\text{Cellulose} + HNO_3 + H_2SO_4 \longrightarrow$$

Cellulose   *Nitric Acid   Sulfuric Acid*

$$\text{Nitrocellulose} + H_2O + H_2SO_4$$

Nitrocellulose   *Water   Sulfuric Acid*

Cellulose in the form of cotton linters (practically pure cellulose) is boiled in a caustic to remove dirt and impurities, then washed. The reaction of nitric and sulfuric acids with the cellulose is carefully and accurately controlled until the desired degree of nitration is reached. The sulfuric acid is used to take up the water formed in the reaction. The reaction product is then washed thoroughly and treated with ethyl, isopropyl or butyl alcohol and adjusted to 30% alcohol and 70% nitrocellulose. A recent addition is the use of water as a dampener in place of alcohol. This results in a product which is safe to handle and ship. Even though it is dampened, one must remember that this cellulosic is a highly flammable material and great care must be exercised in its use at all times. A pamphlet from the Manufacturers Chemist Association, "Handling of Nitrocellulose Manual Sheet N-1," is available, which lists all the safeguards regarding this product.

Types and Viscosity

Nitrocellulose is commercially available in three types based on the degree of nitration of the cellulose, each type having distinctly different solubility characteristics. The nitrogen content and solubilities can be summed up as follows:

TABLE 10.2

Nitrogen Content	Per Cent Nitrogen	Solvent-type Solubility
High	11.8–12.2	Esters, ketones, glycol ethers and ether-alcohol mixtures
Medium	11.2–11.7	Same as high nitrogen content except a greater tolerance for alcohol
Low	10.7–11.2	Considerable tolerance for alcohol

The three types of nitrocellulose are available in a large number of viscosity grades, the viscosity being a function of the modified cellulosic chain. As the length of the polymer chain increases, there is a corresponding increase in viscosity. Thus one can vary the viscosity and keep the solids content constant or vice versa, vary the solids content and keep the viscosity constant.

Formulating Principles of Nitrocellulose Coatings

Volatile or Solvent. As a rule, nitrocellulose is seldom used by itself to produce a coating, but it is normally dispersed in a volatile or solvent mixture to achieve a fluid liquid. The volatile components (active solvents, ketones, esters, glycol ethers), latent solvents (alcohols), and diluents (aromatic and aliphatic hydrocarbons) are blended in a definite ratio to meet a specific end application. Some of these methods of application are the following:

Aerosol	Roller coat, reverse and direct	Spray
Brush	Knife coater	Airless, hot and cold
Dip	Curtain coater	Conventional, hot and cold
Flow coat	Silk screen	Electrostatic
Squeegee	Tumble	Rotogravure

Each of these applications requires a specific volatile or solvent blend. Therefore, it is important to select a volatile or solvent blend with the following considerations in mind:

(1) All the components must be dissolved in the lacquer liquid. Lacquer technology dictates the use of modifiers (both resins and plasticizers) to achieve specific film properties. The compatibility of all the components making up the lacquer is required to assure solubility of these components in the lacquer liquid. Some solvents which are satisfactory for nitrocellulose may not dissolve one or more of the modifying agents.

(2) The rate of volatile evaporation must be balanced to achieve optimum film properties. The volatility of the solvent dictates the evaporation rate of the solvent from the wet film. It is important to have a proper balance of the volatiles to make certain one of the components may not precipitate if sufficient solvency for this component is lost too fast.

(3) The ratio of hydrocarbons to esters or ketones—i.e., diluent to active solvent ratio—must be maintained at high levels for economic reasons. Normally, nitrocellulose can tolerate higher proportions of aromatic than aliphatic hydrocarbons and can be added in excess of 50% by weight of the solvent mixture. Typical solvent balance can be achieved by the following mixtures:

Active solvents (esters, ketones)	30	35
Latent solvents	15	15
Aromatic diluents	55	—
Aliphatic diluents	—	50

Fast-evaporating blends can use higher percentages of diluent. Slow-evaporating solvents have lower solvency.

(4) Blushing, or a white discoloration of the film, will occur if the solvent blend evaporates too fast. The rapid cooling of the film results in the condensation of atmospheric moisture on the wet film trapping it, which causes the nitrocellulose to precipitate from solution. A slowing down of the evaporation rate by a high-boiling active solvent prevents a blush from occurring.

(5) Proper application viscosity is required. As noted above, there are many different methods of application and no two applications utilize the same solvent blend. In conjunction with application method, the drying of the film must also be considered. This includes air drying under different humidity and temperature conditions and force dry by hot air or infrared.

In general, a proper blend of low- to medium-boiling active solvents is required, and at times a high-boiling active solvent is essential.

Nitrocellulose.   The contributing properties of nitrocellulose to a lacquer can be summed up as follows: The higher the viscosity grade, the tougher, more flexible, and more cold-check resistant will the film be. The most common nitrocellulose coatings are produced from the medium-viscosity grades which make possible lacquers of fairly high solids content that maintaining good end properties. The low-viscosity grades result in lacquers of maximum solid content at the expense of some of the end properties.

Plasticizers.   The compatibility of the nonsolvent- and solvent-type plasticizer with nitrocellulose is rather wide. The nonsolvent type includes both raw and blown castor oil, blown soya, treated linseed, tung and other nonoxidizing oils. The solvent-type plasticizers used include the phthalate esters which also serve as a solvent for the modifying resins. The solvent-type plasticizer is more efficient than the nonsolvent; therefore, less is needed to achieve the desired flexibility.

Pigments or Colorants.   Imparting color to a lacquer film can be achieved with organic dyes that are soluble in the volatiles or solvents in the lacquer. These normally give a tint to a transparent film. The other method is to add a pigment to produce an opaque or semiopaque film. Approximately half the lacquers produced are pigmented and are highly decorative. Besides imparting a color, the pigments added are functional since they extend the durability of the lacquer film upon exposure to sunlight. Some pigments are more efficient than others in this respect. Dyes, as a whole, have poor exterior durability and are used primarily for interior purposes.

Pigments of an alkaline nature must be avoided in nitrocellulose lacquers since they react and produce gas-containing nitrogen oxides. When the use of a basic pigment is necessary, tests indicate that small amounts of acids are required to neutralize the alkaline nature of the pigment.

In selecting colorants for lacquers, the following should be considered: (1) high hiding and covering power, (2) lightfastness, (3) bleeding properties, (4) easy dispersibility, (5) chemical resistance, (6) heat resistance and (7) rheological properties.

Formulas

At one point, the different methods of applying lacquers were discussed along with the volatile components which make up the liquid lacquer. These volatile components require a proper balance and a specific viscosity to be able to meet the application properties. The application methods used cover every possible substrate and include:

Metals	Plastics	Foil
Woods	Glass	Fabrics
Paper	Masonry	Rubber
Leather	Plastic films	"Masonite"

These substrates have been coated with many different lacquers because each surface presents different requirements. The task would be endless to present formulas for each since hundreds of formulations are used over one particular substrate. To demonstrate representative formulas in the main areas of use—automotive, furniture and industrial—we can illustrate the difference in the ratios of the film former and exclude the pigmentation.

Furniture Lacquer	*Automotive Lacquer*	*Industrial Lacquer*
33.3% N/C	45% N/C	25% N/C
33.3% Maleic rosin	40% Alkyd	50% Alkyd
33.4% Alkyd	15% Plasticizer	15% Ester gum
		10% Plasticizer

Future Trends

Nitrocellulose lacquers have established themselves in the protective and decorative industry. Many industries prefer these coatings because of the relative ease of application. The lacquer industry has not remained static; it has introduced new coatings and new application techniques which have eliminated many problems while producing better and more durable coatings and increasing production rates. There is still room for new developments, and much work is being done in exploring the many facets of research and development. In view of all the potentialities existing in the coatings industry, there is every reason to believe these nitrocellulose lacquers and their new modifications will keep up with the improvements in other coatings fields.

CELLULOSE ACETATE

Chemistry

The first published reports of the synthesis of cellulose acetate was in 1869 by Schutzenberger. This was accomplished by heating cellulose with acetic anhydride in a sealed tube at 180°C.

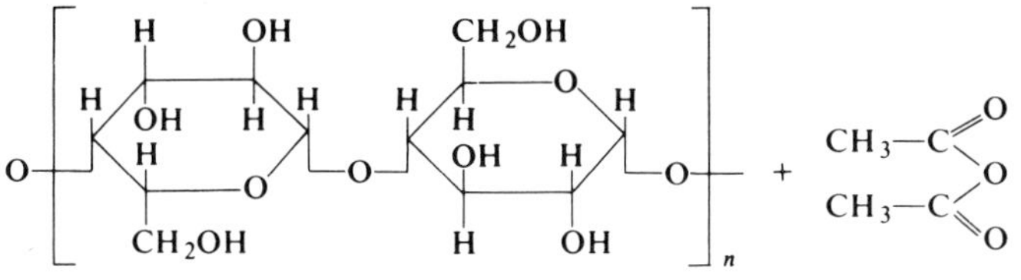

Cellulose

Acetic Anhydride

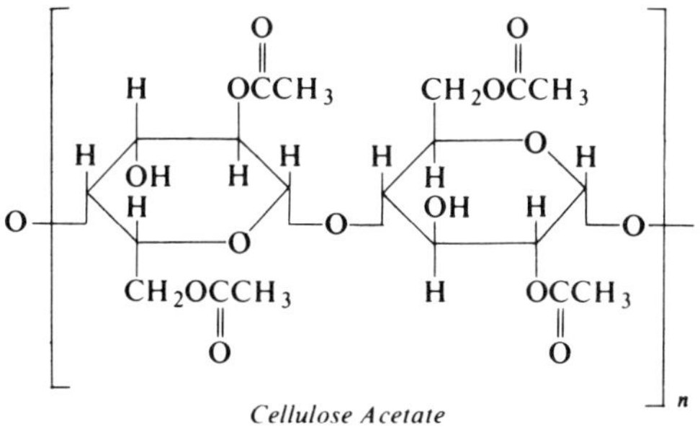

Cellulose Acetate

Today, this reaction is modified to include acetic and sulfuric acids. The reaction proceeds to the triacetate which is then hydrolyzed to reduce the acetyl content, producing a more soluble cellulose acetate. Cellulose triacetate is a rather insoluble polymer and has found practically no use in the coatings field.

Types and Viscosity

Cellulose acetate is available for coatings applications with acetyl content ranging from 38 to 40.5%. These ranges present a cellulosic with the best solubilities and properties. As with nitrocellulose, different viscosity esters are available and the viscosity depends upon the length of the polymer chain. The solubility of the ester increases as the viscosity of the ester decreases.

Formulating Principles

The same formulating principles apply to cellulose acetate as to nitrocellulose. The essential difference is a much lower compatibility for resins and plasticizers as well as special solvent requirements which limit the cellulose acetate to special formulating problems. Due to its high melt index (255°C), it lends itself to use in heat-resistant coatings. It also is used for cable and wire coatings, textiles, leather and paper coatings. Cellulose acetate has poor adhesion to metals and must be properly modified to achieve the desired adhesion. The main use for cellulose acetate is in the production of rayon.

CELLULOSE ACETATE BUTYRATE (CAB)

Chemistry

Because of the limited solubility characteristics of cellulose acetate, modification of cellulose with mixed esters produces a more soluble and more compatible polymer. Acetic and butyric acids and their anhydrides are reacted under controlled conditions to yield cellulose acetate butyrate. The proportions can be varied from a high acetyl content with a small amount of butyryl to a low acetyl, high butyryl content. The fully esterified product can then be partially hydrolyzed to form additional chemical products with a wide variation of the physical properties.

Cellulose Acetate Butyrate

Types and Viscosity

CAB is available in a range from 31% acetyl and 17% butyryl to 6% acetyl and 48% butyryl. Thus one can divide the polymer into two main groups, one with a high acetyl content and the other with a high butyryl content with the dividing range of physical and chemical properties at about 20.5% acetyl and 26.5% butyryl.

As with other cellulosics, there are different viscosity types. With CAB, the viscosity varies within a specified acetyl and butyryl content of the polymer.

Formulating Principles

Because of its numerous modifications, CAB is available to the paint chemist with a variety of physical and chemical properties. At the present time, it is the second most important cellulosic in the coatings industry.

The properties can be summed up as an increase in the butyryl content increases the solubility, tolerance for diluents, compatibility for resins and plasticizers, flexibility, and moisture resistance. As the butyryl content decreases, the tensile strength, hardness and melting point increase.

Two areas of use have been extended to the CAB lacquers: the automotive and furniture industries. The combination of CAB and acrylic resins has produced an excellent nonyellowing, water-white furniture coating. They have also been modified with nonoxidizing or semioxidizing alkyds. Recent advances included modification with urea-formaldehyde resins to produce coatings with exceptional flexibility and adhesion to wood. Coatings of these types can air dry in 10 to 15 seconds when applied in thickness of 0.3 to 0.5 mil with the proper selection of volatiles or solvent blends. The advantage to CAB lacquers is the apparent high build developed by such a coating, along with excellent holdout due to the good bridging properties.

For the last three years, the automotive industry has utilized a low-pigmented, high-gloss CAB acrylic lacquer. To produce the high gloss, the film is force dried at 180°F then reflowed at 325°F to produce a film of unusual smoothness similar to glass. This reduces the need for buffing and polishing the coating, which was required years ago with nitrocellulose types. A coating of this type exhibits exceptional hardness, adhesion and nonyellowing properties.

ETHYLCELLULOSE

Chemistry

The idea of cellulose ethers was first conceived in 1905 by Suide, but it was not until 1926 that L. Lilienfeld received the first patent on synthesis of ethylcellulose. Initially, cotton is swollen in a 12 to 25% solution of sodium hydroxide and then treated with ethyl chloride until a certain percentage of the sodium hydroxide is neutralized.

$$\left[\begin{array}{c} H \quad OC_2H_5 \quad\quad CH_2OC_2H_5 \\ \text{(ethyl cellulose ring structure)} \\ CH_2OC_2H_5 \quad\quad H \quad OC_2H_5 \end{array}\right]_n$$

Ethyl Cellulose

Types and Viscosity

The reaction product has an ethoxy content which varies with the degree of ethylation and yields commercial-grade products which range from 45 to 49.5% ethoxy content. Within these ethoxy types, the viscosity can be widely controlled and, as with other cellulosic polymers, is a function of the polymer length.

Formulating Principles

The most unusual property of ethylcellulose is its low density, which allows greater coverage or greater volume per unit weight than any other cellulosic derivative. It has about 45% more volume than nitrocellulose and about 20% more than cellulose acetate. Costwise on a volume basis it is almost equal to nitrocellulose.

The compatibility properties of ethylcellulose are very wide with regard to plasticizer and resins. Very weak solvent blends (80:20, toluene: alcohol) dissolve ethylcellulose quite readily. Besides being extremely flexible, it is not flammable.

Its main uses are in printing inks, paper coatings and industrial coatings. Recent interest has been shown in high solids gel coatings which are made up in solvent blends that dissolve ethylcellulose only at elevated temperatures. At room temperature these blends have insufficient solvent power. These gel coatings with a single hot dip application have been depositing smooth, uniform film thickness of 15 to 30 mils of dry film.

METHYLCELLULOSE

Chemistry

The dimethyl ether of cellulose, or methylcellulose, is a reaction product similar to ethylcellulose except that methyl chloride is substituted for ethyl chloride. Its basic chemical structure is

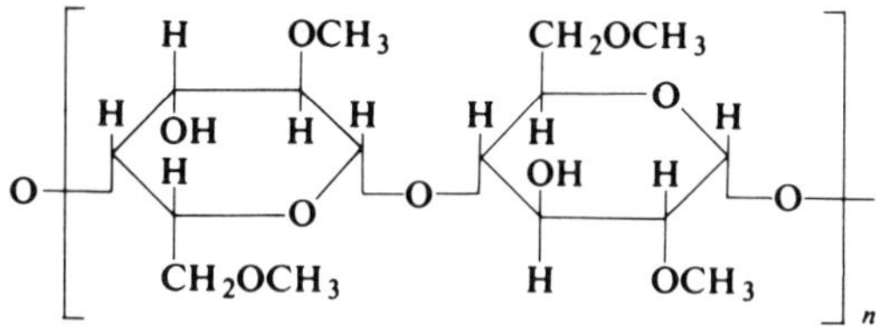

$$\left[\begin{array}{c} \text{H} \quad \text{OCH}_3 \qquad \text{CH}_2\text{OCH}_3 \\ \text{H}\ \text{OH}\ \text{H}\ \text{H}\quad \text{H}\ \text{H}\ \text{O}\ \text{H} \\ \text{O}\ \text{H}\qquad \text{OH}\ \text{H} \\ \text{CH}_2\text{OCH}_3 \qquad \text{H} \quad \text{OCH}_3 \end{array} \right]_n$$

Methyl Cellulose

Types and Viscosity

The methoxy content varies from 27.5 to 32.0%, which gives a product having the maximum water solubility. The viscosity, as with other cellulosics, is controlled by the chain length of the cellulose. This factor makes possible a wide range of viscosities.

Formulating Principles

Methylcellulose by itself is not used as a main film former in the coatings industry, but it has been utilized in water-based systems because of its hydrophilic colloidal nature. Small concentrations are used as a thickener to produce cream-type water-based emulsion paints which are nondrip and nonspatter.

The biggest use of methylcellulose in the last 15 years has been in multicolor paints. Multicolor is a suspension of a colored particle in a colloidal solution of methylcellulose, which acts as a protective colloid around each color particle and prevents them from coagulating. Market survey indicates nitrocellulose to be the main film former and biggest single type of film former used in these paints. The lacquer phase is dispersed in methyl cellulose, and as many as five different colors are used to produce a pleasant decorative coating.   U.S. Patents 2,591,904 and 2,809,119 describe this process.

Today, the multicolor market has reached its peak. Multicolor paints are used very widely in the architectural field, which accounts for the bulk of their sales. The product finishing or industrial field accounts for a large segment of the remaining usage, while a very small percentage is confined to the do-it-yourself market.

ETHYLHYDROXYLETHYLCELLULOSE

Ethylhydroxylethylcellulose is used to speed up the drying of enamels as well as to add some marproofing to the coatings. It has not been used as a main film former because of its high cost.

HYDROXYETHYLCELLULOSE AND CARBOXYMETHYLCELLULOSE

Hydroxyethylcellulose is produced from alkali cellulose by the reaction with ethylene oxide

$$R\text{—}ONa \quad + \; CH_2\text{—}O\text{—}CH_2 \rightarrow RO\text{—}CH_2CH_2OH$$

Alkali Cellulose   *Ethylene Oxide*   *Hydroxyethylcellulose*

It is used as a thickener for emulsion paints.

Carboxymethylcellulose is available as the sodium salt, sodium carboxymethylcellulose. It is prepared from alkali cellulose as follows:

$$R\text{—}ONa \quad + \quad ClCH_2COONa \quad \rightarrow \quad ROCH_2COONa \; + \; NaCl$$

Alkali Cellulose   *Sodium Monochloroacetate*   *Carboxymethylcellulose*

This is also used in emulsion paints.

CONCLUSION

Cellulosic derivatives have been placed in an enviable position as film formers, either modified or by themselves, in the protective and decorative coatings industry. Their many advantages offset the few disadvantages of these film formers. They possess a wide variety of physical and chemical properties found in no other single segment of the coatings industry. In less than half a century, these cellulosic derivatives have increased to a value in excess of a quarter of a billion dollars in a production year, and this growth continues. The future will yield new developments from these film formers which will enhance the numerous products decorated and protected by these cellulosics.

REFERENCES

1. Dow Chemical Co., Midland, Mich., "Methocel Handbook," Form #ML6A-1254, 1953.
2. Eastman Chemical, Division of Eastman Kodak, Kingsport, Tenn., "Eastman Cellulose Esters," Bulletin #T50-848, 1950.
3. Hercules, Inc., Wilmington, Del., "Ethyl Cellulose, Properties and Uses," Form 5M 3-58 7111, 1955.
4. Hercules, Inc., Wilmington, Del., "Nitrocellulose, Chemical and Physical Properties," Form 500-359 6500 2-63 3426, 1963.
5. Otto, E., "High Polymers," Vol. 5, Ch. 8, New York, Interscience Publishing, 1946.
6. Davidson, R. L., and Sittig, M., (editors), "Water Soluble Resins," New York, Reinhold Publishing Corp., 1962.
7. Wertheim, E., "Textbook of Organic Chemistry," p. 381, The Blakiston Co., 1939.

11

*Vinyl Resins for Coatings**

There are many high polymer resins available today which could be classified as vinyl. One system would class any resin a vinyl which has the basic building block group (R—CH=CHX) as part of the molecule. This system would include the 3 billion pound thermoplastics of today (polyethylene, vinyl chloride and copolymers, and polystyrene) and the maleates and acrylics among others. We shall limit this discussion to just one of these classes: vinyl chloride, related copolymers and modified types. We will also limit the discussion to the coatings portion of the vinyl chloride business, which in the United States is about 40 to 100 million pounds, depending on the definition of coatings.

The vinyl chloride class of resins had its beginning in about 1926, when a study was undertaken by Union Carbide Corporation to find uses for ethylene and acetylene gases. This study ultimately resulted in the preparation of vinyl chloride copolymer resins on a commercial scale in 1936. Since that time, the growth of vinyl resins in production and sales has been phenomenal, second only to polyethylene in this respect. Even today, vinyl resins as a class are still expanding into new markets. The vinyl chloride resins have been and remain the most versatile types of materials yet invented by the vast plastics industry. This is especially true in the coatings field.

Vinyl chloride resins are made by polymerizing vinyl monomers, the most important of which are vinyl chloride and vinyl acetate (Figure 11.1). The polymerization is carried out commercially in various ways. Basically, polymerization is a free radical-initiated chemical process wherein monomeric building blocks are converted to long molecular chains by a simple head-to-tail or head-to-head addition process. The monomers retain their identity in the long-chain polymer, except that the

By W. H. McKnight and G. S. Peacock, Chemicals & Plastics Div., Union Carbide Corporation, Bound Brook, N. J.
*Deceased.

VINYL SOLUTION AND DISPERSION RESINS

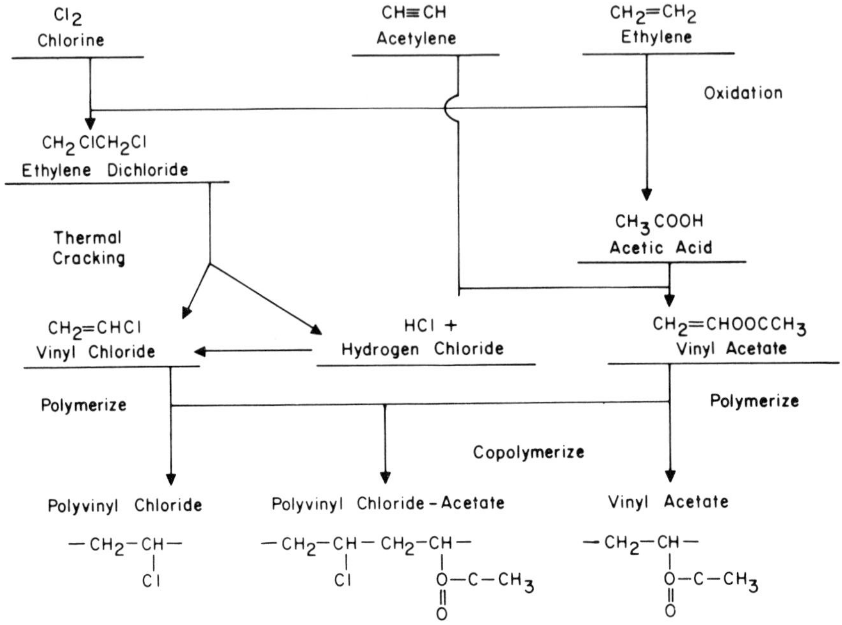

Figure 11.1. Monomer manufacture.

double bonds largely disappear. Thus a vinyl chloride polymer or homopolymer consists of a long chain of vinyl chloride units. Molecular weight or chain length can be controlled in various ways, such as by the time, temperature, and pressure of the reaction, chain-stopping agents, etc. More than one monomer may be used to produce a copolymer having properties suitable for specific uses.

Vinyl chloride and copolymer types are manufactured commercially by one of five basic polymerization processes: (1) solution polymerization, (2) bulk polymerization, (3) suspension polymerization, (4) emulsion polymerization and (5) combination procedures (e.g., polymerization followed by hydrolysis).

POLYMERIZATION METHODS

Bulk Polymerization

Bulk polymerization is probably the oldest and simplest method known. The monomer (or monomers) is simply stirred in an autoclave under proper conditions (with catalyst, etc.). The polymer formed eventually becomes a slurry of flaky or granular resin in the unused monomer. When the proper degree of conversion is reached, the reaction is stopped by discharging the slurry, with a consequent pressure drop. This also serves to strip off the unreacted monomer. Resin made by this process may or may not be washed before the final drying operation. This method poses many problems, not the least of which involves removing the heat of polymerization. Both the resin and the monomers are poor heat conductors; this fact is complicated further by a tendency for layers of polymer to build up on the walls of the equipment. This material is referred to as "ivory." Formation of "ivory" necessitates frequent and costly shutdowns of the equipment for cleaning. Because of the cost of downtime, this method of polymerization is not widely used today for vinyl resin manufacture.

Solution Polymerization

The first copolymer resins manufactured commercially were prepared by the solution polymerization technique—a method that is still in wide use for making low and medium molecular weight resins. Basically, it consists of preparing a solution of monomer (or monomers) in a suitable solvent, which also contains a catalyst. This mixture is stirred in an autoclave, under controlled pressure and temperature conditions. The polymer builds up slowly and is dissolved in the same solvent initially used for the monomers. This results in a solution which gradually increases in viscosity and molecular size. The reaction is stopped when the viscosity reaches a predetermined level. Often additional monomers are added during polymerization, and the process can become quite complicated when monomers having different rates of polymerization are being copolymerized. When the desired viscosity is obtained in the autoclave, indicating a certain molecular weight range, this "autoclave varnish" is passed on to precipitators where the dry resin is recovered by precipitation with nonsolvents. In some cases, the resin may be redissolved and reprecipitated. The wet slurry is usually washed with water to remove traces of catalyst and is finally dried in a hot air blast to produce a fine granular powder. This method of polymerization is costly because of the many steps involved and the use of unique engineering procedures. It is still operated in competition with other polymerization methods. The advantage of this process for coatings applications is that no emulsifiers or other ionic species are deposited with the resin to cause cloudy films and under-film corrosion on exposure.

Suspension Polymerization

Suspension polymerization has largely supplanted bulk polymerization of high molecular weight vinyl resins and is also used to prepare some of the resins which compete commercially with the solution-polymerized type. The greatest tonnage of vinyl resins is produced by this technique. As the name suggests, it involves a suspension of the monomer-polymer in a carrier which is normally water. The system takes advantage of the good heat transfer properties of water which helps control the rate of polymerization. It yields a resin powder consisting of small beads or spheres. The size and shape of the beads may be controlled by the type and rate of agitation in the autoclave and by the use of very small amounts of protective colloids such as polyvinyl alcohol, dissolved in the water phase. The catalyst is normally benzoyl or lauryl peroxide, which are water-insoluble and therefore remain with the resin particles. The engineering of autoclaves and stirring systems for suspension poly-merization is among the vinyl resin manufacturer's most closely guarded secrets. The type and amount of protective colloid also becomes a critical factor which determines the properties of the dry resin. By selecting the proper stirring technique and the judicious use of water-soluble modifiers, resins can be made to have the optimum flowing properties for easy handling as dry powders. This is especially important in preparing resin compounds for automatic extrusion or molding operations. Probably the greatest disadvantage of this polymerization method, once the engineering is mastered, is its tendency to produce a resin whose molecules vary widely in weight or degree of polymerization. The presence of the pro-tective colloid, which is deposited within or upon the resin particles on drying, frequently leads to quality control problems. The catalyst is largely decomposed by the time the polymerization is completed. The suspension polymerization technique is relatively new. As development work progresses, many new types of resins may be expected from this process in the future.

Emulsion Polymerization

Emulsion polymerization differs from suspension polymerization mainly in the degree of dispersion in the autoclave. In this method, water is the carrier or suspending medium. An emulsifier or surfactant is used to obtain a very fine dispersion of particles of monomer in the water. Here again, a water-insoluble catalyst is generally used. The emulsion is often homogenized to produce even smaller particles of monomer in the starting emulsion. The final autoclave product from this polymerization method is a stable suspension of polymer particles in the 0.01 to 1.0 μ size range. This is indeed a very fine dispersion, and the latex or hydrosol produced resembles milk or similar types of stable water dispersions. Resins thus produced may be sold in the latex form, or the dry resin may be recovered and sold as a powder. When recovery of the dry resin is desired, the problems of separating water from such ultrafine particles are complex. Dusting and loss of some of the powder to the atmosphere is common. The method used for recovery is usually spray drying of the latex. Resins of this type, which are sold as dry powder, are used to prepare organic coating dispersions known as plastisols and organosols. These will be discussed later.

Combination Methods

The fifth method is a combination procedure whereby a resin polymerized by one of the foregoing methods is further processed by hydrolysis or some additional chemical reaction. Resins made by combination procedures are always more expensive than the simple polymerized types. These extra process steps justify their cost by adding special properties. Some of the coating resin grades attain very desirable properties such as alkyd compatability through these "double processing" techniques.

VINYL CHLORIDE SOLUTION RESINS

Vinyl solution resins are normally applied as solutions in organic solvents. The chemical and physical properties of polyvinyl chloride are directly dependent on the molecular weight of the polymer. The solubility of polyvinyl chloride of high enough molecular weight to give sufficient toughness or flexibility for coatings is so low that application from organic solvent solution is impractical. Accordingly, copolymers of vinyl chloride and vinyl esters are produced which are considerably more soluble in certain organic solvents. These copolymers contain up to about 20% vinyl ester, usually vinyl acetate. The vinyl ester portion of the comonomer greatly improves solubility and flexibility, but it detracts slightly from the maximum chemical and physical properties obtainable from the vinyl chloride homopolymer.

Copolymer solubility is dependent on molecular weight, comonomer type and content, and uniformity of copolymerization. The latter property is characteristic of the method of polymerization. Solution-polymerized copolymers have polymer molecules that contain uniform but randomly distributed amounts of comonomer. Uniform copolymerization by the solution process is more likely because both monomers are in a continuous phase; hence each polymer molecule is as soluble as the next. Suspension-polymerized copolymers are generally less uniform and contain a considerable proportion of molecules high in vinyl chloride content. Polymerization occurs in the discontinuous phase and the vinyl ester monomers, being relatively soluble in water, are not always readily available for copolymerization. Thus suspension copolymers are less soluble in ketones and less tolerant to aromatic hydrocarbon diluents; added heat is often necessary to prepare solutions. The water solubility of the protective colloid used to maintain the suspension during polymerization tends to increase water sensitivity and decrease the durability of the coating. Despite these apparent shortcomings of suspension-polymerized copolymers, the industry has made rapid strides in polymerization technology, and these resins compete with solution copolymers for many applications. Table 11.1 lists the range of solution resin copolymers available from one manufacturer.

Chemical modification of the vinyl chloride–vinyl ester copolymers notably improves certain properties; adhesion and compatibility are two of these. Air dry adhesion of the unmodified copolymer to smooth, nonporous substrates such as bare metals is so poor that these resins are used to formulate strippable coatings for temporary protection. Copolymerization with 0.5 to 1% of an unsaturated carboxylic acid such as maleic acid gives polymers having excellent air dry adhesion to a wide variety of metallic substrates. Carboxyl-modified polymers are used alone or in blends with unmodified copolymers. Carboxyl-modified copolymers serve as primers for unmodified copolymers and organosols based on polyvinyl chloride dispersion resins.

Hydroxyl modification of the copolymer also improves adhesion to various substrates such as wood, cellulosic materials in general, vinyl butyral wash primer and alkyd resins. A comparison of air dry adhesion of these polymers to a variety of substrates is shown on Table 11.2. Hydroxyl-modified copolymers are produced by preparing the vinyl-ester copolymer followed by partial hydrolysis of the ester to hydroxyl, or by direct copolymerization with a hydroxyl-containing monomer. Finished products generally contain from about 2 to 10% hydroxyl by weight. Hydroxyl modification also greatly increases the compatibility of the vinyl copolymers with a variety of other polymers and modifying resins. Hydroxyl-modified copolymers are compatible with acrylics, alkyds, oleoresinous vehicles, polyamides, polyketones, urethane prepolymers and certain urea- or melamine-formaldehyde resins. Compatibility makes it possible to improve other types of surface-coating materials. As an example, alkyds are often modified to increase speed of drying. In addition, the vinyl increases weather proofness, and water and alkali resistance. Conversely, vinyls can be so modified as to improve their own properties. Melamine-formaldehyde is often thermally cross-linked with a hydroxyl-modified vinyl to increase hardness and solvent resistance. Table 11.3 shows the compatibility of hydroxyl-modified copolymers with a variety of materials.

TABLE 11.1. Vinyl Chloride–Acetate Resins for Solution Coatings[a]

Product	Approximate Chemical Composition (% by wt)			Specific Gravity	Per Cent by Weight	Solvent Ratio	Viscosity at 25°C (cps)	Properties and Uses
	Vinyl Chloride	Vinyl Acetate	Other					
VYHH	86	14	–	1.36	22	1/1 MIBK/toluene	375	Basic coating resin for all coatings uses. Good adhesion upon baking, and good air dry adhesion with addition of VMCH
VYHD	86	14	–	1.36	25	2/1 MIBK/toluene	175	For general coatings use. Similar to VYHH, but has greater solubility and may be used when extreme toughness and durability of VYHH are not required.
VYLF	88	12	—	1.37	30	1/1 MIBK/toluene	188	Use similar to VYHD when even higher solids and greater gloss and guild required, but with much less flexibility and toughness
VYNS	90	10	–	1.36	12	MEK	135	Like VYHH but less soluble and tolerates more plasticizer. Has better heat stability.
VYNW	97	3	–	1.39	8	Tetrahydrofuran	120	Can be highly plasticized to give tough elastic coating, nontacky even at 225°F. Limited solubility
VAGH	91	3	5.7[b]	1.39	20	1/1 MIBK/toluene	300	Like VYHH, but compatible with wide range of other coating materials, including some alkyds. Air dry adhesion to most coating vehicles
VAGD	90	4	6[b]	1.39	20	1/1 MIBK/toluene	150	Like VAGH, but lower viscosity
VMCH	86	13	1[c]	1.35	22	1/1 MIBK/toluene	325	Use alone or blend with other vinyl chloride-acetate resins, such as VYHH, for air dry and low-bake adhesion
VMCC	83	–	0.9[c]	1.34	26	1/4 MIBK/toluene	270	For general coatings use. Much like VMCH but with greater solubility. Used when extreme toughness and durability of VMCH is not required

[a] Union Carbide Corporation—Chemicals & Plastics Div.
[b] Hydroxyl calculated as vinyl alcohol.
[c] Interpolymerized dibasic acid (0.7–0.8 carboxyl)

TABLE 11.2. Air Dry Adhesion of Vinyl Solution Resins

Substrate	Hydroxyl-modified Copolymer	Unmodified Copolymer	Carboxyl-modified Copolymer
Clean, smooth metal	Poor	Poor	Excellent
Glass	Fair	Poor	Excellent
Wood	Fair	Poor	Fair
Paper	Good	Poor	Good
Cloth	Good	Poor	Fair to excellent
Phosphated metal	Fair	Poor	Excellent
Phenolic resin	Good	Poor	Fair
Urea resin	Good	Poor	Fair
Vinyl chloride resin	Excellent	Excellent	Excellent
Chlorinated rubber	Fair	Fair	Fair
Acrylic and methacrylic ester resin	Excellent	Excellent	Excellent
Oleoresinous (varies widely)	Fair to excellent	Poor	Poor
Alkyd resin	Excellent	Poor	Fair
Vinyl butyral resin	Excellent	Poor	Fair
Shellac	Good	Poor	Poor
Concrete (somewhat dependent on type)	Good	Good	Excellent
Plaster (somewhat dependent on type)	Good	Good	Excellent
Nitrocellulose	Poor	Poor	Fair

Solubility

Vinyl solution resins are generally dissolved by stirring in ketones, nitroparaffins, certain esters and chlorinated hydrocarbons. Suspension-polymerized vinyls may require heat to effect a good solution. Solvents such as ketoethers, ether esters, polyethers and certain other types of compounds have varying degrees of solvent action on these vinyl resins. Aromatic hydrocarbons like xylene or toluene are used as diluents and tend to swell, and in some instances partially dissolve, the resins, especially at high temperatures. Water and aliphatic hydrocarbons are strong precipitants for vinyl resins. Alcohol also will precipitate them, except for the hydroxyl-modified type which tolerates alcohols as a small proportion of the diluent. Ketones are the most suitable class of primary solvents for vinyl resins. Compared to other solvents, they will yield higher resin concentration solutions without gelling, and lower solution viscosities at equivalent total solids content; they tolerate greater dilution with non-solvents and exhibit good storage stability.

Some chlorinated hydrocarbons are excellent solvents and are especially useful because of their nonflammability. Methylene chloride, ethylene dichloride or trichloroethylene are generally used in nonflammable solvent mixtures. It should be noted that chlorinated solvents are very selective in their solvent strength for vinyl copolymer resins. Carbon tetrachloride, for example, is a nonsolvent for these resins. These solvents also are highly toxic, and the utmost care should be taken in handling them; they should be used only in applications where effective ventilation is maintained.

Nitroparaffins, nitroethane and nitropropane are also very effective solvents for vinyl resins. They do not give as low a viscosity as ketones, but they have a slower evaporation rate. Because of their low solvent power, esters are generally used in a mixture with another active solvent. In special cases, solutions formulated with esters may result in a slight cost saving over formulas not containing esters. They also have a milder odor then ketones. However, all esters used for vinyl resin solutions should always be of high purity (alcohol free).

The ratio of active solvent to toluene will vary with the solution viscosity and evaporation rate desired for a specific application. Suitable thinner mixtures for spray application may consist of methyl ethyl ketone, methyl isobutyl ketone and methyl isoamyl ketone as solvents, with toluene and xylene as diluents. Methyl ethyl ketone has excellent solvent power and is available at low cost; however, the amount used should be carefully balanced with an appropriate amount of diluent. In a humid atmosphere, a high proportion of methyl ethyl ketone in the thinner will cause blushing and blistering before and during baking in warm weather. Methyl isoamyl ketone tends to eliminate these defects. It retards drying and improves flow-out. Small proportions of isophorone can be used for the same purpose. Other types of solvents and diluents, such as nitroparaffins, small portions of esters or high-solvency naphthas, may be used in certain spraying thinners.

Paper and cloth coatings may be formulated with highly volatile solvents such as acetone and methyl ethyl ketone. Their use in this application presents no problems. The lower heat capacity and rate of heat transfer of paper and cloth diminish the possibility of moisture blushing, and controlled drying conditions prevent blistering. Under these conditions, it is feasible to use mixtures such as acetone and toluene. Even though a large amount of the active solvent portion evaporates soon after application, this loss in solvent power of the mixture is counterbalanced by an increase in solvent power of the toluene at elevated oven temperatures.

Application by roller coaters requires slow-evaporating solvents and diluents. Isophorone has particular merit in such thinners because it combines a slow evaporation rate and an excellent solvent action on vinyls. High-boiling coal tar and hydrogenated petroleum naphthas normally serve as diluents.

TABLE 11.3. Modifiers for Hydroxyl-modified Vinyl Solution Copolymers

Modifier	Typical Examples	Supplier[a]	Remarks
Alkyd	"Beckosol" P-296-70	4	Improves gloss and gloss retention, increases solids. Some sacrifice in chemical and other severe exposure atmospheres.
	"Duraplex" ND-77B	5	
	"Paraplex" D-65-A	5	
Acrylic	"Acryloid" B-66	5	Increases hardness, gloss and gloss retention. Reduces light stability and flexibility. Tends to chalk on prolonged exposure.
	"Acryloid" B-72	5	
	"Acryloid" AT-50	5	
Epoxy	"Bakelite" Epoxy Resin ERL-2772	6	Improves adhesion and light stability, and flexibilizes the epoxy resin. Excessive chalking on exposure.
Bisphenol A type	"Bakelite" Epoxy Resin ERL-2774	6	
	"Bakelite" Epoxy Resin ERL-2795	6	
Peracetic acid type	"Unox" Epoxide 4221	6	
	"Unox" Epoxide 4289	6	
Butoxy	"Buton" 100	2	Improves adhesion and increases solids at some sacrifice in flexibility and color retention.
	"Buton" 200	2	
	"Buton" 300	2	
Phenolic	"Bakelite" Phenolic Resin BKR-2620	6	Increases adhesion, solids and hardness. Suggested for baking primers for vinyl top coats.
	"Bakelite" Phenolic Laminating material CLS-3112	6	
Wax	"Cardis" polymer	7	Improves chip and mar resistance. Used in small quantities for paper coatings.
	"Bakelite" Polyethylene Resin DYDT paraffin	6	

Polyketone	"Bakelite" Polyketone Resin 251	6	Increases solids, compatibility, adhesion and hardness. Some sacrifice in flexibility.
	"Bakelite" Polyketone Resin 252	6	
Amine	"Cymel" 300	1	For baking types. Increases gloss, gloss retention and hardness. Slight loss in flexibility on aging.
	"Resimene" 872	3	
	"Resimene" U-933	3	
	"Uformite" F-240	5	

[a] Key to suppliers:
(1) American Cyanamid Co., Plastics and Resins Division.
(2) Enjay Chemical Company.
(3) Monsanto Co.
(4) Reichhold Chemicals, Inc.
(5) Rohm & Haas Co.
(6) Union Carbide Corporation, Chemicals & Plastics Div.
(7) Warwick Chemicals Co.

Thinners for brush application also call for relatively slow-evaporating solvents, such as glycol ether esters, methyl isoamyl ketone, mesityl oxide and cyclohexanone, to obtain ease of application and good flow-out. The active solvent proportions should be carefully balanced with a sufficient amount of diluent to minimize the tendency of top coats to redissolve and "pick up" undercoats.

Typical viscosities are shown in Figure 11.2. The solubility of a lower molecular weight copolymer in various solvents and solvent-diluent mixtures is shown in Figures 11.3 and 11.4.

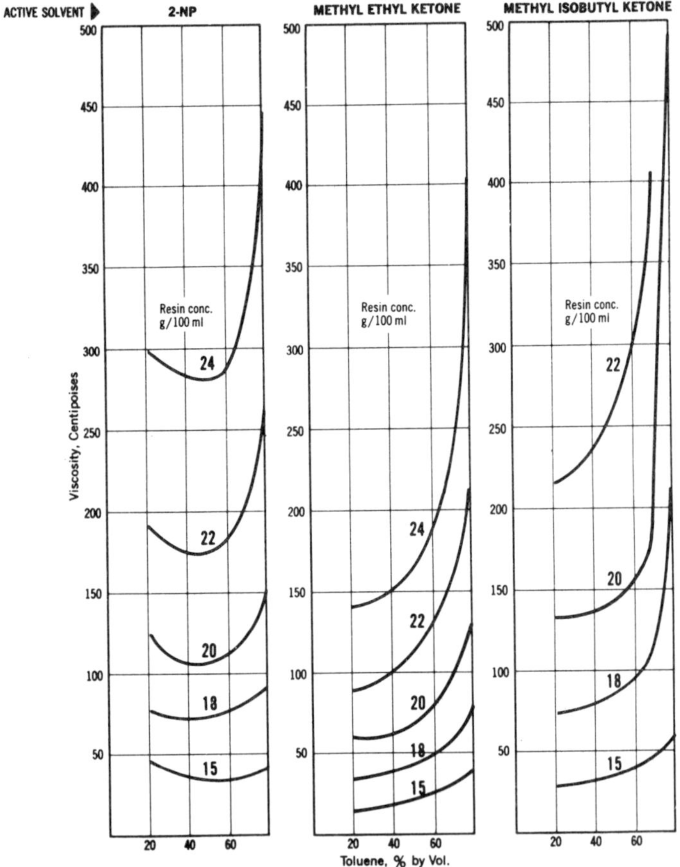

Figure 11.2. Viscosity of vinyl resin VYHH in various solvent mixtures. (*Courtesy of Commercial Solvents Corporation*, "Better Vinyl Printing Inks and Solvents with 2-NP")

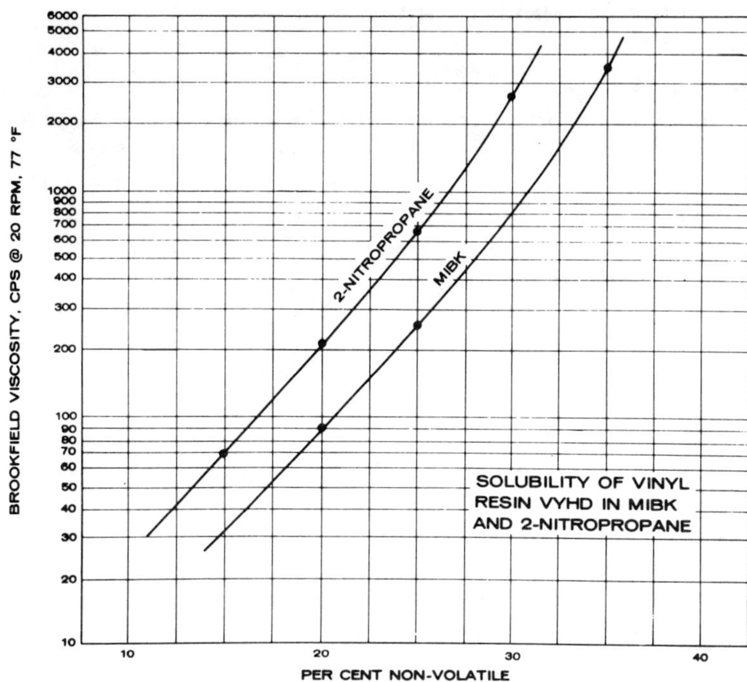

Figure 11.3.

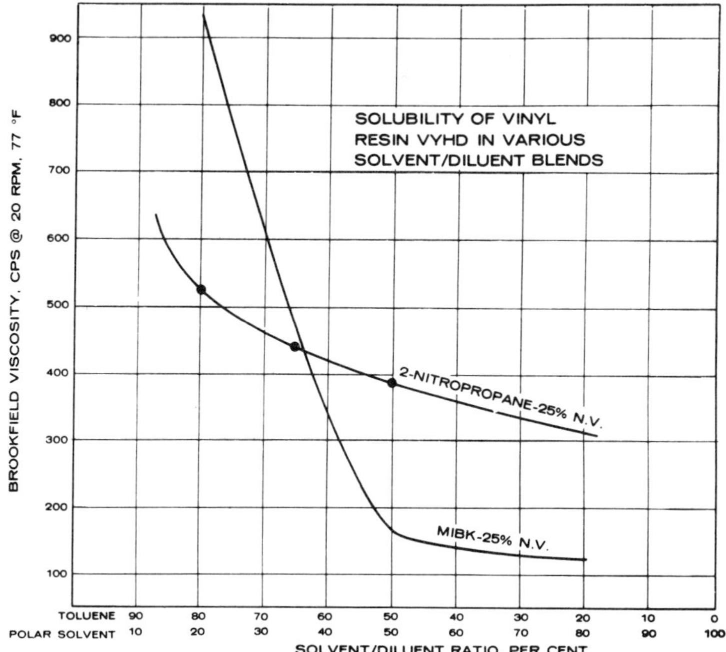

Figure 11.4.

Plasticizers

Plasticizers are an important consideration in formulating vinyl coatings by improving flexibility and impact resistance. A large selection of monomeric and polymeric plasticizers is commercially available. The plasticizer used should be good solvents for the resin, to minimize exudation or migration on aging. Use of a solvating-type plasticizer will also increase solvent release from the coating. An unplasticized vinyl copolymer coating may retain as much as 4 to 5% solvent in the film even after a long air dry or a forced dry at temperatures to about 175°F. Solvent retention is more prevalent where fast-evaporating solvents are used. It should be noted that pigmentation will also aid solvent release.

Vinyl solution resin copolymers are most compatible with monomeric plasticizers and have fair compatibility with most polymeric types. Blends of plasticizers are commonly used to impart specific properties. Plasticizer contents range from about 5 to 30% based on the resin weight. As plasticizer content increases, flexibility increases, while hardness and chemical and water resistance decrease. Other characteristics such as toxicity, flammability, color, odor and taste must be carefully considered before selecting a suitable plasticizer for a particular application. A partial list of plasticizer compatibility with vinyl copolymers is shown in Table 11.4.

TABLE 11.4.   Vinyl Solution Resin Compatibility with Plasticizers

	Weight	Vinyl:	Weight	Plasticizer[c]
	9:1	4:1	2:1	1:1
Di(2-ethylhexyl) phthalate	C	C	C	C
Ethylhexyl isodecyl phthalate	C	C	C	C
Diisodecyl phthalate	C	C	C	C
Tri(2-ethylhexyl) phosphate	C	C	C	C
Di(2-ethylhexyl) adipate	C	C	C	C
Triethylene glycol di(2-ethylhexoate)	C	C	C	C
Acetyl tributyl citrate	C	C	C	C
Acetyl triethyl citrate	C	C	C	C
"Aroclor" 1242 chlorinated diphenyl[a]	C	C	C	C
"Aroclor" 1254 chlorinated diphenyl[a]	C	C	C	C
"Paraplex" 6-60-polyester[b]	C	C	C	C
"Paraplex" 6-62-polyester[b]	C	C	C	C
Tricresyl phosphate	C	C	C	C
Butyl stearate	I	I	I	I
Butyl acetate ricinoleate	C	I	I	I

[a] Monsanto Co.
[b] Rohm & Haas Co.
[c] C = compatible; I = incompatible.

Heat and Light Stabilizers

Vinyl solution resins are degraded by prolonged exposure to UV light or by elevated temperatures. The degradation of vinyl copolymer films is inhibited by the addition of suitable stabilizers on an adequate quantity of certain pigments. The amount and type of stabilizer depends on the vinyl resin itself, the plasticizer and pigment, the baking schedule, and the end use application of the coating. The best heat stabilizers for vinyl copolymer resins are compounds that are slightly basic in nature and insoluble in water. For clear films, the stabilizer must be colorless. In such cases, some of the organic compounds of calcium, barium, cadmium, tin and lead are effective heat stabilizers. Epoxy resins in combination with organometallic derivatives of these metals provide maximum stability. Small amounts of urea-formaldehyde or melamine-formaldehyde resins (between 1 and 3 phr) also effectively stabilize clear films. Such films tend to develop checking or other discontinuities during baking, but these can be prevented by adding to the coating small amounts (approximately 1 phr) of long oil, drying-type alkyds, or non-drying or semi-drying oils. Substrates such as zinc, iron and tin plate accelerate decomposition of the coating during baking. Therefore, prior to application of the coating, the surface of these metals should be chemically passivated by phosphate treatment or the use of a varnish or oil-type primer.

Pigmentation

Pigment requirements for vinyl copolymer solution resins are similar to those of the higher molecular weight dispersion resins, which will be discussed later. General pigment characteristics that must be considered are hiding power, UV light protection, purity, ease of wetting in the vehicle, and resistance to solvent bleed, chemicals and heat. After selecting a suitable pigment, the pigment-to-binder ratio must be established for maximum hiding power and durability. For maximum exterior durability, optimum pigment concentration is somewhat higher than that required for hiding. Increasing pigment content to too high a level, however, will cause premature chalking. Finishes not requiring exceptional resistance to exterior exposure have fewer limitations on the degree and type of pigmentation. The slightly higher pigment contents in exterior formulations protects the vinyl resin from ultraviolet degradation. A list of typical pigments used with vinyl resins is given in Table 11.5. The amounts suggested are based on the quantity of each pigment necessary to provide good hiding and adequate UV light protection.

Rutile titanium dioxides are good pigments for vinyl resins for whites and tints. Where maximum exterior durability is desired, this pigment,

TABLE 11.5. Pigments for Vinyl Resins

Pigment	Parts per 100 Parts Vinyl Resin by Weight
Aluminum powder	40–55
Titanium dioxide	65–120
Titanium dioxide/antimony oxide (9:1)	65–120
Titanium dioxide/zinc oxide/antimony oxide (3:2:1)	65–120
Phthalocyanine green	15–25
Phthalocyanine blue	15–25
Chromic oxide green	65–100
Carbon black	7
Blue lead sulfate	100–170
White lead sulfate	100–170
Red lead	135–170
Iron oxide yellow (synthetic)[a]	55–100
Iron oxide red (synthetic)[a]	55–100
Iron oxide brown (synthetic)[a]	55–100
Iron oxide black (synthetic)[a]	55–100
Lead chromate	100–170
Strontium chromate	100–135
Cadmium reds	100–135
Chrome orange	100–135
Cadmium yellow	100–135

[a] All iron oxides require careful attention to heat stabilization for baking temperatures above 250°F. However, through the selection of proper stabilization systems, these pigments become very useful for exterior coatings. Natural oxides are seldom satisfactory.

as made by the sulfate process,[2] has given the best results. Replacing 10% of the titanium dioxide with antimony oxide will improve gloss retention and retard chalking. Antimony oxide is particularly effective in baking finishes. It functions in two ways: it acts as a UV light screening agent and as a stabilizer. Fillers or extenders with low chalking rates may be included in formulations for economy. High-purity grades of calcium carbonate, mica and other nonfibrous extenders are often used. Small amounts of flatting agents, such as silica gel, may be added where desired.

Traces of natural iron contamination will increase the chalking rate and discoloration of vinyl films. Such contamination from any source should be avoided. Pigments containing unstable forms of natural iron or zinc will also adversely affect the heat stability of vinyl resins. Synthetic iron oxides should be used in place of natural oxides. Iron or Prussian blue should be avoided for the same reason. This restriction also includes chrome greens which are blends of Prussian blue and lead chromate. Chromic oxide green, however, is an excellent pigment.

The use of basic pigments, such as zinc oxide, with the carboxyl-modified solution copolymer or blends of the carboxyl-modified copolymer with other vinyl copolymer resins should generally be avoided, because the chemical interaction between the two causes gelation. This is aggravated by small amounts of water which may be contained in the pigment or solvent. Addition of small percentages of an organic acid such as maleic acid (dissolved in solvent) may sometimes break light gels.

Dispersion of Pigments.   Pigments can be dispersed in vinyl resins by all the commonly used methods, such as two-roll milling, ball-mill grinding, sand grinding and high-speed impeller dispersion. Suitable wetting agents improve the grinding of many pigments. Two wetting agents particularly suitable for vinyl copolymer solution resins are "Nuodex" NA* and "Lexinol" AC-1.** These improve dispersion and gloss, and decrease paste viscosity and grinding time. Pigment dispersion in the relatively nonpolar vinyl chloride–vinyl ester copolymers is more difficult than in the carboxyl or hydroxyl modified copolymer unless sufficient shear can be attained to wet the pigment thoroughly. This can be obtained by two-roll milling the pigment in the resin and subsequently dissolving the "chips" in solvents. When ball mill grinding or high-speed impeller dispersions are made, best results are obtained by grinding in the carboxyl or hydroxyl copolymers or blends of these with the unmodified copolymers.

Vinyl solution resins may be applied by conventional application methods, such as brushing, spraying, dipping and knife coating. Films may also be applied by more specialized equipment such as roll coaters, gelatin roll metal coaters, forward and reverse roll coaters, and curtain coaters. All coating methods require careful adjustment of viscosity and solvent evaporation rates to fit the particular requirements of the operation.

Metal Surface Preparation and Adhesion

Over most surfaces maximum adhesion of vinyl coatings is obtained after baking at a temperature high enough to drive out residual solvent and to actually fuse the resin to the surface. Baking markedly improves adhesion on smooth nonporous substrates such as metals, glass and dense paperboards. Over more porous substrates such as concrete, paper and cloth, mechanical adhesion is usually sufficient. Air dry adhesion of the vinyl copolymers has been shown in Table 11.2.

Proper surface preparation is important to good coating performance. The coating can develop good adhesion only to a surface free of rust, grease or other foreign materials. The cleaning method used to prepare a surface will depend on the particular substrate. Some of the most com-

*Nuodex Products Co.
**American Lecithin Co.

mon cleaning methods for metals are solvent wash, hand and power tool cleaning, vapor degreasing, chemical and flame treatment, white metal blast, and commercial blast cleaning. Sandblasting to white metal is the most preferred procedure for steel, especially for corrosive environments. Metals such as steel, aluminum and copper alloys often require different treatments.

Metal surfaces should be passivated before baking vinyl finishes are applied. This may be done by chemical treatment such as phosphate salt coating or by the use of vinyl butyral-based wash primers or other primers. Primers containing maleic acid-modified vinyl solution copolymer resins have shown excellent coating performance. A carboxyl-modified vinyl should be used where direct adhesion to bare metal is required. Other commercial primers based on various alkyds, types of drying oils or varnishes also perform satisfactorily. These primers must be used with a top coat based on hydroxyl-modified copolymers or blends of the hydroxyl-modified with the unmodified copolymers to assure good intercoat adhesion. Hydroxyl-modified resins blended with alkyds, urea-formaldehyde or polyketone resins make excellent wood seal primers for many wood coating lacquers.

Uses of Vinyl Solution Copolymers

The oldest large use of vinyl solution resins is as a lining for cans, tubes and closures. These resins are tasteless, odorless and are sanctioned by FDA for contact with food. Hydroxyl- and carboxyl-modified copolymers are especially suited for this application because of excellent flexibility and adhesion to metal or various types of primers.

The effectiveness of maintenance paints based on vinyl solution resins has been established. The inert nature of the vinyl polymer aids resistance to attack by chemicals, chemical fumes, and both fresh and salt water. In addition, vinyl coatings have sufficient adhesion and flexibility to withstand the forces encountered during expansion and contraction of the metal during seasonal changes. Vinyl maintenance systems are generally applied in several coats for maximum protection. There are many government specifications describing such coatings. The classical vinyl maintenance system is based on a vinyl butyral wash primer, WP-1 (MIL-C-15328B) for saltwater or freshwater exposure. The wash primer is overcoated with a primer based on a hydroxyl-modified vinyl copolymer and top-coated with an unmodified vinyl copolymer. Such systems have been in regular use for over 20 years on locks and dams on inland waterways, in large chemical plants, and on ships and offshore oil-drilling rigs.

The flexibility of plasticized vinyl solution copolymers had led to considerable use in coil coatings on aluminum, galvanized and cold-rolled steel. Vinyl-based coatings will withstand extreme drawing, stamping, bending and other forming operation without cracking or loss of adhesion over the draw.

Use of vinyl chloride solution copolymer coatings for factory-applied wood coatings is a relatively new trend in the industry and is rapidly expanding. The coated wood is fabricated into panels for exterior house siding and interior decorative wood paneling. These coatings are used as furniture finishes. Hydroxyl-modified vinyl copolymers are formulated with nonoxidizing alkyds and melamine resins to form unpigmented coatings for decorative wood paneling, printed hardboard and chipboard. These vinyl finishes have excellent sanding and polishing characteristics, good cold check resistance, and good resistance to a variety of chemicals and household preparations. Vinyl systems have shown promise as exterior finishes for medium-density phenolic overlaid plywood. One such system is based on a primer consisting of a hydroxyl-modified copolymer and a urethane prepolymer and is top-coated with a blend of hydroxyl-modified and unmodified copolymer.

Coatings based on vinyl resins find wide use on aluminum and paper, both as coatings and as inks. Vinyl coatings have good flexibility and clarity, and they are heat sealable. These coatings can be used in food packaging because they are odor and taste free and have good water, alkali and grease resistance.

VINYL DISPERSION RESINS

Vinyl dispersion resins are very high molecular weight polymers based on vinyl chloride or copolymers of vinyl chloride and minor amounts of a modifying comonomer. Vinyl acetate, various olefins and vinylidene chloride are typical of comonomers used to impart specific properties. Dispersion vinyl polymers are produced by the emulsion polymerization process. Certain special types may be made by the suspension process. The emulsion process involves the preparation of an emulsion of vinyl chloride in water, using a suitable surfactant, and the polymerization of monomer into polymer. The dry resin is recovered from the latex by a spray-drying process which results in agglomerates of the very fine ultimate particles. These agglomerates make up the dry powder as it is sold.

Organosols

In a dispersion of vinyl resins, there is a suspension of discrete particles of resin in a liquid. When this suspending liquid contains volatile solvents, the dispersion is called an "organosol." If the liquid phase consists only of plasticizers, the dispersion is called a "plastisol." A plastisol by definition is changed to an organosol upon the addition of volatile solvents.

High viscosities are avoided in all these dispersions since the resin particles are not dissolved. However, viscosity of a suspension depends on (1) packing effects, (2) the volume of suspended matter in proportion to the volume of liquids, (3) the size and shape of the suspended particles, (4) the solvating effect of the liquids on the suspended particles and (5) the viscosity of the suspending liquids. The solids content can thus be quite high and the cost of volatiles low because of the small amount of liquid required and the elimination of the need for active solvents.

The commercial dispersion resins range in particle size from 0.05 to 1 μ in diameter. The smaller sizes are preferred in organosols because they exhibit less tendency to settle, whereas the large particle sizes are preferred for plastisols since greater fluidity and more stable viscosity are obtained with equivalent proportion of plasticizer. Typical properties of some commercially available dispersion resins are shown in Table 11.6.

TABLE 11.6. Typical Properties of Commercial Vinyl Dispersion Resins

Code	Vinyl Chloride (%)	Specific Viscosity	Ultimate Particle Size (μ)[a]		
			Lower Range	Probable Range	Upper Range
1	99 –	0.25	0.20	0.30	0.7
2	97	0.24	0.10	0.80	1.3
3	96	0.29	0.05	0.12	0.6
4	91	0.26	0.05	0.10	0.4
5	99 –	0.30	0.10	0.75	0.8
6	95	0.20	0.08	0.15	0.6

[a] Determined from electron micrographs.

The liquid constituents of organosols are classified into two types: dispersants and diluents. Dispersants are polar compounds which have affinity for the resins, aiding in the wetting and dispersing them. Plasticizers and volatile liquids, such as ketones and glycol ethers, are typical dispersants. Diluents are usually aromatic or aliphatic hydrocarbons or a blend of these. They are used to balance and modify the wetting and swelling characteristics of the dispersants and to lower the cost and viscosity of the liquid medium. Figure 11.5 shows the effect of thinner balance on organosols. If the organosol is overbalanced with diluent (i.e., contains excess diluent), the dispersion tends to agglomerate and flocculate, showing false body and poor stability on aging. However, if the organosol is oversolvated (i.e., contains an excess of dispersant), the dispersion shows high viscosity, resin swelling and a considerable viscosity increase on aging. A moderate excess of dispersant gives better viscosity stability of aging.

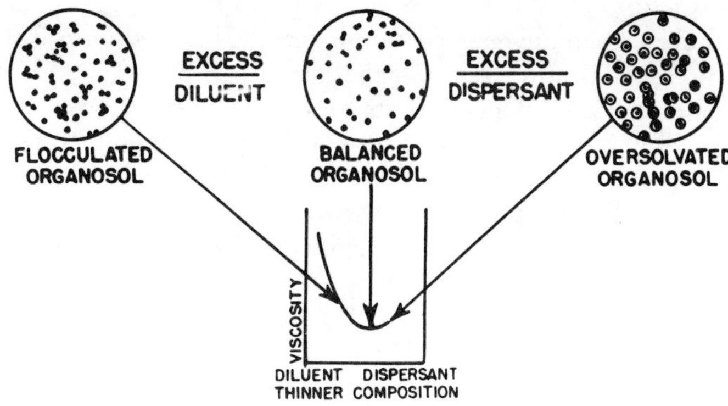

Figure 11.5. Effect of thinner balance on organosols.

The optimum composition of the liquid phase, i.e., a proper balance of dispersant and diluent to give low viscosities and storage stability, is determined by a study of viscosity as a function of composition. This is done by making several dispersions in which the ratio of dispersant to diluent is systematically varied and the viscosity of each measured. Figure 11.6 shows that the viscosity passes through a minimum as the dispersant-diluent ratio is changed, with constant resin content. This figure also

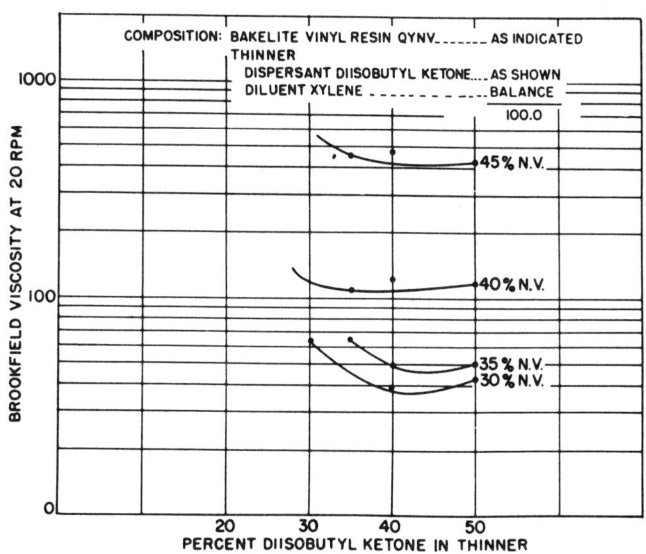

Figure 11.6. Organosol viscosity-composition diagram.

shows that even when the resin content is varied, the optimum dispersant-diluent ratio remains approximately the same.

Many combinations of dispersants and diluents have been studied as formulating aids. Diisobutyl ketone is a good dispersant and is widely used. It has a low solvating effect on the resin, yields low viscosities and has a relatively wide latitude in blends with hydrocarbon at the optimum viscosity. More active solvents such as methyl isobutyl ketone have been used, but they have a greater solvating effect on the resin, and therefore more care is necessary in the choice of diluents to counteract this effect. Less active solvents such as trimethylnonanone can also be used.

Combinations of aromatic and aliphatic hydrocarbons are generally used as diluents. Aromatic hydrocarbons swell the resin particles and therefore cause high viscosities. However, this swelling action aids fusion of the resin and is especially desirable in organosols of low plasticizer content. Aliphatic hydrocarbons, on the other hand, do not exhibit this swelling action and thus give lower viscosities at higher solids content. However, more dispersant is necessary to wet and disperse the resin. Figure 11.7 shows the effect of increasing the aliphatic hydrocarbon portion in the diluent. Note that with increased aliphatic content, the viscosity decreases, but the per cent dispersant needed for optimum viscosity increases. Figures 11.8 and 11.9 show the effect on organosol viscosity and the dispersant per cent of aromatic hydrocarbons ("Solvesso" #100) and of aliphatic hydrocarbons ("Apcothinner") when used as the sole diluent.

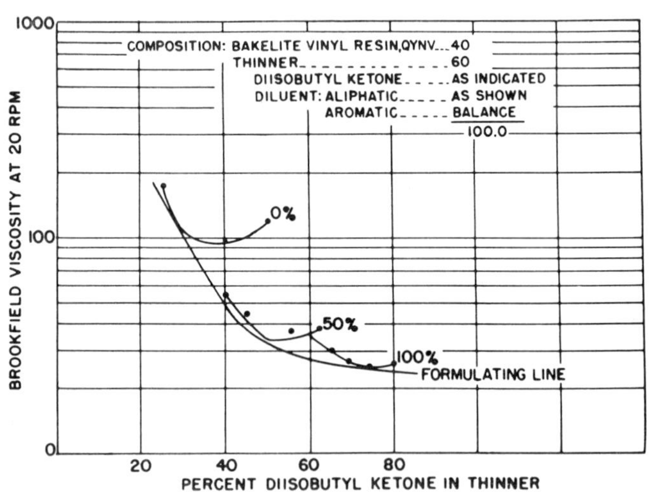

Figure 11.7.   Organosol viscosity with varying blends of diluents.

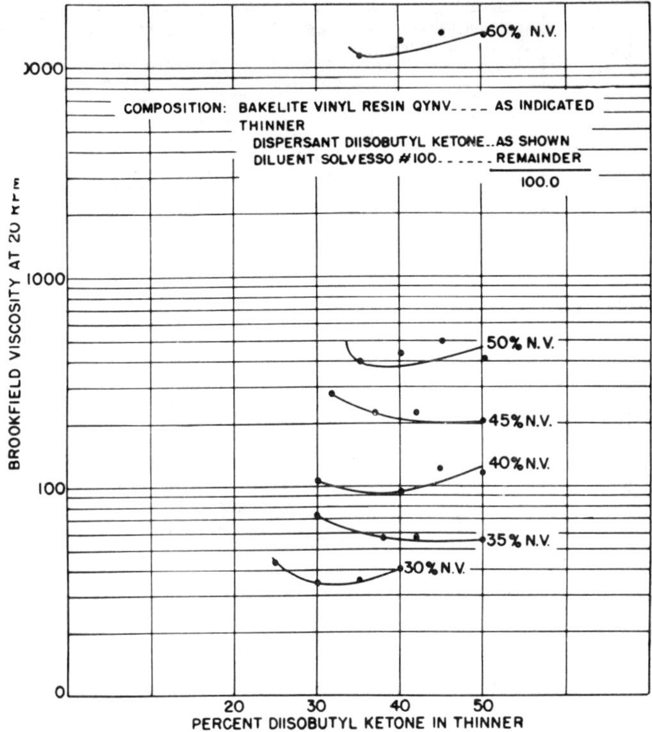

Figure 11.8. Organosol viscosity-composition diagram.

Plasticizers such as DOP [di-(2-ethylhexyl) phthalate] have little effect on organosol viscosity when used to replace portions of the volatile dispersant. Thus a plasticizer is an excellent dispersant, and since it does not volatilize during the fusing of the resin, the amount included will depend on the properties desired in the final film. Figure 11.10 illustrates this effect. Except for the initial increase in viscosity due to the higher viscosity of the plasticizer itself, no difference is observed in the viscosity of an organosol containing from 5 to 30 parts plasticizer per 100 parts resin.

The choice of plasticizer will depend on the properties desired in the final film. The effect of the various plasticizers on organosol viscosity can be determined by the methods described. The better plasticizers such as DOP, diisodecyl phthalate etc., are preferred dispersants in organosols. In compositions containing 30 or more parts plasticizer per 100 parts resin, the plasticizer usually serves as the sole dispersant and no volatile dispersant is needed. While the actual viscosity level, the proportion and the type of diluent used will vary with the plasticizer, the plasticizers perform equally well with all the common vinyl chloride dispersion resins.

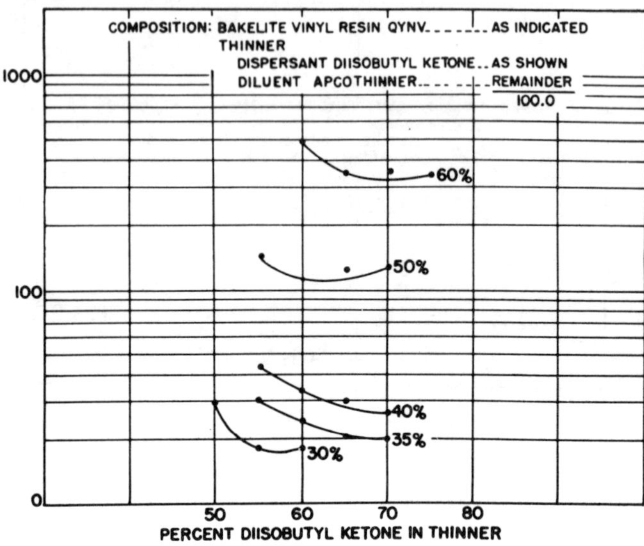

Figure 11.9. Organosol viscosity-composition curve.

Organosol Viscosity - Composition Curve

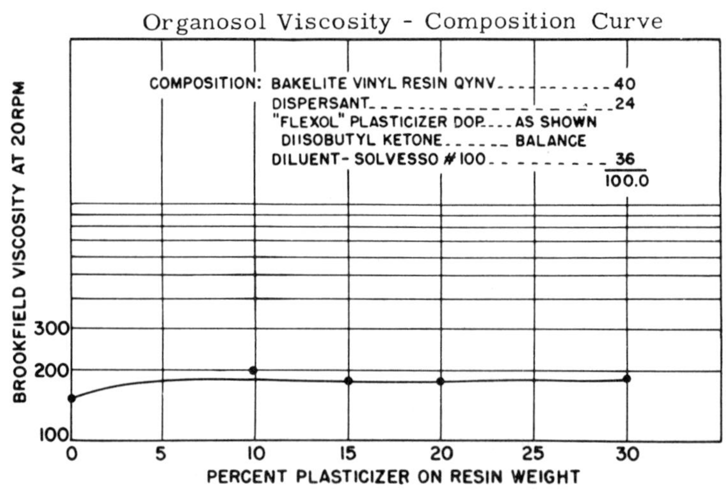

Figure 11.10. Effect of plasticizer on organosol viscosity.

Plastisols

Although the liquid medium of plastisols is solely nonvolatile plasticizers, the same optimum viscosities described for organosols are observed when primary plasticizers (dispersants) are blended with hydrocarbon-type secondary plasticizers (diluents). Since the viscosity of plastisols is more dependent on the resin-to-plasticizer ratio, the choice of plasticizer is more important than in organosol formulating. Generally plasticizers are divided into two classes, monomeric and resinous. Monomeric plasticizers are typified by such compounds as phthalate esters, the phosphate esters, adipate esters, sebacate esters, etc. These are high-boiling liquids having a relatively low viscosity. Resinous plasticizers are compatible resins, usually of the linear polyester type, and are generally used where migration and extraction of the plasticizer must be avoided.

Formulation

Basically a dispersion coating consists of the vinyl dispersion resin, a liquid suspending medium, appropriate pigments and/or fillers, and stabilizers. Other additives may be readily incorporated as desired to obtain any particular final film property. In an organosol, the ingredients are charged to a pebble mill and ground from 4 to 30 hours depending on the degree of dispersion desired. A plastisol is usually made by the stir-in technique, i.e., agitating by an appropriate means until a smooth, free-flowing dispersion is formed.

The amount and type of the above additives will be dictated by the end use. However, the formulator of organosol metal coatings must also consider behavior during drying. An improperly formulated organosol on drying may "mud crack." This is caused by a decrease in volume and a lack of appreciable binding action on the pigment and resin particles as the volatile portion leaves the film. Cracking can be readily avoided (1) by the use of stronger dispersants, (2) by heating the film rapidly to fuse particles together before an appreciable volume of the liquid is volatilized or (3) by using compatible resins that are soluble in the organic liquids present and will coalesce the film at room temperature. Examples of (3) are the solution-grade vinyl chloride–acetate resins. However, these resins should be dissolved in dispersant-diluent type solvents before adding to the organosol. If the dispersant-diluent balance of this solution is similar to that of the organosol, the organosol balance will not be greatly affected.

Fusing of organosols and plastisols occurs when the temperature is raised to the point where the dispersant becomes an active solvent for the resin. This temperature will vary with the resin and plasticizer used. Conventional polyvinyl chloride homopolymers fuse at 325 to 350°F.

Typical industrial bake schedules may range from 5 to 15 minutes at 350°F to 45 seconds at 500°F. Note that the time requirements of these bake schedules are substantially dictated by the mass and thermal conductivity of the substrate being heated and the heat transfer characteristics of the oven being used. Newer low-fusing vinyl chloride copolymers coalesce at temperatures as low as about 275°F. Tensile strength measurements of organosol films are one indication of how completely the dispersion resin has been fused. Figure 11.11 shows how tensile strength increase to a maximum with time at a given temperature.

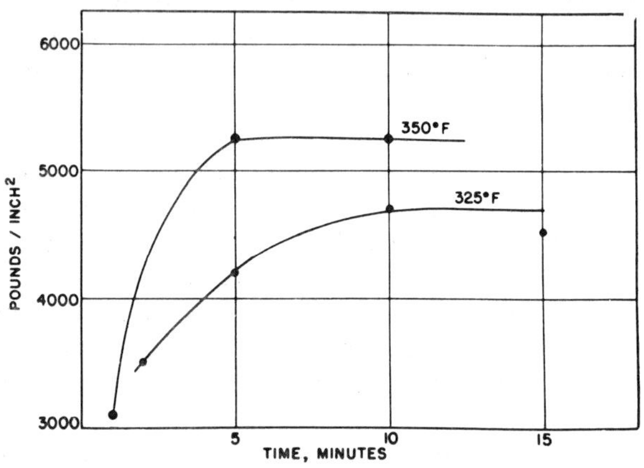

Figure 11.11. Effect of baking time on tensile strength of organosol films resin-QYNV.

Pigmentation

The same general rules of pigmentation apply to vinyl dispersion resins that have been discussed previously for vinyl solution copolymers. Iron and zinc pigments should be avoided since they accelerate thermal degradation of the resin. Properly pigmented dispersion resin coatings show excellent long-term weathering resistance. Pigments can be dispersed in conventional mixing equipment. Pigmented organosols are normally prepared by ball-mill grinding or high-speed impeller mixing. Pigmentation of plastisols is usually accomplished by grinding the pigment in plasticizer to form a concentrated paste which is let down while dispersing the resin in a high-shear mixer.

Heat Stabilizers

Also necessary in organosol and plastisol coatings are heat stabilizers. Under the influence of heat, vinyl chloride resins release hydrogen chloride which results in discoloration, embrittlement and film breakdown. Stabilization against heat is effected by the addition of compounds capable of reacting with the HCl as it is formed, thus preventing further autocatalytic breakdown. Inorganic oxides, complex organometallic compounds and various glycidyl compounds are used as stabilizers. Organic compounds are preferred as stabilizers for unpigmented vinyl coatings. Metallic compounds such as pigments and soaps are very effective in pigmented systems.

Application

Organosols and plastisols can be applied to primed metal surfaces by a variety of methods including spraying, knife coating, roller coating and extrusion. Spraying is the most widely used of these methods. Solids content of 60% and higher are common. Plastisols are sometimes applied by spraying; however, in most cases a small amount (4 to 10%) of volatile diluent must be added before reasonable spray viscosities are obtained. The use of pressure-fed spray equipment further facilitates the efficient application of plastisols and high solids organosols by spraying. One-coat film buildup as high as 5 to 8 mils may be achieved by such systems. However, most spray applications or organosols involve a total coating thickness of 1 to 2 mils.

Reverse roll coating and knife coating have found wide use as methods of applying plastisols and high solids organosols to metal surfaces. In the reverse roll coater, a roll running in a direction opposite to that of the metal substrate applies the coating. For versatility of coating weight in combination with high operating speeds, the reverse roll coater is generally preferred. One-coat deposits as high as 20 mils may be obtained with this type of equipment. In the knife coater, a stationary coating knife is used to meter the coating as the metal substrate passes under it. The lower cost of knife coating equipment makes it attractive for metal coating applications in which a maximum one-pass deposit of 4 mills is desired.

Extrusion coating differs from the above methods in that no post-baking is required to fuse the plastisol. The plastisol is fed to a conventional extruder and heated to the fusion temperature before being extruded through a flat film die. The extruded film is then laminated to the substrate by bringing them in contact between two rolls. Reduced floor space requirements, more efficient heating systems and broader viscosity limits are some of the advantages that make the process attractive.

Primer Formulation

Because of the inert nature of vinyl chloride-type resins, adhesion is one of the most important formulating problems in coatings technology.

In the area of solution coating, this problem has been solved largely by modifying the base resins to make them self-adhering. However, in the field of vinyl dispersions, self-adhering resins have not yet been developed. Because of this, activity has been directed toward the formulation of primers, specifically for organosols and plastisols. For plastisols, primer development has been complicated by the fact that the primer must not only bond the plastisol to the substrate but must also resist migration of the plasticizer into the primer. This migration results in softening of the primer and loss of adhesion. For this reason, plastisol primers must possess a combination of thermoset, or chemically reacted, properties which gives resistance to plasticizer migration, as well as some compatibility, which permits the plastisol to bite into the primer for full adhesion. Several good plastisol primers are commercially available. They are generally based on mixtures of phenolic, epoxy and vinyl resins.

Formulation of one-coat, self-adhering organosols is possible by blending the dispersion vinyl resin with the carboxyl-modified vinyl chloride–vinyl ester solution copolymers. Generally one part of carboxyl-modified solution copolymer is blended with one part of dispersion vinyl resin to give the adhesion necessary for severe forming operations. Higher-application solids can be obtained by using the lower molecular weight, more soluble solution copolymer.

Uses of Organosols and Plastisols

There is virtually a limitless range of formulations for vinyl dispersion resins. A formula may be compounded to meet almost any requirement in product finishes. Some areas where market potential for these resins is high include coatings for office furniture, automobile dashboards, business machines and computers, exposed ducts and vent work, school and industrial lockers, outdoor furniture, automobile roofs, soft edge glass shelving, stairway bannister runners, tape for industry, ladder treads, "easy grip" handles for scissors, tools, levers and brake wheels, electric motor coils, work gloves, textured bottles, nonskid clothes hangers, table model television sets, dishwasher interiors and disk racks, and countless others. No matter where they are applied, however, vinyl dispersion finishes will provide a long-lasting, scratch- and abrasion-resistant surface that will bear up under the hardest wear. For example, the plastisol coating on electroplating racks contributes to their longer service life because of the chemical resistance imparted. Also, the smooth plastisol finish permits the surface to drain rapidly. Flexibility, durability and excellent electrical properties are other features that coaters can count on when plastisols are used.

The advantages of vinyl dispersions are as unlimited as the number of applications. To illustrate, organosols usually contain from 5 to 60%

volatile thinner. This makes them more attractive than plastisols when coatings of lower viscosity are desired. Organosols provide:

(1) Easier handling with conventional spray equipment due to lower viscosity.

(2) A greater range of film hardness.

(3) Higher pigmentation and a greater range of possible modifiers, due to lower viscosity.

(4) Thinner films than may be produced with plastisols.

(5) Film flexibility that may be varied over a considerable range.

Plastisols are preferred for coatings where maximum film thickness is desired. They provide:

(1) Freedom from objectionable volatiles. This minimizes one major hazard, flammability.

(2) Application at 100% solids content for maximum film build.

(3) Films with excellent flexibility.

(4) Tough films of high impact resistance and tensile strength.

(5) Ease of compounding, requiring only simple equipment.

POLYVINYL ACETAL RESINS

Polyvinyl acetals are reaction products of vinyl alcohol and various aldehydes. In practice, they are prepared by solution polymerization of vinyl acetate followed by partial hydrolysis to polyvinyl alcohol and subsequent reaction with aldehyde. Those products containing 15 to 30% polyvinyl acetate are classified as partially hydrolyzed; those containing 0 to 5% polyvinyl acetate are described as completely hydrolyzed. A variety of both types in several molecular weight grades is used in production of the acetals. In addition, polymers of varying acetal content are offered. Thus, these polymers contain polyvinyl acetal, acetate and alcohol groups in controlled proportions, but randomly distributed in the polymer chain. Polyvinyl butyral was originally developed specifically as an interlayer for glass to make nonshattering or "safety" glass. The composition worked out originally, has stood the test of time and is still used in auto windshields. Besides "safety" glass, butyral resin is used as the base for "wash primer," a World War II development of the U. S. Navy in cooperation with industry. This primer is still widely used today for priming ship hulls and for steel structures. The first and still the most extensive use of polyvinyl formal is in the manufacture of high-performance wire enamels. Wire enamels made with polyvinyl formal cross-linked with various phenolic resins produce coatings with a high level of toughness, adhesion, and heat and abrasion resistance that is maintained during high-speed winding and in service.

General Properties

The general properties of polyvinyl formal offered by one supplier are shown in Table 11.7. Properties of typical polyvinyl butyral resins are shown in Table 11.8.

TABLE 11.7.   Properties of Vinyl Formal Resins[a]

Type	*"Formvar"* 7/70	*"Formvar"* 12/85	*"Formvar"* 7/95S	*"Formvar"* 7/95E	*"Formvar"* 15/95S	*"Formvar"* 15/95E
Form			White, free-flowing powder			
Volatiles, % max	1.5	1.5	1.5	1.0	1.5	1.0
Molecular weight (weight average)	19,000– 23,000	26,000– 34,000	16,000– 20,000	16,000– 20,000	24,000– 44,000	24,000– 40,000
Hydroxyl content, % polyvinyl alcohol	5.0–6.5	5.5–7.0	7.0–9.0	5.0–6.5	7.0–9.0	5.0–6.0
Acetate content, % polyvinyl acetate	40–50	22–30	9.5–13.0	9.5–13.0	9.5–13.0	9.5–13.0
Formal content, % polyvinyl formal (approximate)	50	70	80	82	80	82
Specific gravity	1.214	1.219	1.229	1.227	1.229	1.227
Viscosity, cps[b]	7–11	18–22	15–22	15–22	50–80	40–60
Viscosity, cps[c]	100–140	500–600	300–500	300–500	3,000– 4,500	3,000– 4,500

[a] Shawinigan Resins Corp.
[b] Viscosity determined with 5 g resin made to 100 ml with ethylene chloride at 20°C using an Oswald Viscometer.
[c] Viscosity of a 15% solution in 60:40 toluene/ethanol at 25°C using a Brookfield Viscometer.

TABLE 11.8.   Properties of Vinyl Butyral Resins[a]

	XYHL	*XYSG*
Form	White powder	White powder
Intrinsic viscosity	0.81	1.16
Specific gravity	1.12	1.12
Composition (approximate)		
Vinyl butyral resin, %	80.7	80.7
Vinyl alcohol resin, %	19	19
Vinyl acetate resin, %	0.3	0.3
Tensile strength, psi		8000–8500
Modulus of elasticity, psi		350,000–400,000
Modulus of rupture, psi		11,400
Izod impact strength, ft-lb (notched specimen)		0.4–0.6
Softening point, °F		135–140

[a] Union Carbide Corporation, Chemical & Plastics Division.

Polyvinyl acetals are highly resistant to heat and light. They have excellent adhesion to a variety of substrates such as glass, most metals, polar plastics and cellulosic materials. When immersed, these resins readily absorb 5 to 8% water and become softer   Prolonged immersion does not increase the amount of water absorbed and, more importantly, does not affect the bond between the resin and the substrate. A unique property of the vinyl acetals is this ability to undergo many wet and dry cycles without disrupting the adhesive bond. Substitution of butyral or formal groups generally results in a more hydrophobic polymer that is tougher and has better adhesion and a higher heat distortion temperature. Polyvinyl acetals are thermoplastic and may be cross-linked with various thermosetting resins such as phenolic, epoxy, diisocyanate, urea-formaldehyde and melamine-formaldehyde resins. Acetal films have high resistance to aliphatic hydrocarbons, animal, vegetable and mineral oils. They permit unusually high pigment loadings or blending with large amounts of low molecular weight resins such as rosin and ester gum without undue loss of film strength.

Solubility and Plasticization

Polyvinyl butyral resins are soluble in alcohols, alcohol-aromatic blends, glycol ethers, and some ketones and esters. The resins are generally more soluble in mixed solvents. Small amounts of water are especially effective in decreasing the viscosity of solvent mixtures containing water. Plasticizers may be added to butyral resins coatings to increase flexibility and to reduce the viscosity of hot melts. The acetal resins are compatible with a number of commercial plasticizers which can be used where greater flexibility is desired.

Compatibility and Cross-linking

Vinyl acetal resins are compatible with many natural resins, a few oils and certain classes of phenolic resins. Compatibility with nitrocellulose is borderline, differing with the grade of pyroxylin and the solvents used. The same is true with urea-formaldehyde resins. Despite the borderline compatibility, many commercial products are prepared by the addition of plasticizers and mutually compatible resins to increase the formulating latitude. Vinyl acetal resins are often blended with large amounts of low molecular weight resins, such as rosin, ester gum or coumarone types, in the formulation of hot melt adhesives.

The free hydroxyl groups in vinyl butyral resins present a point of chemical reactivity through which the resin may be insolubilized. Both resinous materials and chemical reagents are effective. Typical of the resinous materials are phenolic and urea-formaldehyde resins. Phenolic resins have been used to increase the water and solvent resistance and raise the softening point of the vinyl butyral resins. Flexibility is improved with increasing quantities of the vinyl butyral resin. Solvent resistance increases as the proportion of phenolic is increased. Chemical agents that react with the hydroxyl group may be used to insolubilize vinyl acetal resins. Glyoxal, when added to a vinyl butyral resin solution, acts as a curing agent during the process of air drying to produce a solvent-resistant film. Since the reaction is reversible in the presence of water, the films are not suited for water immersion. Cupric ions, especially in the presence of ammonia, also insolubize the vinyl acetal resins, as do diisocyanates when the solvents used do not react with the diisocyanates.

Uses

Wire Enamels. Polyvinyl formal cross-linked with 50 parts of alkyl phenolic resin in a naphtha–cresylic acid solvent mixture can be considered a basic formulation for magnet wire enamels.

Adhesives. The excellent adhesive qualities of vinyl butyral resin, evidenced by its extensive use as a safety glass interlayer, have been instrumental in encouraging evaluation of these resins in strictly adhesive application. Vinyl butyral–phenolic resin adhesives may be used for bonding rubber, cork, asbestos board, wood, glass or ceramic parts, cloth, paper, or metals to plastics of the thermosetting type. After the bond has been cured, it will be stable to at least 212°F and will not be softened readily by the action of water or solvents. In order to bond two metal pieces together, the surfaces, which should be grease free, are coated and dried to remove solvent. Drying may be accomplished at room temperature or at low bake not above a temperature of 250°F for about 5 minutes. Temperatures higher than 250°F cause excessive hardening of the adhesive and make subsequent bonding difficult. After the solvent has been completely removed, the surfaces are pressed together and heated. The time and temperature required will depend on the materials being bonded; from 15 minutes at 275°F to a few seconds at 400°F will suffice.

Adhesives based on vinyl butyral–phenolic resin combinations are also used in the lamination of plywood. The vinyl butyral resin acts as a "plasticizer" or fortifier for the phenolic resins, serving to increase the shock resistance and improve the adhesion of the phenolic to wood, especially under conditions of high humidity.

While the ratio of vinyl butyral resin to phenolic resin will vary for different applications, depending on the materials being bonded and the nature of the application, increasing the proportion of vinyl butyral resin will have the following general effects:

(1) Increase the tensile strength,
(2) Increase the impact strength,

(3) Decrease creep resistance,

(4) Decrease boiling water resistance.

Vinyl butyral resins can be used in hot-melt form if mixed with compatible plasticizers and resins. Hydroabietyl alcohol and vinyl butyral resin in equal parts will make a melt that is pourable at 350°F, yet remains tough and shock resistant at room temperature. Hot melts based on vinyl butyral resins are used commercially for high-speed bookbinding.

Wood Finishes. Vinyl butyral resin has proved to be a valuable component of wood sealers and finishes. It imparts toughness to the film and enables the coating to maintain adhesion under a wide variety of conditions. When a vinyl butyral resin is used in a sealer under pyroxylin lacquers, the coatings are more resistant to marring and pressure marking as a consequence of the better adhesion to the wood. Normally, the vinyl butyral resin is modified with phenolic resins, shellac or nitrocellulose for wood coatings in order to improve the water resistance. Compositions with shellac have better durability than either resin alone on exterior exposure. Seemingly, the vinyl butyral resin reinforces the shellac, minimizing cracking and loss of adhesion. These same conditions are useful as sanding-sealers under pyroxylin and hydroxyl-modified vinyl solution copolymers.

The excellent adhesion and sealing qualities of the resin are illustrated by its use in knot sealers. Knots, even though they do not loose, often exude pitch and volatile materials which destroy the adhesion of paint films, discolor them and make them brittle. This leads to cracking and peeling over the knot area long before the clear portion of the wood needs repainting.

The following formulation for knot sealers meets the MIL-S-12935 (CE) specification:

	Parts by Weight	*Per Cent by Weight* *(Approximate)*
Phenolic resin varnish (60% nonvolatiles)	5.0	33.3
Vinyl butyral resin	0.5	3.3
95% alcohol (denatured)	9.5	63.4
	15.0	100.0

Vinyl butyral resin is dissolved by adding it to the alcohol under agitation. The phenolic resin is then added to the alcohol solution with thorough agitation.

Compositions containing approximately 50% vinyl butyral resin are useful as sealers over asphaltic materials to prevent bleeding into top coats. They also serve as excellent sealers on porous woods such as redwood. These same coatings have given excellent service as clear outdoor finishes on wood, although they present some difficulty in applying at adequate film thicknesses because of the viscosity of the vinyl butyral resin solution.

Metal Conditioners

The most important application of vinyl butyral resins in the coatings field is in the production of metal conditioners, or pretreatments, more widely known as wash primers. These wash primers are solutions of vinyl butyral resin containing rust inhibitors. They are applied as thin coatings of between .0001 and .0005 inch and merely require air drying to establish a firm bond to metal surfaces. This inhibitive film serves to prevent corrosion and undercutting of the film by rust as well as providing a firm anchor for subsequent paint coats. Wash primers do not replace conventional primers and shop coats due to the thickness customarily applied, but must be classed with industrial phosphate metal treatments. They provide a temporary protection against atmospheric rusting and must be protected by top coats before exposure to the weather. These top coats can be applied within 15 minutes of the application of the wash primer since the vinyl butyral resin solution dries entirely by solvent evaporation. If desired, this drying can be speeded by heat.

Phenolic, oleoresinous, and alkyd paints adhere to the wash primed metal. Coatings based on hydroxyl-modified vinyl solution resins develop excellent adhesion. Most nitrocellulose lacquers adhere to wash primers as a result of modifiers such as alkyds contained in the lacquer. Nitrocellulose itself has relatively little adhesion to the wash primer film. In general, coatings based on vinyl chloride copolymers (except hydroxyl-modified vinyl resins), vinylidene chloride, and acrylic and methacrylic resins do not adhere to the wash primers. Just as alkyd resins improve the adhesion of nitrocellulose lacquers to wash primers, suitable blends of hydroxyl-modified vinyl resins with other vinyl resins will develop the required adhesion.

Application of Metal Conditioners. The proper application techniques play an important role in the eventual performance of paint systems, particularly the more resistant systems such as the vinyl types. Grease, oils and other contaminants should be removed from the surface before applying the wash primer, as should any mill scale and rust present. Grit or sandblasting has proved more effective than wire brushing or flame descaling in providing an acceptable surface for the wash primer. Acid washes, such as those based on acetic acid which are frequently suggested as a pretreatment for galvanized iron, should not be used, since salts left on the surface may be a source of osmotic blisters. The performance of wash primers over rusty steel is always poorer than over clean steel. When a wash primer is applied to a surface covered with mill scale, it is possible that the galvanic action between the steel and the mill scale can be accelerated leading to more rapid failure. This latter effect is more often observed in underwater exposures. Wash primers may be applied by conventional methods such as spraying, dipping, brushing, or roller coating. It is essential, however, that they be applied as a wet coat in order to cover the surface and penetrate cavities and corrosion pits. Since a heavy coating of wash primer may detract from the performance, the primer should be diluted with additional alcohol. This achieves the desired fluidity without applying a dry film thicker than the preferred maximum of approximately .0005 inch. Precautions should be taken to avoid conventional paint thinners for the dilution, since even small amounts of high-boiling aliphatic hydrocarbons may precipitate the butyral resin during drying and ruin performance. Although not good practice, the wash primers may be applied over wet metal. In this case, it is essential that the thinner contain adequate n-butyl alcohol to ensure against resin precipitation. Usually 30% n-butyl alcohol in the thinner will give good performance. On wet or rough surfaces, brush application is preferred since this ensures that the wash primer penetrates any pits in the surface.

Preparation and Storage of Metal Conditioners. Pigmented wash primers are made in conventional paint equipment, precautions being taken to prevent contamination by incompatible ingredients. The vinyl butyral resin is dissolved in the solvents, care being taken that the resin is added slowly to prevent the formation of large resin clumps. The pigment base should be prepared from the base solution by dispersing the pigment in a pebble mill, grinding to a Hegman gauge reading of 7. Steel mills and equipment should be avoided, since the alcohol solvents used often lead to rapid rusting and the iron contamination which is introduced into the wash primer detracts from the adhesion. For the same reason, the containers used for storage should be lined with a phenolic resin-baking or vinyl resin coating or should be of a nonreactive metal such as terneplate. The acid diluent should be packaged in glass or polyethylene containers or in containers having a suitable baked phenolic or vinyl resin coating. The same precautions are required if the diluent and base are combined in a one-package wash primer. Table 11.9 lists suggested pigment grinding bases and diluent compositions for several representative wash primers.

TABLE 11.9. Composition of Wash Primers

	MIL-C-15328A, Parts by Weight	*MIL-C-15328B, Parts by Weight*
Base Grind		
Vinyl butyral resin	7.2	7.35
Basic zinc chromate pigment (insoluble)	6.9	7.08
Magnesium silicate (talc)	1.0	1.04
Lampblack	0.1	0.08
Isopropanol, 99%	—	46.25
Ethyl alcohol, 95%	48.8	—
Butanol	16.1	16.38
Water	—	2.00
	80.1	80.18
Acid Diluent		
Phosphoric acid, 85%	3.6	3.66
Water	3.2	3.28
Isopropanol, 99%	—	12.88
Ethyl alcohol, 95%	13.1	—
	19.9	19.82
Total	100.0	100.0

Basic Zinc Chromate Metal Conditioners.   Basic zinc chromate metal conditioners are the most widely used wash primers, and they have been subjected to the most extensive testing. Both MIL-C-15328A and MIL-C-15328B have been found to give outstanding adhesion to the widest number of surfaces such as:

Steel	Cadmium	Galvanized iron
Stainless steel	Tin	Magnesium
Zinc	Aluminum	Glass

These formulations are the preferred primers for use where exposure to salt water or salt spray is involved. These metal conditioners give maximum protection where the top coat is a relatively permeable type such as alkyd, oleoresinous or emulsion paint.

For outstanding performance, these wash primers should be top-coated with a coating based on hydroxyl-modified vinyl solution resin. This is an especially effective system in seawater and corrosive atmospheres. On freshwater immersion, or under conditions of high humidity, this system may develop osmotic blisters when the temperature is above 100°F. This occurs when the top coat has relatively little ionic permeability as is the case with coatings based on vinyl and phenolic resins. Pigmentation of the top coats with leafing aluminum pigments may minimize the osmotic blisters. In general, however, where higher temperatures and humidity are encountered, it is advisable to use one of the wash primers containing fewer soluble ions.   The rust-inhibitive action of wash primers is enhanced as the quality of the pigment dispersion is improved   This improved dispersion is best obtained by extended grinding rather than by the use of grinding aids, since most surface-active agents detract from the protection afforded by the wash primer.   In the mixing of wash primers, add the diluent to the base slowly, with agitation, to prevent local gelation as a result of a momentary imbalance of the thinner.   These primers should be used within eight hours after mixing.   After this time, any remaining primer should be discarded, since there is a gradual decline in the adhesion of films applied after the mixed primer has aged about eight hours.   Eventually the mixed material may gel, but the usefulness of the primer is spent long before this occurs.

The proportion of acid diluent is an important factor in the performance of wash primers.   While a variation of 10 to 20% in the volume of acid diluent does not bring about noticeable changes in the performance of the wash primer, larger variations should be avoided.   If further reduction in viscosity is required, isopropanol or *n*-butyl alcohol should be used.   On certain surfaces, such as clean magnesium, it may be preferable to reduce the proportion of acid diluent by 75%, but the necessity for this should be established by experiment.

POLYVINYL ACETATE

Vinyl acetate is the principal monomer copolymerized with vinyl chloride to improve solubility and lower the softening range.   The homopolymer of vinyl acetate is quite soluble and soft, and finds little use alone in high-performance coatings and inks.   Polyvinyl acetate is a thermoplastic which is colorless, odorless and tasteless.   The polymer has good compatibility with a wide variety of materials and is used in blends to impart adhesion, gloss, heat and light resistance.   Like the vinyl acetals, polyvinyl acetate tends to absorb water and become softer and more flexible.   However, unlike the vinyl acetals, the polyvinyl acetate bond tends to fail in the presence of moisture if the bond is placed under strain.   If dried, bond strength will return to original level.   The general properties of several commercial grades of polyvinyl acetate are shown in Table 11.10.

Solubility

In general, the vinyl acetate resins are soluble in ketones, esters, chlorinated hydrocarbons, nitroparaffins and lower-boiling aromatic hydrocarbons. With the exception of methanol, they are insoluble in anhydrous

TABLE 11.10. Properties of Vinyl Acetate Resins[a]

	Types				
	AYAC	*AYAB*[b]	*AYAA*	*AYAF*	*AYAT*
Viscosity[c]	1.5	2.5	7	15	23
Inherent viscosity[d]	0.12	—	0.42	0.52	0.68
Softening point,[e] °F	89.6	111.2	150.8	170.6	187
Equivalent fluidity temperature, °F	131	—	239	257	266
Form	Solid Blocks	—	Granular	Granular	Granular
Color	Water White	Water White	Water White	Water White	Water White
Burning rate	Slow	Slow	Slow	Slow	Slow
Effect of aging	Unaffected	Unaffected	Unaffected	Unaffected	Unaffected
Effect of sunlight	Unaffected	Unaffected	Unaffected	Unaffected	Unaffected
Heat stability, hr at 135°C (275°F)	1.15	1.15	No appreciable color change after 4.5 to 5 hr		
Water absorption, %					
[at 25°C (77°F)] 16 hr	2.4	2.0	1.6	1.4	1.4
144 hr	8.3	7.3	4.0	3.6	3.6
Specific gravity [20/20°C (68/68°F)]	1.18	1.18	1.18	1.18	1.18
Tensile strength, psi	—[f]	—[f]	1,500	2,600	4,200
Refractive index, n_D [20°C (68°F)]	1.4665	1.4665	1.4665	1.4665	1.4665
Per cent solids in methyl isobutyl ketone at 100 ± 10 cps and 20°C (68°F)	51.0	41.5	27.5	21.0	18.0
Water immersion[g]	Swells, becomes pliable			Swells slightly	Only a surface attack

[a] Union Carbide Corporation, Chemical & Plastics Division.

[b] Available only in solution, designated A-70.

[c] Viscosity in cps of 86.1 g plus benzene to make 1 liter of solution. (A rough approximation of molecular weight may be obtained by multiplying by 1000.)

[d] ASTM D-1243-58T; Procedure A.

[e] Approximate values determined by modified ball and ring method.

[f] Impossible to break because of tendency to cold flow.

[g] Surface of all films immersed in water was unaffected after drying.

alcohols, but are readily dissolved in lower-boiling alcohols containing a small percentage of water. In the presence of an active solvent, a number of less active solvents may be used as diluents with vinyl acetate resin. These include xylene and hydrogenated naphthas.

Preparation of Emulsions

When desired, lacquer-phase emulsions of vinyl acetate resins can be prepared quite readily. The essential technique of preparation consists in making a solution of the resin in a suitable water-immiscible solvent mixture, incorporating an emulsifying agent, and adding this to distilled or low mineral content water with sufficient agitation to cause the resin solution to be broken up into extremely fine droplets. For maximum stability, this crude emulsion should immediately be run through a homogenizer or colloid mill to break the droplets into even finer-sized particles.

One of the most satisfactory emulsifying agents is ammonium oleate, prepared by adding oleic acid to the lacquer-phase emulsion and an excess of ammonia to the water phase. Triisopropanolamine oleate, prepared in a similar manner, can be used if the volatility of ammonia is undesirable. Sodium lauryl sulfate may also be used and is dissolved entirely in the water phase. The amount of emulsifying agent required is of the order of 0.5 to 5% of the weight of resin present, the exact amount varying for specific agents and specific types of emulsions.

Because of the presence of emulsifying agents, the water resistance of films laid down from emulsions is generally less than those prepared from solutions. Where ammonium oleate is used as the emulsifying agent, however, this difficulty is minimized due to the liberation of ammonia upon drying. A small amount (0.5% of resin weight) of bentonite clay dispersed in a little water and then added to the finished emulsion appreciably improves the storage stability. Bentonite clay should not be added to the water phase before emulsification. A ball mill is suggested as a suitable means of preparing the preliminary bentonite dispersion.

A sample emulsion formula is given below for guidance. It is obvious that any number of formulation modifications are possible, and that specific applications will call for individual variations.

Plasticized Vinyl Acetate Emulsion

Parts by Weight

82.0%	lacquer phase	50.0	Vinyl acetate
		5.0	Tricresyl phosphate
		43.5	Toluene
		1.5	Oleic acid
		100.0	
		92.0	Distilled water
18.0%	water phase	8.0	28% Aqua ammonia
100.0%		100.0	

Prepare each phase separately, stir the water phase, add the lacquer phase slowly, and homogenize. If both phases are warmed first, emulsification is facilitated. If thinned about 15% with distilled water, this emulsion may be sprayed readily, even with an apparent viscosity of 200 to 300 cps.

Uses

Vinyl Acetate Resins in Inks. The clarity, extremely low acid number, stability, ready solubility, flexibility and adhesiveness of vinyl acetate resins make them extremely suitable for use in several types of inks, particularly metallic inks and lacquers. One great advantage that these resins have in such inks is their high "leafing" power for metallic pigments, i.e., their ability to permit the metallic particles to lie one over the other in such a way that the appearance of metal foil is approached. An additional advantage is that the extremely low acid number of the resin prevents the ink from tarnishing or gelling on long storage. The complete absence of color in vinyl acetate resins imparts a more brilliant appearance to the ink, while the excellent light resistance withstands yellowing on aging. In formulating these inks, an aromatic hydrocarbon, such as toluene, is generally used. The metallic pigment is first wet down thoroughly by the solvent, and a solution of the resin is then added gradually to the pigment-solvent mix. Generally, no plasticizer is added. These inks or coatings may be applied from either a gravure or an aniline press, but not from a flat-bed press because of the possibility of the ink drying out and "stringing" due to solvent evaporation. Inks similar to these, containing aniline dyes or pigments, are widely used in printing on cellophane and glassine paper, where they exhibit excellent adhesion. A similar type of ink, containing a mixture of nitrocellulose and vinyl acetate resin, is used where the resistance of vinyl acetate to mild alkalies, oils and greases is important.

Vinyl Acetate Resins in Paper Coatings.   The many desirable properties of the vinyl acetate resins suggest their inclusion in high-gloss protective overlacquers for use on printed paper labels, posters, and decorative or protective papers. The complete absence of color, taste or odor, the excellent light stability, good adhesion, good film strength, and ready solubility of these resins make them of special interest in this field. The ease with which such coated papers can be sealed by heat or solvents is an added advantage. In order to raise the softening point of such overlacquers, the vinyl acetate resins are frequently blended with varying amounts of nitrocellulose or hydroxyl-modified vinyl resin. In such blends, the exact ratios of the ingredients are determined by the properties desired in the final coatings, but 10 to 25% of the modifying agent is enough to raise the softening point appreciably. Vinyl acetate resins are also used in the manufacture of laminated papers for special purposes. In these cases, the vinyl acetate resin is modified only enough to raise its softening temperature to that required for easy handling.

Hydrolyzed Polyvinyl Acetate

Polyvinyl alcohol is a well known water-soluble polymer used for paper sizing, latex thickening and solvent-resistant sheeting, among other things. Since vinyl alcohol does not exist as a monomeric material (isomerizes

100% to acetaldehyde), this polymer is made by back-hydrolyzing polyvinyl acetate. Alkaline catalyst is generally used. This reaction may be carried out to various degrees, so that 100% polyvinyl alcohol may be made, or various partially hydrolyzed grades may be preferred. Complete water solubility and great resistance to organic solvents is the most important attribute of the 100% hydrolyzed product, or any polymer with not more than about 25% residual vinyl acetate polymer. Resins with less vinyl alcohol are not water soluble, but they do become soluble in ketones and alcohols. These materials show improved adhesion to polar and cellulosic surfaces as compared to polyvinyl acetate, and the hydroxyl groups present many interesting cross-linking possibilities.

REFERENCES

1. Hardman, D. E., and Brezinski, J. J., *Offic. Dig. Federation Soc. Paint Technol.*, 36, 18 (1964).
2. Federal Specification TT-P-442, Type III, Grade (A).

12

*Chlorinated Rubber**

Chlorinated rubber is a solid, film-forming resin manufactured by the chlorination of rubber in solution. The main reaction is addition of chlorine to the unsaturated double bonds of the isoprene units $(C_5H_8)_n$. The commercial product contains about 65% chlorine which is almost the stoichiometric proportion. There is evidence of some cyclization and substitution. According to Parker,[7] the structure is considered to be that shown in Figure 12.1.

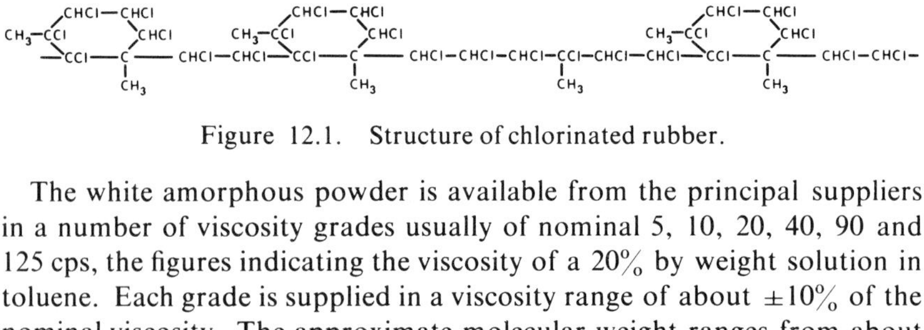

Figure 12.1. Structure of chlorinated rubber.

The white amorphous powder is available from the principal suppliers in a number of viscosity grades usually of nominal 5, 10, 20, 40, 90 and 125 cps, the figures indicating the viscosity of a 20% by weight solution in toluene. Each grade is supplied in a viscosity range of about ±10% of the nominal viscosity. The approximate molecular weight ranges from about 3500 for the 5 cps solution to about 20,000 for the 125 cps one.

Chlorinated rubber is sold under several trademarks, of which the principal are "Alloprene" (Imperial Chemical Industries Limited—I.C.I.) "Parlon" (Hercules Incorporated) and "Pergut" (Farbenfabriken Bayer).

Chlorinated rubber is used either as the major binder in air-drying coatings or as a modifying resin to impart its properties to blends with other resins. The distinguishing characteristics of films of chlorinated rubber are high resistance to attack by acids, alkalies, and oxidation; low permeability to water and water vapor; nonflammability and high electrical resistance. In addition, a suitably formulated film is resistant to

*By D. L. Davies, Imperial Chemical Industries Ltd., Cheshire, England.

attack by molds, fungi and other micro-organisms. Chlorinated rubber has no melting point, but it begins to decompose at 125°C.

Although the solid resin was in process of commercial development in the United States and Europe before 1939, the main development has taken place since 1946. Since 1960 there has been a great increase in usage, and two of the major producers have announced plans for expansion which will result in a world production of about 50 million pounds a year.

In addition to the long established use of chlorinated rubber as a chemical-resistant coating for severe industrial conditions, recognition of the great durability of the coatings has led to renewed interest in their use as long-term protective coatings for steel and other surfaces. See section on uses (p. 196).

PRINCIPLES OF FORMULATION

Plasticization

An unplasticized film of chlorinated rubber deposited from a volatile solvent gives a hard but brittle film. For use in surface coatings, chlorinated rubber must therefore be plasticized.

The choice of plasticizer depends in the first instance on the intended use of the coating; if chemical resistance is required, nonsaponifiable plasticizers are employed, usually of the chlorinated paraffin or chlorinated diphenyl types. The amount of plasticizer is very important, since it affects the properties of the film. If underplasticized, the film will be harder but more brittle, and its adhesion may be lower. If overplasticized, the film will be softer and more thermoplastic, and consequently will suffer more dirt retention. The permeability of the film is also affected (see section dealing with effects of Plasticizers on Permeability, p. 191).

The results of work carried out in I.C.I. laboratories on detached, unpigmented films of 20-cps chlorinated rubber with varying plasticizer contents are shown graphically in Figure 12.2. The ordinate, plotted on a convenient scale, represents the work done in extending the film under a given load and is obtained by integrating the area beneath an Instron trace. It is therefore a combination of tensile strength and modulus of elasticity. The maxima on the curves indicate that the optimum plasticization is in the region 70/30 chlorinated rubber: plasticizer for "Cereclor" S52, "Cereclor" 42 and di-2-ethylhexyl phthalate (DOP), but in the region 60/40 for chlorinated diphenyl ("Aroclor" 1254).

It should be stressed that these curves relate to unpigmented films. The plasticization ratios so obtained, however, are in fair agreement with results of normal film tests, e.g., bend, scratch, hardness, etc., on pig-

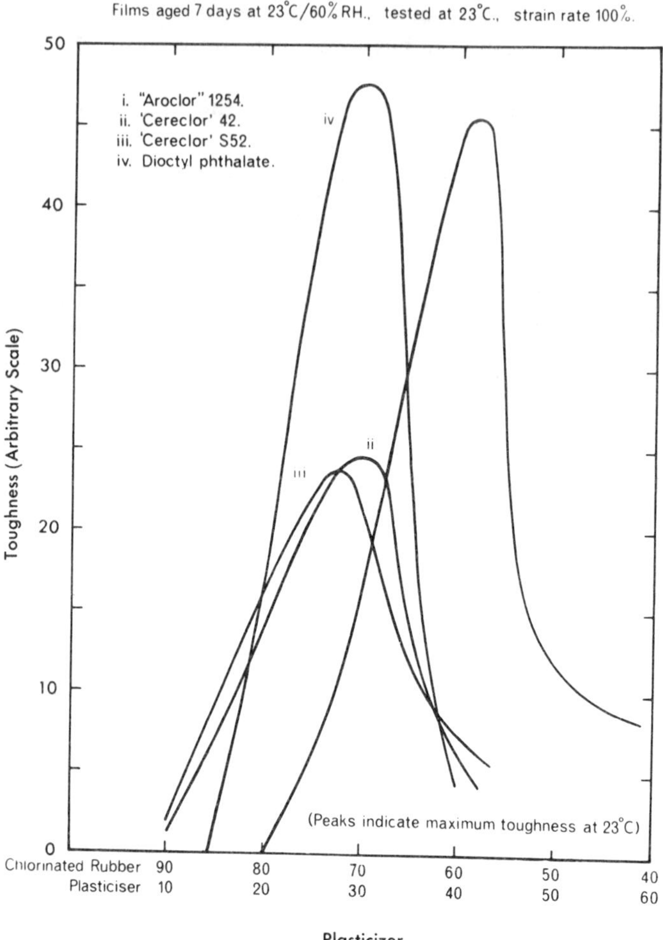

Figure 12.2.   Optimum plasticization of chlorinated rubber.

mented films. A number of plasticizers compatible with chlorinated rubber are given in Table 12.1.

Selection of Plasticizer

For paints with maximum chemical resistance, chlorinated paraffins or chlorinated diphenyls are the normal choice. Where chemical resistance is not required, or for other special reasons, saponifiable plasticizers may be considered, e.g., phthalate esters give films with better light resistance than phosphate esters. Sebacate esters give useful low-temperature flexibility.

TABLE 12.1.   Plasticizers Compatible with Chlorinated Rubber

Chlorinated paraffins	e.g., "Chlorowax" 40, "Chlorafin" 40, "Cereclor" 42 or S52
Chlorinated diphenyl	e.g., "Aroclor" 1254, 1260; "Clophen" A60
Phosphate esters	Tricresyl phosphate
	Triphenyl phosphate
Phthalate esters	Dibutyl phthalate
	Dioctyl phthalate (and diisooctyl phthalate)
	Butyl benzyl phthalate
	Cyclohexanyl and methyl cyclohexanyl phthalate
Miscellaneous	Castor oil (raw, blown or dehydrated)
	Linseed oil
	Tung oil
	Dibutyl and dioctyl sebacate
	Methyl abietate ("Abalyn")
	Hydrogenated methyl abietate ("Hercolyn")

Effect of Plasticizer on Permeability

Besides affecting the mechanical properties of the film, the nature and amount of plasticizer can affect the permeability to water vapor. Some measurements made in I.C.I. laboratories are given in Table 12.2. These

TABLE 12.2.   Effect of Plasticizer on Water Vapor Permeability of Clear Films

Plasticizer	*Ratio Chlorinated Rubber/Plasticizer*	*Water Vapor Permeability ($g/m^2/24\ hr/mil$)*
Chlorinated paraffin, 42% Cl	90/10	26
("Cereclor" 42)	80/20	25
	70/30	33
	65/35	53
	60/40	84
Chlorinated paraffin, 54% Cl	80/20	29
("Cereclor" 54)	70/30	28
	65/35	30
	60/40	32
Chlorinated diphenyl	80/20	19
(e.g., "Aroclor" 1254)	70/30	17
	65/35	17
	60/40	18
Dioctyl phthalate	90/10	26
	80/20	30
	70/30	88
	65/35	156
Dibutyl phthalate	80/20	185
	70/30	186
	60/40	215

results show that overplasticization increases permeability, and that at equal plasticization, permeability decreases with increasing molecular weight of a given type of plasticizer. Comparisons of permeability to oxygen and water vapor have recently been published by Rudram and Sherwood,[11] quoting work at the Paint Research Station, England, which show that clear films of chlorinated rubber had water vapor permeabilities of the same order as epoxy/polyamide and polyurethane films, and that these were all lower than films of linseed stand oil or tung oil phenolic media. The oxygen permeability was also very much lower than that of the traditional coatings.

In comparisons made in I.C.I. laboratories of chlorinated rubber finish paints with conventional enamels based on linseed oil/pentaerythritol alkyds, it was found that the water vapor permeability of the chlorinated rubber paint was approximately one-tenth that of the alkyd.

Clearly, fundamental properties of this kind have important practical implications, e.g., in protection of steel structures from corrosion. Another practical application is the formulation of paints for radioactive areas. Walker[15] has studied the effect of various formulation parameters on the retention of radioactive contaminants and has shown that the retention value is lower with chlorinated paraffin of 54% chlorine content than with chlorinated paraffin of 42% chlorine content, and is still less with chlorinated diphenyl of 54% chlorine content. On the other hand, the retention value with dibutyl phthalate is much higher. The order of merit of these plasticizers is the same as indicated for the water vapor permeability in Table 12.2. The retention value was also shown to depend on the amount of plasticizer present.

Although plasticization has a marked effect on film permeability, there is little effect of using different viscosity grades of chlorinated rubber. Thus, there is no significant difference in permeability of films of equal thickness made from 20- and 90-cps chlorinated rubber, at the same level of plasticizer content. However, the mechanical properties, e.g., tensile strength of films of 90 cps are significantly greater than those of 20 cps films at equal plasticization.

Pigmentation of Chlorinated Rubber Paints

Titanium dioxide is valuable not only for whites but also in colored paints because of its excellent protective properties. Zinc oxide, on the other hand, should not be employed in major amounts as a pigment in finishes since not only does it result in poor weathering properties but it has also been reported[3] that chlorinated rubber can react violently when heated at temperatures of 216 to 220°C. This pigment should therefore not be used in stoving paints, nor should solid chlorinated rubber be dry milled with zinc oxide.

Selection of pigments will be based on normal principles; for chemical-resistant paints, the pigments should be resistant to acids, alkalies and solvents. Most inorganic pigments are satisfactory except Prussian blue, Brunswick green and ultramarine, which are likely to cause flocculation.

Of the organic pigments, arylamide reds and yellows and phthalocyanine blues and greens are satisfactory.

Metallic pigments, especially aluminum powder, require special care as gelation can occur under certain circumstances. Aluminum powders vary widely in this regard, probably due to a number of factors such as composition and method of manufacture, including the use of additives. A mixture of 0.5 to 1% each of red iron oxide and light magnesium oxide on the total weight of the paint has been found effective in extending the can storage stability by a substantial factor.

Milling Techniques

Pebble mills with stone or steatite balls are advisable in preference to steel balls or mills since with the latter, free iron can be abraded, especially with hard pigments, and free iron can induce gelation in the presence of moisture and strongly polar solvents. Although pigments are available in "chip" form in chlorinated rubber, and these give good dispersion with high gloss in printing inks, their use is not normally necessary in paint formulation. With carbon black and low-density organic pigments, a satisfactory milling technique resulting in maximum gloss is to charge the mill with all the pigment, two-thirds of the solvent and one-half of the solid chlorinated rubber, and then mill overnight. The pigment becomes wetted, and as the chlorinated rubber dissolves, the viscosity of the charge rises with consequent increase in shearing action of the pebbles. The balance of the chlorinated rubber, dissolved in remaining solvent and the plasticizer, is then added to the mill base and milling is continued to the fineness of grind desired.

Pigment Volume

With the commonly used pigments, e.g., titanium dioxide in a simple chlorinated rubber plasticized lacquer, pigment volume concentrations for finish paints are usually in the range of 17 to 20%, when gloss ratings (45 or 60°) in the range 90 to 100% are required.

Measurements of water vapor permeability with 20-cps chlorinated rubber, plasticized in the ratio 70/30 with chlorinated diphenyl (54% Cl) and pigmented with titanium dioxide to varying pigment volume concentrations are shown in Table 12.3.

TABLE 12.3. Effect of Pigment Volume Concentration on Water Vapor Permeability

% PVC	Mean Water Vapor Permeability $(g/m^2/24\ hr/mil)$
16.8	26
21.2	27
25.2	36

Solvents

Chlorinated rubber is directly soluble in aromatic hydrocarbons, chlorinated hydrocarbons, esters and ketones higher than acetone. It is insoluble in aliphatic hydrocarbons and alcohols. A proportion of non-solvent can frequently be used as diluent in the presence of a strong solvent, and there are technical as well as commercial advantages in using these blends.

In terms of solubility parameter, it can be said that chlorinated rubber is soluble in solvents of weak hydrogen bonding having solubility parameters in the range 8.4 to 10.6, and in medium hydrogen-bonding solvents with solubility parameters of 7.8 to 10.7. The higher viscosity types may be less soluble in solvents at the higher end of these values.

Besides the normal aromatic solvents, toluene, xylene and higher homologues such as tri-/tetramethylbenzenes (e.g., "Aromasol" H), chlorinated rubber is also soluble in petroleum-derived solvents with a KB value of 80 or over, e.g., "Solvesso" 100 and 150; "Shellsol" A or E; "AMSCO Solv" B90, D80 and F80; "Cosol" No. 2. At equal concentrations, the viscosity of solutions of chlorinated rubber in xylene is a little

TABLE 12.4. Solvent Combinations for Brush Paints

Constituents	Solvents (wt %)							
Xylene	45	38	—	—	—	—	—	—
Solvent naphtha	35	—	—	—	—	—	—	—
"Aromasol" H	—	—	75	—	45	—	—	—
Mineral spirits	—	22	25	22	40	25	20	—
"Solvesso" 100	—	—	—	—	—	—	80	80
Diethyl carbonate	20	—	—	—	—	—	—	—
Ethyl acetate	—	10	—	24	—	—	—	—
Butyl acetate	—	30	—	54	—	—	—	—
"Cellosolve" acetate	—	—	—	—	15	—	—	—
Turpentine	—	—	—	—	—	7	—	20
"Amsco" D	—	—	—	—	—	68	—	—

higher than in toluene; the viscosity increases with the trimethylbenzenes and is considerably higher with trichloroethylene or "Cellosolve" acetate.

Numerous solvent blends are possible of which the following is a selection given by Parker;[7] Table 12.4 gives brush, and Table 12.5 spray (air-assisted), formulations.

TABLE 12.5.   Solvent Combinations for Air-assisted Spray Paints

Constituents	Solvents (wt %)									
Toluene	25	45	30	—	—	—	—	—	—	—
Xylene	—	45	—	—	50	90	80	—	55	46
V.M.&P. naphtha	—	—	30	—	—	—	20	—	—	—
"Aromasol" H	40	—	—	—	—	10	—	15	—	—
Mineral spirits	15	10	—	34	18	—	—	—	—	22
"Cellosolve" acetate	15	—	—	—	—	—	—	—	—	—
Acetone	—	—	40	—	—	—	—	—	—	—
Methyl isobutyl ketone	—	—	—	66	32	—	—	—	—	—
n-Butanol	5	—	—	—	—	—	—	—	—	—
Trichloroethylene	—	—	—	—	—	—	—	85 [a]	—	—
Hi-flash naphtha	—	—	—	—	—	—	—	—	35	—
Turpentine	—	—	—	—	—	—	—	—	10	—
Butyl acetate	—	—	—	—	—	—	—	—	—	32

[a] When spraying trichloroethylene paints, a pencil spray is preferred, using the nozzle without auxiliary atomizing air.

Blends of Chlorinated Rubber with Other Resins

Chlorinated rubber is compatible with many natural and synthetic resins of different types, and blends may be made with the object of achieving a certain combination of properties. Thus, methacrylate resins may be used in blends to take advantage of their outstanding lightfastness, or chlorinated rubber may be added to certain alkyd resins to increase their hardness or reduce their drying time, e.g., in quick air-drying industrial enamels.

Compatibility may be complete or in limited proportions; even with the latter, some useful combinations are possible. With medium to short oil length alkyds, compatibility is greater with low-viscosity chlorinated rubber. On this account, and since variations between apparently similar resins can affect the compatibility, it is not practicable to give comprehensive and precise information in a condensed form. Reference should be made to manufacturers' literature.[1,9]

The following list is indicative only of some of the main types which can be used in blends.

Alkyd resins—medium and long oil   Coumarone-indene
Alkyd resins—short oil (modified)   Natural resins
Acrylic resins   Modified natural resins
Chlorinated paraffin   Melamine- and urea-formaldehyde
Chlorinated diphenyl   Phenolic and terpenic resins

The compatibility can be conveniently determined by preparing separate 25% by weight solutions in an aromatic solvent and mixing the solution in the ratios desired. The mixed resin solutions are then spun on to glass panels and dried at a constant temperature and humidity, since these variables can affect the grading given. The use of a climatic cabinet is advisable. Compatibility is assessed visually or instrumentally on the light transmission through the film. High transmission, or absence of opacity, is taken as a measure of compatibility.

USES OF CHLORINATED RUBBER PAINTS

The major fields of application of chlorinated rubber-based paints are: chemical- and corrosion-resistant coatings; marine coatings; building, masonry and swimming pool paints. Chlorinated rubber is also used, either alone or with other resins, in special types of paints, e.g., road-marking paints; fire-retardant and mold-resistant paints, and primers or finishes for numerous special applications.

Chemical- and Corrosion-resistant Coatings

The simplest type of chlorinated rubber paint, of brush consistency, is as follows:

	% by wt
Chlorinated rubber (20 cps)	20
Inert plasticizer	13
Inert pigment	16
Solvent	51

This type of coating will resist spillage even by a 98% H_2SO_4, concentrated nitric acid and sulfuric acid/nitric acid 3:1 (nitration acid), as well as solutions of caustic alkalies such as sodium and potassium hydroxide, or strongly alkaline oxidizing solutions such as sodium hypochlorite. The maximum dry film thickness which can be applied on vertical surfaces without sagging with this type of formulation is about 1 mil (25 μ).

However, it is accepted by the best authorities on prevention of corrosion of steel that coating thicknesses should be at least 5 mils (125 μ). With the classic type of formulation this would therefore require a minimum five-coat system. High-build types of chlorinated rubber paints have been developed which enable single coats of up to 5 mils to be applied in a single coat.[2,4,8]

This type of formulation is particularly suited for maintenance painting of plants in severely corrosive atmospheres or where occasional spillage occurs. The effect of increasing thickness by increasing the number of coats is shown by Figure 12.3. These panels show classical-type chlorinated rubber finishes brush applied over chipped and wire-brushed pre-rusted steel surfaces, after exposure for $2\frac{1}{2}$ years at a site where the corrosion rate of mild steel is 65 μ/yr. The primer was red lead in chlorinated rubber.

(A) AT A THICKNESS OF
50 MICRONS.

(B) AT A THICKNESS OF
105 MICRONS.

(C) AT A THICKNESS OF
150 MICRONS.

(D) AT A THICKNESS OF
200 MICRONS.

Figure 12.3.   Atmospheric corrosion resistance of chlorinated rubber paints of increasing thickness.   Improvement in performance of chlorinated rubber paint system with increasing thickness of paint film. *System*: 1 primer/various number of top coats. *Time of Exposure*: $2\frac{1}{2}$ years. *Surface Preparation*:   Chipped and wirebrushed. *Location*: Corrosion rate of mild steel 65 μ per year.

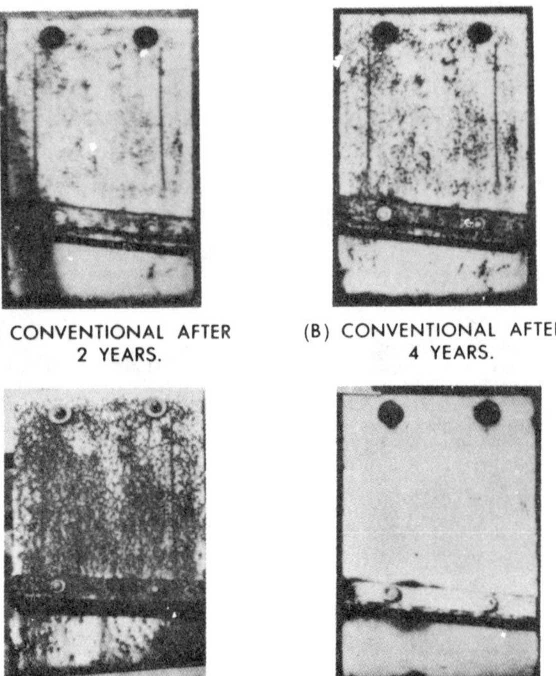

Figure 12.4. Atmospheric corrosion resistance of conventional and thixotropic chlorinated rubber systems. *Systems*: A, B, C. Conventional. 1 Primer/1, Undercoat/2 Finish. D. Thixotropic 1 Primer/2, Thixotropic Coating/1 Finish. *Surface Preparation*: Chipped and Wirebrushed. *Location*: Fairly severe site where corrosion rate of mild steel is 95 μ per year.

Figure 12.4 compares the performance of conventional thickness four-coat brush systems with a four-coat system incorporating two high-build coatings over periods up to six years at a fairly severe site where the corrosion rate of mild steel is 95 μ/yr. As before, the steel was prerusted, and the preparation was chipping and wire brushing. On A, B and C, the systems were primer, undercoat and two finish coats. On D, the system was primer, two thixotropic coats and one finish coat. The value of the thixotropic systems is obvious, and their use for maintenance painting is superseding the conventional thickness type. Moreover, the high labor cost of brush application can be very much reduced by using "airless" spray methods. The formulation of paints suitable for this method must be modified in order to obtain solid films free of voids. Typical brush and spray formulations are given in Table 12.6, which use either modified hydrogenated castor oil or "Bentone" 34 as the thixotropic agents. It is also possible to use both types of gellant in the same formulation.

TABLE 12.6.   High-build Chlorinated Rubber Paint Formulations

Gellant	Modified Hydrogenated Castor Oil		"Bentone"	
Color	Red	Silver Grey	Red	Silver Grey
Chlorinated rubber (10 cps)	17.0	16.4	16.2	12.5
Chlorinated paraffin (70% Cl)	11.3	11.0	9.1	7.9
(42% Cl)	5.7	5.4	6.9	4.5
Titanium dioxide	—	11.0	—	13.6
Red iron oxide	9.5	—	8.4	—
Barytes	14.1	11.0	—	13.6
Vegetable black	—	0.2	—	0.2
Modified hydrogenated castor oil, e.g., "Thixatrol" ST	1.8	1.8	—	—
"Bentone" 34	—	—	2.8	3.4
Xylene	40.6	43.2	41.3	35.0
"Cellosolve" acetate	—	—	—	8.3
Ethanol	—	—	0.8	1.0
Epoxidized vegetable oil	—	—	0.8	—
Blanc fixe	—	—	13.7	—
Application by	Brush	Brush	Brush	Airless spray

It will be noted that these are all based on 10-cps chlorinated rubber, to secure a more rapid rate of dry. The modified hydrogenated castor oil-type gellant must be dispersed warm (about 80°C), but the gel with "Bentone" 34 is developed by the use of ethanol at ambient temperature. For "airless spray" formulation, not only is the chlorinated rubber content reduced, but a proportion of higher boiling point solvent is included. Since the wetting power of these high-build paints is less than that of the conventional type, it is advisable to use a suitable primer. Good results have been obtained with thick chlorinated rubber coatings applied over chlorinated rubber-based primers and over other types of primer. Table 12.7 gives formulations for several types of chlorinated rubber primers. In the zinc-rich type, this formulation contains the maximum amount of zinc which can be tolerated.

In addition to these primers, thick chlorinated rubber coatings can be used over red lead/alkyd, zinc-rich epoxy or inorganic zinc silicate primers. Similarly, primers using calcium plumbate or zinc chromate are readily formulated.

TABLE 12.7. Primers Based on Chlorinated Rubber

	Composition (% by wt)			
	Zinc Rich	Metallic Lead	Aluminum	Red Lead
Chlorinated rubber (20 cps)	—	16.2	15.0	14.5
(90 cps)	3.9	—	—	—
Chlorinated paraffin (42% Cl)	—	7.0	8.2	—
(48% Cl)	2.6	—	—	5.3
Red lead (nonsetting)	—	—	—	25.4
Zinc dust (superfine)	74.8	—	—	—
Metallic lead (90% paste in plasticizer)	—	30.0	—	—
Aluminum powder (nonleafing)	—	—	16.8	—
Titanium dioxide	—	—	3.6	—
Extender, e.g., graphite, silica	—	13.3	7.9	12.7
Barytes	—	—	10.5	—
Red iron oxide[a]	—	—	0.9	—
Magnesium oxide[a]	—	—	0.3	—
Modified hydrogenated castor oil	0.2	0.5	—	0.5
Epoxidized vegetable oil[a]	1.0	1.0	1.0	—
Mineral spirits	—	8.0	—	10.4
Aromatic solvent, e.g., "Solvesso" 100	17.5	24.0	35.8	31.2

[a] Present as stabilizers (see section on pigmentation, p. 193).

Marine Uses

In Europe, chlorinated rubber has been recognized for many years as a valuable medium for underwater protection of steel.[10,12–14] A number of large vessels have had their hulls painted with chlorinated rubber media usually with the conventional thickness-type formulation in multicoat systems. Francis, Preiser and Cook[15] showed that in seawater immersion over a period of 22 months, the performance of a five-coat chlorinated rubber system was very similar to that of a seven-coat vinyl system.

Following the development of high-build chlorinated rubber formulations, interest is now turning to their use in underwater hull paints, especially for application under winter conditions when it is difficult or impossible to use curing systems.

Figure 12.5 shows two steel panels coated with primer and two thick coatings, system thickness 12 mil, after four years immersion in polluted canal water. Surface preparation was chipping and wire brushing in one case, and grit blasting in the other. The grit-blasted panel is somewhat better, but even on the wire-brushed panel the paint system is in good condition. The primer in both cases was a zinc-rich type. Other experience indicates that the metallic lead primer in chlorinated rubber gives

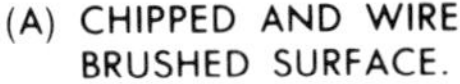
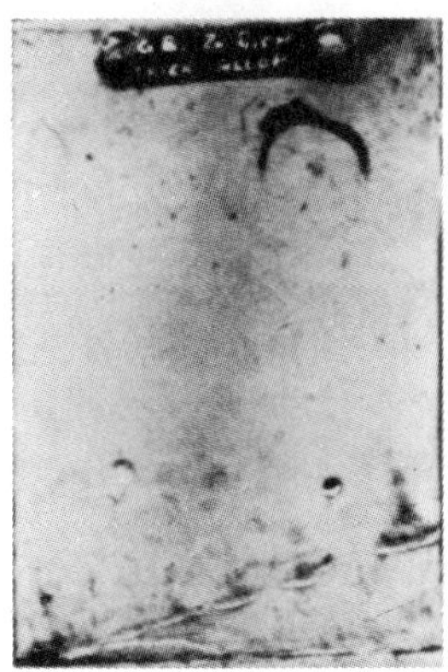

<table>
<tr><td>(A) CHIPPED AND WIRE
BRUSHED SURFACE.</td><td>(B) GRIT BLASTED
SURFACE.</td></tr>
</table>

Figure 12.5.   Thixotropic chlorinated rubber paint after four years immersion in canal water. System: 1 primer/2 thixotropic coatings.

better results below water than zinc-rich primers, though the latter are very satisfactory for above-water use.

Besides their use on underwater hulls, chlorinated rubber paints are used in boot topping and for topside paints, e.g., in alkyd blends. Orgonas and Delahunt[6] have reported favorably on trials of chlorinated rubber top coats over inorganic zinc silicate for topsides and weather decks of tanker fleets.

Building, Masonry and Swimming Pool Paints

Chlorinated rubber paints of conventional types were established early and are widely used for coating asbestos-cement sheeting. These coatings were shown to have excellent durability for periods over 10 years. Use on concrete structures is also well established, with the object of giving long-lasting decoration to the building as well as preserving the surface from spalling.

In some countries, this durability is now attracting attention for the long-term decoration of the outside of brick houses. The formulation listed in Table 12.8 has given excellent results over six years when used as a two-coat system on asbestos cement.

Whiteley[16] refers to exterior masonry paints based on chlorinated rubber being popular in the West Indies and the United States. After two-year test on an experimental wall under tropical conditions at Accra, West Africa, two chlorinated rubber external masonry paints were in excellent condition.

High-build chlorinated rubber paints suitable for application by "airless" spray have been used on large concrete surfaces to provide a

TABLE 12.8. **Chlorinated Rubber Paint for Asbestos
Cement and Concrete**

	% by wt
Chlorinated rubber (20 cps)	14.1
Chlorinated diphenyl ("Aroclor" 1254)	9.4
Titanium dioxide	15.0
China Clay (kaolin)	16.2
"Bentone" 34	0.5
Methanol	0.17
Aromatic solvent, e.g., "Aromasol" H 3	
Mineral spirits 1	44.63
Pigment volume concentration, %	37.5
Viscosity, poise	5.6
Weight pergallon, lb	10

smoother surface which is less subject to dirt retention. Chlorinated rubber is widely used as a coating for concrete swimming pools, using standard-type formulations, although there is some divergence of opinion on the optimum level of pigment volume concentration, which sometimes is in the region of 20%, while other formulations recommended are about 35%.

A number of factors can affect performance such as hydrostatic pressure due to the height of the water table and the detailed construction of the pool.

Road Line Paints

Traffic paints must be quick drying and must have a good durability and visibility. Paints with "no pickup" times of 10 to 20 minutes by the ASTM D-711-55 test can be formulated by using plasticized chlorinated rubber alone or in blends with alkyd or phenolic resins. Paints of this kind are in use in a number of countries of Europe, i.e., Sweden, Norway, Belgium and West Germany. In the United States, in addition to Federal Specifications TT-P-1106 and 115C there are state specifications for New Jersey (Types II and III) and California Division of Highways (May 1960). A formulation based on the latter is shown in Table 12.9.

Fire-retardant Paint

Because of their high chlorine content, chlorinated rubber and chlorinated plasticizers will not burn, and when these compounds are used in the presence of antimony oxide, paints can be formulated which reduce the rate of spread of flame. For use on flammable substrates, depending on the nature of the surface and the degree of protection desired, it may

TABLE 12.9. Traffic Paint Based on Chlorinated Rubber and Phenolic Resin

	% by wt
Chlorinated rubber (10 cps)	6.60
Chlorinated paraffin (42% Cl)	3.18
20-gal tung oil phenolic varnish[a]	18.90
Rutile titanium dioxide	5.15
Titanium calcium pigment (30% TiO_2)	25.70
Asbestine	4.64
Celite	7.30
Mica	5.15
Cobalt naphthenate	0.13
Epichlorohydrin	0.20
Mineral spirits	3.78
Toluene	19.27
Wt/gal/lb	12.0
ASTM D-711-55	
No pick-up time	20 min
Smear-free time	35 min

[a] 50% nonvolatile.

also be necessary to employ paints capable of producing a thermally insulating barrier when strongly heated. "Intumescent" coatings of this kind usually contain special additives such as monoammonium phosphate, pentaerythritol and dicyandiamide. Numerous combinations are mentioned in the patent literature. Nonflammable paints containing chlorinated rubber have been described in specifications MIL-P-17970 to 17974 for use on metal surfaces in ships of the U.S. Navy.

Miscellaneous Special Paints

When buried steel pipelines are to be coated with hot-applied thick pitch enamels, primers must be used to ensure adhesion to the steel. Primers to meet the requirements of American Water-Works Specification C203-62 for Type B are formulated on chlorinated rubber.

Articles of steel after casting in sand molds have a rough surface finish which readily rusts if not protected. The Ford Motor Company Specifications M99J43 and M99J52 relate to casting sealers to be based on chlorinated rubber which will be capable of machining without chipping or will serve as a rust-preventive coating.

Walker's work[15] referred to earlier (see p. 192) showed that correctly formulated chlorinated rubber paints are outstanding in their value for use in radioactive areas for low retention value and for ease of decontamination.

REFERENCES

1. "'Alloprene'—Chlorinated Rubber," I.C.I. Mond Division, Runcorn, Cheshire, England; Chemical Manufacturing Co. Ltd., 444 Madison Avenue, New York.
2. Burgess, T. W., *F.A.T.I.P.E.C. Congr., 8^e*, 185 (June 1966).
3. *Chemical Trade Journal*, 826 (May 24, 1963).
4. Davies, D. L., and Street, D. M., *Comptes Rendus du Centre Belge d'Etude de la Corrosion 1965* (in press).
5. Francis, Preiser, and Cook, *American Society of Naval Engineers Journal*, 401 (August 1958).
6. Orgonas, R. P., and Delahunt, J. F., *Paint Varnish Prod.*, **61** (June 1965).
7. Parker, H. E., (Long, J. S., and Miles, R., editors) in "Treatise on Coatings," Vol. II, New York, Marcel Dekker.
8. Parker, H. E., *J. Paint Technol.*, **38**, 497, 357 (June 1966).
9. "'Parlon'—Properties and Uses," Hercules Incorporated, Wilmington, Del.
10. "Report of Joint Technical Panel," British Iron & Steel Research Association, London, 1950.
11. Rudram, A. T. S., and Sherwood, A. F., *F.A.T.I.P.E.C. Congr. 8^e*, 137 (June 1966).
12. Schroeder, Th., Congress de la Federation Europeene de la Corrosion, Brussels, June 1963.
13. Schroeder, Th., and Kleinschmidt, *Paint Manuf.*, **34**, No. 5, 59 (1964).
14. Talen, H. W., *et al.*, "Report of Corrosion Committee V," *J. Oil Colour Chemists' Assoc.*, **45**, No. 6, 416 (1962).
15. Walker P., *J. Oil Colour Chemists' Assoc.*, **49**, No. 2, 126 (1966).
16. Whiteley, P., *Paint Manuf.*, **33** (Feb. 2, 1963).

13

*Urethane Coatings**

Urethane is now the accepted description for a group of polymers that are sometimes called polyurethanes, isocyanates or polyisocyanates. These polymers are not derived by polymerizing a monomeric urethane molecule, nor are they polymers containing primarily urethane groups. Urethanes are the reaction products of isocyanates with materials possessing hydroxyl groups and simply contain a significant number of urethane groups regardless of what the rest of the molecule may be.

The basic chemistry of isocyanates and urethanes has been known for over a hundred years. In 1848, Wurtz prepared methyl and ethyl isocyanates by reacting potassium cyanate and alkyl iodides. He also found that ethyl isocyanate reacted with ethyl alcohol to form ethyl carbamate which was later named urethane.[82,87]

$$C_2H_5NCO + C_2H_5OH \rightarrow C_2H_5NH \cdot CO \cdot OC_2H_5$$

Polymer formation from reactants each having two or more functional groups is illustrated between a diisocyanate, $R(NCO)_2$ and a polyol material with three reactive hydroxyl groups (a triol), $R(OH)_3$, as follows:

$$3R(NCO)_2 + 2R'(OH)_3 \rightarrow$$

$$\left[\begin{array}{l} -O-R'-OCO \cdot NH \cdot R \cdot NH \cdot CO \cdot O-R'-OCO \cdot NH \cdot R \cdot NH \cdot CO- \\ \qquad | \qquad\qquad\qquad\qquad\qquad\qquad\qquad\qquad\qquad | \\ \qquad O \qquad\qquad\qquad\qquad\qquad\qquad\qquad\qquad O \cdot CO \cdot NH \cdot R \cdot NH \cdot CO- \end{array} \right]_n$$

The resulting polymer will contain urethane groups linking the initial reactant species in chains and cross-links and may also contain unreacted pendant —NCO and —OH groups, depending on the stoichiometry and conditions of the reaction. The isocyanate reaction with hydroxyl groups has been termed the "polyaddition" process by O. Bayer,[11] since addition to a double bond occurs, but without a chain reaction being involved.

*By E. R. Wells, Mobay Chemical Co., Pittsburgh, Pa.

Urethanes therefore belong to the step reaction or condensation polymer classification discussed by Billmeyer.[15]

The modern industrial development of urethane polymers stems largely from the pioneering work of Professor Otto Bayer in Germany since 1937. The history, chemistry, applications and commercial development of the urethanes have been discussed by numerous authors.[9,11,17,29,30,34,43,68,69,84,92,93,99] A comprehensive source of recent information is authoritatively detailed in the two volumes by Saunders and Frisch.[82]

RAW MATERIALS

Isocyanates

Isocyanates, the basic raw materials, are produced in a number of different forms but by far the most widely used is tolylenediisocyanate (TDI), which is supplied commercially in the United States as an 80:20 mixture of the 2,4- and 2,6-isomers as shown:

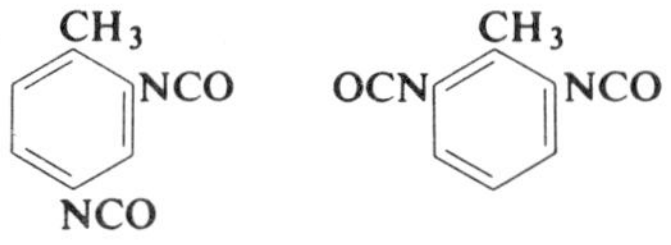

$$CH_3 \quad\quad\quad CH_3$$

(structures of 2,4-TDI and 2,6-TDI)

80%   20%

TDI has a noticeable vapor pressure and has an irritant effect on the mucous membranes and so requires care in handling. It is produced by several suppliers at moderately low lost.

After TDI the so-called prepolymers (p. 211) are the most important sources of —NCO groups for coatings use.[5,21,29,71,75]

Another diisocyanate of commercial importance is 4,4'-diphenylmethane diisocyanate (MDI):

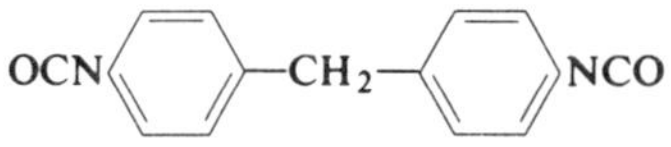

$$OCN—\!\!\!\bigcirc\!\!\!—CH_2—\!\!\!\bigcirc\!\!\!—NCO$$

MDI is a solid at ordinary temperatures, rendering handling less convenient than liquid isocyanates. Its principal use is in the manufacture of elastomers.[93]

More important than pure MDI for coatings use are the lower-cost undistilled varieties of MDI. These dark-colored liquid isocyanates remain fluid well below room temperatures, have low viscosity and vapor pressure and are relatively free from respiratory irritation effects. Approximately 50% of the liquid mixture consists of difunctional isocyanates, the remaining monomers have functionalities of three or more in diminishing proportion.[54,56]

Although these polyaryl isocyanates are higher in cost than TDI they are of increasing value where monomeric TDI cannot be used and where pale initial color is not mandatory. Urethane foam has long been made in the "one-shot" process in commercial operations where all the ingredients of the foam are simultaneously brought together in a mixing head and then ejected onto conveyors. The TDI vapor produced is easily vented from work areas during polymerization. TDI cannot be used in uncontrolled operations where its vapor might permeate the air over large areas, and coatings are therefore not applied by a one-shot operation with monomeric TDI. The one-shot technique recently developed with the polymeric isocyanates in coating applications will be discussed more fully in a later section.[28,29,63]

Polyols

Reactive hydroxyl materials in common use are glycerol, glycols, trimethylpropane (TMP), pentaerythritol, phenols, polyglycols, polyethers, polyesters, alkyds, castor oil, glucosides, sucrose derivatives, etc. Many such polyols combined with just a few commercial isocyanates in various ways to be described indicate the innumerable possible structures of urethane polymers.[23,28,31,41,47,70,73,83,93,96,99,104]

COATING VEHICLE INTERMEDIATES

TDI is processed from its monomeric state into polymeric forms more suitable for coating vehicles by two principal methods depending on the proportion of TDI used in preparation of the adduct. One method utilizes drying oils and polyols and the proportion of TDI used is stoichiometrically equivalent to the total of the nonvolatile ingredients. The second method employs adducts in which TDI is used in considerable excess of the stoichiometric equivalent amount. The resulting prepolymer contains terminal —NCO groups that are utilized in subsequent reactions with ambient moisture, with amines, or with additional polyols, etc., and are packaged as single-component or two-component coating systems.

CHEMISTRY

Isocyanate groups will react with any material containing an active hydrogen, e.g., one attached to an oxygen or nitrogen atom. In coatings technology, the reactions with the hydroxyl group and with water are of chief interest.

$$\sim\sim OH \; + \; OCN\!-\!\!\bigcirc\!\!-\!NCO \; + \; HO\sim\sim \quad\longrightarrow\quad \sim\sim O\!\cdot\!\overset{O}{\overset{\|}{C}}\!\cdot\!\overset{H}{\overset{|}{N}}\diagdown\diagup\overset{H}{\overset{|}{N}}\!\cdot\!\overset{O}{\overset{\|}{C}}\!\cdot\!O\sim\sim$$

$$OCN\!-\!\!\bigcirc\!\!\underset{CH_3}{-}\!NCO \; + \; HO\sim\sim OH \; + \; OCN\!-\!\!\bigcirc\!\!\underset{CH_3}{-}\!NCO \;\longrightarrow$$

$$\bigcirc\!\!-\!\!\overset{H}{\overset{|}{N}}\!-\!\overset{O}{\overset{\|}{C}}\!-\!O\sim\sim O\!-\!\overset{O}{\overset{\|}{C}}\!-\!\overset{H}{\overset{|}{N}}\!-\!\!\underset{CH_3}{\bigcirc}\!\!-\!NCO$$

In the first equation, the two reactants are linked by the formation of urethane groups. In the second equation showing an excess of diisocyanate, the new product contains two urethane groups plus two unreacted —NCO groups illustrating the formation of an —NCO terminated prepolymer.

The isocyanate reaction with water may be considered to proceed first with the formation of an amine and evolution of carbon dioxide, and then to a urea as shown:

$$R\cdot NCO + H\cdot OH \rightarrow R\cdot NH_2 + CO_2\uparrow$$

$$R\cdot NCO + R\cdot NH_2 \rightarrow R\cdot \overset{H}{\overset{|}{N}}\cdot\overset{O}{\overset{\|}{C}}\cdot\overset{H}{\overset{|}{N}}\cdot R\cdot$$

Combining these reactions:

$$2R\cdot NCO + H\cdot OH \rightarrow R\cdot \overset{H}{\overset{|}{N}}\cdot\overset{O}{\overset{\|}{C}}\cdot\overset{H}{\overset{|}{N}}\cdot R + CO_2\uparrow$$

These reactions occur in rapid succession, the amine first produced catalyzing the second reaction and being consumed itself in the process. The CO_2 diffuses from thin films leaving them bubble free when dry, but in the presence of excessive moisture CO_2 causes bubbling and foams in thick layers.

The above reactions occur readily at room temperature and are the important ones to be considered in air dry coating systems. The following reactions occur principally at temperatures above 100°C and usually need be considered only for bake-type coatings.[80,82]

$$R\cdot NCO + R'NHCOOR'' \rightarrow RNHCON\!\underset{\overset{|}{R'}}{}\!COOR''$$

a urethane   an allophanate

$$R \cdot NCO + R'NHCONHR'' \rightarrow RNHCONCONHR''$$
$$|$$
$$R$$

a urethane   *a biuret*

$$R \cdot NCO + R'NHCOR'' \rightarrow RNHCONCOR''$$
$$|$$
$$R$$

an amide   *an acylurea*

Functionality

The concept of functionality in polymer formation has been discussed in a previous chapter (see Chapter 4). Isocyanates and polyols used in the preparation of urethane coatings are usually selected from materials having functionality ($f°$) of two or three. Components having three or more functional groups are "branched." The separate components must all be initially soluble in the solvents used in the coating. Difunctional reactants will yield only linear or chain-like polymers that will continue to exhibit solubility in solvent after the polymer has been formed. If either of the reactants has $f° \geq 3$, the resulting polymer will form a cross-linked polyurethane that may swell in certain solvents but is no longer capable of complete solution.

When the branch points are relatively close together, the resulting polymers will form a tight network that is rigid and highly resistant to solvent and chemical attack. Linear polymers without cross-links are usually soft and flexible and far less resistant to destructive agents. A few standardized di- and trifunctional isocyanates used with a wide variety of polyols available in commerce produce a multitude of polymers for innumerable applications.

Monofunctional materials can play no part in polymer growth and act as "chain-stoppers." Aliphatic alcohols are used for specific purposes, e.g., for removing traces of unwanted TDI at the end of a reaction[76,86] and thus stabilizing the product, as well as for reducing the average functionality of a mixed polyisocyanate and thereby aiding compatibility.[54] Phenols are used as temporary "blocking" agents of —NCO groups in heat-reactive systems.[82,99]

A most rewarding feature of studying the urethanes is the opportunities they afford for instruction in polymer chemistry. Urethane polymer reactions may be followed at normal temperatures and pressures in simple containers, in contrast to the high temperatures and pressures needed for synthesis of many other types of polymers.

CLASSIFICATION OF COATINGS

In 1959 a Committee of ASTM (D-1, Sub. IX, Group 12) was convened to recommend definitions and descriptions of the various types of urethane coatings to avoid the confusion that had become apparent in discussions. Their recommendations were made available in 1960.

Table 13.1 gives broad descriptions of the five distinct types that comprise the rather wide variants within each group. These variations are explained in the text.

TABLE 13.1. Urethane Coatings—General Characteristics

	One-package			Two-package	
	ASTM-1	*ASTM-2*	*ASTM-3*	*ASTM-4*	*ASTM-5*
Description	*Air-cured*	*Moisture-cured*	*Heat-cured*	*Catalyzed Prepolymer*	*Two-can Polyester/ Polyisocyanate*
Free monomer,[a] %	0	3–10	0	3–10	0–1.5
Air dry or cure, hr	0.4–4	0.2–8[b]	30 min at 300°F	0.1–2[b]	2–16
Hardness (Sward)	20–40	20–60	40–85	25–70	30–60
Chemical resistance	Poor-Good	Fair-very good	Excellent	Fair-very good	Good-excellent
Weather durability	Good-very good	Poor-very good	Excellent	Poor-very good	Good-outstanding
Pigmentation	Conventional	Slurry grind (13)	Conventional	Slurry grind	Pigment polyol part

[a] Of types constituting possible vapor hazard.
[b] Relative humidity not less than 30%.

DRYING OIL MODIFIED URETHANES

These comprise the ASTM-1 types and were originally called "urethane oils."[1,32,77] From their analogy with orthodox alkyd resins, in which phthalic anhydride has been replaced by TDI for linking two monoglycerides (from the alcoholysis of drying oils), the term "urethane alkyd" seemed more appropriate, or "uralkyd" as was later suggested.[98]

where $\sim\!\!=\!\!\sim\!\!=\!\!\sim\!\!=\!\!\sim$ is $CH_3 \cdot CH_2 \cdot CH{=}CH \cdot CH_2 \cdot CH{=} CH \cdot CH_2 \cdot CH{=}CH(CH_2)_7 \cdot COO$— as in linseed oil.

The earliest reports[1,77] of these resins were not encouraging to rapid development, but with improvements in raw materials and processing their promise became apparent. Linseed oil was used with TDI until significant yellowing after exposure, mainly a result of the isocyanate component, suggested the use of safflower oil for marked improvement in color stability.[91] Still further improvement was obtained with the addition of UV absorbers (1 to 5%).[52,100]

The uralkyds give properties unobtainable with phthalic anhydride, notably in speed of cure when catalyzed with a trace of drier, greater hardness even at considerably greater oil lengths, excellent flexibility, and significantly superior abrasion, solvent and chemical resistance.[52,90,91,100–102] In exterior exposure there have been reports of earlier loss of gloss compared to the best-quality alkyds,[77,91] which also maintain initial color better, and uralkyd costs were naturally higher. The cost disadvantage has markedly lessened in recent years especially as the maximum amount of TDI that can be used in any stable formulation is approximately 30%. Other studies reported that the uralkyds examined had all-round superior properties when compared to commercial epoxy esters.[90,100,101] Costs also favored the uralkyds. Further developments, e.g., the use of nonyellowing isocyanates for uralkyds, are being pursued.

PREPOLYMERS

Prepolymers have played important parts in practically every branch of urethane polymer technology and in some applications continue as indispensable components. They are adducts of polyols with isocyanates, usually monomeric, in which either of the components is in considerable excess of the other. The resulting intermediate is a medium molecular weight polymer having reactive —OH or, more usually, —NCO groups. Prepolymers serve several functions, principally:

(1) They provide a source of reactive groups which, by dilution, may be more easily controlled in subsequent polymer building, giving products that are more uniform from batch to batch.

(2) They provide a means of avoiding the use of elaborate metering equipment sometimes necessary for one-shot techniques, e.g., in commercial foam operations.

(3) They provide a means of safe handling of monomers having noticeable vapor pressure.

Prepolymers require careful preparation for uniformity, especially where the presence of unreacted monomer in the finished product is undesirable.

Consider a polyglycol HO—G—OH, where G = —$(CH_2)_n$— or —$(C_3H_6O)_nC_2H_4$— which are common polyol groupings. For NCO termination, two molecules of TDI will be required for each molecule of glycol.

$$\text{where} \quad U = -\overset{\displaystyle H}{\underset{\displaystyle |}{N}} \cdot \overset{\displaystyle O}{\underset{\displaystyle \|}{C}} \cdot O-.$$

	Molecules	$f°$	*Equivalents*	
Glycol	1	2	2 —OH	NCO/OH = 2.0
TDI	2	2	4 —NCO	

However, the unreacted —NCO groups may also react further with glycol molecules to form chains as shown:

$$\text{OCN}\!-\!\underset{CH_3}{\bigcirc}\!-\!U\cdot G\cdot U\!-\!\underset{CH_3}{\bigcirc}\!-\!U\cdot G\cdot U\!-\!\underset{CH_3}{\bigcirc}\!-\!U\cdot G\cdot U\!-\!\underset{CH_3}{\bigcirc}\!-\!U\cdot G\cdot U\!-\!\underset{CH_3}{\bigcirc}\!-\!\text{NCO} \;+\; 3\;\text{OCN}\!-\!\underset{CH_3}{\bigcirc}\!-\!\text{NCO}$$

For —NCO termination, where there are x molecules of glycol in such a chain, $x + 1$ molecules of TDI are consumed. But for x molecules of glycol, $2x$ molecules of TDI were provided by stoichiometry. Hence, there will remain in such a system a minimum of $x - 1$ molecules of unreacted TDI.

Conditions for complete absence of unreacted TDI are when $x = 1$ molecule ($x - 1 = 0$), or when one —NCO group of each TDI molecule can be completely prevented from reacting. Since no practical way has yet been found to achieve the latter, chain formation and hence the presence of some unreacted TDI upon completion of prepolymer formation are inevitable. The handling of such a product would be greatly preferred to the handling of an equivalent quantity of liquid TDI; nevertheless any free TDI remaining in the prepolymer will exert its vapor pressure and produce discomfort in unvented operations. Free TDI may be minimized by ordered addition at normal temperatures and avoidance of catalysts,[50] and may be removed from polymers by distillation and solvent extraction.[45]

The amount of unreacted monomer in any prepolymer should preferably be below 1% of the total product if the amount of TDI vapor released to the air by a large area of coated surface is not greater than 0.02%, the level deemed to be safe by the most recent toxicity studies.[43,82]

Safe prepolymers may also be prepared by reacting TDI with a glycol at NCO/OH < 1.0 until all isocyanate is consumed and then "capping" with MDI[93] or one of its lower-cost homologs to a final NCO/OH $\simeq$ 2.0.[56]

Prepolymers continue to be used extensively for urethane coatings. They serve as one-can (ASTM-2) coatings that cure in thin films simply by exposure to moist air. Polyols have molecular weights between 200 and 2000, and functionalities between 2 and 3. A minimum of about 30% relative humidity is required to cure the ASTM-2 prepolymers in acceptable times at normal room temperatures.[5]   Excessively high humidity levels are also liable to produce bubbled coatings by overstimulation of the moisture reaction. The partial use of slowly evaporating solvents such as butyl "Cellosolve"® acetate is suggested to alleviate this condition.

For general use, a prepolymer should not be too highly branched to ensure good can stability, as traces of moisture admitted could cause early gelation. Neither should a prepolymer for moisture cure be completely linear, since that would result in prolonged tack times, dirt collection, and poor solvent and heat resistance.

An example of a flexible prepolymer suggested for an exterior varnish on redwood[75] was prepared from a diol and a triol with TDI as shown:

	Moles	$f°$	Equivalents	NCO/OH
TMP	1	3	3 —OH	
Polypropylene glycol				
(molecular weight = 1000)	1	2	2 —OH	8/5 = 1.6
TDI	4	2	8 —NCO	

$$
\text{OCN}-\!\!\!\bigcirc\!\!\!-N\cdot C\cdot O(C_3H_6O)_n\cdot C_3H_6\cdot O\cdot C\cdot N-\!\!\!\bigcirc\!\!\!-N\cdot C\cdot O\cdot CH_2\cdot C(CH_2\cdot O\cdot C\cdot N-\!\!\!\bigcirc)_3
$$

The can stability of this prepolymer is improved by raising the NCO/OH to 1.8, but the level of free TDI is thereby increased as discussed above.

Prepolymers designed for subsequent blends with additional polyols (ASTM-5) such as polyesters, castor oils, polyethers, etc., are moderately low in molecular weight so that they have low solution viscosities and provide maximum versatility in formulations.[45,71]

BLOCKED ISOCYANATES

The active —NCO groups of prepolymers may be rendered inactive by reaction with certain compounds such as phenol. The "blocked" isocyanate so formed is stable indefinitely at room temperature but is reactivated upon heating above 150°C. This allows such blocked isocyanates to be packaged with hydroxyl-terminated resin solution in a single can. After application, the coating is baked at 150 to 200°C to release the phenol.

®registered trademark—Union Carbide Corp.

Thermal regeneration of the isocyanate is the obvious but not the essential route to the ultimate cured urethane. Carbonyl addition similar to transesterification is believed to be more probable.[82]

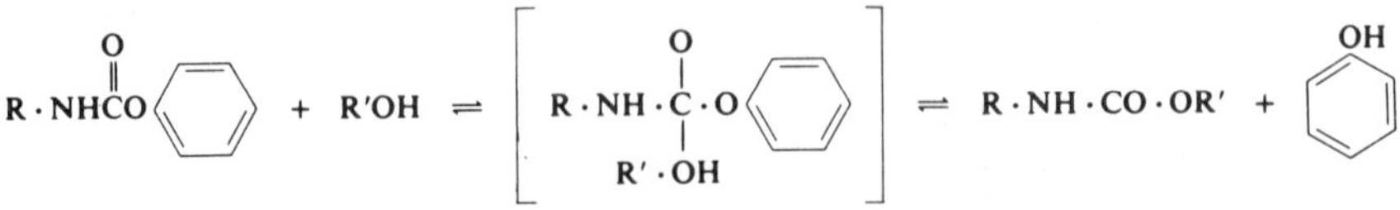

$$R \cdot NHCO\langle\bigcirc\rangle \; + \; R'OH \; \rightleftharpoons \; \left[R \cdot NH \cdot \underset{R' \cdot OH}{\overset{O}{C}} \cdot O\langle\bigcirc\rangle \right] \; \rightleftharpoons \; R \cdot NH \cdot CO \cdot OR' \; + \; \langle\bigcirc\rangle OH$$

The phenol-blocked prepolymers gained immediate acceptance in the electrical industries principally for the coating of magnet wire in insulated windings. Although the urethane single-package low solids (*ca.* 30% nonvolatiles) enamels were higher in cost than established oleoresinous varnishes, they had the unique advantage of being solderable without removal from the ends of the wires to be joined. Solderability facilitated production on electronics production lines and was therefore economically advantageous. Where higher heat resistance is required, as in heavy-duty motor windings, a blocked polyisocyanate based on a trimerized isocyanate, such as is shown below, has been successfully used, together with a polyester based on terephthalic acid. The improved heat resistance is due to the isocyanurate as shown below. Cresylic acids are used for both blocking and resin solution in single-package ASTM-3 type systems.[53,79,82]

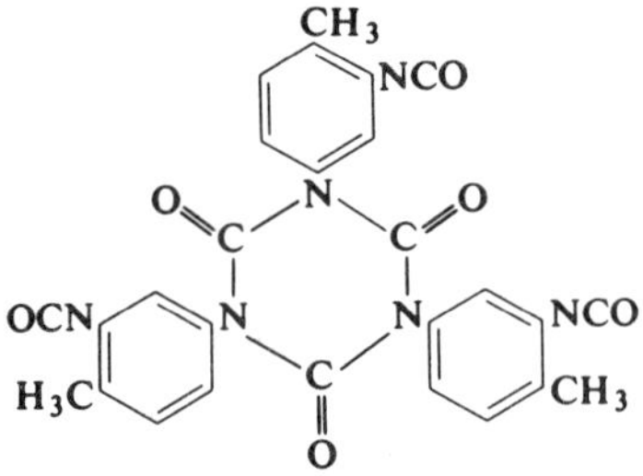

TDI Trimer

These blocked systems failed to gain acceptance in other applications where baked enamels are much used because of their higher cost compared to established alkyd/amine-formaldehyde blends which also have superior color stability.

Polyamide resins unblock phenol-blocked isocyanates at room temperature, causing immediate gelation. However, storage-stable isocyanate/polyamide combinations are possible if an aliphatic alcohol such as amyl alcohol is used as the blocking agent. The greater stability results in a higher unblocking temperature, and catalysts are generally required. These resulting polyurea coatings should gain increasing attention for various applications at the lowered costs of TDI.[39,82]

The uralkyds (ASTM-1) and the two-component polyester-polyisocyanate blends (ASTM-5) also give films of enhanced hardness, flexibility, and chemical and solvent resistance when baked.[82] The moisture-cured prepolymers (ASTM-2 and -4) do not respond so favorably to baking because of the low humidities prevailing in ovens at elevated temperatures.

TWO-PACKAGE URETHANE COATINGS ASTM-4

The use of small amounts of catalyst to assist the cure of prepolymers based on castor oil proved necessary in early formulations owing to the sluggish reaction of such adducts with ambient moisture.[93] These ASTM-4 systems generally resemble the single-can moisture types in their limitations regarding ease of pigmentation and storage stability.[13] Only moderately branched polyols can be used to prepare prepolymers with low monomeric isocyanate content and with adequate storage life, and this restriction limits their chemical resistance. Prepolymers intended for moisture cure at ambient temperatures whether ultimately catalyzed (ASTM-4) or not (ASTM-2) may be conveniently made from TDI with polyether triols having molecular weights in the 500 to 1000 range.[23,51,97] This rather narrow range is imposed by the several requirements of optimum cure time (a few hours maximum), minimal monomer content, convenient solution viscosity and nonvolatile solids content, adequate shelf-stability, and reasonable cost. The physical properties of the moisture-cured films approximate those of the uralkyds, their chemical and water resistances are usually superior, but their exterior durability when not pigmented is often inferior to that of the oil-modified types. Polyethers, unless strongly stabilized, are prone to oxidative degradation especially when exposed to UV radiation.[82,96]

POLYESTER/POLYISOCYANATE TWO-COMPONENT SYSTEMS, ASTM-5

The two-package systems lack the convenience of single-can urethane coatings, but they do permit greater opportunities for formula variation, particularly the polyester-polyisocyanate ASTM-5 types which may also be readily pigmented in any color.[97] By using a range of polyols of varying degrees of complexity with a single isocyanate, an unlimited array of coatings is possible. Even with only three polyesters—linear, moderately branched and highly branched—polymers of optimum properties for any application on rigid or flexible substrates may be formulated with a minimum resin inventory. The versatility of the ASTM-5 systems is consequently far superior to that of any single-can system. This combination of desirable properties is normally only encountered in certain baked coatings and was demonstrated for the first time in high degree by the urethanes. For example, alkyd-amine resin blends when baked have Sward hardness in the 50 to 60 range and often withstand only a few inch-pounds of standard impact. ASTM-5 urethanes of the same hardness will withstand impacts of greater than 160 inch-pounds.[82,99]

The polyisocyanates used are usually made from low molecular weight glycols and triols to give maximum performance, rather than from the polyethers frequently used in the lowest cost moisture-curing prepolymers. Great care is taken to reduce unreacted monomer to an insignificant level.

Commercially available alkyd resins based on nondrying oils and having suitable degrees of —OH functionality may sometimes be substituted for the recommended polyesters, and low-cost polyethers may also be used in part or total replacement.[23,82,99] One of the better low-cost systems employs blown castor oil as the polyol partner to take advantage of the superior durability performance of the "preshrunk" vehicle advocated with justification by Long,[42] and confirmed by Wells[96] for urethane coatings on wood.

COMPARISON OF URETHANE COATINGS WITH COMPETITIVE COATINGS

Limitations of space preclude detailing the comparisons of each of the five urethane types with established orthodox types such as alkyds, epoxies, vinyls, lacquers, etc.; the interested reader should consult the following references: 6, 9, 17a, 22, 25, 58, 60, 67, 72, 78, 82, 90, 91, 99, 100, 101 and 103. Certain qualities appear to be common to the urethanes generally: faster rates of air drying or cure, hardness without sacrifice of flexibility, outstanding abrasion resistance, generally excellent water resistance, and from fair to excellent adhesion, chemical resistance and durability, depending upon the type of formulations being considered.

The success of new coatings depends on outstanding properties and straightforward application. Where costs are significantly higher than those of established coatings, a new system must offer unique advantages to succeed in volume sales. The higher-cost uralkyds succeeded in penetrating alkyd markets by their faster curing and better durability in demanding applications so that they were actually of better value than the alkyds, e.g., in floor and spar varnishes. Similarly the higher-cost blocked isocyanate/polyester types easily succeeded as electrical coatings by virtue of their solderability and their electrical and physical properties.

Although the two-package ASTM-5 coatings displayed the optimum combinations of properties, sales have been principally limited by the two-package feature, the after-yellowing upon exposure and high costs. Considerable research has been directed to solutions of these problems.[1,29,43,57,82,85,94]

IMPROVED COLOR STABILITY

The yellowing of urethanes based on TDI may be significantly diminished by the addition of UV absorbers (1 to 5%),[36,100] and by greater care in the preparation of polyesters and polyols generally. Fast-drying safflower oil is being used in preference to linseed for improved nonyellowing.[91] Trimerization of TDI to give the isocyanurate structure has been employed in Europe for several years to improve both drying speed and color retention of prepolymer types.[30,43] An adduct derived from the nonyellowing hexamethylene diisocyanate is manufactured by Farbenfabriken Bayer and is available from Mobay Chemical Co. in the United States. This new coatings isocyanate, when used with a highly branched polyester has shown outstanding color and gloss retention even after two years exposure in Florida.[29,57]

LOWER-COST URETHANES

The introduction of dark-color varieties of MDI, e.g., liquid polymeric isocyanates,[28,29,56] has opened up important and hitherto unattainable possibilities for low-cost coatings having high-build and protective capabilities. These solvent-free liquid isocyanates of low viscosity contain about 30% —NCO and, being essentially free from irritant vapors, may be applied by one-shot or machine-applied techniques. With this new isocyanate and low-viscosity polyols, urethane coatings may be formulated entirely free from solvent hazards, thus realizing a long-sought goal of the coatings industry.[63]

Coatings having solids up to 90% nonvolatile in clear and pigmented types may be formulated to be sprayed through standard spray equipment without difficulty. At the higher solids content, potlife is reduced to a few hours.[56]

Completely solvent-free systems are usually best applied through twin-stream spray guns equipped with metering to give the mixture in the correct ratio. To obtain rapid gelation of the mixture on vertical surfaces without sagging in thick films, catalysts and/or thixotropes are usually necessary. In the solvent-free types, the coating may be built up to any desired thickness, thus greatly reducing application costs. These coatings have excellent edge coverage and adhesion, and relatively little shrinkage. The dark color of the isocyanate component does not prelude formulation of brilliant shades of color with modern pigments. Only a few pastel shades and pure white are unobtainable with this dark-color isocyanate. Bubble formation from pigment moisture may be controlled by molecular sieves[66] or anhydrous calcium sulfate.[46] The new low-cost systems yellow on exposure, but for many industrial applications this is not important. Such systems make admirable base coats for a nonyellowing urethane system described above. A further advantage is their low cost, in many cases no greater than those of alkyds, epoxy esters, etc. A drawback to wider use of the solvent-free types of coating is the lack of sufficient foolproof dual-stream spray equipment. With the considerable research now proceeding, this is expected to be of only temporary duration.

Since there is frequently a large choice of polyols available for use in urethanes a selection may be made on a cost basis where performances are approximately equal. For example, a highly branched polyether costing 35 cents/lb appears at first sight cheaper to use than a penta-functional resin costing 47 cents/lb that owes its rigidity partly to an aromatic structure.[31] The relative hydroxyl equivalents were 140 and 300, respectively. Consequently the —NCO demand of the aromatic resin is less than half that of the polyether. The cost of using the aromatic resin is therefore only about 60% that of the polyether for a similar urethane product. Such considerations are of concern only with high-cost isocyanates.

With the reduction of the cost of TDI to new low levels, the isocyanate is often no longer the most costly portion of a urethane system. The relative costs of purchasing the —NCO group for three different iso-cyanates early in 1966 were:

TDI monomer	10
Polymeric	16–25
TDI prepolymers (ASTM-5)	90–100

TABLE 13.2.   Isocyanate Usage for Coatings

				Adducts with Hydroxylic Materials		
					Polyols	
					(Glycols, Ethers, Esters)	
	One-shot	$1 - f°$ Alcohols	Mono-glycerides	Low Molecular Weight	Medium Molecular Weight	High Molecular Weight
TDI	(Foams only)	Polyurea (ASTM-3)	Uralkyds (ASTM-1)	Adducts (ASTM-5) Blocked adducts (ASTM-3)	Moisture cured (ASTM-2, -4)	—
Polymeric	One-shot, low-cost coatings (ASTM-5)	Compatibility improve-ment (ASTM-5)	—	—	Capping of TDI/pre-polymers (ASTM-2)	—
MDI	H_2O scavenge, "slurry grind"	—	—	Elastomers	Elastomers	—
Saturated types	—	—	Experimental	Experimental	—	—

Application usages are not always interchangeable. The selection here indicates the polymeric alternative where TDI cannot be used, e.g., in one-shot and some other ASTM-5 type applications.

Actual and potential applications of commercial and experimental isocyanates are shown in Table 13.2.

CONCLUSION

The growth of the urethane coating systems is reflected in marketing reports[17,92] and in the literature, a part of which appears below. References that have appeared since the 168 items listed in reference 82 are 2, 6, 12, 13a, 14, 21, 24, 25, 26, 27, 29, 31, 36, 40, 41, 43, 46, 50, 51, 52, 53, 54, 55, 56, 57, 58, 63, 65, 78, 84, 86, 89 and 92.

Among the various topics discussed, apart from general surveys (references 9, 11, 17, 29, 30, 34, 43, 69, 82, 84, 93 and 99), are formulation details (references 17, 51, 53, 56, 57, 70, 97, 100 and 102), physical properties (references 8, 29, 34, 35, 82, 93, 99, 100 and 101), polymer structural relations (references 8, 15, 20, 23, 24, 55, 81 and 82), resistance to destructive agencies such as the weather, chemicals, solvents, etc. (references 2, 6, 12, 14, 22, 25, 29, 42, 48, 58, 60, 61, 72, 88, 99, 100 and 101), and specific applications such as for maintenance paints (references 17, 17a, 27, 60, 61, 62 and 99), tiles and masonry (references 13a, 17, 49 and 99), wooden substrates (references 14, 89 and 96), flexible substrates (references 13a, 21, 37, 64, 82, 99 and 103). Analytical methods are given in references 3, 10, 24a and 43, and safety precautions are discussed in references 82 and 95.

REFERENCES

1. Armitage, F., and Hammond, W. T. C., *Chem. & Ind.*, 1082 (1951).
2. Arnstein, L. R., *Paint Varnish Prod.*, **54,** No. 8, 29 (1964).
3. ASTM D-1638-60T.
4. Bailey, M. E., *Offic. Dig. Federation Paint Varnish Prod. Clubs*, **31,** 116 (1959).
5. Bailey, M. E., *Offic. Dig. Federation Soc. Paint Technol.*, **32,** 197 (1960).
6. Bailey, M. E., *Sanitary Maintenance*, 88 (September 1963).
7. Bailey, M. E., *et al.*, *Offic. Dig. Federation Soc. Paint Technol.*, **32,** 486 (1960).
8. Bailey, M. E., *et al.*, *ibid.*, 984 (1960).
9. Baldin, E. J., *et al.*, *Offic. Dig. Federation Paint Varnish Prod. Clubs*, **30,** 1070 (1958).
10. Baumann, G. F., and Steingiser, S., *J. Appl. Polymer Sci.*, **1,** 251 (1959).
11. Bayer, O., *Angew. Chem.*, **A59,** 257 (1947).
12. Berger, A. J., and Cizek, A. W., *Offic. Dig. Federation Soc. Paint Technol.*, **36,** 1145 (1964).
13. Bieneman, R. A., *et al.*, *Offic. Dig. Federation Soc. Paint Technol.*, **32,** 273 (1960).
13a. *Ibid.*, **37,** 937 (1965).
14. *Ibid.*, 542 (1965).
15. Billmeyer, F. W., "Textbook of Polymer Chemistry," Second ed., New York, Interscience Publishers, 1962.

16. Bolin, R. E., *et al.*, *J. Chem. Eng. Data*, **4**, 261 (1959).
17. Bristol, F. A., *Paint Varnish Prod.*, **52**, No. 11, 71 (1962).
17a. *Ibid.*, **55**, No. 10, 105 (1965).
18. Britain, J. W., *Ind. Eng. Chem.*, *Prod. Res. Dev.*, **1**, 261 (1962).
19. Britain, J. W., and Gemeinhardt, P. G., *J. Appl. Polymer Sci.*, **4**, 207 (1960).
20. Burrell, H., *Offic. Dig. Federation Soc. Paint Technol.*, **34**, 131 (1962).
21. Chenicek, A. G., *ibid.*, **37**, 565 (1965).
22. Cushing, J. W., *et al.*, *Corrosion*, **17**, No. 12, 28 (1961).
23. Damusis, A., *et al.*, *Offic. Dig. Federation Soc. Paint Technol.*, **32**, 251 (1960).
24. Darr, W. C., *et al.*, *Ind. Eng. Chem.*, *Prod. Res. Dev.*, **2**, 194 (1963).
24a. David, D. J., *Anal. Chem.*, **35**, 37 (1963).
25. Drisko, R. W., *Offic. Dig. Federation Soc. Paint Technol.*, **36**, 767 (1964).
26. Freeborn, A. S., *J. Oil Colour Chemists' Assoc.*, **48**, 539 (1965).
27. Griffith, J. R., and Rohl, G. E., *Offic. Dig. Federation Soc. Paint Technol.*, **36**, 1172, 1225 (1964).
28. Gruber, H., *F.A.T.I.P.E.C. 6^e*, *Wiesbaden*, 306 (1962).
29. Gruber, H., *J. Oil Colour Chemists' Assoc.*, **48**, 1069 (1965).
30. Hampton, H. A., *et al.*, *J. Oil Colour Chemists' Assoc.*, **43**, 96 (1960).
31. Hahn, F. J., *et al.*, *Offic. Dig. Federation Soc. Paint Technol.*, **37**, 1251, 1279 (1965).
32. Hauge, H. M., and Pawlak, J. A. (to Spencer Kellogg), U.S. Patent 2,970,062 (June 1, 1961).
33. Hebermehl, R., *F.A.T.I.P.E.C. 4^e*, *Lucerne*, 85 (1957).
34. Heiss, H. L., *et al.*, *Ind. Eng. Chem.*, **46**, 1498 (1954).
35. Heiss, H. L., *et al.*, *ibid.*, **51**, 929 (1959).
36. Hill, H. E., *Offic. Dig. Federation Soc. Paint Technol.*, **36**, 64 (1964).
37. Hill, H. E., *Rubber Age*, **89**, No. 6, 979 (1961).
38. Hudson, G. A., and Wells, E. R., *Rubber Age*, **91**, No. 3, 419 (1962).
39. Hudson, G. A., *et al.*, *Offic. Dig. Federation Soc. Paint Technol.*, **32**, 213 (1960).
40. Kaatz, E. J., *Offic. Dig. Federation Soc. Paint Technol.*, **37**, 1153 (1965).
41. Kennedy, H. M., and Gibbons, J. P., *Paint Varnish Prod.*, **55**, No. 4, 47 (1965).
42. Long, J. S., *Offic. Dig. Federation Soc. Paint Technol.*, **32**, 1119 (1960).
43. Lowe, A., *J. Oil Colour Chemists' Assoc.*, **46**, 820 (1963).
44. Mattice, J. J., *Offic. Dig. Federation Soc. Paint Technol.*, **34**, 603 (1962).
45. McElroy, W. R. (to Mobay Chemical Co.), U.S. Patent 2,969,386 (Jan. 24, 1961).
46. Menard, A. J., and Williams, T. H. (to Timesaver Products Co.), U.S. Patent 3,196,026 (July 20, 1965).
47. Metz, H. M., *et al.*, *Paint Oil Chem. Rev.*, **121**, No. 8, 6 (1958).
48. Miller, R., *Offic. Dig. Federation Soc. Paint Technol.*, **33**, 1286 (1961).
49. Mobay Chemical Co., reprint, "Floors from Cans."
50. Mobay Chemical Co., bulletin, "The Techniques of Prepolymer Preparation."
50. Mobay Chemical Co., Technical Information Bulletin, 69-C18.
52. *Ibid.*, 70-C19.
53. *Ibid.*, 71-C20.
54. *Ibid.*, 77-E23.
55. *Ibid.*, 88-F33.
56. *Ibid.*, 91-E27.
57. *Ibid.*, 92-C21.
58. Montreal Soc. Paint Technol., *Offic. Dig. Federation Soc. Paint Technol.*, **35**, 1134 (1963).
59. Mort, F., *J. Oil Colour Chemists' Assoc.*, **45**, 95 (1962).
60. Natl. Assoc. Corrosion Engrs., *Materials Protection*, **1**, No. 6, 97 (1962).

61. *Ibid.*, 105 (1962).
62. *Ibid.*, No. 9, 95 (1962).
63. New York Soc. Paint Technol., *Offic. Dig. Federation Soc. Paint Technol.*, **35,** 1143 (1963).
64. Northwestern Soc. Paint Technol., *Offic. Dig. Federation Soc. Paint Technol.*, **32,** 1449 (1960).
65. O'Brien, R. M., and Swindlehurst, R. N., *Offic. Dig. Federation Soc. Paint Technol.*, **35,** 1041 (1963).
66. O'Connor, F. M., and Waythomas, D. J., *Offic. Dig. Federation Soc. Paint Technol.*, **34,** 162 (1962).
67. Panel discussion, *Offic. Dig. Federation Paint Varnish Prod. Clubs*, **31,** 115 (1959).
68. Pansing, H. E., *Offic. Dig. Federation Paint Varnish Prod. Clubs*, **30,** 37 (1958).
69. Patton, T. C., *Offic. Dig. Federation Soc. Paint Technol.*, **34,** 342, 348 (1962).
70. Patton, T. C., and Metz, H. M., *Offic. Dig. Federation Soc. Paint Technol.*, **32,** 222 (1960).
71. Pflueger, E., *F.A.T.I.P.E.C. 6^e*, *Wiesbaden*, 293 (1962).
72. Pittsburgh Soc. Paint Technol., *Offic. Dig. Federation Soc. Paint Technol.*, **32,** 1463 (1960).
73. Poswick, J., and Dramais, C., *Offic. Dig. Federation Paint Varnish Prod. Clubs*, **30,** 1431 (1958).
74. Pratt, B. C., and Rothrock, H. S. (to E. I. du Pont de Nemours & Co.), U.S. Patent 2,358,475 (May 23, 1944).
75. Remington, W. J., and Athey, R. J., *Offic. Dig. Federation Paint Varnish Prod. Clubs*, **31,** 612 (1959).
76. Rhodes, M. S., and Spaunburgh, R. G. (to Allied Chemical Corp.), U.S. Patent 2,970,123 (Jan. 31, 1961).
77. Robinson, E. B., and Waters, R. B., *J. Oil Colour Chemists' Assoc.*, **34,** 361 (1951).
78. Saint Louis Soc. Paint Technol., *Offic. Dig. Federation Soc. Paint Technol.*, **35,** 1188 (1963).
79. Saums, H. L., and Pendleton, W. W., *Elec. Mfg.*, **60,** 130 (1957).
80. Saunders, J. H., *Rubber Chem. Technol.*, **32,** 337 (1959).
81. *Ibid.*, **33,** 1259 (1960).
82. Saunders, J. H., and Frisch, K. C., "Polyurethanes: Chemistry and Technology," Pt. I and II, New York, Interscience Publishers, 1962, 1964.
83. Saunderson, F. T., *Am. Paint J.*, **50,** No. 24, 80 (1961).
84. Sempert, R. E., *Offic. Dig. Federation Soc. Paint Technol.*, **36,** No. 475 (Pt. 2), 78 (1964).
85. Son, Chu Pham Ngoc, University of Delaware, Thesis, 1962.
86. Spitzer, W. C., *Offic. Dig. Federation Soc. Paint Technol.*, **36,** No. 475 (Pt. 2), 16 (1964).
87. Staff report, *Paint Varnish Prod.*, **45,** No. 12, 25 (1955); No. 13, 25 (1955).
88. Staff report, *Petroleum Chem. Transporter* (August 1960).
89. Stamm, A. J., *Offic. Dig. Federation Soc. Paint Technol.*, **37,** 654, 707 (1965).
90. Stanton, J. M., *J. Am. Oil Chemists' Soc.*, **36,** 503 (1959).
91. Stanton, J. M., *Paint Varnish Prod.*, **52,** No. 11, 59 (1962).
92. Tenhoor, R. E., *Chem. Eng. News*, **41,** No. 5, 94 (1963).
93. Toone, G. C., and Wooster, G. S., *Offic. Dig. Federation Soc. Paint Technol.*, **32,** 230 (1960).
94. Wagner, K., and Mennicken, G., *F.A.T.I.P.E.C. 6^e*, *Wiesbaden*, 289 (1962).
95. Walpole, A. L., and Williams, M. H. C., *J. Oil Colour Chemists' Assoc.*, **42,** 694 (1959).

96. Wells, E. R., *Paint Varnish Prod.*, **51,** No. 3, 41 (1961).
97. Wells, E. R., and Hudson, G. A., *Paint Oil Chem. Rev.*, **122,** No. 20, 8 (1959).
98. Wells, E. R., and Hixenbaugh, J. C., *Am. Paint J.*, **46,** No. 47, 88 (1962).
99. Wells, E. R., *et al.*, *Offic. Dig. Federation Paint Varnish Prod. Clubs*, **31,** 1181 (1959).
100. Wells, E. R., *et al.*, *Paint Varnish Prod.*, **53,** No. 8, 33 (1963).
101. Welsted, H. J., *Paint Varnish Prod.*, **52,** No. 11, 50 (1962).
102. Wilson, G., and Stanton, J. M., *Offic. Dig. Federation Soc. Paint Technol.*, **32,** 242 (1960).
103. Wyatt, H. J., *Rubber Age*, **91,** No. 3, 422 (1962).
104. Wyart, J. W., and Vona, J. A., *Paint Varnish Prod.*, **52,** No. 11, 65 (1962).

14

*Silicones**

Silicones, by virtue of their high heat resistance, occupy a very significant place in modern protective coatings, and their entry into conventional finishes, to impart markedly improved weather durability, portends further growth at an accelerated rate. Since their advent in the World War II era, silicone fluids have been used in ever-increasing volume as additives in various types of finishes to improve flow, to prevent pigment flotation, and to correct silking and other surface defects in films.

The first silicone heat-resistant paint dates back to about the same time; it consisted of a silicone resin pigmented with aluminum powder. Although the original technology is not entirely outmoded, many modern paints represent more sophisticated formulations. Several varieties of aluminum and zinc dust pigmented coatings based on both pure silicone resin and silicone-organic resin blends are available. An even greater variety of white and colored highly temperature-resistant coatings for industrial equipment, appliances, space heaters, electrical resistors, etc., is being produced and consumed in substantial volume.

Later, in the coatings area, silicone resins and certain silicone chemicals began to be utilized very successfully as masonry water repellents. This use, too, has seen considerable expansion since its inception.

In addition to these specialty applications which are utilizing increasing volumes of silicone resins, silicone fluids, and chemicals, silicones are now moving rapidly into the conventional protective coatings area. By copolymerization of silicone resinous intermediates with such other resins as alkyds, epoxies and acrylics, and to a lesser extent by cold blending of compatible silicone resins with the conventional organics, markedly improved heat resistance and greatly increased weather resistance are obtained.

*By Harold L. Cahn, General Electric Co., Silicone Products Dept., Waterford, N.Y.

FUNDAMENTAL CHEMISTRY

With all previous coatings technology based on carbon chemistry, it is important to note the departure, to some extent, from this principle in considering the chemistry of the silicones. Just as carbon atoms are building blocks in conventional organic chemistry, silicon atoms assume this role in silicone chemistry. Of course, ordinary carbon side chains are essential in order to build into the various silicones the multiplicity of properties they possess.

Being the first two members of the fourth group in the periodic table of elements, carbon and silicon compounds have certain similarities, but there are also many differences between them—differences that are, in fact, responsible for the entire silicone technology. For instance, there are CH_4 (methane) and SiH_4 (silane). Each is the first member of its own series, but while there is virtually no limit to the carbon chain length in the methane series, the longest analogous silicone compound is hexasilane, Si_6H_{14}. In addition, the silanes, halosilanes and most other silicone compounds are much more reactive than their carbon analogs.

By virtue of this ready reactivity, halosilanes are easily hydrolyzed to the corresponding silanol, which in turn can be condensed to produce the siloxane bond—the skeleton on which all silicones are built.

$$R_2SiCl_2 + 2H_2O \rightarrow R_2Si(OH)_2 + 2HCl$$

$$[R_2Si(OH)_2]_n \rightarrow (R_2Si\underset{\overset{|}{OH}}{-}O\underset{\overset{|}{OH}}{-}Si-R_2)_{n/2} \rightarrow (R_2SiO)_n$$

At this point, then, one can define a silicone. A silicone is any chemical compound comprising the siloxane bond, $-\underset{|}{\overset{|}{Si}}-O-\underset{|}{\overset{|}{Si}}-$, to which various organic side chains are attached to produce characteristic properties in each individual compound. Under proper conditions halosilanes may be alcoholyzed (instead of hydrolyzed), or partially alcoholyzed, to produce kindred, but different end products.

Thus, the entire organosilicon industry is built around three prime factors:

(1) The ability to readily hydrolyze halosilicon chemicals and then to condense them to produce the siloxane bond.

(2). The additional ability to control the hydrolysis and under certain conditions to induce an alcoholysis.

(3). The variation in nature and degree of reactivity of the different silicone chemicals.

MANUFACTURE OF SILICONES

Silicones are manufactured primarily by three methods, all of which are in current use. Each has its limitations from the point of view of yield, products most readily manufactured, or complexity and peculiarity of procedure.

The first and perhaps the simplest method from the point of view of of chemistry involved is the *direct process*. As the name implies, the principle of this method is the direct conversion of silicon metal to chlorosilanes. (Under certain circumstances other halides may be utilized, but chlorides are by far the most widely used because they are the most economical.)[1] Although this method is generally used for the production of methyl and phenyl silicones, other alkyl silicones can be made in this manner.

While the entire gamut of reaction products is derived from the silicon/organic halide reaction, operating conditions can be adjusted to produce that which is most needed and most economical. The direct process is most economical for the production of disubstituted chlorosilanes. A certain amount of trichloro- and monochlorosilane is unavoidably produced, but economics dictates the adjustment of conditions to maximize the yield of dichlorosilanes. The trichlorosilanes are essential, however, and can be fractionated from the yield and used in resin production or can be further methylated by the Grignard method to produce more of the disubstituted product. This cannot be accomplished by recycling in the direct process.[2]

For the production of methyl chlorosilanes, powdered copper serves as a catalyst in the reaction between powdered silicon metal and methyl chloride. As was pointed out, this method yields a wide range of chlorosilanes and even some silicon tetrachloride and silicon tetramethyl. Under proper conditions, however, the yield of dimethyldichlorosilane is maximized, with methyl trichlorosilane next. The reaction and its products may be represented as follows, assuming that R represents the methyl group:

$$RCl + Si \xrightarrow{\text{Cu}} \begin{array}{l} SiCl_4 \\ SiHCl_3 \\ RSiCl_3 \\ R_2SiCl_2 \\ RHSiCl_2 \\ R_3SiCl \\ R_4Si \end{array}$$

If the R is a phenyl group, the catalyst used is powdered silver, and the reactions yield the analogous phenyl chlorosilanes.

In either case, the next step is a tedious and exacting one—the quantitative separation of the chlorosilanes. This is essential so that in the subsequent synthesis of silicones, only the proper quantity of each chlorosilane desired will become a part of the reaction. Some idea of the magnitude of this task can be obtained from an examination of the boiling points and densities of some typical chlorosilanes given in Table 14.1.

TABLE 14.1. Constants of Typical Chlorosilanes

Chlorosilane	Boiling Point (°C)	Density
CH_3SiCl_3	66	1.271
$(CH_3)_2SiCl_2$	70	1.062
$(CH_3)_3SiCl$	57.6	0.846
$C_6H_5SiCl_3$	199	1.280
$(C_6H_5)_2SiCl_2$	303	1.190
$CH_3C_6H_5SiCl_2$	205	1.188
$(C_2H_5)_2SiCl_2$	129	1.106
$C_2H_5SiCl_3$	100	1.238

Another widely used method of manufacture of silicones is the *Grignard process*. This is the most versatile way of producing chlorosilanes in that any organic chlorosilane in any degree of halogen substitution can be made by it. However, the necessity of conducting this reaction in anhydrous ether required that extreme caution be exercised in view of the explosion hazard involved. Since no other method displayed the versatility of the Grignard reaction, its use was continued while efforts were made to improve it. This was accomplished from the point of view of economics and chemistry, although no relaxation of safety measures could be allowed. It was found that tetrahydrofuran could replace the ether, and with it came better and easier solubility of the Grignard reagent. The degree of halogen substitution is dependent upon, and inversely proportional to, the amount of Grignard reagent that is used. The following chemical equations will illustrate this point.

$$SiCl_4 + CH_3MgCl \rightarrow CH_3SiCl_3 + MgCl_2$$

$$CH_3SiCl_3 + CH_3MgCl \rightarrow (CH_3)_2SiCl_2 + MgCl_2$$

$$(CH_3)_2SiCl_2 + CH_3MgCl \rightarrow (CH_3)_3SiCl + MgCl_2$$

Finally, the *olefin addition method* is a very handy tool to produce chlorosilanes with any desired alkyl group attached. The reactants are a silane containing an —Si—H bond and an olefin such as ethylene, $H_2C{=}CH_2$. Although this reaction can be utilized for several compounds, it is used primarily to yield the very reactive trichlorinated products, principally of the alkyl type. The simplest illustration of this reaction is the preparation of ehtyltrichlorosilane from trichlorosilane and ethylene

$$SiHCl_3 + H_2C\!\!=\!\!CH_2 \xrightarrow{\text{Pt}} CH_3CH_2SiCl_3$$

This reaction proceeds without catalysis, but it goes better when catalyzed with platinum.[3] Another example, with a long-chain hydrocarbon is the preparation n-octyl trichlorosilane from trichlorosilane and 1-octene:[4]

$$SiHCl_3 + C_6H_{13}\!\!-\!\!CH\!\!=\!\!CH_2 \rightarrow C_6H_{13}\!\!-\!\!CH_2\!\!-\!\!CH_2SiCl_3$$

Thus, by these methods, and with the starting materials discussed, many different chlorosilanes are available; these are employed in the preparation of the abundance of silicone polymers in use today.

Hydrolysis of Mono-, Di- and Trisubstituted Silanes

The hydrolysis of a dichlorosilane was shown briefly in the discussion of fundamental chemistry to illustrate the subject matter of this paper. A fuller treatment of the hydrolysis is in order, following the discussion of manufacturing methods of the basic intermediates—the chlorosilanes— in order to maintain the proper chronology of the conversion of silicon metal (and other chemicals) into silicone polymers.

One of the principal differences between chlorosilanes and related carbon compounds, it was pointed out, was the ready reactivity of the former with moisture. Now, following the separation of the mono-, di- and trisubstituted materials, regardless of the method by which these intermediates were produced, they are recombined in very specific proportions (as determined through untiring research) and reacted with water (hydrolysis) to yield the corresponding silanols. In these compounds, the chlorine atoms are replaced by hydroxyl groups. When resins are being prepared, the hydrolysis is performed in solvent solution, usually aromatic hydrocarbon. Each of the three types of silanols from the three (mono-, di- and tri-) chlorosilanes reacts differently in the subsequent condensation (polymerization) process.

The least reactive, monosilanol, is capable of dimerizing

$$2R_3SiOH \rightarrow R_3Si\!\!-\!\!O\!\!-\!\!SiR_3 + H_2O$$

and of reacting with the other silanols. It is apparent, therefore, that trialkyl (or aryl) silanol has the properties of a chain stopper, and it is utilized as such.

Dimethylsilanediol is reactive, even with itself, and it is only with difficulty that it is isolated as a monomer. It begins to condense into a polymer instantly upon formation.

$$n[(CH_3)_2Si(OH)_2] \rightarrow [(CH_3)_2SiO]_n + nH_2O$$

In more expanded form, if $n = 5$, $(R_2SiO)_n$ becomes

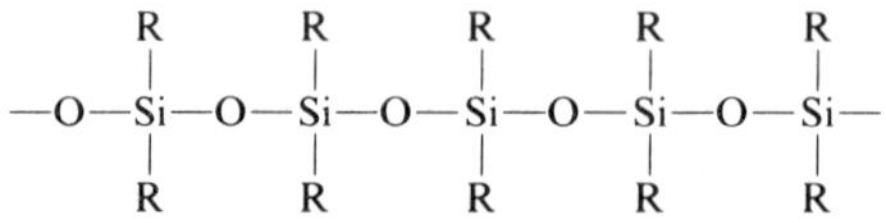

$$-O-\underset{\underset{R}{|}}{\overset{\overset{R}{|}}{Si}}-O-\underset{\underset{R}{|}}{\overset{\overset{R}{|}}{Si}}-O-\underset{\underset{R}{|}}{\overset{\overset{R}{|}}{Si}}-O-\underset{\underset{R}{|}}{\overset{\overset{R}{|}}{Si}}-O-\underset{\underset{R}{|}}{\overset{\overset{R}{|}}{Si}}-$$

Recalling now the dimerization of the monosilanol, and the usage of this reaction as a chain stopper, the five-membered, open-ended chain shown above can be terminated as follows:

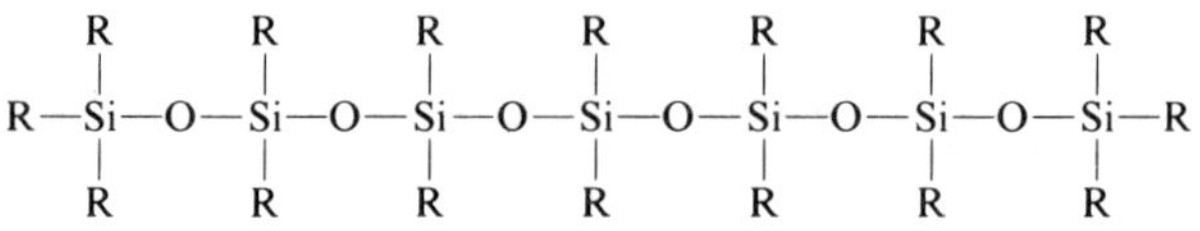

$$R-\underset{\underset{R}{|}}{\overset{\overset{R}{|}}{Si}}-O-\underset{\underset{R}{|}}{\overset{\overset{R}{|}}{Si}}-O-\underset{\underset{R}{|}}{\overset{\overset{R}{|}}{Si}}-O-\underset{\underset{R}{|}}{\overset{\overset{R}{|}}{Si}}-O-\underset{\underset{R}{|}}{\overset{\overset{R}{|}}{Si}}-O-\underset{\underset{R}{|}}{\overset{\overset{R}{|}}{Si}}-O-\underset{\underset{R}{|}}{\overset{\overset{R}{|}}{Si}}-R$$

In contrast to the instability of dimethylsilanediol as a monomer, and its ultimate polymerization to a silicone fluid (or rubber), diphenylsilanediol *is* stable as a white crystalline compound in monomeric form.

$$(C_6H_5)_2SiCl_2 + 2H_2O \rightarrow (C_6H_5)_2Si(OH)_2$$

The polymerization of dimethylsilanediol, illustrated in the above reaction, is a linear one, producing two-dimensional polymers of varying chain lengths, depending upon the extent to which the polymerization is carried. Lower polymers of this type are the familiar, low-viscosity silicone oils or fluids. As the polymer grows in length, the viscosity increases. The ultimate product, as this operation is continued, is a silicone gum which is compounded into many grades of silicone rubber for fabrication by such processes as molding, extrusion, etc. Of course, for various end purposes, many different varieties of fluids and rubbers are formed by modifying the dimethyl silicone with other organic groups. Thus, there are methyl phenyl oils:

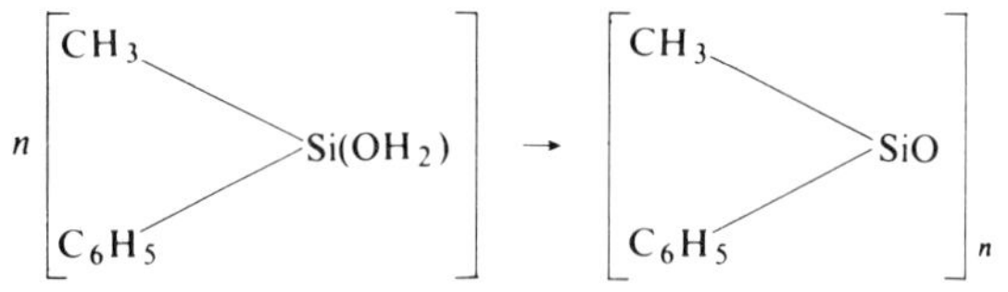

$$n\left[\begin{matrix} CH_3 \\ \diagdown \\ \diagup \\ C_6H_5 \end{matrix} Si(OH_2)\right] \rightarrow \left[\begin{matrix} CH_3 \\ \diagdown \\ \diagup \\ C_6H_5 \end{matrix} SiO\right]_n$$

and methyl hydrogen oils:

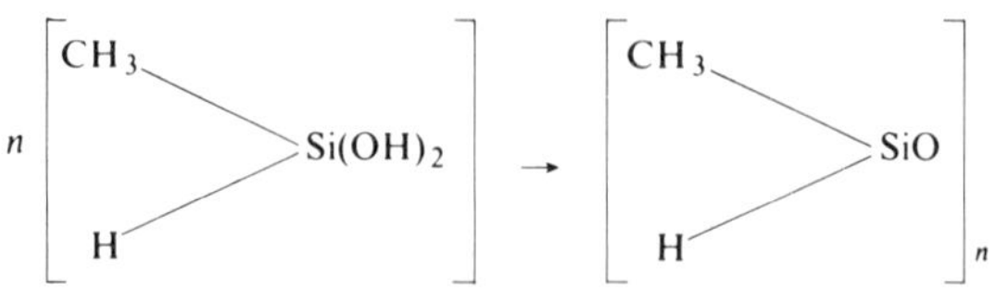

$$n\left[\begin{matrix} CH_3 \\ \diagdown \\ \diagup \\ H \end{matrix} Si(OH)_2\right] \rightarrow \left[\begin{matrix} CH_3 \\ \diagdown \\ \diagup \\ H \end{matrix} SiO\right]_n$$

and others.

In the protective coatings field, large quantities of dimethyl oils and smaller quantities of the other varieties are consumed. These will be more fully discussed in a later section on applications of silicones in protective coatings.

Before leaving the silanediol compounds, it must be pointed out that a possible complicating feature is the formation of cyclic low polymers, which terminates the polymerization. An example of this, as contrasted with the linear polymerization of a silanediol, is the cyclic trimerization of the same material:

$$[R_2SiO]_3 \xleftarrow{\quad} 3R_2Si(OH)_2 \xrightarrow{\quad} R_2Si\underset{O}{\overset{O}{\diagup\diagdown}}SiR_2 \quad (\text{ring with } SiR_2)$$

This is typical of different cyclic low polymers that may be formed. Generally this reaction can be controlled and is held to a minimum when a high degree of polymerization is desired. The cyclics formed in such cases are stripped off by distillation. On the other hand, under certain conditions, low polymers are desired to facilitate purification, after which, by adjustment of reaction conditions, the ring is broken and linear polymerization is effected.

The trisubstituted silanes or, when hydrolyzed, the silanetriols constitute the key characteristics of the three-dimensional silicone resin polymers. These silanetriols alone have little application as vehicle end products in protective coatings, but for purposes of illustration, the condensation of a silanetriol will be shown:

$$4RSi(OH)_3 \rightarrow HO-\underset{\underset{OH}{|}}{\overset{\overset{R}{|}}{Si}}-O-\underset{\underset{OH}{|}}{\overset{\overset{R}{|}}{Si}}-O-\underset{\underset{OH}{|}}{\overset{\overset{R}{|}}{Si}}-O-\underset{\underset{OH}{|}}{\overset{\overset{R}{|}}{Si}}-OH$$

Continued condensation leaves the following:

$$-\underset{\underset{O}{|}}{\overset{\overset{R}{|}}{Si}}-O-\underset{\underset{}{|}}{\overset{\overset{R}{|}}{Si}}-O-\underset{\underset{O}{|}}{\overset{\overset{R}{|}}{Si}}-O-\underset{\underset{}{|}}{\overset{\overset{R}{|}}{Si}}-O- \quad + \quad 6H_2O$$

which can be abbreviated as $(RSiO_{1.5})_4$.

Inasmuch as we know that silicone resins are, in actuality, copolymers of both di- and trifunctional silanols, the condensation product might be represented in this manner:

$$\begin{array}{ccccccccc}
\overset{R}{\underset{|}{|}} & & \overset{R}{\underset{|}{|}} & & \overset{R}{\underset{|}{|}} & & \overset{R}{\underset{|}{|}} & & \overset{R}{\underset{|}{|}} \\
-Si-O- & & Si-O- & & Si-O- & & Si-O- & & Si-O- \\
\end{array}$$

In each of these illustrations of the condensation of silanols, there will be

noted the formation of the siloxane bond $-\overset{|}{\underset{|}{Si}}-O-\overset{|}{\underset{|}{Si}}-$, which was dis-

cussed earlier as the backbone or skeleton of all silicone compounds.

Copolymerization through Hydroxyl or Alkoxy Groups

By virtue of their terminal hydroxyl groups, or alkoxy groups, as the case may be, unbodied, resinous, silicone intermediates are capable of copolymerizing with other organic resins that are similarly terminated. This will be discussed in further detail in a later section of this chapter, since it is the newest and most active area in current silicone coatings technology. It was, however, timely to make mention of it at this time in connection with a discussion of the condensation polymerization of silicone resins.

For the present, this copolymerization can be illustrated with an alkyd resin, where there is a characteristic excess of polyol beyond the stoichiometric requirement for the esterification reaction. The functional points in both the alkyd and silicone intermediate are the hydroxyl groups, and by condensation at these sites, the two constituents are chemically bound through the oxygen atom.

$$(Alkyd\ Polymer)-OH\ +\ HO-(Silicone\ Intermediate)$$

$$\downarrow -H_2O$$

$$Alkyd-O-Silicone$$

APPLICATION OF PURE SILICONE AND HIGH SILICONE CONTENT RESINS

Principal Properties

The principal property by which silicone resins are best and most widely recognized is heat resistance. Because these resins are relatively nonoxidizing, they have excellent color and gloss retention during continuous exposure to temperatures in the 500 to 600 °F range. There is some variation in degree to which this property is exhibited by different silicone resins, just as there is even more variation in the degree to which they will withstand extremely low temperatures. In addition to their stability to color and gloss degradation, they have remarkable retention of film integrity.

While color and gloss retention are proportional (although not directly) to the ratio of silicone resin to non-silicone in a copolymer or cold blend, the retention of film integrity after high-temperature exposure is extremely good, even when the silicone resin represents only 25% of the total vehicle solids. Thus, if protection of the substrate is the prime factor, a relatively low-cost coating can be formulated in a medium to dark color (with heat-resistant pigments) and with extenders incorporated to yield a semigloss to flat finish. These expedients will serve to mask the effects of temperature on the high content of non-silicone vehicles. Film integrity is retained, providing good protection of the substrate at a moderate cost. Conversely, the higher the silicone content (from 50% up), the better will the gloss and color retention be after high-temperature exposure.

The classic example of a high heat-resistant coating is silicone/aluminum paint. Since the early formulations were developed, however, much has been learned about techniques and cost conservation. For instance, a pure silicone or very high silicone content vehicle for an aluminum paint is used today only if temperatures are to be maintained in the 500 to 800°F range. Here it is necessary to maintain integrity via the silicone vehicle, as lesser binders might be destroyed before fusion of the metallic pigment to the steel substrate is accomplished. On the other hand, where temperatures are expected to rise to, and be maintained at, 1000 to 1200°F, silicone resin contents in the vehicle may be as low as 10 to 25% of the total vehicle solids. At those temperatures, substantial amounts of silicone are volatilized, but enough is left, possibly in the form of a binder, while the metal pigment fuses to the substrate.

Silicone resin films are moderately chemical resistant, although, other things being equal, a silicone would not be chosen purely for this property. Likewise, when exposed to solvents, a well-cured film will be quite resistant to aliphatic hydrocarbons, but more vulnerable to aromatics.

Perhaps *the* new, or rather newly recognized and exploited property of the silicone resins, is their high resistance to deterioration by weathering. It has long been known that silicones, being nonoxidizing, are highly resistant to the deteriorating effects of an oxidizing atmosphere. This characteristic is, of course, responsible for their stability at elevated temperatures. Years ago, pure silicone vehicle-based enamels, prepared with chalking-type titanium dioxide, were given an optimum bake for their vehicles and exposed in southern Florida. Contrary to what is exhibited by other organic coatings, these films were unaffected. Not only was there no chalking or other deterioration, but there was little, if any, detectable loss of gloss. Yet fifteen or more years passed before it was realized that some of this outstanding weather resistance could be imparted to conventional coatings by cold blending and, more recently, even more effectively by copolymerizing silicone and organic constituents. Even air-drying finishes can be prepared with outstanding durability as compared to their non-silicone-modified counterparts.

Catalysts

Silicone resin coatings must be catalyzed for optimum performance. High silicone content vehicles require baking at relatively high temperatures to reach an adequate stage of cure. While a fairly adequate cure can be obtained without catalysts, their incorporation serves two important functions: acceleration of cure and enhancement of heat life.

Although many materials such as oil-soluble organometallic compounds, acids, bases and amines catalyze the curing of silicone resins, relatively few are practical for this use. The chief objections are their contribution toward poor package stability and/or the color imparted to white or pastel finishes.

The most useful catalyst is zinc octoate, which imparts good shelf life and heat-aging stability without discoloration. Zinc naphthenate is almost as good, but it lacks the better color properties of the octoate. Where color is not a factor, cobalt and manganese are useful. Color consideration is important with these two metals because the quantities required to catalyze the silicone resins are from 10 to 50 times the quantities required as driers for conventional oxidizing coatings. Here the octoates and naphthenates may be used interchangeably.

Inasmuch as the overall quantity requirements are somewhat higher than for oxidizing coatings, the staining effects of cobalt and manganese detract from their widespread use, except in darker colors. Therefore, zinc is generally used, except in aluminum paints where it is detrimental to aluminum leaf retention and with certain copolymers where instability occurs in the presence of zinc.

Catalyst contents which have resulted in good heat stability for 500 hours at 482°F are:

Zinc	0.5% metal based on silicone resin solids
Cobalt	0.2–0.5% metal based on silicone resin solids
Manganese	0.1–0.5% metal based on silicone resin solids

In aluminum paints, 0.05% to 0.1% cobalt and 0.01% iron are satisfactory.
Iron octoate (or naphthenate) has much more staining effect than any

of the others and is also too active to be the sole catalyst. Appreciable quantities seriously reduce the heat life of films; 0.01 to 0.02% iron metal may be used as a surface hardener where color will permit. Tin, chromium, calcium and lead compounds should be avoided as built-in catalysts. They are very reactive and bring about rapid viscosity increase and subsequent gelation.

Where conditions permit the addition of catalyst on the job, a more reactive catalyst such as lead naphthenate or octoate may be incorporated just prior to use. Lead catalyst usually produces a "hard-to-handle" film sooner or at a lower temperature. Through-cure may also be speeded significantly. It is well to mix it in about an hour or so before use, and the coating will be entirely stable at least throughout the day. Calcium is also reported to be comparable with lead in promoting package instability, but its activity in accelerating cure is not in the same class, and so serves little if any purpose.

Thus, the decreasing order of activity is Pb (lead), Fe (iron), Co (cobalt), Mn (manganese), Zn (zinc). The differences between the last three are so slight that they are often interchangeable.

Finally, silicone resin coatings should be packaged in lined containers. This precaution is necessary since in soldered, unlined metal containers, the lead in the solder can act catalytically to promote gelation of the silicone-based material during storage.

High silicone content copolymers may vary in their catalyst requirements, but it is a fact that certain ones have their idiosyncrasies. To be more specific, in the General Electric product line, there are SR-120 and SR-119 at 75% and 87.5% silicone content, respectively. In spite of the successful wide usage of zinc catalyst with the pure silicone resins, these two copolymers will not tolerate zinc unless it is introduced just shortly before the application of the coating. Upon continued exposure of these resins to zinc, in solution, an incompatibility becomes evident within 24 to 48 hours, resulting in the deposition of films of somewhat diminished gloss.

In these instances, manganese octoate and naphthenate have been found extremely useful as catalysts; 0.1% metal based on resin solids works very satisfactorily. In white, this has a trace of staining effect, which, if too much for certain applications, can be counteracted by bluing the white slightly. If there is any tinting at all, the effect of the manganese color is almost nil. Another alternative in the case of pure whites is to use cobalt, but in individual cases, it will have to be determined whether or not an effective amount will also have a staining effect. In the majority of cases, it has been possible to utilize manganese. Again, where color is not a factor, 0.01 to 0.02% iron may be used along with other curing catalysts for surface hardening.

A complex silicone amine has been found useful in curing silicone resin films in one to two days without heat or in less time if some heat (under the usual silicone baking temperatures) is applied. Since the chemical nature of the product is proprietary, the commercial designation SC-3900 is used to identify it. A concentration of 5% of this catalyst based on resin solids is effective, but in various applications lower percentages may be adequate. Available temperature for baking (if any) and nature of anticipated exposure are factors which will have a bearing on the value of this catalyst. Much more must still be learned about all the factors governing the wide usage of this type of product, as it is of relatively recent origin.

It is known that in many silicone resins gelation will occur quite quickly if the pure catalyst is added. Therefore, it is suggested that a 20% solution of the catalyst in *n*-butanol be used to introduce it. Stability for a day or so is thus imparted, but this will still not permit building in the catalyst when a coating is prepared. On the other hand, when used with certain resins that contain considerable *n*-butanol as a part of the resin solution, this catalyst can be built in to the coating which will remain stable for several weeks. Thus, there is no rule of thumb for silicones in general, although each manufacturer may be able to provide specific data regarding his own products.

Pigmentation

Conventional methods, with certain variations, are used in the pigmentation of silicone coatings. As a group, silicone resins are of low viscosity. They wet pigments easily and permit fine grinds in a minimum of time. To obtain any degree of gloss, however, consideration must be given to pigment volume concentration and oil absorption of pigment. Generally speaking, for a glossy white, TiO_2 enamel, the pigment volume should be in the range of 10 to 15% and the pigment should be the lowest oil absorption type available. In addition to providing gloss, low pigment volume concentrations (PVC) of prime pigments help prolong heat life during extended high-temperature exposure. Extended pigments such as titanium-calcium or siliceous titanium dioxide pigmentations can be tolerated up to about 25% PVC with enhanced heat life, but they result in a sharp reduction in gloss. As has been indicated, semigloss and flat enamels are no problem, since such finishes are easily attainable with relatively low pigmentation.

One of the problems accompanying the low viscosity of silicone resins is settling. In most cases, however, this settling is soft, and remixing to a smooth uniform enamel is easy. High-bulking extenders would help here, but at some expense to the gloss. Some of the vehicle-type viscosity builders are serviceable if they are either heat stable or will volatilize under high heat without disrupting the film. More work is being done in this area.

In the pigmentation of silicone resins with aluminum, an entirely different approach from the aluminum pigmentation of conventional organic resins is needed. While most silicone resins require a high percentage of aromatics as solvents, the aliphatic solvents in aluminum pastes will be tolerated, thus obviating the need for either special pastes or powder.

In contrast to white and color pigmentation, aluminum should be incorporated in relatively high quantities for best results. Whereas $1\frac{3}{4}$ to 2 pounds of paste per gallon of vehicle is used in high-grade conventional paints, $2\frac{1}{2}$ or 3 pounds is in order for silicone aluminum paints intended for very high-temperature exposure (in the range of 1000°F). This is particularly true if weather exposure is also involved. The reason for the high pigmentation is that at such temperatures we are depending solely on the aluminum to provide the barrier between the primed substrate and the corrosive atmosphere. Experience has shown that simultaneous exposure to weather and such temperatures as 600 to 1200°F requires the use of a good high-temperature primer, such as zinc dust. Both silicone-based and all-inorganic, zinc-rich coatings have been used successfully in such applications.

It has been observed that the coarse types of aluminum flake have given consistently good results. The normal 325-mesh grades are usually good for this service also. The general field of application does not seem to warrant the extra-fine lining grades.

Zinc dust may be used in place of aluminum to form durable, heat-resistant paints. These paints, when used as top coats, provide satisfactory service in temperature environments as high as 800°F. Such coatings, in the form of zinc dust–zinc oxide primers overcoated with silicone aluminum paints, are capable of withstanding exterior exposure at temperatures in the range of 1000 to 1200°F. Since zinc is quite reactive, such paints should be packaged in two-component containers to obviate any "gassing" problems.

At one time, there were strict limitations on color for high-temperature-resistant nonmetallic coatings. Now, with additional data on conventional pigments and more complete lines of inorganic colors especially designed for heat resistance, one may produce heat-resistant finishes in any color. Of course, some high-temperature colors may be a bit more costly, but the important thing is that color is no longer a limitation.

To be specific about certain typical colors, carbon black is good for temperatures up to about 500°F. Beyond that, it will lose color, but here a mixed oxide color can take over. These are not as jet as good carbon blacks but are, nevertheless, satisfactory. One of these is a blend of copper, manganese and chromium oxides. Ordinary black iron oxide is not at all satisfactory, as it turns red below 500°F. Cadmium colors for all shades of maroon, red, orange and yellow are very satisfactory. Various shades of high oxide-type red iron oxide are entirely stable. A wide variety of shades can be prepared with these colors in full strength and in tints with white. Chromium oxide green and phthalocyanine green are very serviceable for these colors. Regular chrome greens are not suitable, as they lose color. Similarly, iron blues are very poor. Phthalocyanine blue will take the heat up to 500°F but has some tendency to fade on exposure after high heat. Cobalt blues and ultramarine blue are quite satisfactory.

Colors to be avoided are yellow, brown and black iron oxides, as they all tend to turn red. Some lead colors such as chrome, yellows and molybdate oranges are heat and craze resistant but are usually avoided because of the lead. Since lead compounds are quite active catalysts for silicone resins, lead-pigmented coatings have a tendency to be very unstable, although some of the newer ones are reported to be fairly resistant to elevated temperatures.

Corrosion-inhibitive pigments for primers, which must also withstand high heat, will be a consideration. Zinc yellow itself is not especially good for this purpose, but combined with red iron oxide, it has been very satisfactory up to 450°F. For higher temperatures, strontium chromate yellow is reported to be color stable up to 1100°F.

Among the white pigments, titanium dioxide is the first choice for the same reasons it is of prime importance in conventional coatings. In addition, it has excellent heat stability. Other whites which are satisfactory in silicone coatings are lithopone, zinc sulfide, zinc oxide and antimony oxide. One observation: In some cases, titanium enamels, in which about 5% of the pigment is zinc oxide, have exhibited slightly lowered gloss when a zinc catalyst is incorporated.

Extenders may be used as freely as in any other type of coating with regard to high-temperature stability. One must only remember that gloss reduction is accomplished somewhat more readily in silicone coatings than in others. Typical extenders are magnesium and aluminum silicates, diatomaceous silica, air-floated silica, mica, calcium carbonate, and barytes.

Typical Coatings

Many metallic pigmented paints are used in various applications. Some of these will be illustrated here.

#597   Heat-Resistant Aluminum Paint

Material	*Pounds per 100 Gallons*
G-E silicone resin SR-82 (60% in xylene)	61.6
"Aroplaz" 7323 (60% in xylene)	91.5
Hercules N-22 ethyl cellulose solution (10% nonvolatiles in 95/5 xylene/*n*-butanol)	137.3
Xylene	279.0
Alcoa aluminum paste #205 or Reynolds #30	310.6
	880.0

Test	
Total solids	35%
Pigment	23%
Vehicle solids	12%
Silica in vehicle solids	18%
Viscosity, #4 Ford Cup	17 sec
Drying time Dust-free	Within 15 min
Dry-through	Within 60 min
Full hardness, baked at 325°F	Within 30 min
Heat resistance	Passes test
Salt spray resistance	Passes test
Hydrocarbon resistance	Passes test
Water resistance	Passes test

#1023 Heat-Resistant Aluminum Paint

Material	*Pounds per 100 Gallons*
G-E silicone resin SR-112 (50%)	279.0
Ethyl cellulose solution (5.5%)[a]	126.5
6% manganese naphthenate	2.3
"Solvesso" 100	178.2
Alcoa aluminum paste #206 or Reynolds #32	310.0
	896.0

Paint Properties	
Weight per gallon	9.0 lb
Viscosity (25°C)	61 KU
Total solids	39%
Solids in vehicle	25%
Aluminum paste per gallon of vehicle	3.9 lb
Reduction for spray	5:1 with xylene or Solvesso 100
Reduced viscosity	20–21 sec, #4 Ford Cup
Aluminum paste per gallon at spray viscosity	3.1 lb
Fineness of grind	6
Storage stability (part full container)	Excellent (passes spec. test)
Brushing properties	Excellent
Spray properties (reduced)	Excellent

Paint Properties

Storage stability	Satisfactory
Film properties Set to touch	Within 60 min
Dry hard	Within 2 hr
Specification heat test	Passes (programmed heating to and including 8 hr at 1200° F)

[a] Hercules T-100 (T-type, 100-cps grade), in "Solvesso" 100.

#1332 Heat-Resistant Aluminum Paint

Material	*Pounds*	*Gallons*
"Pliolite" S-5B solution (50% in xylene)	207.9	26.50
"Aroclor" 5460 solution (60% in xylene)	93.6	9.24
"Aroclor" 1254	47.8	3.73
G-E silicone resin SR-82 (60% in xylene)	38.5	4.32
Xylene	140.8	19.55
V.M. & P. naphtha	113.1	17.40
Aluminum paste (Reynolds #30 or Alcoa #205)	240.0	19.68
	881.7	100.42

This formulation contains 10% silicone resin in the vehicle solids. Considerable effort has been made to minimize the raw materials cost, while still meeting the specification requirements. The same formulation, with the silicone eliminated, failed the specified heat resistance test. The composition breakdown, and other properties, as observed in actual tests, compared to specification requirements, are as follows:

Paint Properties and Constants	*Found*	*Specified*
Silicone solids, % of total vehicle solids	10	—
Total solids in paint, %	44	40 (min)
Solids in vehicle (exclusive of solvent in aluminum paste), %	36	—
Weight per gallon of paint, lb	8.8	—
Viscosity, KU	62–64	60–85
Aluminum paste content, lb./gal of vehicle	3	—
Reduction ratio for spraying (with 60/40 xylene/ V.M. & P. naphtha)	2.5 to 1	—
Reduced viscosity, sec in #4 Ford Cup	17–18	—
Fineness of grind	5	5 (min)
Brushing properties	Excellent	Satisfactory
Spraying properties (reduced)	Excellent	Satisfactory
Storage stability	Satisfactory	Satisfactory

Film Properties of #1332 Formulation	*Found*	*Specified*
Set to touch	Within 1 hr	2 hr (max)
Dry hard	Within 2 hr	6 hr (max)
Programmed heat test, including 8 hr at 1200° F	Excellent	Satisfactory

Film Properties of #1332 Formulation	*Found*	*Specified*
Salt spray resistance, 24 hr after programmed		
Heating to 500°F	Excellent	Satisfactory
Heating to 600°F	Excellent	Satisfactory
Heating to 900°F	Excellent	Satisfactory

#407 Heat Resistant Aluminum Paint

(Starting formulation for test against MIL-P-14276B and TT-P-28d)

	Pounds	*Gallons*
25% Silicone/alkyd copolymer 249–393[a]	302.0	40.00
Xylene	169.0	23.35
VM & P naphtha	100.0	15.30
6% Cobalt octoate	1.2	0.16
6% Manganese octoate	0.5	0.07
"Exkin" #2	2.7	0.35
Aluminum paste (Reynolds #40 or Alcoa #1570)	312.0	25.27
	887.4	104.50

[a]25% Silicone/Alkyd Copolymer, 249–393

Alkyd Portion	*Parts by Weight*
Soya Fatty Acids	50.00
Phthalic Anhydride	27.80
Monopentaerythritol	13.24
Glycerol (99.5%)	8.96
	100.00
Ethyl Benzene (for azeotropic distillation)	3.00
Amsco Mineral Spirits 3704	50.45

Charge alkyd solids constituents into kettle equipped for solvent reflux cooking and raise temperature slowly to 230°C. Cook to acid value of 6–8 and viscosity of $Z-Z_2$ at 60% solids in Amsco 3704. Reduce batch to 60% solids with indicated solvent.

Copolymer Formulation	*Wt. %* *(solids)*	*Wt. %* *(solution)*
Alkyd Intermediate (60%)	75	78.9
SR-174 (75% in Amsco 3704)	25	21.1
	100	100.0
Amsco mineral spirits 3704 (to adjust to 50% solids)		25.3

Charge alkyd and silicone portions to same kettle used for alkyd preparation and bring to reflux temperature (171°C). Continue cooking to $Z-Z_2$ viscosity at 50% solids in indicated solvent, removing any additional water formed. At conclusion of cook, dilute to 50% solids in Amsco Mineral Spirits 3704 and filter.

Composition and Tests	*Found*	*Specified*
Total solids, %	40	40 (min)
Pigment, % by wt of paint	23	23–27
Vehicle solids, % by wt of paint	17	13 (min)
Silica (SiO_2), % by wt of vehicle nonvolatiles	11	11 (min)
Aluminum paste, lb/gal. of vehicle	3.95	—
Viscosity, KU	61	54–67
Heat resistance (400–1200°F on black iron)	Passes test	Pass test
Heat resistance (1000°F on cold-rolled steel)	Passes test	Pass test
Salt spray resistance (heated on black iron)	Passes test	Pass test
Salt spray resistance (unheated, cold-rolled steel)	Passes test	Pass test
Water Resistance	Passes test	Pass test
Hydrocarbon resistance	Passes test	Pass test
Drying time (all tests)	Passes tests	Pass test

A similar paint has been prepared, complying completely with the Anti-air pollution Rule 66 by replacing the xylene and the VM & P Naphtha with the Amsco Mineral Spirits #3704 or equivalent. Inasmuch as the latter solvent is somewhat lower in specific gravity than those which it replaces, the volume yield is larger and the weight per gallon is lower.

Undoubtedly a similar vehicle, equally satisfactory for this application, could be prepared from other fatty acids such as linseed or safflower.

#125 Zinc-Dust, Zinc-Oxide Primer

Material	*Pounds per 100 Gallons*
"Asarco" #1 zinc dust	312.5
"XX-601" zinc oxide	150.0
#1132 graphite	50.0
Diatomaceous silica	43.8
G-E silicone resin SR-112 (50%)	462.5
"Solvesso" 100[a]	231.3
	1250.1

[a] For spray gun application, xylene may be substituted for the slower solvent.

Zinc dust should be stirred into the paint base shortly before use. Volume yield of paint base is 94.7 gallons. Zinc dust is added at the rate of 3.3 lb/gal of paint base. For priming stacks and processing equipment which may be simultaneously exposed to the elements and to high temperatures, this primer would be useful.

A normal high-temperature primer for silicone enamels and coatings is illustrated in the next formulation.

G-E silicone resin SR-120 has excellent adhesion to most metals. To build substantial film thickness, or where service in a corrosive atmosphere is anticipated, priming may be needed. A typical formulation for a hard, flexible, impact-resistant metal primer is:

#465 Heat-Resistant Metal Primer

Material	Pounds per 100 Gallons
Imperial X-883 zinc yellow	215.5
R-C #1094 indian red[a]	161.3
Micro Velva A	161.3
G-E silicone resin SR-120 (65%)	445.2
70:30 xylene/n-butanol	222.2
	1205.8

Properties

Weight per gallon	12.1 lb
Viscosity	66 KU
Reduction ratio	5:1 with same solvent used in formulation
Reduction viscosity	25 sec, #4 Ford cup

[a] C. K. Williams #8098 red iron oxide may be used in place of R-C #1094 on equal weight basis.

0.1% manganese will provide tight cure.

This primer formulation bakes very satisfactorily in 30 minutes at 440°F without catalyst. The higher pigmentation appears to bring about a faster cure than is possible with an enamel. However, the metallic catalyst system may be used to provide even faster cure.

Other high-temperature coatings covering the broad spectrum of types of finishes that are properly formulated with silicone resins are shown in the next group of formulations:

#148 White Missile Coating

(High Infrared Reflectance)

Material	Pounds per 100 Gallons
Zinc sulfide	650
G-E silicone resin SR-112 (50%)	292
G-E silicone resin SR-82 (60%)	129
"Acryloid" B-66 (40%)	183
"Nuogel" AO	9
Xylene	54
	1317

This will air dry to handle within half an hour. It requires typical silicone baking temperatures for full cure.

#209 Black Heat-Resistant Coating

Material	Pounds per 100 Gallons
Ferro F-2302 black	56.7
#1132 graphite	113.4
"Micalith" G	56.7

Material	*Pounds per 100 Gallons*
7% ethyl cellulose T-200 in toluene	410.1
G-E SC-3900, 20% in *n*-butanol	9.5
G-E silicone resin SR-82 (60%)	94.5
"Aroplaz" 7323 (60%)	68.0
6% cobalt octoate	0.8
6% manganese octoate	0.5
Xylene	88.2

Properties

Weight per gallon	9.0 lb
Viscosity	84 K U

This coating passed the following heat test exposures:

    15 minutes at 800° F followed by 60° F water quench.

    4 hours at 1000° F followed by air quench.

    2 hours at 1200° F followed by air quench.

After 30 minutes bake at 400° F and 16 hours air aging, the film withstands $\frac{1}{8}$-inch bend and 24 hours immersion in gasoline.

#297 Cocoa Brown High-Temperature Appliance Coating (Baking Type)

Material	*Pounds per 100 Gallons*
TiO_2 RANC	57.5
Ferro F-6112 red brown	86.0
"Bentone" 11	14.3
G-E silicone resin SR-120 (65%)	689.0
"Cymel" 301	78.8
Catalyst 1010	5.4
6% manganese naphthenate	7.2
6% iron naphthenate	1.4
70/30 xylene/*n*-butanol	44.8
	984.4

Properties

Weight per gallon	9.85 lb
Viscosity	63 K U

Reduced 5:1 by volume with the solvent blend shown above, this enamel sprays excellently. After a 1 hour bake at 500° F, the film exhibits a 4H hardness while taking both direct and reverse impacts of at least 72 inch-pounds. These tests were made on cold-rolled steel Q panels.

#637 Light Brown Resistor Coating

Material	*Pounds per 100 Gallons*
Ferro F-6109 light yellow brown	25.2
Ferro F-6112 red brown	25.2

Material	*Pounds per 100 Gallons*
325-mesh mica	149.2
Antimony oxide KR	28.2
Zinc oxide XX-4	16.3
"Santocel" CS	. 8.9
"Bentone" 38[a]	7.7
Denatured ethyl alcohol (95%)[a]	3.4
G-E silicone resin SR-112 (50%)	180.3
G-E silicone resin SR-125 (50%)	180.3
6% manganese naphthenate	3.0
Xylene	332.5
	960.2

Properties

Weight per gallon	9.5–9.7 lb
Viscosity	61–63 KU

[a] "Bentone" 38 is dispersed along with the pigments and the alcohol is added in the final letdown.

Dip coating will pick up a 15 to 20 mil dry film in three coats. Although in most cases it is highly desirable to apply thick coatings with a minimum of operations, it is also necessary to program the heating of each coat to some extent in order to avoid solvent entrapment. Final baking can be done at 250°C, but 275°C will accomplish the curing in a shorter time. Final baking at 350°C can be done also if a short period at 250 to 275°C is provided.

ADDITIVES IN COATINGS

Silicone resins, particularly fluids, impart some remarkable surface properties to protective coatings. In many instances, the addition of one of these can "make" a finish, whereas without it the coating would be inferior.

As additives to paint formulations, either silicone resins or silicone fluids can be utilized, depending on the specific needs in individual coatings applications. This utilization in varying concentrations will result in at least one or a combination of benefits for the end product in terms of processing, application and/or performance.

For depressing foam in the manufacture of natural gum and synthetic resin varnishes, silicone fluids or antifoam compounds may be used. The exact type and amount will be determined by the chemical characteristics of the material to be defoamed. In some cases, a fluid will serve adequately while in others, an antifoam compound will be best. The latter type of product contains a dispersing agent to increase the efficiency of the defoamer.

In protective coatings, silicone fluids are utilized to prevent pigment flotation, silking, orange peel and other surface defects. A good example of their usefulness is a multipigment pastel pink enamel in which the red pigment was floating, thus preventing uniform tinting. The use of 100 ppm of a low-viscosity silicone fluid quickly corrected this problem.

A high-viscosity fluid in dipping enamels can provide a defect-free surface. Without the additive, the turbulence in the dip tank can cause bubbles which sometimes come out on the work and remain through the bake cycle, thereby necessitating repair. The fluid in the coating minimizes bubble formation and causes the bubbles to break quickly when they do come out on a coated part. Similarly, these products depress extensive bubbling that occurs on roller application of some paints, thus making it possible to roller coat the finish.

Certain specific fluids act as flow control agents in polyester coatings, providing for smooth, uniform finishes which the coating does not have without the silicone additive. A similar effect is obtained in epoxy enamels and clear coatings when a certain silicone resin is incorporated. Without that resin additive, cratering is quite prevalent.

A silicone resin that was originally designed as a base for masonry water repellents has the peculiar property in very small amounts of acting as an antiflooding agent in polyurethane finishes. It is introduced as a very dilute solution and is effective in quantities as small as 20 ppm of the silicone resin, based on the total weight of the formulation.

The same product or, in this application, newer products designed as masonry water repellents have a most interesting effect on very highly pigmented masonry paints. Normally, such coatings are highly porous because of their high pigment loading. By incorporation of 5% of silicone resin solids, based on vehicle solids of the paint, all the desirable characteristics of this coating that high pigmentation imparts are retained, while the sponge-like absorbency of water is eliminated. This gives less deterioration of the paint film and, above all, provides dry walls with less risk of interior wall damage.

Similar improvements have been observed in many latex paints when a small amount (2 to 3%) of a water-repellent chemical, sodium methyl siliconate, is added to the latex coating. Particularly noteworthy were the shorter period until the coating becomes washable and the improved resistance to water spotting.

The newest and very dramatic effect that can be obtained through a small addition of a specific type silicone is the production of a hammertone finish from almost any type of vehicle. A variety of sizes of patterns and textures can be obtained by varying the quantity of additive, the gun pressure and the distance of the gun from the work. Still greater variety is possible if the application solids content of the coating is changed.

SILICONE MASONRY WATER REPELLENTS

The most widely used silicone masonry water repellents are the resinous types in hydrocarbon solvents.   On limestone and other non-siliceous stone and masonry, a water-soluble alkaline silicone chemical must be used, as the resin-solvent type is not fully effective on calcareous masonry. To some extent, the aqueous type is used on concrete—poured, precast and prestressed—but it is rarely used on brick masonry. The solvent type serves this market very adequately.

The principal reasons for the widespread use of silicone masonry water repellents are:

(1) They are clear and colorless, preserving the original appearance of the masonry after treatment.

(2) Their non-film-forming characteristics leave the pores of masonry open, thus permitting the walls to "breathe" and discharge moisture-laden air to the outside atmosphere.

(3) Being in the pores (forming a lining) rather than on the surface of the masonry, the treatment is highly durable and will remain effective for several years. All these very desirable properties represent a bonus in that they are realized in addition to obtaining the expected benefits of minimized efflorescence, spalling, and cracking, and the maintenance of a stain-free masonry surface.

The resin-solvent type product is a hydrocarbon solution of a partially hydrolyzed, partially alcoholyzed silicone resin. Different products are characterized by being derivatives of varying combinations of alkyl di- and trichlorosilanes. By being partially alcoholyzed with one of the lower alcohols (usually n-butyl), the end products are easily cured by (hydrolysis) reaction with moisture of the air, liberating the alcohol.

The solvent-based materials may vary in solids content and in solvent, as produced, but all are optimized at 5% nonvolatile with mineral spirits for application. These products are packaged at application solids content and are very stable under normal storage conditions, providing their containers are tightly closed.

The sodium methyl siliconate is a reaction product of aqueous caustic soda and a fully condensed resinous silicone material comprising a preponderance of trisubstituted and a lesser amount of disubstituted silicone. This is usually produced at a total solids (sodium methyl siliconate) content of about 30%, although the active silicone portion is only about two-thirds that amount. To formulate this material, it is reduced from 20% to 2 or 4% silicone solids. The widest usage of this product is found in connection with the limestone industry and in limestone structures. Nothing else is as effective in making such masonry water repellent and in preserving its cleanliness.

The curing mechanism of sodium methyl siliconate is via cleavage of the molecule to liberate the highly insoluble fully condensed silicone from which the product originated.   The sodium reacts with CO_2 from the

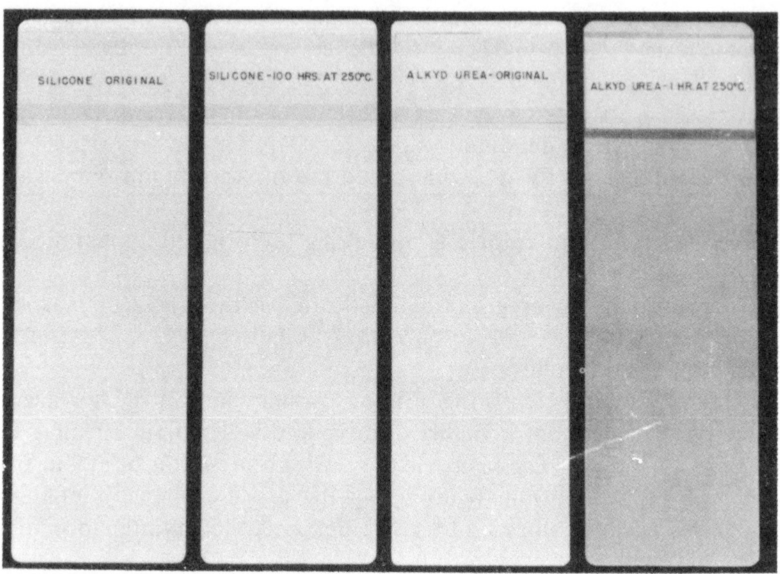

Figure 14.1.   Silicone and alkyd-urea enamels after heat aging.   Note that the silicone exposure was for 100 hours versus only 1 hour for the alkyd urea.

Figure 14.2.   All-alkyd (left) versus 25/75 silicone/alkyd-based olive drab paint following 24-hour exposure to 480°F.   The unmodified coating has lost its integrity as evidenced by fracturing of the film and powdering of the pigment, while the silicone-modified coating is intact, even in the bend area.

atmosphere to form soluble sodium carbonate which is ultimately washed away.

A newer product employing modified techniques has recently appeared on the market. It is available at 50% solids in methanol and is reduced to 5% solids by volume, with water, at the time of application. Its instability in reduced form precludes its being formulated to application solids in advance. Since it is new, its impact on the market is yet to be seen.

It is important in any discussion of these products, to emphasize one word in their generic name—repellents. These are water *repellents,* not water*proofers.* They merely line the pores of masonry, but do not fill them. They are, therefore, not water barriers and will not contain water under pressure or hydrostatic head. For this reason, the silicone masonry water repellents are useful only for above-grade applications.

Figure 14.3.   Hot oil heater and stack coated with silicone-based aluminum paint withstands high temperatures and outdoor weathering.

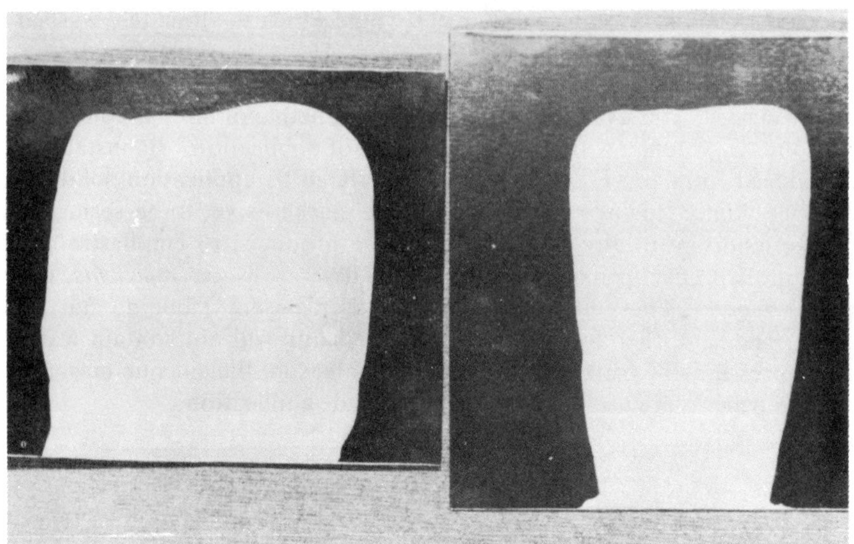

Figure 14.4.   Bad pigment float, at left, corrected by a very small addition of silicone fluid.

Figure 14.5.   Panel at right illustrates the smooth, uniform appearance of an ordinary alkyd resin-based metallic enamel.   The next two panels illustrate a textured pattern (center) and a hammered pattern (left) made from the same enamel modified with a specific silicone additive.   Both patterns were produced by the same addition of the silicone, but with different air pressures at the gun (50 psi for the texture and 80 psi for the hammer).

Figure 14.6. Concrete stadium under construction, in which all precast sections were treated with aqueous silicone masonry water repellent prior to being installed.

Figure 14.7. Droplets of transparent dyed water remain standing on silicone masonry water repellent-treated brick, while absorption into the untreated brick (right) is instantaneous.

Figure 14.8. The unsightly efflorescence shown on this brick wall can be essentially, if not entirely, eliminated by a silicone masonry water-repellent treatment.

Figure 14.9. Ordinary exterior white paint after about 1500 hours weatherometer exposure. Note free chalking.

Figure 14.10. Silicone-modified, very long oil alkyd-based house paint after about 1500 hours weatherometer exposure. Note absence of chalking.

Figure 14.11.   Exterior storage tanks protected by silicone/alkyd-based industrial maintenance enamel.

SILICONE RESINOUS INTERMEDIATES IN CONVENTIONAL COATINGS

It was pointed out earlier that unbodied, resinous, silicone intermediates are capable of copolymerizing with organic resins which have reactive hydroxyl groups. For the most part, these reactions are carried out with specifically prepared intermediates which would have little value in other applications. However, certain silicone resins which are useful *per se* in protective coatings formulation also qualify by definition for use as intermediates in copolymer work. They are unbodied and have terminal hydroxyl groups in sufficient quantity to react with organic resin constituents such as alkyds, epoxies, etc. The economics of the operation usually dictate the selection of the specially prepared silicone intermediates for the practice of this technology, but academically the unbodied resins are potential reactants.

It will be recalled that one certain characteristic of all silicone resins is the presence of some trisubstituted silicone—the most reactive type—which condenses to a three-dimensional polymer. Similarly in the copolymer resins, the silicone portion must possess silicone resin qualifications. Putting these data in terms of chlorosilanes that will comprise the blend, an aryl trichlorosilane is almost certain to be present because its hydrolysis product will aid materially in providing compatibility with the organic constituent. Other chlorosilanes might include dichlorosilanes whose hydrolysis products would contribute flexibility, and one or more alkyl trichlorosilanes, or perhaps just the alkyl trichlorosilane(s) without any dichlorosilanes. Further variation in physical properties is controlled by the selection of alkyl trichlorosilanes. As the carbon chain increases in length, the end product becomes softer—varying from a friable resin to a waxy material to a soft, resinous mass to a liquid.

Somewhat analogous intermediate products are prepared by alcoholyzing (usually with methanol) instead of hydrolyzing the selected blend of chlorosilanes. Alternatively, certain patents[5,6] teach the utilization of silicone intermediates which are neither entirely hydrolyzed nor entirely alcoholyzed, but partially reacted by both procedures.

Earlier, a schematic design was shown of the copolymerization of a characteristically hydroxy functional alkyd resin and an hydroxy functional silicone intermediate. Had the intermediate been prepared by alcoholysis with methanol, the copolymerization reaction would have been represented as:

$$\text{Alk—OH} + \text{CH}_3\text{O—Si} \rightarrow \text{Alk—O—Si} + \text{CH}_3\text{OH}$$

Frequent use is made of this reaction, and of the one with a hydrolyzed silicone intermediate, in the preparation of silicone/alkyd resins (oil modified) and silicone/polyester copolymers.

Although there is probably more utilization of these reactions than of some other types, copolymerization of silicone intermediates is by no means limited to these reactions. Any polymer with reactive, terminal hydroxyl groups is a potential copolymer constituent. The epoxy resins based on bisphenol A are essentially linear polyalcohols with terminal oxirane rings. The structures may be schematically diagrammed as follows:[7]

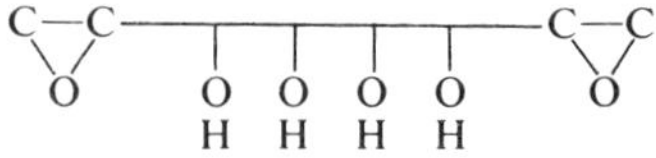

where there are four "monoglycidyl-ether-of-bisphenol A" groups between the terminal glycidyl groups. Each of these hydroxyl groups is a potential reaction site for copolymerization with a silicone intermediate, after which the modified epoxy resin may be used as a vehicle constituent, just as it was prior to the introduction of the silicone; or it may be esterified to form a silicone-modified epoxy ester vehicle.

Another important group of vehicles that may be successfully and beneficially modified with silicones is the thermosetting acrylic resins. One approach is to use the silicone-modified epoxy resin as the curing agent in an acid (acrylic acid) containing acrylic polymer.

Another type of thermosetting acrylic resin contains reactive hydroxyl groups. These may be introduced by copolymerizing with any of several hydroxyl-containing monomers,[8] although the simplest, for illustrative purposes, would be a hydroxyl-modified ester of acrylic or methacrylic acid. These acrylic polymers are quite amenable to further copolymerization with silicone intermediates through the hydroxyl group. Like their counterparts that are not further modified with silicones, these thermosetting silicone/acrylic resins may be cured by incorporating melamine- or urea-formaldehyde resins into the finish system.

Still another type of thermosetting acrylic is that which contains methylolated acrylamide. The preparation and chemistry of these resins are detailed by Brown and Miranda.[8] The methylolacrylamide which characterizes this polymer has several hydroxyl reaction sites. In spite of the etherification of at least 50% of the hydroxyls by lower alcohols to insure stability of the resin,[8] there are remaining sufficient sites for further copolymerization with silicone intermediates such as have been described. Although certain catalysts may be used to speed the cure of these resins or bring it about at lower temperatures, this group of resins is very interesting in that they can be cured by heat alone. Temperatures from 250 to 300°F will suffice with acidic catalysts, but it is necessary to go to 350°F to cure the uncatalyzed copolymer in 30 minutes. In an increasingly important application of these resins, including those that are silicone modified—coil coatings—baking schedules such as those just cited are quite obsolete. Schedules counted in seconds, for the most part, at temperatures in the range of 500 to 700°F are more common.

Where, then, do these silicone modified alkyds, epoxies, acrylics, etc., fit into the current coatings picture? They are contenders for vehicle selections for the entire gamut of exterior coatings—architectural, industrial maintenance and product finish.

Wood trim paints based on silicone-alkyd copolymer vehicles have proved their superior gloss retention and preservation of film integrity. Very long oil alkyd-silicone copolymers are being evaluated in house paints and appear to be proving their merit in that application. Industrial maintenance coatings based on such copolymer vehicles are exhibiting unprecedented performance in withstanding weather and corrosive atmospheres. Coil sheet and strip coatings on metal are utilizing silicone-modified copolymers in coatings that are guaranteed for years against fading and general deterioration.

Although much of the developmental work in this area is relatively new and proprietary in nature, the following will illustrate the preparation of a silicone/alkyd copolymer and an industrial maintenance coating based on that vehicle:

30% Silicone/Alkyd Copolymer

Alkyd Portion	*Parts by Weight*
Soya fatty acids	36.3
Phthalic anhydride	13.0
Monopentaerythritol	12.4
Mineral spirits	38.3
Xylene (for azeotropic distillation)	5–10% of kettle charge

Add all ingredients, except mineral spirits, to the kettle and maintain heat between 425 and 450°F. Remove the water azeotropically as it is formed. Continue cooking until the acid number is down to a range of 5 to 10. Reduce to 60% solids with the mineral spirits.

Silicone Alkyd Copolymer	*Parts by Weight*
Alkyd resin from above (60% nonvolatiles)	70
Silicone intermediate SR-187 (75% nonvolatiles)	24
Mineral spirits	6

Charge the alkyd resin and the silicone intermediate to a kettle equipped with a reflux condenser, water trap, thermometer and agitator. Heat to a steady reflux which should occur at about 340°F and hold until a clear thin pill is formed when a drop of resin is put on a glass slide or panel. Cooking may be continued from that point, to the desired viscosity of the end product. When cooking is completed, remove the heat and reduce to 60% nonvolatile with mineral spirits. The SR-187 shown in this vehicle is the hydroxy functional type.

#242 Dark Green Exterior Maintenance Enamel

	Pounds per 100 Gallons
Imperial X-3163 iron blue	16.8
Imperial X-1285 Iroquois blue	45.0
Imperial X-2846 empress yellow	6.9
30% silicone/alkyd copolymer[a]	526.0
Modaflow	3.1
"Nuosperse" 657	2.5
"Nuogel" AO	3.0
"Exkin" #2	4.6
6% cobalt octoate	2.6
6% manganese octoate	1.6
5% calcium octoate	3.1
Mineral spirits	170.0
	785.2

[a] Vehicle is that described immediately above.

This coating performs very well and retains its high gloss through continuous outdoor exposure in a chemical manufacturing plant atmosphere.

(It is protecting multi-thousand gallon storage tanks at General Electric Silicone Products Department plant.)

Another typical copolymer vehicle is the following short oil baking-type modified alkyd resin prepared by a one-stage cook.

262-209 Short Oil Silicone-modified Copolymer

	Parts by Weight
Coconut fatty acids	28.7
Phthalic anhydride	23.3
Glycerol	23.0
Silicone intermediate SR-187 (75% nonvolatiles)	33.3
Xylene (for azeotropic distillation)	

Charge the ingredients to a kettle equipped for a solvent reflux cook, and raise the temperature slowly to 230°C (about 3 to 5 hours). Cook to an acid number 2 to 3, remove the heat and thin to 60% solids with xylene.

This vehicle, alone or blended with another silicone-modified longer alkyd resin, can be formulated with thermosetting amine resins in exactly the same manner as would be used with unmodified alkyd resins.

The methoxy functional silicone intermediate is represented by the General Electric product SR-191 which contains 15% methoxyl. A silicone/alkyd resin vehicle prepared from that product would be formulated as follows:

30% Siloxane-modified Long Oil Alkyd Resin

Alkyd Portion	Parts by Weight
Soya fatty acids	59.0
Phthalic anhydride	17.6
Pentaerythritol	23.4
Triphenyl phosphite	0.25
Xylene (for azeotropic distillation)	3.0

Charge all ingredients into a kettle equipped for solvent reflux cooking. Bring the temperature slowly to 235°C (in about 2 hours), collecting water in the trap as formed. Cook to a low acid number (about 5 hours) and reduce to 75% solids with mineral spirits.

Copolymer	Parts Solids	Parts Solution
Alkyd resin (75% nonvolatiles)	70	93.3
SR-191 (15% methoxyl)	30[a]	33.8[a]
2-Ethylhexoic acid		1.0

[a] SR-191 is 100% solids, but 33.8 parts are necessary to yield 30 parts of siloxane which will remain in the copolymer after liberation of CH_3OH.

Charge the alkyd into a kettle and dry by azeotropic distillation for 30 minutes. Set up packed column for fractionation to remove methanol, and add premixed SR-191 and 2-ethylhexoic acid. Cook until 90 to 95% of the theoretical amount of methanol is removed, over a period of about 5 hours, being guided by the boiling point of the alcohol and cure of the resin on a 200°C cure plate. The viscosity will rise during cooking, necessitating gradual dilution with mineral spirits, down to 60% solids at the conclusion of the cook, at which point the viscosity should be U-W.

As shown above, 2-ethylhexoic acid can serve as a copolymerization catalyst. Certain patents[5,6] pertaining to this subject teach the preparation of silicone copolymer vehicles utilizing both alkyd and oil-modified alkyd resins in the presence of a titanate catalyst.

A typical starting formulation for evaluation as a semigloss coil or strip coating based on a silicone/polyester vehicle would be the following:

#258 Coil or Strip Coating

	Parts by Weight
Titanium dioxide (nonchalking)	294
Magnesium silicate (325 mesh)	26
Magnesium silicate (extra fine)	101
Silicone/polyester vehicle (50% nonvolatiles)	553
Hexamethoxymethyl melamine resin	31
Acid catalyst	3
"Solvesso" 150	150

Here, too, the great bulk of data in this relatively new activity is proprietary information. Some starting formulations, as guides for further development, are available in the commercial literature but these are, admittedly, intended to be further developed.

The potential in both vehicle and end product development is gigantic, and the early, more or less crude tools illustrated here can serve as useful guidelines in the research and development of silicone copolymer coatings, not only for coil and strip metal, but for all exterior applications where maximum durability is the primary objective.

RAW MATERIALS SOURCES

Acryloid Resins	Rohm & Haas Co.
Alcoa Aluminum Pastes	Aluminum Co. of America
Amsco Solvents	American Mineral Spirits
Antimony Oxide	The Harshaw Chemical Co.
Aroclor Resins	Monsanto Co.
Aroplaz Resins	Archer Daniels Midland Co.
Bentone 11, 38	National Lead Co.
Catalyst 1010	American Cyanamid Co.
Cymel 301	American Cyanamid Co.
Ethylcellulose	Hercules Powder Co.

Exkin #2	Nuodex Products Div., Tenneco Chemicals
Ferro Colors	Ferro Corporation
G-E Silicones	General Electric Co., Silicone Products Dept.
#1132 Graphite	Joseph Dixon Crucible Co.
Imperial Color Pigments	Imperial Color Div., Hercules Powder Co.
Mica, 325-mesh	English Mica Co.
Micalith G	English Mica Co.
MicroVelva A	Carbola Chemical Div., International Talc
Modaflow	Monsanto Co.
Nuogel AO	Nuodex Products Div., Tenneco Chemicals
Nuosperse 657	Nuodex Products Div., Tenneco Chemicals
Pliolite Resins	The Goodyear Tire & Rubber Co., Chemical Div.
R-C Iron Oxides	Reichard-Coulston Company
Santocel CS	Monsanto Co.
Solvesso Solvents	Humble Oil & Refining Co.
Titanium Dioxide Pigments	Titanium Pigments Div., National Lead Co.
Zinc Oxide Pigments	The New Jersey Zinc Co.
Zinc Dust, #1	American Smelting & Refining Co.
Q-Panels	The Q-Panel Co.

REFERENCES

1. Rochow, E. G., "An Introduction to the Chemistry of the Silicones," p. 148, New York, John Wiley & Sons, 1951.
2. *Ibid.,* p. 150.
3. Wagner, G. H., and Strother, C. O., U.S. Patent 2,632,013 (March 17, 1953).
4. Rochow, E. G., "An Introduction to the Chemistry of the Silicones," p. 143, New York, John Wiley & Sons, 1951.
5. Rauner, L. A., and Tyler, L. J., U.S. Patent 3,015,637 (Jan. 2, 1962).
6. British Patent 909,366 (Oct. 31, 1962).
7. Spitzer, W. C., "Official Digest," *Journal of Paint Technology and Engineering,* **36,** 52 (1964).
8. Brown, W. H., and Miranda, T. J., "Official Digest," *Journal of Paint Technology and Engineering,* **36,** 92 (1964).

ACKNOWLEDGMENT

The author wishes to acknowledge with thanks the critical review of the sections on Chemistry and Manufacture of Silicones by Dr. R. N. Meals, and of the section on Silicone Resinous Intermediates by Dr. G. F. Roedel, both of the Research and Development Staff of Silicone Products Department, General Electric Company.

15

*Miscellaneous Resins**

In recent years many new resin finishes have appeared on the market, however the use of varnishes in coatings continues to be large. One of the prime considerations of the varnish craftsman down through the ages was the choice of suitable resins needed to produce the type of protective coating desired. This search for better resinous materials has continued to the present day and still exists.

Within the last half century, the coatings industry has experienced a period of extraordinary growth brought about primarily by the availability of synthetic resins and the introduction of scientific methods into the field of varnish and paint making.

Varnish resins are divided into two general categories—natural resins and synthetic resins. Natural resins include only those used in preparing coatings without modification by chemical processes. Synthetic resins would cover chemical modifications of natural resins and resinous products made from nonresinous chemicals.

This chapter is concerned with miscellaneous resins such as natural resins, rosin, shellac and asphalt used in the coatings industry. Resins such as the acrylics, alkyds, amino resins, cellulosics, chlorinated rubber, epoxies, hydrocarbons, phenolics, polyurethanes, silicones, vinyls, etc., have been covered in earlier chapters.

NATURAL RESINS

The use of natural resins in the preparation of decorative and protective coatings goes back to very early times. These resins which are hardened exudations from trees of many different genera and species are classified as fossil or recent types. "Recent types" refers to a resin extracted by tapping a living tree, whereas "fossil types" are dug from the

*By F. J. Martinek, The Sherwin-Williams Co., Cleveland, Ohio.

earth or found in beds of streams, where they were deposited by vegetation of past geological periods and covered by the shifting of land masses.

Natural resins are highly complex chemical compounds whose exact composition is still a matter of uncertainty. Analysis shows that they all consist mainly of carbon and hydrogen with a small proportion of oxygen. The compounds which may be found in them are:

(1) Resenes (complex saturated organic compounds),

(2) Essential oils such as terpenes,

(3) Complex aromatic acids of high molecular weight.

The resins formed in a tree in pockets or passages between cells exist in the form of a solution of the acids and resenes in essential oil. These solutions exude as viscous liquids or oleoresins. Upon exposure to air, the solution hardens to form the resin partly as a result of evaporation of the essential oil and partly by oxidation and polymerization. The natural resins are not subject to depletion since they are products of trees which themselves are capable of infinite renewal and increase. From the viewpoint of terminology, rosin and shellac are not included within the term "natural resins" although basically they do fall into this category.

The coatings industry usually refers to the natural resins as gums. This is a misnomer since gums are related to sugars and carbohydrates. Gums, being soluble in water, form viscous solutions, whereas they are insoluble in drying oils and organic solvents. When heated, they decompose completely without melting. However, the natural resins described in this chapter are insoluble in water. They are somewhat soluble in organic solvents and vegetable oils, and are chemically related to the terpenes or the essential oils. When heated, the resins melt and volatile oils are distilled off. The residue known by varnish formulators as "run" gum or resin is soluble in solvents or oils, but in all cases is totally insoluble in water.

In general, natural resins are characterized by good chemical stability and inertness; they are not readily attacked by either acids or alkalies. The classification as to type of natural resins is outlined in Table 15.1.

Congo

Congo is a hard fossil resin originating in the Belgian Congo in Africa. In its natural state, Congo resin does not dissolve in any organic solvent, but it does soften in some alcohols and hot solvents to form gelatinous liquids or transparent, viscous gels. Many of these jellylike masses are solutions of the solvents in the resin.

When thermally processed or run (cracked), Congo is soluble in a wide range of petroleum solvents of all varieties—coal tar solvents, alcohols such as butanol, terpene solvents, ketones and ethers, as well as the fatty

TABLE 15.1.   Classification of Natural Resins

(I) Dammars—Low acid number resins of recent origin; solvent and oil soluble	

(I)   Dammars—Low acid number resins of recent origin; solvent and oil soluble
    (A) Batavia
    (B) Singapore
(II) East Indias—Semifossil or semirecent resins related to the dammars; solvent and oil soluble
    (A) Batu
    (B) Black
    (C) Pale East India Singapore (Rasak)
    (D) Pale East India Macassar (Hiroe)
(III)   Copals—In general, higher acid numbers than dammars
    (A) Manila
        (1) Melengket or soft resins of recent origin; spirit soluble
        (2) Loba or half-hard of semirecent origin; spirit soluble
        (3) Philippine Manilas (Almaciga) of semirecent origin; spirit and oil soluble
        (4) Pontianak, semifossil hard; spirit and oil soluble
        (5) Boea, fossil hard; oil soluble
    (B) Congo, African fossil hard; oil soluble
        (1) White, ivory, straw, pale and amber
    (C) Kauri, New Zealand fossil hard; oil soluble
        (1) Pale and brown
        (2) Bush
(IV)   Miscellaneous resins
    (A) Accroides, red gum, or grass gum, of recent origin; spirit soluble
    (B) Elemi, soft of recent origin, balsam type; spirit soluble
    (C) Mastic, of recent origin; spirit soluble
    (D) Sandarac, of recent origin; spirit soluble

acids and the vegetable oils. One suggested running operation recommends that the resin be placed in a kettle having a lid with a vent for fumes and openings for a thermometer and a stirrer. The temperature is first raised to 600°F and maintained until the lumps have disintegrated to a plastic mass and foaming begins to subside. The temperature is then raised to 650°F and held at that point until the resin becomes liquid and gives a clear drip from a stirring rod when removed from the kettle for sampling. A fine melt may require as long as 60 or 80 minutes at 650°F.

In this procedure, the resin loses anywhere from 20 to 35% of its weight depending upon its hardness. The distillate which comes off during the running operation, if condensed, is an oily liquid, dark in color, with good solvency (sometimes referred to as copal oil) used in insecticide sprays, shoe polish, wood preserving substances and putty. It is not practical to give a precise procedure for running such resins since shipments are seldom duplicated.

After running, the resin is compatible with ethylcellulose, nitrocellulose, a wide variety of other resins of a natural and synthetic nature, waxes, putties and asphalt. Varnishes prepared with it are equal or superior to those containing modified phenolic resins in durability, elasticity, weather resistance, color retention and drying time.

Because of gloss and "depth of film" associated with Congo varnishes, this resin has long been the standard rubbing varnish resin. Thermally processed Congo imparts excellent adherence to varnishes and finishing compositions when these are applied to metal surfaces. Congo was the first resin to be used in the preparation of so-called wrinkle finishes.

Because of its high acid number, Congo cannot be used in vehicles for paints or enamels containing reactive pigments. Insoluble "soaps" are formed due to a neutralization reaction. This phenomenon is known as "livering" and occurs whenever the acidity of a resin is great enough to cause a reaction with basic pigments. Livering may result in the formation of a "cheesey" mass or merely in an increase of viscosity. Congo esterified with glycerin or other polyhydric alcohols forms Congo esters, sometimes referred to as copal esters. Such esters have low acid values and are free from "livering" tendencies. They also possess good heat, moisture, dilute acid and alkali resistance, as well as color retention at high baking temperatures and bleaching of the varnish film upon aging; they are nontoxic and do not impart taste or odor to aqueous liquids and solutions with which they come in contact. These processed copals can be dissolved in oil by heating with the oils to approximately 500° F. They are also soluble in a wide variety of solvents.

When used with tung oil, Congo resin gives quick-drying varnishes of good durability. The alkali resistance of Congo-tung oil varnishes is good but can be made better by the addition of about 10% of a phenol-formaldehyde resin.

In addition to its esterification with polyhydric alcohols and fatty acid glycerides, Congo is capable of being acetylated to yield a product known as "acetocopal" which could be used for varnish manufacture and as a lacquer plasticizer. By treatment with butyric acid, "butyrocopal," which possesses kindred properties and uses, can be obtained.

Kauri

Kauri is produced by *Agathis Australis*, a conifer, the largest forest tree of the North Island of New Zealand. The tree is an evergreen and is believed to reach an age of nearly 3600 years. Kauri is a fossil copal, of the same general class as Congo and Boea. It is found underground in hilly clay ranges and Kauri peat swamp. The trees which produced the Kauri resin are no longer in existence. Some minor amounts of Kauri of the soft type are obtained by tapping living trees. The bulk of Kauri, however, is fossilized resin of considerable age.

Kauri has the lowest acid value of the fossil copals (60 to 80) and, unlike Congo, shows solubility in alcohols and ketones as well as in some lacquer solvents. In its original form, it is compatible with a wide range of vegetable, mineral and petroleum waxes, as well as stearic acid, but the temperatures of combination are such that thermal processing of the kauri takes place. However, it must be cracked before it can be used in oleoresineous varnishes. After the cracking operation, which is conducted in a manner similar to that for Congo, Kauri is compatible with all animal and vegetable oils. The heat treatment necessary to make it oil soluble is not as severe as that for Congo. Over a period of about $1\frac{1}{4}$ hours, it is heated to 600°F and then held at that temperature for about a half hour.

Running losses commonly are of the order of 20 to 25%. When used in conjunction with tung oil, there is no difficulty in preparing satisfactory varnishes of high gloss, adhesion and durability, which dry in less than four hours.

Originally, Kauri was used in varnishes for furniture. The old varnishes referred to as piano, furniture and carriage finishing varnishes, were made with Kauri and linseed oil. They took much longer to dry than present-day coating materials, but they would polish to a much higher gloss than the latter.

Kauri is used in varnishes of widely varying oil length, japans, lacquers, paint, cements, linoleum and oilcloth. It finds some small application in jewelry as an amber substitute.

Manilas

Manila resins are the exudation from Agathis Alba trees found throughout the Philippines and the Netherlands East Indies, with the exception of Java. The Manilas are obtained by deliberate tapping of the tree and from accidental wounds. In tapping, the bark is cut off horizontally down to the wood and the wood underneath is then cleaned to provide a surface upon which the resin collects. After every collection of resin a thin slice of bark is cut off to provide a fresh opening for the resin ducts. Average production is 26 to 27 lb/yr per tree.

The softest grades are gathered between one and three weeks after tapping and called Melengket by the natives. The finest grade, PWS (prime white soft), is nearly water white in color and almost entirely free from impurities. WS (white soft) is very clean and quite light in color. MA (Melengket A) may contain some bark and wood fiber. MB (Melengket B) may contain considerable amounts of woody impurities. A slightly older and harder grade is called E and is marketed in admixture with the C grade of the Loba-type manila as Loba CNE.

Half-hard manila, or Manila Loba, is the designation of material gathered from one to three months or more after tapping. The resin has become hard enough so that the pieces do not block together as does the soft manila. It is scraped free of surface impurities and graded according to size of piece from A (over 4 cm) to D (less than 1 cm) and dust. The FAX pale nubs are cleaner and lighter than the DK or small dark chips, which are quite dirty, or the CNE which contains considerable crusty resin. These are Netherlands East Indies gradings.

Singapore Manila, which originates in Borneo, is of the same type. Philippine Manila is also a resin of about the same age; it is sorted according to both color and size of piece, and is graded into bold, pale, scraped, bold chips, small chips, sorts, seeds and dust. The fossil grade of Manila is derived from the Netherlands East Indies and is known as Boea or hard Manila. The light-colored variety is found where it has exuded in the crotches of trees still standing. The amber and dark grades are found buried in the ground, often in masses so huge that the botanists are completely at a loss to account for their formation. Manilla CBB and Manila DBB are grades consisting of a mixture of Boea and Loba. The excellence of Boea as a varnish resin is outstanding.

A closely related resin used for oil varnishes is Pontianak, also a fossil resin. It behaves more like Kauri in the varnish kettle, but otherwise its properties and the properties of its varnishes make it almost indistinguishable from Boea.

Manila copal is used in oil and spirit varnishes and paints, and also finds some application in the manufacture of lacquers, sizing materials, plastics, japans, driers, linoleum, oilcloth, waterproofing compositions, printing inks, adhesives and other miscellaneous uses.

The Melengkets and Lobas are not compatible with oils, but may be rendered so by thermal cracking. They are readily compatible with cellulose derivatives such as nitrocellulose and ethylcellulose, but are incompatible with cellulose acetate in acetone. They may be combined with vegetable and mineral waxes and stearic acid at temperatures approximating those encountered in thermal processing.

The soft Manilas, however, show only partial compatibility with Japan, Ozokerite, and petroleum waxes, while the Lobas show compatibility with Japan wax but not with Ozokerite or petroleum waxes.

In general, the Manilas are not readily miscible with petroleum asphalt or stearin pitch, but form homogeneous mixtures with coal tar and vegetable pitches. After thermal processing, the Manilas may be combined readily with a wide variety of asphalts and pitches.

The Melengkets and the Lobas are compatible with a wide range of synthetic resins, except for the paracoumarones in which they form cloudy dispersions. After thermal processing, they show compatibility with all of the phenol aldehydes in the 100%, modified, heat-reactive forms, with rosin, ester gum, the maleic-rosin materials, and the paracoumarones. Also, as might be expected from their relatively high acid values, the Manilas can be esterified with glycerol or other polyhydric alcohols. Other chemical alterations are also possible which produce resinous materials of markedly different characteristics when compared to the original copal. The relationships between the various classes of Manilas are shown in Figure 15.1.

Dammar

Among the recent resins, the Dammars are used most widely. Dammar is obtained from *Shorea* and *Hopia* trees which have been tapped with holes. The resin which exudes from the holes is collected when it is hard enough to handle. The Dammars are of the hydrocarbon solubility type, being soluble in petroleum and coal tar solvents, hydrogenated naphthas, terpenes and chlorinated solvents. They are only partially soluble in alcohols. Compatible mixtures are formed with vegetable oils by simple heating; no running operation is required.

Several types of coatings can be made using Dammar resin. Films of colorless Dammar varnishes do not yellow on aging; in fact, any color they may contain is bleached when exposed to sunlight. Dammar resin has been used to produce overprint varnishes with many outstanding qualities. It was also used for a long time as a cold-cut resin, in solvents blended with oils, in the making of enamels and mill whites. For that application, it was cut with mineral thinner or a combination of turpentine and mineral thinner.

In cellulosic lacquers, the light-colored Dammars impart the necessary adhesive qualities in addition to improving the gloss and hardness and preventing the occurrence of after-yellowing. For this purpose, the Dammar must undergo a "dewaxing" operation since it contains between 10 and 12% wax. The dewaxing consists of dissolving the dammar in coal tar petroleum solvent, and then precipitating the wax with alcohol. By allowing the solution to stand, the wax and alcohol separate from the gum solution, and then the two layers can be separated by decanting.

Dammars are modifying agents for alkyd resins, whether in clear or pigmented coatings, and have proved to be equal to the maleic anhydride resins used for that purpose. The inclusion of dammar in alkyd baking enamels markedly improves color retention and often gloss and hardness as well. In addition, Dammar has the advantage of suitability for use with reactive pigments for which maleics are not satisfactory. It is not necessary that the Dammar be dewaxed for this purpose.

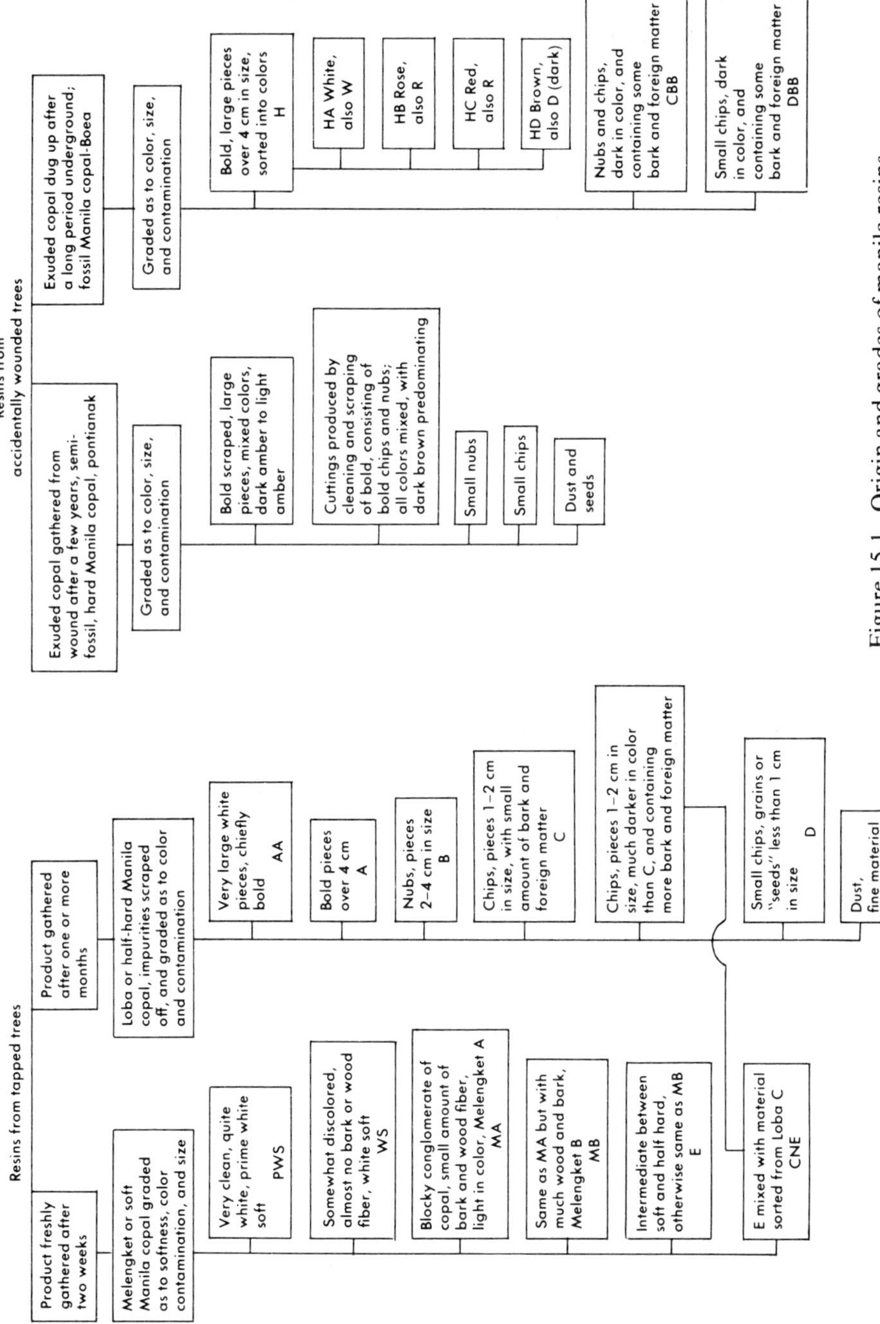

Figure 15.1. Origin and grades of manila resins.

East India

The resins which constitute the East India group are Batu, Black East India and Pale East India. The East Indias are generally similar to each other in properties and resemble the Dammars quite closely. The East India resins are not tapped from the trees, but are collected where found as a result of accidental wounding or normal exudation. Their range of solubility is less broad than that of the Dammars, particularly in petroleum solvents. Like the Dammars, they require no running operation to make them oil compatible. East India resins are used in both spirit and oleoresinous varnishes.

If one overlooks differences in color and viscosity, almost any one of the East India resins can be used in the same formula. They are used in floor varnishes, grinding vehicles, priming coats for metal, enamels, sanding sealers, mill whites and as vehicles for oil-base traffic paints. In floor varnishes and traffic paints, they show outstanding resistance to abrasion and chipping. Excellent, inexpensive spar varnishes can be formulated with these resins. East India vehicles show the best leafing properties of all the natural resins when used in aluminum paints.

Accroides

Accroides is a dark-colored resin known also as Yacca, grass tree gum or red gum and obtained from various members of the species *Xanthorikoea* of the natural order Juncaceae which occurs throughout Australia and Tasmania. The resin appears in a yellow and in a red form. The tree has a thick jacket of old leaves which surround the trunk. Resin collects around the bases of the dead leaves which are long and similar to rushes. The resin is collected by beating the stem of the tree. As a result, the resin is powdered and falls to the ground from which it is gathered. Accroides contains several per cent of paracoumaric acid, cinnamic acid and their esters. Some 80 to 85% of the resin is resinotannol.

Accroides differs markedly from the other natural resins in that it is the only material which, as marketed, is heat reactive in a manner analogous to that of the heat-reactive phenol-formaldehyde resins.

When heated to thermal processing temperatures, the resin is converted into hard, completely insoluble, and usually resistant films. Because of its heat-reactive properties, Accroides is not adapted to the preparation of oil varnishes. The resin is compatible with cellulose acetate and cellulose nitrate but not with ethyl acetate.

The low price and heat-reactive properties have a definite influence upon its commercial use. One application is as a binder in the preparation of wallboard. It is a substitute for rosin in paper coating and for shellac in spirit varnishes, lacquers and printing inks. The resin is also used in the preparation of finishes for cellulosic materials, metals, metal foil and glass.

Elemi

Elemi is the resinous product obtained from the tree *Canarium luzonicum*, a member of the Burseraceae family located in the Philippine Islands.

It is a soft, balsam-like, plastic resin which shows solubility in coal tar hydrocarbons and esters but not in petroleum solvents, alcohols and ketones.

The softening and melting points of commercial Elemi are below room temperature. Acid number of the resin shows a range of 30 to 35, and its saponification number is of the order of 20 to 40.

Elemi is compatible with vegetable, fish and animal oils of the drying and nondrying types, waxes of the mineral and vegetable variety, stearic acid, and with cellulose derivatives such as nitrocellulose and ethylcellulose, although it shows a slight incompatibility with cellulose acetate. It is readily compatible with the phenol aldehyde resins of the 100%, modified and heat-reactive type, as well as rosin, ester gum, maleic rosin resins, and the paracoumarone materials. It is readily miscible with petroleum asphalts, petroleum, coal tar, and vegetable pitches, as well as fatty acids. In resin-wax combinations, it often serves as a mutual solvent.

Elemi finds application as a plasticizer, confering increased adhesion to lacquers applied to metal surfaces, in adhesives and cements, wax compositions, printing inks, surface coatings applied to textiles and paper, linoleum and oilcloth, as a base in perfumery, an ingredient in waterproofing compositions, as well as in engraving and lithography.

When added in small quantities to lacquers and spirit varnishes, it markedly improves the ease of brushing as well as adhesion and flexibility. Elemi contains 20 to 25% of essential oils, 13 to 19% acids, about 30 to 35% resenes and 20 to 25% alcoholic resinols, terpenic in nature.

ROSIN

Rosin is obtained from several species of the pine tree indigenous to the southeastern United States. There are two types of commercial rosin—Gum Rosin and Wood Rosin.

Gum rosin is obtained by tapping living trees, whereby a sticky mixture, called oleoresin, is exuded from the wound. The distillation of this oleoresin yields gum turpentine as the distillate and Gum Rosin as the residue.

Rosin is graded according to color. American grades in order of decreasing color are B, D, E, G, H, I, K, M, N, WG and WW. In actual practice, most of the Gum Rosin consumed is graded WG or WW. Since light color is important in most varnishes, demand is greater for the WG and WW grades.

Wood rosin is extracted by solvents from pine stumps which have been in the ground for a number of years after the tree has been cut down. The crude rosin after extraction from the stumps is refined by treatment with a pair of immiscible solvents such as furfural and a petroleum hydro-carbon.

Crude rosin is ruby red in color and is graded FF. Because of its color and poor solubility, FF wood rosin has limited use in surface coatings, but a large proportion is refined by various methods to give pale wood rosin grades as indicated above for gum rosin. For most purposes, wood rosin and gum rosin can be used interchangeably, although in some products special treatment is required to produce identical results. Rosin is approximately 85% resin acids and 15% neutral material. The acids are monobasic and unsaturated, and these two characteristics are the basis of the various rosin modifications. The principal acid is abietic acid, which may be isomerized to *levo*-pimaric acid by heat.

Abietic Acid   *levo-Pimaric Acid*

While the carboxyl group of rosin is capable of undergoing the usual reactions of organic acids, such as salt formation and esterification, the reactivity of the group is greatly influenced by the fact that it is attached to a tertiary carbon. This steric hindrance factor is particularly enhanced by the fact that the tertiary carbon is a member of the bulky ring system. As a result, esterification is difficult, high temperatures being required, and the esters are correspondingly resistant to saponification. Because most of the unsaturation is in the form of a conjugated diene, it permits a wide variety of reactions such as the addition of maleic anhydride, hydrogen, oxygen or phenol-formaldehyde condensates, as well as rendering the resin acids susceptible to polymerization and isomerization.

The most important resin-forming reaction of rosin is esterification. The carboxyl group may be esterified with mono-, di- or polyhydric alcohols to give nonresinous, resinous nonconvertible, and resinous con-vertible products, respectively.

TABLE 15.2.   Esters of Resin Acids from Monohydric, Dihydric and Polyhydric Alcohols

Alcohol	Ester	Drop Melting Point (°C)
Methanol	$R \cdot COO-CH_3$	Liquid at room temperature
Diethylene glycol	$R \cdot COO-CH_2-CH_2$ $\quad\quad\quad O$ $R \cdot COO-CH_2-CH_2$	45
Ethylene glycol	$R \cdot COO-CH_2$ $R \cdot COO-CH_2$	60
Glycerol	$R \cdot COO-CH_2$ $R \cdot COO-CH$ $R \cdot COO-CH_2$	92
Pentaerythritol	$R \cdot COO-CH_2$ $\quad$ $CH_2-OOC \cdot R$ $\quad\quad\quad C$ $R \cdot COO-CH_2$ $\quad$ $CH_2-OOC \cdot R$	115

The carboxyl group may also be neutralized with lime or zinc oxide to produce the widely used limed rosin and zinc resinate.   The lead, cobalt and manganese resinates serve as driers for oil and varnish.

SHELLAC

Shellac rightfully belongs in the family of natural resins and is used to a great degree in the coatings field.   It is a resinous material obtained almost entirely from India, although it is also produced in comparatively small quantities in the Asiatic countries east of India.

Shellac is produced by the lac bug, *Laccifer (Tachordia) lacca*, a tiny red insect, not larger than the smallest apple seed.   It is synthesized in glands in the skin of these insects which cluster together on twigs of the kusum, ber, ghont and palas trees.

The "scale" of the lac insect consists of an excretion which acts as a protective coating for the insect's body.   This is an amber-colored resinous substance known as "lac," the raw material of shellac manufacture.   The insect lives most of its life under this resinous shield, sucking the sap of the tree to which it has fixed itself.

The lac is removed from the trees by cutting off the twigs or by twisting the branches so that the resin falls from the trees.   It is then taken to a central place where the wood is removed.   It is washed to remove the wood pulp, refuse and dye which float to the top, and then is ground up into seed lac.   The next step is the removal of the remaining impurities by melting and filtering through cloth. Melted Shellac is flowed onto a flat surface where it is picked up before becoming solid and stretched into a thin sheet. Finally the sheets are broken up into flakes, in which form Shellac is generally available on the market. This is known as "Orange Shellac."

Orange Shellac has five grades which are usually classified under several general names according to quality, ranging from the lowest, basic grade, up. The larger proportion goes into the cheapest grade which is known as TN. It contains more residue and dirt than the finer grades and is used where color is not a factor and where cheapness is required. The TN mark is a standard of the trade and is nonproprietary. It has been in existence for so long that its origin cannot be traced, however in the United States it is suggested that it stands for "truly native."

Although TN is a universally recognized standard, the quality is not fixed, but varies with the quality of the stick lac crops from which it is manufactured. TN is manufactured only from Bysaki stick lac and other inferior lacs.

Other grades of Orange Shellac are known as fine, superfine, heart and superior. These better grades, generally lighter in color and cleaner, have many distinguishing or proprietary marks among which are DC, VSO, ASO and others. These grades are manufactured from the very best Bysacki lac, the finest grades being made from Kushmi.

Commercial Shellac comes in two general grades or classifications: Orange Shellac, manufactured as described previously, and dewaxed lac. The latter is manufactured to meet the demand for a shellac yielding a transparent solution. The naturally occurring shellac wax is removed during manufacture so that its alcoholic solution is transparent. Frequently a colorless product is required, in which case the Orange Shellac must go through a process of bleaching before it is usable. As the first step in the bleaching process, the Orange Shellac is dissolved in a hot-water solution of borax and soda; chlorine gas, or a similar bleaching agent, is then introduced into this solution until the color has been changed from orange to white. The Shellac is then precipitated out of the solution with sulfuric acid, broken or crushed into small pieces while wet, and dried.

Bleached and dewaxed Shellac is the most popular type in the United States at the present time. The liquid Shellac commonly sold is a 4- or 5-pound cut, meaning that 4 or 5 pounds of Shellac have been dissolved in 1 gallon of denatured ethyl alcohol.

At the 4-pound cut level, there is an equivalent of 35% Shellac by weight whereas a 5-pound cut contains 40%.

Shellac is used for a wide variety of coating purposes—as a clear sealer coat under varnish or as a finishing coat for floors, bowling alleys, furniture and many other wooden articles because of its fast drying, hardness, toughness, transparency and wear resistance. It is elastic in that it will dent but not crack under hammer blows.

Some of the disadvantages are its high price and limited package stability of about six months. The acids in the shellac react with the alcohol to form esters and then kick out of solution. These esters also attack the interiors of cans to form iron salts which darken the shellac. For this reason, shellac is usually packaged in glass containers or terneplate lined cans. Shellac has poor water and humidity resistance.

ASPHALTS AND PITCHES

Asphalts and pitches are dark film-forming compounds used to a large degree in the surface coatings industry, as well as for the more common uses of wood surfacing and roofing.

Asphalts are classified as natural and petroleum types. The natural types are believed to have originated from decomposed prehistoric vegetable matter and were probably produced by the evaporation of the volatile constituents of asphaltic crudes at or near the surface of the ground. Chemically, they seem to be saturated hydrocarbon chains containing varying amounts of free carbon, sulfur, oxygen and nitrogen.

Gilsonite, a hard brittle resin with a softening point of 270 to 400°F, mined in the states of Utah and Colorado, is probably the most important natural asphalt in the United States. It is found in vertical veins from a fraction of an inch to 18 feet wide, several miles in length and up to 200 feet deep. This asphalt has been pushed up between the rock formations by great pressure.

There are three grades of gilsonite, namely, jet, select and second. The jet grade comes from the middle of a vein called "cowboy" and it is more or less free from mineral matter. The "second" grade includes all that is not suitable for the other two grades and may also contain some mineral matter, because it generally comes from the sides of vein next to the rock and shale. The selects are the softest, and the jet grade is the hardest.

Gilsonite is generally melted and cooked into oils to manufacture black varnish in the same way as other hard resins are used in the coatings industry, since it is a hardening agent for the oil.

Gilsonite is low in cost and used in high-gloss, black and dark-colored surface coatings in which color is less important than cost.

Glance pitch is a dense, black, natural asphalt containing 20 to 30% free carbon used principally in dull gloss, eggshell black and dark-colored surface coatings because it produces finishes much lower in gloss than those formulated with gilsonite.

Both gilsonite and glance pitch have a tendency to bleed through light-colored top coats, and their use therefore is limited.

Petroleum asphalts are produced as still residues in the distillation of asphaltic-base crude petroleum. They have largely replaced the natural asphalts because of better uniformity, fewer impurities and lower transportation costs since much of the natural asphalt has to be hauled from distant places, particularly Trinidad.

Petroleum asphalts are widely used in acid- and alkali-resistant coatings, roofing materials, underground coatings, waterproofing compounds, and the impregnation of paper and cloth. They are available in three forms, namely solid or semisolid, solution or cutback in solvent, and water emulsion.

It is necessary to melt solid asphalts before they can be applied and therefore they are only practical for use on roads and roofs.

Solution asphalts prepared by using low-cost, off-color petroleum fractions similar to kerosene are used in chemical-resistant coatings, roof coatings and waterproofing compounds.

Emulsified asphalts consist of minute particles of asphalt suspended in water, together with emulsifying agents and inert fillers such as clay. When applied, the water evaporates and the particles of asphalt coalesce into continuous films. The clay prevents excessive running in warm weather.

Coal tar pitch, the still residue from coal tar distillation, is somewhat different chemically from the true asphalts in that it consists mostly of aromatic and cyclic compounds. Aromatic solvents must be used in cutting back coal tar pitch. Its chief uses include ship bottom as well as flat roof coatings.

REFERENCES

1. American Gum Importers Assn. Inc., "Natural Resins Handbook," American Gum Importers Assn., Brooklyn, N. Y., 1939.
2. Barry, T. H., "Natural Varnish Resins," London, E. Benn Ltd., 1932.
3. Hicks, E., "Shellac, Its Origin and Application," New York, Chemical Publishing Co., 1961.
4. Mantell, C. L., Kopf, C. W., *et al.*, "Technology of Natural Resins," New York, John Wiley & Sons, 1942.
5. Payne, H. F., "Organic Coating Technology," Vol. I: "Oils, Resins, Varnishes and Polymers," New York, John Wiley & Sons, 1954.
6. Von Fisher, W., "Paint and Varnish Technology," Vols. VII and VIII, New York, Reinhold Publishing Corp., 1948.

16

*Plasticizers**

A plasticizer is a substance incorporated into a polymer to increase its workability, flexibility, toughness or distensibility. When these effects are achieved by chemical modification of the polymer molecule, e.g., through copolymerization, the resin is said to be internally plasticized. A plasticizer may reduce the melt viscosity, lower the second-order transition temperature or lower the modulus. Plasticizers are used with thermoplastic polymers (see Table 16.1). Additional information on plasticizers for specific resins is included in chapters on these resins. Plasticizers are referred to as primary or secondary plasticizers. Primary plasticizers are solvents for the polymer. Secondary plasticizers cannot be used by themselves but are used as extenders for primary plasticizers.

**TABLE 16.1. Protective Coatings Resins
Commonly Plasticized**

Acrylic	Chlorinated rubber
Ethylcellulose	Chlorinated polypropylene
Nitrocellulose	Styrene-butadiene copolymers
Cellulose acetate	Polyamides
Cellulose acetate butyrate	Zein
Polyvinyl acetate	Silicones
Polyvinyl chloride	Shellac

HISTORY

The earliest commercial application of a plasticizer was made by the Hyatt Brothers around 1870 when they incorporated camphor with nitrocellulose. Next tricresyl phosphate came into use as a plasticizer followed by the phthalate esters in 1920. About 1930, Dr. W. Semon of B. F. Goodrich discovered that rigid, insoluble polyvinyl chloride polymer could be dissolved in a number of high-boiling solvents when heated to

*By C. R. Martens, The Sherwin-Williams Co., Cleveland, Ohio.

300° F and that the viscous solutions thus formed would set to rubbery gels.

Today plasticizers for polyvinyl chloride account for 65% of all plasticizer production. In 1965 plasticizer production reached 1 billion pounds annually.

THEORY

Thermoplastic polymers are composed of long-chain molecules held together by secondary valences or van der Waals forces. The incorporation of a plasticizer into the polymer by solvent, heat or mechanical means reduces the intermolecular forces in the polymer. The attractive forces between polymer chains vary with different polymers. Some resins with strong polar groups such as cellulose acetate are strongly bonded together, while others such as polyethylene exhibit little attraction between the polymer chains. The stronger the bonding forces between the polymer chains, the more polar must the plasticizer be.

Plasticizers contain polar or solvating groups which are capable of reducing van der Waals forces, by physical separation of polymer molecules, thereby reducing the intensity of the attractive forces between chains, and by replacing the polymer-polymer bonds with polymer-plasticizer bonds, thus reducing the effective number of bonds available for intermolecular attraction. The solvating groups determine the compatability of a plasticizer and they must have enough attraction for the resin to keep the plasticizer from being forced out or exuded. Thermosetting polymers which consist of three-dimensional networks connected through primary valence bonds are not usually susceptible to softening by external plasticizers.

PROPERTIES

Plasticizers used in protective coatings can improve such important characteristics as flow, leveling, brushability, impact resistance, drawability, gloss and luster.

Compatability

The maximum amount of plasticizer which can be added to a polymer without causing phase separation determines its degree of compatability. It is important that the plasticizer be compatible with the resin as well as with other components in the system, such as stabilizers, fillers, etc.

The plasticizer must remain compatible with the resin over the operating temperature. A plasticizer's affinity for a resin, as determined by swelling tests and the use of the solubility parameter, can provide a useful guide in determining whether a plasticizer will or will not be compatible.

The incompatibility of a plasticizer is shown by exudation from a film on long standing or upon exposure to air, light, certain liquids and low temperatures. This exudation or spewing of plasticizer is a sign of phase separation and appears as a wet, oil or tacky surface on the surface.

Functional groups in plasticizer molecules which are usually polar in nature contribute compatibility. Some of the more important are ketone, ester, amine, chlorine, nitro and sulfonic groups. The functional groups in plasticizers that impart compatibility are the same groups that make organic liquids act as solvents.

Efficiency

Efficiency is sometimes expressed as the concentration of a given plasticizer necessary to produce the standard modulus in a given resin. In general, the plasticizer that will produce the desired performance in a given surface coating is considered the most effective. However, the most efficient plasticizers are only the better choice when they are more expensive on a pound-volume basis than the other ingredients. On the other hand it may be more economical to use a larger amount of a plasticizer of lower effectiveness if it is less expensive than the other ingredients. Efficiency ratings of plasticizers for coatings must take into account many different properties such as low-temperature flexibility, hardeners, etc.

Permanence

Plasticizers for use in films should have low volatility. While the vapor pressure of the plasticizer is important, the volatile losses are determined not on the plasticizer alone but on the plasticizer resin combination. Plasticizers of limited compatability may exhibit unexpected high volatile losses. The volatility of the members of any homologous series decreases as the molecular weight increases. Polymeric plasticizers are more permanent than chemical-type plasticizers.

Stability

Plasticizers should be resistant to moisture, oxygen, light, heat and in many cases oil, chemicals and flame. Plasticizers containing oxygen linkage or hydroxyl groups tend to have poor moisture resistance. Aromatic groups have better water resistance than aliphatic groups. Flame resistance is imparted by plasticizers containing phosphates or chlorinated type.

Odor, Taste, Toxicity, Color, etc.

Plasticizers used in food containers, clothing, surgical appliance, sheeting, etc., should be free from odor, taste and toxicity. Dark-colored plasticizers are usually limited in their utility. Table 16.2 shows plasticizers sanctioned by the Food and Drug Administration.

TABLE 16.2.   Plasticizers Sanctioned by Food and Drug Administration[a]

Acetyl tributyl citrate
Acetyl triethyl citrate
p-tert-Butylphenyl salicylate
Butyl stearate
Butylphthalylbutyl glycolate
Dibutyl sebacate
Diethyl phthalate
Diisobutyl adipate
Diphenyl 2-ethylhexyl phosphate
Epoxidized soybean oil (max iodine value 6, min oxirane oxygen 6.0%)
Ethyl phthalyl ethyl glycolate
Glycerol monooleate
Monoisopropyl citrate
Mono-, di- and tristearyl citrate
Triacetin
Triethyl citrate
3-(2-Xenoyl)-1,2,-epoxypropane

[a]*Federal Register* (Feb. 18, 1966). 31 F.R. 2897 Title 21, Ch. 1, Subchapter B, Part 121, Food additives Subpart E, p. 1.

Performance of various plasticizers in coating films are shown in Table 16.3, 16.4, 16.5 & 16.6.

TABLE 16.3.   Performance of Plasticizers in Cellulose Acetate Butyrate[a] Film

Plasticizer (33 1/3 phr)	Sward Hardness	Tensile Strength (psi)	Elongation (%)
Dibutyl phthalate	38	1800	12
Dioctyl phthalate	44	2900	19
Dioctyl adipate	28	2100	12
Tricresyl phosphate	54	3000	20
Triphenyl phosphate	40	2500	10

[a] 38% butyral.

TABLE 16.4.   Performance of Plasticizer in Vinyl Chloride Film

Plasticizer (66 2/3 phr)	Tensile Strength (psi)	Elongation (%)	Hardness, Durante A	Volatile Loss 72 hr at 82°C (%)	Brittle Temperature (°C)
Dioctyl phthalate	2000	400	74	0.25	−36
Tricresyl phosphate	2500	360	72	1.00	−10

TABLE 16.5.   Performance of Plasticizers in Nitrocellulose Film

Plasticizer (50 phr)	Tensile Strength (psi)	Elongation (%)	Moisture Permeability	Flexibility, Schaffer Fold Cycle
Control (none)	8660	6	100%	20
Dibutyl phthalate	2740	4	65	10
Triphenyl phosphate	4480	8	56	24
Tricresyl phosphate	1230	24	28	—

TABLE 16.6.   Performance of Plasticizers in Ethylcellulose[a] Film

Plasticizer (15 phr)	Tensile Strength (psi)	Elongation (%)	Hardener (relative)
Control	8960	30	100
Dibutyl phthalate	5120	38	35
Diphenyl phthalate	7680	38	80
Tricresyl phosphate	7465	48	70
Triphenyl phosphate	6685	30	70

[a] 48–49.5% ethoxy.

CLASSIFICATION OF PLASTICIZERS

Plasticizers can be classified as follows:
(1) Chemical,
(2) Polymeric,
(3) Natural oils and modifications.

These are general classifications, and there is some overlapping between the classes. A recent publication lists some 500 different plasticizers. Within these classes, plasticizers can also be classified as primary and secondary. A primary plasticizer is completely compatible with the resin. Primary plasticizers are active solvents for the resin. A secondary plasticizer is a material which is relatively insoluble in a resin and has limited compatibility with it. Secondary plasticizers are used in conjunction with, or as a partial substitute for, primary plasticizers. They are low in cost and usually of a hydrocarbon type: Physical properties of some common plasticizers shown in Table 16.7.

Monomeric Chemical Plasticizers

These are relatively low molecular weight materials, range 200 to 400. About twelve of these materials account for more than 60% of all of the plasticizer consumption in this country. The classes are as follows:

TABLE 16.7 Physical Properties of Plasticizers

Compound	Molecular Weight	Average Pounds per U.S. Gallons, 20°C	Refractive Index, n_D^{25}	Viscosity at 20°C (cps)	Flash Point, Cleveland Open Cup (°F)	Vapor Pressure (mm Hg at 150°C)
Butyl stearate	340	7.15+	1.442	7.9	370	0.11
Castor oil (refined)	926	7.98	1.4773	680	555	—
Castor oil (blown)	—	8.40	1.4832	15,000	450	—
Dibutyl phthalate	278	8.74	1.490	20	340	1.1
Dibutyl sebacate	314	7.79	1.439	8	356	—
Diethyl phthalate	222	9.32	1.5002	14	305	0.003
Diisodecyl phthalate	446	8.05	1.483	108	425	0.02
Dioctyl adipate	370.56	7.72	1.446	—	377	0.15
Dioctyl phthalate	390	8.21	1.484	81.4	426	1.3
Methyl acetylricinoleate	346	7.80	1.4545	20	375	—
Octyl decyl phthalate	408	8.13	1.482	45	435	0.01
Tributyl phosphate	266	8.14	1.423	3.7	380	7.3
Tricresyl phosphate	368	9.70	1.555	120	500	0.02
Triphenyl phosphate	326	—	1.550	—	437	—

Phthalates. These are the most important group of chemical plasticizers. They are esters of phthalic anhydride and various alcohols. They are low in cost, have good compatibility with most resins and have good stability. One of the most common is dibutyl phthalate

$$\text{(benzene ring)}\begin{array}{l} -\overset{\displaystyle O}{\overset{\|}{C}}-O-C_4H_9 \\ -\underset{\displaystyle O}{\underset{\|}{C}}-O-C_4H_9 \end{array}$$

Dibutyl phthalate is used primarily with nitrocellulose and polyvinyl acetates. Dioctyl, dicapryl and dimethoxyethyl phthalate are used

TABLE 16.8 Plasticizer Compatibility

	Cellulose Acetate	Cellulose Acetate Butyrate	Cellulose Nitrate	Ethyl Cellulose	Cellulose Acetate Propionate	Chlorinated Rubber	High Styrene Butadiene
Butyl stearate	I	I	LC	LC	—	C	—
Dibutyl phthalate	C	C	C	C	C	C	C
Diethyl phthalate	C	C	C	C	C	C	C
Dioctyl adipate	I	C	C	C	C	C	C
Dioctyl phthalate	LC	C	C	C	—	C	—
Diphenyl phthalate	I	C	C	C	LC	C	C
Tributyl phosphate	C	C	C	C	—	C	—
Tricresyl phosphate	LC	C	C	C	C	C	C
Triphenyl phosphate	C	C	C	C	C	C	C

principally with vinyl resins. Dimethyl, diethyl and di(methoxyethyl) phthalates are used with cellulose acetate.

Phosphates. The phosphates, second to the phthalates in volume used, low-temperature flexibility. They are higher in cost than phthalates. (*meta* and *para*)

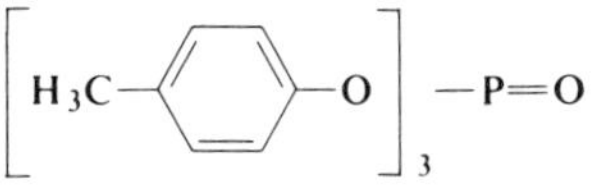

$$\left[H_3C-\left\langle\bigcirc\right\rangle-O \right]_3 -P{=}O$$

is used with nitrocellulose and vinyl chloride. Others in this series are diphenyl cresyl, diphenyl 2-ethylhexyl, triphenyl, tri(2-ethylhexyl), and tri(butoxyethyl) phosphate.

Adipates. The adipates are adipic acid esters and are used to impart low-temperature flexibility. They are higher in cost than phthalates. They are used primarily with vinyl resin

$$C_8H_{17}-O-\overset{\overset{\displaystyle O}{\|}}{C}-(CH_2)_4-\overset{\overset{\displaystyle O}{\|}}{C}-O-C_8H_{17}$$

Dioctyl Adipate

Adipic plasticizers are di(2-ethylhexyl), diisooctyl, octyl decyl, dibutoxyethyl, diisooctyl decyl and diisodecyl adipate.

Sebacates. Sebacic acid esters are premium grade plasticizers with good low-temperature flexibility, low volatility and good resistance. Their principal use is in vinyls.

$$C_4H_9-O-\overset{\overset{\displaystyle O}{\|}}{C}-(CH_2)_8-\overset{\overset{\displaystyle O}{\|}}{C}-O-C_4H_9$$

Dibutyl Sebacate

Dibutyl and di(2-ethylhexyl) sebacate are two plasticizers in this class.

Chlorinated Biphenyls. Chlorinated biphenyls ("Aroclor"—Monsanto) are used as plasticizers for the cellulosics vinyls and chlorinated

TABLE 16.8 continued

Protein Compounds	Shellac	Acrylic-type Resins	Polyamides	Polyvinyl Acetate	Polyvinyl Butyral	Polyvinylidene Chloride	Polyvinyl Chloride	Polystyrene	Nitrile and Neoprene Rubbers
—	I	—	—	LC	LC	LC	LC	LC	C
C	I	C	C	C	C	C	C	C	C
C	LC	C	I	C	C	C	I	C	C
I	I	C	C	I	C	C	C	C	C
—	I	—	—	LC	C	LC	C	C	C
I	I	C	LC	C	C	C	C	C	C
—	C	—	—	C	C	C	C	C	C
C	I	C	C	C	C	C	C	LC	C
LC	I	C	LC	C	C	C	C	C	C

rubbers. They are chemically inert, being resistant to both acid and alkali.

Polymeric

These are high molecular weight polyesters derived from esterifying dibasic acids and dihydric alcohols. They have excellent permanance properties: low volatility, low extraction by oils and good resistance to migration. Polymeric or resinous plasticizers are used with both cellulosic and vinyl polymers. The molecular weights of these materials are in the range of 800 to 8000. They vary from viscous liquids to soft resins.

Oils and Fatty Acid Derivatives

Drying oils and nondrying oils, when cooked in oleoresinous varnishes, alkyds and epoxies, combine chemically to internally plasticize the composition. Linseed, soya, dehydrated castor, safflower and fish oils are not compatible with most cellulosic, vinyl and other thermoplastic resins. However, blown oils have good compatibility with nitrocellulose and other resins, and are used as plasticizers.

Castor oil because of its hydroxyl group is compatible with resins such as nitrocellulose. Blown castor oils in various viscosities are used as plasticizers in lacquers. Esters such as acetylated castor oil are used as plasticizers for nitrocellulose and vinyls. Esters of castor fatty acids (ricinoleic) are efficient plasticizers for nitrocellulose and vinyls.

Epoxidized oils are used as plasticizers. They are produced by epoxidizing the double bond of drying oils.

$$-\underset{\underset{H}{|}}{C}=\underset{\underset{H}{|}}{C}- \; + \; O \; \rightarrow \; -\underset{\underset{H}{|}}{\overset{\overset{O}{\diagup\diagdown}}{C}}-\underset{\underset{H}{|}}{C}-$$

These are used as primary plasticizers for vinyl resins where they also act as stabilizers against degradation by heat and light. They act as acceptors for any hydrochloric acid released. They have low volatility and excellent low-temperature flexibility.

Compatability of common plasticizers shown in Table 16.8.

REFERENCES

1. Bruins, P. F., "Plasticizer Technology," New York, Reinhold Publishing Corp., 1965.
2. "Modern Plastics Encyclopedia," 1967.
3. "Plasticizers," pp. 418, 443–450, New York, McGraw-Hill Book Co., 1966.

17

*Hydrocarbon Solvents**

Hydrocarbon solvents and thinners used in surface coatings are liquids at normal atmospheric pressures and temperatures. They may be pure chemical compounds of carbon and hydrogen, or mixtures of many compounds obtained as distillate fractions from petroleum, coal tar or pine trees. Turpentine is a mixture of terpene hydrocarbons and at one time had extensive use in coatings. In recent years, turpentine has been almost totally replaced with less-expensive, petroleum-derived solvents.

According to the U.S. Tariff Commission,[20] the total sales of hydrocarbon solvents in the United States in 1963 amounted to approximately 1 billion gallons. It is estimated that two-thirds of this amount goes into the production of paints, varnish, lacquers and inks. In 1965, the paint varnish and lacquer industry estimated approximately \$2,200,000,000 in factory sales.[1] Of this amount, about one dollar in seven went for automotive finishes, including trucks. Hydrocarbon solvents, particularly aromatics, comprised a major portion of the thinners for these coatings. Over 90% of all hydrocarbon solvent production came from petroleum.

Solvents are the temporal components of surface coatings, but during this existence they must perform the important functions of solubilizing the resins, reducing the viscosity and controlling the evaporation rate of the applied coating. Each of these functions will be discussed in some detail.

CLASSIFICATION OF SOLVENTS

Hydrocarbon solvents may be classified in a number of different ways, e.g., according to molecular structure, volatility, viscosity, boiling range, odor and the like. Petroleum solvents are often divided into two principal categories—aliphatic hydrocarbons and aromatic hydrocarbons. Com-

*By E. C. Larson, Shell Development, Emeryville, Calif.

mercial solvents in each of these groups may be subdivided into (1) close cut, (2) medium range and (3) wide range with respect to boiling point. Since usage has become more and more specialized, the trend is to produce solvent fractions as building blocks for greater flexibility in composition and volatility. Thus, future production will go toward close cut and medium range rather than wide boiling range solvents.

Normal paraffins (straight chain), isoparaffins (branched chain), and naphthenes (cycloparaffins) are included among the aliphatic hydrocarbons, and it is not uncommon for commercial aliphatic solvents to contain as much as 20 to 30% aromatics. Similarly, the commercial "aromatic" solvents may contain sizable amounts of paraffinic hydrocarbons. Limited quantities of olefinic (unsaturated) hydrocarbons are available, but they have such specialized use that they will be omitted from this discussion.

It is also quite common for the paint chemist to refer to the normal and isoparaffins as one group, the naphthenes as a second, and aromatics as a third group, and this discussion will be considered primarily in this sphere of reference. In general, the paraffins are the weakest solvents with least odor; the isoparaffins used in the production of odorless thinners are completely odorless. Naphthenes have intermediate solvency and odor, and the aromatics are the strongest solvents and have the most odor. These properties are illustrated schematically in Figure 17.1.

Each commercial solvent, exclusive of such pure hydrocarbons as *n*-hexane, *n*-heptane, isooctane, cyclohexane, methylcyclohexane, benzene, toluene and the xylenes, is comprised of a large number of isomeric hydrocarbons. Gas chromatographic analyses on solvents in the mineral spirits boiling range (300 to 400°F) reveal more than 100 separate hydrocarbons

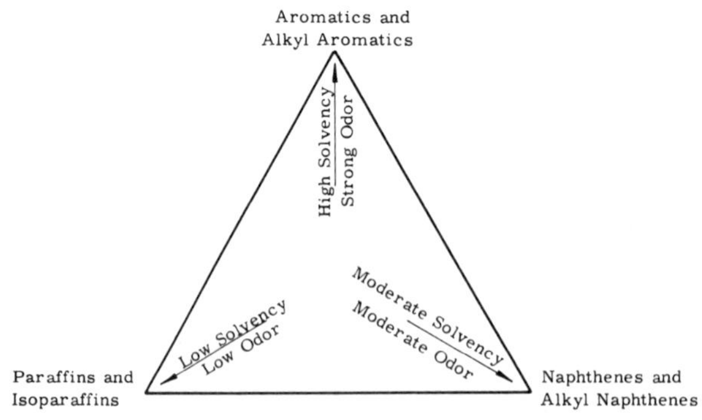

Figure 17.1.   Interrelation of solvent properties.

in some aliphatic solvents and as many as 50 isomers in the aromatic solvents. Consequently, the general properties of the solvents are a summation of the properties of the individual hydrocarbons present in them. The varied sources of crude oil supply and minor variations which occur during manufacture make it impossible to duplicate exactly the type and amount of each and every hydrocarbon from batch to batch or from day to day. However, the variations are relatively minor, and the average performance properties can be duplicated with amazing consistency. A general list of some commercial solvent types is given in Table 17.1. A comprehensive listing of individual solvents is given in the Raw Materials Index, Solvents Section, published by the National Paint, Varnish and Lacquer Association, Inc., Washington, D. C.

From the standpoint of volatility, solvents are often classified as low, medium and high boilers.

TABLE 17.1.   Approximate Properties of Various Types of Commercial Hydrocarbon Solvents[a]

Solvent	Boiling Range (°F)				Flash Point, Tag Closed Cup (°F)	Specific Gravity at 60°F/ 60°F	Weight per Gallon (lb)	Aniline Point (°F)	Kauri-butanol Value
	Initial Boiling Point	50%	Dry Point	End Point					
Hexanes	140	152	160	—	<0	0.68	5.66	146	30
Heptanes	200	207	225	—	20	0.72	6.00	135	35
Lacquer diluent	170	207	225	—	0	0.74	6.15	120	40
Naphtha (light)	205	225	250	—	25	0.76	6.40	115	40
Naphtha (heavy)	240	265	300	—	50	0.75	6.25	125	38
Naphtha (V.M. & P.)	210	250	300	—	25	0.74	6.20	130	38
Mineral spirits (light)	310	325	340	360	102	0.76	6.35	135	36
Mineral spirits (regular)	310	340	—	395	105	0.78	6.50	134	37
Mineral spirits (low odor, light)	310	320	—	340	101	0.76	6.35	142	33
Mineral spirits (low odor, heavy)	360	370	—	410	140	0.78	6.50	150	31
Odorless thinner (light)	345	360	—	395	125	0.76	6.35	183	27
Odorless thinner (heavy)	372	415	—	490	150	0.77	6.42	175	27
Toluene	230	231	232	—	40	0.87	7.25	50[b]	105
Xylene	280	285	293	—	80	0.87	7.25	52[b]	98
Aromatic solvent (light)	315	327	—	350	110	0.87	7.25	60[b]	92
Aromatic solvent (medium solvency)	315	340	—	390	110	0.85	7.08	85[b]	70
Aromatic solvent (medium heavy)	350	370	400	410	140	0.87	7.25	62[b]	90
Aromatic solvent (heavy)	400	460	—	530	200[c]	0.95	7.90	58[b]	95

[a] A comprehensive list of commercial solvents and their properties is available in the Raw Materials Index, Solvent Section, published by the National Paint, Varnish and Lacquer Association, Inc., Washington, D. C.

[b] Mixed aniline point.

[c] Cleveland open cup.

Low Boilers (140 to 180°F)

These solvents are used where fast evaporation is desired. They are usually weak aliphatic solvents with perhaps very small quantities of naturally occurring benzene. Although benzene is a strong aromatic solvent in this range, it is prohibited in many states because of its relatively high toxicity. Even so, its high solvency for resins allows it to find some use in paint and varnish removers.

Medium Boilers (200 to 300°F)

Lacquer diluents composed of C_7 and C_8 hydrocarbons (200 to 250°F) and V.M. & P. naphtha (240 to 300°F) are examples of this group. They are predominately aliphatic but are often fortified with varying amounts of toluene and xylene to increase solvency. Lacquer diluents are also used to reduce the viscosity of gravure inks. V.M. & P. naphthas are used extensively in sprayable paint coatings.

High Boilers (300 to 500°F)

The term "mineral spirits" is applied to the many distillate fractions of paraffinic and naphthenic solvents that boil in the 300 to 400°F range and have flash points greater than 100°F. Mineral spirits comprises about three-fourths of all hydrocarbon solvents used in the paint industry and about 90% of the solvents used in architectural finishes. There are three general categories of mineral spirits: (1) regular, (2) low odor and (3) odorless. Each of these may be subdivided according to boiling range to give a fast- and a slow-evaporating fraction. The general characterization is based on solvency and odor as contributed by aromatics. Regular mineral spirits contains 15 to 25% aromatics, low-odor mineral spirits contains 2 to 10% aromatics, and odorless mineral spirits is essentially isoparaffinic and with no aromatics. Mineral spirits is an excellent solvent for house paints and many architectural enamels based on oleoresinous and alkyd resins. Brushability and wet-edge control are attained by selection of solvents with the proper volatility. Slower-evaporating solvents lengthen wet-edge time and generally improve brushability but also lengthen the drying time.

Odorless mineral spirits is an excellent solvent for interior architectural enamels, but its low inherent solvency necessitates careful resin design to ensure compatibility.

Aromatic solvents in the mineral spirits boiling range are used to increase the solvency of mineral spirits and as the complete solvent in baking finishes. The latter utilize rather difficultly soluble short oil alkyds, amine alkyds, and thermosetting acrylics for industrial coatings.

Limited quantities of kerosene-type solvents (400 to 500°F) composed of aliphatic and some aromatic hydrocarbons are used to improve brushability, maintain wet-edge and control flow properties.

IMPORTANCE OF SOLVENCY

In order for resins to be dissolved there must be an interaction between the resin and solvent molecules. The extent of this interaction determines the solubility or compatibility of the molecules. So-called strong solvents are required for difficultly soluble resins, and weak solvents are adequate for easily soluble resins. It is apparent that the "natural" attractive forces of resin molecules must be "neutralized" or disrupted by the solvent molecules and remain in a uniformly dispersed state throughout a wide range of concentration for practical application. Complete molecular dispersion of the resin is never achieved, and indeed is unnecessary. Even with strong aromatic solvents and easily soluble alkyd resins it has been shown[18] that the molecular aggregates may be 40 to 800 Å in diameter and composed of 40 primary alkyd molecules.

Solute-solvent interactions may be classified into two groups according to the energy involved in the interaction.[10]

(1) Strong "specific" interactions result in the formation of more or less stable complexes of polar solutes with polar liquids. This kind of interaction is usually of the donor-acceptor type, in many cases of the hydrogen-bonding character, and has been the subject of numerous investigations.[15] The formation of such complexes involves a rather large interaction energy which is approximately equal to the energy of the donor-acceptor bond formed.

(2) Weak "nonspecific" interactions between polar solutes and nonpolar, or weakly polar, liquids have also been studied. Some investigators refer to these short-lived entities as "collision complexes." This is in line with the lattice theories of the liquid state[4] which assume that a molecule in a liquid spends most of its time vibrating slowly around potential minima in the vicinity of other molecules. In some solutes, the polar group is shielded by bulky substituents which create a potential barrier for the solvent molecules; such solutes are capable of forming simultaneously both donor-acceptor type complexes and collision complexes with highly polar solvents.

Since hydrocarbon solvents are essentially nonpolar but are used with resins which are significantly polar, our interest is primarily directed toward the weak nonspecific interactions. The process of formation of the primary collision complex involves only a small interaction energy of the dipole-induced dipole type. It is postulated that the solvent molecule entering the collision complex enters the primary solvation site and the collision complex is solvated by several additional solvent molecules in the secondary solvation shell.

Many analytical techniques have been recommended and used for measuring solvent strength: kauri-butanol value, aniline point, dimethyl sulfate test, solubility parameter, dilution limit. Historically, kauri-butanol and aniline point are probably the oldest and most widely used. All the methods have certain limitations, which will be discussed in some detail.

Solubility Parameter

Hildebrand's[9] solubility parameter concept as applied by Burrell[6] has been quite helpful in assigning a numerical magnitude to the strength of solvents. The solubility parameter, S, equals $(E_v/V_1)^{1/2}$ where E_v is the isothermal energy of vaporization at zero pressure in calories per mole and V_1 is the molal volume in cubic centimeters per mole. The term (E_v/V_1) represents the mutual attraction of the compound's molecules

TABLE 17.2.   Solubility Parameter of Solvents (Poorly Hydrogen Bonded)

Group I Solvents	Solubility Parameter	Group I Solvents	Solubility Parameter
Silicones	5.5	p-Xylene	8.7
2,2-Dimethybutane	6.7	p-Diethylbenzene	8.7
Mineral spirits	6.9	m-Diethylbenzene	8.7
Isooctane	6.9	m-Xylene	8.8
n-Hexane	7.3	Mesitylene	8.8
n-Heptane	7.4	Ethylbenzene	8.8
"Freon" 113	7.4	Ethyl chloride	8.8
V.M. & P. naphtha	7.6	Toluene	8.9
n-Nonane	7.7	o-Diethylbenzene	8.9
n-Decane	7.7	o-Xylene	9.0
Methylcyclohexane	7.8	Benzene	9.2
n-Dodecane	7.8	Styrene	9.3
n-Tetradecane	7.9	Chloroform	9.3
n-Hexadecane	8.1	Tetrachloroethylene	9.4
Turpentine	8.1	"Tetralin"	9.5
Cyclohexane	8.2	1,1,2-Trichloroethane	9.6
Amyl chloride	8.4	m-Chlorotoluene	9.7
Benzonitrile	8.4	Nitrobenzene	10.0
Dodecylbenzene	8.5	Carbon disulfide	10.0
Propylbenzene	8.6	Bromobenzene	10.3
Pine oil	8.6	l-Nitropropane	10.7
Decalin	8.6	Nitroethane	11.1
Carbon tetrachloride	8.6	Acetonitrile	11.9
Butylbenzene	8.6	Nitromethane	12.7

and is called the cohesive energy density. Energy is required to change a substance from a liquid to a vapor. This energy is proportional to the size of the molecule and is equivalent to the energy necessary to separate like molecules from each other. Consequently, when the energy values of solvent and resin are equal, or nearly so, the molecules of each can commingle. This results in solubility or miscibility.

The heat of vaporization is available in the literature on most solvents, so the cohesive energy density and solubility parameter may be calculated. The solubility parameter of a resin is generally determined experimentally and a value assigned on the basis of the midpoint of the solubility parameter range for the solvents in which it is soluble.

Polar compounds such as alcohols, esters, ketones and paint resins associate into clusters of molecules due to hydrogen bonding. If two

TABLE 17.3.   Solubility Parameter of Solvents (Moderately Hydrogen Bonded)

Group II Solvents	Solubility Parameter	Group II Solvents	Solubility Parameter
Diisopropyl ether	7.0	Isophorone	9.1
Diethyl ether	7.4	Ethyl acetate	9.1
Diisobutyl ketone	7.8	Propylene oxide	9.2
Methyl butyrate	8.0	Ethylene glycol monomethyl ether acetate	9.2
Methylamyl acetate	8.0		
Methyltetrahydrofuran	8.1	Methyl ethyl ketone	9.3
sec-Butyl acetate	8.2	Dibutyl phthalate	9.4
Isobutyl acetate	8.3	Methyl acetate	9.6
sec-Amyl acetate	8.3	Diethylene glycol monoethyl ether	9.6
Methyl isobutyl ketone	8.4		
Isopropyl acetate	8.4	Benzyl acetate	9.8
Methyl amyl ketone	8.5	Dioxane	9.9
Butyl acetate	8.5	Diethyl ketone	9.9
Amyl formate	8.5	Cyclohexanone	9.9
Amyl acetate	8.5	Ethylene glycol monoethyl ether	9.9
Methyl propyl ketone	8.7	Ethyl lactate	10.0
Ethylene glycol monoethyl ether acetate	8.7	Acetone	10.0
		Methyl formate	10.2
Tetrahydrofuran	8.8	Cyclopentanone	10.4
Propyl acetate	8.8	Acetaldehyde	10.4
Dimethyl ether	8.8	Pyridine	10.7
Methyl isopropyl ketone	8.9	Ethylene glycol monomethyl ether	10.8
Ethylene glycol monobutyl ether	8.9		
		Ethyl acetamide	12.3
Mesitylene oxide	9.0	Ethyl formamide	13.9
Butyraldehyde	9.0	Methyl acetamide	14.6
Methyl n-butyl ketone	9.1	Ethylene carbonate	14.7
		Methyl formamide	16.1

substances differ widely in their hydrogen bonding characteristics, they may not be miscible even though they may have similar solubility parameters. Burrell and CDIC Paint Varnish and Production Club[7] classified solvents and resins according to their hydrogen-bonding tendencies:

(1) Poorly hydrogen bonded,
(2) Moderately hydrogen bonded,
(3) Highly hydrogen bonded.

Petroleum solvents fall in the first class, ketones and esters in the second and alcohols in the third. A list of some of the important hydrocarbons and petroleum solvents and their solubility parameters is given in Table 17.2. For comparison Tables 17.3 and 17.4 contain values for moderately and highly hydrogen-bonded materials. Hydrocarbons and petroleum solvents fall in the rather narrow range of about 7 to 9 solubility parameter.

TABLE 17.4. Solubility Parameter of Solvents (Highly Hydrogen Bonded)

Group III Solvents	Solubility Parameter
Diethylene glycol	9.1
Diacetone alcohol	9.2
Ethylhexanol	9.5
Methyisobutyl carbinol	10.0
n-Octyl alcohol	10.3
2-Ethylbutane	10.5
Methoxybutanol	10.6
n-Heptyl alcohol	10.6
n-Hexyl alcohol	10.7
sec-Butyl alcohol	10.8
Amyl alcohol	10.9
Acetic acid	10.9
Isobutyl alcohol (primary)	11.1
Cyclohexanol	11.4
n-Butyl alcohol	11.4
Methylbenzyl alcohol	11.5
Isopropyl alcohol	11.5
n-Propyl alcohol	11.9
Benzyl alcohol	12.1
Furfural alcohol	12.5
Ethyl alcohol	12.7
Ethylene glycol	14.2
Methyl alcohol	14.5
Glycerol	16.5
Water	23.4

A list of typical resins and their approximate solubility parameters as assigned by CDIC[7] is shown in Table 17.5. Several of the resins are soluble in solvents in each of the hydrogen-bonding categories. Burrell proposed that a mixed solvent would perform as a single solvent having an average solubility parameter and hydrogen-bonding character proportional to the volume fraction of the components in the mixture. For example, a 50:50 mixture of ethyl alcohol ($S = 12.7$, highly hydrogen bonded) and toluene ($S = 8.9$, poorly hydrogen bonded) will have the same solvency, $S = 10.8$, as ethylene glycol monomethyl ether, which also has a solubility parameter of 10.8 and is classed as moderately hydrogen bonded. This concept explains the success of the common practice of using a mixture of nonsolvent or diluent like toluene with a true solvent like ethylene glycol monomethyl ether to dissolve nitrocellulose. It is economically attractive to use as much diluent as possible. Toluene can be added until the average solubility parameter and hydrogen-bonding capacity passes the lower limit of the range for nitrocellulose (Table 17.5).

TABLE 17.5. Solubility Parameter of Polymers and Plasticizers

	Solubility Parameter		
Polymer	*Poorly Hydrogen-bonded Solvents*	*Moderately Hydrogen-bonded Solvents*	*Strongly Hydrogen-bonded Solvents*
Nitrocellulose	11.9 ± 0.8	11.2 ± 3.4	Insoluble
Ethylcellulose T-10	9.0 ± 0.5	8.8 ± 1.0	10.4 ± 1
45% soya-glycerol-phthalic alkyd	9.1 ± 2.1	9.7 ± 2.3	10.7 ± 1.2
30% soya-glycerol-phthalic alkyd	10.4 ± 1.9	11.6 ± 3.7	Insoluble
Soya oil	9.0 ± 2.0	9.7 ± 2.3	10.7 ± 1.2
45% linseed-glycerol-phthalic alkyd	9.5 ± 2.4	9.7 ± 2.2	10.7 ± 2.2
Polyvinyl chloride-acetate (IYHH)	10.2 ± 0.9	10.6 ± 2.8	Insoluble
Polyvinyl acetate (AYAA)	10.8 ± 1.9	11.6 ± 3.1	Insoluble
Polystyrene KT PL-A	9.3 ± 1.3	9.0 ± 0.9	Insoluble
Ester gum	8.8 ± 1.8	9.1 ± 1.7	10.2 ± 0.7
Phenolic resin, "Durez" 220	9.5 ± 1.1	8.8 ± 1	10.5 ± 1.0
"Epon" 1001	10.8 ± 0.2	11.1 ± 1.1	Insoluble
Chlorinated rubber	9.5 ± 1.0	9.3 ± 1.5	Insoluble

The present concept, thus far, suggests that a resin will be soluble in all proportions in solvents which fall within the variance limits suggested in Table 17.5. However, practical experience shows that this is not necessarily true. Larson and Low[11] called attention to the fact that a resin may be compatible at high concentrations in a solvent but incompatible at low concentrations, and thus precipitate from solution. The point at which the resin "kicks out" of solution is called the "dilution limit" and is expressed in terms of the weight per cent of resin in the solution at incipient kick-out. Long oil alkyds are not critical to solvency and are soluble in all proportions in weak isoparaffinic solvents. On the other hand, short oil alkyds have critical solvency and require highly aromatic solvents for solubility in all proportions (i.e., near zero dilution limit). A simple measurement of dilution limit of a resin and candidate solvent is the best assessment of compatibility.

In general, solvency increases as the solubility parameter increases. However, Reynolds and Larson[18] showed that reversals often occur as in the case of normal paraffins and an alkyd resin ("Beckosol" 7) in which the increasing molecular volume of the solvent overshadows the solvency predicted by the increased solubility parameter.

Kauri-butanol Value (ASTM D-1133-61)

This method of test is designed to measure the relative solvent power of hydrocarbon solvents and is suitable for solvents within the initial boiling range of 40°C (104°F) to 300°C (572°F). The kauri-butanol value of a solvent is the volume in milliliters at 77°F of the solvent, corrected to a defined standard, required to produce a defined degree of turbidity when added to 20 grams of a standard solution of kauri resin in n-butyl alcohol. For kauri-butanol values of 60 and over, the standard is toluene and has an assigned value of 105. For values below 60, the standard is a blend of 75% n-heptane and 25% toluene and has an assigned value of 40. High values are indicative of strong solvents, and low values of weak solvents. (A similar test known as "heptane number"—ASTM D-1132-53—is also used to determine the relative solvent power of high-solvency hydrocarbons in the presence of certain resins.)

Aniline Point (ASTM D-1012-62)

Aniline point (also called aniline cloud point) values are indicative of the relative solvent power of hydrocarbon solvents. Aniline point is the minimum equilibrium solution temperature for equal volumes of aniline and solvents. For strong aromatic solvents, the equilibrium is below room temperature and the test is modified to give mixed aniline points. Mixed aniline point is the minimum equilibrium solution temperature of a mixture of two volumes of aniline, one volume of sample, and one volume of n-heptane of specified purity. Aliphatic hydrocarbons have the highest aniline points (158 to 185°F), naphthenes are intermediate (86 to 131°F), and pure aromatics have the lowest (below -22°F).

INTERRELATION OF SOLUBILITY PARAMETER, KAURI-BUTANOL VALUE, ANILINE CLOUD POINT AND DILUTION LIMIT

An interesting comparison of the various solvency concepts for several solvents is given in Table 17.6 along with the measured dilution limits on a commercial melamine-formaldehyde resin. For example, a 40/60 weight blend of toluene and hexane with its lower solubility parameter and higher molecular volume gives a lower dilution limit, so it is a better solvent than cyclohexane, which has about the same aniline point but higher kauri-butanol value than the blend. These results can only be explained on the basis of specific interaction between the aromatic hydrocarbon and the resin molecules. Very probably a dipole-dipole interaction occurs between the unsymmetrical force fields of the aromatic hydrocarbons and the oxygenated functional groups of the resin, whereas this does not occur in the cycloparaffinic solvent.

TABLE 17.6.   Solubility of Melamine-Formaldehyde Resin

Solvent	Molar Volume	Solubility Parameter	Aniline Point (°F)	Kauri-butanol Value	Dilution Limit (% by wt resin at kick-out
n-Hexane	132	7.30	156.6	28.0	81.2
20% by wt toluene 80% by hexane	127	7.6	122.0	35.6	53.0
30% by wt toluene 70% by wt hexane	125	7.7	105.0	41.1	24.8
40% by wt toluene 60% by wt hexane	122	7.75	83.0	48.8	11.2
Methylcyclohexane	128	7.85	103.0	50.9	44.9
Cyclohexane	109	8.20 .	86.0	57.9	40.2
Cumene	135	8.50	63.8 mxd	85.6	Miscible
Toluene	106	8.90	50.0 mxd	105	Miscible
Benzene	89	9.15	< -22	107	Miscible
Nitroethane	71	11.1	—	—	Miscible
Acetonitrile	53	11.9	—	—	57.8

It has been shown[18] that a reasonably good relationship exists between dilution limit and aniline point and also dilution limit and kauri-butanol value. The authors also showed that solubility parameter and kauri-butanol values are related according to the equation

$$S = 6.9 + 0.02 \times \text{kauri-butanol value} \tag{1}$$

For those who wish to relate the kauri-butanol value and aniline point of hydrocarbon solvents, Harvey and Mills[8] suggest the following equations. For KB values below 50:

$$KB = 99.6 - 0.806G - 0.177ACP + 0.0755(340 - B) \qquad (2)$$

For KB values of 50 to 75:

$$KB = 117.7 - 1.06G - 0.249ACP + 0.10(340 - B) \qquad (3)$$

where B equals the mid-boiling point in degrees Fahrenheit and G equals the API gravity at 60°F/60°F.

Each of these tests for solvency serves a useful purpose, but each has limitations and may be unreliable when there is appreciable interaction between solvent and resin. When exceptions occur, it is wise to apply the more practical test of dilution limit to the particular solvent and resin in question.

As further evidence that simple solvency tests on solvents are often unreliable, Larson and Low[11] demonstrated that the presence of small amounts of polar substance such as glycerol monooleate, sorbitan monooleate or pentaerythritol monooleate could improve the compatibility of alkyd resins and relatively low solvency solvents. The results shown in Figure 17.2 illustrate this dramatically. The medium oil alkyd is not completely miscible in a low-odor thinner with a kauri-butanol value of 30. The resin had a dilution limit of 5 in regular mineral spirits ($KB = 36$) and is completely miscible only after the kauri-butanol value is increased to 43.6 by introduction of aromatics. Each increase in the solvency with aromatics introduced objectionable odor to the system. On the other

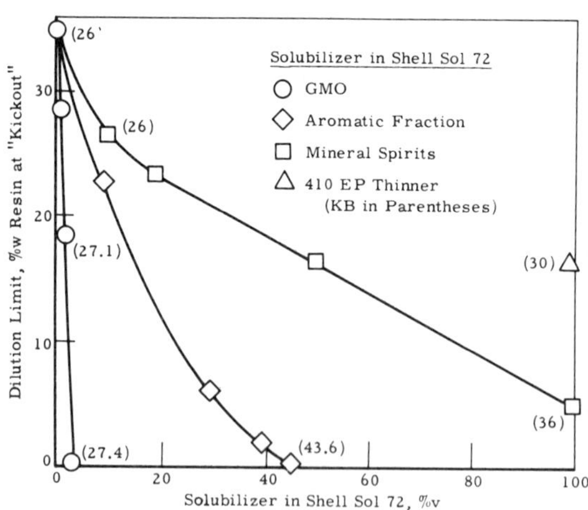

Figure 17.2. Dilution limit versus solubilizer content of isoparaffinic mineral spirits.

hand, the addition of only 3% glycerol monooleate increased the kauri-butanol value of odorless thinner from 26 to 27.4 and gave complete miscibility between resin and solvent without introducing objectionable odor. Apparently the polar sites with their attendant hydrogen bonding promote strong interaction between the resin molecules and cause the resin aggregates to kick out in weak solvents. The presence of a low molecular weight polar additive like glycerol monooleate disrupts the hydrogen bonds and destroys, or reduces, molecular aggregation, thus effecting compatibility between resin and solvent molecules.

IMPORTANCE OF VISCOSITY

When the problem of selecting suitable resins and compatible solvents has been resolved, the viscosity of the solution becomes of prime importance. This property governs the ease of application and to a large extent the appearance of the dried film. It is well known that a different viscosity is required for brushing, spraying, flow coating, dipping and electrostatic techniques. Temperature, pressure (for spray application) and flow-control additives also influence viscosity. The resin manufacturer and paint formulator must supply products which will give quality performance over this rather wide range of conditions.

Although solvents reduce viscosity by solvation, they also possess inherent differences in viscosity which influence the viscosity of resin solutions and paints. The following discussion stresses the importance of solvent selection on viscosity.

Solvents are simple liquids and behave as Newtonian fluids in which the rate of flow divided by the exerted force equals a constant. When resins are added, the system becomes complex, and the flow behavior depends upon the nature and concentration of the resin. The relative increase contributed by the resin is expressed as

$$\text{Relative Viscosity} = \frac{\text{Solution Viscosity}}{\text{Solvent Viscosity}} \tag{4}$$

The dimensionless term specific viscosity is introduced to negate the influence of solvent viscosity on solution viscosity and is represented by the equation

$$\text{Specific Viscosity} = \frac{(\text{Solution Viscosity} - \text{Solvent Viscosity})}{\text{Solvent Viscosity}}$$
$$= \text{Relative Viscosity} - 1 \tag{5}$$

The effect of resin concentration on viscosity may be expressed in terms of reduced viscosity by the equation

$$\text{Reduced Viscosity} = \frac{\text{Specific Viscosity}}{\text{Concentration (in grams per deciliter)}} \tag{6}$$

Reduced viscosity, or viscosity number, has the dimension of deciliters per gram, and thus indicates the apparent size of molecular resin aggregates. The reduced viscosities of extremely dilute solutions form a linear relationship with resin concentration. An extrapolation of this line to zero resin concentration yields a value called *intrinsic viscosity* which represents the apparent size of the individual resin molecule. Intrinsic viscosity also has the practical significance of allowing one to calculate the molecular weight of polymers.

It has been recognized by several investigators[2,13,16,18] that solvent viscosity has an important influence on the viscosity reduction of resins. Reynolds and Gebhart[16] showed that a linear relationship existed between solvent viscosity and solution viscosity of two long oil alkyds, "Aroplaz" 1273 and "Beckosol" 70. Since the solvents included paraffins, naphthenes, and aromatics, they concluded that the interaction with these resins was the same irrespective of hydrocarbon type.

A resin solution is a dispersed system whose viscosity depends upon the concentration of the resin, the degree of aggregation of the resin molecules and the viscosity of the solvent. The viscosity of a simple dispersed system can be represented by the equation

$$A = B(1 + k_1 c_1 + k_2 c_2 + \cdots) \tag{7}$$

where A is the viscosity of the resin solution, B is the viscosity of the solvent, k_1 and k_2 are solvent-resin interaction constants, and c is the volume fraction of the dispersed phase. If it is assumed that a resin dissolved in one solvent has a viscosity A_1, and in another solvent viscosity A_2, and that the interaction is the same in both solvents, then for a given resin concentration

$$\frac{A_1}{A_2} = \frac{B_1}{B_2} \tag{8}$$

where B_1 and B_2 are the viscosities of the solvents. Therefore, the relative ability of the two solvents to reduce the viscosity is proportional to the viscosity of the solvents.

When one compares the viscosity reducing abilities of solvents, the ratio of the viscosities of the solvents should be considered and not the absolute difference. For instance, at 77°F, n-heptane has a viscosity of 0.394 cps, while toluene has a viscosity of 0.557, and methylcyclohexane has a viscosity of 0.686. Methylcyclohexane is 74% more viscous than n-

heptane, and toluene is 41% more viscous than n-heptane.   These are obviously important differences in terms of the previous discussion.   This is shown clearly by three isoparaffinic solvents prepared from odorless thinner (Table 17.7).   Each solvent has the same hydrocarbon type, but the molecular weight varies in accordance with the boiling range, which extends from 336 to 433°F; there are only minor variations in kauri-butanol value and aniline point.   It may be assumed that the interaction between the solvent and the resin is relatively similar for these three solvents (i.e., the interaction constant k is the same).   It would be expected that if resin solutions were prepared with these solvents, the viscosity would be proportional to the viscosity of the solvents.   This was found to be the case as shown in Table 17.8, which gives the viscosity ratios for 40% by weight solutions of a long oil phthalic alkyd.   The viscosity ratios of the resin solutions are in close agreement with the viscosity ratios of the solvents themselves.   The long oil alkyd has little tendency to form a highly aggregated structure.

TABLE 17.7.   Properties of Odorless Solvents

Hydrocarbon composition: All essentially isoparaffinic

Sample	A	B	C
Boiling range, °F	336–371	346–431	375–433
Kauri-butanol value	27.0	26.4	26.0
Aniline cloud point, °F	183	185	189
Viscosity at 77°F, cps	1.2	1.4	1.8
Viscosity ratio	1.00	1.17	1.50

TABLE 17.8.   Resin Solutions with Odorless Solvents

Resin: Long oil phthalic alkyd
Solution nonvolatile content: 40% by wt

Sample	A	B	C
Viscosity at 77°F, cps	250	320	410
Viscosity ratio	1.00	1.28	1.64

Aggregation of resin molecules immobilizes the solvent, thereby greatly increasing the viscosity of the system.   Some paint resins have a strong tendency to aggregate.   The ability of the solvent to solvate the resin molecules and thus break up aggregate structure has pronounced effect on the viscosity of the system.   When the resin is sufficiently dispersed that it no longer immobilizes a significant amount of the solvent, the viscosity of the system is again dependent upon the viscosity of the solvent. Reynolds and Griebel[17] referred to this condition as the "principle of adequate solvency."

Solvent Power and Viscosity as a Function of Shear Rate

The discussion so far has been concerned with the viscosity of resin solutions under relatively low shear conditions. The viscosity of the system under conditions of low shearing stress determines the leveling, sagging and flow-out properties of a paint film. However, brushability[3] is a high shear rate phenomenon with brushing speeds exerting a shear rate of up to 30,000 reciprical seconds (sec^{-1}). In order to evaluate the brushability of a paint, it is necessary to determine viscosity at shear rates approaching these high values.

Larson and Reynolds[18] studied brushability with a modified Stormer Viscosimeter similar to that used by Asbeck and Van Loo[3] and showed that apparent viscosities so obtained and plotted on a log-log plot of viscosity and shear rate could be extrapolated to cover the high shear rate region. They prepared a series of flat enamels with a medium oil alkyd resin ("Rezyl" 405) and a series of low-odor solvents varying widely in aromatic and naphthene contents. The extrapolated viscosities of all the enamels (Figure 17.3) were below 1 poise in the region of high shear (30,000 sec^{-1}), and all had excellent brushability. A comparison at 500 sec^{-1} shows that the enamel made with Solvent 4 had twice the viscosity of the one prepared with Solvent 1; however, at 30,000 sec^{-1} the viscosities were essentially equal. This can be explained on the basis of the highly aggregated nature of the systems. At the low shear rates, the

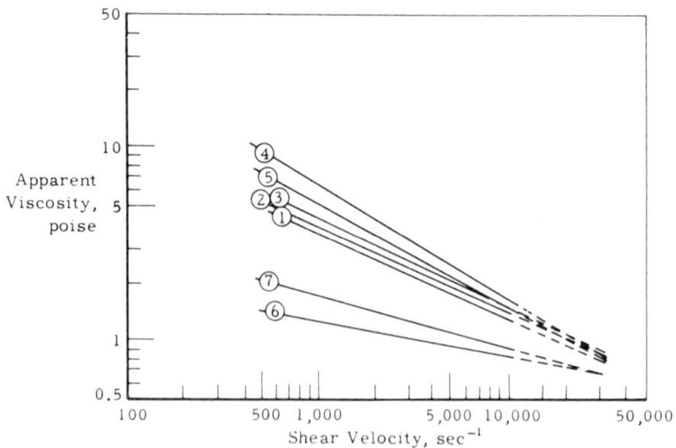

Figure 17.3.   Viscosity characteristics of flat enamels.

degree of aggregation is strongly influenced by the solvent power of the solvent. Therefore, wide differences in the viscosities of the paints can exist depending upon the solvency of the solvent. At high shear rates, the aggregates are mechanically destroyed and the solvating action of the solvent has relatively little influence on the viscosity of the paint; the inherent residual viscosity of the paint is approached which depends on the hydrodynamics of the system. In other words, at high shear rates the viscosity of a paint will be dependent upon the viscosity of the solvent. Thus, the brushability of a paint is influenced by the viscosity of the solvent. Further analysis of the data showed that aromatic solvents were about four times as effective as naphthenes in solvating the resin molecules.

IMPORTANCE OF VOLATILITY

The volatility of a solvent is the primary controlling factor which governs the rate at which the solvent will escape from a film. The process of changing from a liquid to a vapor is called evaporation and will be discussed in a subsequent section. Solvent evaporation influences such properties as sagging, gloss, leveling, brushability, wet edge or lap time, rate of viscosity build, and drying. It is generally agreed that evaporation is controlled either by the diffusion rate of solvent vapor through a layer of stagnant air above the liquid surface or by an equilibrium-type evaporation similar to ordinary distillation.

Several devices are available for measuring evaporation; the Shell Thin Film Evaporometer[14] seems to be the most precise and has been recommended by the New York Society for Paint Technology. As pointed out by Larson and Sletmoe,[12] the instrument has been modified so that an evaporation curve is plotted automatically during the course of the evaporation. The solvent is evaporated from a filter paper disk under controlled conditions of temperature, humidity and airflow. While the standard test is run at 77°F, evaporations have been made at temperatures as low as 65°F (18°C) and as high as 140°F (60°C) and at relative humidities between zero and 80%. The instrument is applicable for very volatile solvents such as used in gravure inks and lacquers and also for conventional paint thinners and retarder solvents used in paints and enamels.

Reynolds and Gebhardt[18] showed that volatility, as measured by ASTM distillation, and evaporation of *hydrocarbon* solvents could be related in suitable mathematical equations. The equations were developed to calculate evaporation in 10% increments corresponding to the 10% ASTM distillation temperatures. For instance, the calculated value based on the 10% ASTM temperature represents the time required for the first

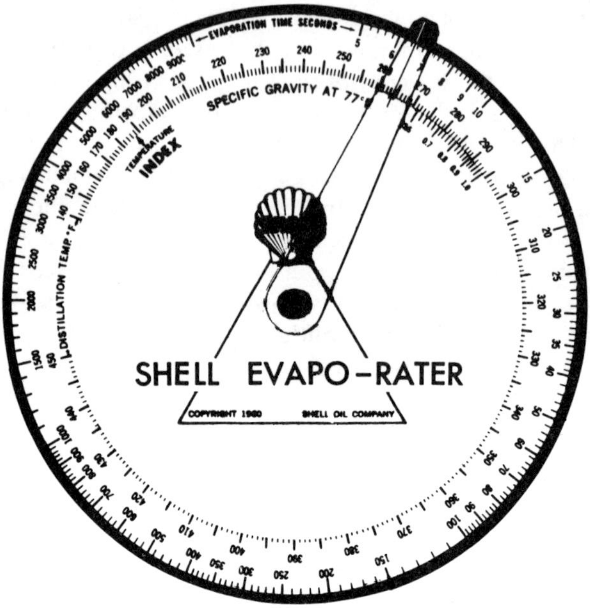

Front View

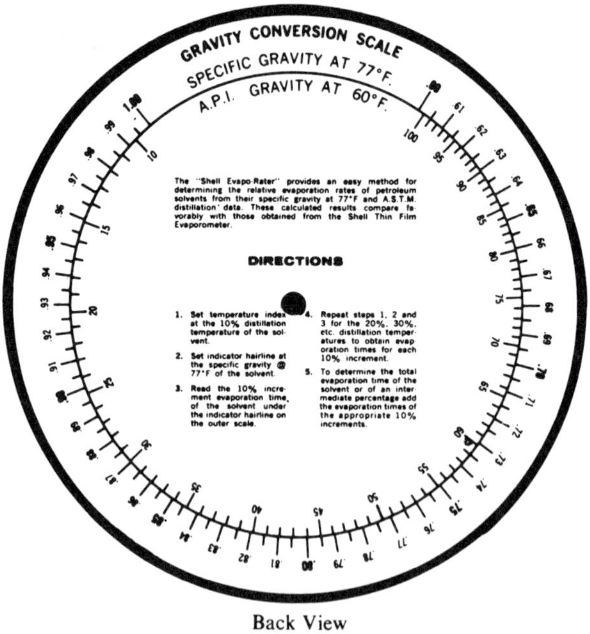

Back View

Figure 17.4. Shell Evapo-Rater.

10% to evaporate. The 20% ASTM temperature yields the evaporation time for the second 10% to evaporate and so on. Thus, the complete evaporation curve may be calculated by a summation of the individual 10% calculations. In a further development, the three equations were incorporated in a circular slide rule called the Shell Evapo-Rater.

The Evapo-Rater permits rapid calculation of the effect of changes in composition of solvents on evaporation from paint films. Stewart and Bewick,[19] and Billmeier and Rittershausen,[5] found that the wet-edge time of varnishes is related to the time for 30% of the solvent to evaporate. Data obtained with the Shell Thin Film Evaporometer and actual wet-edge time measurements agree with these findings. Thus, if it is desired to alter wet-edge time, it is only necessary to consider the alteration in terms of percentage change in the 30% evaporation time of the solvent. Suppose a paint is reduced with mineral spirits and the wet-edge time is to be increased 50% with (1) 140 Solvent or (2) Insecticide Base. Table 17.9 shows that the 30% evaporation time of mineral spirits is 925 seconds and the corresponding value for 140 Solvent is 2910 seconds. The 30% evaporation time of the desired blend would be 1388 seconds and would be equivalent to a blend composed of 76.7% mineral spirits and 23.3% 140 Solvent. By similar calculations, only 6.2% of the heavy solvent would effect the same change in wet-edge time. But the heavy solvent would extend the set touch, through-dry and final evaporation far beyond that of 140 Solvent as noted from the much slower evaporation time of this solvent blend.

The set-touch drying time appears to be related to the time at which 60% of the thinner has evaporated, or only 40% remains in the film. This period can be shortened by the addition of a more volatile solvent or lengthened by adding a less volatile solvent. As implied in Table 17.9, the set-touch time, based on the 60% evaporation time of a film thinned with 140 Solvent could be shortened by one-third by substituting a blend composed of (1) 49.3% 140 Solvent plus 50.7% mineral spirits, or (2) 65.4% 140 Solvent plus 34.6% Super V.M. & P. naphtha. The latter blend would have greater viscosity reducing power because of the lower viscosity of Super V.M. & P. naphtha and would be advantageous when greater film build is desired. On the other hand, if film build and hiding power in the case of pigmented coatings are not critical, blend (1) would give the more economical formulation because more solvent could be used per gallon.

Evaporation of Commercial Solvents

Several factors affect the evaporation rate of solvents. Larson and Sletmoe[12] called particular attention to (1) temperature, (2) evaporative

TABLE 17.9. Sample Calculations of Evaporation Times for Petroleum Solvents[a]

%	Mineral Spirits (Specific Gravity at 77°F = 0.78)		Evaporation Time for 76.7% Mineral Spirits, 23.3% 140 Solvent (sec)	% Slower than Mineral Spirits	Evaporation Time for 93.8% Mineral Spirits, 6.2% Insecticide Base (sec)	% Slower than Mineral Spirits
	ASTM Distillation (°F)	Evaporation Time (sec)				
10	328	285	431	51	410	44
20	330	585	887	52	865	48
30	334	925	1388	50	1,388	50
40	336	1285	1920	49	1,955	52
50	340	1695	2491	47	2,601	53
60	344	2155	3116	45	3,342	55
70	348	2675	3823	43	4,208	57
80	354	3315	4640	40	5,316	60
90	366	4225	5711	35	7,051	67
100	390	6125	7739	26	11,313	85

%	140 Solvent (Specific Gravity at 77°F = 0.78)		Evaporation Time for 49.3% 140 Solvent, 50.7% Mineral Spirits (sec)	% Faster than 140 Solvent	Evaporation Time for 65.4% 140 Solvent, 34.6% Super V.M. & P. Naphtha (sec)	% Faster than Shell 140 Solvent
	ASTM Distillation (°F)	Evaporation Time (sec)				
10	366	910	593	35	607	33
20	368	1,880	1224	35	1255	33
30	370	2,910	1904	35	1941	33
40	372	4,010	2628	34	2673	33
50	372	5,110	3379	34	3407	33
60	374	6,280	4187	33	4186	33
70	378	7,600	5103	33	5063	33
80	380	9,000	6118	32	5994	33
90	384	10,600	7368	30	7057	33
100	398	13,050	9539	27	8683	33

%	Insecticide Base (Specific Gravity at 77°F = 0.78)		Super V.M. & P. Naphtha (Specific Gravity at 77°F = 0.75)	
	ASTM Distillation (°F)	Evaporation Time (sec)	ASTM Distillation (°F)	Evaporation Time (sec)
10	396	2,300	250	35
20	403	5,100	252	71
30	408	8,400	254	109
40	412	12,100	254	147
50	416	16,300	256	187
60	421	21,300	256	227
70	428	27,400	258	269
80	438	35,600	260	313
90	456	49,800	264	361
100	490	89,800	278	428

[a]Evaporation times calculated with the Shell Evapo-Rater.

cooling and (3) humidity.   They pointed out that since pure solvents evaporate at a constant rate at constant temperature and pressure, a single rate value would be suitable representation of the relative volatility of the solvent. It has been rather common practice irrespective of test method to express evaporation in absolute units of time for complete evaporation and intervals thereof; these values also allow arbitrary comparison of volatility.

Since petroleum solvents are usually composed of mixtures of hydrocarbons, it is common for them to deviate somewhat from a linear evaporation rate. Thus, solvent mixtures show a greater time differential than pure solvents for the last few per cent to evaporate.   Furthermore, there is an apparent retardation in rate as evaporation nears completion.   This is commonly referred to as a "tail" on the evaporation curve.   The tail is caused by such factors as the test method, impurities in the solvent and hydrogen bonding. From a practical standpoint, the user is interested in the drying rate of the major portion of the solvent and will ignore final traces or even the final 5% which may linger in the "dry" film. Hence, the average rate for the major portion of the solvent is of greatest interest. Evaporation curves are nearly linear between 10 and 90% by weight evaporated, so these points are used to calculate an average evaporation rate. The values are then expressed in grams evaporated per square centimeter per second. For convenience in interpretation and to obtain whole numbers, the values are multiplied by 10^8. Average evaporation rates on several hydrocarbon solvents and some oxygenated solvents are compared in Table 17.10.   These data show that the evaporation rates range from 80 for mineral spirits to 5560 for hexane. According to this system, a high value indicates fast evaporation and a low value indicates a slow evaporation rate. With a little practice, the formulator or painter should be able to select the solvents whose evaporation rates are most appropriate for each type of application. Because evaporations are expressed on a weight basis, if two solvents have the same indicated rate, 1 pound of each will evaporate in the same time under identical conditions.   However, if their specific gravities differ, 1 gallon of the solvent of lower specific gravity will evaporate faster than 1 gallon of the other solvent. To convert the evaporation rate data in Table 17.10 to a volume basis, divide by the specific gravity of the solvent. For certain practical applications, it may be desirable to express evaporation in terms of pounds evaporated per square foot per hour. In such instances, the values given in the table may be multiplied by the factor 7.39×10^{-3}. On this basis, mineral spirits has an evaporation rate of 5.9×10^{-3} lb/sq ft/hr at 77°F.

**TABLE 17.10.   Comparison of Average Evaporation Rates on Hydrocarbon
and Oxygenated Solvents**

	Evaporation Rate[a]	Boiling Range	
		(°C)	(°F)
Hydrocarbon Solvents			
Hexanes	5560	66–70	150–158
Fast diluent naphtha	5290	60–82	140–180
n-Heptane	2860	98–99	208–209
Lacquer diluent	2500	93–116	200–240
n-Octane	1095	125–126	257–258
V.M. & P. naphtha	990	121–149	250–300
Mineral spirits	80	154–204	310–400
Toluene	1900	110–111	230–232
Xylene	650	135–143	275–290
Oxygenated Solvents			
Isopropyl alcohol	1305	81–83	178–181
n-Butyl alcohol	338	116–119	241–246
Ethyl acetate	4440	72–80	162–176
n-Propyl acetate	2250	95–103	203–217
Amyl acetate	474	120–150	248–302
Ethylene glycol monoethyl ether	330	132–137	270–279
Acetone	5830	56–57	133–135
Methyl ethyl ketone	3620	78–81	172–178
Cyclohexanone	290	130–173	266–343

[a]Grams evaporated per square centimeter per second × 10^8. Measurements were made with the Shell Thin Film Evaporometer.[12]

Effect of Temperature on Evaporation

While most printing and architectural painting is done at room temperature, industrial coatings are generally dried at elevated temperatures. Evaporation rates for a few hydrocarbons and polar solvents measured[12] at 77, 104 and 140°F are shown in Table 17.11. The rate is much greater at the higher temperatures and is essentially linear with temperature—about four times faster at 140 than at 77°F. Furthermore, the increase in rate is greater for high-boiling solvents. Evaporative cooling is greater for low than for high molecular weight solvents, and this fact contributes to the smaller effect of temperature on the evaporation rate of the former. The higher heat of vaporization of high-boiling solvents contributes to a faster increase in vapor pressure with temperature and hence a proportionately faster increase in evaporation rate than with lower molecular weight solvents.

TABLE 17.11. Effect of Temperature on Evaporation Rate

	Evaporation Rate at Indicated Temperature[a]			
	77°F	104°F	140°F	Relative Rate[b]
Hydrocarbon Solvents				
n-Heptane	2860	4960	7440	2.6
n-Octane	1095	2320	4430	4.0
V.M. & P. naphtha	990	—	4350	4.4
Toluene	1900	—	5640	3.0
Xylene	650	1475	2990	4.6
Oxygenated Solvents				
Isopropyl alcohol	1305	2350	3240	2.5
n-Butyl alcohol	338	840	1830	5.4
Ethyl acetate	4440	6310	8500	1.9
n-Propyl acetate	2250	3860	7020	3.1
Amyl acetate	474	—	2630	5.5
Ethylene glycol monoethyl ether	330	760	1840	5.6
Methyl ethyl ketone	3620	5560	7450	2.1
Cyclohexanone	290	—	1730	6.0

[a]Grams evaporated per square centimeter per second $\times 10^8$.
[b]Evaporation rate at 140°F divided by evaporation rate at 77°F.

Effect of Evaporative Cooling on Evaporation

The molecules in a liquid are in a constant state of motion, and the collisions brought about by this motion increase the energy level of a few of the colliding molecules to a point where they can no longer remain in the liquid but escape into the atmosphere. This change in state is called evaporation. The remaining molecules have, on the average, a little less energy than before, and this is manifested by a reduction in temperature of the liquid during evaporation. This cooling effect is to some degree offset by the surroundings, which tend to restore the original temperature. If evaporation is slow, the loss of energy from the system is almost completely compensated by the heat input from the surroundings. If evaporation is fast, the heat input lags behind the evaporative heat loss and there is a net cooling effect. The extent of the temperature drop is a function of the evaporation rate, the quantity of solvent being evaporated, the rate of airflow, and the composition and mass of the substrate.

The extent of evaporative cooling for several solvents at 77°F and 140°F is shown in Table 17.12. Because evaporation rates at 140°F are much faster than at 77°F, the rate of heat loss is much greater than at the lower temperature. However, the rate of heat input from the sur-

TABLE 17.12.   Temperature Drop Caused by Evaporative Cooling[a]

	$\Delta t\ (°F)$	
Solvent	*Ambient Temperature, 77°F*	*Ambient Temperature, 140°F*
n-Hexane	27	60
Methylcyclopentane	27	—
Cyclohexane	22	—
Toluene	8	31
n-Octane	—	20
V.M. & P. naphtha	—	17
o-Xylene	—	15
n-Decane	—	9

[a]Shell Thin Film Evaporometer.

roundings is not sufficiently increased to compensate for the heat loss. Therefore, the drop in temperature is much greater at 140 than at 77°F.

It is impossible to predict what effect evaporative cooling would have on applied printing inks, lacquers and paints, but it is interesting to note that the greater share of solvents used in these applications produce some degree of evaporative cooling. The phenomenon could become a problem when the temperature of the substrate is lowered to, or below, the dew point and water condenses on the film. This often occurs when lacquer and varnish films are applied under humid conditions; moisture condenses on the film as a result of evaporative cooling. The film defect may be temporary or permanent, and is called moisture blush.

Effect of Humidity on Evaporation

When the relative humidity is 50% at 77°F, the dew point occurs at 57°F; when the relative humidity is 75% at 77°F, it occurs at 68°F. This indicates that many low-boiling solvents could promote moisture condensation on the cooled surfaces when the relative humidity is high at the time of application of ink or paint films.

It has been demonstrated[12] by measurements on (1) nitrocellulose–methyl isobutyl ketone and (2) "Lucite"-toluene solutions that the evaporation rate of the solvent is not affected by the relative humidity below about 75%. At a threshold region of 76 to 80% relative humidity and above, evaporation was affected by blooming caused by moisture condensation on the films. Thus, it was concluded that relative humidity has little influence on evaporation rate or solvent release when the temperature of the substrate is above the dew point for the relative humidity in question.

Effect of Film Thickness on Evaporation

Paint and lacquer films are applied in thicknesses ranging between 0.1 and 2.5 mils in single or multiple coats, depending on such factors as the kind of coating, method of application, type of cure, degree of protection required, position of coating and type of substrate. Because of their low solids content and relatively high viscosity, lacquers are applied in very thin films and require several coats to reach a durable thickness. In contrast, thixotropic paints and mastics have high solids content and often furnish a 5 to 25 mil coat in one application.

Reynolds and Griebel[17] showed that solvents evaporated at a constant rate from medium oil glycerol phthalic alkyd solutions applied at 2 to 8 mil wet film thickness. In the thicker films, a greater amount of solvent is applied and the time is extended for a given percentage of the solvent to evaporate. Consequently, if a solvent evaporates in one unit of time from a 1-mil film, it will require two time units to evaporate from a 2-mil film. On the basis of these results it may be concluded that the rate of diffusion of the solvent through the viscous film is adequate to maintain a layer of solvent at the air interface; the solvent evaporates proportionately to the weight of solvent per unit area of evaporating surface.

REFERENCES

1. *Am. Paint J.*, 15 (Aug. 30, 1965).
2. Artel, V., *Offic. Dig. Federation Soc. Paint Technol.*, **36,** No. 474, 753 (1964).
3. Asbeck, W. K., and Van Loo, M., *Ind. Eng. Chem.*, **46,** 1291 (1954).
4. Barker, J. A., "Lattice Theories of the Liquid State," Oxford, Pergamon Press, 1963.
5. Billmeier, R. A., and Rittershausen, E. P., *Offic. Dig. Federation Paint Varnish Prod. Clubs*, **26,** No. 351, 283 (1954).
6. Burrell, H., *Offic. Dig. Federation Paint Varnish Prod. Clubs*, **27,** No. 369, 726 (1955).
7. CDIC Paint and Varnish Production Club, *Offic. Dig. Federation Paint Varnish Prod. Clubs*, **29,** No. 394, 966 (1957).
8. Harvey, W. T., and Mills, I. W., *Anal. Chem.*, **20,** 207 (1948).
9. Hildebrand, J. H., and Scott, R. L., "The Solubility of Nonelectrolytes," Third ed., New York, Reinhold Publishing Corp., 1950.
10. Horak, M., and Pliva, J., *Spectrochim. Acta*, **21,** 911–917 (1965).
11. Larson, E. C., and Low, H., *Offic. Dig. Federation Paint Varnish Prod. Clubs*, **26, ** 45 (1954).
12. Larson, E. C., and Sletmoe, G. M., *Gravure* (August–September 1964).
13. McArdle, E. H., *Am. Paint J.* (June 4, 1945).
14. New York Paint and Varnish Production Club, *Offic. Dig. Federation Paint Varnish Prod. Clubs*, **28,** No. 380, 1060 (1956).
15. Pimentel, G. C., and McClellan, A. L., "The Hydrogen Bond," San Francisco, W. H. Freeman and Co., 1960.
16. Reynolds, W. W., and Gebhart, H. J., Jr., *Offic. Dig. Federation Soc. Paint Technol.*, **32,** No. 428, 1146 (1960).
17. Reynolds, W. W., and Griebel, R. D., *Offic. Dig. Federation Soc. Paint Technol.*, **33,** 921 (1961).

18. Reynolds, W. W., and Larson, E. C., *Offic. Dig. Federation Soc. Paint Technol.*, **34,** 311 (1962).
19. Stewart, R., Jr., and Bewick, H. I., *Ind. Eng. Chem.*, **28,** 940 (1936).
20. U. S. Tariff Commission, "Synthetic Organic Chemicals," U. S. Production and Sales 1963, TC Publication 143, U. S. Government Printing Office, Washington, D. C., 1964.

18

*Oxygenated Solvents**

A wide variety of oxygenated solvents are available to the lacquer chemist for use in the formulation of solvent systems and thinners for lacquers and other coatings. There are numerous factors such as solvency, evaporation rate and cost that must be considered in the final selection and blending of components for a lacquer solvent system. It would be very desirable if there were a universal solvent system that fulfilled all the necessary criteria. However, there is no one simple answer to this problem. In fact, it is possible to have a number of practical formulations that will meet specific performance requirements. The primary task of the lacquer chemist then becomes one of making the best choice between performance and cost.

The oxygenated solvents can be classified according to their functional groups as esters, ketones, glycol ethers, alcohols and other types. The chemical structures of typical solvents are illustrated in Figure 18.1.

The physical properties of some typical oxygenated solvents are given in Table 18.1. While this listing is not complete, it does include some of the more commonly used lacquer solvents. Additional data on these and other solvents can be found in suppliers' literature and other reference sources.

Ester Solvents

Esters, the original solvents for nitrocellulose, are an important class of solvents and are well known in the coatings industry. The ester solvents are available over a wide range of boiling points or evaporation rates and are good solvents for many of the commonly used coatings resins. Some of the more important ester solvents are ethyl acetate, a good solvent for the cellulosic resins; poly(methyl methacrylate); polyvinyl butyrol; polyvinyl acetate; isopropyl acetate, often used as a replacement

*By R. H. Duzy, Chemicals Division, Union Carbide Corporation, New York, N. Y.

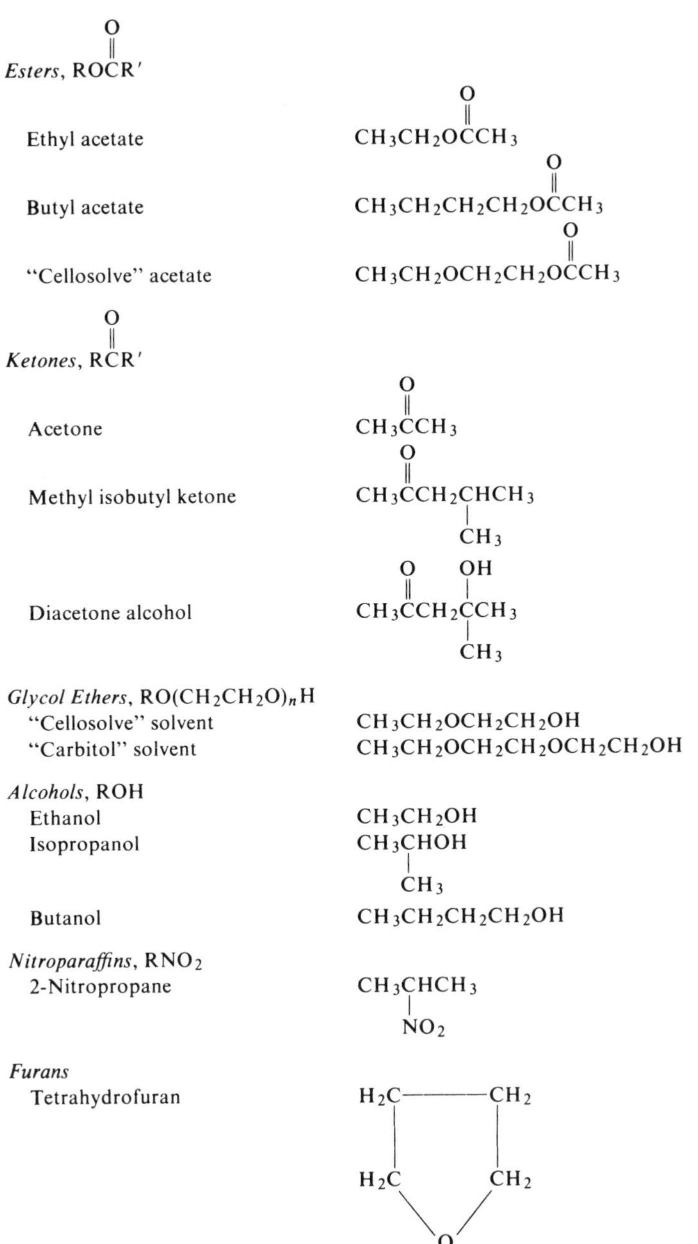

Figure 18.1.   Oxygenated solvents.

TABLE 18.1. Selected Physical Properties of Typical Lacquer Solvents

Solvents	Formula	Molecular Weight Pure Material	Median Specific Gravity, 20/20°C	Average Pounds per Gallon, 20/20°C	Boiling Initial Point (C°)	Range Dry Point (C°)	Flash Point (F°), Open Cup
Esters							
Ethyl acetate (85–90%)	$C_2H_5OCOCH_3$	88.11	0.886	7.37	71	79	44
Isopropyl acetate (95%)	$(CH_3)_2CHOCOCH_3$	102.14	0.870	7.24	85	90	62
Isobutyl acetate (90%)	$(CH_3)_2CHCH_2OCOCH_3$	116.16	0.867	7.22	110	119	83
Butyl acetate (90–92%)	$C_4H_9OCOCH_3$	116.16	0.883	7.28	118	120	90
Methylamyl acetate	$(CH_3)_2CHCH_2CH(CH_3)OCOCH_3$	144.22	0.858	7.14	143	150	110
"Cellosolve" acetate	$C_2H_5OC_2H_4OCOCH_3$	132.16	0.975	8.10	145	165	138
Ketones							
Acetone	CH_3COCH_3	58.08	0.792	6.59	56	57	<20
Methyl ethyl ketone	$CH_3COC_2H_5$	72.10	0.806	6.71	78	81	38
Methyl isobutyl ketone	$CH_3COCH_2CH(CH_3)_2$	100.16	0.802	6.67	114	117	74
Methyl isoamyl ketone	$(CH_3)_2CHCH_2CH_2COCH_3$	114.19	0.815	6.77	141	148	110
Diisobutyl ketone	$(CH_3)_2C_2H_3COC_2H_3(CH_3)_2$	142.24	0.808	6.75	163	173	120
Cyclohexanone	$CH_2(CH_2)_4CO$	98.15	0.948	7.88	153	157	130
Isophorone	$COCHC(CH_3)CH_2C(CH_3)_2CH_2$	138.21	0.922	7.67	207	218	205
Glycol Ethers							
Methyl "Cellosolve"	$CH_3OC_2H_4OH$	76.10	0.966	8.03	124	126	110
"Cellosolve" solvent	$C_2H_5OC_2H_4OH$	90.12	0.930	7.74	134	136	120
Butyl "Cellosolve"	$C_4H_9OC_2H_4OH$	118.18	0.902	7.51	169	173	165
"Carbitol" solvent	$C_2H_5(OC_2H_4)_2OH$	134.18	1.027	8.55	190	205	205
Alcohols							
Ethanol	C_2H_5OH	46.07	0.811	6.76	78	80	75
Isopropanol	$(CH_3)_2CHOH$	60.10	0.786	6.55	81	83	63
Isobutanol	$(CH_3)_2CHCH_2OH$	74.12	0.805	6.68	107	109	95
Butanol	C_4H_9OH	74.12	0.811	6.75	118	119	97
Methylamyl Alcohol	$(CH_3)_2C_3H_4(CH_3)OH$	102.18	0.808	6.72	130	133	106
2-Ethylhexanol	$C_4H_9CH(C_2H_5)CH_2OH$	130.23	0.834	6.94	182	186	185

for ethyl acetate; butyl acetate, one of the more important nitrocellulose solvents and a good solvent for other resins; isobutyl acetate, a somewhat faster evaporating substitute for butyl acetate; amyl acetates, used mainly for nitrocellulose lacquers; "Cellosolve" acetate, a good solvent for many coatings resins.

In general, the ester solvents provide good resin compatibility and contribute good flow-out to lacquer formulations. They are not as strong in solvency as comparable ketone solvents, but they can be utilized in the formulation of lacquers to minimize undercoat softening or lifting problems. The esters, particularly "Cellosolve" acetate, are used in acrylic coatings to impart good flow-out and gloss. Some high-purity esters, having a minimum of reactive hydrogen present to react with isocyanate groups, are available for use in the formulation of solvents systems for urethane coatings. These high-purity esters can also be used in some vinyl solution coatings.

Ketone Solvents

The ketone solvents, as a class, are newer to the coatings industry than the esters. The development of vinyl chloride–vinyl acetate resins in the late 1930's stimulated the need for ketones other than the available acetone and methyl ethyl ketone. Ketones usually exhibit greater solvent strength than do esters of comparable evaporation rate and are solvents for a wider number of coatings resins. The ketones will often tolerate higher dilution with nonsolvents than do esters, thereby allowing the use of larger amounts of these lower-cost materials in the formulation of lacquers. The stronger solvency characteristics of the ketone solvents over many other solvents offer an advantage in the preparation of lacquers having minimum viscosity at fixed solids content or higher solids content at a given application viscosity.

The three most important ketones used by the coatings industry are acetone, methyl ethyl ketone and methyl isobutyl ketone. Acetone and methyl ethyl ketone are solvents for a wide variety of resinous materials including cellulose acetate, mixed cellulose esters, nitrocellulose, vinyl chloride–vinyl acetate resins, epoxies and poly(methyl methacrylate). Methyl isobutyl ketone, slower evaporating than acetone or methyl ethyl ketone, is also widely used in the formulation of coatings based on these resins. Other ketones employed by the coatings industry are methyl isoamyl ketone, ethyl amyl ketone, diacetone alcohol, cyclohexanone and isophorone.

Glycol Ether Solvents

These solvents are generally considered to be slow evaporating, but they exhibit some unique solvency characteristics because of their ether-alcohol functionality. They are normally used in limited amounts in conventional lacquers because of their slow-drying characteristic. Their high tolerance for dilution with hydrocarbons make them useful in nitrocellulose lacquer formulations. Butyl "Cellosolve," because of its ability to increase blush resistance, flow-out and resin compatibility, is of particular value in the formulation of lacquers. The glycol ether solvents are also employed as components of solvent systems for coatings based on epoxies, shellac, cellulose and other resins.

Alcohols

While alcohols are solvents for only a few resins such as ethylcellulose, shellac and polyvinyl butyral, they are generally accepted as components of solvent systems for lacquers based on a much wider variety of resins. Although alcohols, except methanol, are not active solvents for RS-type nitrocellulose, they can be used in significant quantities without sacrificing the solvency of the solvent system or thinner for nitrocellulose lacquers. The use of the faster-evaporating alcohols results in lower formulating costs.

While methanol is a good resin solvent, it is generally used in limited quantities because of its toxicity. In addition to their use in nitrocellulose lacquers, ethanol and isopropanol are used as solvents for shellac and other natural resins. Combinations of these alcohols with aromatic hydrocarbons are useful solvent systems for ethylcellulose, and the further addition of small amounts of other oxygenated solvents allows their use as solvent systems for alcohol soluble-type nitrocellulose. Butanol is employed as a latent solvent for nitrocellulose coatings, and isobutanol, while slightly faster evaporating, can be substituted for butanol in many instances. Other alcohols used in the formulation of lacquers are amyl alcohol, methylamyl alcohol and 2-ethylhexanol.

Other Solvents

Other oxygenated solvents which are used in the formulation of solvent systems or thinners are products such as 2-nitropropane, tetrahydrofuran and dimethylformamide.

Tetrahydrofuran is a very powerful solvent and is used as a component of solvent systems for polyvinyl chloride, polyvinylidene chloride, vinyl chloride–vinyl acetate copolymers, and many other coatings resins or combinations of resins. Its strong solvent power permits the preparation of usable solutions of the more difficultly soluble resins and polymers.

Dimethylformamide and 2-nitropropane are used particularly in solvent systems for vinyl chloride–vinyl acetate copolymer-based lacquers. Among other outstanding properties, they exhibit the ability to tolerate large quantities of aromatic hydrocarbons and still maintain adequate solvency for the vinyl copolymer resins. They are also useful in solvent systems for lacquers based on numerous other film-forming resins.

SOLVENT PROPERTIES

A wide array of oxygenated solvents with varying performance characteristics is available for combining with hydrocarbons in the formulation of practical solvent systems for various types of lacquers. A lacquer can be defined as a solution of nonvolatile resins or a combination of resins in volatile solvents which, on application, forms a continuous film or coating as the volatile components evaporate. True lacquers do not undergo any cross-linking or polymerization.

There are a number of general performance properties determined for the various solvents which the lacquer chemist must consider in the formulation of practical and economical solvent systems or thinners for lacquers. The improper selection of components can lead to problems such as poor flow-out and leveling, blushing, sagging or running. Thus, the proper selection of components for the volatile portion of a lacquer is extremely important since the properties of the dried lacquer films can be affected considerably by the solvent system.

Resin Solubility

One of the prime considerations in determining the utility of a solvent is its ability to solvate the various film-forming resins employed in conventional coatings. The larger the number of resins dissolved by a particular solvent, the more useful will it be to the lacquer chemist who formulates a wide range of coatings. Table 18.2 shows the qualitative solubility of some of the more commonly used lacquer resins in typical solvents. From this listing, prime solvents can be selected to form the basis of the solvent system or thinner for a particular film-forming resin. Solvents showing only partial or slight solubility for a given resin are included in this listing since such solvents can often be used in combination with other active solvents to impart special performance characteristics to the finished formulation.

This listing of resin solubilities can be used in other ways. For example, the solution to a problem of designing a solvent system for a lacquer applied to molded polystyrene objects is to find solvents with suitable solvency for the lacquer resin but with minimum solvency for polystyrene, and thus reduce the tendency of the lacquer to solvate or craze the polystyrene object.

TABLE 18.2. Solubilities of Selected Resins in Typical Solvents[a]

(0.50 g resin per 4.5 ml of solvent)

Solvents	Cellulose Acetate	Cellulose Acetate Butyrate 17% Butyryl	37%	Ethyl-cellulose (N-22)	Poly-styrene	Methyl Meth-acrylate	"Bakelite" Vinyl Resins VYHH	AYAF	XYHL
Esters									
Ethyl acetate (85–90%)	SlS	S	S	S	S	S	S-G	S	S
Isopropyl acetate (95%)	I	PS[b]	S	S	S	S	S-G	S	S
Isobutyl acetate	I	I	S	S	S	PS	S	S	G
Butyl acetate	I	SW	S	S	S	S	S	S	SW
Methylamyl acetate	I	I	SW	S	S	I	S-G	S	SW
"Cellosolve"* acetate	SlS	PS	S	S	S	S	S	S	G
Ketones									
Acetone	S	S	S	S	PS	S	S	S	SW
Methyl ethyl ketone	S	S	S	S	S	S	S	S	G
Methyl isobutyl ketone	I	I	S	S	S	PS	S	S	PS
Methyl isoamyl ketone	I	I	S	S	S	SlS	S	S	SW
Ethyl amyl ketone	I	I	I	S	S	I	S	SlS	SW
Cyclohexanone	S	S	S	S	S	S	S	S	S
Isophorone	S	S	S	S	S	PS	S	S	S
Glycol Ethers									
Methyl "Cellosolve"*	S	I	S	S	I	S	S	PS	S
"Cellosolve"* solvent	I	I	S	S	I	I	S	I	S
Butyl "Cellosolve"	I	I	I	S	I	I	PS	I	S
"Carbitol"* solvent	I	SW	PS	PS	I	I	PS	I	S

Alcohol									
Ethanol (95%)	I	I	I	S	I	I	I	S	S
Isopropanol (anhydrous)	I	I	I	PS[b]	I	I	I	SlS	S[b]
Isobutanol	I	I	I	S	I	I	I	SlS	S[b]
n-Butanol	I	I	I	S	I	I	I	SlS	S[b]
Amyl alcohol, primary	I	I	I	S	I	I	I	I	S[b]
Methylamyl alcohol	I	I	I	S[b]	I	I	I	SW	S[b]
2-Ethylhexanol	I	I	I	S[b]	I	I	I	I	S[b]
Miscellaneous									
Tetrahydrofuran	S	S	S	S	S	S	S	S	S
Dimethylformamide	S	S	S	S	S	S	S	S	S

[a] S = soluble; PS = partly soluble; G = gel; SlS = slightly soluble; SW = swelling; S-G = soluble, tendency to gel; I = insoluble.
[b] Concentration: 0.50 g resin to 9.5 ml of solvent.

* Registered trademark — Union Carbide Corporation.

Evaporation Rate

As indicated earlier, evaporation characteristics play a major role in the selection of components for solvent systems and thinners for lacquers since improper balance can affect the physical properties of the coating. Furthermore, the overall evaporation characteristics of the formulated coating will have to be tailored to meet end use requirements such as spray, dip coat or roller coat methods of application.

TABLE 18.3. General Solvent Properties of Selected Solvents

Solvents	Relative Evaporation Rate (Butyl Acetate = 100)	Nitrocellulose Dilution Ratios			8% Solutions of R.S. 1/2-sec Nitrocellulose	
		Toluene	Naphtha	Xylene	Blush Resistance (% RH at 80°F)	Viscosity
Esters						
Ethyl acetate (85–90%)	615	3.3	1.1		37	16
Isopropyl acetate (95%)	500	2.7	1.1		69	20
Isobutyl acetate	145	2.7	1.1		'80	29
Butyl acetate	100	2.9	1.3	2.7	83	30
Methylamyl acetate	47	1.7	1.0	1.6	91	57
"Cellosolve"[a] acetate	21	2.5	0.9	2.3	94	67
Ketones						
Acetone	1160	4.5	0.7		35	9
Methyl ethyl ketone	572	4.3	0.9		51	10
Methyl isobutyl ketone	165	3.6	1.0	3.2	78	19
Methyl isoamyl ketone	45	3.8	1.1		89	26
Ethyl amyl ketone	26	2.1	0.8		95	42
Cyclohexanone	23	5.7	1.1	4.8	92	79
Isophorone	3	6.2		5.1	96	104
Glycol Ethers						
Methyl "Cellosolve"[a]	47	4.0	Imm.[b]	2.9	42	65
"Cellosolve"[a] solvent	32	4.9	1.1	4.3	59	73
Butyl "Cellosolve"	6	3.5	2.3	3.2	96	110
"Carbitol"[a] solvent	1	1.9	Imm.[b]	1.2	76	300
Alcohols						
Ethanol (95%)	230					
Isopropanol (anhydrous)	230					
Isobutanol	80					
Butanol	45					
Amyl alcohol, primary	26					
Methylamyl alcohol	33					
2-Ethylhexanol	1					
Miscellaneous						
Tetrahydrofuran	800	2.8	1.0		50	15
Dimethylformamide	18	7.7	Imm.[b]		38	17

[a] Registered trademark, Union Carbide Corporation.
[b] Imm. = immiscible.

Evaporation rates are indicated by a number or relative evaporation rate comparing the time necessary for a given amount of a solvent to evaporate with that required for an equal amount of a well-known reference solvent, generally butyl acetate, under identical conditions. The reference solvent is assigned a value of 100 in some instances and 1.0 in others. The relative evaporation rates for some of the more commonly used lacquer solvents are listed in Table 18.3. The components of solvent systems are generally divided into three classes according to evaporation rate:

(1) Low boilers or fast evaporating: Solvents evaporating at a rate greater than twice that of butyl acetate or with relative evaporation rates greater than 200.

(2) Medium boilers or medium evaporating: Solvents evaporating at a rate between 0.8 and twice that of butyl acetate or with relative evaporation rates betwen 80 and 200.

(3) High boilers or slow evaporating: This group includes all solvents evaporating at a rate below 0.8 of butyl acetate or all solvents with relative evaporation rates of less than 80.

Solvent Strength

The strength of solvents for the various coatings resins is usually indicated by solution viscosity measurements and their ability to tolerate non-solvents or diluents. There is a maximum viscosity at which a coating can be applied in any type of finishing operation, and this maximum viscosity limits the practical solids content of the coating. Since, in the interest of economy, it is desirable to apply the coating at the highest possible solids content, the choice of solvents to achieve maximum solids at a particular application viscosity is an important factor. Solution viscosity data as a function of solvent choice are available for a wide variety of coatings resins in the literature of resin and solvent suppliers.

Viscosity data for 8% solutions of nitrocellulose in some of the more important oxygenated solvents are shown in Table 18.3. Although these values provide relatively clear-cut estimates of the viscosity characteristics of low molecular weight esters and ketones, they are not necessarily a good indication of the true solvent strength of the higher molecular weight solvents. Furthermore, the majority of practical lacquers and coatings will often contain latent solvents or couplers and hydrocarbon diluents which can also influence viscosity.

Solvents also differ in their solvency for alkyd and hard resins commonly used to modify nitrocellulose lacquers. It is difficult to study the performance of the many solvents commercially available in the numerous practical combinations of resins and plasticizers in multicomponent nitrocellulose lacquer formulations. Four test lacquers were selected to indicate performance of solvents under conditions or normal use. Relative viscosity data for these lacquers as a function of solvent choice are listed in Table 18.4. Relative viscosity (n-butyl acetate = 100) was used to minimize differences in viscosity resulting from lot-to-lot variations in nitrocellulose and the particular resins used.

As with nitrocellulose lacquers, solution viscosity as a function of solvent choice plays an important part in the selection of solvents for the vinyl chloride–vinyl acetate copolymer resins as well as other coatings resins. Viscosity data for vinyl copolymer resin solutions in some of the more important lacquer solvents are shown in Table 18.5. The use of aromatic hydrocarbons as diluents illustrates the effect these materials can have on solution viscosity, particularly with some of the higher

TABLE 18.4.   Relative Viscosities of Nitrocellulose Lacquers

n-Butyl acetate = 100

Solvent	N	ND	NRAD	NRMD
Acetone	25	40	20	Immiscible
Tetrahydrofuran	50	70	55	60
Ethyl acetate (85–90%)	52	64	46	60
Methyl ethyl ketone	30	52	32	40
Isopropyl acetate	65	76	72	80
Methyl isobutyl ketone	60	80	55	65
Isobutyl acetate	95	95	85	95
Butyl acetate	100	100	100	100
Methyl "Cellosolve"	220	120	60	Immiscible
Methylamyl acetate	190	140	420	190
Methyl isoamyl ketone	85	90	75	85
"Cellosolve" solvent	240	130	80	110
Ethyl amyl ketone	140	120	170	120
Cyclohexanone	260	120	70	130
"Cellosolve" acetate	220	140	120	130
Dimethylformamide	55	45	—	—
Butyl "Cellosolve"	360	170	150	150
Isophorone	340	120	100	150
Formula Components (parts by weight)				
R.S. $\frac{1}{2}$-sec nitrocellulose (dry)	8.0	8.0	8.0	8.0
Nonoxidizing alkyd resin (100% basis)	—	—	12.0	—
Maleic hard resin	—	—	—	12.0
Dibutyl phthalate	—	—	—	4.0
Ethanol	—	4.3	4.3	4.3
Solvent (see above)	92.0	41.7	35.7	33.7
Toluene	—	23.0	20.0	19.0
Xylene	—	23.0	20.0	19.0

**TABLE 18.5.   "Bakelite" Vinyl Resin VYHH
Solution Viscosity**

Viscosity at 68°F[a] (cps)

	A	B
Acetone	90	90
Methyl ethyl ketone	90	130
Methyl isobutyl ketone	230	340
Methyl isoamyl ketone	300	500
Ethyl amyl ketone	870	Gcl
Cyclohexanone	670	360
Isophorone	930	485
Formula Components (parts by weight)		
"Bakelite" vinyl resin VYHH	20	20
Solvent (see above)	80	40
Toluene	—	20
Xylene	—	20

[a] Brookfield viscometer, No. 2 Spindle, at 20 rpm.

molecular weight solvents. For example, significant reductions in solution viscosity are observed when aromatic hydrocarbons are used to replace half of the cyclohexanone or isophorone.

Another method of measuring the strength of solvents for various film-forming resins is by determining the amount of nonsolvents that a resin solution will tolerate before precipitation of the resin occurs. Nitrocellulose dilution ratio values for some typical oxygenated solvents are listed in Table 18.3. These values indicate that the oxygenated solvents will tolerate greater amounts of aromatic than aliphatic hydrocarbons. Dilution ratios for aromatic naphthas, relatively high in aromatic content, will fall between values for aliphatic naphtha and toluene, and generally are roughly proportional to the aromatic content. A primary consideration here is that high dilution ratios generally mean that a larger amount of the low-cost diluent can be used in the lacquer formula.

Flash Point

The flash point of a solvent is the temperature at which enough vapor has evaporated from a material to propagate a flame above its surface when an ignition source is present. Flash point values for oxygenated solvents are normally reported in the suppliers' literature. The Interstate Commerce Commission requires that red warning labels be placed on products having Tagliabue open cup flash points below 80°F. The flash point of solvents or lacquer solutions provide an indication of their flammability or ability to catch fire.

Odor

There are coatings applications such as printing and packaging in which solvent odor is an important consideration since small amounts of retained solvent can be objectionable. In addition, the particular odor of a coating might be offensive to those persons involved in its application in a commercial finishing line. Certainly all solvents have their characteristic odors which may or may not be particularly objectionable depending on the persons involved. Familiarity with an odor will often be a factor in the acceptance of a solvent or solvent blend. Persons used to the fruity odor of esters may object to the different odor of ketones until they become familiar with it. Generally speaking, the lower-boiling or faster-evaporating solvents have less tendency to present problems of odor. Although personal preferences are important and individuals can vary considerably in their reaction to odor, classes of solvents listed in order of increasing degree of odor levels are alcohols, glycol ethers, esters and ketones.

Water Sensitivity

Their water sensitivity is an important criterion in the selection of solvents for use in the production of emulsions or coatings dispersed in water. Solvents exhibiting minimum sensitivity to water are utilized in the production of nitrocellulose or dispersion coatings to achieve maximum stability. Water-sensitive solvents are avoided for this application since they will tend to blend the lacquer phase into the water phase and thereby reduce the stability of the coating. The ability of selected solvents to couple water into solution can, on the other hand, be used to advantage to minimize or prevent problems caused by the unintentional entrapment of water in coatings or films.

Pounds per Gallon

In many instances coatings are prepared on a weight basis and sold on a gallon or volume basis. Since the majority of oxygenated solvents are marketed on a cost per pound basis, the cost per gallon figure becomes an important factor in calculating the raw materials cost of the formulated coating. For example, an ester and a ketone might have the same cost per pound, but the lower weight per gallon of the ketone results in a lower cost per gallon figure than for the ester. Thus, the coatings formulator must constantly be aware of solvent cost on both a weight and gallon basis.

FORMULATION OF SOLVENT SYSTEMS

Nitrocellulose

The solvent system or thinner for nitrocellulose lacquers usually consists of a combination of components which are divided into three classifications.

(1) Active solvents: Ketones, esters, glycol ethers, and other selected solvents required to dissolve the nitrocellulose.

(2) Latent solvents: Generally alcohols except methanol; these solvents will not dissolve nitrocellulose alone but can be used in significant amounts without reducing the solvent power of the system.

(3) Diluents: Aromatic and/or aliphatic hydrocarbons; these are solvents for many of the modifying resins but not nitrocellulose. These materials generally are less expensive than the active and latent solvents and are normally used to improve the economics of formulating nitrocellulose lacquer.

Base lacquers are generally prepared at high solids and viscosity in a solvent system containing a relatively high ratio of active and latent solvents to diluents. This base lacquer must then be reduced to application viscosity with a thinner or reducer. Since the producer of the base lacquer cannot, in many instances, control the quality of the thinner that will be used, it is a general practice to build a safety factor of solvency into the base lacquer.

In comparison, a thinner may often contain as much as 60 to 70% diluent, with the balance being active and latent solvents. When combined at a ratio of approximately one part by volume of base lacquer and one part of thinner, the overall solvent system of the reduced or thinned lacquer will generally consist of equal volumes of active and latent solvents to diluent.

A guide to basic nitrocellulose thinner formulation is shown in Figure 18.2. Although this guide pertains to nitrocellulose lacquers, many of the basic properties and considerations are also relevant to the formulation of thinners for coatings based on other film-forming resins.

There are many practical combinations of components which can be used to prepare solvent systems and thinners for nitrocellulose lacquers. The best choice will depend on current economics, application requirements, and in some instances the personal preference of the formulator. There are two general approaches or techniques to formulating thinners. One is to select components that provide essentially a constant rate of evaporation, while the second utilizes fast-evaporating components and compensates with much slower-evaporating components to average out

Active Solvent — Required to dissolve nitrocellulose

Evaporation Rate	Fast	Medium	Slow	Very Slow
Cost		Increasing →		
Flow	Very Poor	Fair	Good	Excellent
Viscosity		Increasing →		
Drying Rate	Too Fast	Medium	Satisfactory	Too Slow Usually
	Dusting	Orange Peel		Sagging — Force Dry
Blush Resistance		Generally Increases — Are Exceptions →		
	Bad		Good	

Latent Solvents — Can reduce cost without reducing solvency

Evaporation Rate	Fast	Medium	Slow	Very Slow
Cost		Increasing →		
Flow	Very Poor	Fair	Good	Excellent
Viscosity		Increasing →		
Drying Rate	Too Fast	Medium	Satisfactory	Too Slow Usually
	Dusting	Orange Peel		Sagging — Force Dry
Blush Resistance		Increasing →		
Coupling Action		Decreasing →		
		Optimum 20–30%		

Diluents — Reduce cost, often good solvents for modifying

Concentration	Increasing →	90
Cost	Decreasing →	
Flow	Decreasing →	
Solvency	Satisfactory	Rapidly above 55%
Aromatic	65 →	D.R. Phase separation or gelation
Aliphatic	55 →	

Figure 18.2. Guide to basic nitrocellulose thinner formulation.

evaporation characteristics. Both types of formulating approaches are employed in commercial production of lacquers.

While it is not possible to show all of the practical combinations of active solvents, latent solvents and diluents used in the formulation of thinners, the systems in Table 18.6 illustrate some typical formulations.

TABLE 18.6.   Typical Thinner Formulations

	Per Cent by Weight				
	A	B	C	D	E
Methyl ethyl ketone	12	10	10	—	7
Ethyl acetate	—	—	—	10	—
Methyl isobutyl ketone	13	8	10	10	—
Butyl acetate	—	7	10	15	25
Butyl "Cellosolve"	—	—	3	—	3
Isopropanol	10	10	10	6	5
Butanol	—	5	7	9	10
Toluene	65	60	50	50	30
Xylene	—	—	—	—	20

The improper selection of components in the formulation of nitrocellulose lacquers can cause various performance problems:

(1) Moisture blush,
(2) Nitrocellulose or cotton blush,
(3) Resin or gum blush,
(4) Improper flow and leveling.

While the causes of these problems and the formulating approaches necessary to overcome them pertain to nitrocellulose lacquers, similar problems can occur with other types of coatings and the approaches for eliminating these problems also have merit.

Moisture blush is caused by condensation of moisture in a lacquer film during drying, particularly under conditions of high humidity. To eliminate this problem, (a) reformulate and replace part or all of the fast-evaporating solvent with a slower-evaporating solvent having a higher blush resistance value (see Table 18.3); (b) keep fast-evaporating alcohols to a minimum since the blush resistance of a lacquer drops off rapidly beyond 15% concentration of these latent solvents; (c) add a retarder such as butyl "Cellosolve."

Nitrocellulose or cotton blush is caused by improper balance of active solvent to diluent during evaporation and the nitrocellulose is precipitated from solution. To eliminate this situation, (a) reduce diluent content; (b) replace diluent with one having a faster evaporation rate; (c) reformulate using slower-evaporating active solvents or add a small amount of a high-boiling solvent with high diluent tolerance.

Resin or gum blush indicates incompatibility of lacquer solids in the solvents used. When gum blush occurs, check the compatibility of the solids components in equal parts of n-butyl acetate and toluene. If the solids are compatible, check each component of the solvent system and then reduce the concentration of the component causing the blush. Resin blush can frequently be eliminated by the addition of a small amount of butyl "Cellosolve."

Flow and leveling problems often are a function of the volatile portion of a lacquer. When a lacquer is sprayed, the wet film is made up of small particles of lacquer which must flow together to provide a smooth, glossy film. Incomplete fusion of these particles, convection currents set up during drying or a nonuniform shrinkage of the film can cause a rough surface and reduce gloss. Flow and leveling can be improved by formulating the lacquer with slow-evaporating solvents. However, care must be exercised not to use too high a concentration of the slow-evaporating solvents, otherwise sagging and running of the coating may occur or an undue increase in drying time will result.

Vinyl Copolymer Resins

Some of the most important vinyl resins used in solution coatings are the vinyl chloride–vinyl acetate copolymers. The solvents used in formulating coatings based on these copolymer resins can be divided into active solvents and diluents. The most commonly used active solvents are the ketones, with other selected solvents such as tetrahydrofuran, 2-nitropropane and dimethylformamide also employed. The aromatic hydrocarbons are used as the diluents. Alcohols are generally avoided since they tend to precipitate the resin from solution or cause viscosity increase during storage. Aliphatic hydrocarbons can be tolerated only in limited concentrations.

The preferred solvent systems for vinyl lacquers are blends of ketones and aromatic hydrocarbons, with the choice and amounts of these components being dictated by application requirements and current economics. Some typical solvent systems are illustrated in Table 18.7.

TABLE 18.7. Typical Solvent Systems

	Spray			Roll Coat
	A	*B*	*C*	*D*
Methyl ethyl ketone	45	—	—	—
Methyl isobutyl ketone	—	50	—	—
n-Nitropropane	—	—	32	—
Cyclohexanone	5	—	—	—
Isophorone	—	—	—	50
Toluene	50	—	68	—
Xylene	—	50	—	—
"Solvesso 100"	—	—	—	50

Acrylics

Acrylic resins used in protective coatings are either poly(methyl methacrylate) or copolymers of methyl methacrylate with acrylic or methacrylic acids, or their higher esters. These resins can be used alone or with modifying resins such as nitrocellulose, cellulose acetate butyrate or vinyl chloride–vinyl acetate copolymers. The acrylic resins are soluble in a variety of solvents. In general, the solubility of the resin becomes more difficult as the hardness or the methyl methacrylate content of the acrylic resin is increased. Solvents for the acrylic ester resins are ketones, esters, aromatic hydrocarbons and chlorinated hydrocarbons. The aliphatic hydrocarbons and alcohols are nonsolvents for most acrylic lacquer resins, but they may be used in small amounts to impart special performance properties.

Thinners for acrylic lacquers generally consist of low-boiling ketones to provide maximum solids at application viscosity, aromatic hydrocarbons to improve economics and a high-boiling solvent such as "Cellosolve" acetate important to flow-out and leveling of the coating. Some typical thinner formulations for acrylic lacquers are illustrated in Table 18.8.

TABLE 18.8.   Typical Thinner Systems

	Parts by Weight			
Acetone	40	15	—	20
Methyl ethyl ketone	—	15	20	—
Methyl isobutyl ketone	—	—	10	20
"Cellosolve" acetate	20	15	10	15
Ethanol	—	10	10	5
Toluene	40	45	50	40

Epoxy

The selection of solvents in formulating epoxy coatings is influenced by the type of coatings. Epoxy resins are used in surface coatings in three principal ways: esterifying with various vegetable acids and subsequent curing by oxidation or heat polymerization, in combination with urea or phenol-formaldehyde resins as heat-converted coatings, and as coatings converted with amines or polyamide resins.

The solvent systems for epoxy esters are generally aliphatic and/or aromatic hydrocarbons with minor amounts of alcohols sometimes being required with the short oil epoxy esters. Solvent systems used with the urea or phenol-formaldehyde converted epoxy coatings are based on ketones, alcohols and aromatic hydrocarbons. A typical solvent system might consist of equal parts of methyl isobutyl ketone, *n*-butanol, toluene and xylene. The amine-cured epoxy coatings are two-package systems which are mixed just prior to application. Solvents useful in preparing the amine catalyst solution are butanol, "Cellosolve" solvent and aromatic hydrocarbons. A typical solvent system for the epoxy resin portion of the coating could consist of 45 parts methyl isobutyl ketone, 5 parts butyl "Cellosolve" and 50 parts by weight of toluene.

19

*Pigments—General Classification and Description**

*By Robert S. Radcliffe, The Dimlich-Radcliffe Co., Cleveland, Ohio.

Part A. PIGMENT CLASSIFICATION

Definition of Paint

Paint has been described as "a mixture of pigment with vehicle, intended to be spread in thin coats (films) for decoration or protection or both."[1] Such a definition encompasses a wide variety of decorative and protective coatings such as dry powder paints, lacquers, emulsion paints, certain paper coatings, metal primers and anti-corrosion compounds. Fundamentally, all such mixtures are alike in that they contain a pigment —generally a solid, insoluble in the binder and its diluent—and a binder, which can hold the pigment to the surface to be covered and can be spread to a thin film either because it is a liquid or because it can be converted into a liquid-like phase through solution, emulsion or chemical reaction. The dilution agent or solubilizing agent of the binder is such that on exposure in thin films to the air, it will evaporate and leave the film. A paint then is a three-phase system of pigment, binder and diluent.

In the parlance of the paint industry, paint is a three-phase system of pigment, binder and thinner. Because of the great diversity of available pigments, binders and thinners, the student or formulator is well advised to keep in mind the possible physical and chemical changes that obtain as the paint is manufactured, held in the wet state and spread to a film, and as it dries to a hard film and carries out its functions of decoration and protection. Knowledge of the interaction of the three phases during these positions is fundamental to the full appreciation of paint technology. One further point, there are organic coatings now being developed which are applied with both binder and pigment as solids. The film is formed on application of heat.

The Functions of Pigments in Paint

The obvious first function in paint is to decorate or obscure the underlying surface to which the paint is applied. It is this property that distinguishes a paint from a varnish. The second function is to protect the surface from the ravages of time and the elements, including chemical reactions. It is a rare case when the durability of an organic binder is not enhanced by the addition of pigment.

However, there are many other reasons for the incorporation of pigments into the system which are more subtle, but must be known and understood by the student and paint formulator. The first of these is to produce consistency[14] in amount and type adequate to make a paint which may be applied by the procedure selected to form a film of desired specifications. Thus, pigments are selected to insure adequate film thickness with the right flow and application properties. Pigments are also added to control, as desired, the amount of penetration of the vehicle into the substratum.

Paints are manufactured which produce various decorative effects. For example, there are paints whose dried films are of high gloss—enamels—and those with extremely low angular sheen—flat finishes. In between, there are "gloss," "semi-gloss," "eggshell," "flat enamels" and many other finishes. Most of these effects are obtained by the manipulation of the type and amount of pigment incorporated with the binder.

Specific pigments or combinations of pigments can impart to the paint film a degree of chemical resistance; an example is the incorporation of zinc chromate to induce corrosion resistance.

The practical formulator, of course, must maintain a basic interest in the raw materials cost of paint. To do this, he must manipulate the amount, type and cost of his pigmentation to be certain that the desired film can be produced with the most efficient use of pigments of the lowest possible cost.

The reader can imply from the above discussion that while the primary function of pigments in paints is to obscure and protect the surface to which the paint is to be applied, they have a great many other uses in commercial paints. It can also be implied that the solid phase part of paint may require a wide range of types of pigment, whether or not they affect the paint obscuring power, the decorative effect or the durability.

Types of Pigments

Pigments, the solid phase of modern organic protective coatings, fall into three general types: (1) Colors are pigments that absorb certain wavelengths of the light falling on them and reflect others. The wavelengths reflected to the eye produce the sensation of color.[24] (2) The white pigments—light entering them is refracted according to their refractive index and, except for minor absorption, is reflected completely to the eye as white, providing the index of refraction of the pigment is greater than that of the surrounding medium. (3) Metallic powders and dusts constitute the third basic type of pigment.

The colors such as blue, green, yellow and black must exhibit some hiding power. Generally, their insolubility in the binder and thinner of the paint distinguishes them in this respect from dyestuffs.

White pigments are divided into two general classes—*prime* pigments and so-called *supplemental* pigments. Prime pigments display hiding power when completely immersed in normal paint binders. Their index of refraction must be in excess of 1.5. Supplemental pigments appear white when exposed to air with its index of refraction of 1.0, but they do not exhibit hiding power when completely wet by normal paint binders. They have indices of refraction of 1.5 or less.[21]

Principles of Hiding Power

The hiding power of a white pigment is determined by the difference between its index of refraction and the index of refraction of its surrounding medium.[21] Hiding power is proportional to the square of the difference between the index of refraction of the pigment and that of the surrounding medium. The indices of refraction of the common oleoresinous binders are about 1.50. Any white pigment having an index of refraction in this neighborhood or below has no hiding power when completely covered with such a binder.

Calcium carbonate, though a white pigment in air, has an index of refraction of 1.60 and thus exhibits little, if any, hiding power in certain paints. Rutile titanium dioxide, however, has an index of refraction of 2.76 and has the highest hiding power of any white pigment. The common prime white pigments are basic carbonate white lead, basic sulfate white lead, zinc oxide, zinc sulfide, antimony oxide and titanium dioxide. Representative supplemental white pigments are: barium sulfate, calcium carbonate, magnesium silicate, calcium sulfate, aluminum silicate and silica.

While the above principles cover the theoretical basis for the hiding power of white pigments in paint, the actual application of these principles in the formulation of paints is quite involved.

First, it is assumed that pigment is completely covered by the binder. Such a situation exists in all paints in the wet film and is known as "wet" hiding power. If the paint is so formulated that in the dried film, a part of the pigment extends above the binder film into the air, white pigments with low indices of refraction can show hiding power. Such a condition exists in flat paints which show a low angular sheen, and supplemental pigments now show hiding power because the medium in which they exist is air with an index of refraction of 1.0. Such hiding power has been called "dry flat hiding."[29]

In the case of an enamel which has a high angular sheen or gloss by definition, the pigment is generally fully covered by the organic binder in the dried film and the true hiding—or "dry" hiding power—is exhibited. Paint formulators set up their products to show adequate dry and wet hiding to assure the desired coverage of substratum, and they use "dry flat" hiding to produce an economical paint with high light diffusion.

As has been noted, paint must be manufactured, stored in the wet state, applied in the wet state, allowed to harden or dry, and finallly judged for its value as a dry film. Consequently, to obtain efficient hiding power from pigments, they must be dispersed, held in dispersion through the wet states, and not allowed to agglomerate to larger particles exhibiting inefficient hiding power.[9] There is no point in adding more pigment to the paint than is necessary to do a commercial covering job.[16] The film thickness of the paint is generally an index of its durability during exposure to the elements. If the film to be applied is thick or thin, the hiding power of paint should be governed by the amount and kind of hiding pigment to be used in the product. At the present time, a great many research programs are under way to establish information on the effect of pigment particle size and size distribution on the hiding power of prime pigments.[9] The dispersion of the hiding pigment and its maintenance in a dispersed state may call for the use of a supplemental pigment.

The Hiding Power of Paint

Though we have spent considerable time discussing the hiding power of pigments, the important point is the hiding power of the paint. While comparative hiding powers of pigments—whites particularly—have been widely studied and compared, the formulator can fall into critical error by assuming generalized values for hiding power of various white pigments and by assuming fixed hiding power values for a pigment in different paint formulas. Care should so be taken to maintain a desired balance between wet hiding power and dry hiding power. In paints to be tinted, these values should be kept close to each other.[2] Where ultimate use can allow the development of dry flat hiding power, the economical paint formula demands such development.

With this introduction, we shall proceed to a more definite description of pigments. Future chapters will deal with colors and white pigments. Part B of this chapter will discuss supplemental pigments in some detail.

Part B. SUPPLEMENTAL PIGMENTS

In 1948, the author prepared an article on this group of white pigments.[26] He insisted on the term "extender pigments" to designate the group. Previously, these pigments had been referred to as fillers, inerts and extenders. "Filler" was rejected because it implied adulteration. "Inert" was rejected because it implied chemical inertness. "Extender" and "extender pigments" are now rejected because these terms imply that these pigments are used only to extend the value of the prime white pigment or color. In present-day paint formulas, supplemental pigments are actually included in the formulas for specific properties they can add to the paint, including hiding power. The writer has no quarrel with the use of the terms filler, inert, extender or extender pigment, but feels that at this time a more proper and tighter designation should be given to this group of pigments.

Definition

Supplemental pigments are white pigments of low refractive index, reasonably nonreactive chemically, and generally of lower cost, which may be added to paint formulas to enhance the hiding power of white pigments and colors, impart specific paint properties to the paint, improve the durability of the paint film, and when possible reduce the cost of the paint.

As an example, we may consider barium sulfate. Years ago, it was used as filler to adulterate basic carbonate white lead because of its high specific gravity. Because of its chemical inertness, it was considered a representative inert. But, in its precipitated form, blanc fixe, the pigment's tendency to carry with it occluded water-soluble salts, makes it a questionable inert. There is no question that in lithopone, it enhances the hiding power of zinc sulfide and therefore is an extender. Today, the largest use of barium sulfate in paints is in automotive primers where its unique property of imparting excellent sanding characteristics to the primer becomes the main reason for its use. Thus, it is a supplemental pigment.

History

Prior to about 1920, i.e., before the advent of light-stable lithopone and extended titanium pigments, the fundamental white pigments were basic carbonate white lead, basic sulfate white lead and zinc oxide. Supplemental pigments were considered adulterants and frowned on for use in quality paints.

In 1911, Gardner[12] published an extensive work on the durability of paints which had been adulterated with supplemental pigments and showed that such adulteration had actually improved the quality of the paints. The New Jersey Zinc Co. carried out extensive research on outside house paints and suggested in 1924 a formula of 40% lithopone, 40% zinc oxide and 20% talc.[12] In 1930, Krebs Pigment and Chemical Co. promoted a formula having a basic pigment combination of 40% lithopone, 45% leaded zinc oxide, and 15% talc.[15] By 1935, the results of numerous house paint exposure studies, started in the late 1920's or early 1930's, began to be published.[8] The many studies which have followed have examined the place of white pigments and supplemental pigments in outside house paints for 30 years.*

With the advent of light-stable lithopone—a pigment about 30% zinc sulfide and 70% barium sulfate—the progress toward white pigments of higher hiding power began. About the same time, barium-based titanium pigment—25% titanium dioxide, 75% barium sulfate—was brought out. These pigments were followed by calcium-based titanium pigment (30% TiO_2, 70% $CaSO_4$), titanated lithopone (15% TiO_2, 85% lithopone), 50% lithopone (50% ZnS, 50% $BaSO_4$), and much later 50% calcium-based titanium pigment (50% TiO_2, 50% $CaSO_4$). Pure zinc sulfide as a pigment followed the development of light-stable lithopone.

During this period from about 1920 until 1930, pure titanium dioxide was being slowly developed into a satisfactory commercial product. From 30% lithopone to pure (anatase) titanium dioxide, there was about a five-fold increase of hiding power.[18] And, it should be noted that supplemental pigments, namely, barium sulfate and calcium sulfate, were part of the all the prime pigments based on zinc sulfide and titanium dioxide until these two high-hiding power pigments were available as pure products commercially.

The obvious cause for the development of white prime pigments with increasingly higher hiding powers was to obtain paint films which could obscure the background with thinner films of paint (or fewer coats of paint) which led to shorter production time in application and drying and therefore lower cost of painting. Strangely enough, more durable paints were to be made with these new pigments.

As shown by Madsen,[18] the sales growth of pure titanium dioxide began about 1930 and grew very rapidly. With the introduction of pure titanium dioxide, the paint formulator was faced with handling a pigment of three to ten times the hiding power of the whites he had used before. He wanted to take advantage of the much greater hiding power to make products which could not be made with pigments of lower hiding power, and he

---

*The space allotted to supplemental pigments in this volume is inadequate to even list, let alone develop, the specific values attributed to such pigments in outside house paints. For background literature, the reader should consult references 20, 27, 30, 31 and 32.

wanted to develop the potential savings titanium dioxide could afford. He therefore turned to the study of supplemental pigments which could allow the use of titanium dioxide. With the development of rutile titanium dioxide and the more recent chloride process rutile pigments, progress toward this end has continued.[9]

The history of developments in the field of supplemental pigments has been a continuous search for better quality control, improved fineness, more easily dispersed pigments and pigments to answer specific paint problems. These same improvements have constituted the basis for studies by prime pigment producers. However, as titanium dioxide became more popular compared to its own extended pigment and that of zinc sulfide, the supplemental pigments were forced to become better in quality and far more diverse in type.

Around 1933, two products were placed on the market which foreshadowed the development of supplemental pigments to meet the new needs. Both were calcium carbonate pigments produced by two different manufacturers. One manufacturer approached the problem of producing a quality calcium carbonate pigment by water grinding and separation of calcite.[17] A pigment was developed with a maximum particle size of 15μ, and a mean particle size of 2.5μ. The second product was produced in an ingenious way from precipitated calcium carbonate by organic surface treatment.[10]

Shortly thereafter, the first of the "ultrafine" supplemental pigments appeared, first a calcium carbonate pigment,[6] then silica gel pigments.[35] With the advent of these products, low price per pound was no longer criterion for supplemental pigments. Although these new pigments approached the prime white pigments in price and in some cases commanded higher prices, their intrinsic value has made them commercial products of wide use.

With the arrival of the titanium dioxide pigments — both pure and extended — many studies were carried out to establish the relative values of these new pigments versus the older ones with regard to their formulating properties, durability status and commercial economy. When pure titanium dioxide appeared on the market, more emphasis was placed on the study of supplemental pigments to accompany titanium dioxide in paint formulas in order to obtain efficient use of the prime pigment and establish the paint characteristics of the supplemental pigments. No discussion of white pigments would be complete without a brief history of methods for evaluating them.

White Pigment Evaluation Methods

Obviously, the substitution of one pigment for another by weight in paint was the simplest method of comparison. Since the specific gravity or volume yield of white pigments varies considerably (Table 19B.1), a weight substitution did not produce the same volume of paint and, as a consequence, equal volume substitution followed. However, it was found that even when pigments were compared on an equal volume basis they produced substantially different properties such as consistency, durability, enamel holdout, permeability and others. Consequently, a great many studies were made which compared white pigments, pure and extended, with supplemental pigments and which led to the collection of a great mass of data about paint formulation and the behavior of prime and supplemental pigments.

Armstrong and Madsen[2] used "enamel holdout" as their criterion for paint film equality when one pigment was substituted for another. They developed the term "saturation value" using oil absorptions and pigment volume relations to arrive at equality.

Vannoy[32] for his outside house paint studies developed "equal free binder" using equal paint consistency as the criterion for pigment comparison. Asbeck, with Van Loo and Laiderman, developed "critical pigment volume" for comparison. In this work, film permeability was used as the criterion of comparison.[4,5] Asbeck's work was particularly enlightening, since he studied pigment packing, dispersion, fineness of paint grinding and pigment agglomeration. Color uniformity, as a criterion of paint quality, added another approach to obtaining efficient use of the prime pigment and the selection of the best supplemental pigment to be used with it.[11]

The preceding discussion and the literature cited have dealt with solvent-thinned paints. A great deal of evaluation work of a specific nature has been done on supplemental pigments in water-based paints. The basic facts developed above are applicable to emulsion, water-dispersed or soluble resin paints. However, the use of pigments must be accompanied by full knowledge that such paints cannot employ slightly water-soluble pigments and pigments of high electrolyte content, and the pigment carrying properties of the binder[25] must always be considered.

General Properties of Supplemental Pigments

To give detailed properties of the available pigments of low refractive index is beyond the scope of this work. Instead, the basic properties of this group of materials will be presented and those properties which distinguish the individual pigments will be noted.

Refractive Index. The common commercial supplemental pigments have indices of refraction from 1.50 to 1.64 (see Table 19B.1). Prime white pigments have indices of refraction ranging 1.93 to 2.76.[21]

Color. The dry brightness (when compared as dry pigments surrounded by air) varies commercially from about 70 to 98%. Comparisons in the dry state are much preferred over comparisons in oil or other binder since the index of refraction of air is lower.

As pointed out previously, all white pigments do absorb some wavelengths of light. This absorption, caused largely by impurities, produces yellow to brown or blue to gray undertones. Consequently, to define the color of a pigment, the brightness is obtained and values of undertone are established. McCleary[19] used a Hunter Multi-Purpose Reflectometer: The reading for brightness was the green reading; the tint or undertone, the difference between the amber and blue filter readings.

Specific Gravity.   With the exception of barium sulfate, the common supplemental pigments have specific gravities of approximately 2.0 to 3.0. Since paint is sold on the basis of volume, specific gravity is the index of the volume a given weight of pigment will produce. Paint formulators are most likely to use a derivation from specific gravity, namely bulking value. Bulking value is expressed in two ways:

(1) Pounds of pigment needed to produce a gallon, or the specific gravity of the pigment multiplied by 8.33 pounds—the weight of 1 gallon of water.

(2) Gallons produced by each pound of pigment—the reciprocal of pounds per gallon.

One can readily see that pigments of low specific gravity produce higher volumes of paint than pigments of higher specific gravity. This greater yield of paint by supplemental pigments is one of the reasons for their economy.

Particle Size, Distribution and Shape.   It is not the function of this chapter to pursue in detail the fundamental concepts of pigment particle size and its effect on paint formulation and paint properties, nor should this section pursue the effect of such phenomena as dispersion, flocculation, pigment packing, etc. However, certain other aspects of particle size, because they apply more readily to supplemental pigments, must of necessity be discussed.

Pigments of the supplemental group vary in particle size from those which contain a considerable percentage above $44\,\mu$—held on a 325-mesh screen—to those with particle size of $0.04\,\mu$.

As we shall see later, the pigments are made by two general methods: grinding of minerals or ores, and chemical precipitation. Both methods are open to critical control and sophisticated manipulation. Generally speaking, chemical precipitation produces pigments with narrower ranges of particle size. Figure 19B.1 will show how one producer of ground ore pigments has commercially classified the product of a grinding system into a group of products, graded by particle size.

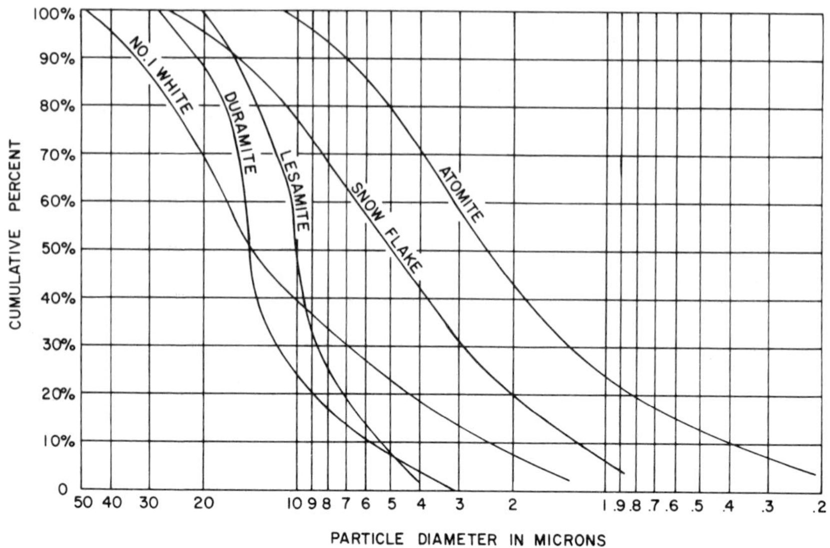

Figure 19B.1.   Typical particle size distribution curves. (*Courtesy of Thompson, Weinman & Co.*)

In references cited earlier, the student and formulator is advised to consider supplemental pigments for use in paint not only by average particle size, medium particle size or ultimate particle size, but by particle size distribution, since this distribution may be vital to obtaining the paint and film properties desired.

With regard to shape, supplemental pigments are spheroid, micaceous, needlelike or fibrous. It should also be stated that some pigments are relatively porous in the size distributions at which they are sold commercially. Some are tight crystal conformations while others are relatively amorphous.

Reactivity. Reactivity with binders is an important property of all white pigments. Since a wide range of binders—both solvent thinned and water thinned—is involved, a careful scrutiny of reactivity must be made.

Reactivity is derived either as part of the pigment or through impurities present. The reactivity of materials such as MgO or CaO with oleoresinous vehicles is obvious. Trace amounts of ionizable salts such as sulfates, chlorides, silicates and phosphates can have decided effects on the stability of emulsions, the water resistance of paint films, mixing and other properties.

It is impractical for the paint formulator to analyze for such trace impurities, but information about the order of magnitude of such impurities can be helpful. pH determinations of water slurries of the pigment are common testing procedures. McCleary[19] developed a conductivity test of basic importance.

Reactivity in itself does not eliminate a pigment, but use without knowledge of its reactivity can be dangerous, particularly when supplemental pigments are involved. Similarly, reactivity can be used to advantage in proper formulations.

Binder Demand. In the historical and pigment evaluation sections above, it was shown that supplemental pigments have assumed primary importance in paint formulation because of the increasing use of pure titanium dioxide as the source of white prime pigment. It has also been shown that extended tests have been conducted by numerous investigators to compare these pigments, particularly when combined with pure titanium dioxide, with regard to the qualities they produce in paints and paint films. A perusal of the literature indicates that these workers were seeking a quick, reliable and simple test for comparing the amount of binder these pigments require to make a given paint of specified characteristics. To date, no satisfactory test to establish this "binder demand" (which will be applicable to solvent-thinned and water-thinned paints) has been developed.

Oil Absorption. Of the tests which have been considered, oil absorption[13] is by far the most widely used and remains probably the most useful. The two tests most commonly employed both require a specified linseed oil and explicit methods of determining the amount of oil a given weight of pigment will absorb under specific hand manipulation. Since both tests use as an "end point" a certain paste condition, the procedure does aid in developing the amount of linseed oil to wet the pigment.

The oil absorption is expressed in pounds of oil necessary to wet 100 pounds of pigment, but it is probably better expressed by a volume relationship, i.e., volume of oil to wet a volume of pigment.

Oil absorption can be used quite well to predict the consistency a given pigment will produce in an oleoresinous paint. This is particularly true when pigments of the same chemical composition and method of manufacture are involved, e.g., pigments made by grinding calcite calcium carbonate. Armstrong and Madsen[3] have pointed out the necessity of running the tests on the composite mixture of the pigments to be used rather than adding the individual oil absorptions to arrive at a figure for the composite. Patton[23] has calculated an interesting set of results to establish the general apparent pigment packing status of pigments at the end point of oil absorption. This "loose packing" factor has essentially the same character as the pigment dispersions on which the modified Stormer viscometer method of determining consistency was based.*

A great many workers have shown that the relationship of oil absorption test to critical pigment volume concentration, while tenuous, does warrant the assumption that the test may be used as a first trial index of binder demand.[3-5] Thus oil absorption can indicate the binder needed by comparative pigments to establish relative factors of consistency, "enamel holdout," gloss, sheen, water penetrability, uniformity of tints over porous surface and other properties of the paint film.

On an economic basis, this test can indicate how much pigment may be crowded into the binder to give a cheaper raw material paint. Many supplemental pigments are cheaper per gallon of paint made than binder. Conversely, the high oil absorption pigments produce paints of thicker consistency, allowing larger amounts of less expensive thinner to be added to the paint for application.

Importance of the Test Paint Grind. While oil absorption has been used with reasonable success, it does have decided weaknesses. Reproducible results are hard to obtain because of the difference of pressure used from one operator to another. This test measures the linseed oil absorption but does not measure the amount of the actual binder to be used. It works well with hydrophobic pigments but not with hydrophilic ones. With water-based or water-thinned paint systems it has little value. It does not take into account the solvent in the paint system. In many cases, the binder cannot be used without solvent. In some cases, the thinner can produce wetting. More important, this test gives no indication of the amount of energy, as mixing and grinding, needed to produce a satisfactory paint.

For basic comparative results, there is no shortcut to avoid the test paint mix and grind. The test paint should be made in test equipment and by procedures which can be duplicated in plant equipment. It is believed that no matter how well the test paint appears to follow predetermined data and meet the specifications required, at least two further test paints should be made at lower and higher pigment volumes. Plant errors, economy of pigmentation, economy of binder, stability of consistency and film properties can be brought to light by these test paints which may, in the long run, be of inestimable value.

---

*The writer is familiar with the original work by which the Stormer was modified to determine consistency. Paints were tested by spatula stirring for body. Ten paints representing the thinnest body through the thickest body found on the commercial market were selected. All paints were pigmented by the same amount and type of pigment and the same vehicle, and all had what appeared to be true "plastic" flow. The term "consistency" was selected to denote Bingham[7] plastic flow as differentiated from Newtonian flow or viscosity.

Manufacture of Supplemental Pigments

There are two basic processes for the manufacture of supplemental pigments, namely, grinding of ores and chemical precipitation. As indicated earlier, in recent years these processes have become more and more sophisticated as the quality of such pigments has been improved to meet the requirements of the paint industry. The basic methods of production have been maintained, but process changes brought about by new equipment and new quality control methods have led to improved pigments. Some pigments such as calcium carbonate, barium sulfate and calcium sulfate are made by both grinding and precipitation.

In recent years, a highly sophisticated silica pigment has been placed on the market.[34] This pigment is made by the oxidation of silicon tetrachloride by gaseous reaction. This pigment has developed a sound market in the coatings and allied industries despite a relative high price.

One other processing step is used with some pigments, namely, calcination. The pigment at some stage of manufacture is heated to high temperatures. The calcination process tends, generally, to produce some crystal growth or tightening of the crystal structure. This generally lowers the oil absorption of the pigment and lowers the tendency of the particle to absorb to its surface the normal air-moisture coating, leading to easier mixing with binders.

Pigments from Ores. The basic operations used in the production of supplemental pigments from ores are mining or quarrying of the ore, followed by crushing, selection for quality and benefaction, grinding and separation of finely ground pigment, and packaging. Two general processes are recognized—based on the method of collecting the finely ground pigment—namely, water grinding and separation, and air grinding and separation. Both methods are employed and can produce highly satisfactory materials.

In the dry or air grinding and separation process, the benefeted crushed ore is placed dry in a mill which is exhausted by a controlled vacuum. As the ore reaches a predetermined particle size, the vacuum lifts the particle from the mill to a collection system from which it is packaged. In the water grinding, the benefeted crushed ore is ground in a rotating mill with water. The pigment in slurry form is conducted to an elutriation settling system or a centrifugal system which separates the pigment, when ground to a specified fineness, from the coarser material which is drawn off and returned to the mill for further grinding. After grinding and separation, the pigment is dried, disintegrated and packaged.

Both processes have been highly improved and every year become more specifically controlled to obtain uniform quality products. The use of water in grinding has the advantages of removing water-soluble salts and reactive products from the finished materials. Also a sharper cutoff of top size particles is possible by water grinding. At commercial production levels, air grinding has not yet been able to produce a pigment as tightly controlled. For example, a water-ground and separated calcium carbonate pigment with a top size of 25μ and a mean size of 5μ can be produced with no particles above 25μ. But, when air grinding and separation are used, a residue of 0.2 to 0.5% above 25μ remains although the mean particle size is 5μ. The formulator is advised to remember that the most sophisticated methods of air grinding and separation may still produce products with a very small but definite amount of material remaining on a 325-mesh screen.

Pigments by Precipitation.   The manufacture of supplemental pigments by precipitation is a straightforward chemical reaction followed by filtering, washing, drying, disintegration and packaging. The precipitation is highly controlled to obtain a definite particle size or type. The filtering and washing are tightly controlled to assure low reactivity and water-soluble salts. The drying and disintegration are regulated to maintain a dry state condition and to keep all particles below the specified top size desired. Where calcination is required or used, further restraints are set up. So much technology and quality control go into the manufacture of modern supplemental pigments that the formulator should not assume that competitive products will perform alike in all paint formulas; wherever possible or practical, comparison of test paints should be made. Each year, pigment vendors increase the amount of information available on the manufacturing of their products, indicating the changes that improve the properties of these products. This literature should be closely studied and consulted.

Surface Treatments.   Surface treatments—either organic or inorganic —have not been widely adopted by the manufacturers of supplemental pigments. Such treatments are decidedly specific, and pigments thus treated should be used with full knowledge that the treatment limits their possible use. Such treatment is applied to wet-ground, dry-ground or precipitated products.

The Major Supplemental Pigments

Barium sulfate appears in the paint industry in two forms, namely, as ground barytes and the precipitated form blanc fixe. Barytes as a ground ore product is sold both as a crude ore and as a "bleached" product. In bleaching, the ore is treated with acid to remove certain impurities such as iron compounds, thereby improving the color of the product. Recent advances in grinding procedures have led to pigments of good fineness and good color, compared to older "crude" or "bleached" products.

Blanc fixe, while a pigment of excellent color and small particle size, has never been used in large tonnages in the paint industry as a distinct pigment. Its greatest use has been in the extended pigments of zinc sulfide (lithopone) and titanium dioxide (barium-based titanium pigment). In these cases, its good color and fine particle size enhance the hiding of the basic prime pigment.

The high specific gravity and, in recent years, the high price of barium sulfate pigments have curtailed their use. Nevertheless, their low binder demand, softness and good exterior durability do recommend them. Perhaps their greatest use now is in the formulation of sanding primers, particularly automotive types.

Calcium carbonate pigments (known as whiting, calcite, ground limestone and chalk) represent the largest group of supplemental pigments of a single chemical composition. Products from grindings of ore and precipitation are produced in a wide range of particle sizes and brightness, for specific uses. A whole family of products made by grinding calcite mineral to oolitic limestone is available. Selected calcite products of 95.0% brightness and top particle size of 15μ are sold. Coarse grades well above 44μ and free of fine particles are made and used. Despite their origin, many of these products are at least 95% calcium carbonate.

The precipitated group of calcium carbonate pigments is almost as varied as the natural types. Here, control of the precipitation reaction develops pigments of varied particle size and excellent brightness, as well as binder demands over a wide range.

Surface treatments can be applied practically to these pigments. Since many of the pigments are made from by-products of other chemical reactions, their selling prices are generally low. Since no composite pigment has ever been made with the prime·white pigments, such as the extended pigments of barium sulfate and calcium sulfate, very little has been published concerning the ability of the calcium carbonate pigments to enhance the hiding power of the prime pigments.

Calcium carbonates are the basic supplemental pigments for practically all types of paint from dry powder paints through emulsion paints to house paints. They are basic ingredients for putty, caulking compounds and sealants. In outdoor house paints, they act to inhibit mold growth and aid in the maintenance of color in tinted paints. They chalk satisfactorily when not used in excessive amounts and when the relative coarse grades of the natural-type product are used. They are economical because of low binder demand in the natural types and the higher oil absorption of the precipitated grades aided by the addition of more thinner. Their low specific gravity (2.71), low price and high brightness make them attractive.

Calcium sulfate occurs as gypsum ($CaSO_4 \cdot 2H_2O$), plaster of Paris ($CaSO_4 \cdot 1/2H_2O$) and anhydrite ($CaSO_4$). Gypsum is found as a natural ore and the other forms can be made from it. Anhydrite ($CaSO_4$) is generally produced by precipitation. In any form, a high-brightness, low specific gravity pigment can be produced. As a supplemental pigment, it has not been widely used in any form. Its greatest use has been in composite pigments, such as titanium pigment calcium base or Venetian red. In systems carrying water, it tends to go from the anhydrite form to the hydrated gypsum form with some additional crystal growth. In certain oeleoresinous systems, a small amount of water added to the paint system will flocculate the pigment with the resulting development of desirable false body which leads to easy brushing. In outdoor house paints, anhydrite is quite durable, but it has the general failings of a fine particle size pigment. The hemihydrate has been used to absorb moisture and avoid gas formation in metallic paints.

The Silicate Pigments. Silica and the silicates comprise the largest group of supplemental pigments used by the paint industry. Most of these pigments are obtained by grinding naturally occurring minerals, but a few are chemically prepared. Their use in paint formulating is derived from their particle size and shape, their aid in acting as stress relief points in otherwise cracking-type paint films and their ability to produce mat or flat finish paints. The list presented in Table 19B.1 is only an outline of these pigments. To discuss each type at any length would require more space than is available. We shall merely present some very general background material. However, it should be remembered that despite their mineral origin and low cost, these pigments are manufactured to high-quality specifications and represent a real part of present-day paint pigmentation.

The *clays* are essentially hydrated aluminum silicates and are sold in forms from simple, air-dried China clay to highly sophisticated water-washed, bleached and calcined products. Clays are hydrophilic, and thus the largest increase in their use occurred with the arrival of the emulsion or water-thinned paints. They have extremely fine ultimate particle size of platelike structure. In the case of the bentonite-type clays, the particle size is so small that such clays are used largely for the production of thixotropy, consistency increase and as nonsettling agents. The color of the clay depends on its origin and on the sophistication of its benefaction and manufacture. In oleoresinous and outdoor house paints, clays have never been extremely popular, due largely to their more difficult mixing and their inability to perform functions of unique character. Pyrophyllite, an aluminum silicate of larger particle size than clay, has been used as a substitute for talc with very similar durability. While oil absorptions for

TABLE 19B.1. Properties of Supplemental Pigments

Pigment	Approximate Chemical Composition	Refractive Index	Specific Gravity	Bulking Value (gal/100 lb)	Oil Absorption (lbs oil/100 lb of pigment)
Barium sulfate pigments					
Ground barytes	$BaSO_4$	1.64	4.476	2.650	6.0
Blanc fixe	$BaSO_4$	1.64	4.35	2.753	14.0
Calcium Carbonate Pigments					
Precipitated types					
Colloidal	$CaCO_3$	1.63	2.68	4.48	55.0
High oil absorption	$CaCO_3$	1.63	2.68	4.48	40.0
Low oil absorption	$CaCO_3$	1.63	2.68	4.48	17.0
Surface treated	$CaCO_3$	1.63	2.65	4.53	15.0
Natural types					
Calcite					
Water ground	$CaCO_3$	1.60	2.71	4.43	15.0–6.5
Dry ground	$CaCO_3$	1.60	2.71	4.43	9.0
Limestone	$CaCO_3$	1.60	2.71	4.43	10.5
Imported chalk	$CaCO_3$	1.60	2.71	4.43	12.5
Calcium sulfate pigments					
Gypsum	$CaSO_4 \cdot 2H_2O$	1.53	2.35	5.107	21.0
Anhydrite	$CaSO_4$	1.59	2.95	4.070	25.0
Precipitated calcium sulfate	$CaSO_4$	1.59	2.95	4.070	50.0
Silicate pigments					
Silica	SiO_2	1.55	2.6	4.60	25.0
Diatomaceous silica	SiO_2	1.40–1.50	1.95–2.35	6.15–5.10	25.0–150.0
Clay	$Al_2O_3 \cdot 2SiO_2 \cdot 2H_2O$	1.56	2.60	4.60	36.0
Pyrophyllite	$Al_2O_3 \cdot 4SiO_2 \cdot 2H_2O$	1.588	2.85	4.22	
Talc	$2MgO \cdot 4SiO_2 \cdot H_2O$	1.59	2.85	4.212	27.0
Mica					
Phlogopite	$K_2O \cdot 6MgO \cdot Al_2O_3 \cdot 6SiO_2 2H_2O$	1.606	2.5–3.0	4.365	24.0
Muscovite	$K_2O \cdot 3Al_2O_3 \cdot 6SiO_2 \cdot 2H_2O$	1.59	2.5–3.0	4.272	47.5

the clays have been given in Table 19B.1, these values should not be considered valid but merely comparative, since the determination of oil absorption of clays is a questionable procedure.

Diatomaceous silica pigments are made by grinding diatoms, the geological remains of unicellular plants. These pigments induce flatting in paints with small additions. They have a real value where high reflection of light is needed, as in traffic paints. The reader should refer to Armstrong and Madsen's[2] findings as to the binder demand of such pigments when mixed with other supplemental pigments.

Chemically, *mica* is a hydrated potassium aluminum silicate. Two forms are used: muscovite and phylogopite. Both dry- and water-ground types are available. The platelike structure of mica is unique, and consequently it has been widely used to eliminate cracking in outside house paints and other products whose normal failure is of a cracking nature. Mica has also been used in emulsion-type paints to prevent cracking and to give improved brushing characteristics.

The *talcs* are hydrated magnesium silicate pigments. In many ways, they have been the "workhorse" of the supplemental pigment industry. The early outdoor paint work showed the value of these materials in preventing cracking. This factor was particularly important when all the other pigments of the formula were of small particle size or reactive. The flatting effect of talc has been used for years. Talc pigments occur in two particle shapes—micaceous and fibrous. Commercial grades contain some of each shape. The manufacturers of talc have been in the forefront in improving the fineness, mixing and grinding of their products. Magnesium silicate has become the supplemental pigment for white outdoor house paints, but its tendency to chalk with a white chalk has limited its use in tinted outdoor house paints.

Silica in the past was not an important pigment. It was made by grinding quality deposits of quartz. It was used largely in wood fillers and wood undercoaters to give good sanding characteristics and "tooth." However, with the development of the high-purity silica gels, a new group of supplemental pigments became available. These micron-sized pigments have been used as flatting agents and to increase consistency and, at times, thixotropy. While high in price, they nevertheless have developed a real commercial market for themselves. One product is designed to stop "gassing" of metallic paints; another of extremely fine particle size is made by the oxidation in gaseous reaction of silicon tetrachloride.

Specific Supplemental Pigments. We have not attempted to describe or list such specific supplemental pigments as magnesium oxide, magnesium carbonate, barium carbonate, lime and aluminum hydrate which are used as reactants for increased consistency or false body. The stearates (zinc and aluminum) and the waxes (castorbean, japan wax and others) are too specific in their use to be discussed here.

The Future of Supplemental Pigment Technology

In reviewing the developments in the technology of supplemental pigments over the past few years, one is amazed at the problems which have been solved by the paint industry. However, a review of the literature and the industry indicates likely areas for future development. The first of these is a wider study of the new ultrafine supplemental pigments. A gap in paint industry research is the small amount of work done to show the value of supplemental pigments in small quantities. Durability tests have been carried out with paints pigmented with 15% or more supplemental pigments, but little has been done on the durability value of such pigments at low concentrations. The future of supplemental pigments is, of necessity, very bright.

REFERENCES

1. American Society For Testing Materials, Book of Standard, 1936.
2. Armstrong, W. G., and Madsen, W. H., *Offic. Dig. Federation Paint Varnish Prod. Clubs* (June 1947).
3. Armstrong, W. G., and Madsen, W. H., *Ind. Eng. Chem.*, **39,** 944 (August 1947).
4. Asbeck, W. K., and VanLoo, M., *Ind. Eng. Chem.*, **41,** 1470 (July 1949).
5. Asbeck, W. K., Laiderman, D. D., and VanLoo, M., *Offic. Dig. Federation Paint Varnish Prod. Clubs* (March 1952).
6. Bates, W. C., and McClure, R. R., (to Diamond Alkali Co.), U. S. Patent 2, 141, 458 (1938).
7. Bingham, E. C., "Fluidity and Plasticity," New York, McGraw-Hill Book Co., 1924.
8. Broeker, J. F., *Offic. Dig. Federation Paint Varnish Prod. Clubs* (June 1935).
9. Bruehlman, R. J., Thomas, L. W., and Gonick, E., *Offic. Dig. Federation Soc. Paint Technol.* (February 1961).
10. Church, J. W., and McClure, R. R., (to Pure Calcium Products Co.), U. S. Patent 2, 034, 797 (1936).
11. Fremgen, R. D., Melester, M. T., and Hilliard, D. A., *Offic. Dig. Federation Paint Varnish Prod. Clubs* (April 1956).
12. Gardner, H. A., "Paint Technology and Tests," New York, McGraw-Hill Book Co., 1911.
13. Gardner, H. A., "Physical and Chemical Examination of Paints, Varnishes, Lacquers and Colors," New York, McGraw Hill Co., Tenth ed., p. 289, 1946.
14. Green, Henry, and Melsheimer, L. A., "Protective and Decorative Coatings," Vol. IV, p. 106, New York, John Wiley & Sons, 1944.
15. Krebs Pigment and Chemical Co., Formula "K," Bulletin 1, August 1930.
16. Lightbody, A., and Dawson, D. H., *Ind. Eng. Chem.*, **34,** 1452 (December 1942).
17. Lukems, A. R., (to Thompson, Weinman & Co.), U. S. Patents 2,323,550 (1943).
18. Madsen, W. H., "Paint and Varnish Technology," Ch. V, p. 65, New York, Rheinhold Publishing Corp., 1948.
19. McCleary, R. L., *Offic. Dig. Federation Paint Varnish Prod. Clubs* (July 1946).
20. McCleary, R. L., and Kummer, P. E., *Offic. Dig. Federation Paint Varnish Prod. Clubs* (October 1953).

21. McMullen, E. W., and Ritchie, E. J., "Protective and Decorative Coating," Vol. IV, p. 141, New York, John Wiley & Son, 1944.
22. New Jersey Zinc Co., *Paint, Oil, Chem. Rev.*, 77, No. 25 (June 18, 1924).
23. Patton, T. C., "Paint Flow and Pigment Dispersion," p. 184, McGraw-Hill Book Co., 1964.
24. Paul, M. R., "Protective and Decorative Coatings," New York, John Wiley & Son, Vol. IV, p. 95, 1944.
25. Radcliffe, R. S., "Protective and Decorative Coatings," New York, John Wiley & Son, Vol. III, p. 475, 1943.
26. Radcliffe, R. S., "Paint and Varnish Technology," Ch. IV, New York, Reinhold Publishing Corp., 1948.
27. Ramsay, H. L., *Am. Paint J.* (June 24, 1963).
29. Sawyer, R. H., *Ind. Eng. Chem.*, 6, 113 (March 15, 1934).
30. Vannoy, W. G., "Protective and Decorative Coatings," Vol. III, p. 269, 1943.
31. Vannoy, W. G., *Offic. Dig. Federation Paint Varnish Prod. Clubs*, No. 251, 558 (1945).
32. Vannoy, W. G., *Offic. Dig. Federation Paint Varnish Prod. Clubs*, No. 292, 235 (1949).
33. Vannoy, W. G., *Offic. Dig. Federation Soc. Paint Technol.* (October 1961).
34. Wagner, E., (to DeGussa, Germany), U. S. Patent 2,990,249 (1961).
35. Young, L. O., (to W. R. Grace & Co.), U. S. Patent 2,856,268 (1958).

20

*White Pigments**

The white pigments which are the subject of this chapter are distinguished from the nearly colorless extenders discussed in Chapter 19 by their opacity when dispersed in typical paint vehicles. As large single crystals they are actually colorless and transparent. The characteristic "whiteness" which the eye sees is the result of the diffuse scattering and reflectance of light by a myriad of fine particles and by the fact that they do not selectively absorb any of the wavelengths of visible light. Under red light they appear red; under blue light, blue; under yellow light, yellow. Only when the light by which they are illuminated contains all wavelengths of the visible spectrum can they appear white.

The importance of white pigments results from the fact that they are used not only in white paints but in most colored finishes as well. Many colored pigments do not reflect light at all. They are actually transparent and produce opacity only by virtue of the light that they absorb. As a consequence, all but the darkest-colored paints must contain some white pigment to produce opacity and brightness. Most contain more white than color.

The opacity of white pigments when dispersed in a paint film is due to two factors—particle size and the difference between their refractive indices and those of the paint vehicles in which they are dispersed; the greater the difference, the greater is the opacity produced. If the refractive index of a colorless pigment is essentially the same as that of typical paint vehicles (see Table 20.1), it becomes transparent when dispersed and is called an *extender*—not a white pigment.

The relationship between reflectance and the refractive index of pigment and vehicle is described by an equation which bears the name of its discoverer, Fresnel. In its simplest form this equation may be expressed as:

$$R = \frac{(n_p - n_v)^2}{(n_p + n_v)^2}$$

*F. B. Stieg, Titanium Pigment Corporation, New York, N. Y.

where

R = relative reflectance,
n_p = refractive index of pigment,
n_v = refractive index of vehicle.

The refractive indices for some of the white pigments included in this chapter and those of a few typical extender pigments and paint vehicle components appear in Table 20.1.

TABLE 20.1. Indices of Refraction of Some Common Paint Materials

Material	Refractive Index
Rutile titanium dioxide	2.76
Anatase titanium dioxide	2.55
Zinc sulfide	2.37
Antimony oxide	2.09
Zinc oxide	2.02
Basic lead carbonate	2.00
Basic lead sulfate	1.93
Barytes	1.64
Calcium sulfate (anhydrite)	1.59
Magnesium silicate	1.59
Calcium carbonate	1.57
China clay	1.56
Silica	1.55
Phenolic resins	1.55–1.68
Melamine resins	1.55–1.68
Urea-formaldehyde resins	1.55–1.60
Alkyd resins	1.50–1.60
Natural resins	1.50–1.55
China wood oil	1.52
Linseed oil	1.48
Soya bean oil	1.48

High refractive index is not enough to guarantee opacity, however. The refraction and reflection of light by a white pigment take place at the pigment-vehicle interface, but many interfaces are required to produce opacity in a paint film.

The more interfaces present in a given volume of film, the greater is the opacity of the film. This is why particle size is important to the performance of a white pigment. There is a limit, however, to the process of reducing the size of the individual pigment particles to develop more hiding power. If the particles become too small, they are incapable of scattering light of any visible wavelength.

Generally accepted optical theory has suggested that the minimum particle size should not be less than one-half the wavelength of visible light, or about 0.3 μ. This theory, however, was based upon the assumption of random particle distribution and on a minimum particle separation of 50 wavelengths so that interference effects between particles would not exist. In practice, such separations do not exist between pigment particles,[1] and the optimum particle size for maximum white opacity is more probably in the 0.15 to 0.20 μ range.

The materials in Table 20.1 are listed essentially in the order of their refractive indices. All the white pigments therefore appear in the first grouping. The extender grouping which follows is noticeably in the same range as the vehicle components in the lower half of the table.

OPACITY

The opacity of a white pigment cannot be reported in terms of its refractive index, however, and have any meaning to the paint chemist. He is accustomed to rating such pigments in terms of the number of square feet of "hiding power" that each pound contributes to the coverage of a paint product, or in terms of the "tinting strength," or whitening power, they exhibit when used in conjunction with dark colors such as blue or black.[2] The numbers given in Table 20.2 have been generally accepted in the paint industry for many years,[3] but are actually far from precise, since the opacifying power of any white pigment will vary with its concentration in the paint film.[4]

TABLE 20.2.   Tinting Strength and Hiding Power of White Pigments

Pigment	Tinting Strength	Hiding Power (sq ft/lb)
Rutile titanium dioxide (PSC)	1850	157
Rutile titanium dioxide (conventional)	1750	147
Anatase titanium dioxide	1250	115
50% rutile calcium-base	880	82
Zinc sulfide	640	58
30% rutile calcium-base	600	57
Lithopone	280	27
Antimony oxide	300	22
Dibasic lead phosphite	250	20
Zinc oxide	210	20
35% leaded zinc oxide	175	20
Basic carbonate white lead	160	18
Basic sulfate white lead	120	14
Basic silicate white lead	80	12

Although these hiding power numbers once assigned to white pigments are now recognized as imprecise, they still provide a logical explanation for dividing the group of white pigments into two categories: reactive and nonreactive.

The individual white pigments vary not only in opacity, but also in specific gravity and cost per pound. When listed in the order of the hiding power they are capable of producing *per dollar of cost*, as in Table 20.3, it becomes apparent that the use of at least half of them cannot be justified on the basis of their opacity or whitening power alone. This lower-hiding category contains the *reactive* white pigments; they are employed in paint formulations because of the reactions they undergo with paint vehicles, and/or the properties other than hiding power which they impart to a paint film. The remainder, used solely because of their white opacity, are the *nonreactive* white pigments.

TABLE 20.3. Hiding Power of White Pigments as a Function of Cost

Pigment	Pounds per Gallon	Cost[a] per Pounds (cents)	(A) Cost per Gallon (dollars)	(B) Hiding Power per Gallon	(C) Hiding Power per Dollar
30% rutile calcium	27.1	.09125	2.47	1545	625
Rutile TiO_2 (PSC)	35.0	.27	9.45	5845	619
50% rutile calcium	29.0	.14125	4.10	2378	580
Rutile TiO_2 (conventional)	35.0	.27	9.45	5145	544
Anatase TiO_2	32.5	.25	8.13	3738	460
Lithopone	35.8	.0875	3.13	967	309
Zinc sulfide	33.3	.2875	9.56	1931	202
Zinc oxide	47.1	.1475	6.95	942	135
35% leaded zinc	48.1	.15375	7.30	962	132
B.C.W.L.	56.2	.205	11.42	1001	88
B.S.W.L.	53.0	.19	10.07	770	77
"45X"	33.3	.175	5.83	400	69
Antimony oxide	47.9	.475	22.75	1054	46
"Dyphos" PG	55.5	.62	34.41	1110	32

[a]Approximate.

THE REACTIVE WHITE PIGMENTS

The reactive white pigments are, with one exception, lead or zinc compounds. They are, as a class, the oldest of the white pigments and once held a far more important position in the industry than they do now. This should in no way suggest that they do not fulfill a useful function; it is merely a reflection of the fact that the science of paint manufacture has progressed to the point that pigments of various types are used today to fill only those roles for which they are best suited. A paste white lead house paint in which basic carbonate white lead served 50 years ago to produce opacity, to react with fatty acids and the acid degradation products of the linseed oil vehicle, and to impart a desirable consistency, has been replaced by the so-called ready-mixed house paints in which these three functions may be performed (in that order) by titanium dioxide, basic silicate white lead plus zinc oxide, and extender.

Basic Carbonate White Lead

Physical Properties

Specific gravity	6.7–6.8
Refractive index	2.00
Tinting strength	160
Oil absorption	8–15
Fume proof	no

Basic carbonate white lead is the oldest of all white pigments used by the paint industry; the first known reference to its manufacture dates back to 400 B.C.

Its composition is generally represented by the chemical formula $2PbCO_3 \cdot Pb(OH)_2$, but commercial white lead may vary considerably from this, principally in the amount of normal lead carbonate present. Fundamentally, it is produced by the oxidation of metallic lead to lead oxide, the reaction of the lead oxide with acetic acid to form normal lead acetate, the combination of the normal lead acetate with additional lead oxide and water to produce basic lead acetate, and finally the conversion of this basic lead acetate to basic lead carbonate through treatment with carbon dioxide gas.

The early Greeks performed this process by simply suspending metallic lead in an earthenware pot over vinegar and waiting for the white lead to form, the carbon dioxide being obtained solely from the air. This was naturally a long process with a highly unsatisfactory yield. All the improvements that have taken place in the intervening 23 centuries have been concerned with speeding up the process and improving the end product as a white pigment:

(1) The Dutch Process (seventeenth century) produced carbon dioxide and heat first from the fermentation of manure, and later of tanbark, in an enclosed space (or "stack") surrounding the lead and vinegar (later, acetic acid) containers, shortening the process to about three months for an 80% conversion of the metallic lead to white lead.

(2) The Carter Process (1885) reduced the time required to 12 days by tumbling finely divided lead (produced by atomizing the molten metal) in a revolving wooden cylinder in the presence of acetic acid, water and carbon dioxide gas.

(3) The Euston Process separated the various steps of the process into individual operations, making more precise control possible—first oxidizing the lead in air, then dissolving off the lead oxide in a normal lead acetate solution, and finally precipitating white lead from the basic lead acetate solution formed by carbonating with carbon dioxide gas.

(4) The Sperry Process was the first to exhibit any great originality and is one of several electrolytic processes for manufacturing white lead. The process is carried out in a concrete cell having a lead anode and an iron cathode separated by a porous diaphragm. The lead anode is surrounded by a solution of sodium acetate, the iron cathode by a solution of basic sodium carbonate. When a current is passed through the cell, lead is dissolved from the anode, forming lead acetate, and hydrogen and hydroxide ions are liberated at the surface of the cathode. The hydroxide ions and carbonate ions pass through the porous diaphragm from the cathode to the anode portions of the cell, where they react with the lead acetate to form white lead. This is a continuous process, and the reaction products are continuously being removed.

(5) The Thompson-Stewart Process is essentially a modification of the earlier Carter Process in which precise control of the reactants is made possible. This process is capable of producing a white lead of lower lead carbonate content than any previously used method, but it may also be varied to produce any desired composition within the range of $4PbCo_3 \cdot 2Pb(OH)_2 \cdot PbO$ and $2PbCO_3 \cdot Pb(OH)_2$.

Basic carbonate white lead is used today almost entirely in exterior house paint formulations, but seldom to the extent of more than 35% of the total pigmentation. It reacts with the fatty acids present in linseed and other oils to form complex lead soaps which contribute to the adhesion, toughness and elasticity of the paint film. It also reacts with the acidic degradation products of these oils to stabilize the film and increase durability. Various grades are available which vary essentially in particle size, the finer-particle grades, as might be expected, having the highest reactivity and the greatest opacity.

Basic Sulfate White Lead

Physical Properties

Specific gravity	6.3–6.4
Refractive index	1.93
Tinting strength	120
Oil absorption	10–14
Fume proof	No

Basic sulfate white lead is a more recent development than is the basic carbonate. It was first produced in 1855 in Birmingham, Pennsylvania, by E. O. Bartlett, although commercial production did not begin until 1876. Its theoretical chemical composition may be represented by the formula $2PbSO_4 \cdot PbO$, but, as with the basic carbonate, considerable variation from this composition may occur in the various grades and types.

Basic sulfate white lead is manufactured by two types of processes, the fume processes and the chemical process:

(1) The original fume process comprised the high-temperature roasting of a selected lead sulfide ore (galena) in an oxidizing atmosphere. In a more recent version of the process, molten lead is sprayed into burning fuel gas in a special furnace into which sulfur dioxide is also fed. In both processes, basic sulfate white lead is produced as a fume and is collected as the white pigment by filtering through large cylindical cloth bags in a "bag house."

(2) The chemical process is somewhat similar to the Carter Process for producing the basic carbonate in that it starts with the finely divided metallic lead, or a mixture of lead and lead oxide, which is treated with dilute sulfuric acid to precipitate the basic sulfate white lead. This process lends itself to somewhat more precise control than the fume processes and yields pigment particles that are needle-like, or acicular, in shape.

Basic sulfate white lead closely resembles the basic carbonate in its properties and uses in the paint industry. White slightly lower in cost and in specific gravity (see Table 20.3), it is also lower in opacity and is not as effective in enhancing the performance of exterior white house paints. Its substitution for the basic carbonate generally results in slightly increased erosion of the paint film.

Basic Silicate White Lead

Physical Properties

Specific gravity	4.0
Refractive index	1.83 (estimated)
Tinting strength	80
Oil absorption	15
Fume proof	No

Basic silicate white lead is a comparatively recent development in lead pigments and represents an entirely new concept of reactive white pigments. It consists of a core of silica which is surrounded by a surface layer of monobasic lead silicate and monobasic lead sulfate. It is produced in a "dry-phase" reaction by heating finely divided silica, lead oxide and basic lead sulfate together in the proper proportions. While no theoretical formula has been proposed, the resulting white pigment is known to analyze from 47 to 48% lead monoxide, 47 to 48% silica, and approximately 4% sulfur trioxide.

Basic silicate white lead was developed at a time when it was apparent that only the *reactivity* of white lead would justify its continued use by the paint industry. As may be seen by a comparison of the positions of white lead and the titanium pigments in column C of Table 20.3, it had been hopelessly outclassed as an opacifying white pigment. Since the lead soap-forming reactions which are the desirable feature of white lead pigments take place only on the surface of the pigment particles, a far more efficient use of the elementary lead is made possible by concentrating the lead compounds on the surface of an inert, low specific gravity silica particle. While lower in opacity than either of the previously described white lead pigments, an equal volume of basic silicate white lead is equally as effective from a reactivity standpoint, and its lower volume cost (see column A of Table 20.3) makes it much less expensive in normal paint formulation.

Dibasic Lead Phosphite

Physical Properties

Specific gravity	6.7
Refractive index	2.25
Tinting strength	250
Oil absorption	14
Fume proof	No

The dibasic lead salt of phosphorous acid is the most recently developed of the lead pigments and has the average formula: $2PbO \cdot PbHPO_3 \cdot \frac{1}{2}H_2O$. It is the only white pigment to possess anticorrosive properties and, although low in opacity, finds considerable application in metal primers and in top coats exposed to a saline environment. It is also an efficient stabilizer for chlorine-containing resins, functioning as an acid acceptor which retards degradation of the coating with age.

Zinc Oxide

Physical Properties

Specific gravity	5.6
Refractive index	2.02
Tinting strength	210
Oil absorption	12–25
Fume proof	Yes

Zinc oxide, like basic carbonate white lead, was mentioned in ancient literature, but does not appear to have been proposed as a white pigment for paint until 1770, at which time Courtois in France noted that it was not discolored by sulfur-bearing gases—a major disadvantage of the basic carbonate white lead which was the only available white pigment at that time. Successful painting tests were made on the interior of a French man-of-war in 1786, but commercial production of the pigment did not begin until 1835.

Two principal methods, the French Process and the American Process, are used to manufacture zinc oxide. The French Process is an indirect one which requires the recovery of pure zinc metal from zinc sulfide ores (zinc blende or sphalerite). The metal is then melted, vaporized and oxidized under controlled conditions.

The American Process, or "direct process," eliminates the step of obtaining the zinc as a metal. A hot furnace is charged with a layer of anthracite, which ignites, and a mixture of coal and ore is then spread over the bed of burning coal. Careful control of reaction times and temperatures and of the cooling rate of the emitted fumes make possible the production of a pigmentary zinc oxide.

Zinc oxide is basic in nature and readily forms zinc soaps with the acid constituents of paint vehicles. This reactivity is utilized by the paint chemist to increase the hardening of the paint film, to increase the consistency of the fluid paint, to aid mixing and grinding of pigments (because of better wetting), to reduce after-yellowing, and to improve the self-cleaning and mildew resistance of exterior house paints. The latter two effects are apparently the result of the slight water solubility of zinc soaps.

Zinc oxide is available in a number of grades, varying principally in particle size, and consequently in opacity and reactivity. In exterior house paints, the less-reactive types are generally preferred, to avoid excessive hardening and embrittlement of the film, while the finest variety (and the most reactive) is commonly used as 5 to 10% of the pigmentation of white architectural finishes to minimize after-yellowing.

Leaded Zinc Oxide

Physical Properties (35% Grade)

Specific gravity	5.9
Refractive index	1.99 (estimated)
Tinting strength	175
Oil absorption	11–15
Fume proof	No

Leaded zinc oxides may contain almost any amount of lead sulfate in the form of mixed normal and basic sulfates. The common commercial grades are 5, 35 and 50% leaded, the 35% grade being most widely used. Because of its greater economy (see column A of Table 20.3), as compared to an equivalent mixture of pure oxide and basic sulfate white lead, this type of zinc oxide is an extremely popular reactive white pigment for exterior house paints.

Leaded zinc oxides are produced by two methods, one being the co-fuming of mixed zinc sulfite and lead sulfide ores in an oxidizing atmosphere. This method produces a leaded zinc oxide of somewhat poorer color than the second method, which consists of simply blending the lead and zinc components after producing them in separate processes.

Antimony Oxide

Physical Properties

Specific gravity	5.7
Refractive index	2.09
Tinting strength	300
Oil absorption	12
Fume proof	No

Although there are three possible oxides of antimony, only the trioxide (Sb_2O_3) is used as a white pigment. It was first manufactured in Europe about 1920 by a process similar to the production of French Process zinc oxide. Either the pure metal or the sulfide may be vaporized in an oxidizing atmosphere to form an antimony trioxide fume, which is cooled and collected in bag houses. The available grades of antimony oxide differ primarily in their content of antimony tetroxide and other impurities, and consequently in color.

Antimony oxide is the only one of the "reactive white pigments" that is not a lead or zinc compound. As a matter of fact, it would not be classed as a "reactive" pigment at all, were it not for the fact that the definition includes those pigments employed in paint formulations because of "properties—other than hiding power—which they impart to the paint film." It may be used in automotive enamels and other types of synthetic exterior finishes to improve chalk resistance. Before the development of the modern, nonchalking grades of rutile titanium dioxide, antimony oxide was quite widely used for this purpose, but today finds application primarily in fire-retardant paints.

THE NONREACTIVE WHITE PIGMENTS

The nonreactive white pigments are used in paint formulations for the production of whiteness in whites, brightness in tints and opacity in all types of finishes. The degree to which they develop these properties in relation to their cost (see column C of Table 20.3, therefore, generally determines their usefulness to the paint chemist. In certain types of formulations in which maximum opacity must be obtained with the minimum amount of pigment, their hiding power per gallon (column B of Table 20.3) may become the deciding factor. Of the two general types

—zinc sulfide pigments and titanium pigments—only the latter have any foreseeable future in the paint industry. It is more than probable that no chemist entering the paint industry today will ever be called upon to formulate a zinc sulfide-containing paint, one major producer after another having withdrawn from the field under the pressure of competition from titanium pigments. The zinc sulfide pigments are included in this classification only because they are still being manufactured in limited quantity and have not yet been replaced in all formulations.

Zinc Sulfide

Physical Properties

Specific gravity	4.0
Refractive index	2.37
Tinting strength	640
Oil absorption	13
Fume proof	Yes

Zinc sulfide was first produced in 1783 in France and used for the first time in paint around 1862. Its full-scale commercial production was delayed, however, until after 1920, when the demand for a pigment having greater opacity than lithopone resulted in renewed interest in pure zinc sulfide as a white pigment. It has been produced by precipitation processes using zinc chloride and either sodium or barium sulfide, by heating mixtures of zinc oxide and sulfur, and by methods involving precipitation from solutions of soluble salts with hydrogen sulfide. This latter process is the one still in limited use today. After precipitation, the zinc sulfide is muffled at about 1330°F, in the absence of air, to develop the crystals to optimum pigmentary size (0.2 to 0.3 μ).

Lithopone

Physical Properties

Specific gravity	4.3
Refractive index	1.84 (estimated)
Tinting strength	280
Oil absorption	12–15
Fume proof	Yes

Lithopone is a composite pigment of zinc sulfide crystals coprecipitated upon a barium sulfate or calcium sulfate base. It was first produced in 1847 in France by a very simple process involving the plunging of barium sulfide into a solution of zinc sulfate. A later step, added in 1874, required roasting of the washed precipitate to increase whiteness, and lithopone had become one of the most important white pigments in the paint industry by 1900.

Lithopone is the first known application of the principle of increasing the efficiency of a white pigment by extending it with an inert, non-opaque base. The most commonly used form of lithopone contained 28% of zinc sulfide and 72% of barium sulfate, yet was able to produce almost 50% as much hiding power as the pure zinc sulfide (see Tables 20.2 and 20.3). This greater "hiding efficiency" of lithopone was actually responsible for delaying the commercial development of pure zinc sulfide for many years, but eventually the demand for more concentrated opacity (see column B, Table 20.3) could no longer be satisfied by a pigment of this type.

Titanium Pigments

Physical Properties (Rutile)

Specific gravity	4.2
Refractive index	2.70
Tinting strength	1750–1850
Oil absorption	16–36
Fume proof	Yes

Physical Properties (Anatase)

Specific gravity	3.9
Refractive index	2.55
Tinting strength	1250
Oil absorption	16–26
Fume proof	Yes

The titanium dioxide pigments are the newest and the most important white pigments in the paint industry. Titanium dioxide was first produced commercially in 1920 in Niagara Falls, New York, but, paralleling the history of zinc sulfide and lithopone, the pure oxide was initially felt to be too expensive and the first widely used titanium pigment was a composite of 30% titanium dioxide coprecipitated with 70% of a barium sulfate base. This pigment, sold under the trademark "Titanox,"* was able to compete successfully with lithopone because of the low cost of the barium sulfate base and because the "built-in" dilution of the inert base enabled it to produce 40% as much hiding power as the pure titanium dioxide.

The second titanium pigment to appear was a 30% "calcium-base" pigment having even greater opacity. This was the result of precipitating the titanium dioxide on an anhydrous calcium sulfate base. The lower specific gravity of the calcium sulfate, as compared to barium sulfate, produced a greater separation of the titanium dioxide crystals in the composite pigment, and thus increased hiding power as the result of the

---

*Registered trademark, National Lead Co.

previously described "dilution effect." This higher hiding pigment made great inroads into the use of both lithopone and the earlier barium-base titanium pigments.

By 1932, pure titanium dioxide had also been established as a practical white pigment; this was followed by a number of modified grades of increasing chalk resistance in exterior finishes, poor chalk resistance having been the greatest weakness of the original titanium dioxide pigment.

All these early titanium pigments were of the anatase crystal-line structure, although it was known that titanium dioxide also existed in a second crystal form usable as a white pigment—rutile. (A third crystal form of titanium dioxide, brookite, also exists, but it has no commercial significance). Although chemically identical, these different crystal forms result in different specific gravities and different refractive indices.

As would be expected from the previous discussion of the hiding power of white pigments, the higher refractive index of the rutile crystal form results in increased opacity—20 to 30% more for the rutile titanium dioxide than for the anatase form (see Table 20.2).

The first commercial rutile titanium pigment appeared in 1941 in the form of a coprecipitated rutile calcium-base pigment composed of 30% rutile titanium dioxide precipitated on a calcium sulfate base; this was followed a year later by a pure rutile titanium dioxide. The greater hiding efficiency of the rutile calcium-base pigment quickly eliminated the earlier anatase composite pigments from the market, along with most of the remaining lithopone.

The latest composite titanium pigment to appear was a 50% coprecipitated rutile calcium-base pigment in 1952.

Today, titanium pigments are manufactured by two quite different processes. The original rutile titanium dioxides and all anatase titanium dioxides and rutile calcium-base pigments are manufactured by the sulfuric acid, or "sulfate," process, while a growing percentage of rutile titanium dioxide is being manufactured by a vapor phase, or "chloride," process.

(1) The sulfate process uses a black ore, often considered to be an iron titanate, known as ilmenite. This ore occurs both in rock formations, where it is usually associated with iron ore, and as a black sand, deposits of both types being rather widely distributed throughout the world. The ground, concentrated ore, or the sand, is first dissolved in concentrated sulfuric acid, forming iron sulfate and a complex sulfate of titanium. Metallic iron is added to the digesting tank to reduce the iron to the ferrous state; after removal of all the undissolved ore, ferrous sulfate is precipitated from the solution as a by-product. In the production of the pure titanium dioxides, the purified titanium sulfate solution is concen-

trated, carefully nucleated or "seeded," and then boiled to precipitate the titanium dioxide as a hydrate. This hydrate is then calcined to drive off water, producing crystalline titanium dioxide, and finally ground on various types of mills, depending upon the grade of pigment being manufactured. Variations in the process, principally at the nucleation and calcination steps, determine whether the end product is of the rutile or anatase crystal form.

In the production of the calcium-base composite pigment, the calcium sulfate is first precipitated from sulfuric acid solution. The complex titanium sulfate solution is then added and boiled to precipitate the titanium dioxide hydrate upon the calcium sulfate crystals. Calcination transforms the calcium sulfate to the anhydrite form and the titanium hydrate to titanium dioxide.

(2) The chloride process for manufacturing rutile titanium dioxide involves the conversion of titanium tetrachloride to titanium dioxide by its combustion in a stream of oxygen under carefully controlled conditions in a special burner. Titanium tetrachloride is an intermediate product in the Kroll Process for manufacturing titanium metal and is generally produced by chlorinating rutile ore. With the growth of the titanium metal industry, surplus capacity to produce titanium tetrachloride first made the chloride process of interest to titanium pigment manufacturers, as also has the elimination of the problem of sulfuric acid waste disposal. A high rate of chlorine recovery is necessary for the economic operation of this process.

The particle size distribution produced by the chloride process is somewhat narrower than that of sulfate process pigments, and trace elements are more easily eliminated.

Anatase titanium dioxide is used in the paint industry today only because of its slightly better initial color as compared to the rutile titanium dioxide, which is slightly yellower in tone, and for that property which was once considered to be its greatest disadvantage—its relatively poor chalk resistance. Rutile titanium dioxide is actually so chalk-resistant that paint formulators have found it advisable to introduce one-third to one-half of the nonreactive white pigment used in white exterior house paints as the anatase form in order to obtain sufficiently fast "cleanup." As indicated in Table 20.3 (columns B and C), it is inferior to the rutile titanium dioxide in all aspects of hiding power. Several grades are available, varying slightly in chalk resistance and resistance to after-yellowing in paints.

Rutile titanium dioxide has the greatest opacity of any white pigment ever produced. It is ideally suited for use in high gloss synthetic enamels and lacquers in which pigment volume must be kept at an absolute minimum. While not quite as white initially as anatase titanium dioxide, the rutile pigments produce white finishes which yellow less on aging or baking, and hence are preferred for almost all applications except those in which absolute maximum initial whiteness is required.

Recent improvements in rutile titanium dioxide pigments, resulting in finer ultimate particle size, increased their hiding power and tinting strength by about 6 to 10% (depending upon grade). These newer sulfate process pigments are identified by the abbreviation P.S.C. (particle size controlled). All chloride process pigments are manufactured in this same finer particle size range. Finer particle size has also resulted in bluer tone.

The control of chalk resistance in titanium dioxide pigments is achieved by surface treatments of inorganic hydrates, principally of alumina and silica. Since the untreated titanium dioxide actually contributes to the oxidation of organic binders due to its photochemical reduction under the influence of UV radiation, it is theorized that these surface treatments inhibit the reduction-oxidation cycle by occupying reactive sites on the crystal surface.

These same surface treatments increase the oil absorption of titanium dioxide pigments. A complete series of specialized pigments has been produced for use in latex paints that are fast replacing the older oleo-resinous or alkyd-resin interior and exterior architectural finishes. These newer formulations characteristically require relatively high oil absorption pigments to produce maximum opacity due to the pigment packing which results from a colloidally dispersed, rather than a dissolved, resin system.

The most economical of all white pigments, however, are still the 30% rutile calcium-base pigments (see Table 20.3). Due to the dilution effect provided by the precipitation of the rutile titanium dioxide upon an inert base, these pigments are far more efficient than any blend of the pure rutile titanium dioxide with inert "extenders." Consequently, they are preferred for all applications where maintaining minimum pigment volume is not a factor. The 50% calcium-base pigment has been designed with higher hiding power per gallon (see column B, Table 20-3) to take advantage of the greater hiding efficiency of a composite pigment, while still allowing room in the formulation for functional "extenders." This advantage of the 50% calcium-base pigment over the 30% grade has been responsible for its widespread use in alkyd flat wall paints in recent years. Despite the slight water solubility of their calcium sulfate base, the calcium-base pigments have excellent exterior durability and are widely used as the nonreactive white pigments in house paint formulations and traffic marking paints. They are also being used in some types of high-polymer emulsion or "latex" paints, but they are not recommended for use in paints to be applied to bare metal because of their lack of cor-rosion resistance.

Titanium pigments, due to their substantial hiding power advantage (Table 20.3), are the only important nonreactive white pigments in use by the paint industry today. At the present state of their development, it is improbable that any future increase in their strength is to be expected, but it is likewise improbable that they will be challenged by any new product.

REFERENCES

1. Steig, F. B., *Off. Dig.*, **31** No. 408, 52–64 (1959).
2. Steig, F. B., *Off. Dig.*, **34** No. 453, 1065–79 (1962).
3. "The Handbook," Titanium Pigment Corp., New York, N. Y.
4. "Physical and Chemical Examination Paints, Varnishes, Lacquers, and Colors," 12th Edition, 1962, Gardner Laboratories Inc., Bethesda, Md.

21

*Colored Pigments**

In general, colored pigments can be divided into two classifications—organic and inorganic.

Inorganic

Chrome yellows	Chrome greens	Ultramarine blue
Chrome oranges	Iron oxides	Red lead
Molybdate orange	Cadmiums	Siennas
Zinc yellow	Iron blue	Chrome oxides
		Nickel titanate

Organic

Lithols	Benzidine yellow	PTA (phosphotungstic)
Para reds	Nickel azo yellow	PMA (phosphomolybdic)
Toluidine red	Quinacridones	Indanthrones
BON (β-oxynaphthoic acid)	Arylide red	Phthalocyanines
Hansa Yellow	Thioindigoids	Anthraquinones
Toners		

This chapter will touch only slightly on the composition of various pigments but will attempt to cover their advantages and disadvantages, and how and where they are being used. A pigment is a colored material which is essentially insoluble in the medium in which it is to be dispersed. (Dyes are soluble and will not be considered.)

Pigments are produced in a variety of physical forms to satisfy specific working properties or requirements.

Toners: Unextended or full-strength organic pigments having the maximum tinting strength for their type.

C. P. (chemically pure): Undiluted full-strength inorganic.

Resinated toners: Organic pigments treated with barium or calcium resinate which acts as a dispersion agent.

Lakes: Pigments treated with alumina hydrate or some other trans-

*By M. Malaga, The Glidden Co., Cleveland, Ohio.

parent medium to give lower opacity but improve flocculation-resistant qualities and ease of dispersion.

Extended pigments: Physical mixtures of colored prime pigments (organic or inorganic) with colorless extenders.

Another term that should be added to this listing is *particle size pigments* (easy dispersing). The trend over the last few years has been toward dispersing pigments in high-speed equipment to avoid long milling times. Since particle size will be broken down by physical or chemical means to its maximum fineness, less problems will be encountered by over- or undergrinding.

Great strides have been made by the pigment industry toward producing better, more durable and more easily dispersing pigments. This trend will continue until practically all pigments will be stir-in types.

To simplify this chapter, the pigments are listed according to color instead of grouping in organic or inorganic category.

Chrome Yellows

Chrome yellows are clean in color (high in chroma), low in cost, high hiding, nonbleeding and easy to disperse. They have very good durability which varies depending on the shade range. The lighter the shade, the poorer is the durability. The disadvantages of chrome yellows are poor alkali and soap resistance, the presence of toxic metals, and the tendency to darken in masstone on exposure. These pigments cost about 42 cents per pound. They are used in almost all types of finishes except toys, trade sales and automotive. The reasons they are not used in these finishes are poor alkali resistance and darkening tendency on outdoor exposure; also, their lead content makes them too toxic for use in interior enamels and toy enamels.

Chrome yellows can be purchased in various shade ranges from the light greenish yellow known as primrose or lemon to the redder shades commonly known as medium yellow. Dirtier (lower chroma) shades of yellows, known as shading yellows, have better outdoor durability than their counterpart—lemon yellow. These yellows blended with iron or phthalocyanine blue give bright attractive greens.

Chrome yellows vary considerably in chemical composition but all are based on the precipitation of lead chromate from solutions of soluble lead salts and biochromate of soda. C.P. chrome yellow is not an indication that the pigment is chemically pure but that it is free of any inert extenders.

Zinc Yellows

Zinc yellows can be divided into two grades—ordinary zinc yellow or basic zinc chromate. Both of these have low tinting strength and are not desirable for masstone use as far as color is concerned. They have poor lightfastness and poor alkali resistance. In combination with phthalo-cyanine blue, certain clean shades of green can be made with zinc yellow that can only be duplicated with more expensive pigments.

Zinc yellows cost approximately 30 cents/lb. They are used mostly in rust-inhibitive protective coatings, metal primers and government specification materials. Extreme caution should be used in formulating with this type of pigment since it is highly reactive in some vehicle systems and will have a tendency to "liver." Zinc yellows or zinc chromates generally come in only one shade—a rather yellowish green in the primrose yellow range. This pigment, zinc yellow, is a complex zinc potassium chromate which is formed by precipitation.

Basic zinc chromate is a compound of zinc and chromium and differs from zinc yellow in both chemical composition and water solubility. It was used quite extensively during World War II in a new chemical coating applied to the underside of naval vessels. This coating was called "wash primer" and consisted of a solution of polyvinyl butyral resin, basic zinc chromate, magnesium silicate, lampblack, alcohol, water and phosphoric acid.

Zinc yellow plays a very important role in the formulation of primers for aluminum. No lead-containing pigments in primers could be considered because of the stimulating effect of such materials upon the corrosion of aluminum. Zinc yellow will continue to play a very important role in this respect as more and more uses are being found for aluminum.

Strontium Yellow

Strontium yellow is a precipitated strontium chromate. It is nonbleeding, easy grinding, and has good lightfastness in both masstone and tints, as well as good heat resistance. Its disadvantages are very low tint strength (approximately 30% of a chrome yellow, in a primrose shade), poor acid and alkali resistance and high cost in comparison to chrome yellow. The current market price is approximately 50 cents/lb. At present, it is being used in vinyl resin systems for fabric coating and in metal-protective coatings. The outstanding property of strontium chromate is its low viscosity in high acid vehicles, whereas zinc chromate in this same vehicle system would show a very definite increase in body.

Nickel Titanate Yellow

Nickel titanate yellow is commonly referred to as sun yellow or titanium yellow. This pigment is composed essentially of the oxides of titanium, nickel and antimony. The color is developed by the thermodiffusion of nickel and antimony atoms into the crystal lattice of rutile titanium dioxide. It is calcined at about 1000°C. This pigment is weak in strength but has very good lightfastness, good outdoor weathering durability and good chemical resistance; it can withstand very high temperatures without color change. Recommended uses are for industrial enamels where resistance to chemicals is important, interior or exterior paints, baking enamels, exterior paints for stucco and mortar, and baking enamels used on aluminum siding for the exterior of houses. This pigment will have a tendency to chalk, but the chalk is the same color as the original so it is not too noticeable. Since this is a relatively new pigment, it is difficult to evaluate it accurately.

Nickel Azo Yellow

Nickel azo yellow, commonly referred to as green gold, has excellent lightfastness in masstone and tints, very slight bleed in some vehicle systems and good bake resistance. It is a metal chelate pigment which will have a tendency to change color in some acidic conditions. Extra precaution should be taken to check color stability on aging, especially in lacquer systems. The cost is rather high, approximately $6.15/lb. In masstone this pigment is green, but it produces a greenish yellow tone in tints. It gives quite an attractive appearance when used in polychromatic finishes. It is recommended for automotive and high-grade enamel finishes.

Cadmium Yellow

Cadmium yellows are nonbleeding; they have excellent lightfastness in deep shades, excellent baking resistance and good alkali resistance. Their disadvantages are low tinting strength, poor gloss retention and relatively high cost, approximately $3.10/lb in pure grade and $1.35/lb in lithopone grade. They are used in high heat specialty items such as washing machines, heaters, ovens and appliances. Cadmium yellows in both lithopone and pure grades are easy to disperse. They come in various shade ranges, from primrose to a deep yellow with a red cast. Pure cadmium colors are made by a reaction of cadmium sulfate and hydrogen sulfide. The lithopone-type cadmium colors are a precipitate of cadmium sulfide in which barium sulfate is formed as part of the reaction. In tints, cadmium yellows have a tendency to fade quite badly.

Yellow Iron Oxide

Yellow iron oxide is widely used in the paint industry because of its low cost—5 to 25 cents/lb—and good performance characteristics in paint coatings. It can be purchased as a synthetic or in its natural state. The synthetic is more widely used since it is more uniform and easier to disperse. It is lightfast in both masstone and tints, nonbleeding and free of toxic materials; it also has high hiding power and good chemical resistance. Its disadvantages are that it is dirty in color (low in chroma), has a tendency to settle and has poor gloss retention. In its natural state, this pigment is much cheaper than in the synthetic version. The accepted usage for this pigment is in interior or exterior finishes, primers, enamels, toy enamels and automotive finishes. They are generally nonreactive with most vehicles and are quite stable for shelf goods. In recent years, suppliers have been producing low-opacity oxides for use in transparent or metallic finishes. There are many patented processes for making synthetic iron oxides, all involving the principle of precipitating hydrated ferric oxide from a chemical solution of iron salts. The color may vary from a light cream to a dark buff.

Hansa Yellows

Hansa Yellow, commonly known as toluidine yellow, is an insoluble azo organic pigment obtainable in shade ranges from primrose to medium chrome yellow. It is excellent in masstone lightfastness, has good alkali and acid resistance, is lead free and is easy to grind. Its disadvantages are poor lightfastness in tints, bad bleeding in almost any type of finish, relative transparency, poor heat resistance and high price, approximately $2.25 to $2.50/lb, depending on shade range. Hansas are classified in color groups: Hansa R being the reddest, Hansa G slightly less red, and Hansa 10G primrose or greenish. They are recommended for emulsion paints and toy enamels where toxic metals are not tolerated.

Benzidine Yellows

Benzidine yellows are insoluble azo organic pigments having approximately twice the tinting strength of Hansa Yellow. They are obtainable in two forms. One form, the acetoacetanilide type, is used almost entirely in the printing industry. The other (*o*-toluidide type) is used in the paint industry. They have high tint strength, are free of toxic metals, and have good masstone lightfastness and soft texture. They also have good chemical resistance, excellent bake resistance and good bleed resistance. Their disadvantages are high cost (approximately $2.60/lb for the toluidide type), low hiding power and high oil absorption. Their primary use is in interior lead-free chemical-resistant finishes, toy enamels, floor coverings and rubber products. Although the benzidine yellows are stronger than Hansa Yellow, they are much poorer in lightfastness.

Vat Yellows

Vat yellows are commonly referred to as anthropyrimidine and flavanthrone yellows. They are alkali and acid resistant, nonbleeding and have excellent exterior durability in tints. Their disadvantage is their very high cost, ranging from 15 to 25 dollars/lb in toner and 7 to 10 dollars/lb in lake form. They are relatively new pigments and, because of their high cost, are not used to full advantage. Recommended uses are for automotive enamels, nitrocellulose lacquers, vinyl coatings and emulsion paints. Extra care should be taken when using the lake types as they are extended quite highly with alumina hydrate and will have a tendency to be reactive in some vehicle systems.

Chrome Orange

Chrome oranges are low in cost, nonbleeding and have good light-fastness. They contain lead, darken in masstone, have poor alkali and soap resistance and poor hiding in comparison to the molybdate oranges. They cost approximately 42 cents/lb. Chrome oranges differ chemically from chrome yellows in that they are basic lead chromates. Their usage has dropped considerably during the last few years as formulators are replacing chrome orange with combinations of molybdate orange and chrome yellow. These combinations will match almost any range of chrome orange (metameric only) with better gloss, gloss retention, durability, suspension and increased hiding. Chrome oranges are still being used in some exterior enamels and rust-inhibitive primers.

Molybdate Orange

Molybdate oranges are low in cost (50 cents/lb), non-bleeding, light-fast and easy to disperse; they also have good bake resistance. Their disadvantages include the presence of toxic metals, poor alkali and soap fastness, and a tendency to darken in masstone on exposure. This pigment is used quite extensively in blending with organic red pigments to produce low-cost, high-intensity red finishes, having good hiding power and light-fastness. Various shades of red can be acquired, depending on the organic toning red being used. Combinations of quinacridone and molybdate orange or "Indo" maroon and molybdate orange give very good durable outdoor finishes. This pigment is being used in all types of finishes except where lead-free pigments are required. Molybdate orange, at the present time, can be purchased in various shade ranges. The redder shades have a tendency to be lower in hiding power and poorer in lightfastness than the orange shades. This pigment is a coprecipitated mixed crystal of lead chromate, lead sulfate and lead molybdate.

Cadmium Orange

Cadmium orange, like cadmium yellow, is also available in both the C.P. and lithopone types. Its use is basically the same as cadmium yellow; it is high in cost and low in tint strength. The orange tone is obtained by substituting cadmium selenide for some of the cadmium sulfide during processing. The approximate cost of this pigment is $4.10/lb in toner type and $1.65/lb in lithopone grade. This pigment is used in high heat specialty finishes and industrial enamels.

"Mercadium" Orange

"Mercadium" Orange is a cadmium mercury pigment also available in both C.P. and lithopone form. Instead of cadmium selenide, mercury sulfide is used with cadmium sulfide to obtain this product. The current market price of this pigment is the same as that for cadmium orange. Claims are that the "Mercadium" pigments are slightly cleaner and brighter than cadmium pigments. Their recommended uses are the same as cadmium.

Benzidine Orange

Benzidine orange, commonly referred to as pyrazolone orange, is a diazo pigment dyestuff prepared by coupling diazotized dichlorbenzidine with two molecules of phenylmethylpyrazolone. It has high color intensity, high tinting strength, is free of toxic metals, has good chemical resistance and is easy to disperse. It is high in cost (approximately $3.30/lb) it has poor lightfastness, poor bleed resistance and low hiding power. Recommended uses are in lead-free paints such as toy enamels.

Dinitraniline Orange

Dinitraniline orange is an (mono) azo pigment dyestuff prepared by coupling diazotized dinitroaniline to β-naphthol. It has high tint strength, is free of toxic metals, has good chemical resistance and good masstone lightfastness. Its disadvantages are high cost ($1.60/lb), poor lightfastness and poor bleed resistance. It is used in some latex and specialty finishes.

Vat Dye Oranges

Vat dye oranges are commonly referred to as Orange RK or "Indofast" Orange. They are anthraquinone derivatives. They have excellent lightfastness in pastels and metallics, freedom from toxic metals, excellent chemical resistance and excellent bake resistance. They are very high in cost, ranging from 9 to 25 dollars/lb, depending on whether they are lake or toner types, and have a tendency to show slight bleed in some systems. At present, they are used in exterior baking-type house paints, where maximum durability is required, and also in automotive finishes. Care should be taken, when formulating with the lake types, to check reactivity. The high percentage of alumina hydrate in the lakes has a tendency to be reactive in some systems.

Chrome Greens

Chrome greens are coprecipitated combinations of iron blue and chrome yellow. A variety of greens can be made with this combination, from light to very dark. They are low in cost (45 cents/lb), have high hiding and good bake resistance, and are are nonbleeding. Their disadvantages are poor alkali resistance, moderate lightfastness and a tendency to flood and float. They are used in baking and outdoor enamels. Some formulators develop chrome green colors using combinations of phthalocyanine blue or green with shading yellow to get better lightfastness and alkali resistance. From the standpoint of plant inventory, formulators are using combinations of shading yellow and iron (Milori) blue to keep this to a minimum. There are also some economic advantages to using this combination, but care should be taken to avoid color float.

Chromium Oxide

Chromium oxide is a calcined inorganic green pigment having good heat, light and chemical stability. It is easy grinding and nonreactive in most paint systems. Pure chromium oxide is manufactured by the reduction of alkali chromates with reducing agents at elevated temperatures with subsequent wash, drying and pulverizing. It is low in tint strength, being only about one-fourth the strength of a chrome green in comparable shade.

Pure chromium oxides find wide applications in paints, enamels, building products, rubber, floor coverings and other uses where color permanence is a factor.

Chromium oxides are olive green in shade (low in chroma) and have poor gloss retention on exterior exposure. The cost of this pigment is approximately 46 cents/lb.

Hydrated Chromium Oxide

Hydrated chromium oxide, also known as Guignet's green or C.P. permanent green, is quite different from chromium oxide, and the two should not be confused. Hydrated chromium oxide is prepared by calcining bichromate of soda in the presence of boric acid, with subsequent hydrolysis during washing. It is clean in color and high in chroma; it has excellent outdoor durability and has been used over the years in automotive metallic finishes. It is also used for tinting pastels because of its excellent color stability. This pigment costs about $1.50/lb. Its disadvantages are low tinting strength, difficulty of dispersion, and poor blister resistance.

Copper Phthalocyanine Green

Phthalocyanine green is derived from phthalocyanine blue by chlorination. Chlorination, in addition to producing a green pigment, imparts still greater stability to the molecule. It is available in a number of forms: full strength toner, resinated toner, flushed form, water-dispersable form and reduced form. It is clean in color (high chroma), has high transparency for metallics, excellent chemical resistance, bake resistance, lightfastness, tinting strength and bleed resistance. It is high in cost, except in tints, and dark shades have a tendency to bronze on exposure. It is very seldom, if ever, used as a straight masstone pigment. Current market price for this pigment is $3.50/lb. It is available in a number of shade ranges from a very yellow green to a bluish green. It is used quite extensively in all types of finishes from automotive metallics to trade sales tints, both in bake and air dry quality.

Organic Green Toners

Organic green toners are commonly referred to as PTA (phosphotungstic) or PMA (phosphomolybdic) green. Although it is difficult to distinguish, the PTA types are considered to give better brilliancy and slightly better lightfastness than the PMA. Both types are transparent in masstone, vary high in tint strength and high in chroma. Their disadvantages are high cost, poor outdoor durability and poor bleed resistance. Current market price is between $4.00 and $6.50/lb. They have very little usage in the paint industry, but are used in printing inks and foil lacquers.

Iron Blues

Iron blues have high tinting strength, are low in cost—60 cents/lb— and have good durability and lightfastness in masstone and dark tints. Common names are Chinese, Milori and Prussian blue. Their disadvantages are poor lightfastness in light tints, poor can storage stability and poor alkali resistance. They are used in all types of finishes, including automotive finishes in conjunction with phthalocyanine blue. The mixture of the two blues give characteristics that cannot be obtained with one alone. Phthalocyanine blue improves the alkali resistance and durability; iron blue reduces cost, helps overcome bronzing and increases jetness. Prussian blues are usually referred to as the red-shade iron blues; Chinese and Milori, the green-shade. Nickel-treated iron blues have a tendency to be better for nonbronzing qualities. Iron blues have been considered to be very hard grinding, but in recent years improvements have been made regarding their dispersability. In *light* tints, iron blues have a tendency to fade either in the wet state or on outdoor weathered panels. In some formulations there is a definite cost advantage in using iron blue instead of phthalocyanine blue.

Copper Phthalocyanine Blues

Phthalocyanine blues were discovered in Great Britain about 1927; and their manufacture in the United States began about 1936. The phthalocyanine pigments are generally regarded as representing the highest standards among organic pigments for lightfastness, heat stability and chemical resistance. When first manufactured they had a tendency to crystallize and were hard to disperse, but over the years much has been done to improve these qualities. They are very high in tint strength, have excellent lightfastness even in light tints and excellent bleed resistance. They are high in cost, except in light tints and tend to bronze in dark shades. The current market price is about \$3.20/lb. They can be purchased in a red shade or green shade and are used in all types of finishes—industrial, automotive, bake or air dry, trade sales alkyds or oils, latex, etc.

Ultramarine Blue

Ultramarine blue is a deep reddish blue made by firing a combination of sulfur, alkali, clay and a reducing agent at elevated temperatures. It is produced in a variety of tones, all of which are very brilliant. It is low in cost (30 to 40 cents/lb), has good alkali resistance and excellent heat resistance. Its disadvantages are very poor acid resistance, low hiding power, poor tinting strength, poor outdoor durability and poor wettability. It is used in the manufacture of machinery and toy enamels and also for shading baking whites. Another area of use is in special applications where high infrared reflecting properties are required. Recently developed ultramarines have tinting strength considerably higher than those obtained a few years ago. They are used quite extensively in many industries other than the paint industry.

Organic Blue Toners

Organic blue toners fall in the same category as the organic green toners, PTA (phosphotungstic) and PMA (phosphomolybdic). Their properties are essentially the same as those listed for the organic green toners. The current price of these pigments is from \$4.50 to \$5.50/lb, depending on shade range. In spite of their deficiencies (poor bleed resistance and poor lightfastness on outdoor exposure), the brilliance and high tinting strength of these pigments make them suitable for various interior paints of more or less special character, such as poster paints, foil lacquers and some toy enamels.

Indanthrone Blue

Indanthrone, or "Indo" blue, is classified as a vat dye pigment. It has good outdoor durability in all depths of shade, has good chemical resistance, good bake resistance and good bleed resistance. Its disadvantages is its high cost. In tone form, this pigment is approximately three times as expensive as phthalocyanine blue. The cost of toners is $16.50/lb; of lakes, $7.50/lb. Indanthrone blue is considerably redder in hue than phthalocyanine blue and is far superior in bronze resistance. When used in automotive metallic finishes, no other pigment or combinations of pigments could duplicate the tone acquired with indanthrone blue. Because of its high cost and hard texture, indanthrone blue is rather limited in use in the paint industry. In addition to automotive finishes, it is being used in exotic specialty finishes and the tinting of white baking enamels.

Carbazole Dioxazine Violet

Carbazole dioxazine violet is also known as "Indofast" Violet, Hostaperm Vat Violet, "Monarch" Violet, Calbazole Violet, etc. It is an opaque organic toner with excellent lightfastness, bleed resistance, and alkali and acid resistance. Its disadvantages are extremely high cost and poor dispersion. It can be purchased in lake form at 10 dollars/lb and $16.50/lb for toner form. It is used as a lightfast violet for tinting systems, for reddening phthalocyanine blues and bluing white baking enamels. It is used in automotive finishes and industrial white baking enamels.

Organic Violet Toner

Both PTA (phosphotungstic) and PMA (phosphomolybdic) pigments are being used for the printing ink industry. They have high tinting strength and high color purity. The disadvantages are high cost ($3.00 to $4.50/lb), poor outdoor durability and poor bleed resistance. The PMA violets are more transparent and bluer than the PTA violets.

Mineral Violet

Mineral violet, an organic pigment, is sometimes referred to as manganese violet. It has excellent bake resistance, excellent acid resistance, excellent bleed resistance and low cost (approximately $1.65/lb). Its disadvantages are very low tinting strength, poor alkali resistance, hard grinding and poor suspension. The color has excellent lightfastness for interior exposure but is not recommended for exteriors. It is used for tinting white interior baking enamels.

Quinacridone Violet

Quinacridone violet, commonly referred to as "Quindo" and Monastral Violet is alkali and acid resistant, nonbleeding, light-fast in masstone and tints, and heat resistant; it also has high hiding power. These pigments are high in cost, approximately 11 to 12 dollars/lb except in tints. They are used quite extensively in combination with molybdate orange to give brilliant low-cost reds with good durability. They are also used in automotive finishes, enamels, nitrocellulose lacquers, vinyl products, etc. These violets are especially suited for high bake finishes or plastics where high temperature is required. They do not fade or change color when baked at elevated temperatures when used in masstone or tints. Silicone finishes baked at 500°F for long periods of time show practically no color change when quinacridone pigments are used.

Lithols

Lithol reds are the most economical organic reds available (approximately 1 dollar/lb), have good bleed resistance and are easy grinding. They are widely used where brilliance, hiding and lost cost are of primary importance, as in toy enamels, dipping enamels and general industrial applications where good durability is not required. They are recommended for use on interior work only, since they check and fade badly on exterior exposure. They can be purchased in various shade ranges in both resinated and nonresinated types. The light shades are sodium lithols, the medium shades barium lithols, the deeper shades calcium lithols. The oil absorption of this pigment increases as the shade gets deeper.

The acid resistance of barium and calcium lithols is generally considered good. Alkali resistance is only fair. The sodium lithols have poor alkali and acid resistance. Lithols, although not considered bleeders in enamels, will bleed in some solvents and lacquers. The extent of bleed is not as severe as in toluidine or para red. Lithols have good hiding power in nonresinated form but are rahter transparent in resinated types. The deep lithols are used with molybdate orange for producing economical, bright red shades for industrial enamels.

Para Reds

Para reds are relatively low in cost, having good hiding power, ease of grind, good acid and alkali resistance, and good deeptone lightfastness. They are not recommended for tints because of very bad bleed and poor lightfastness, or for baking finishes since they turn brown with heat. They are used in cheap air dry exterior finishes such as farm implements, lawn mowers or spreaders. The current cost is $1.25/lb. Care should be taken to avoid contamination when using these pigments around other batches. A small amount of para red dust can cause severe bleeding problems in any white batch. These pigments are available in light to very dark dark shades. Chemically, they are couplings of diazotized p-nitraniline with β-naphthol.

Toluidine Reds

Toluidine reds are exceptionally bright reds produced primarily in two shades—light and dark. They are relatively low priced ($1.70/lb), easy-grinding, have excellent deeptone lightfastness and outdoor durability; they also have excellent acid and alkali resistance but very poor bleed resistance. They are not recommended for tinting light tints and have a tendency to bloom in urea- or melamine-modified baking enamels. They are used in the manufacture of air dry or bake enamel finishes where good outdoor durability is required and where bleeding is of no concern. These pigments have been used in automotive finishes where the finish was not over-stripped or two toned, and are still being used on some truck finishes where bleeding is not detrimental. Toluidine reds represent the azo coupling of diazotized *m*-nitro-*p*-toluidine with β-naphthol. A definite study was made regarding the bloom characteristics of toluidine reds in baking enamels and evidence was obtained that the bloom is definitely related to the solubility of the pigment in the enamel vehicle constituents.

BON (β-Oxynaphthoic) Reds and Maroons

BON reds and maroons are commonly referred to as rubine reds and are chemically similar to lithol rubine, being treated with calcium or manganese to improve masstone lightfastness. These pigments range in shade from relatively light red to very dark maroon. The darker the shade, the harder they are to disperse. They have very good masstone lightfastness and excellent bleed resistance. Their disadvantages are poor alkali and soap resistance and poor tint lightfastness. The approximate cost of these pigments is $2.00 to $2.25/lb, depending on color range. They are used in high-quality bake enamels and in some automotive finishes. Extra care should be taken when working with manganese BON to avoid skinning or wrinkling due to the manganese. These colors should not be used in highly acid or alkaline systems.

Lithol Rubine

Lithol rubine is a deep masstone red (high in chroma) with poor hiding power. It is used in blending with molybdate orange for low-cost reds. Because of its very clean bluish tone, many bright shades of red can be obtained. Lithol rubine comes mostly in resinated form, has good bleed resistance, easy grinding and good bake resistance. It has poor lightfastness, and poor alkali and soap resistance. Its primary uses are in industrial enamels where better bleed, baking and lightfastness properties are required. The cost of this pigment is approximately $1.50/lb. Lithol rubines are based on compounds made from various amines and β-oxynaphthoic acid.

Chlorinated Para Red

Chlorinated para red is commonly known as tanager, fire toner, blazing red, permaton red, etc., and is closely related to para red. Its properties are similar to those of para red, except it has better tint lightfastness. The approximate cost of this pigment is $1.40/lb. It can be obtained in two types—the o-chloro-p-nitroaniline coupled with β-naphthol and the p-chloro-o-nitroaniline. The former product is superior to the latter in masstone and extension lightfastness; in fact, it is even more lightfast than toluidine red.

Quinacridone Reds and Maroons

Quinacridone reds and maroons are commonly known as MONAS-TRAL® and "Quindo" colors in this country. They represent newly developed pigmentary forms of complex organic compounds. They have exceptional color brightness, outstanding lightfastness in tints, excellent resistance to chemicals, acids, alkali and soap. They are nonbleeders and have excellent resistance to heat. They are expensive—approximately $11.75 to $13.75/lb, depending on the shade range. They are used in all types of finishes where the best durability is required. With aluminums, they make very attractive automobile coatings. Other uses are for vinyls, baking enamels, printing inks, floor and wall coverings, etc.

Red Iron Oxide

Red iron oxides have relatively the same characteristics as the yellow iron oxides. There are three principal types of pure red oxides and each is manufactured by a different process. Let us start with synthetic yellow oxide to produce a red. The yellow oxide is passed slowly through a rotary kiln under controlled heat. This drives off the water of hydration and results in the formation of a red oxide. The copperas red type is made by passing ferrous sulfate slowly through a rotary kiln under controlled heat. The ferrous sulfate, or copperas as it is called, first loses its water of crystallization and then oxidizes to ferric oxide with the liberation of sulfur trioxide. The third type is made by direct precipation, and the ferric oxide content of this type is in the 97% range.

Pure red iron oxides in synthetic form are easy to disperse and range in color from a light salmon to a very rich blue. These types are used quite extensively in tinting pastel shades where good color retention is required. There is one precaution that should be pointed out regarding the use of deep shades of the calcined yellow and the precipitated types. Both are very soft pigments and sensitive to overgrinding in a steel ball mill. Overgrinding causes the shade to lighten and then redden. High-speed dispersers or sand mills are ideal for this type of pigment. Other types of red oxide are manufactured in the 93 to 95% range which are lower in cost than the pure iron oxides.

Red iron oxides can be purchased in synthetic or natural form. The synthetic form is carefully standardized for color, tinting strength, hiding

^(R)—Trademark—E. I. Dupont De Nemours & Co.

power, ease of dispersion, particle size, oil absorption, composition and many other properties which are essential to quality paint performance.

Natural types are still available and are being used in floor and deck enamels, primers, and barn paints where price is an advantage.   The synthetic types are used in all kinds of enamels and lacquers including epoxy and silicone resin finishes and emulsion paints.

Costs of this pigment vary from 5 to 25 cents/lb, depending on whether it is natural or synthetic; the cost will also vary with shade.

Besides being low in cost, iron oxides have high hiding, excellent light-fastness, good chemical resistance and excellent bake resistance.   Their disadvantages are dirty color (low in chroma), poor gloss retention and fair suspension characteristics.

Cadmium Red and Maroons

Cadmium reds have the same characteristics as the cadmium yellows except for durability, since they are much better and can be purchased in pure or lithopone grades.   The reds are coprecipitated, cocalcined mixtures of cadmium sulfide and cadmium selenide; those which contain large quantities of barium sulfate are known as cadmium lithopones.   As the depth of the red increases, the tint lightfastness increases.   These pigments are high hiding, nonbleeding, very lightfast in deep shades and very bake resistant.   Their disadvantages are low tint strength, poor gloss retention on outdoor exposure and poor lightfastness in tints.   The cost of the pures ranges from \$4.65 to \$6.00/lb, depending on the shade, the lithopone types vary from \$2.00 to \$2.75/lb.   They are widely used in coated textiles, automobile finishes, water paints and special finishes that require unusually good heat and chemical resistance.

"Mercadium" Reds and Maroons

"Mercadium" Reds and Maroons are calcined pigments made by combining sulfides of mercury and cadmium.   Their characteristics and properties are almost equal to the cadmiums, and they might be slightly brighter in masstone and cleaner in tints.

Red Lead

Red lead is a very ancient pigment, low in hiding, nonbleeding, and easily dispersible. It is very high in specific gravity and low in oil absorption.   It comes in various lead content grades—85, 95, 97 and 98%.   It has excellent film integrity on exterior exposure but has a tendency to fade.   It is used primarily for undercoats on steel bridges and iron structures.   It has been used for many years to protect iron and steel from corrosion, and is still being used because of its rust-inhibitive properties.

Red lead is manufactured by heating litharge in a reverberatory oven at 900 to 950°F for approximately 24 hours to obtain the 85% grade. Longer times are required for the higher-percentage grades.

Thioindigo Reds and Maroons

Thioindigo's are commonly referred to as "Indo" or "Thiofast" pigments. They have excellent chemical resistance, excellent light-fastness in dark shades and good bake resistance. They are high in cost—approximately $5.85 to $18.00/lb, depending on the shade and special characteristics; bronze in masstone and some varieties have a tendency to bleed. These pigments are alkali and acid resistant, and cannot be qualified for tints as the shade affects the lightfastness. They can be rated from poor to very good. They have been used quite extensively in the automotive and vinyl industry; they are also used in alkyd resin enamels, nitrocellulose lacquers, plastics and emulsion paints.

Arylide Maroons

Arylide maroons, commonly referred to as toluidine maroon, "Naphthanil"[*] and naphthol maroons, are azo pigments based on arylides of β-hydroxynaphthoic acid. These colors have a yellowish undertone which is unique for maroons. These pigments bleed severely and are very hard to disperse. Lightfastness in masstone depth is considered very good, and the pigments show no color change when baked for one-half hour at 250°F. Arylide maroons have excellent acid and alkali resistance. They are low in hiding power and relatively expensive. Costs range from $3.50 to $4.25/lb. They have been used in some industrial and specialty finishes.

Siennas, Ochers and Umbers

The following pigments have been classified in one group since they are all earth colors; ochers, raw siennas, raw umber, burnt sienna, burnt umber and Van Dyke brown. They are all rather weak in strength but do give warm, clean tones that are not obtainable with any other pigments. There are two types of umbers—crude turkey umber mined on the island of Cyprus, and domestic raw umber. The domestic type is stronger from a tint standpoint than the turkey umber. Both types can be used in emulsion vehicle systems. If raw umber is desired in a baked finish, only the domestic grade should be used because of its color stability. Turkey umber becomes reddish-brown at high temperatures. The burnt umber is made by calcining the raw crude umber, and the domestic type is made by blending domestic crude and some black.

Ochers and raw siennas have a yellowish-brown masstone similar to ferrite yellow but slightly richer in tint.

Raw sienna is a natural, hydrated yellow iron oxide, with the best grades imported from Italy. Burnt sienna is made by calcining the pro-

[*]Trademark—E. I. Dupont De Nemours & Co.

cessed, but unground, raw sienna. The calcining drives off the water of hydration and results in various yellowish and brownish reds of varying shades, depending on the temperature and the chemical content of the crude.

Burnt siennas have a reddish-brown masstone and a yellowish-red tint tone. Burnt umber has a very dark brown masstone and brownish-gray undertone. Van Dyke brown is similar to burnt umber but has a purplish-brown undertone. These pigments have all been used by artists for many years, the pigments being ground in linseed oil. Other uses for these pigments are in stains and the tinting of pastels.

Carbon Blacks, Lampblacks and Bone Blacks

The latest statistics show that there are some 25 grades of carbon black manufactured in the United States and more than a dozen carbon black producers employed in their manufacture. Including the 25 carbon blacks, a formulator has a choice of over 30 different grades or types of black he can use for formulation work. His choice can be made from furnace blacks, lampblacks, thermal blacks, acetylene blacks, or bone black, each having its own specific use.

The oldest of the carbon blacks are lampblacks. The difference between true carbon blacks and lampblacks is their method of manufacture. Carbon blacks are manufactured from raw gas, whereas lampblacks are made from various oils. Acetylene black and thermal blacks are very seldom used in the paint industry.

Bone blacks are made from cattle bones that are carefully cleaned and selected. These bones are then placed on trays and charred in the same manner used in making industrial coke. The bone char is then ground to pigment fineness for use in the coatings industry.

Over 95% of the world's supply of carbon black is produced from natural gas.

The various types of furnace blacks are available in both powdered or beaded form. The powdered form is normally used on roller or disk-type mills where it will disperse more readily. The beaded form is generally used in steel or pebble ball mills and considered advantageous from the standpoint of easier and cleaner loading.

An instrument used quite extensively in the pigment industry to evaluate jetness or color intensity of blacks is called a Nigrometer. This instrument assigns a numerical value to each black—the higher the number, the blacker (or jetter) is the color. For example, a carbon black in the 255 range would be considered a very jet black; a black in the 60 range would be considered a gray black. As the blackness index varies, so does the particle diameter—the smaller the particle diameter, the blacker is the color.

TABLE 21.1.   Properties of Colored Pigments[a]

Pigment	Advantages	Disadvantages
	Yellows	
Yellow iron oxide	Low cost. High hiding, nonbleeding. Excellent lightfastness. Good chemical resistance. Free of toxic metals.	Low in chroma. Low in tinting strength. Poor gloss retention. Fair suspension characteristics. Fair bake resistance.
Lead chromate yellow	Low cost. High hiding, nonbleeding. Good lightfastness in pastels. Excellent outdoor durability. Good bake resistance. Soft texture.	Contains toxic metal. Discolors in presence of sulfides. Darkens in masstone on exposure. Poor alkali and soap resistance. Tends to be reactive in highly acid vehicles.
Zinc yellow	Low cost. Metal protective.	Low tinting strength. Poor lightfastness as a yellow. Poor alkali resistance.
Basic zinc chromate	Specific use in metal-protective wash primer.	No color value.
Cadmium yellow	High hiding. Nonbleeding. Excellent lightfastness—deep shades. Excellent bake resistance.	Low tinting strength. Low color intensity. Poor gloss retention. Fair weathering characteristics. Poor tint lightfastness. Sensitive to discoloration due to metal sulfides. Relatively high cost.
Hansa Yellow	High tinting strength. Free of toxic metals. Excellent lightfastness deep shades. Good lightfastness in pastels. Soft texture. Good chemical resistance.	High cost. High oil absorption. Poor bleed resistance. Low hiding power.
Benzidine yellow	High tinting strength. Free of toxic metals. Good masstone lightfastness. Soft texture. Good chemical resistance. Excellent bake resistance. Good bleed resistance.	High cost. Low hiding power. High oil absorption.
Green gold	Excellent lightfastness in masstone and tint. High color intensity in tints and green blends. High transparency in metallics. Good bleed resistance. Good bake resistance.	High cost. Has very slight bleed in some systems.

Oranges

Pigment	Advantages	Disadvantages
Benzidine orange	High tinting strength. Free of toxic metals. Good chemical resistance. Soft texture.	High cost. Poor lightfastness. Poor bleed resistance. Low hiding power.
Dinitraniline orange	High tinting strength. Free of toxic metals. Good chemical resistance. Good masstone lightfastness. Soft texture.	High cost. Poor tinting lightfastness. Poor bleed resistance. Low hiding.
Orange R K (vat dye)	Excellent lightfastness, even in pastels and metallics. Free of toxic metals. Excellent chemical resistance. High transparency for metallized polychromatic finishes. Excellent bake resistance.	High cost. Moderate color intensity. Fair bleed resistance.
Molybdate orange	Low cost. High hiding. Nonbleeding. Good lightfastness. Excellent outdoor durability. Good bake resistance. Easy grinding. Blends well with reds.	Contains lead. Discolors in presence of sulfides. Darkens in masstone on exposure. Poor alkali and soap resistance in pastels.
Chrome orange	Low cost. Nonbleeding. Good bake resistance. Good lightfastness.	Contains lead. Discolors in presence of sulfides. Darkens in masstone on exposure. Poor alkali and soap resistance in pastels. Sensitive to overgrinding. Poor gloss retention.

Greens

Pigment	Advantages	Disadvantages
C. P. chrome green	Low cost. High hiding. Easy grinding. Nonbleeding. Good bake resistance.	Poor alkali resistance. Moderate lightfastness. Tend to flood and float.
C. P. chromium oxide	Excellent lightfastness. Excellent chemical resistance. Excellent bake resistance. Nonbleeding. High hiding power.	Low tinting strength. Low chroma. Somewhat abrasive. Poor gloss retention on exterior exposure.
Hydrated chromium oxide	High chroma. High transparency for metallics. Excellent chemical resistance. Excellent bleed resistance. Excellent lightfastness.	Relatively difficult to disperse. Low tinting strength. Low hiding. Poor blister resistance.
Phthalocyanine green	High chroma. High transparency for metallics. Excellent chemical resistance. Excellent bake resistance. Excellent lightfastness. High tinting strength. Excellent bleed resistance.	High cost except in tints. Dark shades bronze on exposure.

TABLE 21.1 (continued). **Properties of Colored Pigments**[a]

Pigment	Advantages	Disadvantages
	Greens	
PTA–PMA (Phosphotungstic-phosphomolybdic)	Very high tinting strength. Very high chroma.	High cost except in tints. Poor outdoor durability. Poor bleed resistance.
	Blues	
Iron blues	Low cost. High tinting strength. Good durability and lightfastness in masstone and dark tints. Excellent bleed resistance.	Fair lightfastness in light tints. Poor can storage color stability in light tints in presence of oxidizable vehicles. Poor alkali resistance.
Phthalocyanine blue	Very high tinting strength. High degree of color intensity. Excellent lightfastness even in light tints. Excellent chemical resistance. Excellent baking resistance. Excellent bleed resistance.	High cost except in light tints. Tendency to bronze in dark shades. Special forms required for flocculation resistance.
Ultramarine blue	Low cost. Good alkali resistance, except lime. Excellent heat resistance.	Very poor acid resistance. Low hiding power. Low tinting strength. Poor outdoor durability in all shades. Poor wetting in organic vehicles.
Indanthrone blue	Good outdoor durability in all depths of shade. Good chemical resistance. Good bake resistance. Good bleed resistance.	High cost except in light tints. Moderate color intensity in tints.
PTA blue	High tinting strength. High color intensity.	High cost except in light tints. Poor outdoor durability, especially in tints. Poor bleed resistance.
	Violets	
Dibenzanthrone violet (vat dye)	High tinting strength. Good lightfastness at high dilution. Excellent acid and alkali resistance. Excellent bake resistance except in whites.	High cost. Fair bleed resistance. Poor high bake resistance in white finishes.

Quinacridone violet	Have exceptional color brightness. Outstanding lightfastness in tints. Excellent resistance to chemicals, acids, alkali and soap. Excellent resistance to bleeding. Excellent resistance to heat.	High cost.
Thioindigoid maroon	Excellent chemical resistance. Excellent lightfastness in dark shades. Good bake resistance. Good lightfastness at high dilution.	High cost. Masstone bronze. Some varieties bleed in lacquer and aromatic solvents.
Mineral violet	Excellent lightfastness. Excellent bake resistance. Excellent acid resistance. Excellent bleed resistance. Low cost.	Very low tinting strength. Poor alkali resistance. Hard grinding.

Reds

Red iron oxide	Low cost. High hiding. Nonbleeding. Excellent lightfastness. Good chemical resistance. Excellent bake resistance.	Low in chroma. Poor gloss retention. Fair suspension characteristics.
Cadmium red	High hiding. Nonbleeding. Excellent bake resistance. Excellent lightfastness in deep shades.	Low tinting strength. Fairly low in chroma. Poor gloss retention on outdoor exposure, poor polish back. Sensitive to discoloration due to metal sulfides. Poor lightfastness in pastels—outdoor exposure.
Toluidine red	Relatively low cost. High color intensity. Excellent deeptone lightfastness and outdoor durability. Easy grinding. Excellent acid and alkali resistance.	Poor bleed resistance. Danger of bloom in urea- and melamine-modified baking enamels. Moderate hiding power. Fair tint lightfastness.
Para red	Low cost. Good hiding power. Good acid and alkali resistance. Good deeptone lightfastness.	Moderate color intensity. Poor bake resistance. Poor bleed resistance. Poor tint lightfastness.
Chlorinated para red	Good deeptone lightfastness. Moderate cost. Excellent acid resistance. Good alkali resistance. Easy grinding.	Poor bleed resistance. Poor tint lightfastness.
Naphthol red	Excellent acid and alkali resistance. High chroma. Easy grinding.	Low hiding power. Fair bleed resistance. Poor outdoor durability. Relatively expensive.

TABLE 21.1 (continued). **Properties of Colored Pigments**[a]

Pigment	Advantages	Disadvantages
	Reds	
Lithol red	Low cost. Good bleed resistance. Easy grinding. High chroma.	Poor lightfastness. Poor soap resistance. Fair bake resistance.
Lithol rubine	High chroma. Blends well with molybdate orange for low-cost reds. Good bleed resistance. Good bake resistance. Easy grinding.	Low hiding power. Poor lightfastness. Poor alkali and soap resistance.
BON red	Very good masstone lightfastness. Excellent bleed resistance.	Poor alkali and soap resistance. Fairly hard grinding.
Pyrazolone red	Good bleed resistance. Excellent acid and alkali resistance. Easy grinding. Good bake resistance. Good masstone lightfastness.	Low hiding. Poor lightfastness in tints. Relatively expensive.
Quinacridone red	Excellent lightfastness. Excellent bleed, heat and chemical resistance.	High cost except in tints.
Thioindigo red	Excellent lightfastness. Excellent acid and alkali resistance.	High cost.
PTA–PMA reds	Good acid resistance. High chroma. Easy grinding.	Poor alcohol bleed. High cost. Poor tint lightfastness. Fair bake resistance.
"Mercadiums"	Brighter in masstone, stronger and cleaner in tints than cadmiums. Cheaper than cadmium and easier to disperse. Nonbleeding.	Outdoor durability not as good as cadmium. Low tinting strength.
	Maroons	
Alizarine maroon (dyestuff)	Very good lightfastness. Good acid and alkali resistance. Excellent bleed resistance.	Low hiding. Relatively low chroma. Fair bake resistance. Poor durability.
"Helio" Bordeaux Maroons (calcium toners)	Very good lightfastness at masstone depth and is considered as possessing good permanency in relatively strong tints. Nonbleeding in enamel. Good hiding power.	Suitable for interior exposure only.

Cadmium maroon	High hiding. Nonbleeding. Excellent lightfastness—deep shades. Excellent bake resistance.	Low tinting strength. Fairly low in chroma. Poor gloss retention on outdoor exposure. Poor lightfastness in pastels—outdoor exposure.
BON maroon	Very good masstone lightfastness. Excellent bleed resistance.	Poor alkali and soap resistance. Fairly hard grinding.
Toliudine maroon	Relatively low cost. Excellent deeptone lightfastness and outdoor durability. Easy grinding. Excellent acid and alkali resistance.	Poor bleed resistance. Danger of bloom in baking finishes. Moderate hiding power. Fair tint lightfastness.
"Mercadium"	Same as "Mercadium" Reds.	Same as "Mercadium" Reds.

[a] This information was compiled from literature obtained from E. I. du Pont de Nemours & Co., Imperial Chemical Industries, Ltd., and Harmon Color, and reprinted with their consent.

The high-color blacks are commonly referred to as high-color channel blacks, and are indeed jet black. They are very high in cost, ranging from $.60 to $1.70/lb, depending on jetness. In this case, the jet black would be considered the finest particle size black. This type of black is used in automotive finishes, high-color industrials, high-color shelf goods and specialties where optimum blackness is required.

The medium-color channel blacks are not as jet in color and are used for utility finishes, medium-color industrials, medium-color shelf goods, and formulations where supreme blackness is not required but a good jet black is essential. The approximate cost for this type of pigment is from 35 to 45 cents/lb.

The regular color channel blacks are used for trim paints, drum paints, chassis paints, ship paints and low-cost industrials. Channel blacks of this type are used for tinting and shading where a brownish black is required. The approximate cost of this pigment varies from 15 to 25 cents/lb.

Thermal blacks are very weak in strength and are being used in some rubber finishes where special characteristics are desired.

Acetylene blacks are high in oil absorption and specifically used in wire coatings because of their electrical conductivity features.

Lampblacks are used for tinting purposes, as are some of the large particle size furnace blacks.

Often one type of black will work effectively in a formulation whereas another will not, and vice versa. However, by choosing between the furnace or lampblacks, good tinting can be obtained with minimum float.

Bone blacks have high color value but are very weak in strength. They are used in combination with high-grade channel blacks for lacquers and enamels and are also used for silk screen paints. Their chief use is for tinting purposes.

TINTING PROPERTIES OF COLORED PIGMENTS

One can refer to the definition of tints as interpreted by American Society for Testing Materials as "a color produced by the mixture of white pigment or paint in predominating amount (50% or more) with a color pigment or paint."

To clarify this point, it should be noted that in most cases a nonchalking rutile-type titanium dioxide is used in order to get a true picture of fade. The results listed below are based on an overall picture and vary in different vehicle systems, due to additives or processing by suppliers which may be subject to slight changes.

Chrome Yellows
Primrose or lemon yellows are considered poor for tints, as they fade quite rapidly.
Light chrome yellows are considered fair.

Medium chrome yellows are considered good.
Shading yellows—dirtier shade lemon yellows, are considered fair.
Zinc Yellows
Zinc yellows are considered fair to good.
Basic Zinc Chromate
Basic zinc chromate is not recommended for tints; recommended for primer use only.
Strontium Yellow
Strontium yellow has very good lightfastness in tints.
Nickel Titanate Yellow
Nickel titanate yellow or titanium yellow has very good lightfastness in tints.
Nickel Azo Yellow
This pigment has excellent lightfastness in tints.
Cadmium Yellow
Excellent lightfastness in tints.
Yellow Iron Oxide
Excellent lightfastness in tints.
Hansa Yellow
Very good but recommended for interior use only.
Benzidine Yellow
Fair and recommended for interior use only.
Vat Yellows
Excellent lightfastness in tints.
Chrome Orange
Light chrome orange is considered good but has a tendency to darken.
Medium chrome orange is considered good but has a tendency to darken.
Deep chrome orange is considered good but has a tendency to darken.
Molybdate Orange
Molybdate oranges rated from fair to good.
Cadmium Orange
Light cadmium orange considered excellent.
Deep cadmium orange considered excellent.
"Mercadium" Orange
Considered excellent for tints.
Benzidine Orange
Rated fair and recommended for interior use only.
Dinitraniline Orange
Rated good but recommended for interior use only.
Vat Dye Orange
Anthraquinone types have excellent fade resistance.
Chrome Greens
Light chrome green considered good to fair.
Medium chrome green considered good.
Deep chrome green considered good.
Chromium Oxide
Considered excellent for tints.
Hydrated Chromium Oxide
Considered excellent for tints.
Copper Phthalocyanine Green
Considered excellent for tints.

Organic Green Toners
PTA and PMA have only fair durability and are recommended for interior use only.
Iron Blues
Considered good to fair depending on treatment.
Copper Phthalocyanine Blue
Considered excellent in any tint shade range.
Ultramarine Blue
Considered poor to good depending on vehicle system.
Organic Blue Toners
PTA and PMA blue toners recommended for interior use only; rated poor to fair.
Indanthrene Blue
Considered excellent in tints of all ranges.
Carbazole Dioxazine Violet
Considered excellent except in *high* bake white finishes.
Organic Violet Toners
PTA and PMA violets not recommended for exterior use. Fair in tints for interiors.
Mineral Violet
Not recommended for exterior. Excellent for interior finishes only.
Lithols
Rated fair and for interior use only.
BON (β-Oxynapthoic)
Rated as poor to fair in tints.
Toluidine Reds
Rated as fair but would not recommend use for tints in white due to bleeding character-
istics.
Para Reds
Rated as poor for tints.
Lithol Rubines
Rated as good but for interior use only.
Chlorinated Para Red
Rated as good but for interior use only.
Quinacridone Reds and Maroons
Rated as excellent.
Red Iron Oxide
Rated as excellent.
Cadmium Reds and Maroons
Rated as excellent.
"Mercadium" Reds and Maroons
Rated as excellent.
Red Lead
Not recommended as a tinting color.
Thioindigo Reds and Maroons
They can be rated from poor to very good depending on type. "Thiosafast" and "Thio-
fast" are rated good. Most others rated poor in light tints.
Arylide Maroons
Rated as good for interior use only. Would not recommend as a tinting color of whites
due to bleeding characteristics.
Siennas and Umbers
Rated as excellent for tinting.

REFERENCES

1. E. I. du Pont de Nemours & Co., *Tech. Bull.* (April 1956).
2. Sherman, L. R., Hercules Powder Co., Imperial Color and Chemical, "Colored Pigments for Chemical Coatings," March 1962.
3. Spengeman, W., E. I. du Pont de Nemours & Co., "Tinting Colors."
4. Treade, M., Reichard-Coulston, Inc., "Iron Oxide Pigments."
5. Venuto, L. J., Columbian Carbon Co., 1960.
6. Wormald, G., E. I. du Pont de Nemours & Co., "Pigment Colors for Paints."

22

*Metallic Soaps: Driers, Suspending Agents, Flow Modifiers, Flatting Agents and Sanding Aids**

METALLIC SOAPS

Metallic soaps are the metal salts of higher monocarboxylic organic acids. A generic name for this family of chemical compounds is *metallic carboxylates*.

The significant chemical effects of metallic soaps or carboxylates in coating applications are imparted by the metallic or cationic section of their individual molecules, the organic radical portion of the respective molecules functioning as a carrier or means of obtaining compatibility with coating vehicles. To illustrate, the metallic cations of driers exert the catalytic activity which accelerates or controls the oxidative polymerization mechanism of oleoresinous film hardening.

The significant physical effects of metallic carboxylates in coatings originate in both the organic radical and the cationic portions of their molecules. The relative importance of each of these factors in the realization of physical effects depends upon the molecular composition of specific metallic carboxylate compounds. Examples of physical effects are the pigment suspension, flow modification and gloss reduction obtainable by using specific metallic soaps of fatty acids in oleoresinous coating sys-

*By Siegfried Meinstein and H. Russell Spielman, Witco Chemical Co., Inc., Organics Div., Chicago, Ill. The authors dedicate this chapter to Edward F. Wagner, Ph.D. (Ch.E.), who, by his own example, is a continuous source of inspiration and motivation to his associates.

tems. As will be shown later, the ability of a metallic soap to suspend pigments or to modify flow (rheological) properties is governed by its balance of polar versus nonpolar tendencies. The flatting effect of an appropriate metallic soap depends upon its particle size and solubility characteristics in coating vehicles. Other benefits are directly related to the inherent water repellency and unctuousness of certain metallic carboxylates.

The following tabulation indicates the principal metals and organic acids that are combined selectively to form individual metallic carboxylates useful to the coating industry:

Metals				*Organic Acids*	
Aluminum	Cobalt	Copper	Stearic	Naphthenic	Oleic
Calcium	Lead	Mercury	Palmitic	Octoic	Linoleic
				(2-ethyl-	
Zinc	Manganese			hexoic)	
Magnesium	Zirconium			Tall oil	
	Iron			fatty acid	

Let us discuss briefly the methods and reactions involved in the manufacture of metallic carboxylates. Two processes, *fusion* and *precipitation,* are used in general industrial practice; often both are used by a single manufacturer. The choice of method depends upon both the product properties attainable and the degree of process feasibility for any specific metallic carboxylate product.

In the fusion process, a metallic oxide, metallic hydroxide or salt of a weak acid is reacted directly with organic acid at elevated temperature according to the following general equation:

$$2RCOOH + MeO \longrightarrow Me(RCOO)_2 + H_2O$$

The precipitation process involves a double decomposition reaction in aqueous medium. In this method, a dilute aqueous solution of neutral sodium soap is initially prepared by reaction of caustic soda with the desired organic acid. An inorganic salt solution of the desired metal is prepared separately and is then added to the sodium soap solution to precipitate the metallic soap product. The resulting sodium salt by-product must be removed by thorough water washing. These reactions proceed as indicated:

$$RCOOH + NaOH \longrightarrow RCOONa + H_2O$$

$$2RCOONa + Me^{++} \longrightarrow Me(RCOO)_2 + 2Na^+$$

Metallic carboxylates useful in coatings appear commercially both in finely ground powder form and as liquids having viscosities low enough to permit easy handling in coating production operations. The powders are typified by the metallic stearates and allied fatty acid soaps which are used primarily for their physical effects. Driers are typical liquid carboxylates.

Powder-form metallic carboxylates for coating applications are preferably produced by the precipitation process. This method is suitable because it offers excellent control of particle size and provides an in-process flexibility which permits useful modifications of procedures and reaction conditions. These process changes help to attain carboxylate products that have specifically desired performance properties. Examples of powders produced by precipitation are zinc stearate, calcium stearate, magnesium stearate and various compositions of aluminum stearate.

Liquid-form metallic carboxylate products, depending upon type, grade and intended use, can be either intrinsic fluids or solutions. They are produced by fusion or precipitation, the method varying with the individual products. The standard drier grades are solutions of respective carboxylates in hydrocarbon solvent. In the precipitation of driers, a two-phase process medium (water-hydrocarbon solvent) can be used.[18,19] As precipitation progresses, the metallic soap reaction product dissolves and remains in the solvent phase. The resulting drier solution is washed, thoroughly dehydrated, precisely adjusted to prescribed metal content, filtered and then packaged.

A third process, the *direct metal reaction method*,[12] is employed in the manufacture of some driers, as typified by the reaction yielding cobalt naphthenate directly from cobalt metal and naphthenic acid:

$$2Co + 4RCOOH + O_2 \longrightarrow 2Co(RCOO)_2 + 2H_2O$$

Metallic carboxylate molecules have polar properties resulting from their metallic (cationic) groups, as well as nonpolar properties originating from their hydrocarbon groups. Especially with powdered metallic soap products, the qualities of polar-nonpolar balance specific to the various product compositions permit controllable surface-active and colloidal phenomena. A range of useful functions for coating applications has been made possible through judicious adjustment of the metal and organic acid compositional variables by manufacturers of metallic carboxylates.

To illustrate how the type of metal can affect metallic soap performance, under proper incorporation conditions an aluminum distearate can be caused to disperse colloidally in a hydrocarbon fluid medium and subsequently impart marked modification of the medium's rheological properties. In opposition, a zinc distearate made from the same fatty acid and incorporated under identical conditions into the same hydrocarbon fluid would simply disperse like an inert pigment. Nevertheless, these opposite effects are both employed in coating technology.

The equal importance of the influence exerted by the type of organic acid can be demonstrated by the ability of aluminum octoate to form gels rapidly with low-boiling hydrocarbon solvents at room temperature, whereas any aluminum stearate mixed in the same manner with the same solvent would require heating, then cooling, to achieve similar results.

Further, the polar-nonpolar balance of certain metallic carboxylates can be affected for application purposes by varying the ratio of metal to organic acid in their molecular compositions. Aluminum mono-, di- and tristearates each contain different amounts of metal; accordingly, each causes a different degree of flow modification to occur after incorporation in a given hydrocarbon fluid when all other conditions are held constant.

The following list illustrates the major physical functions of the family of powdered metallic carboxylates in coating systems:

> Gloss reduction (flatting)
> Pigment suspension and control of settling
> Modification of rheological properties
> Pigment-grinding assistance
> Improvement of film water resistance and washability
> Reduction of coating penetration into porous substrates
> Improvement of pigment tinting effects
> Facilitation of lacquer film sanding

The large variety of metallic carboxylate compositions available in powdered form enables a wide choice of characteristic properties such as solubility, gelling power and particle size. Each grade of metallic soap is prepared from an organic acid or blend of organic acids selected to yield a product of predetermined functionality. In the manufacture of these soaps, correct choice of raw materials, close tolerances in conditions of precipitation, thorough washing and drying of precipitates, proper grinding, avoidance of contamination and rigid analytical control[23] of raw materials and finished soaps are required to yield quality products of reliable performance.

Liquid metallic carboxylates, in addition to their major function of drying oleoresinous coatings, are used as antifouling additives in marine paints (copper soaps); water displacers in the manufacture of flushed colors (zinc and lead naphthenates); polymerization catalysts and promoters for epoxy, unsaturated polyester, silicone and urethane resins (tin, zinc, cobalt and lead octoates); fungicides and mildewcides (copper, zinc, phenylmercury and organotin carboxylates), and as alcoholysis catalysts in the manufacture of alkyd resins (lithium soaps).

In the manufacture of the liquid carboxylates, the proper choice of raw materials, exacting quality control, diligent production care and thorough removal of reaction by-products such as moisture and inorganic impurities are necessary for the realization of stable, efficient products. In the development of the solution grades, special attention must be given to the stability of these products at the various ambient temperatures likely to be encountered in their transport, storage and use.

DRIERS

A drier is an additive to coatings which accelerates or controls the hardening of applied films by catalyzing the oxidative polymerization of oleoresinous vehicle components. To be effective, driers must be miscible in unpolymerized and polymerized drying oils and resins, in mixtures of these, and in solvent thinners normally used in coating formulations.

The liquid-form metallic salts of organic acids are of greatest convenience and are the most widely accepted for this purpose. Liquid driers are manufactured to $\pm 0.1\%$ metal concentration tolerance and at the highest practical metal content consistent with pourability and handling ease for the user. Drier products of the lightest possible color are necessary to minimize potential tinting effects in coatings.

Mechanism of Drying

As indicated previously, the drying activity of metallic carboxylates is due to their metallic cations. In the absence of any basic experimental results on the specific chemical nature of drier function, it has generally been thought that the essential processes by which driers accelerate the hardening of films are those of oxidation and polymerization or a combination thereof. Cobalt and manganese, which readily undergo reversible valence changes, have been considered to be oxidation promoters. It has been supposed that lead accentuates the rate of polymerization.

While the full scope of chemical reaction mechanisms and the precise manner of drier involvement in film formation of oleoresinous compositions have still not been completely elucidated, considerable understanding has been gained in recent years.

It has been shown that the process by which applied drying oils or resins are transformed into hard, continuous, protective coatings is one of autoxidative polymerization.[1] Atmospheric oxygen is absorbed at activated alpha-methylene groups adjacent to one or more nonconjugated double bonds of a fatty acid portion of the oil or resin molecule. The result is formation of the hydroperoxide:

$$-CH-CH=CH-$$
$$|$$
$$OOH$$

Double bonds, particularly when unconjugated, activate adjacent methylene groups. The hydroperoxide may then isomerize to the conjugated form, and polymerization will proceed via free radicals as the unstable hydroperoxide dissociates.

If we depict an oleoresinous molecule as RH, a polyunsaturated fatty acid ester, then as S. R. W. Martin has shown,[10] the hydroperoxide, ROOH, may dissociate into the free radicals: $\cdot$RO, R$\cdot$ and $\cdot$OH, which may lead to the following linkages:*

RO—OR   (peroxide)
R—O—R   (ether)
R—R   (carbon-carbon)
R—OH   (hydroxylated ester)

Water, oxygen and hydrogen peroxide form as by-products. In addition to the autoxidative polymerization, certain other oxidation reactions will occur.[14] These will frequently involve chain scission and yield volatile and nonvolatile decomposition products such as formic and acetic acids, aldehydes and related compounds.

Studies of the chronology of film formation[11] show that after a variable induction period of a number of hours, during which no weight changes occur, an oxygen intake of some 14% takes place. Upon completion of drying, about 5 to 8% of oxygen, based on the original weight of the oil film, is retained as part of the polymer.

To accelerate and guide the film-forming free radical reactions, metallic driers must be soluble in the oil or resin and remain so throughout the polymerization process. Since this type of mechanism normally involves homogeneous catalysis, the reactive oxygen should also be readily available in the homogeneous liquid medium.

As will be shown, driers should help satisfy this latter requirement.

Further, to perform as a catalyst, the drier metal should either be capable of existing in two valence states, the lower one being the more stable, or the metal should be capable of forming coordination compounds. It should be present as available cations rather than stable complexes.

Driers are believed to increase reaction rates of the autoxidative polymerization of oleoresinous vehicles catalytically in several ways:

(1) They practically eliminate the induction period by overcoming the inhibiting influence of small amounts of natural antioxidants normally present in oils (i.e., by promoting their destruction via oxidation).

---

*Cobaltous and manganous soaps are used as accelerators in the polymerization of unsaturated polyesters. Cobaltous compounds also serve as catalysts in the manufacture of adipic acid from cyclohexane. Both types of reactions are of a free radical nature and involve hydroperoxide formation.

(2) Cobaltous and manganous driers are believed to act as oxygen carriers which make the oxygen available in the proper physical state for homogeneous reaction at reactive sites.

(3) Lead driers are thought to accelerate through-drying at the interiors and bottoms of films largely by promoting hydroperoxide decomposition and also by influencing the polymerization via formation of intermediate coordination compounds.

(4) The rate of oxygen absorption is definitely increased by the presence of driers.

(5) Total oxygen retained in dried films is less when driers are used.[15] This would indicate a greater amount of more stable carbon-carbon bonding than carbon-oxygen linking (ether, ester and alcohol structures). In substantiation of this, the resistance of films to water and alkali was found to improve on aging when optimum drier amounts were used, as compared to film formation without driers or proper drier combinations.[4,16]

(6) Driers are thought to increase the rate of cross-linking by promoting hydroperoxide decomposition.[16]

Classification of Driers by Type of Acid

The organic acids from which driers are produced are selected to insure that the acid radical part of the metallic carboxylate molecule acts as an effective means of physical transport for the catalytic metal portion of the molecule in all of the organic media into which driers are likely to be introduced. This miscibility requirement narrows the field of "candidate" acids considerably. Classified according to type of organic acid, only three groups of metallic carboxylates are currently of proved and established suitability: the metallic naphthenates, octoates (2-ethylhexoates) and tallates. Concurrently with this writing, driers made from neodecanoic acid are appearing commercially and are being evaluated by coating manufacturers.

Naphthenate driers remain the most widely employed and offer excellent overall properties such as good stability, light color, low viscosity, high oxidation resistance and very good compatibility with solvents, oils and resins.

Octoate driers, prepared from 2-ethylhexoic acid, a synthetic acid of precise and uniform composition, also give excellent performance. The octoates have the best color, lowest odor and lowest nonvolatile content of the three drier groups. In some alkyd systems, octoates function more efficiently than naphthenates. Octoates are especially effective under drying conditions of high atmospheric humidity.

Tallate driers, in many applications, can replace naphthenates on a direct pound-for-pound basis. Although tallates can occasionally demonstrate less stability and lower oxidation resistance than naphthenates and octoates, their overall good performance has established them as highly acceptable materials. Through constant developmental effort and progressive improvement in available tall oil acid raw materials, drier producers have consistently enhanced the color and stability of tallate driers over the past decade.

Metals

Table 22.1 lists driers in general use, their usual metal concentrations as supplied and ranges in weight percentages of metal normally employed in the drying of oleoresinous vehicles.

TABLE 22.1

Driers[a]	Per Cent Metal Concentration Used[b]
6% cobalt	.005–0.2
24% lead	.1 –5.0
6% manganese	.005–0.2
4% calcium	.01 –0.5
8% zinc	.01 –1.0
6% iron	.02 –0.1

[a] The percentages represent the metal content of the respective driers themselves as they are normally commercially available in the form of naphthenate, octoate or tallate.

[b] The figures represent the weight percentage of metals added to formulations and are calculated on the basis of vehicle solids. Metals are generally added in concentration ranges shown. The optimum amount of each metal will depend on specific vehicles in formulations.

Cobalt is the most potent of the commercially available drier metals and is known as a top-drying metal because it is operative at the film surface and induces the initial tack-free state of film hardening. Although cobalt is probably the most generally useful drier metal available to the coating formulator, it must be used with due caution. The introduction of excessive concentrations may cause overactivity at the film surface and a resultant wrinkling or reduction of film durability. Because of the high activity of all cobalt driers, they should be added to paints and enamels after pigment grinding has been completed. Additions to varnishes should be at the lowest practical temperature. Thorough incorporation of cobalt driers into vehicles is necessary in order to avoid surface "skinning" tendencies and to insure uniform surface drying.

Lead is described as a through-drying metal in that it preferentially promotes the hardening of the interior and bottom of the film to a durable, insoluble, yet flexible state. In coating formulations, lead is used in larger amounts than any other drier metal. Apart from drying activity, lead driers often improve the water and salt spray resistance of films. Lead naphthenate possesses the property of improving the pigment wettability of some vehicles during grinding, and it is also an effective rust inhibitor. Lead is rarely used alone, but rather in combination with cobalt, manganese and possibly an auxiliary metal. Lead driers can react directly with the free carboxylic groups in alkyd resins, thus for this type of vehicle it is preferable to incorporate lead at room temperature. Lead driers should not be added to aluminum paints because they interfere with leafing, nor should lead be used in sulfide-resistant coatings and in low-toxicity finishes such as those for baby furniture and toys. Lead has the inherent disadvantage of being the least compatible of all drier metals in alkyd and varnish vehicles because of its occasional reaction with such vehicles to form insoluble lead compounds. Calcium drier, an auxiliary, is commonly co-added to troublesome vehicles to improve the stability of lead.

Manganese is a highly useful drier metal in that it offers drying characteristics similar to those of both cobalt and lead, specifically a combination of top- and through-drying. When used alone, manganese can cause excessive film brittleness which can be modified by the use of an auxiliary drier. Manganese can replace cobalt in formulations containing lead drier. It finds much application in baking finishes. A combination of manganese and cobalt results in a film hardness not obtainable with cobalt alone.

Iron driers are used for selected purposes where effective drying is required. Iron has a mild catalytic effect at room temperature, but very pronounced activity at elevated temperatures. Iron driers are therefore extensively used in baking finishes. Because of the tinting effect resulting from the natural deep-brown color of all iron driers, their use is generally restricted to darker-colored finishes. Iron naphthenate improves the dispersion of carbon black pigments and thereby helps to give glossier films.

Calcium is a highly versatile auxiliary drier metal. For lead-free finishes where cobalt or manganese is present, the lead drier requirement can be replaced by calcium without sacrifice of drying time. Each 1% of lead metal in the coating can be replaced by approximately 0.25 to 0.33% of calcium metal. Calcium driers improve the hardness and gloss of soft oil varnish films. Calcium driers also stabilize lead driers against precipitation in various reactive vehicles and exert a beneficial solubilizing action on all driers. Calcium naphthenate is a useful wetting agent and dispersing aid for pigment grinding.

Zinc driers are auxiliaries and are true driers almost in name only. Their effects are nevertheless highly advantageous. Zinc, in combination with cobalt, is used to prevent film wrinkling. This is accomplished by the ability of zinc to hold the film surface open longer and thus bring about a more uniform drying rate throughout the entire film cross section. When used for this purpose, zinc retards the initial hardening of the surface, but the overall drying time is generally not extensively affected. In combination with cobalt, zinc (similarly to manganese) yields film hardnesses which would not be attainable with cobalt alone. In zinc-cobalt formulations, the amount of zinc metal used is generally two to three times the amount of cobalt present. Zinc naphthenate is an important wetting and dispersing aid for pigment grinding and often prevents flooding and floating of colors. Zinc is also known to contribute higher gloss and better color retention to enamels.

Zirconium driers[8,9] offer a synergistic action with cobalt and manganese which enhances drying efficiency. Zirconium also supplements the auxiliary activity of calcium and can replace a portion of the cobalt and manganese requirement. Lead, however, can be replaced completely by zirconium. Zirconium driers are claimed to have no adverse effect on the leafing of aluminum pigments.

Drier Selection

The importance of a proper balance of drier metals in coating formulations should not be underestimated. Several factors influence the selection of a proper drier combination for any individual coating application:

The type of drying oil or resin is the most important factor because each has its own appropriate drying requirement. In general, the more saturated the oil or resin, the greater are the quantity requirements for drier metal. However, since in the drying process the nature and stability of peroxide intermediates varies with the ratio of conjugated to unconjugated double bonds, the degree of conjugation will also have to be considered in making the best drier selection.[1] Raw oils may require more drier than heat-bodied oils of the same viscosity. Hard oil varnishes customarily contain a 1:10 cobalt-to-lead ratio; for softer oils a higher ratio is desirable. Linseed oil varnishes can contain typically a cobalt drier or a combination of manganese, lead and calcium for good stability and durable film characteristics. Iron can remove residual tack in fish oil paints. A combination of cobalt, lead and calcium will reduce tack-free time in tall oil ester varnishes. In short oil varnishes, the type of resin plays an important part only when the weight of resin approaches or exceeds the weight of oil. In such varnishes, the proportion of top drier to lead can be increased without incurring wrinkling. Maleic baking varnishes typically contain small amounts of cobalt in combination with zinc.

Pigments of some kinds are troublesome in that they can cause loss of drying activity by adsorption of driers. Additional comments on this problem will appear subsequently in this chapter. Separately, pigments reactive with vehicles can affect drier requirements.

Performance requirements and drying conditions have as much influence on the selection of driers as does the composition of the coating. Considerations such as desirability of "wet edge" for brushing, requirements for rapid drying from a service or utility standpoint, the applied film thickness, the availability of sunlight and ventilation, and the ambient atmospheric temperature and humidity all have a role in establishing the drier requirements for individual types of coating formulations. For example, the thicker wet films need less top drier to avoid wrinkling; as mentioned, zinc can also be used to advantage in such cases. High-humidity conditions require relatively more top drier, whereas high temperatures reduce the overall amount of drier needed.

General Suggestions for Handling and Testing of Driers

Suffcient time must be allowed for the drier to "sweat" into an experimental coating formulation. A minimum of overnight standing is necessary before testing for drying. If drying is unsatisfactory, the tests should be repeated after 24 hours of additional standing.

Driers should be added to varnishes with the use of good agitation after the batch has been thinned and cooled. The batch should be packaged promptly to minimize any possible "skinning" tendency.

Top driers should be added to paints and enamels after pigment grinding, during reduction of the paste. Uniform and complete distribution of these driers through efficient mixing is mandatory for optimum activity.

Drier Dosages and Calculations

The drier dosage recommendations for a variety of vehicles will be found in Tables 22.2 through 22.6.[24] These data have been compiled primarily from various vehicle manufacturers' reports and can serve as starting point information. Specific suggestions cannot be made because each coating formulation must be treated individually; all of the foregoing factors must be weighed carefully.

As shown in these tables, drier additions are conventionally expressed as *per cent metal*, based on the weight of the vehicle solids. All commercially available driers are identified as to the per cent metal contained. The following formula serves as a convenient means of calculating drier additions:

$$\text{Pounds of Liquid Drier Required} =$$

$$\frac{(\text{Pounds of Vehicle Solids}) \times (\text{Concentration of Metal Required})}{(\text{Per Cent Metal Concentration of Drier Used})}$$

Driers in Water-based Paints

Oil- or alkyd-modified latex (polyvinyl acetate, acrylic or butadiene-styrene) systems and emulsified alkyd or oil paints depend on driers for

TABLE 22.2. Drier Recommendations for Air-Drying Oils—Basic Starting Points

| Type of Oil | Driers (%) Metal on Oil | | | Remarks |
	Co	Pb	Mn	
Dehydrated castor, bodied	0.03–0.10	0.2–0.5	0.0–0.02	—
Fish, bodied	0.04–0.08	0.5–0.8	—	—
	0.05	—	0.05	+0.3% Fe for Dark Paints
Fish, pressed	0.05–0.10	0.2–0.5	—	For high stearin content oil use low Pb, high Co
Fish, refined	0.05	0.5	—	—
Linseed, blown or heat-bodied	0.03	0.25	—	—
Linseed, raw	0.02–0.03	0.20–0.25	—	—
	—	0.20–0.25	0.02–0.03	—
	0.03–0.04	—	—	—
	—	0.15	0.01	"Boiled Oil" 16 hr dry
	0.015	—	0.02	"Boiled Oil" 16 hr dry
	0.01–0.03	0.3–0.6	0.02–0.04	Exterior white paints
	0.025–0.05	—	0.025–0.05	Iron oxide paints
Linseed, refined	0.02	0.15–0.20	—	—
	—	0.15–0.20	0.02	—
	—	0.10	0.01	"Boiled Oil" 16 hr dry
	0.01	—	0.02	"Boiled Oil" 16 hr dry
Oiticica	0.03–0.10	0.3	0.0–0.03	—
Perilla	0.02	0.15	—	—
Safflower	0.05	0.5	0.02	—
Soyabean, blown	0.05–0.10	—	—	—
Soyabean, heat-bodied	0.05–0.10	—	—	—
Soyabean, raw	0.02–0.03	0.4–0.5	0.03–0.07	—
Tall oil	0.01–0.05	0.5–1.0	—	—
Tung	0.02–0.04	0.15–0.40	—	—

TABLE 22.3. Drier Recommendations for Tung Oil Varnishes and Paints— Basic Starting Points

| Type of Resin | Driers (%) Metal on Oil | | |
	Co	Pb	Mn
Coumarone	0.03–0.05	0.3–0.5	—
Maleic	0.02–0.05	0.3–0.7	—
Natural	0.03–0.05	0.3–0.5	—
Phenolics (modified)	0.05–0.10	0.3–0.5	—
Phenolics (pure), heat reactive	0.03–0.10	0.3–0.5	0.00–0.02
Phenolics (pure), nonreactive	0.01–0.10	0.1–0.5	—
Rosin esters	0.01–0.10	0.3–0.7	0.00–0.03

TABLE 22.4. Drier Recommendations for Dehydrated Castor Oil Varnishes and Paints—Basic Starting Points

Type of Resin	*Driers (%) Metal on Oil*			
	Ca	*Co*	*Pb*	*Mn*
Coumarone	—	0.03–0.05	0.5	—
Maleic	—	0.07–0.10	0.3–0.5	0.0–0.02
Natural	0.0–0.1	0.03–0.10	0.2–0.5	0.0–0.01
Petroleum	—	0.05–0.10	0.5	0.0–0.05
Phenolics (modified)	—	0.05–0.10	0.5	0.0–0.05
Phenolics (pure), heat reactive	—	0.03–0.10	0.3–0.5	0.0–0.02
Phenolics (pure), nonreactive	—	0.03–0.10	0.3–0.5	0.0–0.02
Rosin esters	—	0.05–0.10	0.2–0.5	0.0–0.02

TABLE 22.5. Drier Recommendations for Linseed Oil Varnishes and Paints—Basic Starting Points

Type of Resin	*Driers (%) Metal on Oil*			
	Co	*Pb*	*Mn*	*Remarks*
Coumarone	0.03–0.08	0.2–0.5	—	—
Maleic	0.05–0.10	0.5–1.0	—	—
Natural	0.02–0.10	0.2–0.5	—	0.0–0.2% Ca
Petroleum	0.00–0.10	0.3–0.5	0.0–0.10	Baking 0.2% Mn, 0.2% Fe
Phenolics (modified)	0.02–0.10	0.2–0.7	0.0–0.05	To eliminate hazing add 0.02–0.1% Ca
Phenolics (pure), heat reactive	0.03–0.10	0.3–0.5	0.0–0.05	—
Phenolics (pure), nonreactive	0.02–0.10	0.2–0.5	0.0–0.02	—
Rosin esters	0.04–0.15	0.2–0.7	—	0.05–0.2% Ca

In aluminum vehicles, eliminate lead, increase cobalt.
For baking, 0.02–0.03% manganese or cobalt.
Long oil varnishes will usually require the upper limits of drier values given above.

TABLE 22.6. Drier Recommendations for Alkyds—Basic Starting Points

Type of Oil	*Length*	*Driers (%) Metal on Nonvolatile Content of Vehicle*			
		Ca	*Co*	*Pb*	*Mn*
Dehydrated castor	Short	—	0.05	0.3–0.6	0.0–0.03
	Medium	0.1	0.05–0.07	0.4–0.7	—
	Long	—	0.05–0.10	0.5–0.8	—
Linseed	Short	—	0.05	0.3–0.5	0.03
	Medium	—	0.03–0.05	0.3–0.5	0.03
	Long	—	0.02–0.05	0.5–0.8	0.02–0.03
Soyabean	Short	0.1	0.05	0.1	0.03
	Medium	0.1	0.01–0.05	0.2–0.3	0.0–0.03
	Long	0.1	0.03–0.07	0.3–0.6	0.0–0.03
Tall oil	Medium	0.0–0.5	0.05–0.10	0.0–0.2	0.0–0.02

For baking, 0.01–0.02% cobalt

proper film formation. The amount of drier required is calculated on the basis of the total oil or alkyd component in the same manner as that shown above for oleoresinous vehicles. In many instances, the conventional drier grades can be employed in these types of paints; however, care must be exercised to obtain uniform and complete dispersion of these driers.

Emulsifiable grades of driers are offered specifically for ready incorporation into water-based systems by simple mixing procedures. Stable and uniform drier dispersions are easily attained with these grades. A special application of emulsifiable cobalt drier is found in unmodified butadiene-styrene latex paints. In these systems, cobalt accelerates the cure of the film and permits earlier washability.

Other Driers

Other types of driers and drier promoters exist and will be mentioned briefly here. These are not as commonly used as the previously discussed materials because their applications are more specialized. o-Phenanthroline[21,22] solution, which is not a metallic carboxylate, forms weak chelating complexes with cobalt and manganese drier metals and increases their activity. The presence of chemically available zinc, which may be present as driers, or as zinc oxide pigment in acidulous vehicles, is detrimental to the catalytic activity of this additive. Vanadium has been reported by Klebsattel[5] to be among the most active drier metals, but it remains experimental. Carboxylate driers based on rare-earth metals, such as cerium and cerium-lanthanum combinations, have found some application both in baked finishes and in fume-proof paints in which these metals contribute good color retention.

Formulation Problems

Mention has already been made of the problem of decreased drying activity which results from adsorption of driers by pigments. Certain pigments, notably carbon blacks, iron oxide, titanium dioxide and selected inorganic colors, can be particularly troublesome. Adsorption of driers is evidenced by a progressive loss of drying activity as a packaged coating formulation shelf-ages. The presence of calcium, zinc, lead or iron naphthenate in the pigment grind helps to minimize drying loss. When tolerable, litharge added to the pigment before grinding is an old but sometimes effective remedy. "Feeder" driers (which are specialty additives containing drier metals in semi-soluble form), *ortho*-phenanthroline solution and zirconium drier have been found to be beneficial in individual situations. Rebalancing of drier ratios often helps, but the temptation to increase metal concentrations beyond normal as a compensation can be hazardous. Coating formulations which have lost drying activity usually can be restored to original drying rate by regrinding. As with many other formulating problems, it is best to approach each drying loss situation on an individual basis.

"Seeding," which is the occurrence of an unexpected and undesirable rough-textured appearance of the coating surface after drying, is caused by an incompatibility of ordinarily soluble formulation components. This condition, although occasionally caused by reaction of lead driers with acidulous vehicles, is more often brought about by other sources such as the presence of highly polymerized particles of vehicle or moisture contamination.

"Skinning" is a term describing the formation of a solid or gelatinous skin on the surface of packaged coatings during normal storage. This condition is caused by oxidative polymerization of the vehicle at the air-liquid interface. The problem can be created by incorrect balance of driers, especially excess cobalt. Vehicles which have a high degree of pre-polymerization are highly susceptible. In addition to drier adjustments such as reduction of cobalt content and addition of auxiliary metals, special antiskinning additives can be included in the coating formulation to inhibit surface polymerization. Compounds offered for this purpose include phenolic derivatives and oximes. Their effectiveness is quite specific for individual vehicles, and therefore careful trials are necessary. Inclusion of more powerful solvents such as aromatics in the volatile portion of the vehicle will help to alleviate skinning in some instances.

The mechanism of "gas checking," a surface deformation of dried films exposed to noxious atmospheric or baking oven gases, is still the subject of investigation. Vehicle-drier relationships are known to be involved with its occurrence, and the presence of manganese drier has been demonstrated to initiate or aggravate this condition in outdoor-exposed varnish films containing cobalt drier.[2] The use of calcium drier is often effective in inhibiting gas checking. When encountered in baked coatings, changes in drier metals and amounts, consistent with good color retention, can be tried.

In some cases, a permanent surface "hazing" or dulling of dried paints, and in others the development of a removable "bloom" on the dried paint surface, can be related to drier balance as well as to other sources. Presence of excess cobalt in pentaerythritol-based alkyds has been found to encourage blooming.[3] When minimum cobalt was used with these alkyds, the addition of calcium and lead was reported to assist in controlling blooming.[3] Drier balance is only one of many possible sources of surface hazing. When choice of driers is implicated in this problem, a general reduction of drier levels is indicated.

SUSPENDING AGENTS

A suspending agent is any material added to a coating system in small amounts for the purpose of improving the stability of pigment dispersion and limiting the degree, as well as controlling the nature, of pigment settling during prolonged storage of the coating system. Suspending agents practically prevent the settling of pigment particles in some formulations and retard settling to reasonable and practical rates in others. Of equal importance, these agents also prevent the hard caking or packing of any pigment particles which may eventually settle out and thus impart a "soft settling" or easy resuspendibility property to such particles in packaged coatings.

Powder-form metallic soaps of fatty acids find application for suspension purposes in oleoresinous coatings. These soaps offer three separate modes of action in the mechanism of suspension.[17]

First, the metallic carboxylates aid in pigment wetting and dispersion during the grinding operation through the surface-active functionality of their molecular structures. The respective metallic or polar sections of individual metallic soap molecules are strongly attracted and held to surfaces of pigment particles. The long nonpolar organic chains of the soap molecules are thereby oriented outward toward the vehicle medium. This imparts a lipophilic nature to the pigment particle surfaces and thus improves particle wettability by the vehicle and facilitates dispersion of the particles in the vehicle as discrete units.

After pigment wetting and dispersion have been accomplished, a limited degree of pigment flocculation* is required to preclude hard caking at the bottoms of containers and to provide easy resuspendibility to particles which may settle. This is especially necessary when working with vehicles of high pigment wetting power or with dispersed pigments which have been highly wetted by the vehicle. Flocculated pigment particles in dispersion are loosely bonded to each other and thereby form soft clusters having relatively low packing tendency. The metallic soap molecules attached to the pigment particle exteriors provide a desirable degree of flocculation by formation of polar-nonpolar bridges from one pigment particle to another.

Separately, specific metallic carboxylates promote pigment suspension by modifying the rheological properties of the coating system. Flow modification as a major function of metallic soaps will be treated separately in this chapter. However, a salient feature of selected soaps in

---

*An important distinction must be realized between limited and excessive degrees of pigment flocculation. The former is generally desirable from the standpoint of settling characteristics, whereas the latter is objectionable and indicates an overt condition of dispersion instability.

relation to suspension is their ability to establish a three-dimensional network of colloidal gel structure throughout the vehicle when they are properly incorporated therein. The reinforcing effect of this structure increases the consistency of the coating, hinders its free flow and counteracts the pull of gravity on suspended pigment particles.

The various types of carboxylates useful for suspension effects are numerous, yet they are rather specific for individually desired degrees of effectiveness and separate application categories. The fatty acid soaps of aluminum, zinc and calcium are of major utility. Aluminum mono-, di- and tristearates, intermediate compositions of aluminum stearate, and aluminum palmitate are all employed for suspension. The commonly used zinc soaps include zinc stearate and zinc palmitate. The preferred calcium carboxylate is the stearate.

The choice of metallic carboxylate type depends upon vehicle composition, pigments used, and the desirability or nondesirability of simultaneously imparting other properties coexistent with pigment suspension. The derivable corollary effect of greatest utility is flow modification.

Metallic soaps for suspension may be incorporated into coating compositions in several ways. When possible, the addition of a soap-solvent pre-gel is usually preferable, since this procedure is best for raw material economy and uniformity of results. This method is applicable primarily to the aluminum soaps, because these can be solubilized in most nonpolar vehicle thinners to yield stable gels. A "cooked" pre-gel masterbatch can be prepared by mixing an aluminum soap at approximately 5% concentration by weight into a hydrocarbon solvent, heating the mixture with agitation to 180°F to solubilize the soap and then allowing the preparation to cool to room temperature. The result is a gel. Alternatively, in situations where cooking is not desirable or practical, a pebble mill or intensive shear dispersion equipment may be used for preparation of aluminum soap pre-gel masterbatches. Portions of pre-gel masterbatch are added to coating formulations in proper amounts as desired.

Another satisfactory method of incorporation is the direct co-dispersion of metallic soap and pigment in the vehicle by means of a grinding mill or an intensive shear apparatus. This latter technique is applicable to all of the soaps generally used for suspension purposes.

Like the choice of soap type, the amount of soap to be used also depends upon the composition of the system and the desired primary and secondary effects. The typical use concentration of aluminum stearates, as an example, is about 2% based on the total pigment weight, or 3 to 8 pounds per 100 gallons of coating.

The coating formulations shown on p. 411 illustrate the use of metallic stearates for the combined purposes of pigment suspension and flow modification.

FLOW MODIFIERS

As previously mentioned, certain metallic carboxylates can bring about rheological changes in oleoresionous coatings. The coating formulator takes excellent advantage of these available modifications of flow properties. Metallic soap additives for this purpose are usually incorporated into formulations by use of the same methods as indicated above for pigment suspension agents. Often the desired result is a combination of both suspension and flow modification.

With the use of proper processing techniques, the applicable metallic soaps will establish colloidal gel structure throughout the vehicle of an oleoresinous coating. The reinforcing action of this structural network in the vehicle modifies the flow properties of the coating, as explained in the foregoing discussion of suspending agents.

Most systems containing the metallic carboxylates of general utility require either heat and agitation or the application of high shear in order to incorporate the soap and enable gelation to occur. Heat may be externally applied as in the instance of aluminum soap pre-gels prepared by cooking. Mills or intensive dispersion machines provide a combination of shear and internally developed heat, although the greater significance here is the effect of shear. Gel formation can be obtained at room temperature and with simple agitation in a few unique systems comprised of selected solvents and special-purpose soaps.

Whenever elevated temperatures are involved in the incorporation of metallic soap, the gel strength progressively increases while the system is allowed to cool. Gel strength attains a maximum after the system has returned to room temperature.

Although a coating formulation without a flow-modifying additive may already exhibit to some extent a non-Newtonian flow characteristic (i.e., a nonlinear relationship between shear* stress and rate of shear) resulting from interaction of pigments and vehicle alone, the incorporation of a suitable metallic carboxylate such as an aluminum soap can impart or enhance, one type of non-Newtonian flow known as *thixotropy*.

The consistency of a thixotropic system varies according to both the time duration of applied shear and the degree of shear rate. More specifically, a thixotropic system decreases in consistency when subjected to a period of shearing force (e.g., stirring, agitation or brushing); with the system held at a constant rate of shear over an elapse of time, the consistency decreases to a certain minimum value. The system subsequently recovers

---

*The discussion of shear here and subsequently pertains to the rheology (deformation and flow) of preestablished systems indicated. The reader will therefore differentiate these references from the preceding mention that the application of shear is one basic means of incorporating metallic soaps, or in effect, of establishing certain systems.

consistency when at rest after the shearing force ceases. This is a repeatable cycle involving dynamic interrelationships by which the structural quality of the system changes from gel toward sol, then back toward gel. Relative properties such as the required energy input and the rate or amount of consistency decrease or recovery vary with the compositions of specific systems.

Through the addition of a metallic carboxylate and the resultant formation (or alteration) of structure, the consistency of an oleoresinous coating system at the beginning or at the completion of such a cycle is typically greater than if no soap were added. If all other variables could be excluded from consideration the relative amount of bodying effect obtained would depend upon the composition and quantity of metallic soap added.

The practical benefits obtained from developing an appropriate degree of mild thixotropy and the resulting coexistence of a reasonably higher consistency under low shear or quasi-static conditions are readily evident in the application of wet coating films. The brush application of paint containing a thixotropic additive serves as a primary illustration. The act of brushing applies a shearing force to the liquid coating and consequently decreases its consistency as it flows from the brush. A sufficient rate and amount of decrease are necessary to permit easy brushability; sufficient flow is necessary to allow leveling and disappearance of bristle streaks; finally, a sufficient rate and amount of consistency recovery are necessary to prevent sagging of the wet film. Related advantages will be apparent for coatings applied by spray, roller or dip methods.

Specific types of powdered-form metallic carboxylates are used for flow modification in oleoresinous coatings. The aluminum stearates are by far the most useful, because they are solubilizable in a wide variety of vehicle liquids and are available in a large number of product grades from which a broad range of gel strengths can be derived. Aluminum octoate (2-ethylhexoate), the most readily soluble and the most potent gel former of the aluminum soaps, should be used with caution as a direct additive. However, it can be employed with common aliphatic, aromatic and chlorinated hydrocarbon solvents to make special-purpose gels with minimum or no heating. Aluminum palmitate can be used for purposes similar to those of the aluminum stearates. Calcium stearate finds application when limited bodying effect is desired.

Table 22.7 illustrates the gelling behavior of several aluminum soap compositions in mineral spirits solvent. The systems were prepared by adding 4% soap to mineral spirits at room temperature with agitation and then heating the mixture to 180°F. Mechanical stirring was maintained until incorporation of the soap was complete. The systems were allowed to cool and develop gel strength overnight. Gel consistencies, expressed in Kreb units (KU), were then measured at room temperature with the Stormer Viscosimeter.

It should be noted that, of the fatty acid types listed in Table 22.7, neither the hydrogenated tallow nor the hydrogenated marine is compositionally a single fatty acid compound. Since they are directly derived from natural sources, both of these types are mixtures of various individual fatty acid compounds. They differ distinctly in their respective quantitative distributions of individual fatty acid components. Thus, hydrogenated tallow-type fatty acid normally consists of approximately 60% stearic, 30% palmitic and possibly 3 to 5% oleic acids; the remainder consists of shorter-chain acids (e.g., myristic, lauric) and small amounts of longer-chain acids. In contrast, hydrogenated marine-type fatty acid usually contains only about 25% each of palmitic and stearic acids, with arachidic and behenic (saturated C_{20} and C_{22}) acids composing possibly another 35%. Unsaturated acids, mainly oleic, account for some 5% of

Figure 22.1.   A gas chromatograph instrument for analysis of fatty acid compositions. (*Courtesy Witco Chemical Co., Inc., Organic Div.*)

TABLE 22.7. Aluminum Soap Gel Consistencies—Effects of Organic Acid Type and Soap Type

Aluminum Soap Type	Fatty Acid Type	Gel Consistency (KU)	Gel Characteristic
Di-soap "A"	Hydrogenated tallow, with peptizing additive[a]	44	Fluid
Tri-soap "A"	Hydrogenated tallow-hydrogenated marine blend	47	Buttery
High tri-, low di-soap	Hydrogenated tallow-hydrogenated marine blend	51	Buttery
Di-soap "B"	Hydrogenated marine	51	Buttery
Tri-soap "B"	Hydrogenated tallow	54	Buttery
High di-, low tri-soap	Hydrogenated tallow-hydrogenated marine blend	56	Buttery
Mono-soap	Hydrogenated tallow	57	Buttery
Di-soap "C"	Hydrogenated polymeric blend[b]	90	Rubbery
Di-soap "D"	Octoic (2-ethylhexoic)	101	Rubbery

[a] Additive incorporated during manufacture of the metallic soap.
[b] U.S. Patent 2,699,428

the total; myristic, lauric and perhaps some of the odd-numbered carbon chain acids comprise the balance.

Excepting the aluminum octoate, all of the soaps listed in Table 22.7 would be commercially known as "aluminum stearates."

When viewed in relation to each other, results shown in this table indicate a softening effect of the hydrogenated marine-type acid on the gel (tri-soap "A" versus tri-soap "B"). The peptizing additive fluidizes the system, whereas the polymeric acid increases gel strength markedly. Rubbery gels such as formed by the aluminum soaps of polymeric or octoic acid can be troublesome in paint systems. Hence, soaps of such very high gelling power are generally not used as the sole or major gelling component of coating systems.

Gelling behavior differences in other solvents such as aromatics are even more significant than those observed in mineral spirits. Yet, in longer-chain aliphatic hydrocarbon solvents, the hydrogenated marine-type fatty acid will have a strengthening rather than softening effect. When the pre-gel is incorporated into the remainder of the coating formulation, this same type of acid will usually exert a stabilizing influence.

If other factors are not variable, an increasing thickening trend will be experienced in mineral spirits as the gellant composition is changed from the aluminum tristearate toward the aluminum monostearate.

Under the proper incorporation conditions, aluminum soaps will form gel structures in all usual nonpolar aliphatic and aromatic hydrocarbon solvents comprising coating vehicles. These soaps are more easily solubilized in aromatic solvents than in the aliphatics and thus will form gels with the aromatics at lower incorporation temperatures than with the aliphatics. Aluminum soaps are practically insoluble in polar solvents such as alcohols and lower ketones. When an aluminum soap is incorporated into a blend of polar and hydrocarbon solvents, the soap's capacity to form gel structure is reduced; thus the use of polar-nonpolar solvent blends affords a means of controlling the amount of bodying effect obtained from metallic soaps. In practice, excessive proportions of polar solvents in such mixtures can cause separation of soap from vehicles and consequent seeding problems.

Also, the type of vehicle resin or oil bears a relationship to final viscosity when metallic soaps are used. In a three-component system consisting of a given resin or oil, a nonpolar solvent thinner and a gel-forming grade of metallic soap, when all other conditions are considered to be constant, relative factors pertaining to the resulting viscosity are the activity of the soap in the resin or oil as compared to the activity of the soap in the solvent. An example of differences under moderate temperature conditions are the mild bodying effect of an aluminum stearate in some straight linseed oil-solvent vehicles, as contrasted to the drastic thickening capability of the same aluminum stearate in selected varnishes containing the same types of solvent.

The following alkyd coating formulations illustrate the use of metallic stearates for the combined purposes of flow modification and pigment suspension:

Architectural Gloss White Enamel[a]

Pounds	Gallons	
300.0	8.57	Rutile titanium dioxide
15.0	0.32	Zinc oxide
8.0	0.92	Calcium stearate
5.2	0.67	4% calcium naphthenate
140.0	17.50	Long oil oxidizing alkyd resin
420.0	52.50	Long oil oxidizing alkyd resin
145.0	22.31	Mineral spirits
3.4	0.43	6% cobalt naphthenate
1.0	0.13	Antiskinning agent
1037.6	103.35	
Pounds per gallon	10.0	
Vehicle nonvolatile content	55.6%	
Pigment volume concentration	16.7%	
60° gloss value	93–95	
Consistency	80–90 K U	

[a] Adapted from Formulary No. I-47, ADM Chemical, Div. Ashland Oil & Refining Co., Minneapolis, Minn. (see authors' note, p. 420).

Automotive Primer for Air Dry or Baking[a]

Pounds	Gallons	
84	2.86	Zinc yellow
84	3.52	Magnesium silicate
168	4.75	Iron oxide
40	5.46	10% aluminum stearate (di-tri soap) pre-gel in xylene
1.3	0.17	Maleic anhydride
67	7.97	Short to medium oil oxidizing alkyd resin
148	20.59	Xylene
1	0.13	Antiskinning agent
362	43.04	Short to medium oil oxidizing alkyd resin
40	5.56	Xylene
4	0.42	24% lead naphthenate
1	0.13	6% cobalt naphthenate
1000.3	94.60	

Grind: Pebble or ball mill to desired fineness.

Pounds per gallon	10.55
Vehicle nonvolatile content	41.65%
Pigment volume concentration	29.7%
Consistency, No. 4 Ford cup	125 sec

[a] Adapted from Formulary No. Q-407, ADM Chemical, Div. Ashland Oil & Refining Co., Minneapolis, Minn. (see authors' note, p. 420).

Special gel-forming zinc soaps, selected grades of aluminum stearate, calcium stearate and aluminum stearate–calcium stearate blends have been used (often with highly specialized techniques) to produce the thixotropic gels of heavy buttery consistency required for silk screen pigment extender bases.[6,7] This application calls for smooth gels having no stringiness or tack.

Supplementary benefits available from metallic carboxylate additives are the improvement of water repellency and water washability of dried films, enhanced development and depth of colors, inhibition of coating penetration into substrates and, in some formulations an assistance in the prevention of pigment flooding.

FLATTING AGENTS

A flatting agent is an additive to paint, varnish, enamel or lacquer which reduces the gloss or angular sheen of the dried film. The flatting effect contributed by powdered-form metallic soaps is attributable to their presence within an oleoresinous or lacquer vehicle as small, dispersed particles which cause microscopic irregularities at the dried film surface. The amount of alteration in the specular characteristics of the film is directly related to the type and amount of metallic soap incorporated into the coating system. Optically, the result is a reduction in the regular reflection and an increase in the diffuse reflection of incident light.

Major requisites for metallic carboxylate flatting additives are that their particles be readily wettable by the vehicle and that the particles be dispersible as stable, separate units. The particles must reduce specular reflection without creating an objectionable surface roughness.

In dead flat pigmented paints, wherein no film luster whatsoever is desired, the type and the quantity of major pigment constituents are the predominant factors in achieving this extreme degree of flatting. In other types of paints and in enamels in which relatively less, but more precisely controlled, gloss reduction is sought, the use of flatting additives becomes preferable. In lacquers and varnishes, additives find widespread use to achieve flatting effects. In clear films there is no substitute for flatting additives in reducing sheen.

When depending primarily upon major pigment constituents to provide dead flatness, suitable low-gloss pigment grades are employed at relatively high pigment-to-binder ratios. A metallic carboxylate such as an aluminum stearate or calcium stearate is commonly employed in these highly filled systems for the purpose of suspending pigments rather than flatting. The inclusion of metallic soap in this type of formulation will also improve film washability which would otherwise suffer from the highly increased pigment volume concentration.

In clear varnishes or lacquers and in finishes of relatively low pigment volume concentration, a metallic soap may be used in combination with a silica or other inorganic flatting agent for the dual purpose of suspending the inorganic additive and augmenting the degree of flatting.

For a proper flatting effect and a desirable film surface appearance, the particles of metallic carboxylate flatting agent must have both an optimum average size and a relatively narrow size distribution range. Particles which are too small yield undesirably glossy films because such particles do not impart sufficient microscopic film surface irregularity to give the necessary amount of light diffusion. At the opposite extreme, particles which are too large have a low surface area per unit weight and can create unattractively rough films.

Overall, the choice of metallic soap for purely flatting purposes is dictated primarily by the factors of low or *manageable* solubility in the vehicle, correct average particle size and the absence of outsize particles resistant to attrition during dispersion operations. For flatting of paints, enamels, varnishes and lacquers, many of the same types of powdered me-

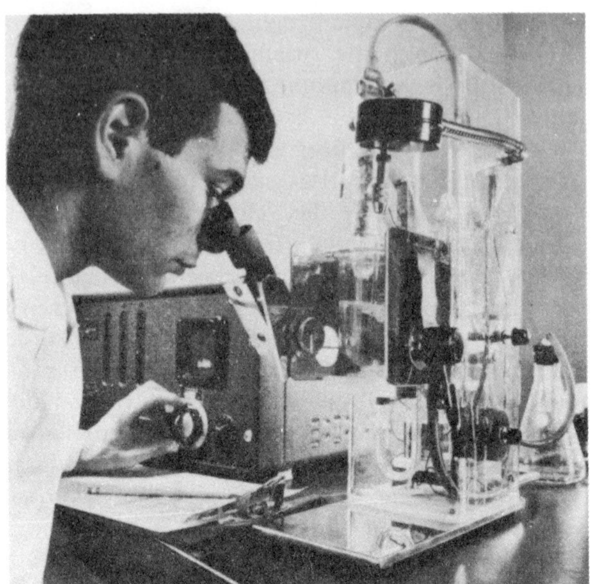

Figure 22.2. Determining metallic stearate particle size with the Coulter counter. (*Courtesy Witco Chemical Co., Inc., Organics Div.*)

tallic carboxylates discussed under suspending agents are also suitable. These include the previously mentioned aluminum stearates, calcium stearate and zinc stearate, but with the addition of magnesium stearate.

Determinations using the Coulter Counter apparatus have shown that the typical average *initial* particle diameters of various flatting grade metallic stearates (i.e., the typical particle sizes of these stearates as commercially supplied) range up to 20 μ (microns), with some grades averaging in the 3 to 5 μ range.*

The average initial particle sizes of aluminum stearates are known to be considerably larger than those of zinc, calcium or magnesium stearates. Two factors could be responsible for this: the *ultimate* (*individual crystal*) *size*** of the soap particles and the *degree of agglomeration* of these ultimate particles. Experimental evidence indicates that the extent of agglomeration constitutes a major factor.* Observations reveal that the initial particles of aluminum stearates are agglomerates of relatively high

---

*The information reported here is derived from laboratory studies performed by the technical staff of Witco Chemical Co., Inc., Organics Div.

**The ultimate particle size of a precipitated, powdered metallic soap product is governed by the conditions of reaction during manufacture.

order. Similar investigations show that those of zinc, calcium and magnesium stearates are also agglomerates, but are of decidedly lower order.*

In the production of flatted coatings, the grinding of powdered metallic soap flatting agents, whether directly into coating formulations or in separate mill bases, serves to reduce the soap particles from their initial size to units having a lower order of agglomeration and to disperse these units uniformly.

Whereas for flow modification a degree of solubility of aluminum stearates is necessary, the opposite is sought in the application of these particular soaps exclusively for flatting purposes. Therefore, when aluminum stearates are used for flatting, extra attention to processing details is required. The greater the degree to which an aluminum stearate is solubilized in the vehicle, the less will the flatting effect be because of the resulting reduction in the quantity of available stearate in particulate form. The solubility of aluminum stearates increases in proportion to the elevation in the temperature of grinding and the extension in the time span of grinding (fineness of dispersion). The concentration and solubilizing power of solvents employed are additional important factors bearing upon the solubility of aluminum stearates in mill bases.

When using a pebble mill, aluminum stearates should be ground with a part of the vehicle at temperatures below 130°F to avoid excessive solution of stearate particles and to minimize the development of higher consistency in the vehicle. A 5 to 10% concentration of aluminum stearate generally will require 3 to 5 hours of grinding. The mill base at the completion of the grinding period is typically a paste, which can then be added to the remainder of the coating system with no further grinding. When a roll mill must be used, or when higher grinding temperatures will be developed in any type of dispersion process, a grade of aluminum stearate having low gelling power can be used. Alternatively, carefully determined increments of polar solvents such as alcohol or ethyl acetate can be included in the mill base to inhibit development of gel structure.

Table 22.8 shows the relative flatting efficiency of several aluminum stearate grades in a varnish formulation. The Photo-Volt 60° Glossmeter, Model 610, was used for readings on the dried films.

When dispersing calcium, magnesium or zinc stearate in a mill base, 20 parts by weight of stearate with 80 parts of straight solvent or mixed solvents would be a typical grinding formulation. In contrast to the situation with aluminum stearates, the temperature developed during grinding of calcium, magnesium or zinc stearate in the presence of nonpolar solvents is not highly critical because of the low tendency of these latter

*The information reported here is derived from laboratory studies performed by the technical staff of Witco Chemical Co., Inc., Organics Div.

TABLE 22.8. Flatting Effects of Aluminum Stearates in Spar Varnish (Mill Base Consistencies Are Also Shown)

Aluminum Stearate Type	Mill Base Consistency (KU)[a,b]	60° Gloss Value of Dried Film[c]
None (varnish without stearate)	—	96
Tri-soap "A"	61	53
High tri-, low di-soap	53	62
Di-soap "B"	64	53
Tri-soap "B"	61	66
High di-, low tri-soap	56	62
Mono-soap	61	80
Di-soap "C"	54	80

[a] 20% stearate in mineral spirits, ball-milled for 18 hr. Stormer Viscosimeter used for determination of consistencies at room temperature.

[b] Stearate compositions are the same as those correspondingly shown in Table 22.7. While stearate concentration in mill-bases was five times greater than that used in the gels of Table 22.7, consistencies were not commensurately higher. Incorporation via ball milling avoids heat, minimizes solubility, preserves flatting properties of individual stearate particles.

[c] 10% mill base in varnish (2% stearate in total).

soaps to dissolve. Under highly intensive grinding conditions, calcium stearate and magnesium stearate can yield a mild bodying effect. However, zinc stearate will not increase the consistency significantly.

The effective amount of aluminum, calcium, magnesium or zinc stearate used for flatting applications varies between 2 and 6%, based upon the weight of the vehicle solids. Excessive concentrations can cause harmful film softening.

"Stir-in" types of zinc stearate for use in lacquers are available. These grades are dispersible with mere mechanical stirring and thus eliminate the need for mill grinding (see p. 418).

Table 22.9 shows the flatting effect of several lacquer-grade zinc stearates. It will be noted that, despite the longer than normal (18-hour) ball-milling time for each of the mill bases, very little increase in consistency occurs. Flatting will be more pronounced at a conventional 3 to 5 hour milling cycle. Although two grades designed for stir-in incorporation were included in this study, all of the zinc stearates were subjected to identical ball-mill grinds.

A study by Parker and Wendt[13] demonstrated the effect of mill base grinding time on the individual flatting efficiencies of aluminum mono-, di- and tristearates, calcium stearate and zinc stearate in clear lacquer films. It was shown that with all of the aluminum stearates evaluated, the film gloss increased as the duration of grinding increased. These findings were partially ascribed to the enhanced solubility of the aluminum soap particles in the mill base solvents as a consequence of the progressive re-

TABLE 22.9.   Flatting Effects of Zinc Stearates in Nitrocellulose Lacquer

Zinc Stearate	Typical Average Particle Size as Supplied (μ)	Designed Mode of Incorporation	Mill Base Consistency (KU)[a]	60° Gloss Value of Dried Film[b]
None (lacquer without stearate)	—	—	—	96
"A" (basic soap)	3	Stir-in	<40	72
"B"	4	Stir-in	<40	57
"C"	4	Mill grind	<40	66
"D"	5	Mill grind	51	58
"E"	10	Mill grind	<40	57

[a]Consistency of mill base at 20% zinc stearate concentration in methylisobutylketone after 18 hr ball milling. Determined at room temperature with the Stormer Viscosimeter.

[b]Lacquer (25% solids) reduced with 10% mill base; 5% zinc stearate concentration on total solids.

duction of the particles from their large initial size to smaller, more readily soluble sizes. In contrast, calcium stearate and zinc stearate, which are, respectively, moderately soluble and barely soluble in such solvents regardless of their small initial particle size, tended to yield maximum film gloss after a relatively short grinding time.

The refractive indices of the metallic soap flatting additives are important when these materials are used in clear finishes. The refractive indices of the appropriate metallic stearates typically range between 1.35 and 1.55, which is intermediate between those of the major pigments and vehicles. Aluminum stearates have the lowest refractive indices of the applicable soaps and yield excellent film transparency while imparting flatness to varnishes and lacquers.

SANDING AIDS FOR LACQUERS—ZINC STEARATES

The following discussion of zinc stearate sanding aids for lacquers is related to the foregoing discussion on metallic carboxylate flatting agents because the lacquer formulator employs zinc stearate to perform both of these functions, sometimes simultaneously, in nitrocellulose- and ethylcellulose-based formulations. In lacquer systems, most of the currently available lacquer-grade zinc stearates can be used for sanding, flatting or for both purposes.

In the lacquer finishing of wood surfaces the purpose of sanding sealer coatings is to close the pores and provide a sufficiently thick layer of easily sandable film which will serve as a base for the subsequent top coats.

Zinc stearate is incorporated in these sealers to impart the required sanding qualities to dried films.

At the beginning of this chapter, it was stated that application benefits are related to the inherent unctuousness of certain metallic carboxylates. This property, also known as dry lubricity or "slip," is characteristic of numerous metallic stearate powders and is one of the prominent features of zinc stearate in particular. The lubricating effect originates mainly from the fatty acid component of the soap molecule, but the type of metal involved bears a strong modifying influence. Zinc stearate exhibits a high level of lubricity which can be demonstrated by simply rubbing a small amount of it between the fingers and noting the very slippery feel.

Zinc stearate yields excellent results as a sanding aid. It disperses well in lacquer systems and is not dissolved to any practical extent by lacquer thinners. Upon drying of an applied sealer coat, the zinc stearate particles form a micro-irregular film surface as previously described for flatting additives. If a cross section of a dried sealer film containing zinc stearate were viewed under magnification, the minute surface irregularities would appear as "hills and valleys" covered by an outer layer of dried vehicle. The interiors of the "hills" are particles of zinc stearate.

Under sanding conditions, these embedded stearate particles lubricate the interface between the coating surface and the abrasive particles of the moving sandpaper, as each "hill" is worn away. This effectively reduces friction and lowers the amount of work associated with sanding the surface to a smooth state. Also, since less frictional heat is developed during sanding, the concurrent amount of thermoplastic softening of the applied film is relatively less. This results in a much lower tendency toward clogging or "loading" of abrasive sheets.

Zinc stearate does not discolor lacquer vehicles; its refractive index is conducive to good transparency in clear films.

The usual concentration of zinc stearate used in lacquer sanding sealers ranges from 2 to 10%, based on the weight of the vehicle solids. A concentration range of 5 to 8% zinc stearate is generally optimum for nitrocellulose vehicles, whereas ethylcellulose vehicles require approximately 10%. The following sanding sealer formulation demonstrates the use of zinc stearate.

Although intensive grinding of zinc stearates in mill concentrates to attain the desired degree of dispersion had been a necessary general practice by lacquer manufacturers for many years, the advent of the more recently developed stir-in lacquer-grade zinc stearates has enabled this grinding step to be eliminated in the production of clear sanding sealers and clear flat lacquers, if so desired. Stir-in type zinc stearates, which as of this writing have been adopted for use by a majority of lacquer manufacturers in the United States, can be dispersed easily in lacquer bases

Water-White Lacquer Sealer for Spray Application[a]

Pounds	Gallons	Pounds Nonvolatile	
114	12.00	74.0	$\frac{1}{2}$-sec RS nitrocellulose (65%)
43.5	5.40	37.0	Vegetable oil lacquer plasticizer
22.2	2.67	11.1	Short oil nondrying coconut alkyd resin
5.9	0.65	5.9	Zinc stearate, lacquer grade
546.4	79.28	—	Lacquer thinner
732.0	100.00	128.0	

Nonvolatile content: 17.5%

Lacquer thinner, % by wt:

Methyl ethyl ketone	10
n-Butyl acetate	20
n-Butyl alcohol	10
Denatured ethyl alcohol	10
Toluene	35
Petroleum naphtha (195–250°F boiling range; 38.5 KB value)	15
	100

[a]Adapted from Formulary No. R-11, ADM Chemical, Div. Ashland Oil & Refining Co., Minneapolis, Minn. (see authors' note, p. 420).

with the use of simple mechanical stirring. The dispersions obtainable are smooth, pourable, foam free and readily reducible with clear lacquer. Obviously, stir-in zinc stearate grades must be exceptionally free-flowing powders having an unusually low order of particulate agglomeration.

"Nonblooming" grades of zinc stearate* for sanding sealers have been developed to resist the attack of acid catalysts resins present in some of the varnish top coat formulations. Conventional zinc stearates when used in sealers under such top coats are decomposed at the sealer–top coat interface by the acid catalysts. The resulting liberated stearic acid, being incompatible with the coating film, eventually migrates to the outer surface and forms an objectionable "bloom" or "frost." A non-blooming lacquer-grade zinc stearate can be described chemically as a highly basic soap, specifically a zinc hydroxide–stearate or mono-di soap of relatively high mono to di ratio. The zinc hydroxide portion of the compound preferentially combines with available acid and thereby preserves the zinc–fatty acid portion of the compound. The currently available nonblooming-type zinc stearates are all stir-in grades; however, all stir-in grades are not necessarily nonblooming.

*See zinc stearate "A," Table 22.9.

Separately from, but akin to, its function in the sanding of undercoats, zinc stearate has been shown to be an effective rubbing aid[13] when employed for its flatting effect in lacquer top coats. The lubricating property of zinc stearate enables selected areas of the flat finish to be rubbed to a high gloss when desired for decorative highlighting.

Lithium stearate has been suggested as potentially useful in hot-sprayed lacquer sanding sealers because of its unusually high (approximately 212°C) softening point.[20]

Authors' Note: For the purposes of this presentation, we have altered the designations of various components of all coating formulations appearing in this chapter from the original specific trademark form to a generalized descriptive form.   As set forth herein, these formulations are meant for illustrative purposes only; because of the aforementioned modifications they should not be taken as shown, if considered for intended applications. Counterpart original (or superseding) formulations, specific as to individual components, are available as information recommended or suggested for laboratory evaluation, without guarantee or representation as to results, from the source indicated in the appropriate footnotes.

REFERENCES

1. Allan, L. H., *Paint Technology*, **22**, No. 248, 161 (1958).
2. Atherton, D., in "Paint Technology Manuals," Part 2, p. 46, London, Chapman and Hall, Ltd., 1961; New York, Reinhold Publishing Corp., 1961.
3. *Ibid.*, p. 49.
4. Gregg, G. W., and Anderson, L. E., *Offic. Dig. Federation Paint Varnish Prod. Clubs*, No. 270, 424 (1947).
5. Klebsattel, C. A., in (Mattiello, J. J., editor) "Protective and Decorative Coatings," Vol. 1, p. 531, New York, John Wiley & Sons, 1941.
6. Licata, F. J., *Offic. Dig. Federation Paint Varnish Prod. Clubs*, No. 291, 165 (1949).
7. *Ibid.*, No. 388, 485 (1957).
8. Mack, G. P., and Parker, E. (to Carlisle Chemical Works, Inc.), U.S. Patent 2,739,902 (March 27, 1956).
9. Mack, G. P., and Parker, E. (to Carlisle Chemical Works, Inc.), U.S. Patent 2,739,905 (March 27, 1956).
10. Martin, S.R.W., in "Paint Technology Manuals," Part 3, p. 33, London, Chapman and Hall, Ltd., 1962; New York, Reinhold Publishing Corp., 1962.
11. *Ibid.*, p. 28.
12. Nowak, M., and Fischer, A. (to Nuodex Products Co., Inc.), U.S. Patent 2,584,041 (Jan. 29, 1952).
13. Parker, K., and Wendt, R. E., *Offic. Dig. Federation Paint Varnish Prod. Clubs*, No. 328, 311 (1952).
14. Payne, H. F., "Organic Coating Technology," Vol. 1, p. 111, New York, John Wiley & Sons, 1954.
15. *Ibid.*, p. 238.
16. *Ibid.*, p. 27.
17. Pilpel, N., *Paint Technol.*, **27**, No. 11, 16 (1963).

18. Roon, L., and Gotham, W. (to Nuodex Products Co., Inc.), U.S. Patent 2,113,496 (April 5, 1938).
19. Roon, L. (to Nuodex Products Co., Inc.), U.S. Patent 2,139,134 (Dec. 6, 1938).
20. Wendt, R. E., *Offic. Dig. Federation Paint Varnish Prod. Clubs*, No. 344, 604 (1953).
21. Wheeler, G. K. (to R. T. Vanderbilt Co., Inc.), U.S. Patent 2,526,718 (Oct. 24, 1950).
22. Wheeler, G. K. (to R. T. Vanderbilt Co., Inc.), U.S. Patent 2,565,897 (Aug. 28, 1951).
23. Witco Chemical Co., Inc., Organics Div., New York, Tech. Bulletin 55-4R, "Metallic Stearates: Their Properties and Uses," p. 28, 1963.
24. Witco Chemical Co., Inc., Organics Div., New York, Tech. Bulletin, "Metallic Carboxylates as Catalysts, Lubricant Additives and Paint Driers," p. 20, 1966.

23

*Preservatives and Fungicides**

The manufacturer of coatings must contend with problems of micro-organism growth control for materials in process, finished formulations and final film performance. Chemical additives are used for the preservation of aqueous systems from bacterial spoilage and suppression of fungi, which attack films derived from all types of paints. Often, a single agent of broad antimicrobial activity is incorporated in aqueous coatings at sufficient concentration to give dual protection.

MICROORGANISMS

Bacteria and fungi, the causative organisms, are actually simpler members of the plant kingdom which lack roots, stems, leaves or flowers. They are grouped together in a subphylum, Fungi, to differentiate them from Algae and Lichens, which contain chlorophyll.

Bacteria comprise one class of Fungi which is unicellular and reproduces by fission (Schizomycetes); four additional classes include mildews (Ascomycetes), parasitically living on plant life, and molds, familiar to most as mushrooms and various growths easily observable by eye on foodstuffs, leather, paint films, plastics and wood. After further separation into Family and Tribe, they are finally referred to by genus (always capitalized), species and strain; for example, the fungus commonly found growing on exteriorly exposed paint films is *Pullularia pullulans* in a variety of strains and mutations. A variant used to evaluate paint film fungicidal properties is designated, American Type Culture Collection #9348. Other variants, from field isolates, differ markedly in resistance to the same fungicide.[1]

Many different kinds of bacteria live in air and water; their species number in the thousands. It is not surprising that aqueous paints become

*By Dr. Nathaniel Grier, Merck Sharp and Dohme Research Laboratories Div., Merck & Co., Inc., Rahway, N.J.

inoculated during manufacture and, unless properly preserved, promote their survival and multiplication.

Unique respiratory processes enable bacteria to tolerate most environmental conditions. One group, autotrophs, derives carbon for cell synthesis solely from carbon dioxide: heterotrophic types metabolize organic compounds as carbon sources either with oxygen from air (aerobic), in the absence of oxygen (anaerobic) or in either circumstance (facultative). Some are extremely resistant to elevated temperatures; nearly 100 hours at just below 100°C is required to kill *Bacillus subtilis* variant *niger*.

More than ten genera have been isolated from a can of spoiled latex paint, and one laboratory test for assaying preservative capabilities employs eleven in a single inoculum including *Pseudomonas, Achromobacter, Paracolobactrum, Aerobacter, Proteus, Flavobacterium* and *Bacillus*, together with a fungus.[5]

A similar variety, found on exteriorly exposed paint films at air-film and film-wood interfaces, may be partly responsible for peeling and adhesion failures.[10]

PRESERVATION OF AQUEOUS SYSTEMS

Water serves uniquely as a favorable and necessary medium for microbial growth; its presence suffices even as a very thin condensed phase on solids or as minute droplets in immiscible liquids.

Additional essential factors are readily at hand in formulating aqueous coatings. Practically all ingredients function as nutrients for bacteria commonly associated with in-can deterioration, including the primary synthetic resin vehicles of acrylic, butadiene-styrene, and polyvinyl acetate types.[11] Residual monomers, e.g., vinyl acetate, are claimed to be slightly inhibitory. Copolymers and many resin and oil modifiers such as alkyds, epoxies, polyesters, polyamides, linseed oil and tung oil are biodegradable in water dispersion. Binders made water soluble by incorporating salt-forming acid groups in molecules of oils or synthetic polymers also supply carbon sources.

Asphalt emulsified in water supports rapid bacterial growth, equivalent in efficiency to a rich nutrient, aqueous dextrose solution. Researchers hope to provide low-cost protein food for an expanding world population using this bacterial conversion of petroleum components. Fungi can metabolize asphalt by surface growth; Figure 23.1, a 60X microphotograph of granule-imbedded asphalt roofing shingle shows colonies of *Aspergillus niger* thriving on an untreated section (A) compared to an area protected with fungicide previously mixed in the asphalt (B).

All thickeners are biodegradable and, in addition, suffer marked viscosity loss when attacked chemically by extracellular enzymes. The

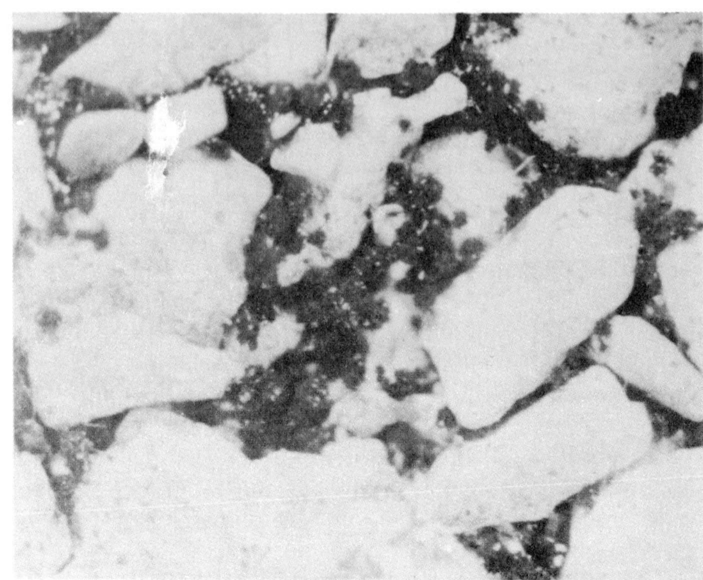

Figure 23.1(A).   *Aspergillus niger* growing on asphalt of roofing shingle.

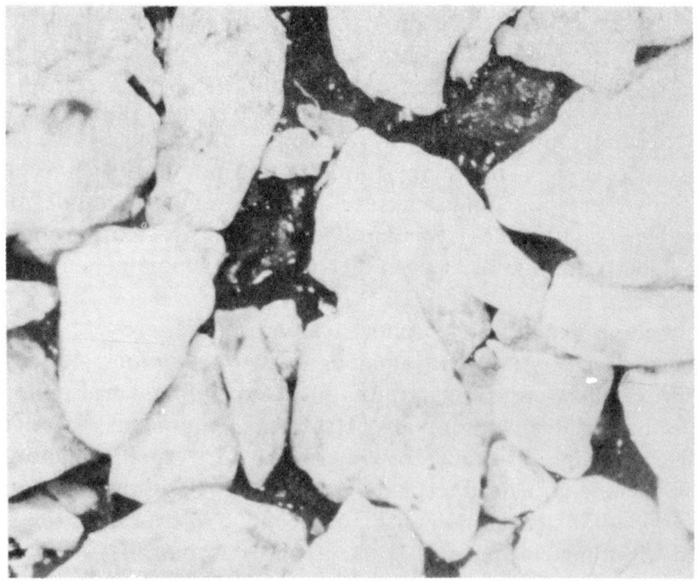

Figure 23.1(B).   No growth—asphalt treated with fungicide.

widely used cellulose derivatives do not appear to differ in susceptibility. It has been claimed that these polymers of glucose, joined by 1,4-glycosidic linkages and partially etherified to yield methyl, carboxymethyl or hydroxyethyl ethers, become less vulnerable to microbial breakdown as the degree of etherification per anhydroglucose unit increases.

A most perplexing problem faced by paint manufacturers was the sharp drop in viscosity of aqueous paints when stored for periods as short as six months, unaccompanied by familiar signs of spoilage like obnoxious odor, colloid disruption or gas formation. Repeated effort by competent microbiological laboratories failed to isolate organisms from these protected batches; sterility had been achieved. Enzymatic breakdown of thickener proved to be the cause. Bacteria and fungi, along with other forms of life, are able to synthesize enzymes, organic chemical catalysts, for use within or outside their cells, to accelerate the conversion of substrate complexes to simpler molecules in many instances. Cellulase, an enzyme secreted extracellularly, continues to cause cellulosic fission long after the death of all microorganisms has occurred, the polymer chains being opened at the β-1,4-glycosidic linkages. Duplication of the observed viscosity fall is easily accomplished in less than 4 hours at room temperature by the addition of 1 mg of cellulase, available commercially as a dry powder, to a liter of sterile cellulosic-thickened coatings.

Cellulases function optimally at pH 4 to 6 but can cause fragmentation of cellulosic polymers at alkaline pH too. As is typical of enzymes, it is inactivated by heating at above 70°C for several hours or with chemical agents including heavy metals, 8-hydroxyquinoline, organomercurials, chlorinated phenols and sulfur derivatives, at higher levels than used to stem microbial growth. The enzyme has been isolated as a contaminant from commercial lots of dry methylcellulose.

Other thickeners include lecithin, casein, alginates, starches, polyvinyl alcohol, polyacrylates, natural gums, complex polysaccharides, soybean protein and gelatin—with none being resistant to microbial or enzyme conversions. Proteinaceous types more closely resemble reactive sites of microorganisms and compete for microbiocides; casein, for example, contains thiol groups available as inactivators of metallic inhibitors. Chemical complexing occurs with thickener amino or hydroxyl substituents removing formaldehyde-based preservatives. Increased concentrations of inhibitor will offset these adverse effects. In some instances, gradual reaction between biocide and viscosity agent leads to increased bodying of coatings during storage.

Surfactants, too, are metabolized; as a result, in spoiled paints colloid systems collapse, made obvious by emulsion breakdown, pigment settling and other separations. The so-called hard or microbiologically resistant detergents are in reality biodegradable; under sewage treatment conditions, 20% of surfactant escapes destruction to cause effluent foaming.

Anionic compounds having sulfonic acid or sulfate salt substituents support anaerobic sulfate-reducing bacteria, e.g., *Desulfovibrio desulfuricans*. Conditions of an oxygen-free atmosphere occur during coatings manufacture when little or no air is mixed into aqueous liquids. Hydrogen sulfide production signifies their presence at stagnant points of plant piping, in unagitated "heels" left in tanks and in sealed cans of paint. They multiply readily with the assist of oxygen-scavenging aerobes in quiescent open systems. Sanitation measures, elimination of piping blind spots and inhibitor addition in the early steps of paint making effectively control these organisms as well.[4]

Surface-active agents of the quaternary ammonium type can increase the antimicrobial efficiency of organometallic preservatives and broaden their spectrum of activity. In contrast, sequestrants, occasionally used in coatings, cause marked lowering of inhibitory properties. Ethylenediaminetetraacetic acid chelates phenylmercury compounds reducing antimicrobial potency to one-twentieth of the normal level. Polyphosphates do not interfere.

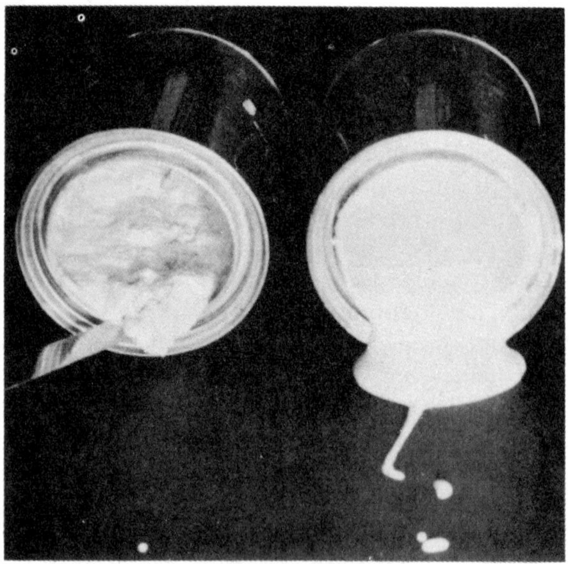

Figure 23.2. *Left can*: Coagulated latex paint with solubilized phenylmercuric acetate, one freeze-thaw cycle. *Right can*: Same latex paint with equivalent loading of phenylmercuric propionate powder, five freeze-thaw cycles.

Ammonia, primary aliphatic amines and their carboxylic acid salts, added to aqueous coatings for pH adjustment, solubilize phenylmercury compounds as ammonium complexes. Although this increases toxicant availability, the dissolved phenylmercurial is reactive with anionic surfactants causing decreased freeze-thaw stability. Figure 23.2 illustrates such an effect: a latex paint treated with a commercial ammoniacal phenylmercuric acetate solution has become a pasty mass after but one freeze-thaw cycle; the same coating protected with an equivalent loading of dispersed phenylmercury propionate powder is unaffected after five freeze-thaw cycles.

Additional components of aqueous formulations help microorganisms thrive. Film coalescing agents, whether synthetic compounds like hexylene glycol and carbitol acetate, or natural oils such as pine oil, readily undergo bacterial utilization. Inorganic salts, originating from polymerization processes, surfactants, buffers and other sources, furnish vitally needed mineral elements. Very few additives contribute anti-microbial properties; lower alcohols, glycols and other freeze-thaw stabilizers are not usually present in sufficient concentrations to be inhibitory. High pH in acrylic and butadiene-styrene paints eliminates some bacteria, but *Pseudomonads* and others multiply readily at alkaline pH. Ambient temperatures of 25 to 30°C in paint factories and storage areas encourage growth; below 10°C most bacteria remain static.

PRESERVATIVE TESTING

Many microbiological assay methods have been studied to develop a practical means for estimating the long-term stability of aqueous coatings. Variables given consideration include the kinds and amounts of organisms to be used as inocula, the frequency of introduction to test samples, periods of incubation, time-rates of kill determinations, etc.

Current procedures utilize a mixed bacterial inoculum, predominant in *Pseudomonas aeruginosa*. Weekly dosing of test paints is made for periods up to ten weeks, and sampling for viable bacteria follows each inoculation at intervals from 1 hour up to seven days. The rate at which sterility returns after challenging with bacteria is a measure of inhibitor efficiency; 48 hours is considered satisfactory. Long-term storage is required to determine chemical stability of toxicant in paints. For accelerated aging of formulations, a month's residence at 60°C is deemed equivalent to a year of shelf life.

PRESERVATIVES

Organomercurial salts are the most active antimicrobial agents in use for the preservation of aqueous systems.

Phenylmercury Salts

Direct mercuration of benzene with mercuric acetate in benzene-acetic acid at 80°C produces phenylmercuric acetate:

$$C_6H_6 + Hg(CH_3COO)_2 \rightarrow C_6H_5HgOOCCH_3 + CH_3COOH$$

Other salts are prepared by conversion of the acetate to the hydroxide with sodium hydroxide and neutralization with acids or by metathesis using inorganic or organic acids. For the preservation of practically all aqueous coatings, satisfactory performance can be obtained with 2 to 12 ounces (avoir.) per 100 gallons of total formulation; phenylmercury salts and solutions vary considerably in mercury content, the dosage is based upon 60% mercury metal equivalency. In systems containing protein-aceous materials, twice the normal amount should be used to compensate for partial inactivation.

Phenylmercurials are commercially available as water-miscible solutions, water-dispersible powders or water-emulsifiable solvent solutions. The solid salts are represented typically by the acetate, propionate, benzoate, 2-ethylhexoate, di(phenylmercury)ammonium propionate, and in dry physical mixture with water-solubilizing tris(hydroxymethyl)amino-methane. Inorganic salts include the borate, chloride and nitrate. Aqueous solutions of all but the chloride can be prepared with ammonium hydroxide or primary aliphatic amines; they are most stable at pH 7.5 to 9.

Water-emulsifiable solutions of the oleate, propionate and di(phenyl-mercury)dodecenylsuccinate generally consist of water-immiscible solvents, e.g., high-boiling petroleum naphtha, toluene or xylene, coupling agents like ethyleneglycol and oil-soluble surfactants. They are "universally" applied, either by dispersion in aqueous coatings or solution in oleoresinous systems. Phenylmercurials are incompatible with such reducing agents as formaldehyde, sulfides and nonferrous metals. Oxidized forms of sulfur, i.e., sulfates and sulfonates, do not affect activity. Certain plant species are extremely sensitive to mercury vapor; paints containing mercurials should not be used in greenhouses.

Drs. L. J. Goldwater and M. B. Jacobs[3] conducted a study of mercury exposure to humans from the use of a mercury-containing paint inside homes. The latex paint employed in their tests contained approximately 0.2 pound of mercury metal (in the form of a phenylmercury salt) per 100 gallons. They observed slight volatilization of mercury vapor from coated surfaces during the first 24 hours, with a decrease of mercury concentration in air to an insignificant level thereafter. No evidence was found of mercury exposure or absorption to a degree which would constitute a health hazard to the painter or occupants of the painted room.

Many millions of gallons of aqueous paints have been preserved successfully with phenylmercury salts in the past decade. Very few adverse reactions to humans exposed to these have been reported in the medical literature.

Only one other type of organomercurial inhibitor is offered for coatings. One member, chloromethoxypropylmercuric acetate, synthesized by the addition of mercuric acetate in methyl alcohol to allyl chloride, is purportedly a more efficient bacteriostat then phenylmercurials. Its water-miscible solution, containing active ingredient equivalent to 10% mercury, prevents spoilage at concentrations of $\frac{1}{2}$ to 1 pound per 100 gallons of paint.

Sulfur Compounds

A most active nonmercurial preservative is 3,5-dimethyl-1,3,5,2H-tetrahydrothiadiazine-2-thione, the condensation product of methylamine, formaldehyde and carbon disulfide. At levels of 1 pound per 100 gallons it affords excellent protection. However, strongly colored complexes form in the presence of metallic ions; copper turns black. Further, in moderately warm water (45°C), gradual hydrolysis liberates initial reactants.

Other carbon disulfide-derived compounds, e.g., sodium and zinc dimethyldithiocarbamates and tetramethylthiuram disulfide, are offered commercially for the preservation of proteinaceous thickener solutions and starch coatings at suggested use concentrations of 5 to 10 pounds per 100 gallons. Potential discoloration problems have curtailed their applications to other coatings. Under acidic conditions, the salts rapidly release unstable dithiocarbamic acids.

Two chlorinated alkylthio compounds, N-trichloromethylmercapto-4-cyclohexene-1,2-dicarboximide and N-trichloromethylthiophthalimide are claimed to be effective bacteriostats at loadings of 2.5 to 5 pounds per 100 gallons. Their nontoxicity to humans is of great advantage, but the use cost per gallon of paint is high. There is also reluctance to employ sulfur compounds potentially capable of forming colored metallic sulfides.

Phenols

Halogenated phenols, especially those containing three to five chlorine substituents per mole of phenol, represent a group of the most widely used industrial antimicrobial agents. These also include chlorinated cresols, xylenols and phenylphenols. However, use in coatings is limited because of their odor, conversion to pink and red quinones on exposure to light and air, and formation of colored complexes with metallic ions. Concentrations of 5 to 10 pounds per 100 gallons are necessary for adequate protection against bacterial deterioration. In alkaline coating systems, phenolate ions form; these are less biocidal than nonionized phenols.

Halogenated phenol-formaldehyde condensates, 2,2'-methylenebis(4-chlorophenol) and 2,2'-methylenebis(2,4,5-trichlorophenol), are lower in mammalian toxicity and odor, with approximately the same antibacterial activity as the chlorinated phenols. 3,4',5-Tribromosalicylanilide is also less toxic and equally efficient; treatment costs per gallon of paint cannot meet the low mark set by organomercurials.

Other Compounds

The newest product offered commercially is a quaternary salt of 1,3-dichloropropene with hexamethylenetetramine, namely 1-(3-chloroallyl)-3,5,7-triaza-1-azoniaadamantane chloride; 1 to 2 pounds per 100 gallons of total formulation is proposed for control of microbial growth. Its mode of action is formaldehyde release, independent of pH.

Another formaldehyde condensate, 2,6-bis(dimethylaminoethyl)cyclo-hexanone, in resin emulsions has offered substantial bacterial inhibition at loadings of one-half pound per 100 gallons. The antibacterial spectrum and high activity indicate it does not function by formaldehyde release extracellularly.

2,4-Dimethyl-6-acetoxy-1,3-dioxane at concentrations of $\frac{1}{2}$ to 1 pound per 100 gallons is claimed to be effective over a broad pH range against both Gram-negative and Gram-positive bacteria. It is not compatible with systems containing amides or amines, since discolorations may result.

Barium metaborate, for in-can preservation and inhibition of can corrosion, is used in polyvinyl acetate and acrylic coatings at 10 to 20 pounds per 100 gallons.

ADDITIONAL MEASURES

Positive effort to insure clean equipment is essential; thorough flushing of all units immediately after use simplifies maintenance. Mixing tanks which contain accumulations splattered on walls and cover should be scraped to metal; piping with self-draining features and round-bottom tanks which empty completely are easier to keep free from high bacterial growths. Water for thickener solution preparation and pigment grind slurries should be dosed with preservative initially. Measuring equipment, transfer tanks and other accessory units, after flushing, can be filled with treated water to maintain sterility.

Most enzymes are denatured above 60°C, and nearly all are inactivated by heat at 80 to 100°C. Wherever feasible, heat is an efficient way to remove enzyme contamination from formulation components; 15 to 30 minutes at inactivation temperature suffices.

Selection of paint ingredients based on ease of microbial inhibition is worthwhile, especially when comparable materials are available. It has already been noted that certain thickeners create a need for twice the level of inhibitor otherwise effective in a formulation.

PAINT FILM DETERIORATION

Fungi pose minimal problems in the preservation of aqueous paints but are a prime cause of disfigurement and premature failure of all types of paint films. Ross[10] has listed microflora associated with painted surfaces from six widely separated test locations in the United States (see Tables 23.1 and 23.2).

TABLE 23.1. Microorganisms Isolated from Oil Paint Films Exposed on Test Fences at Six Different Locations in the United States[a]

Fungi	Bacteria
Alternaria dianthicola	*Alcaligenes recti*
Aspergillus flavus	*Alcaligenes* sp.
Botryodiplodia malorum	*Bacillus cereus*
Botrytis cinerea	*Bacillus mycoides*
Cladosporium sphaerospermum	*Bacillus sphericus*
Cladosporium sp.	*Bacillus* sp.
Cephalosporium carpogenum	*Flavobacterium invisibile*
Fusarium flocciferum	*Flavobacterium marinum*
Helminthosporium spiciferum	*Micrococcus albus*
Paecilomyces varioti	*Micrococcus candidus*
Penicillium oxalicum	*Micrococcus ureae*
Phoma glomerata (*Peyronellaea*)	*Sarcina flava*
Pullularia pullulans	
Stemphylium consortiale	

[a] From *Advan. Appl. Microbiol.*, **5**, 220 (1963).

Although more than a dozen genera were isolated, only four appear to be major contributors to film deterioration: *Pullularia pullulans, Phoma glomerata, Cladosporium* and *Alternaria* species. *Pullularia* was isolated from all sites and undoubtedly is most capable of utilizing dry paint films as nutrients.

Practically pure cultures have been seen growing over an entire exterior wall surface of wooden buildings; Figure 23.3 is a photomicrograph of a typical single colony. High humidity, moderate ambient temperatures and shielding from direct sunlight promote the growth of microorganisms. Fungal spores, spread in air, soil and water, survive under the most difficult conditions and inoculate wet or dry paint films. Surfaces particularly vulnerable are exteriors facing north or shaded by eaves, overhangs and

TABLE 23.2.   Microorganisms Isolated from Emulsion Paint Films Exposed on Test Fences at Six Different Locations in the United States[a]

Fungi	*Bacteria*
Alternaria dianthicola	*Bacillus mycoides*
Aspergillus flavus	*Flavobacterium marinum*
Cladosporium sphaerospermum	*Sarcina flava*
Cladosporium sp.	
Cephalosporium carpogenum	
Helminthosporium spiciferum	
Penicillium oxalicum	
Phoma glomerata (*Peyronellaea*)	
Pullularia pullulans	
Stemphylium consortiale	
Torula nigra	
Unknown yeasts	

[a]From *Advan. Appl. Microbiol.*, **5**, 220 (1963).

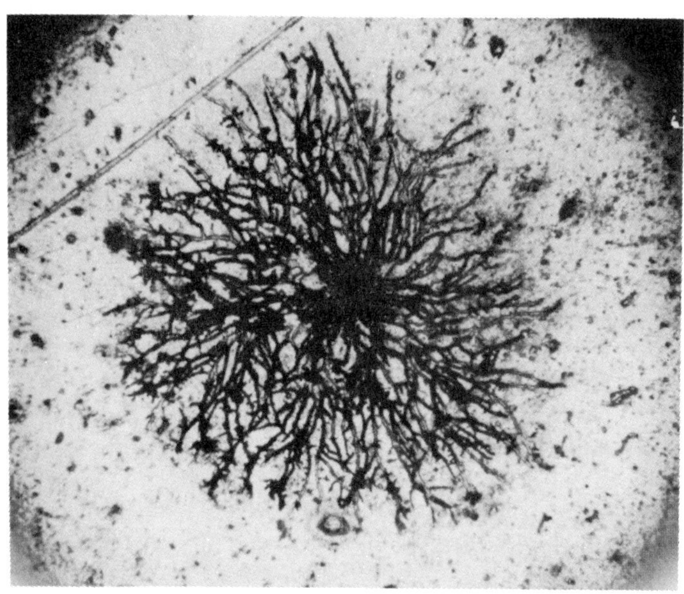

Figure 23.3.   Fungal colony, *Pullularia pullulans.*

porch roofs. Figure 23.4 illustrates a wall exposed to the south; heavy mold attack predominates in the cooler area kept in shadow by the porch roof. Interiors, especially of breweries, food packing plants, bathrooms and kitchens, are also favorable sites. Figure 23.5 depicts heavy growth

Figure 23.4.   Fungal attack on shaded section of wall facing south.

on polyvinyl acetate-covered ceiling and support surfaces of a shower room and swimming pool within three months after painting.

Black pigmentation of microorganisms has often been mistakenly identified as dirt. Examination with a low-power hand lens helps distinguish filamentous structures of microbes. Their organic pigments can be bleached with 5% aqueous sodium hypochlorite solution; color will not be restored on sequential treatment with 3 to 5% hydrogen peroxide solution as happens with lead, mercury and other metal sulfide coating stains. Darkening of painted surfaces by metal sulfide formation has been attributed erroneously at times to mold growth. An excellent study on film sulfide formation and testing aspects has been made by H. C. Wohlstein and M. Feldstein.[13]

Discolorations produced by fungi can range over the entire color spectrum; confirmation of origin can be obtained microbiologically by subculture techniques applied to minute samples carefully removed from affected surfaces.

Substrates of wood, hardboard, plaster and other porous materials

Figure 23.5.   Heavy microbial growth on ceiling and support of shower room.

let moisture penetrate films, permitting fungal extension through prime and top coats from either direction.   In contrast, weathering of hard, factory-baked finishes on aluminum, steel and other nonporous siding creates surface micropockets; dirt accumulation and moisture condensation in these tiny craters allow fungi to degrade the film from the outer surface inward.

Before paint is applied to any fungus-contaminated surface, mechanical cleaning to bare substrate is advisable.   This should be followed with a medium-soft brush scrub using disinfectant solution consisting of: 3 ounces of trisodium phosphate, 1 ounce of detergent, 1 quart of 5% sodium hypochlorite and 3 quarts of warm water or enough to make 1 gallon, followed by clean rinsing with fresh water.[8]

Constituents of paints vary in their ability to support fungal life, and details of these findings are available in literature reviews.[2]

For example, more resistant plasticizers include phosphoric acid esters; tricresyl and tributyl phosphates, dibutyl and di(2-ethylhexyl)phthalates, triethyleneglycol di(2-ethylhexoate) and di(2-ethylbutyrate); chlorinated paraffins; toluenesulfonic acid derivatives; diphenyl, diphenyl sulfone, silicone oils and decalin. Among the more susceptible are glyceryl and other esters of lauric, stearic, oleic, azelaic, sebacic, succinic and ricinoleic acids, castor and cottonseed oils.[12]

In general, molds do not thrive in alkaline media; formulations which include calcium carbonate and other basic salts impart significant film protection.

WATER PASSAGE AND PAINT FILMS

A major factor in coatings susceptibility is water permeability, as liquid and vapor. Microorganisms live only in a water environment; spore forms can survive but not multiply in its absence. Water passage into and through films is dependent upon at least two properties: *solubility*: the number of molecules of penetrant a unit volume of film can accommodate under the existing conditions; *diffusivity*: the ease with which the penetrant can move within the film.[6]

Water solubility in polymer films is influenced by their physical structure. In amorphous, rubbery, linear polymers typical of many coatings, liquid water dissolves at a slow increase with temperature. Under constant relative humidity this temperature effect disappears; however, water vapor solubility decreases rapidly with temperature. Thus, sun-warmed coated surfaces are kept completely dry.

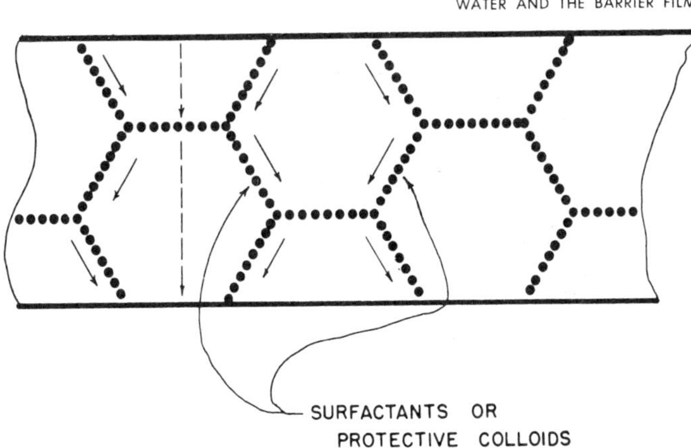

Figure 23.6. Paths for water permeation in latex-deposited films. From "Water and the Barrier Film," by Alan S. Michaels, of Massachusetts Institute of Technology. Published in *Official Digest*, Vol. 37, No. 485 (1965).

Micro-voids in glassy polymers and improperly formed films can fill with water and serve as reservoirs of high, localized microbial degradation. In latex paints, the microstructure of the final dried film and the types of minor constituents remaining in the coating affect moisture-barrier properties to a major degree. Water-soluble surfactants and hydrocolloids concentrate between the latex particles on film drying, forming an intercellular passage for water as graphically outlined in Figure 23.6 by Michaels.[7] He recommends that surfactants and other ingredients be selected which become chemically altered on drying to water-insensitive products; volatile ammonium salts, for example, lose ammonia, leaving water-insoluble acids.

Fillers and pigments must be well dispersed and completely polymer wetted to act as moisture barriers in dried films. Some of the paths for water transport which may otherwise result are outlined in Figure 23.7. A proposed solution is tailored copolymers, prepared from a relatively low-polarity monomer and a small quantity of high-polarity monomer, the latter to aid in wetting filler and pigment particle surfaces with resin. At the same time low-polarity polymer, with low water absorption characteristics, interspersed between them could impede water movement.

Water diffusivity, and hence vulnerability, of films is controlled by the following polymer structural features: (1) polarity, (2) crystallinity, (3) chain stiffness and (4) cross-linkage. Very high cross-linkage is needed to arrest the movement of so small a molecule as water—cured long oil

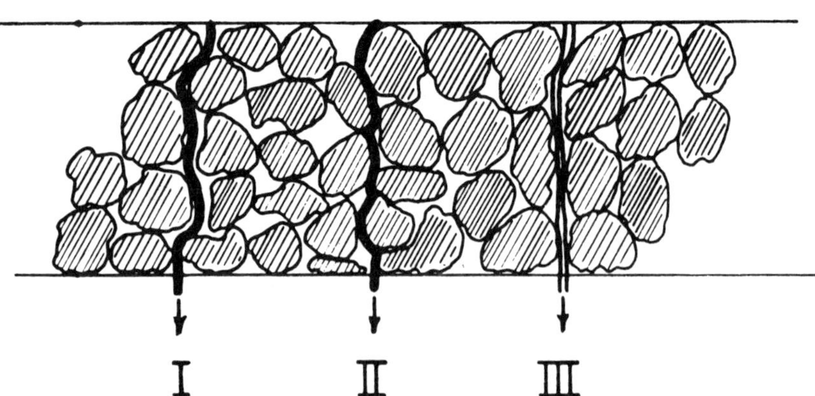

I. Diffusion in polymer matrix; II. Diffusion along filler surfaces; and III. Bulk flow through microcapillaries (poor resin-filler adhesion)

Figure 23.7.   Paths for moisture transport in filled or pigmented films. From "Water and the Barrier Film," by Alan S. Michaels, of Massachusetts Institute of Technology. Published in *Official Digest,* Vol. 37, No. 485 (1965).

alkyds have an insufficient number, but phenol-formaldehyde resins do prevent water ingress because of high cross-linkage.

Plasticizers greatly increase water diffusivity by separating chains in polymer structures of latex paints and thereby decrease fungal resistance, in addition to providing nutrient.

The extreme difficulty in protecting blister-resistant alkyd films may be attributable more to water permeance than resin. The same coating, made impermeable by zinc oxide, is much less susceptible. Our laboratory studies indicate that zinc oxide at high concentration is poorly antifungal and functions primarily by a hardening action.

In formulating coatings, a balance must be obtained between desired properties of film flexibility, blister resistance and water permeability. The degree of water transport in and through films parallels the difficulty in protecting them against premature failure by microbial action. Presently available fungicides will not furnish resistance to poorly constituted films at any loadings.

SCREENING OF FUNGICIDES

The search for practical paint fungicides begins with microbiological assay of candidate chemicals in water-nutrient agar media against *Pullularia pullulans* and other fungi associated with film degradation.[14] Compounds found active are studied at much higher concentrations in paints to compensate for loss of mobility and decreased availability when locked in dried films. Coated wood, paper, glass, metal, plaster and cord specimens—the substrate selection varying with investigator—are weathered artificially in cyclic exposures to light, heat, water and dust erosion. Finally, they are inoculated with fungi and maintained in incubators or tropical test chambers under optimal growth conditions of 90 to 100% relative humidity and 28 to 30°C up to four weeks.

A relatively simple adaptation, which uses a two-coat filter paper substrate, heat-water leach weathering and pure cultures of *Pullularia pullulans*, shows the ability to discriminate between subthreshold and adequate concentrations of inhibitor even for very small differences in loadings (see Figure 23.8). Except for extremely hard films, a useful guide to expected field performance is realizable for many types of coatings.[9]

Coated wood panels are then exposed outdoors on test fences in areas free from industrial fumes and conducive to fungal growth. Panels facing north at an angle of 45° to vertical promote attack much more than counterparts in opposite direction or staked vertically. A typical test specimen from our New Jersey panel farm is shown in Figure 23.9; the left half, relatively fungus-free, was painted with the identical alkyd formula-

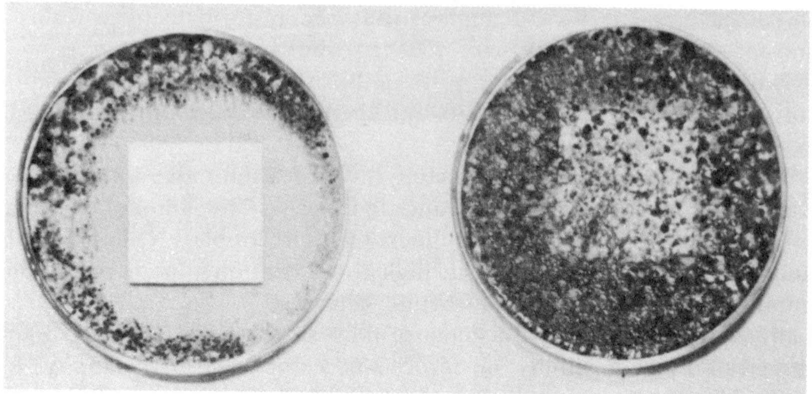

Figure 23.8.   *Left dish*: Test paint containing adequate fungicide loading. *Right dish*: Same paint with subthreshold concentration—both "weathered" with a heat-leach cycle.

Figure 23.9.   *Left half test panel*: Alkyd paint with 2 pounds of phenylmercury propionate per 100 gallons. *Right half*: Same paint, no inhibitor; three years exposure, north, 45° to vertical.

tion as the right but in addition contained 2 pounds of phenylmercury propionate per 100 gallons. Exposure was north, 45° to the vertical, for three years.

FUNGICIDES

Phenylmercury salts are also the major fungal inhibitors used for film protection. In the past few years, average consumption of mercury metal for this purpose amounted to 400,000 pounds annually in the United States.

Four compounds represent approximately 90% of the sales volume; the acetate, di(phenylmercury)dodecenylsuccinate, oleate and propionate. Dosage recommendations are: service in mild fungal growth locations, 1 to 2 pounds of phenylmercury salt* per 100 gallons of paint; for severe areas, 2 to 4 pounds* per 100 gallons of paint.

For blister-resistant alkyd coatings, the lower toxicant loadings suffice if 1 to 2 pounds of zinc oxide is included. This level will also protect linseed oil films having high zinc oxide content.

In latex systems, solid phenylmercury acetate and propionate are less reactive than their water-solubilized preparations toward surfactants and must be added to the pigment grind for good dispersion. They neither affect freeze-thaw stability (Figure 23.2) nor lower water resistance of dry films, being free of solvents and solubilizing agents.

Oil-soluble phenylmercury salts, e.g., oleate and dodecenylsuccinate, originally introduced for oleoresinous systems, find use in latex paints in emulsified form. The propionate has unique solubilities in oil-water phases and more efficiently distributes to water from a resin film layer without undue leaching.

Inhibitor additions to finished latex paint are best made using water-immiscible solutions of fungicide. Pigment flocculation or colloid disruption is avoided, yet both in-can preservation and film protection is attainable.

Phenylmercury salts have vapor pressures at 20°C of 10^{-6} mm, or one-thousandth that of mercury itself.

Other Fungicides

Halogenated phenols were once proposed as fungicides for paints; pentachlorophenol loadings of 10 to 50 pounds per 100 gallons, needed for film protection, caused serious odor and color problems. Wood presently impregnated with the same agent to eliminate rot suffers in its paintability properties. Substitute treatments are actively being investigated.

4-Chloro-3,5-dimethylphenol is offered as a fungicide for aqueous and oil coatings at a proposed loading of 3 pounds per 100 gallons.

Organotin compounds have failed to gain a significant market share. Bis(tri-*n*-butyltin) oxide at loadings up to 10 pounds per 100 gallons is ineffective in oil-type systems, whereas in polyvinyl acetate paints 5 pounds per 100 gallons is claimed to be satisfactory. Levels of 1 to 3 pounds per 100 gallons in acrylic and other latex systems are suggested.

---

*Equivalent to 60% mercury content.

Bis(tri-*n*-propyltin) oxide has been reported as appreciably more anti-fungal, but it has not as yet attained commercial use in paints.

Very few nontoxic (to humans) fungicides have proved practical for coatings. Generally, the levels for mold inhibition are so high as to be detrimental to film properties. N-Trichloromethylmercapto-4-cyclo-hexene-1,2-dicarboximide and the corresponding phthalimide find some application in relatively hard, nonpermeable finishes for food plant interiors at loadings of 5 to 15 pounds per 100 gallons. They have been incorporated in calcium carbonate-containing exterior oil paints at similar levels.

Copper 8-quinolinolate, a khaki-colored nontoxic antimicrobial agent, has found limited use in food plant coatings at concentrations ranging from 10 to 50 pounds per 100 gallons. It is stable to light and heat, completely insoluble in water and has no measurable vapor pressure—all characteristics of an "ideal" toxicant. However, its color and moderate inhibitory strength are major drawbacks to a wider utility.

For interior maintenance paints, barium metaborate loadings of 200 to 250 pounds per 100 gallons are advocated. In prime and top coats of oil, alkyd, vinyl and metal paints, 50 to 250 pounds per 100 gallons must be added. The borate is additionally claimed effective as an antimicrobial for all drying oils and alkyd resins. Each formulation must be specially designed to accept such high loadings of inhibitor.

A quaternary compound, *p*-ethylbenzyl dimethylammonium cyclohexl-sulfamate, is recommended at minimum loadings of 11 pounds per 100 gallons in latex paints. No effectiveness is evident for oleoresinous systems.

Some efforts have been made to produce fungi-inert resin systems using monomer components having antifungal properties. Undecylenic acid, vinyloxazolidine and tris(hydroxymethyl)aminomethane are a few re-actants reported to yield polymers with a degree of resistance to microbial degradation. Another approach is to give films a self-cleansing action by controlled chalking rates with anatase and rutile titanium dioxide blends.

Antimicrobial agents have contributed to maintaining and extending quality, performance and service life of coatings. Although paint formulators look forward to gradual elimination of additive upon additive in their products, the most unusual adaptive capabilities of microbes serve to prevent this.

REFERENCES

1. Baniecki, J. R., and Hanan, S. E., *Am. Chem. Soc., Div. Org. Coatings and Plastics Chem. Preprint*, **25**, No. 2, 13 (1965).
2. Chicago Paint Production Club, *Am. Paint J.*, **41**, No. 6-A 8 (1956).

3. Goldwater, L. J., and Jacobs, M. B., "Mercury Exposure From the Use of a Mercury-bearing Paint," *Natl. Paint Varnish Lacquer Assoc. Bull.* (Nov. 9, 1964).
4. Grier, N., and Ramp, J. A., *Chem. Specialties Mfrs. Assoc. Proc.*, **49**, 110 (December 1962).
5. Kent, M. M., *Am. Paint J.*, **5**, 21 (1966).
6. Lowrey, E. J., and Broome, T. T., *Offic. Dig. Federation Soc. Paint Technol.*, **38**, No. 495, 228 (1966).
7. Michaels, A. S., *Offic. Dig. Federation Soc. Paint Technol.*, **37**, 638 (1965).
8. *Natl Paint Varnish Lacquer Assoc., Circular*, **786**, 5 (1960).
9. Ramp, J. S., and Grier, N., *Offic. Dig. Federation Soc. Paint Technol.*, **33**, 1073 (1961).
10. Ross, R. T., *Advan. Appl. Microbiol.*, **5**, 219 (1963).
11. Ross, R. T., and Buckman, S. J., *Offic. Dig. Federation Paint Varnish Prod. Clubs*, **31**, 276 (1959).
12. Wessel, C. J., *SPE Trans.*, 193 (July 1964).
13. Wohlstein, H. C., and Feldstein, M., *J. Air Pollution Control Assoc.*, **16**, 19 (1966).
14. Wolf, P. A., *et al.*, *Am. Chem. Soc., Div. Org. Coatings and Plastics Chem., Preprint*, **25**, No. 2, 23 (1965).

24

*Testing of Raw Materials for Coatings**

The evolution of the protective coating manufacturing industry from the applied "black arts" period to the present state of a reasonable basis in scientific fact has accelerated in the last 20 years. A contributing factor toward the direction of fundamentals has been the *meaningful* testing of raw materials used by the industry.

The availability to the average raw materials laboratory of modern analytical techniques, once considered the domain of the industrial giants and the graduate schools of the larger colleges and universities, is today commonplace. Rapid modernization of methods has been fostered by the marketing of dependable and comparatively low-cost instrumentation, especially in recent years.**

Conclusions should not be drawn that "wet chemistry" and physical testing (viscosities, melting points, etc.) are outmoded. Far from it. The combination of instrumental techniques with the older methods has been a powerful tool in the hands of competent technicians and has fostered a new pride in the data output of raw materials laboratories.

It is not the province of this chapter to explain the theory of operation of spectrophotometers, gas chromatographs, etc., but rather to discuss their use in specific applications. However, the references cited usually go into more detail with regard to actual steps and theory.

Before proceeding, we are assuming that such important considerations as the reasons for testing of raw materials and the techniques and equipment for proper sampling are part of the professional background of the reader.

---

*By Morton L. Levy, Crobaugh Laboratories, Cleveland, Ohio.
**An excellent example of the foregoing was the announcement in March 1966 of an infrared spectrophotometer for less than $3000.

The broad classifications of the raw materials whose testing will be discussed here are:

> Pigments
> Oils
> Resins, Polymers, Plasticizers
> Solvents

PIGMENTS

Specifications for this class of materials have become more stringent due to the greater use of emission spectrographic analysis and x-ray diffraction procedures. The former is most useful in the area of trace or contaminant analysis, whereas the latter is an excellent tool for the determination of basic crystal structure, e.g., classification of titanium dioxide as anatase or rutile.[8] Infrared spectrophotometry is also useful for the study of some pigments.[2,9,16]

As a class, pigments should give the analytical laboratory the least trouble. The inorganic analyses are usually straightforward. The organic pigments and toners can be more troublesome. If a laboratory has much concern with the latter, instrumentation should be available.

Chemically, pigments are usually analyzed for the percentage of active metals, i.e., Pb, Cu, Cr, Ba, Zn, Sb, etc., and inerts. The methods are covered in the ASTM Standards[1] and represent no unusual chemistry or equipment requirements.

Other useful determinations required are oil absorption, color and tinting strength, and particles retained by a 325-mesh screen (coarse particles).

Oil Absorption

This is defined as the pounds of oil required per 100 pounds of pigment to maintain a stiff and putty-like consistency of the combination, i.e., completely wet the pigment. In the laboratory, oil absorption is determined by adding to a weighed amount of pigment (usually 1 gram or any multiple thereof) on a glass plate, foot-free raw linseed oil (acid value of 1 to 3) from a burette and incorporating the oil into the pigment after each drop by rubbing up with a sharp-edged steel spatula. The test is complete when enough oil has been added to the pigment to produce a very stiff, putty-like paste, which does not break or separate (bleed). The calculation is simple: the milliliters of linseed oil used multiplied by its specific gravity gives the number of grams. Exact details are given in ASTM D-281 (Part 20).

Color

This property is a subjective one. The color is judged by comparing a pigment paste in oil versus a standard or referee pigment paste in oil. The oil is light, refined linseed oil with an acid number of 4 to 6. The pastes are prepared by first working up the required weight of pigments and volume of oil with a spatula, followed by mulling with either a Hoover muller or a glass muller with a grinding face of from $2\frac{3}{4}$ to 3 inches in diameter. The pigment pastes to be compared are placed in juxtaposition on a bright tin or clear glass panel and then drawn down or leveled off simultaneously with a scraper blade so as to present both daubs on an even plane. The color tones should be judged immediately. The weights of pigments to use, volumes of oil and the number of mulls are rigorously defined in tables which can be found in ASTM D-387 (Part 20). These variables depend on the type of pigment or toner to be tested.

Tinting Strength of Colored Pigments

The tinting strength of a colored pigment is measured by mulling pastes prepared above with a standard white paste in the weight ratios of 1:20 or 1:50. These mulls are "drawn down" as in the color test. If the tinting strength of the sample is unequal to that of the standard, the operation is repeated using smaller weighed amounts of the stronger paste (sample or standard) until the reduction of the sample paste appears equivalent in color strength to that of the standard paste. The tinting strength is calculated by taking the ratio of the weight of standard paste to that of sample paste used to obtain equality of strength. Multiplying this ratio by 100 will express the result in per cent.

The preparation of the standard white paste used in the reduction is described in ASTM D-387 (Part 20) as well as the ratios to be used with the colored pigments under examination. The number of mull cycles is also standardized and is given in the same tables.

Tinting Strength of White Pigments

The mechanics of this test are similar to the two given above. ASTM D-332 (Part 20) governs the conditions for the testing of the tinting strength of white pigments, i.e., weight of white pigment, weight of standard ultramarine blue, volume of oil and the number of muller strokes. Here, of course, the white pigment under test is compared to a standard agreed upon by buyer and seller. The prepared pastes are "drawn down" and compared. If the sample is lighter than the reference standard it has greater tinting strength (the reverse notation for the tinting strength of a colored pigment).

Coarse Particles

This usually refers to the residue left in a 325-mesh screen (44 μ) after flushing a given weight of pigment with water.   Though this test is standard and relatively easy to perform, it is time consuming.   Some pigments need as much as 1 hour for flushing the required amount through the fine mesh screen (see ASTM D-185, Part 20).

Instruments have recently been developed which not only precisely measure particles larger than 325 mesh, but also give the particle size distribution of the entire pigment down to 0.5-μ diameter.   If the only information required is the amount of pigment greater than 325 mesh, a sample can be run every 10 minutes.   A complete particle size distribution run can be made in an hour.[3]

OILS

The paint industry, including its raw materials suppliers, has been a prime mover in developing and refining many of the methods used in testing triglycerides, fatty acids, rosin and tall oils.   As the largest consumer of drying oils, it has steadily moved the frontiers of knowledge ahead so that the analysis and testing of fatty oils supplies not only empirical data for quality control but fundamental structural information as well.

The chemical and physical testing of oils as a group (both drying and nondrying) usually calls for specific gravity, acid number, saponification number, iodine number, unsaponifiable, peroxide value, moisture, insolubles, ash, color and refining loss.   In many plants, all the preceding tests are usually coordinated with instrumental characterization involving ultraviolet and infrared spectrophotometry and gas chromatography.   The latter technique has for the first time made available rapid, inexpensive and accurate fatty acid distributions of any particular oil sample to guide production and purchasing personnel.

The individual determinations listed above are described in detail as industry standards in "The Official and Tentative Methods of the American Oil Chemists' Society."* The AOCS methods are brought up to date by frequent revisions of older methods and the insertion of new techniques from time to time.

The American Society for Testing Materials (ASTM) in Part 20 also publishes details of standard tests applying to the paint industry.   The ASTM methods are now published every year in book form.   Fortunately, the methods for oils and fats as standardized by either group are identical; this is in part due to the cross-fertilization of the committees from both of these important groups.

*American Oil Chemists' Society (AOCS), 35 East Wacker Drive, Chicago, Ill.

The *specific gravity* of an oil is not reliable for identification purposes. It is important for commodity trading purposes (drying oils are sold by the pound, not by volume), plant storage and gauging, and to the formulator for cost and performance extensions. The specific gravity of oils varies over the range of 0.90 to 0.99 and is measured by using a pycnometer inserted in a constant-temperature bath. Complete details, equipment, etc., are given in AOCS Official Method Cc 10a-25. Briefly, the weight of the oil in the pycnometer at 25°C is divided by the weight of the water occupying the same pycnometer at the same temperature.

Acid number or acid value of a fatty oil is defined as the number of milligrams of potassium hydroxide required to neutralize the free fatty acids in 1 gram of sample. In fats or drying oils, the acid number will give some indication of the freshness, care of manufacture or degree of refining; i.e., the greater the acid number, the lower is the quality likely to be. In fatty acids, the acid value is an indication of average chain length and is used as a control measure in the manufacture of alkyd resins, soaps, etc. The test (AOCS Official Method Ka 2-58) is accomplished by dissolving a specified weight of fatty oil in an isopropyl alcohol–toluene solvent. The titrant is sodium hydroxide in methanol, but the results are calculated to potassium hydroxide.

Saponification number or saponification value is defined as the number of milligrams of potassium hydroxide that will react with 1 gram of sample. The weighed sample (AOCS Official Method Ka 8-48) is refluxed for 1 hour in an excess of standard alcoholic potassium hydroxide. The excess base is titrated with standard acid.

Iodine number is defined as the percentage of iodine absorbed by the sample, whether or not the halogen used is actually iodine. The iodine number is a function of the degree of unsaturation in the carbon-carbon bonding of the chain. Fatty oils may be classified as to their drying ability by the iodine number:

Drying oils	160 and up
Semidrying oils	120 to 160
Nondrying oils	Below 120

The method for iodine number in favor in the United States is the Wijs modification. The absorbing halogen is iodine monochloride in glacial acetic acid. The iodine number determination is highly empirical and details are important (AOCS Official Method Ka 9-51). Some of the important factors in determining iodine number are:

(a) An excess of ICl must be used, the excess depending on the iodine value expected and the type of unsaturation, i.e., conjugated or unconjugated. Tung oil and dehydrated castor oil are particularly "touchy" to analyze for iodine number.

(b) The temperature of the reaction is usually 25 ± 5°C.

(c) The reaction time is usually 1 hour.

Moisture, insolubles and unsaponifiables are three analyses which indicate the quality of an oil. These three determinations are usually referred to in trading circles as M.I.U. and deductions from the selling price are made if they exceed 1 to 2%, depending on the contract stipulations.

Methods for *moisture* are legion. They include drying in an oven, drying for a short time on a hot plate, low-temperature vacuum oven technique, Karl Fischer titration and distillation with an immiscible solvent. All these methods are permissible, but only the latter gives accurate results on drying and/or highly conjugated oils, since the other methods involve some reaction with the active unsaturation. The AOCS Official Methods for moisture in fatty oils are:

Distillation method	Ca 2a-45
Hot plate method	Ca 2b-38
Air oven method	Ca 2c-25
Vacuum oven method	Ca 2d-25
Karl Fischer titration	Ca 2e-55

Insolubles covers dirt, scale, meal and other foreign substances insoluble in kerosene and petroleum ether. The details are covered in AOCS Official Method Ca 3-46.

Unsaponifiable matter includes those substances frequently found dissolved in fats and oils which cannot be saponified by the caustic alkalies but are soluble in ordinary fat solvents. Examples of unsaponifiable matter are the higher aliphatic alcohols, sterols, pigments and hydrocarbons. The method (AOCS Official Method Ca 6a-40) involves saponifying a known quantity of the oil and extracting the unsaponifiables in petroleum ether.

Peroxide value is a measure of *active oxygen* in an oil and is defined as the number of equivalents of peroxide in 1000 grams of oil. It is used as an indication of the quality of oil and the presence of incipient rancidity when over specified limits. Peroxide values are also useful in monitoring the progress of oils being blown or heat bodied.

Ash is not a significant factor in oils but is usually run as a gross check on impurities. A quantity of sample is ignited in a crucible with a Bunsen flame and allowed to burn quietly until nearly consumed. The residue is then ignited in a muffle at 600°C until all the carbon is consumed. The remainder is cooled, weighed and reported as per cent ash.

Color determination of an oil is not without controversy. There are several systems for its determination, but the most common and useful is covered by AOCS Official Method Ka 3-58. It is comparatively crude but has the advantage of speed and simplicity. The oil sample whose color is to be determined is compared with 18 standards tubes containing potassium chloroplatinate, ferric chloride and cobalt chloride in solution. The standards are darker as one proceeds through the series. To use this method, the sample and standards must be similar in color density. If an oil has a green or red color, the comparison cannot be easily made.*

Refining loss of an oil is the weight of material lost as settlings or "foots" when a sample is neutralized by an amount of alkali used according to the original free fatty acid content. The test is rigorously defined in AOCS Official Method Ca 92-52.

Mono-, Di- and Triglycerides

The chemical determination of any of these in the presence of the other two has always been a challenge to the oil chemist. The reaction of periodic acid selectively with alpha-monoglycerides is subject to the interference of free glycerine which must be extracted with water before the beginning of the test (AOCS Official Method Cd 11-57). Furthermore, periodic acid will not react with beta-monoglycerides. An excellent method using column chromatography[22] with silica gel as the column material has been devised. The separations are quantitative. The triglyceride is eluted from the column with benzene and diglyceride with 10% ethyl ether in benzene. Monoglyceride is eluted with ethyl ether. Also, the fraction of eluted monoglycerides can be reacted with periodic acid to determine the ratio of alpha-monoglycerides to beta-monoglycerides.

Instrumental Analysis of Oils

Ultraviolet spectrophotometric methods in fatty oil analysis are based on the absorption of the conjugated double bonds in the UV region. The characteristic wavelengths of the absorptions are 233 μ for dienes and 268 μ for trienes. The tetraenes and more highly unsaturated acids absorb at higher wavelengths. AOCS Official Method Cd 7-58 gives the methodology for measuring the conjugated diene, triene, tetraene and pentaene groups. The percentages of linoleic, linolenic, arachadonic and pentaenoic acids can then be calculated.

There are many infrared spectrophotometric methods in oil analysis.[20] However, identification of natural drying oils by infrared is not practical because of the similarity of spectra. Oils such as soybean and linseed are indistinguishable. The spectra of oils with conjugate unsaturation, tung, oiticica and dehydrated castor oil show minor differences in the region

---

*The standards for this color comparison can be purchased from Gardner Laboratories, 4723 Elm Street, Bethesda, Md.

near 10 μ. Castor oil has a strong absorption at 3 μ because of the hydroxyl group.

Infrared spectroscopy has more value in the identification of modified oils containing dicyclopentadiene, vinyl toluene, styrene and other materials which may be copolymerized or reacted with the drying oils.

Gas Chromatography

The oil chemist has always been looking for a single test to characterize the triglycerides with which he worked. Tests which were helpful such as saponification value, iodine number, titer, etc., always left something to be desired when decisions had to be made on unknowns. Gas chromatography is a method that supplies the characterization in a single test.

The fatty acid composition and distribution from any oil can be determined in a short time by gas chromatography. With this data, an experienced chemist can decide which oil is being studied, or more to the point, can give the proportions of each oil in a mixture. Gas chromatographic techniques are routine for the fatty acid distribution in a whole host of materials. Both AOCS and ASTM publish tentative standards for running these determinations, i.e., AOCS Tentative Method Ce 1-62, "Fatty Acid Composition by Gas Chromatography"; ASTM Tentative Method D 1983, "Fatty Acid Composition By Gas-Liquid Chromatography of Methyl Esters"; ASTM Tentative Method D-2245, "Identification of Oils and Oil Acids Separated From Solvent-Type Paints (Gas Chromatography)."

Most of the work done with gas chromatography in elucidating fatty acid composition of oils has depended upon the formation of the relatively volatile methyl esters to send through the column. The fatty acids themselves, because of a combination of high polarity and low volatility, behave difficultly as far as gas chromatography is concerned. One attempt to use fatty acids directly in gas chromatography is referenced.[14] Apparently it was successful, but the method does not seem to have been pursued. At any rate, the formation of the methyl esters is easily accomplished in about 2 minutes by the use of boron trifluoride catalyst in methyl alcohol.* The method for this esterification technique is referenced[15] and contains instructions for the preparation of the catalyst.

Until very recently, an unsolved gas chromatography problem was the difficulty of esterification of resin acids in rosin and tall oil. It is relatively easy to esterify the tall oil fatty acids with the use of the boron trifluoride catalyst, but the resin acids are not affected. It turned out that a relatively mild (though highly toxic) methylating agent, diazomethane, would com-

---

*This catalyst can be purchased all ready for use from Eastman Organic Chemicals Dept., Rochester, N. Y., or Applied Science Laboratories, State College, Pa.

pletely esterify the resin acids. As a result, the resin acids can now be identified and determined easily.[18,21] The diazomethane is generated from *n*-methyl-*n*-nitroso-*p*-toluenesulfonamide (MNSA).* Redistilled "Carbitol" is used as the reaction solvent. Diazomethane is generated from MNSA with a solution containing 6 ml of potassium hydroxide in 10 ml of water.

The apparatus for the methylation with diazomethane consists of three test tubes with side arms connected in series. A stream of nitrogen is saturated with ether in the first tube (16 × 150 mm) and carries diazomethane generated in the second tube into a third tube where esterification takes place. The dimensions of the latter tubes are 15 × 85 mm. The side arms (0.7 cm o.d.) of the tubes are bent downward, and each almost reaches the bottom of the following tube. The ends are drawn out to about 1 mm o.d. Rubber stoppers are used for connections. The flow of nitrogen through the ether contained in tube 1 is adjusted to 6 ml/min. 0.7 ml of "Carbitol" and 1 ml of a solution of 6 grams of KOH in 10 ml of water are placed in tube 2. The resin acids and/or fatty acids are dissolved in 2 to 3 ml of ether containing 10% methanol and placed in tube 3. MNSA (2 meq per 1 meq of resin acid) dissolved in 1 ml of ether is added to tube 2. The tubes are connected immediately, and the diazomethane generated in tube 2 is swept into tube 3. As soon as a tinge of yellow appears in tube 3 when it is viewed against a white background, it is disconnected and the slight excess of reagent is destroyed by adding a dilute solution of acetic acid in ether, or the excess diazomethane can be removed in a stream of nitrogen. The procedure requires 10 to 12 minutes. *Diazomethane is very toxic and must be handled with caution.*

The column substrate used in fatty acid gas chromatography is usually diethylene glycol succinate or butanediol succinate. The temperature mode of operation may be isothermal (290°C) or programmed. The latter is preferable when short-chain length fatty acids are expected as they will not all elute at one time at the lower temperatures occurring at the beginning of the temperature program. Temperature programming, column materials, flow rates, identification and quantitation are critically discussed in a paper which should be in the library of every laboratory doing work in the fatty acid field.[11]

Molecular weight determinations are very useful in a quality control laboratory. The classical procedures for obtaining this data never allowed good precision unless performed by the microanalytical laboratory. There is available an instrument that gives molecular weight data rapidly and with a precision of ±1% up to a number average of approximately 8000.[13]

---

*This compound is available commercially under the trade name "Diazald" (Aldrich Chemical Co.).

This instrument can be operated in any macro laboratory by technicians with little previous training. Molecular weight data are seldom asked for, but they should be used in characterizing resins, plasticizers, etc. Furthermore, molecular weight data on raw materials where applicable will pinpoint possible troubles and prevent them.

The molecular weight instrument referred to operates on the principle that the vapor pressure of a solution is lowered from that of the pure solvent in proportion to the molar concentration of the solute. When the weight concentration of a solution is known, a measurement of the vapor pressure lowering allows an easy calculation of molecular weight of the solute. In operation, a single drop of the solution is placed on a thermistor bead in a thermostated chamber saturated with solvent vapor. Because the solution has a lower vapor pressure, condensation takes place on it and the solution droplet becomes heated by the release of the latent heat of vaporization of the solvent. This slight increase in temperature as opposed to a bead of solvent alone on a similar thermistor is measured by an electrically balanced bridge circuit. The instrument will accurately detect changes in temperature as little as $0.0001°C$.

RESINS, POLYMERS, PLASTICIZERS

The compositions of matter falling under the classification of resins, polymers and plasticizers are myriad. Their testing covers every imaginable chemical reaction and physical property. The obvious tests such as specific gravity, viscosity, volatile matter, color, acid number, odor, taste, distillation range and flash point are well covered in the ASTM Standards, Parts 20 and 21. The important point is that the raw materials chemist must know with what he is working and the application to which it is being put. Common sense will then dictate the type of chemical and physical testing required. It is almost impossible to work with this class of materials in any kind of volume without an IR spectrophotometer and a gas chromatograph with several good choices of columns on hand.

Several good texts are available which will aid the raw materials chemist in almost any problem that may come up with this group of materials. These have been carefully screened and are referenced.[4,10,12] Excellent industrial compilations of IR spectra are available for identification of unknowns. A booklet containing 125 spectra of plastics and resins is available[19] as is a booklet containing 189 spectra of plasticizers.[5] Typical of the methods used to chemically "tear apart" plastics for examination by IR spectrophotometry is one advanced by Stafford and Shay.[23] It beautifully demonstrates the combination of wet chemistry and instrumental analysis to solve a quality control problem. Stafford and Shay have solved the problem of identifying the components in polyesters, i.e., the dibasic acids and the polyols. The chemistry involved is the ammonolysis of the polyester with dibenzylamine, thereby forming the amide of the dibasic acid and releasing the polyol in its original form. The acid amides are identified by comparing with standard spectra, as are the polyols after extraction in alcohol.

The technique of Stafford and Shay has now been developed into a gas chromatographic method. The polyester is subjected to ammonolysis by a suitable amine such as butylamine, with the resulting polyols being acetylated *in situ*. The reaction mixture is injected into a gas chromatograph with a polyester column operated isothermally at about 275°C. The dibasic acids and the polyols which were used in the original polyester can now be identified from the chromatogram by the retention times, and they can be quantitated by the areas under the curve or by comparison with internal standards which may be purposely added to the unknown.

Alkyd resins can also be subjected to ammonolysis reactions. The dibasic acids and the polyols are analyzed by gas chromatography as indicated in the above paragraph, while the monobasic acids can be recovered by hydrolysis, methyl esterified, and identified and determined as indicated previously.

G. G. Esposito, in a publication[6] from the Aberdeen Proving Ground, gives an excellent method for the identification and determination of plasticizers in lacquers. The technique involves the separation of the resins by a chemical route and then injecting the plasticizers into the chromatograph for separation. Identification and quantitation are possible at the same time.

Infrared spectrophotometry is used to estimate the urea-formaldehyde-melamine resin ratio in a paint vehicle comprising both of these in admixture with an alkyd resin.[17] The method is based on infrared absorbance measurements at 5.8, 6.1, and 12.25 μ on thin vehicle films. Though this work describes a rapid means for estimating the individual resin components in typical two- and three-component alkyd-nitrogen resin blends, it is indicative of other methods which can be worked out by first working with known materials.

SOLVENTS

The general testing of solvents dramatically illustrates the transition of a few instrumental analyses replacing older tests which were very general in nature and, when taken as a whole, did not thoroughly characterize the solvent under examination. More reliance is being placed on IR spectrophotometric identification together with gas chromatographic analysis since characterization is almost absolute. However, tradition is not easily displaced, and the general tests such as appearance, color, specific gravity, distillation range, water, odor and acidity have not been removed from specifications and remain as part of the work load in most raw materials laboratories. No comment will be made with regard to such descriptive tests as general appearance, color and odor. These tests are well defined in the ASTM Standards, Part 20.

Specific gravity is measured at a specified temperature, usually 15.56°C, 20°C, 25°C, or 60°F. The actual temperature used in determining specific gravity is, of course, noted with the result. The specific gravity may be determined by a calibrated hydrometer or by the use of a pycnometer.

Distillation range characterizes a solvent more completely than almost any other general test. Its value has been enhanced by the fact that very little has been left to the judgement of the technician performing the test. Every detail has been standardized so that reproducible results are easy to obtain on both an interlaboratory and intralaboratory basis. ASTM Standards, Part 20, is the source for the details in the distillation range test. ASTM D-86 covers the distillation of petroleum products, ASTM D-850 details the distillation test for industrial aromatic hydrocarbons, and ASTM D-1078 contains instructions for the distillation range of volatile organic liquids.

Nonvolatile matter is determined by volatilizing a specified amount of solvent, usually 100 ml, and weighting back the residue.

Instrumental analysis of solvents usually involves IR spectrophotometry and gas chromatography. Because of the volatility of most solvents, their IR spectrophotometric examination is usually done with sealed cells, 0.025 mm thick for characterization, and 0.1 mm thick, or greater, when trace or minor contamination is being studied in a particular solvent.

The gas chromatograph is usually operated isothermally when testing solvents. The column material receiving the largest use in gas chromatographic analysis of solvents is "Carbowax" with an average molecular weight of 20,000. Almost equally popular as a column material for solvent studies with gas chromatography is silicone rubber.

Gas chromatography has found an invaluable niche in the characterization of solvents consisting of blends of aliphatic and aromatic materials. Blends of lacquer solvents are almost impossible to analyze properly unless gas chromatographic equipment is at hand.

One of the problems of gas chromatographic analysis of solvents, especially complex blends, is that it gives so many peaks that it becomes burdensome to identify them or to use them as "finger print data." By the use of a little ingenuity, i.e., combining wet chemistry with gas chromatography, the problem of complex data is solved.

A generous portion of the solvent sample is extracted with 10% brine, then with water, followed by an extraction with 85% sulfuric acid and finally, by an extraction with concentrated sulfuric acid. A portion of each of the insoluble solvents from each of the treatments is saved and run through the gas chromatograph. The 10% brine extract washes out the lower alcohols such as methyl and ethyl. The water extract removes water-soluble alcohols, many ketones and most of the glycol ethers. The 85% sulfuric acid treatment removes additional esters and additional alcohols, as well as ketones and polar solvents. The concentrated sulfuric wash removes most of the aromatic hydrocarbons and unsaturates.

By looking at the runs of each of the insolubles of the above treatments, as well as a run of the sample without any treatment, an experienced technician can supply very nearly complete identification of all peaks and thereby characterize the blend. This analytical scheme combining gas chromatography and extractions with varying media can be found in reference 17.

Moisture testing of solvents is done in large part by the use of the Karl Fischer titration. The method is capable of good accuracy and is sensitive to traces of water. The Karl Fischer method should not be used in the presence of mercaptans, peroxides, or appreciable quantities of aldehydes or amines.

Karl Fischer reagent can be purchased ready for use from laboratory supply houses. The reagent has good stability but should be standardized daily. A visual end point can be used with light-colored solvents, and an electrometric end point can be used with darker materials. Accuracy is increased by using the electrometric end point. It is not necessary to purchase expensive equipment in order to set up for the electrometric end point. Low-cost kits are available from the manufacturers of pH meters, and they have the advantage of allowing the pH meter to be used for other types of electrometric end points. ASTM D-1364 (Part 20) contains all the instructions needed to carry out this test competently.

REFERENCES

1. American Society for Testing Materials, Part 20, 1966, "Paint Materials Specifications and Tests; Naval Stores; Aromatic Hydrocarbons."
2. Chicago Society for Paint Technology, "Infrared Spectroscopy, Its Use As An Analytical Tool In The Field of Paints and Coatings," *Offic. Dig. Federation Soc. Paint Technol.*, **33**, (March 1961).
3. Coulter Electronic Laboratories, 5227 N. Kenmore Street, Chicago, Ill.
4. Critchfield, F. E., "Organic Functional Group Analysis," New York, The Macmillan Co., 1963.
5. DuVall, R. B., "Infrared Spectra of Plasticizers and Other Additives," The Dow Chemical Co., 1962.

6. Esposito, G. G., "Identification and Determination of Plasticizers in Lacquers by Programmed Temperature Gas Chromatography," CCL Rept. No. 141, Aberdeen Proving Ground, Md.

7. Haken, J. K., and McKay, T. R., *J. Oil Colour Chemists' Assoc.*, **47**, 517 (1964).

8. Hannawalt, Rinn and Frevel, *Anal. Chem.*, **10**, 457, (1938).

9. Harkins, T. R., Harris, J. T., and Shreve, O. D., *Anal. Chem.*, **31**, 541 (1959).

10. Haslam, J., and Willis, H. A., "Identification and Analysis of Plastics," Princeton, N. J., D. Van Nostrand Co., 1965.

11. Horning, E. C., *et al.*, *J. Lipid Res.*, **5**, 20 (January 1964).

12. Kline, G. M., "Analytical Chemistry of Polymers, Part I, Monomers and Polymeric Materials," New York, Interscience Publishers, 1963.

13. Mechrolab-Vapor Pressure Osmometer, F & M Scientific Co., Division of Hewlett-Packard Co., Avondale, Pennsylvania.

14. Metcalfe, L. D., "Direct Analysis of Fatty Acids," *Facts and Methods for Scientific Research*, **2**, No. 1, (Spring 1961).

15. Metcalfe, L. D., and Schmitz, A. A., *Anal. Chem.*, **33**, 363 (1961).

16. Miller, F. A., and Wilkens, C. H., *Anal. Chem.*, **24**, 1253 (1952).

17. Miller, C. D., and Shreve, O. D., *Anal. Chem.*, **28**, 200 (1956).

18. Nestler, F. H. M., and Zinkel, D. F., *Anal. Chem.*, **35**, 1747 (1963).

19. Nyquist, R. A., "Infrared Spectra of Plastics and Resins," The Dow Chemical Co., 1960.

20. O'Connor, R. T., *J. Am. Oil Chemists Soc.*, **33**, 1 (1956).

21. Weaver, John L., General Tire and Rubber Co., Akron, Ohio, private communication (Nov. 3, 1964).

22. Quinlin, P., and Weiser, H. J., Jr., *J. Am. Oil Chemists' Soc.*, **35**, 325, (1958).

23. Stafford, R. W., and Shay, J. F., *Ind. Eng. Chem.*, **46**, 1625 (1954).

25

*Color in Paint**

Color means many things to many people. It is often used to describe objects, e.g., "the man was driving a red car," or "The girl was wearing a blue dress." It is also used to denote a person's reaction to a situation, as when we say, "He was red with anger." Most people will define color depending upon their area of interest, and they could all be correct.

We are concerned here with color as related to paint finishes so we will consider it in this manner.

Color is a psychophysical reaction as seen by the eye and interpreted by the brain. There is much written on how this amazing process takes place. This is a function of nature over which we have little or no control.

The eye-to-brain communication is a science in itself, but we will be concerned only with how the normal process works for those people who have to make color decision or judgments. They must, by reason of necessity, have normal color vision. This is sometimes referred to as having satisfactory color aptitude. There are several recognized tests, such as the Federation of Societies of Paint Technology Color Aptitude Test designed by Carl Foss and Forest Dimmick, to check an observer's ability to see color difference.

VISIBLE SPECTRUM

We are aware of color as light when the retina of our eye is exposed to the action of electromagnetic radiation having a wavelength of between 400 and 700 mμ. The wavelength of the color we see ranges from violet (400 to 430 mμ), blue (430 to 485 mμ), green (485 to 570 mμ), yellow (570 to 585 mμ), orange (585 to 610 mμ) to red above 610 mμ. This range composes what is known as the visible spectrum of colors (Figure 25.1).

This range of color can also be recorded and shown by the use of an instrument known as a spectrophotometer. The recorded range of color is

*By S. J. Huey, The Sherwin-Williams Co., Cleveland, Ohio.

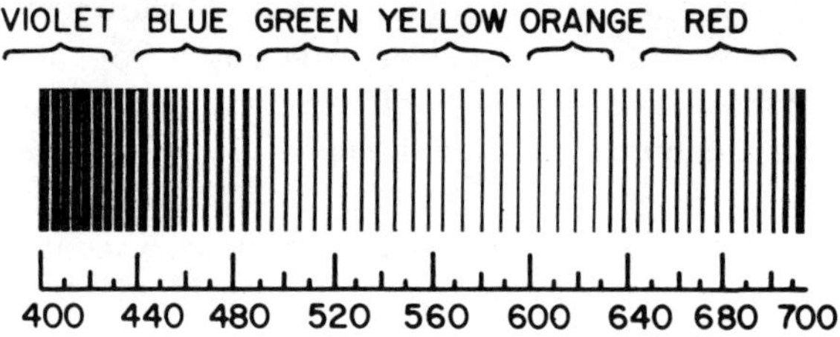

Figure 25.1.   The visible spectrum.

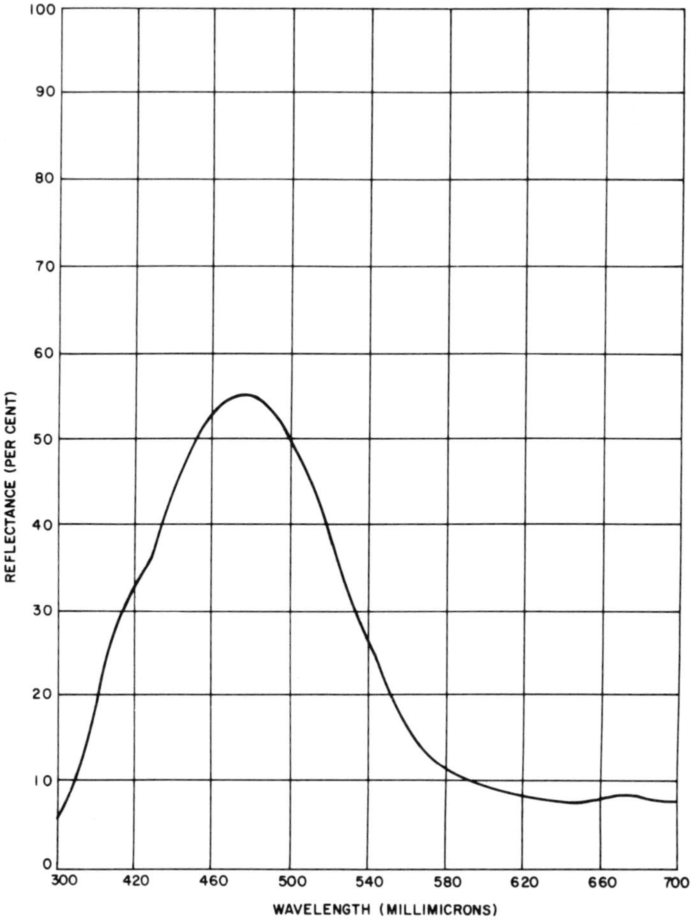

Figure 25.2.   Spectrophotometric curve of letdown of phthalo-cyanine blue.

known as a spectrophotometric curve. The curve shows us the amount of light reflected from a sample at a given wavelength from 400 to 700 mμ.

Figure 25.2 shows a spectrophotometric curve of a let-down of Phthalocyanine Blue in titanium. The sample has a high reflectance in the 430 to 458 mμ region of the spectrum; therefore, it appears blue. Toluldine has very little reflectance in any part of the spectrum except the red region, consequently it appears red (Figure 25.3). If a sample has a fairly equal distribution of light across the spectrum, it will appear white (Figure 25.4)

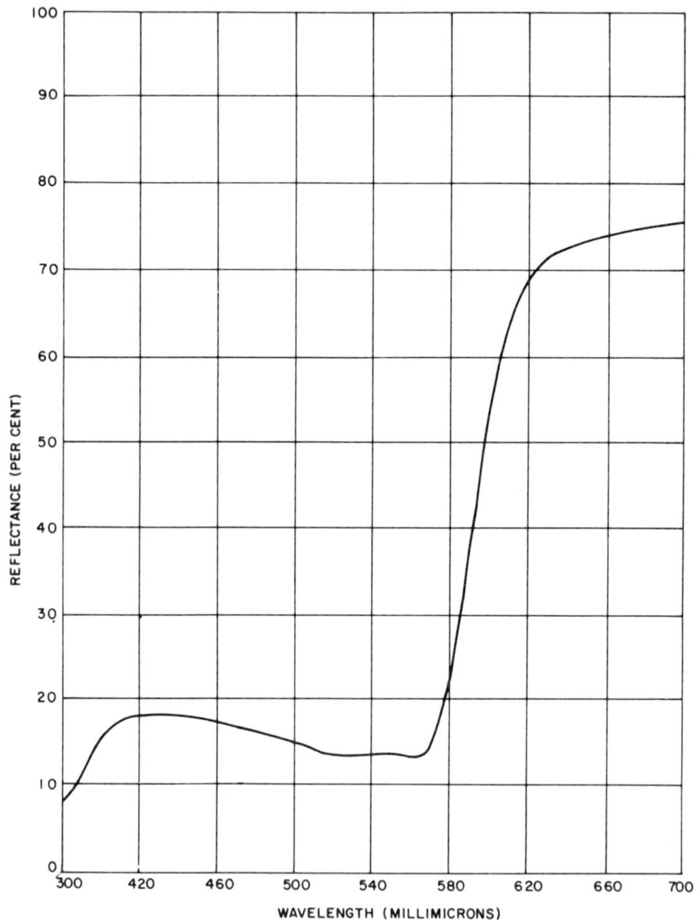

Figure 25.3.   Spectrophotometric curve of letdown of toludine red in TiO$_2$.

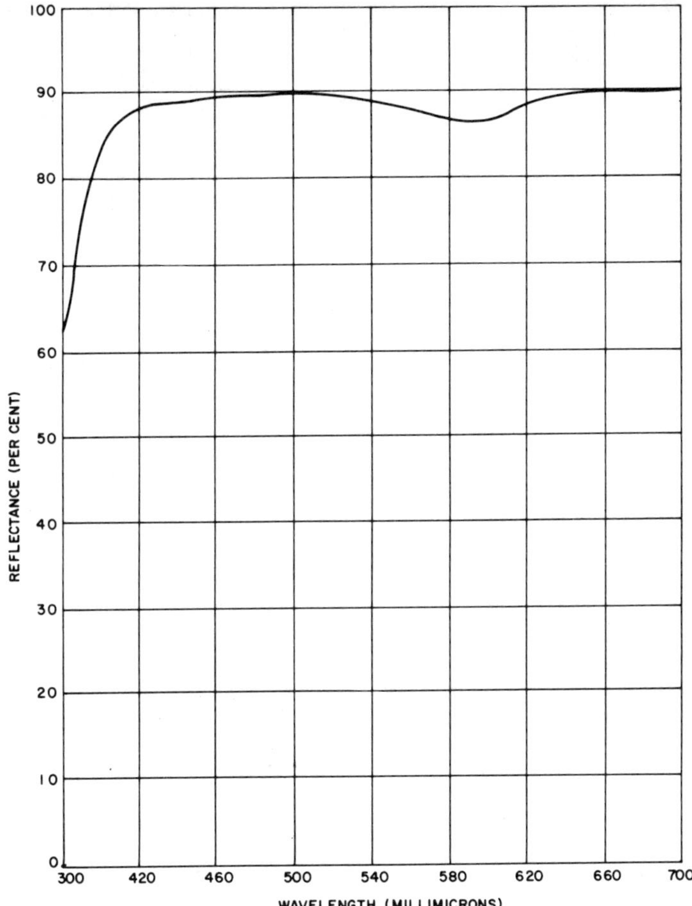

Figure 25.4.   Spectrophotometric curve of a white enamel.

or various degrees of gray (Figure 25.5), depending upon the total light reflectance.

The curve of a given color has sometimes been referred to as the fingerprint of the color. Spectrophotometric curves have many useful functions in colorimetry.

THREE DIMENSIONS OF COLOR

Too often, we find ourselves trying to describe color by a single word or characteristic. We would not attempt to describe a varnish by a single

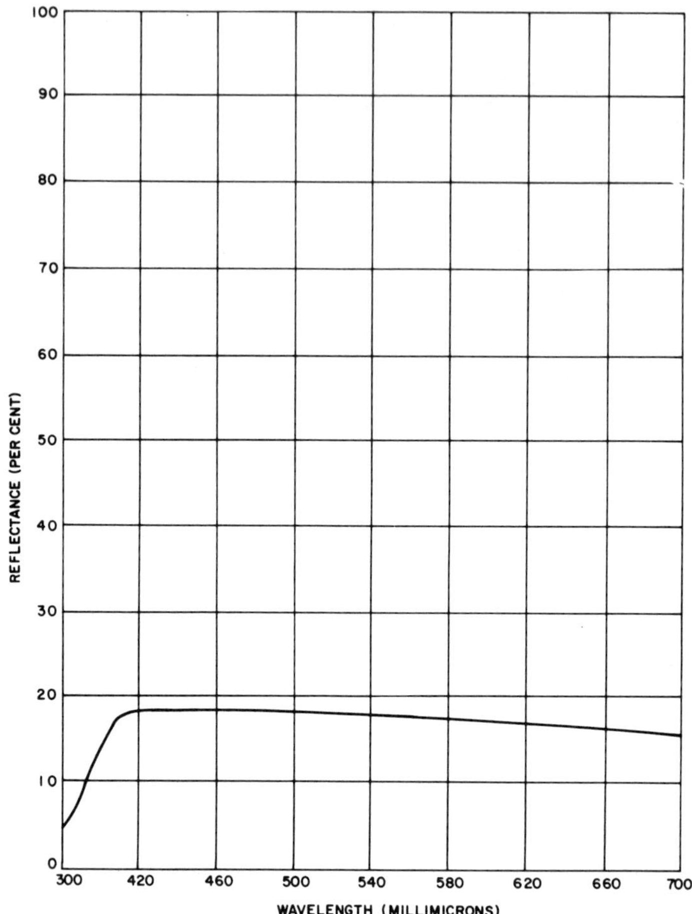

Figure 25.5.   Spectrophotometric curve of medium-gray enamel.

word, such as saying, "The varnish has an acid number of 4.5." Such an explanation of varnish would leave us with much wanted information.

Often in an attempt to broaden a description of color, we use such simple terms as depth of color, strength and character. These are names which are known only to the paint industry and would be quite confusing to anyone else. There is considerable doubt as to whether the individuals using the terms understand each other or transfer the correct information from one to another.

Paint shaders or tinters sometime make the following type of statement, "I will increase the depth of this batch." What does he mean? Does he mean that he must add more red or more black to increase the depth?

With color being such an important attribute of an organic coating, we should define it correctly.

When we say that a piece of lumber is 10 feet long, we have defined one dimension of it, but is the board narrow, wide or thick? If the description says the board is 10 feet long and 1 foot wide, we have a better mental picture of it, but the picture is incomplete until we know its thickness. When we learn that the board is 10 feet long, 1 foot wide and 1 inch thick, then and only then is its linear description complete, leaving nothing else to the imagination.

So it is with color; color by its definition has dimension. Since anything that has dimension must occupy space, we speak of color as to where it lies in color space. To describe color adequately, all three dimensions must be characterized in the same manner as we described a length of lumber. If we think in such terms, we will find our understanding of the colorimetric data much clearer and more meaningsul to us.

The first dimension of color is known as reflectance, also defined as value or lightness. There is a specification that states, "The color of a school room can be any color so long as the reflectance is 65% or above." From this, we can see that the reflectance of paint film is the amount of light that it reflects, regardless of its color. By this we mean that green, red, blue or yellow can have the same light reflectance. We can say that the perfect white reflects 100% of the light that strikes it and perfect black would reflect no light, having a zero reflectance. All other colors fall in between this scale, as illustrated by Figure 25.6.

The second dimension of color is known as hue. Hue is the attribute of color that distinguishes green from red and blue from yellow. It is commonly referred to as the color of the object.

The third dimension of color is saturation or chroma. This dimension is the most difficult to understand and describe of the three dimensions. Saturation is the variable of color that denotes its departure from grayness. Let us assume we have a gray and a red which both have the same reflectance. Now if we take this red and add to it 20% of the gray, we have a mixture of 80% red and 20% gray. The red will appear less brilliant or less red. It has not become less orange or blue but less red. It is now less saturated. If we make a mixture of 50% red and 50% gray, it will appear more gray, and so on with additional amounts of gray until the red is diluted to the point where the sample no longer appears red but gray. At this point, it would have 100% gray and no red so it would have zero saturation. The change in saturation has been obtained without any increase or decrease in reflectance.

We cannot assume that whenever a color is added to another color only one dimension will be affected at a time. Sometimes only two dimensions are affected, but usually all three are involved. If to the red we add 20%

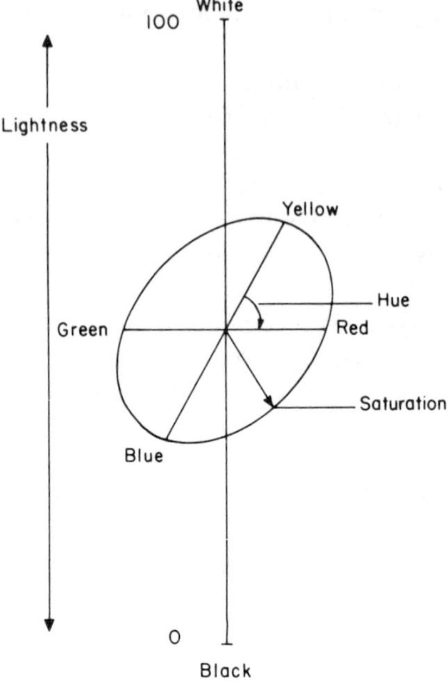

Figure 25.6. Relation of the three dimensions of color to each other.

black instead of gray with the same reflectance, we will immediately dilute the red by 20%. Therefore, it will move closer to the center because now our mixture is 80% red and 20% black, which is less saturated. When we add a material which has a lower reflectance than the red, the resulting mixture will also have a lower reflectance, so it will drop further down in color space. The sample will move down and in, by the addition of the black. If we add white to the sample, the same thing will happen, only in the reverse direction. The red is now more diluted. In other words, it is not as brilliant or as saturated and we have added material which has a higher reflectance than the red. So now the color moves further up in color space and also into the center. White has the same effect on saturation as gray, i.e., moving it into the center of color space.

At this point, the hue of our color has not been affected because all we have added to it has been white, gray and black. Now let us take the red and add to it a yellow, which has the same degree of saturation and the same reflectance. Of course, yellow will have a different hue or dominant wavelength than red. This mixture of two colors of different hues results in a third color, in this case orange. As long as the two colors which we mix have the same reflectance and the same saturation, the resulting color changes only in hue. However, if the yellow had a higher reflectance than the red and was less saturated, the mixture would then change in all three dimensions.

This is an example of what usually happens when colors are mixed in normal shading operations. When we view the results from mixtures of several colors, we sum up in our minds all the steps that we have described and usually refer to it as a color change. However, color measuring instruments resolve each step in a logical order and assign numerical values to them. For this reason, it is important that we have a clear knowledge of color space so that we may properly interpret each dimension as obtained from color measuring instruments.

It is important to understand and remember the three dimensions of color: (1) lightness, (2) hue and (3) saturation.

Tristimulus Values X, Y, Z

One of the basic working tools for colorimetry is what is known as the tristimulus X, Y and Z. The Commission Internationale d'Eclairage (CIE) tristimulus values are obtained mathematically by evaluating the effect on the standard observer's eye of special flux, either directly from a light source or as it is reflected or transmitted by an object.

CIE tristimulus values are defined in the following manner:

$$X = \int_0^\infty \bar{x}(\lambda)\, P(\lambda)\, p(\lambda)\, d\lambda$$

$$Y = \int_0^\infty \bar{y}(\lambda)\, P(\lambda)\, p(\lambda)\, d\lambda$$

$$Z = \int_0^\infty \bar{z}(\lambda)\, P(\lambda)\, p(\lambda)\, d\lambda$$

Here, $\bar{x}$, $\bar{y}$ and $\bar{z}$ represent the amounts of the CIE primaries required by the standard observer to color match bands of wavelengths throughout the visible spectrum with equal energy present for all bands (Figure 25.7). P is the spectral radiant flux from the light source illuminating an object. λ represents the wavelength. These values can be also obtained directly from the automatic integrater attached to the spectrophotometer. Approximate values can also be obtained from filter colorimeters.

Chromaticity Coordinates

The X, Y, Z values of a color have been defined as the relative amounts of each of the three unreal primaries needed to make a color match under

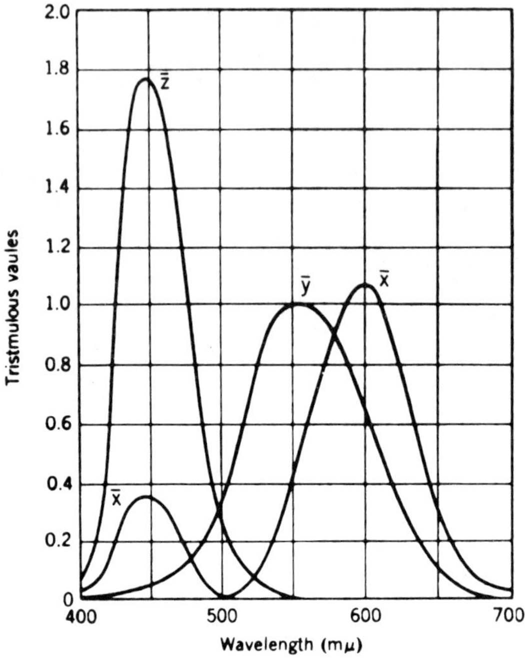

Figure 25.7. CIE tristimulus values $\bar{x}\,\bar{y}\,\bar{c}$ for the spectrum colors.

certain conditions. The chromaticity coordinates, often known as little x and y, are calculated in the following manner.

$$x = \frac{X}{X + Y + Z}$$

$$y = \frac{Y}{X + Y + Z}$$

Small z is seldom calculated because by definition

$$x + y + z = 1$$

Using the chromaticity coordinates x and y, the spectral colors can be plotted on a rectangular coordinate system known as the CIE chromaticity diagram.

Chromaticity Diagram

The chromaticity diagram is sometimes referred to as a road map in color space. This being so, it should be able to be used to locate colors

and the distance between colors (Figure 25.8). This diagram could, therefore, be used to specify colors.

When the chromaticities of colored lights (x and y) are plotted in the
chromaticity diagram and the points connected, they form a curved line
known as the spectrum locus. A straight line is drawn to connect the two
ends (longest and shortest) to form a figure within whose boundaries all
real colors will plot. We should think of it as a plane in color space at a
given reflectance level. The so-called center of the chromaticity diagram
is determined by what illuminant is used. Illuminant C is the one, more
often referred to, whose chromaticity coordinates are x = .3101 and
y = .3162.

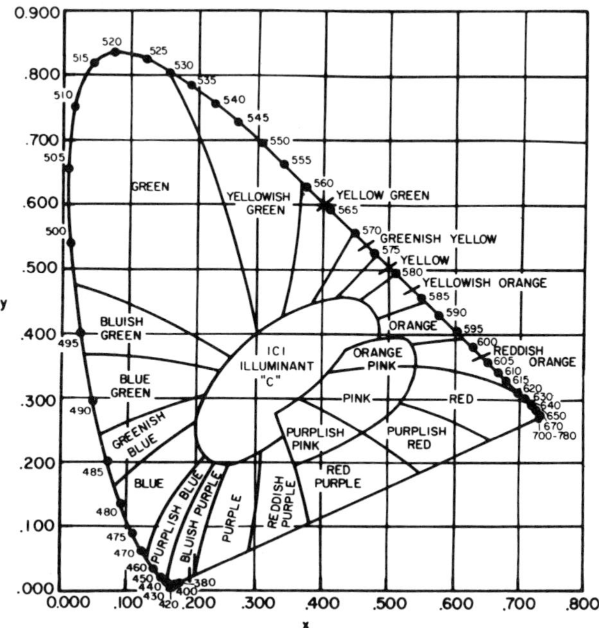

Figure 25.8. Chromaticity diagram and relation of the different hues to each other on the diagram.

The chromaticity diagram can be used to specify colors in terms of
dominant wavelength (hue) and purity (saturation).

Dominant Wavelength

The dominant wavelength of a sample is determined first by obtaining
its chromaticity coordinates. They are then plotted on the diagram. A

line is drawn from point C, which is illuminant C, through the point of the sample and extending the line until it intersects the spectrum locus. The wavelength at which the line intersects the spectrum locus is called the dominant wavelength.

In Figure 25.9, the line intersects the spectrum locus at 550 mμ so the dominant wavelength is 550 mμ.

The purity or saturation is the ratio of the distance between illuminant C and that of the sample to the distance between C and the point on the spectrum locus representing the dominant wavelength of the sample. In Figure 25.9, this is calculated to be 76.92. In the case of the nonspectral colors (those which lie along the bottom of the chromaticity diagram, and are called complementary colors), reference to their dominant wavelength and purity is followed by subscript c. The dominant wavelength of a com-

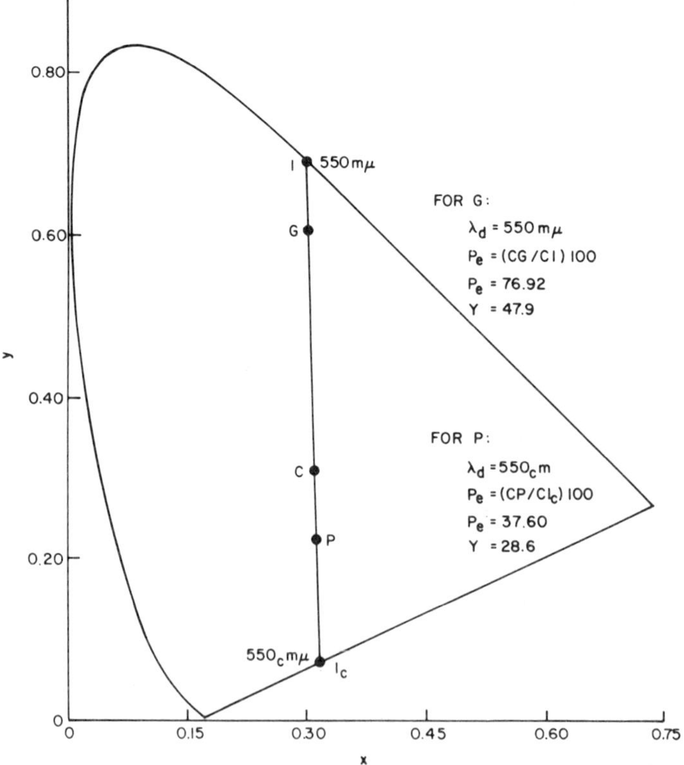

Figure 25.9. Determination of dominant wavelength and purity on the chromaticity diagram.

plementary color is determined by drawing a line from the point represented by illuminant C, through the plotted point of the sample and extending it to the boundary of the diagram in the same manner as with spectral colors. The line is then extended in the opposite direction through the spectrum locus. The wavelength, where it bisects the line, is the dominant wavelength with a subscript c notation. The present purity is calculated in the same manner as with spectral colors.

COLOR ORDER SYSTEMS

There are many possible colors in color space and they can be arranged in an orderly manner. The Munsell System is one such arrangement. It is essentially a scientific concept for describing and analyzing color in terms of three attributes, identified in this system Hue, Value and Chroma. The method of color notation developed by A. H. Munsell, as the principal feature of this system, arranges the three attributes of color into orderly scales of equal visual steps, so that the attributes become dimensions or parameters by which color may be analyzed and described accurately under standard conditions of illumination.

Chromatic colors in the Munsell System of color notation are divided into five principal classes which are given the Hue names of red, yellow, green, blue and purple. A further division yields the intermediate Hue names of yellow-red, green-yellow, blue-green, purple-blue, and red-purple, these being combinations of the five principal Hues. Hence the Hue notation of any color indicates its relation to the five principal and intermediate Hues or any of their subdivisions. Capitalized initials such as "R" for red, or "YR" for yellow-red, are used as symbols for the Hue names. When finer subdivisions are needed, the ten Hue names or symbols may again be combined to produce such combinations as red-yellow-red, which is symbolized "R-YR." For even finer divisions, the Hues may be divided into ten steps each (1R to 10R and 1YR to 10YR), thus increasing the Hue notation to 100.

The Value notation indicates the degree of lightness or darkness of a color in relation to a neutral gray scale, which extends from a theoretically pure white symbolized as 10 to theoretically pure black symbolized as 1. A gray or a chromatic color that appears visually half-way in lightness between pure black and pure white has a Value notation of 5/. Lighter colors are indicated by numbers above five.

The Chroma notation of a color indicates the strength (saturation) or degree of departure of a particular Hue from a neutral gray of the same Value. The scales of Chroma extend from /0 for a neutral gray out to /10, /12, /14 or farther, depending upon the strength or saturation of the individual color. A color classified popularly as "vermilion" might have a Chroma as strong as /12, while another color of the same Hue and Value classified popularly as "rose" might have a Chroma as weak as /4.

The complete Munsell notation for any chromatic color is written Hue Value/Chroma, or symbolically H V/C. A particular sample of vermilion might then have a Munsell notation of 5R 5/12, while a particular sample of rose might have a notation of 5R 5/4.

Whenever a finer division is needed for any of the three attributes, decimals may be used, such as 2.5R 4.5/2.4.

The notation for a neutral gray is written N V/. A very dark neutral (black) would be written N 1/0, and N 9/0 would be the notation for a very light neutral (white). For grays of slight chromaticity, the notation is written N V/(H,C), using only the symbols for the ten major Hues to indicate the Hue; thus a gray of a slightly yellowish appearance is written N 8/(Y,0.4).

The Munsell System of Color Notation can be thought of in terms of a color solid or color space in which the Neutral Value scale runs vertically from a theoretically pure black at the bottom to a theoretically pure white at the top, while the various Hues are located around it, describing an approximately cylindrical shape with the Neutral Value scale in its center. The Chroma scales radiate from the vertical Neutral scale in the center to the periphery of the color solid in equal visual steps.

The Ostwald system was developed by Wilhelm Ostwald about 40 years ago and has been utilized by the Container Corporation to form what is commonly known as the Color Harmony Manual. The manual consists of 28 charts which represent a fairly wide sampling of color space. On each chart the colors of chips are arranged in the form of a triangle (Figure 25.10). The triangle has white (a), black (p) and the particular full color (f). From (a) to (p) are the letters of the alphabet. (a) to (p) represent equal departure from white. Letters (p) to (a) represent equal departure from black. These two sets of ordinates are superimposed and a two-letter notation is derived. The first letter is the white departure and the second is the black departure.

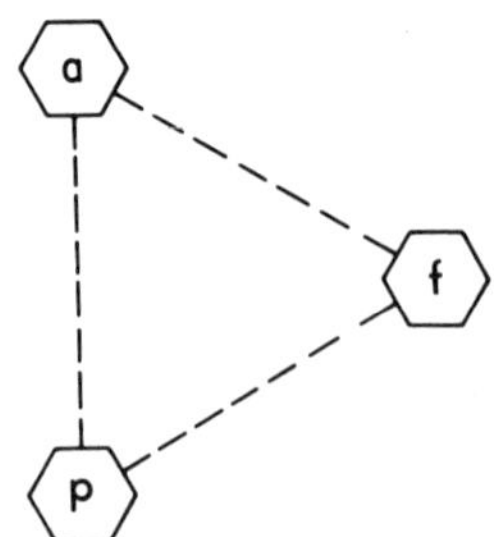

Figure 25.10. Ostwald color triangle: a (white),
p (black), and f (particular full color).

A triangle represented in the Color Harmony Manual is shown in Figure 25.11. What was referred to as the full color now has the specific notation pa and all the other two letter notations now label various colors between the three points a, p and pa. The Color Harmony Manual consists of 28 triangles, such as Figure 25.11, using different hues. In some regions of certain hue charts, extra color samples seem to be desirable and special charts are available.

```
a
  ca
c    ea
  ec  ga
e   gc  ia
  ge  ic  la
g   ie  lc  na
  ig  le  nc  pa     Figure 25.11.   This abridged monochromatic triangle is the
i    lg  ne  pc      plan used for the development of the basic charts of the
  li  ng  pe         Color Harmony Manual.
l    ni  pg
  nl  pi
n    pl
  pn
p
```

The Color Harmony Manual has been used extensively by designers and decorators. By the use of the unique numbering system, it is possible to select the colors of cloth in clothing in a catalog, for example, that will not clash and will be pleasing to the prospective buyer's eye.

COLOR MEASURING INSTRUMENTS

Many instruments are capable of measuring color and, as might be expected, they vary in accuracy. The spectrophotometer is recognized as the basic instrument for absolute measurement of color. Tristimulus colorimeters are used for color difference measurement and therefore have many useful applications in the paint industry. We can say that the two classes of instruments, which find wide acceptance in the coating industry are spectrophotometers and tristimulus colorimeters.

Spectrophotometers

There are numerous spectrophotometers on the market, but we are concerned only with those which are capable of making readings in the visible range. Spectrophotometers give a more complete description of a color than do tristimulus colorimeters. They yield a curve of a sample which represents the reflectance at any given wavelength from 400 to 700 mμ. This can be used for color matching and for determining what type of pigments are in a sample. The latter is done by comparing the curve of the sample to curves of known pigments. Pigments have certain "curve shapes" which are not lost when mixtures of the pigments are made. The usefulness of a spectrophotometer can be increased if an integrater is used in conjunction with it so as to obtain the tristimulus values X, Y and Z. Spectrophotometers have also been used successfully to "tie down" standards. Physical samples change with age, but spectrophotometric data can be used as a permanent record of the color.

When spectrophotometers are used to obtain absolute data, special care must be taken to be sure that proper standardization procedures are used. Some of the more prominent spectrophotometers are manufactured by General Electric, Baush and Lomb, Cary, and Beckman. At the present time the General Electric Spectrophotometer is probably the most extensively used in the paint industry.

Tristimulus Colorimeters

Tristimulus colorimeters have had wide acceptance in the coatings field for several reasons. When properly used, the data obtained from them is very reliable; they are also rugged, easy to operate and relatively inexpensive. The accuracy with which these instruments can make absolute measurements depends upon how well the source-filter-photocell combination matches the CIE quantities of Ex, Ey, and Ez for a given standard illuminant.

If one were to plot a series of colors with constant chroma and brightness, it would be found that with a given instrument the blues may be better spaced than the yellows. With another instrument, the reverse may be true. What this means is that they cannot be used for absolute measurement or to obtain accurate results from one instrument to another, and often from instruments of the same model. However, colorimeters can be used very successfully for color difference measurements. For such measurements, an observer is not as concerned with how accurately the colors are located in color space as with the difference between the two. After all, this is what he does when he compares a sample to a standard.

The data obtained from these instruments must of necessity be the results of these measurements, lightness and the chromaticity parameters. Most of the instruments are manually operated, with filters being inserted and removed from the light beam by some mechanical or electrical means. This may be done automatically and the circuit is then balanced out for each filter. The actual operation of the colorimeters varies from one type to another, but the overall principle is the same.

As stated before, tristimulus colorimeters function best as color difference instruments, but they can be used for many other functions in colorimetry. Their usefulness can be extended if direct comparison of the data between two instruments is not needed and a given instrument is used only for internal control, without reference to data of other instruments in laboratories.

Four of the most commonly used tristimulus colorimeters in the United States are: (1) Gardner Color and Color Difference Meter, (2) Colormaster Differential Colorimeter, (3) Color-Eye and (4) Hunterlab. There are several European instruments that are also used but not to the extent of the four already mentioned.

Colorimeters are currently employed by the coatings industry for maintaining tolerances and determining what colors are necessary to match a sample batch qualitatively and quantitatively. They can also determine what color changes take place between two samples upon aging, weathering, etc. They do all these things successfully because this is essentially measuring color difference. When quantitative additions are made, a computer is needed to make the calculations from the data.

Sample Preparation for Color Measurement

When samples are prepared for color measurement, extreme care must be taken to make sure that the samples are representative. A simple rule to remember is that a bad sample will give bad results. The samples should have complete hiding so that the instrument will not be including the effect of the substrate in the readings. They should also be uniform, free of finger marks, smudges, dirt and surface imperfections. More effort is needed in preparing samples for instrument readings than for visual observation because a visual observer will taken into consideration certain imperfections, which an instrument cannot do.

TOLERANCE SPECIFICATION

All paint formulators find themselves involved in tolerance specifications. If a given standard could be matched perfectly, and at the same time economically and with ease, tolerances would not be a problem. But this situation is not possible in practice. When specifications for any manufactured item must be met, they must have limits which are called tolerances. The limits cannot exceed ability to manufacture the product.

Color Control by Visual Observation

At the present time there are three ways tolerance can be maintained: visually, by instrumentation, and by a combination of both.

There is no doubt that nothing can compare with the speed with which the human eye can perceive small color differences. There has always been a question, however, as to the reliability of these visual color decisions. It is a well-known fact that observers will vary in their decision from day to day and even from hour to hour.

The reliability of visual observations of color difference can be increased greatly when standard conditions for viewing the samples are maintained. Realizing the need for standard methods, ASTM developed D-1729-62, Recommended Practices for Visual Evaluation of Small Color Difference of Opaque Materials. This method describes the light source, sample size, viewing conditions, etc., which are so important in any color evaluation. Any method of visual evaluation, must of necessity clearly specify the conditions under which the observations are to be made if the results are to be reliable.

There are two basic variables in visual evaluation of color tolerances. The first is the observer and the second is the way in which the tolerance is usually stated. An all too common statement in regard to color tolerance is that it must be a good match to the standard. What is a good match to the standard is determined by the end use of the material. Certainly a good match for floor paint would not be considered a good match for prefinished house siding. Indeed, what is considered a good match for any type of finish will vary with individual observers.

It may appear from all these variables that it is difficult to maintain tolerance by visual observation. This is not the case, as it has been done successfully for many years when the conditions have been correctly controlled. The effectiveness of this method can be greatly improved if physical standards representing the standard and the variation, presented in terms of lightness, hue and chroma, are made available to the observer.

By this system the observer does not have to determine what is good, but only whether or not his sample falls within the prescribed limits. This procedure is not used as extensively as it should be, partly because it requires extra effort in preparing the limits. Those who have done this have found the extra effort well worthwhile.

The Munsell Company has prepared tolerance sets which use this method. For a green standard, a maximum and minimum lightness standard is made. There will also be a blue limit and a yellow limit. To complete the set, maximum and minimum standards for saturation are made. This makes a boundary for all possible variations of a green standard. All the observer is required to do is determine whether the samples fall within these limits.

Color Control by Instrumentation

With instrument designs progressing at a rapid rate, it would be naive to assume that any application is completely outside the scope of instrument control and will always depend on human judgment. One area which will yield readily to instrumental analysis is the maintenance of color tolerance; indeed, this is one of the facets of colorimetry where instruments excel.

If colorimeters can accurately determine numerical values for colors and the colors are slightly different, then the values should tell us how far apart these colors are. This, however, cannot be done successfully from the three sets of values which are usually obtained from colorimeters. These values must be converted in some manner so as to consider the three dimensions of color and their relationship to each other. There have been many attempts to simplify the method of color difference. In many cases they were found to be inadequate because they used short-cut methods and did not take into consideration the three dimensions that are necessary to adequately describe color difference.

Tolerances can be established by instruments in two broad categories, graphical and numerical. Either approach requires data from some sort of colorimeter. There are some who question the ability of colorimeters to police color tolerances successfully. There has been much published condemning them. It is the abuse of the colorimeter that is the fundamental problem. It is true, of course, that these instruments have certain limitations that must be recognized. When the necessary precautions are taken, colorimeters have proved to be very successful in this area and are even superior to visual observers.

There are several well-known equations for determining color difference. One equation uses the coordinates L, a and b. Data in this form are determined directly from the Hunterlab Color Difference Meter and the Gardner Color Difference Meter. The same coordinates can be obtained from the Colormaster by the use of a set of charts. The equation using this data is as follows:

$$\Delta E = \overline{(\Delta L)^2 + (\Delta a)^2 + (\Delta b)^2}$$

The ΔE obtained from the equation is often referred to in NBS units.

This formula for computing color difference has been widely used for several reasons, one being the simplicity of the calculations, and the other the number of instruments in use which give data directly in terms of L, a and b. This, however, would not justify its use. The method must be reliable, which it is.

Before we proceed further, we should have an understanding of what the coordinates L, a and b stand for. They are as follows: L = lightness, $+a$ = red, $-a$ = green, $+b$ = yellow and $-b$ = blue.

What is an NBS unit as obtained from this formula? A just perceptible color difference would be about 0.3 NBS unit. We do not mean an acceptable match, because this would vary with the product. What we mean is that in this range the difference between two samples is just perceptible. Acceptable color matches range from 0.3 NBS unit upward, depending upon how close a match is desired.

Like any one-number system describing color difference, this method has limitations because it tells only the total color difference and not the direction of that difference. In any successful numerical system, the direction of the color difference is usually expressed. This should be agreed upon between the buyer and the seller because usually the buyer has a preference, even though he may not know it; if the direction from the standard can be agreed upon, a workable tolerance system can be adopted. For example, if a specification permits a tolerance of 1 NBS unit, we are implying that it could be 1 NBS unit in all directions from the standard. But very often we do not mean this; Sample A could be 1 NBS unit away from the standard but on the red side of the standard, Sample B is also 1 NBS unit away from the standard but on the green side (Figure 25.12). Both samples meet the required tolerance, but when they are compared, they would be a total of 2 NBS units apart. We may not want to

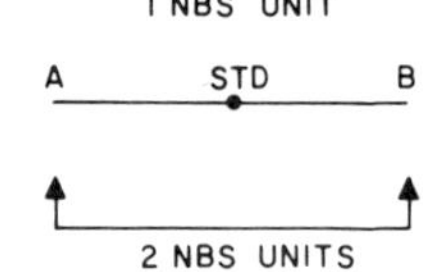

Figure 25.12.   Simple diagram of possible variation of two samples from standard.

permit one sample on the red side of the standard and another on the green side. The batches would be more acceptable if they were all on the red side or the green side; in other words they would have the same hue batch to batch. This is an area where instruments excel because we can assign limits to the numerical values to control the hue variation. A tolerance for a yellow standard could permit a total color difference of 1 NBS unit, but the sample must always have a value for a of zero or $+a$. Thus the sample will always be on the red side of the standard (Figure 25.13).

To obtain an even more sophisticated tolerance, we could control the saturation by stating that the sample could not be on the blue side of the standard, because as blue is added to a yellow, it becomes grayer. This would be done by saying that all the b values must be zero or on the plus

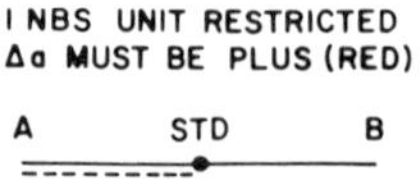

Figure 25.13.   Simple diagram of restricted tolerance.

side. If we said samples could only have a $+a$ and $+b$, they would all be in the red area and would appear clean to the standard. This leaves only one variable that is not restricted—lightness. This can be established by stipulating that the sample must be lighter or darker than the standard. A specification taking all aspects of color into consideration might read as follows: "Total color difference must not exceed 1 NBS unit; Δa must be zero or plus, Δb must be zero or plus, and L must be equal to or larger than the standard." With this type of tolerance, there is no question as to what the manufacturer is supposed to do.

MacAdam Tolerances

It has been said that when one really gets serious about color tolerances, one usually ends up by using the tolerance system devised by MacAdam, or some variation of his original formula.

Davidson and Friede, in their paper "Size of Acceptable Color Difference," compared visual observations with color differences obtained from several equations. The per cent acceptance of each color was plotted against the color difference between the sample and standard calculated according to the equations for color differences under study. If the color difference were the same, a straight line would result. However, it was found that there was quite a bit of scatter, which represented the collection of errors made by the color differences equation. If the best possible curve were drawn from these points, it was found that the least scatter was from the data of the MacAdam equation. Some of the other equations were almost as good, but none was better.

The standard and samples were then arranged so that the sample with the least color difference visually was next to the standard, and so on, the poorest being the fartherest away. The same samples were then arranged according to their color difference as calculated by the particular color difference equation.

If the ranking was the same visually and by calculation, there was a correlation rating of 1. None of the color difference equations gave a correlation of 1, but the MacAdam equation had a correlation of 0.8 to 0.9.

It was concluded that the MacAdam color differences correlated better than an individual observer and that it was the best equation to use for color difference calculations.

If this is the ultimate in tolerance, then why is it not used more extensively? The reason is that it is the most difficult to compute. So complicated is the formula that it would be impossible to make calculations in production control for a large number of samples easily. The formula for computing MacAdam tolerances is as follows:

$$\Delta E = [(1/K)(g_{11}\,\overline{\Delta x}^2 + 2g_{12}\,\Delta x\,\Delta y + g_{22}\,\overline{\Delta y}^2 + G\,\overline{\Delta Y}^2)]^{1/2}$$

To simplify the calculations, a series of charts was developed by Davidson and Hanlon. Later Simon and Goodman developed a similar series, which was a large and improved version of the Davidson and Hanlon charts. These can be obtained from the Union Carbide Corporation. By the use of these charts, calculations can be made more rapidly, but are still time consuming. Recently a method, said to be faster, has been developed by R. S. Foster of Columbus Coated Fabrics Company.

Using ellipses is not new; Lawrence Rudick and George Ingle reported, as far back as 1952, that ellipses were used for tolerance of plastics. They stated that plots showing three views of concentric portions of the color solid are necessary to properly represent the three dimensions of color.

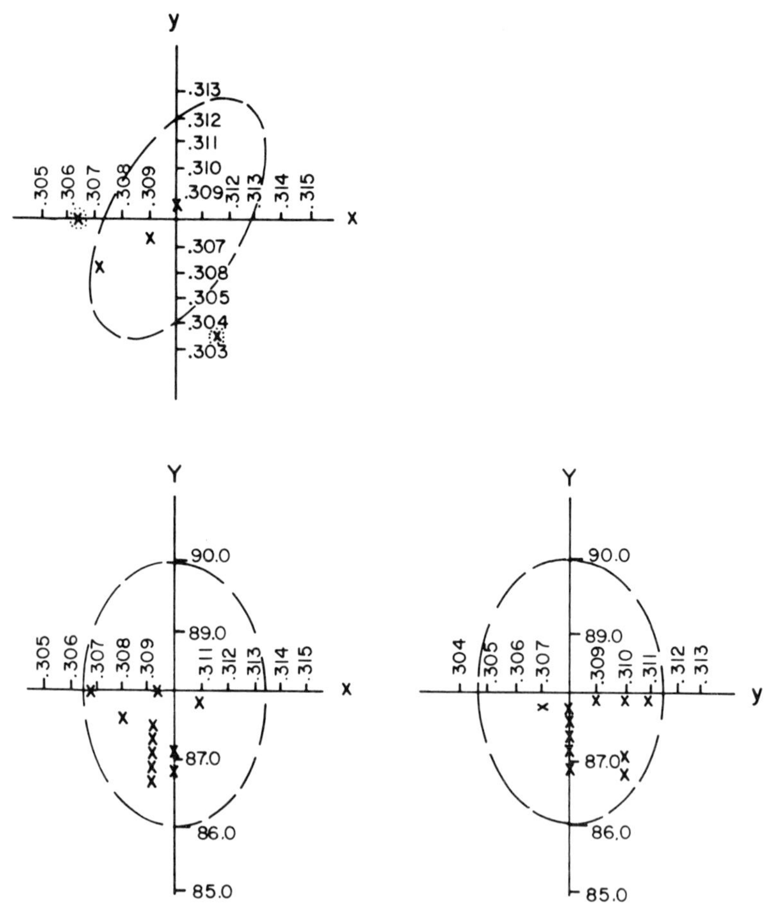

Figure 25.14.   Inspection tolerance, ellipoids for white plastic.

They made tolerances for each color which were plotted concentrically with MacAdam ellipses around the standard point. The size of the solid ellipsoid was fixed by the data from the accumulated physical limit specifications. Figure 25.14 shows that it is first necessary to compute the chromaticity coordinates x and y. Once these are known, the plots then can be made on the three charts. The size of the ellipses was predetermined in terms of MacAdam units. If a sample plots within all three ellipses, then it is said to be satisfactory for color.

This method has one drawback when using the Color-Eye. It is necessary to convert Color-Eye X, Y and Z values to CIE X, Y and Z values, and then convert CIE values to little x, y and cap Y. Even though this could be done as stated by Rudick and Ingle with charts and a slide rule or desk calculator, it still takes time when a large number of production samples are to be checked. Sherwin-Williams developed a method where the coordinates of the ellipses were in terms of Color-Eye ratios. The MacAdam ellipses, which are expressed in the coordinates little x, y and cap Y, were transformed to the X, Y, Z Color-Eye ratio coordinates and projected on three planes. This meant that the operator could go directly from the data obtained on the Color-Eye to the ellipses (Figure 25.15) and determine if the sample was in tolerance. It is necessary to develop an ellipse for each color for which a desired tolerance is needed. However, this system has one drawback. Like the single number specification in the NBS system, it is possible to have one sample on the red side of the standard by 1 MacAdam unit and another on the green side by 1 MacAdam unit; thus they are 2 units apart when compared.

Sherwin-Williams developed a new set of charts in which they changed the true MacAdam ellipses but arrived at a tolerance that was more acceptable by visual comparison (Figure 25.16). As stated earlier in regard to the NBS tolerance, it is desirable to control what side of the standard the sample should be on, and this has been incorporated into the tolerance. The chart stipulates that X must be equal to or lower than Y and that Y must be equal to or greater than Z. The position of the sample in color space in relationship to the standard is now established. This could be done visually, of course, by trying to have a good match and keeping it on the red side, but it would be difficult to do this consistently from batch to batch.

MacAdam color difference can be calculated by the use of analog and digital computers, which are readily available today. A small, relatively inexpensive analog computer is made by Davidson and Hemmendinger, Inc., for rapid calculations of MacAdam tolerances from X, Y and Z data or Color-Eye ratios. The tristimulus values of the standard are set into the computer and then it is balanced out until the chromaticity coordinates x and y are obtained. From these, proper constants and factors

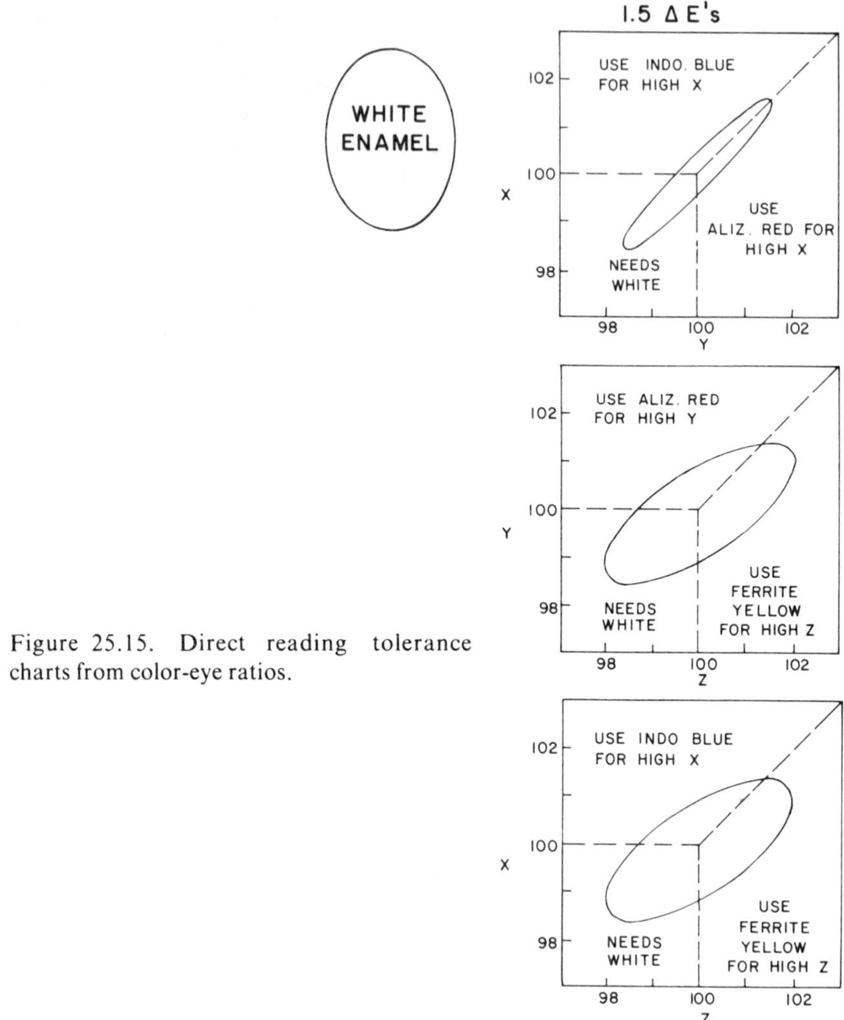

Figure 25.15. Direct reading tolerance charts from color-eye ratios.

are selected from a chart, similar to the one used by Simon and Goodman, and set into the instrument. Then the value of the batch is put in. From this data, the total color difference in terms of MacAdam units is obtained.

A more elaborate color difference computer, manufactured by the Instrument Development Laboratories, not only calculates values but also presents a pictorial view of the sample. Once the sample is located in color space, a dot representing the sample appears on the screen and the instrument is set to 1 MacAdam unit and if the dot is within that area,

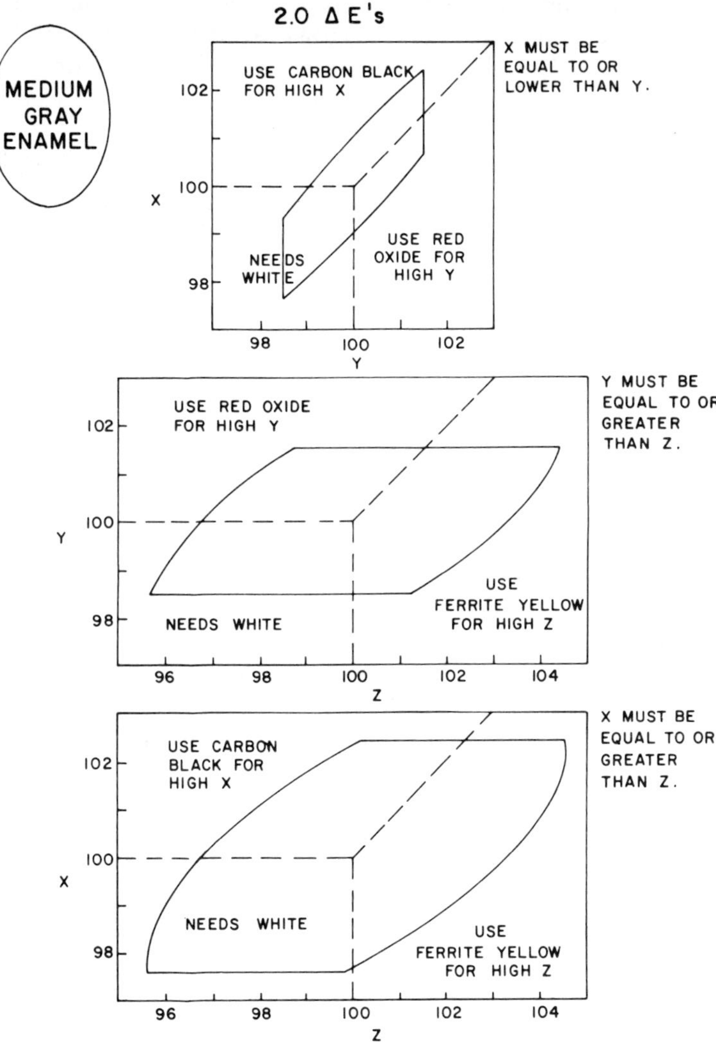

Figure 25.16.   Restricted direct reading tolerance charts.

the sample is 1 MacAdam unit or less. The digital dial readout will give the exact difference. If 2 MacAdam units are desired, then the switch is set to 2 and a larger ellipse appears on the screen. In other words, the sample is stationary and the ellipses are made larger or smaller to determine what the true color difference of the sample is in regard to the standard.

Large digital computers, such as IBM and GE, are being used success-fully to compute color differences using the MacAdam equation. Digital computers are much faster and require very little manipulation so they are convenient to use when large numbers of samples are to be calculated.

Establishing Tolerance Limits

In establishing a color tolerance, consumer acceptability must be used as a basis. Hence, the first step is accumulating physical limit specimens which represent as complete a range of customer acceptability as possible. The specimens should be in all directions colorimetrically from the standard. This may not always be possible because the customer's acceptability may be in one direction from the standard. It is most important to be sure that all samples are spectrally similar to the standard, and extreme care should be exercised to avoid metamerism. Since actual production samples will show chromaticity and reflectance variation from the standard, the difference must be characteristic in three dimensions. The tolerance also must be realistic and obtainable.

The New York Paint Society, Subcommittee 67, made this last point very clear: "When instruments are used for measuring color difference and numerical limits are set, the tendency is to make them more restricted than previous limits. This results in (A) either discouraging of all instru-mental data as impractical, (B) serious delays in production or (C) an un-warranted increase in cost.

Setting up color tolerances between a supplier and a customer can be done by a process of merely agreeing on what is acceptable and what is not, then determining what the tolerance is of these samples and using this to establish the initial tolerance. After experience is gained with the products involved, it will soon be determined if the tolerances are satis-factory. If it is found that the customer is rejecting materials that plot within the tolerance, then the tolerance must be narrowed (Figure 25.17). Care must be used not to include exceptional cases. Unfortunately there is nothing final about a tolerance because the customer has the perfect right to ask for tighter tolerances if they are realistic. The supplier can suggest tolerances that are not so rigid, if production is being held up or if he thinks the limits are beyond his ability to meet due to raw materials or the capability of the instruments.

Instrument Accuracy for Tolerance Limitations

It must be pointed out that there are certain limitations to the accuracy of the instruments. Therefore, it must be assured that tolerances are not established beyond the accuracy of the colorimeters used. If tolerances are set beyond the capability of the instruments, there will be nothing

2 Macadam Units

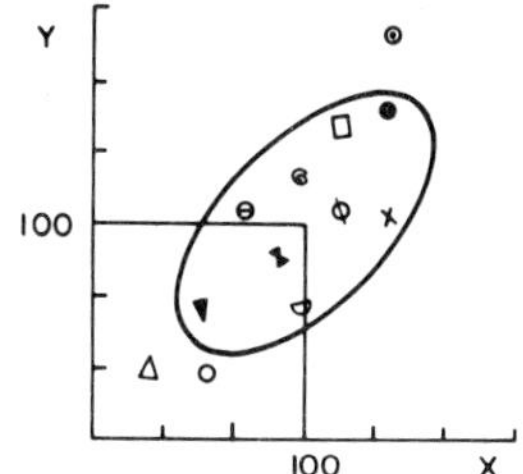

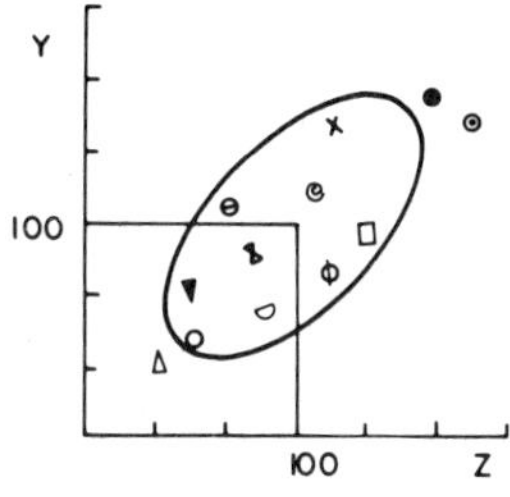

Figure 25.17.   Customer acceptance.   All samples plotting within the three ellipses were acceptable to customer.

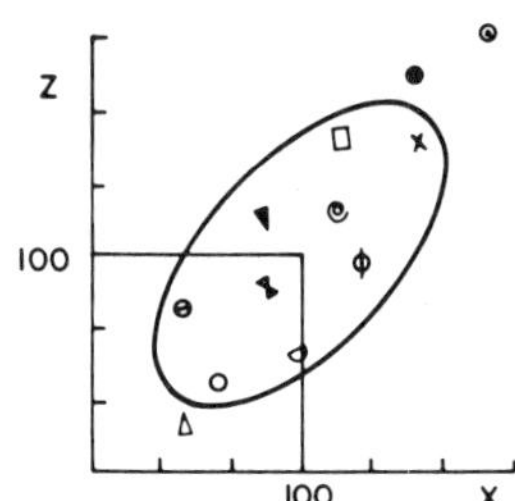

but trouble between supplier and the customers. Much has been published on the accuracy of colorimeters both in relationship to multiple instruments of the same make and also instruments of different makes. Very often all that is of concern is the comparison of a given instrument of a supplier and that of a buyer.

A statistical analysis of the precision and accuracy of the two instruments would be very desirable. However, a simple procedure for comparing the ability of two colorimeters to read differences the same way is the following: Obtain a series of pairs of colors that exhibit small color differences and are in the same general area of color space of the material supplied to the customer. Read these on a colorimeter. They should then be read on a second colorimeter or the customer's colorimeter (Figure 25.18).

Color-Eye Correlation Chart

Sample Pairs	Sellers Color-Eye, ΔE	Customer Color-Eye, ΔE	Difference Between Instruments, ΔE
Green	1.32	1.15	.17
Gray	.42	.38	.04
Blue	1.50	1.32	.18
Brown	1.01	1.23	.22
Yellow	1.58	1.46	.18
Orange	1.37	.97	.40

Figure 25.18. Difference between various pairs of samples as read on two Color-Eyes.

Opaque colored glass or porcelain panels are excellent for this type of work because they are permanent and can be easily cleaned. If a series is run on one colorimeter, the color difference between the pairs should be noted and the same difference should be obtained on the customer's colorimeter. The exact difference will not be obtained; if it is, it would be most unusual. Hence, the difference should be computed between the pairs on one instrument in terms of some system, such as NBS units or MacAdam units, and the same should be done on the second instrument. A difference between the two instruments of say one-half of a MacAdam unit would mean that production could not be held any closer than this because the instruments are not capable of obtaining differences to any greater accuracy. Before a test is run, of course, the instrument must be correctly calibrated to be sure it is in good working order. Also several sets of data should be obtained and an average taken.

A word of caution; direct data from the instrument must not be compared. Data for each instrument must be converted to some type of accepted color difference system, such as the NBS or MacAdam systems. Checking the correlation of the instruments should be done at regular intervals because the accuracy of the instrument can change with time.

In checking the panels, it would be best if they were in the same general range for the hue, chroma and lightness of the samples that will be eventually checked. Also the color difference should be of the same magnitude as the difference that will be expected between batch and standard.

GLOSS AND ITS EFFECT ON COLOR

Gloss has such a profound effect on color that it is impossible to discuss one without the other. This is especially true of colors having low reflectance. A very slight change in gloss will have a noticeable effect on color. This can be shown by instrumental data as well as by visual observation. Any adjustment of a production batch of paint for gloss will usually require a color adjustment. Figure 25.19 shows a spectrophotometric curve of two samples. The only difference between the two is that one has a lower gloss reading than the other. Samples with higher gloss appear deeper when viewed visually. The same effect occurs with instrument data.

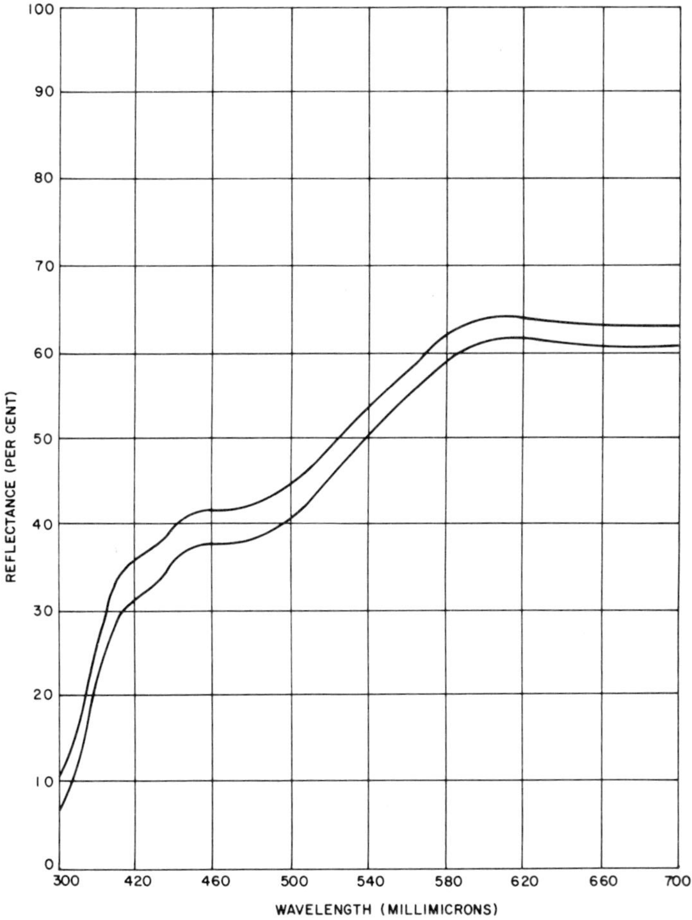

Figure 25.19.   Effect of gloss on color. Both samples are the same color but differ only in gloss. Lower curve represents sample with higher gloss. Upper curve represents sample of lower gloss.

The tristimulus values will also indicate that the sample with the higher gloss is deeper. Color-measuring instruments make their reading at fixed angles, and any variation in gloss between samples will cause a variation in color readings. A paint shader will go through all kinds of gymnastics trying to match a standard and batch that are different in gloss. What he is trying to do is arrive at an angle of viewing where the gloss of both samples will appear the same. If gymnastics won't solve the problem, the shader will moisten the lower gloss sample with his tongue to increase the gloss.

At the present time we do not have a colorimeter that will adjust for gloss in this manner. Nor would we want an instrument that could do so because if a sample looks deep because of gloss, it should read deep. For special situations, some instruments can be changed so as to exclude the effect of gloss.

LIGHT SOURCES FOR COLOR MATCHING

Joseph Joubert, the French moralist, said, "What is true in the lamp light is not always true in the sunshine." He was not talking about color matching but if he were, the words would have been just as meaningful. A physicist says color is not how you see it but how you light it.

In daylight the complete range of visible waves are from 400 to 700 mμ. If this light strikes an opaque body, such as a painted surface, and all the light is reflected back, the surface appears white. The majority of surfaces that we deal with do not reflect all the light that strikes them, but because of special optical properties, absorb a greater fraction of the light of some wavelengths than of others. The reflected light thus lacks the wavelength region of high absorption, and the surface no longer appears white but has a definite color depending upon the degree and spectral location of the absorbed energy. A red object absorbs all the colors in the spectrum except red; that portion of the light is reflected back to the observer, and the object appears red. With a blue object, only the blue light comes back to the observer.

From this example, the importance of the light source in making color judgment is evident. If a light source contains no red or is deficient in red, then samples viewed under it would not appear red, or as red as they should be, depending upon how deficient the light was in the red region.

North daylight has long been considered to be the most satisfactory source of light for color matching, having a fairly even distribution of energy across the complete visible spectrum. North daylight varies depending upon the time of day and, of course, is not available at night. Since many paint factories operate around the clock, color decisions must be made regardless of the time of day. Artificial light sources have, therefore, been developed to reproduce north daylight as closely as possible. Common artificial light sources are tungsten and fluorescent lamps.

The color of light sources can be classified in a system known as degrees Kelvin and is referred to as color temperature. The higher the color temperature, the bluer is the color of the light. A tungsten bulb has a color temperature of about 2800°K. North daylight varies from 6500 to 8000°K. The Inter Society Color Council (ISCC) has selected what is known as CIE Illuminant D, which has a color temperature of 7500°K, as a substitute for north daylight. This is the light source which they recommend for color-matching booths. It was long thought that the correct substitute for north daylight could only be achieved with a filtered tungsten combination. However, recent developments in the manufacture of special fluorescent lamps indicate that such lamps will be able to meet the specifications of Illuminant D (7500°K).

There are requirements other than color temperature for lights used for color matching. The light source should have a conformity index of not less than 92. It should have an overall illumination intensity of approximately 100 foot-candles on the samples being viewed. The light source should be in a booth whose walls are painted a neutral gray with a Munsell value of N7. More detailed specifications for lighting conditions are outlined in ISCC's Standard Practices for Visual Examination of Small Color Difference and ASTM D-1729-60 T, "Visual Evaluation of Color Difference of Opaque Materials."

COLOR FORMULATION

Color formulation is a problem, and there is an old saying that a problem well put is half solved. This becomes more obvious when one states and lists one's objectives and requirements. Naturally, certain goals will be more difficult to meet than others and necessarily will be compromised. With this approach to the problem, a good beginning is a listing of the requirements of the ideal color formula:

(1) It must be able to match the color with the pigments selected.

(2) It must produce matches that are not metameric to one another.

(3) Performance requirements such as chemical resistance, lightfastness, etc., must be met.

(4) It must have the lowest possible cost.

(5) Commonly stocked pigments should be used.

(6) The hiding and tinting strength requirements of the product must be produced consistently.

The four-pigment concept of color formulation states that all colors can be matched with a minimum-maximum of four pigments with one held constant. These pigments must be selected so that the color to be matched lies within their gamut and gives adequate control of the three visual variables of color: hue, saturation and lightness. This can only be done by four pigments. Quoting from Judd, "Alteration of the proportions of the four constituents (pigments) provides the requisite three degrees of freedom for a color match. Thus, one pigment, say white, gives no freedom at all. The colorant layer may be white, but no color is involved. Addition of another color, say black, gives one degree of freedom. The colorant layer may then be any one of the series of bluish grays produced by varying proportions of white and black pigment. Addition of a third pigment, say red, permits the colorant layer to be any one of a two-dimensional series of colors, all essentially of red hue. But addition of a fourth pigment permits three-dimensional variation, so that by proper adjustment of the proportions of the four pigments the colorant layer may be made to match exactly the specimen submitted if its color lies within the gamut of colors producible by these four constituents."

How does one know the color lies within the gamut of the pigments without actually making a trial mixture? It is possible to determine the color gamut of pigments beforehand by making systematic mixtures. The resultant physical specimens could be used as a visual color gamut guide. Given a color to match, by visual examination find the color gamut specimens that bracket the color and use the pigments contained in the specimens.

A better method, however, that provides permanent information is to measure each specimen to determine its tristimulus values X, Y, Z and the chromaticity coordinates x and y. Plotting this data results in color gamut charts (Figure 25.20). A color to be matched is specified in CIE terms. A plot of the coordinates quickly shows the sets of four pigments that can match the color. Also by plotting the contours of equal concentration of the pigments a calibrated gamut, such as Figure 25.21 is obtained (described by Davidson and Hemmendinger in the *Journal of the Optical Society of America*) 415, 216 (1955). Then an estimate of the amount of each pigment can be determined.

There will usually be more than one set of four pigments that will match the color. Selection of the set will depend upon the performance requirements of the product, cost, etc., or whether a metameric match can be tolerated. If a nonmetameric match is wanted, each set would need to be tested by trial and error until metamerism was not evident. However, if the standard contained more than four pigments, the trial-and-error method is virtually an impossible task.

To be nonmetameric when there are more than four pigments, it is not enough to have just the same pigments: The ratio or proportion of the

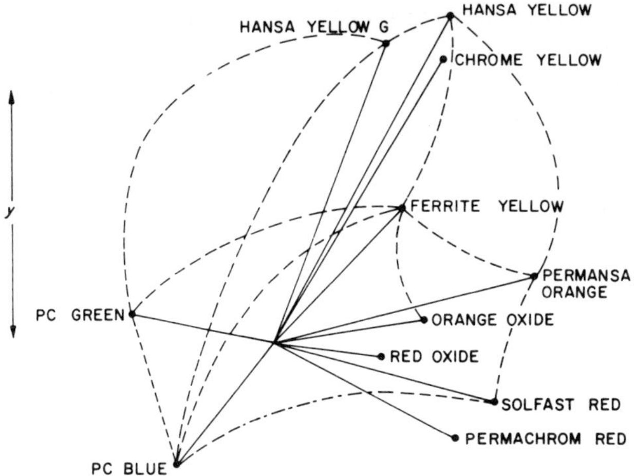

Figure 25.20.   Color gamut chart.

pigments must be the same as in the color being matched. A five or more pigment formula is not unique; an infinite number of proportions of pigments will produce a match, each match being metameric to the other. Therefore, unless a preventive measure is taken, it is possible to have batch-to-batch metamerism.

The four-pigment concept states that one pigment must be held constant. In actual practice, this constant can be a mixture of pigments or, stating it in another way, only three pigments are allowed to vary no matter how many the formula contains. Therefore, when a nonmetameric

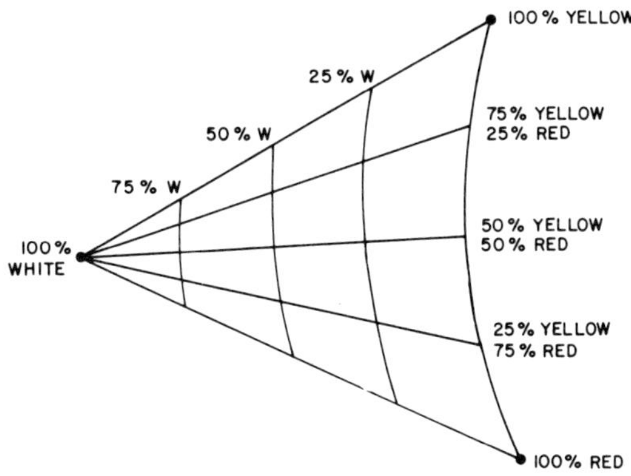

Figure 25.21.   Calibrated pigment gamut.

match requires more than four pigments, by lumping all but the three pigments that are allowed to vary into a single constant, the advantages of the four-pigment concept should be realized and batch-to-batch metamerism will be minimized.

The trial-and-error methods of formulating nonmetameric colors are outmoded by spectrophotometric methods. By comparing spectral curves of knowns to the unknowns, the proper pigments can be chosen. Figure 25.22 is a spectral curve of a sample, and Figures 25.23, 25.24 and 25.25

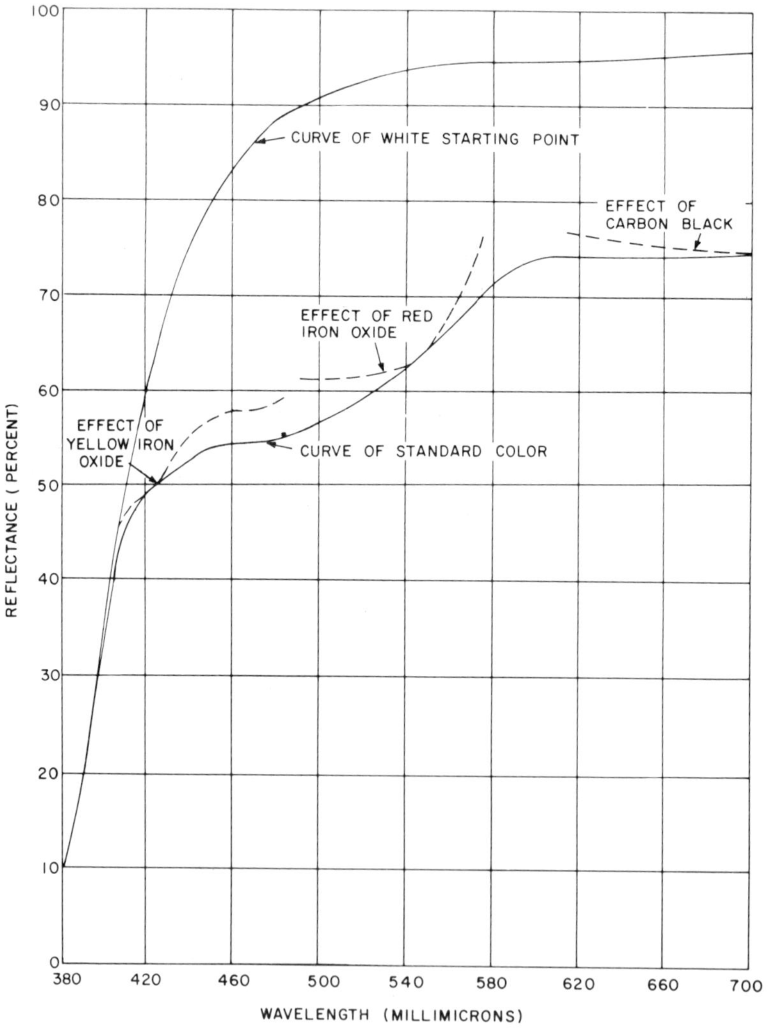

Figure 25.22.   Spectral curve of a sample.

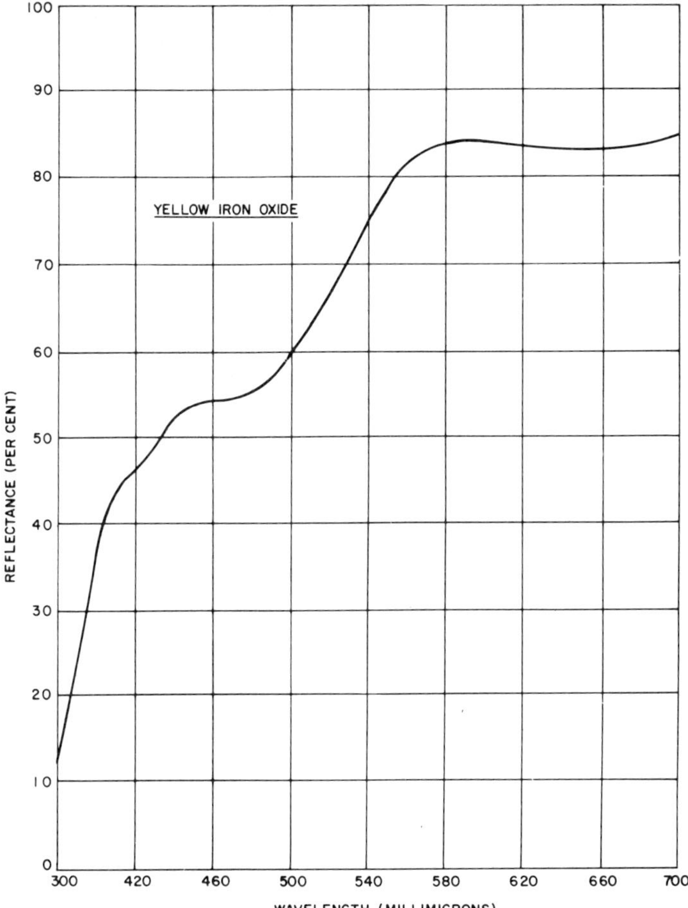

Figure 25.23.   Yellow iron oxide.

are curves from the library of known curves. By examining the absorption characteristics of knowns and comparing them to the unknowns, similarities can be seen. The proper pigments are selected in this manner. Also by applying the Kubelka-Munk analysis, the amount of each pigment can be calculated. Davidson and Hemmendinger's Colorant Mixture Computer and Digital Computer programs can be written to do this.

Dark and/or highly saturated colors that contain little or no white are difficult to formulate and, once formulated, are problems in production. Applying the four-component concept greatly helps these problem colors. One of the most common errors, when the color contains a small percentage of white pigment, is to hold it constant when it should be a variable. The amount is usually critical, and this does not provide enough

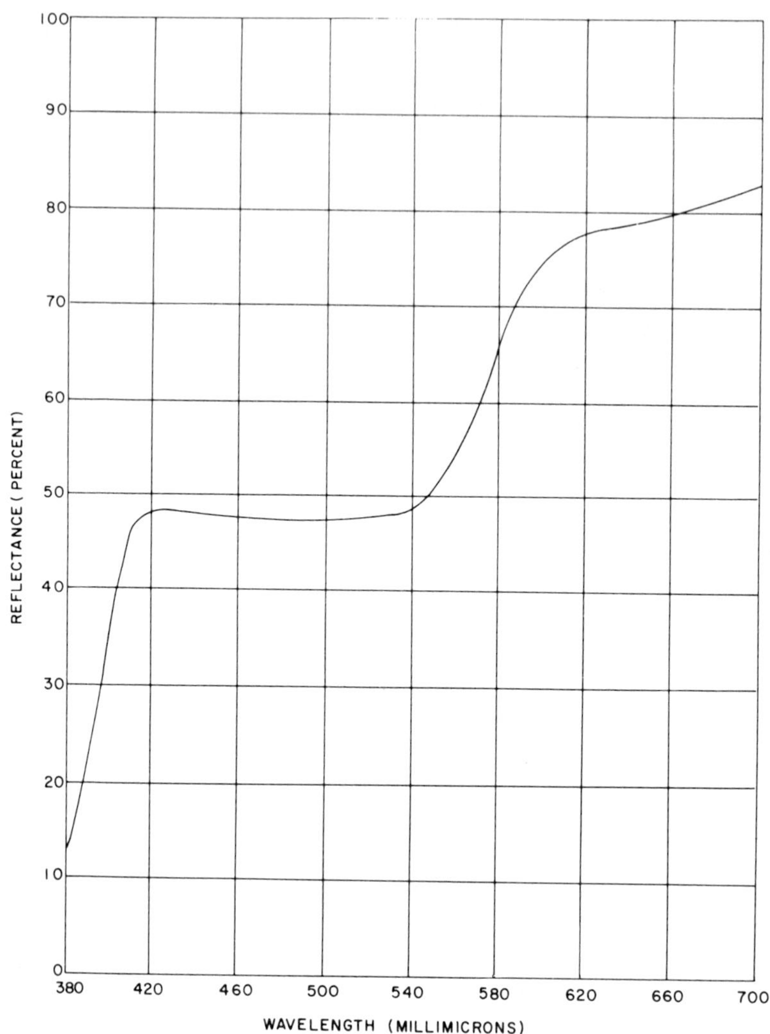

Figure 25.24.   Red iron oxide.

latitude for normal variation. The effect of white in a dark and/or highly saturated color is similar to the effect of black in a light and/or highly saturated color. In colors that contain little or no white, the constant is usually the pigment that is used in the largest amount and/or that which provides excess saturation.

When one pigment is held constant, the hiding or tinting strength will remain uniform, or nearly so, from batch to batch. In dark colors, there sometimes is a tendency to formulate with only three pigments. Unless one remains constant, there will be wide variation in hiding, tinting strength and cost of the product. Full control over a color cannot be expected when there are only three pigments.

Whether or not visual or instrumental computer methods are used to formulate a color, the principles of the four-pigment concept should be used to realize the following benefits:

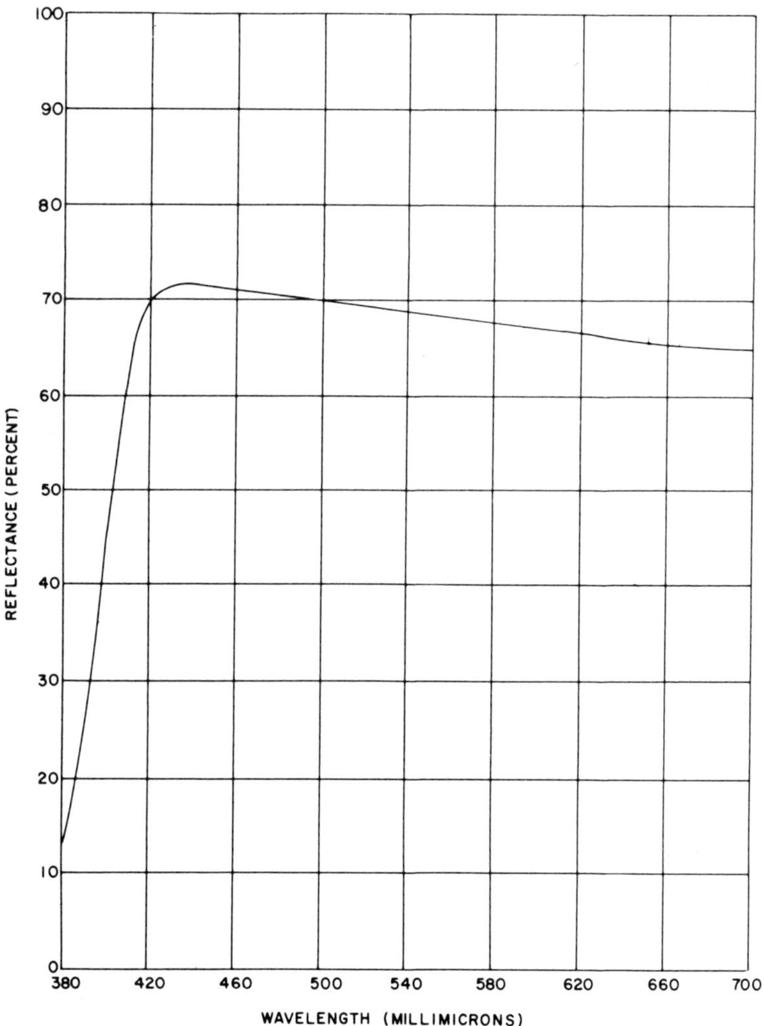

Figure 25.25.   Carbon black.

(1) There will be no question about matching the color with the pigments selected.

(2) Since only three pigments or colorants are allowed to vary to match the color, it is easier for the production shader to decide what and how much to add.

(3) Being a nonmetameric match, visual judgment to determine if the match is close enough will be easier.

(4) The formula will lend itself to control by instrumentation and quantitative shading.

INSTRUMENT COMPUTER COLOR MATCHING

When spectrophotometers and colorimeters first came into accepted use, there was some skepticism as to whether they could even be used to determine quantitatively the amount of pigment necessary to match a given sample. Such early workers as Duncan in England and others had shown that it was possible to do this, but their work was limited in practical application because of the time it took to make the readings and necessary calculations.

Even this limited work gave encouragement to others to investigate various methods which would make quantitative color matching practical. In the United States, there was considerable interest in the textile and plastics industries in this method of color matching.

Real progress in quantitative color matching did not take place until Davidson and Hemmendinger introduced their Colorant Mixture Computer. This was an analog computer which solved the Kubelka-Munk equation using data obtained from a spectrophotometer or colorimeter. The first use of this approach in the paint industry was in 1957 when Davidson and Hemmendinger's first commercial computer was put into operation by The Sherwin-Williams Company for quantitative shading of production batches of paint. These computers are now being extensively used in the United States and Europe. Digital computers are also being used successfully to solve various forms of the Kubelka-Munk equation. The future of the application of computers in paint color formulation and production shading appears to be unlimited. In the next ten years, tremendous progress in this field can be expected.

Instrumental computer color matching begins with the application of the Kubelka-Munk two-constant turbid media theory. Two constants— K, the absorption coefficient, and S, the scattering coefficient—describe the optical properties of a pigment and lead to the following relationship.

$$K/S = \frac{(1 - R)^2}{2R}$$

R is the reflectance of a paint film with total hiding at any specified wavelength in the visible spectrum. Given R, the respective K/S values can be computed, or vice versa. The important property of the K/S value is that it is proportional to the pigment concentration. Also, the K/S value in a mixture of pigments is an independent additive property of the individual pigments. The above is not always true, but it is valid for a wide range of colors where the white constant is high. The following equation represents the process:

$$(K/S)_m = CA\,(K/S)_a + CB\,(K/S)_b + CC\,(K/S)_c + \cdots (K/S)_w \quad (1)$$

At a specified wavelength $(K/S)_m$ is the K/S value of the mixture of pigments, $(K/S)_w$ is the K/S value of the white base, $(K/S)_a$, $(K/S)_b$, $(K/S)_c, \ldots$, are the K/S values per unit concentration of pigments A, B, C, $\ldots$, relative to white and $CA, CB, CC, \ldots$, are the concentrations of pigments A, B, C, $\ldots$, in the mixture and are ratios. With the equation, it is possible to predict the reflectance of a given mixture of pigments at many different wavelengths, thereby predicting the spectrophotometric curve and hence the color. Given the spectral curve of a color, the concentration of pigments A, B, C, $\ldots$, can be calculated, based on the concentration of the white.

The equations are simultaneous linear equations and in their simplest form there are as many equations as unknown pigment concentrations excluding the white. They can be solved by several methods but are tedious and time consuming. To facilitate their solution, special analog computers have been produced and digital computer programs have been written. As mentioned previously, the most widely used computer is the Davidson and Hemmendinger Colorant Mixture Computer.

This computer solves the equations at 16 different wavelengths and for as many as five pigments in addition to white. Digital computer techniques outlined for the paper industry but applicable to paint are described in C. W. Carroll's "Color, Industrial Color Measurement, Laboratory Color Matching, Production Color Control," IBM Corporation, June 26, 1963. The number of wavelengths predicted and the number of pigments used would be determined by the user's individual requirements when writing a digital computer program.

Accuracy of Equation (1) and the entire measuring process is sufficient enough to obtain matches of 1 MacAdam unit of color difference in three corrections or "hits." Of course, sometimes it will take more and sometimes less than three.

When the white content is less than 30 or 40% of the total pigment, Equation (1) is inaccurate making it necessary to use the more fundamental equation:

$$\frac{K_m}{S_m} = \frac{CAK_a + CBK_b + CCK_c \ldots K_w}{CAS_a + CBS_b + CCS_c \ldots S_w} \tag{2}$$

The same additive properties are assumed but here the K's and S's are separate instead of the K/S ratio. More effort is necessary to obtain the separate K and S values relative to white, and a digital computer program is slightly more complex. Color matches equal to using Equation (1) can be obtained.

This color matching process is based on the assumption that the pigments used in the sample are the same as those in the color being matched.

Improvements could be made in the basic Kubelka-Munk equations, and many variations have been tried. Any improvement in the equation may not effect the end result. In production computer shading, there are many more areas that need improving, and as they are improved, they will have more effect on the end result. At this time we should be concerned with our present manufacturing techniques and whether they are suited to computer shading. If they are not, changes should be made.

Better instruments to give more reliable data are needed and good standards must be available. These improvements will be made because computer shading is the best approach to the problem of production color matching. As this becomes apparent to more people, pressure will be exerted in this area to make it possible.

REFERENCES

1. Burhad, R. W., Hanes, R. M., and Bartleson, C. James, "Color: A Guide to Basic Facts and Concepts," New York, John Wiley & Sons, 1963.
2. "Color Measurement," *Bayer Farban Revue*, 3 (February 1964).
3. Davidson, H. R., and Hemmendinger, H., "Colorimetric Calibration of Colorant Systems," *J. Opt. Soc. Am.*, **45**, 216 (1955).
4. Felsher, Hal-Curtis, and Hanaw, Walter J., "Evaluation and Description of Metallic Colors," *SPE J.*, **21**, 12 (December 1965).
5. Huey, S. J., "Low Temperature Storage of Color Standard Panels," *Color Engineering*, 24 (September–October 1965).
6. Huey, S. J., Hunter, R. S., Schreckendgust, J. G., and Hammond, H. K., III, *Symposium on Gloss Measurement*, St. Louis Society of Paint Technology and Mississippi Valley District of ASTM (Jan. 30, 1962).
7. Judd, Deane B., and Wyszecki, Gunter, "Color in Business, Science, and Industry," New York, John Wiley & Sons, 1952.
8. Mackinney, Gordan, and Little, Angela C., "Color of Foods," Westport, Conn., The Avi Publishing Company, 1962.
9. Rudick, Lawrence, and Ingle, George W., "Control of Small Color Difference in Plastic Manufacturing, *ASTM Symposium on Color Difference Specifications*, 10 (1952).
10. Saltzman, Max, and Keay, A. M., "Variables in the Measurement of Colored Samples," *Color Engineering*, 14 (September–October 1965).
11. Simon, F. T., and Goodwin, W. J., "Rapid Graphical Computation of Small Color Difference," *Am. Dyestuff Rept.*, 105 (Feb. 24, 1958).
12. "So You Want to Set Color Tolerances," *New York Society for Paint Technology Official Digest*, 331 (April 1964).

26

*Paint Formulation**

Paint formulation is the art of scientifically compounding properly selected raw materials for efficient manufacture of a competitive product that satisfactorily meets performance requirements.

Several terms are used in this definition of paint formulation and each term should be analyzed individually to give a better insight into the complexities of formulation.

Art

Art as the term is used here means a skill acquired by experience, study and observation, with emphasis on experience. Knowledge of the performance of the raw materials under the given conditions and in relationship to each other is essential. This knowledge is largely dependent on experience. Many good formulators have had no formal technical training but develop excellent products based on years of practical experience. There are no precise scientific principles to apply that would insure a good formulation. Experience is an indispensable factor to the art.

Science

Though formulation is not a precise science, still scientific principles do apply. Several facets of paint formulation lend themselves to chemical and mathematical techniques. Pigment volume concentration is one such important approach. It will be discussed in some detail later in this chapter. With today's more sophisticated coatings such as epoxies, stoichiometric equivalents can be calculated to achieve proper balance between the resin and converter used to get maximum reactivity and eliminate trial-and-error methods. With the rapidly advancing technology of new vehicles and other raw materials, the scientific approach is gaining daily in importance.

*By W. Tomc, The Glidden Co., Cleveland, Ohio.

Raw Materials

The chemical industry is constantly introducing new compounds designed specifically or potentially for use in the paint industry. The alert formulator must keep abreast of these developments. He must know their properties as well as those of established materials. Proper selection of raw materials is essential. Small variations in properties between two otherwise comparable materials can mean the difference between a good or a poor finish. For example, two titanium dioxides may be identical in hiding power, whiteness, tinting strength and oil absorption, but can still result in vastly different performance. One of these may be surface treated to give it chalk resistance and tint retention and, hence, will produce an excellent exterior enamel. The other will chalk, lose its cleanness of tint and show rapid erosion. Knowledge of these small but important differences cannot be overemphasized for sound formulation.

Manufacture

Efficient manufacture is always an important consideration in proper formulation. High-speed dispersion equipment, sand mills, attritors and other modern dispersing machinery are rapidly replacing the older roller, ball and pebble mills. This gives the formulator additional factors to consider: Particle size and particle size distribution of the pigment and the equipment to be used for dispersion now become problems for the formulator.

Cost

The need for a competitive product is almost self-explanatory. If a product is overformulated, i.e., too good for its intended use, it may cost too much and no sales will result. Conversely, if it does not perform well, it will not be repurchased.

Performance

With all these elements affecting the development of a good formulation, it becomes obvious that there is no unique answer for a specific formulation problem but rather that several equally good formulations can evolve. The only real criterion for a formulation is its performance.

For a proper start, there should be a clear definition of what is expected of the projected formulation. With these objectives defined, the selection of raw materials to achieve these objectives can begin. Frequently, a compromise may be necessary. Thorough knowledge of the characteristics of the ingredients and their interrelationship is necessary for a judicious selection.

TABLE 26.1.   Bulking Values of Typical Hiding Pigments[a]

Trademark	Producer	Specific Gravity	wt/ gal	gal/ lb	Type
Azo Dox-11	American Zinc Co.	5.6	46.65	0.02144	Zinc oxide
XX-602	The New Jersey Zinc Co.	5.6	46.7	0.0214	Zinc oxide
EPA #506	The Eagle-Picher Co.	6.06	50.48	0.01981	Leaded zinc oxide
Ozlo 18 M	The Sherwin-Williams Co.	5.72	47.65	0.02099	Leaded zinc oxide
St. Joe #911	St. Joe	5.6	46.7	0.0214	Zinc oxide
Unitane 0–110	American Cyanamid Co.	3.9	32.5	0.0308	Titanium dioxide Anatase
Ti-Pure LW	E.I. du Pont de Nemours & Co.	3.89	32.4	0.0309	''
Horse Head A920	The New Jersey Zinc Co.	3.80	31.65	0.0316	''
Titanox AMO	Titanium Pigment	3.90	32.49		''
Zopaque R. G.	The Glidden Co.	3.9	32.5	0.0308	''
Unitane OR540	American Cyanamid Co.	4.1	34.2	0.0292	Titanium dioxide rutile
TiPure R510	E.I. du Pont de Nemours & Co.	4.11	34.2	0.0293	''
Horse Head R-750	The New Jersey Zinc Co.	4.1	34.15	0.0293	''
Titanox RA	Titanium Pigment	4.2	34.99		''
Zopaque R-88	The Glidden Co.	4.1	34.15	0.0293	''

[a] Raw Materials Index, Pigment Index NPVLA (April 1965).

Principles

The basic prerequisite for good formulation is a broad knowledge of the properties of the raw materials used in coatings. Also essential for a good formulation is a full awareness of the limitations of the raw materials to be used, their interrelationship with the other materials employed in the formula and, of course, the economics of their use.

Many factors, in addition to the basic ingredients, must be considered for a good formulation. These must be clearly defined before the development of a formulation is even attempted. These factors are determined by the anticipated end use of the proposed product. This is often overlooked and can result in much unnecessary laboratory work. The performance requirements, application properties, drying characteristics, color, gloss, hiding, production facilities, availability of raw materials, and selling costs are some of the factors that must be considered before development is begun.

With these requirements carefully outlined, selection of the proper raw materials based on a knowledge of their performance can be made, and the coating can be formulated to meet the stated objectives. Obviously this cannot be learned from a book but is dependent largely on a knowledge of the raw materials and the necessary experience to formulate a good, competitive product.

Although emphasis is placed on the necessity of thorough knowledge of raw materials, there are some basic mechanics common to all formulations.

If it is assumed that the proper selection of raw materials has been made, the performance characteristics of these materials in a given formulation are significantly controlled by the relationship of pigment and vehicle, specifically their respective volumes. The pigment-vehicle volume relationship will determine gloss, drying, stain removal, brushing, flow, holdout, viscosity and other properties.

Pigment-Volume Concentration

This relationship is known as the pigment volume concentration (PVC) or simply pigment volume (PV). It is a percentage expressed as a number and represents the pigment volume divided by the sum of the pigment volume and the vehicle solids volume, multiplied by 100. Thus,

$$\text{PVC} = \frac{\text{Pigment Volume}}{\text{Pigment Volume} + \text{Vehicle Solids Volume}} \times 100$$

The same relationship can be expressed as the pigment-binder ratio or P/B. In this case, the pigment volume is set equal to 1 and the vehicle volume is calculated to its proper value.

$$\text{P/B} = \frac{\text{Pigment Volume}}{\text{Vehicle Volume}} = \frac{1}{x}$$

Critical Pigment Volume Concentration. Still another concept of this relationship is the critical pigment volume concentration (CPVC). This is the specific PVC at which the vehicle demand of the pigment is precisely satisfied; i.e., there are no voids among the pigment particles. There is no excess vehicle present. This concentration is critical because above or below this value the properties of a formulation change dramatically (see Figure 26.1).

In the pigment volume relationship, the formulator has a tool by which he can change the scrubbability, enamel holdout, stain removal, dry hiding, gloss and other properties by varying the PVC. As defined before, the CPVC is the specific PVC at which the pigment is completely surrounded by the vehicle with no excess vehicle left and no pigment voids. If additional vehicle is added, free vehicle is immediately available to enhance the gloss, scrubbability, etc.

As a specific example, if two given latices are compared in a given formulation, one may have a CPVC higher than the other. The higher

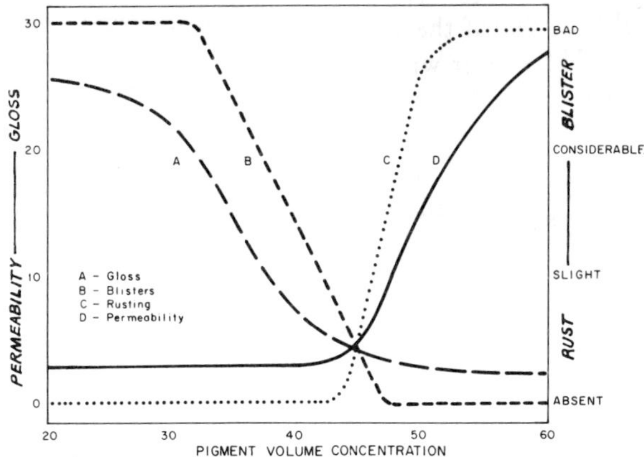

Figure 26.1.   Effect of CPVC on paint characteristics.

CPVC formulation would predictably have better stain and scrubbability characteristics than the lower. If more pigment were added to the latter formulation to raise its CPVC, this would yield properties comparable to those of the first latex. Since pigment usually bulks at lower cost, comparable properties would result at a lower total cost.

It is quite apparent now that the pigment volume is important. Exactly how is pigment volume or bulk determined? Bulk simply means the volume occupied by a given weight of pigment. For pigments, it can be calculated by multiplying the specific gravity by 8.33, the weight of a gallon of water (specific gravity = 1.0). Thus, if a pigment has a specific gravity of 1.5, its bulk will be

$$\text{Bulk} = 1.5 \times 8.33 = 12.495 \text{ lb/gal}$$

or if expressed as the reciprocal

$$\frac{1}{12.495} = .08 \text{ gal/lb}$$

This simply means that every 12.495 pounds of this pigment used in a formulation will yield 1 gallon, or for every pound used, .08 gallon results.

These values are usually expressed as bulk per 100 pounds for ready comparison with the bulk of other pigments. Thus,

$$\text{Bulk} = \frac{100}{12.495} = 8 \text{ gallons}$$

$$.08 \times 100 = 8 \text{ gallons}$$

Hence, if another pigment has a bulking value of 9 gal/100 lb, it is immediately apparent that if the cost and other properties of the two pigments are similar, the latter would be more advantageous because for the same amount we would obtain 1 gallon more yield.

Vehicle volume solids can be readily calculated if the weight per gallon of the vehicle and the total solids are known, as well as the solvent involved and its weight per gallon. Merely subtract the volatile volume to get the vehicle solids volume.

For example, the vehicle under consideration has a weight per gallon of 8.0 and a solids content of 70%, with mineral spirits as the solvent (weight per gallon = 6.5 pounds):

$$\text{Vehicle Solids Volume} = \text{Total Volume} - \text{Volatile Volume}$$

$$8 \times .70 = 5.6 \text{ pounds solids}$$

$$8.0 - 5.6 = 2.4 \text{ pounds volatile}$$

$$\frac{2.4}{6.5} = 0.37 \text{ gallon volatile bulk}$$

$$\text{Vehicle Solids} = 1 - 0.37 = 0.63\% \text{ solids by volume}$$

If 10 pounds of pigment (bulk 0.08 gal/lb) are ground in one gallon of the above vehicle, what would the PVC be?

$$\text{PVC} = \frac{\text{Pigment Volume}}{\text{Pigment Volume} + \text{Vehicle Solids Volume}} \times 100$$

$$\text{Pigment Volume} = .08 \times 10 = .8 \text{ gallon}$$

$$\text{Vehicle Solids Volume} = .63$$

$$\text{PVC} = \frac{.8}{.8 + .63} \times 100 = \frac{.8}{1.43} \times 100 = 55.9\%$$

Critical pigment volume concentration determinations are a bit more complex and in a routine formulation are seldom calculated precisely. Usually a series of coatings at various levels or ladders of PVC are made. An enamel holdout test is run, and the approximate CPVC is established. Enamel holdout is a simple test and is exactly what the name indicates. The coatings under consideration are used as a prime coat and the enamel is applied over it. The concentrations between which the gloss or holdout changes markedly would be the approximate CPVC.

For a detailed discussion of CPVC, the reader is referred to the original work by Asbeck and Van Loo[1] and their subsequent refinement,[2] and also to the work reported by the New England Production Club.[4]

Since the knowledge of raw materials has been stressed, we should now consider how their properties can affect the coating which is being formulated.

Pigments

Prime pigments, i.e., those which provide the hiding or color, are of course essential because of their opacity or hiding power. The selection of a pigment for exterior application would be dictated by its durability. Will it chalk or not? How much will be required to get the proper hiding? For colored pigments, the lightfastness must be considered. From a viscosity standpoint, the oil or vehicle demand must be known. This property could affect the brushing, as well as leveling and gloss. The possible reactivity of the pigment with the proposed vehicle and its effect on package stability are important. Particle size and distribution may have an effect on the type of production equipment to be used in making the product.

In general, the same factors must be considered for the inert pigments. Here, these properties contribute to hardness, mildew resistance, stain removal, viscosity and control of gloss, in fact the same as for the prime pigment selection with the exception of hiding power.

Vehicle

The nature of the vehicle will to a large extent determine the performance of the coating. The vehicle must be selected with continuous awareness of the requirements of the product listed in the original request. In selecting the best vehicle, the following factors must be considered: whether it will be used inside or outside, its own color particularly in whites, its color retention, oil length, drying characteristics, reactivity with the proposed pigments, viscosity, the solvent present and the driers needed. Again there is no easy route to a quick selection of a vehicle except thorough knowledge and experience in its use.

Solvents and Driers

Although solvents are an integral part of most vehicles, additional solvents are frequently added to the formulation to adjust the viscosity, brushing or spraying characteristics. Solvent strength, odor, color, flash point, evaporation time, flash off point and toxicity must all be considered to obtain the desired balance of properties. With driers, the proper combinations of top- and through-dry must be evaluated.

Formulation Example

The following is an example of the calculations required for the formulation of a white house paint to give the desired characteristics. Heatbodied linseed oil is chosen as the vehicle. The pigment is made up of titanium dioxide, leaded zinc oxide and talc. The principal hiding pigment is titanium dioxide and the proper balance is made between anatase (chalking) and rutile (nonchalking). The leaded zinc oxide is included to give mildew resistance, and the talc prevents settling and reinforces the film, thereby aiding resistance to chalking.

White House Paint

Desired Characteristics

Appearance
  (1) Color   White
  (2) Gloss   Approximately 80 on 60° glossmeter.
  (3) Hiding   Complete with one coat
Application
  (1) Viscosity   80–90 KU, generally applied at package consistency
  (2) Flow   Fair to good, must not sag
  (3) Dry   Overnight in air
Durability
  (1) Adhesion   Good to oil-type primers
  (2) Color retention   Good—must resist mildew and chalk enough to remove
                               dirt
  (3) Gloss retention   Fair-good
  (4) Mechanical resistance   Flexible—sufficient to match expansion of wood
                               Hardness—soft film
                               Moisture penetration—slight, film must "breathe"
  (5) Chemical resistance   Resist industrial fumes, generally sulfide
Miscellaneous
  (1) Selling price $6.00/gal = R.M. costs $2.00/gal max
  (2) Package stability   Good—no viscosity increase or settling

Selection of Raw Materials

Vehicle
  Heat-bodied linseed oil   Nonvolatile
  Ph naphthenate   Nonvolatile
  Mn naphthenate   Nonvolatile
  Mineral spirits   Volatile
Pigment
  TiO_2 Mixture of anatase and rutile
  Leaded zinc
  Talc

Quantitative Requirements

PVC   32%
Pounds of oil per gallon of paint   4
Pounds of TiO_2 per gallon of paint   2
Pounds of leaded zinc per gallon
  of paint   3
Pounds of 24% Pb drier $\frac{1}{2}$% × lb of oil/gal
Pounds of 6% Mn drier   .005% × lb of oil/gal
% Mfg. loss (vol)   2%
Total output (vol)   100 gal

Calculations — Vehicle
Input 102 gal
Yield: 100 gal
Loss: 2% by vol

Vehicle NVM	Pounds		Weight per Gallon	Gallons
Heat-bodied linseed, 4 lb/gal, 4 × 102	408	÷	7.87	51.75
Pb naphthenate, $\dfrac{.005 \times 408}{.24}$	8.5	÷	12.5	.68
Mn naphthenate, $\dfrac{0.00005 \times 408}{.06}$	0.34	÷	8.7	0.04
Vehicle solids	416.84			52.47

% vehicle nonvolatile volume	$= 100\% - 32\% = 68\%$
Total gallons nonvolatile volume	$= \dfrac{52.47}{.68} = 71\frac{1}{2}$ gal of solids
Total gallons of volatile	$= 102$ gal $- 77\frac{1}{2}$ gal $= 24.5$ gal
Total pounds of volatile (min sp.)	$= 24.5 \times 6.5 = 160$ lb
Vehicle volatile	160 lb 24.5 gal
Vehicle for 102-gal input	576.84 lb 76.97 gal

Calculations — Pigment

Total volume pigment $= 102$ gal $- 76.97$ gal $= 25.03$ gal

Hiding Pigments	Pounds		BV gal/lb.		Gallons
Anatase TiO_2	100	×	.031	=	3.1
Rutile TiO_2	100	×	.0293	=	2.93
Leaded zinc 12/88	300	×	.0212	=	6.36
Hiding pigment for 102-gal input	500				12.39

Gallons of inert	$= 25.03$ gal $- 12.39$ gal $= 12.64$ gal for 102-gal input
Pounds of inert ("Nytal" 300)	$= \dfrac{12.64}{.0421} = 295$ lb for 102-gal input
Total weight of pigment	$= 500$ lb $+ 300$ lb $= 800$ lb for 102-gal input
Weight per gallon of paint	$= \dfrac{800 + 576.84}{102} = 13.5$ lb/gal
% NVM (wt)	$= \dfrac{800 + 416.85}{1376.84} = 88\frac{3}{8}\%$
% NVM (vol)	$= \dfrac{76.97}{102} = 75.46\%$
Spreading rate gal/mil	$= 1604$ sq ft
Spreading rate ($2\frac{3}{4}$ mils dry film)	$= \dfrac{1604 \times .7546}{2.75} = 480$ sq ft/gal

White House Paint Formula

Raw Material	Pounds	Gallons	Cost per Pound (¢)	Cost ($)
Heat-bodied linseed oil	408	51.75	15.8	64.46
Lead naphthenate	8.5	.625	31	2.64
Mn naphthenate	.34	.375	41	.14
Mineral spirits	160	24.5	2.7	4.32
	576.84	77.25		$71.56

White House Paint Formula (*continued*)

Raw Material	Pounds	Gallons	Cost per Pound (¢)	Cost ($)
Anatase TiO_2	100	3.10	25.3	25.30
Rutile TiO_2	100	2.93	27.3	27.30
12/88 leaded zinc	300	6.36	12.6	37.80
"Nytal" 300	300	12.36	2.5	7.50
	795	24.75		$97.86
	1371.8	102.00		$169.46

$$\text{Total Raw Material Cost} = \frac{71.60 + 97.80}{100} = \$1.694/\text{gal.}$$

Computer

More sophisticated techniques are gradually being introduced into paint formulation. For example, instead of many random experiments to establish the effect of a given set of variables, statistically designed experiments are used. By defining the limits of the variables under consideration and designing the experiments statistically to fit these variables, many needless experiments are eliminated.

Computers find more and more applications in the daily routine of a formulator. Williams and Bacchetta[5] describe one such program for paint formulation. From one set of input data the program is capable of producing complete formulation data for up to five PVC's at a preselected volumetric solids content. In addition, the computer will generate such data as weight per gallon, pounds and gallons of each ingredient per 100 gallon batch, raw materials cost per gallon, pigment/binder ratio (P/B), and other useful information that is most time consuming when done by hand.

Naturally computers are expensive and hence not readily available. However, computer programs are available even for a small-sized company at nominal costs by a time-shared, client access computer system. Briber[3] describes the operation of such a system, and many paint problems can be solved by its use.

REFERENCES

1. Asbeek, W. R., and VanLoo, M., *Ind. Eng. Chem.* **41** 1470 (1949).
2. Asbeek, W. R., Laiderman, D. D., and VanLoo, M., *Offic. Dig. Federation Paint Varnish Prod. Clubs*, No. 326, 156 (1952).
3. Briber, R. M., *J. Paint Technol.*, **39**, 284–289 (1967).
4. New England Production Club, *Offic. Dig. Federation Paint Varnish Prod. Clubs*, No. 370, 803 (1955).
5. Williams, D. M., and Bacchetta, V. L., Jr., *J. Paint Technol.*, **39**, 267–283 (1967).

27

*Pigment Dispersion**

Much has been written about pigment dispersion, and current writers have done an excellent job explaining and outlining the various theories and applications of dispersion. This chapter is meant to serve as a practical, basic guide for both the beginning and the experienced formulator. No new theories or breakthroughs will be presented.

Definition

The definition of dispersion as used in this chapter is: Dispersion is the physical wetting of pigment particles in a liquid medium, to be accomplished for the purpose of obtaining in a film the full physical characteristics of the pigment. This is done without changing the particle size. When the particle size is reduced, the operation is called grinding.

Method

In the past, pigment dispersion has been accomplished by a number of methods, i.e., simple mixing, ball mills, stone mills, attritors, Kady mills, high-speed dispersers, roller mills and sand mills. All of these manufacturing processes require slightly different techniques and all have been successful.

Today the most important of these processes are the high-speed disperser and the sand mill. These two types of equipment process the greatest portion of paint made today and serve as the backbone of all up-to-date facilities. This, however, does not mean the obsolescence of other kinds of equipment; it only means they will be relegated to more specific applications on specific products.

The primary cause of this increased use of the disperser and sand mill is the pigment manufacturer. Pigment technology has advanced to the point that particle size no longer has to be reduced in most grades, and thus grinding is eliminated. Much of this improvement has been accom-

**By R. G. Hummeldorf, The Sherwin-Williams Co., Cleveland, Ohio.*

plished in the last 10 years. This basically means that the formulator by judicious choice can make almost any paint he wants by either the disperser or the sand mill, or a combination of the two. This refinement in pigment technology applies to both the extender pigment manufacturer and the prime and dry color manufacturer.

Equipment

A simple definition of a high-speed disperser would be a single shaft with a circular blade attached. The shaft revolves at such a rate as to give the blade a peripheral speed of between 4500 and 5500 ft/min. This speed will usually give the maximum dispersion in a very short time, provided that the mill base has been properly balanced for dispersion. Dispersers are available from a number of companies, with both fixed and variable speed, and a very wide range of horsepower. The most popular are the variable speed with hydraulic lift which enables the paint manufacturer to operate one disperser with up to four tanks.

A sand mill can be described as a single shaft with more than one circular blade operating in a cylinder. The shaft speed is much lower than that of the high-speed disperser. The cylinder is usually filled with sand, glass or other grinding media. The mill base is then pumped through the mill and into a thin-down tank. The mill base can also be recirculated to obtain a greater degree of dispersion.

All the basic patents on the sand mill are held by E.I. du Pont de Nemours & Co. At the present time, there are several manufacturers producing these mills in wide range of models. The mills are adaptable to both small volume and large volume production.

Mill Base Formulation

One of the most important phases of any dispersion or processing operation is the mill base formulation. An unbalanced mill base formulation leads to all sorts of problems:
    (1) Poor dispersion and flocculation,
    (2) Poor color development,
    (3) Off-standard gloss,
    (4) Viscosity (high or low),
    (5) Settling.

In addition to the technical problems it creates, poor mill base formulation also leads to a poor economic situation since mill time can increase as much as tenfold without producing the desired results. This is immediately reflected in high manufacturing costs, slow equipment turnover and eventually dissatisfied customers.

The above are the chief reasons why proper mill base formulation is so important.

A search through the literature was undertaken to find a quick simple method of determining a good starting point for processing a formula. Some very good data on mill base formulation and processing were found, but not a simple method for the practical paint chemist, who usually does not have the time for a complete study of each formula. Since what we were searching for was unavailable, we took the data obtained and converted it to suit our needs.

This starting point guide was restricted to a single vehicle system (long oil soya alkyd and mineral spirits at 40% vehicle nonvolatile). The pigment variables investigated were:

(1) Oil absorption,

(2) Specific gravity,

(3) Bulking value.

The equipment was set up as recommended by the manufacturer. The only other variable was the peripheral speed of the disperser and this was set at a nominal 5000 ft/min.

A number of these variables were plotted and the only sensible relationship was that between oil absorption and per cent pigment by weight in the dispersion phase. Using this data (Figure 27.1) as a basic, pigments

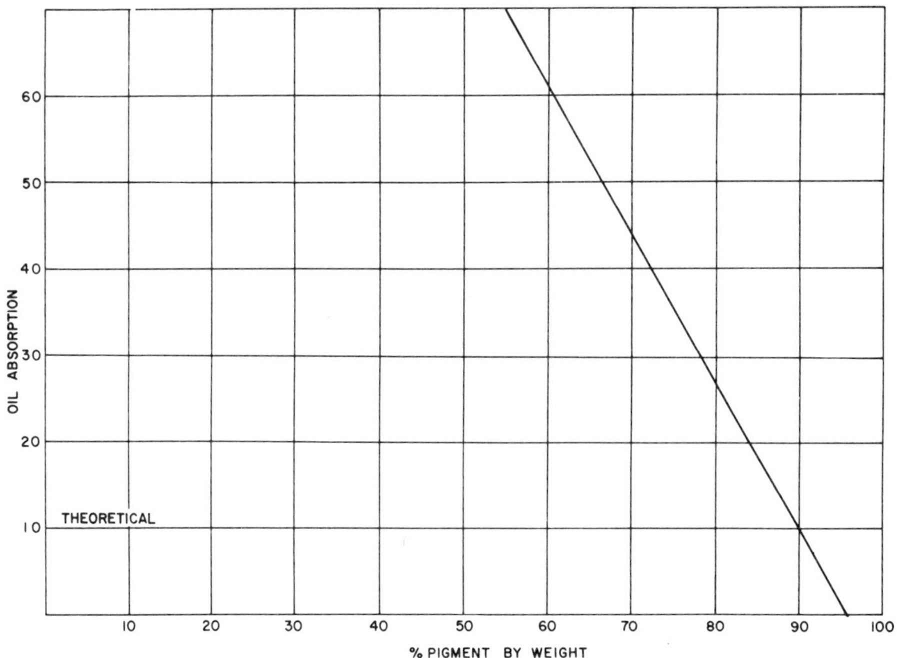

Figure 27.1.   Roller mill—pigment loading for premix.

typical of several classes, were then made into a flowing paste, and the data were plotted as per cent pigment by weight versus oil absorption. This curve (Figure 27.2) followed the same pattern as the theoretical curve. To confirm the data, a number of formulas were checked out and found to fit into the general pattern. Series of laboratory batches were also made using this same chart (Figure 27.2), and these points again fell

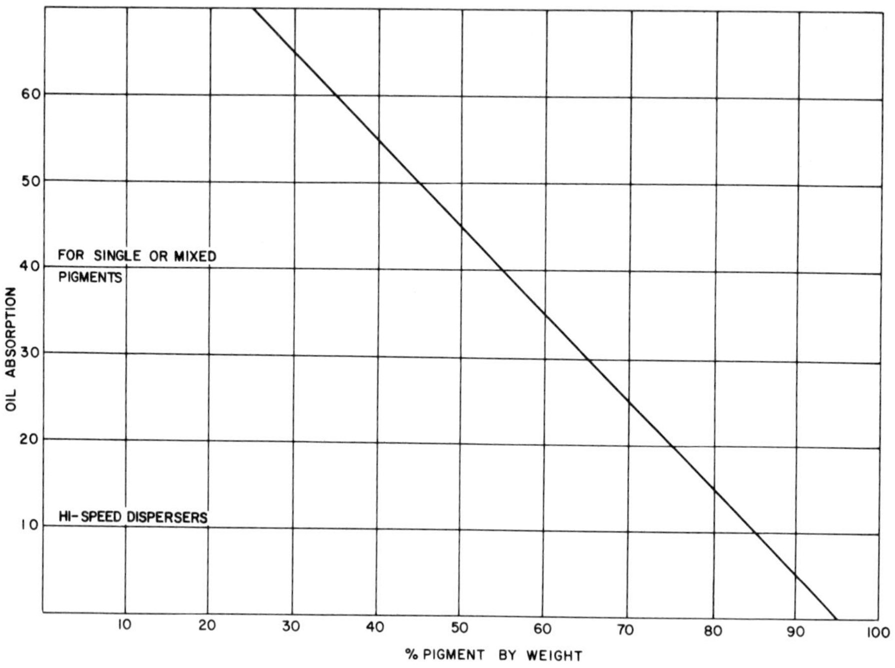

Figure 27.2.   High-speed dispersors—pigment loading for premix.

where they were expected. This was repeated through a series on a high-speed disperser, roller mill premix (paste mixer) and sand mill. The paste mixer and disperser were found to be almost the same, but the sand mill required additional vehicle (Figure 27.3). The data were then used as the starting point in setting up new formulas in the laboratory and were extremely helpful in assuring a workable formulation approximately 90% of the time.

Setting Up a Laboratory Formula

(1) Determine oil absorption of pigment or pigments. Published data is usually sufficient.

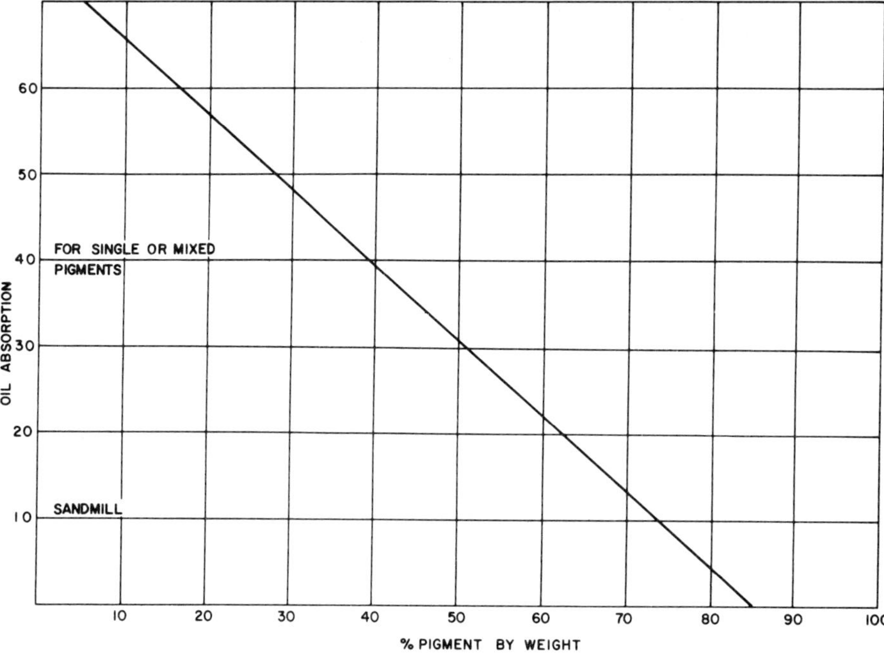

Figure 27.3.   Sandmill—Pigment loading for premix.

(2) From Figure 27.2 or Figure 27.3, depending on the equipment, determine pigment per cent by weight in mill base.

(3) Convert mill base to proper size for the equipment.

(4) Add any wetting agents found in the formula.

(5) Charge mill and run. Electrical load should be constant.

(6) Add either vehicle or pigment to obtain proper flow.

(7) Check dispersion in 10 to 15 minutes.

(8) Complete the batch.

Following the above procedure should minimize laboratory time and also give the plant a formula which can be processed with maximum efficiency.

This method does not take into account certain newer wetting agents or the newer vehicles, some of which are very poor wetters, but it does give the limited technical staff a quick and useful tool and guideline for proper mill base formulation. For more detailed and technical considerations, the works of T. Patton, F. Daniels and Dr. Weisberg should be thoroughly reviewed.

Equipment Setups and Limitations

Since the formula proper mill base is now set up, we are ready to begin manufacture. As previously mentioned, the dispersers come in very wide range of sizes and horsepowers, and are extremely flexible; however, there are some limitations which must be observed to produce the formula successfully.

The equipment variables are as follows:

(1) Horsepower,
(2) Peripheral speed,
(3) Blade size,
(4) Tank diameter,
(5) Paste volume.

A search of the literature on pigment dispersion reveals that the dispersing speed should be 4500 to 5500 ft/min. Since manufacturing experience bears this out, we will eliminate this variable, which then leaves us with four variables.

As a general guideline, the following table (Table 27.1) will provide a good idea of the conditions accompanying a given batch size. These figures are strictly a guide and there are definite exceptions. The table covers only the more useful and common sizes.

TABLE 27.1

1	*2*	*3*		*4*	*5*	*6*
	Blade	*Tank Diameter (in.)*		*Paste*	*Shaft*	*Minimum*
Blade Diameter	*Circumference (ft)*	*Minimum*	*Maximum*	*Volume*	*rpm*	*Horsepower*
8	2.09	20	24	31	2390	5
10	2.62	25	30	61	1910	$7\frac{1}{2}$
12	3.14	30	36	105	1590	10
13	3.40	33	39	132	1470	15
15	3.93	38	45	205	1280	20
16	4.20	40	48	250	1190	25
18	4.70	45	54	350	1065	30
19	4.96	47	57	418	1010	40
20	5.25	50	60	490	950	50
22	5.77	55	66	650	865	75
24	6.30	60	72	910	795	100
28	7.33	70	84	1340	682	125

(1) Blade diameter—Stock blades available from manufacturer.
(2) Blade circumference.
(3) Tank diameter—The minimum figure is 2.5 times the blade diameter and the maximum is three times the blade diameter.
(4) Paste volume—The volume to be filled in a tank at maximum diameter and the liquid height is exactly two blade diameters.
(5) Shaft rpm—The rpm required to give the blade a peripheral speed of 5000 ft/min.
(6) Minimum horsepower—Taken from manufacturer's literature plus personal observation.

Tank Configuration

Dispersers have been used in everything from a 5-gallon can on up. The tanks have had flat bottoms, dished bottoms, sloping bottoms, square sides or whatever else has been available. Of course, not all of these tanks provided the ideal shape. The tank coming closest to the ideal configuration would be a simple round tank with a dished bottom. The dished bottom enables the operator to drain and clean the tank much more easily than any other style.

Now that we have a guideline for picking the most suitable size disperser for our batch of paint, we can proceed to the next step which would be the sand mill. If our paint product is such that it requires the energy available with the sand mill, we now have to consider time as opposed to size. Table 27.2 gives a rough idea of the potential of the various size mills in gallons of finished paint per 8-hour day.

TABLE 27.2[a]

Mill Size (gal)[b]	Gallons per Hour[c]	Gallons per Day[d]
3	40	240
8	120	720
16	240	1440
30	500	3000

[a] Based on running 50% of the finished paint through the mill. The rate will vary up or down depending on the degree of dispersion needed and the ease of dispersion.
[b] Mill size—Volume of mill cylinder as stated by manufacturer.
[c] Gallons per hour—Finished gallons of paint.
[d] Gallons per day—Finished paint based on 6 hr mill time per 8 hr day.

Premixers

The premixers for the sand mill should preferrably be dispersers for two principal reasons: (1) a uniform mix, and (2) the time factor. The premixers should be sized in such a way that the sand mill can be kept busy almost continuously, and they should be in tandem. Table 27.3 gives a breakdown of premix tank ranges. If a disperser is not available for a premixer, any conventional type mixer will suffice.

TABLE 27.3

Mill Size (gal)	Tank Size	Disperser Size (hp)
3	30–60	$7\frac{1}{2}$
8	105–250	25
16	250–500	50
30	500–1000	75–100

Conclusions

If proper care is taken in formulating the mill base for dispersion, and if the equipment is properly sized for processing and economics, then the full properties of the pigment and the formulas will be realized with the lowest possible cost.

REFERENCES

1. Daniel, Frederick K., "Determination of Mill Base Composition for High Speed Dispersers," Journal of Paint Technology *38*, No. 500, 534–542 (Sept. 1966).
2. Fischer, E. K., "Colloidal Dispersions," pp. 261–262, New York, John Wiley & Sons, 1953.
3. Guggenhein, S., *Offic. Dig. Federation Paint Varnish Prod. Clubs*, **30**, No. 406, 729 (1958).
4. Fashinger, *Offic. Dig. Federation Soc. Paint Technol.*, **35**, No. 456 (1963).
5. Patton, Temple C., "Paint Flow and Pigment Dispersion," *J. Paint Technol.*, **38**, No. 500, 534–542 (1966).
6. Patton, Temple C., *Offic. Dig. Federation Soc. Paint Technol.*, **35**, No. 467, 1328 (1963).
7. Stieg, F. B., Jr., *Offic. Dig. Federation Paint Varnish Prod. Clubs,* **31,** 736 (1959).
8. Taylor, J. J., "Fine Pigment Dispersion," Paper to Southern Prod. Club (1954).
9. Taylor, J. J., "Mathematical Approach/Proper Pigment Vehicle Ratios," 1956.
10. Weisberg, H. E., *Offic. Dig. Federation Soc. Paint Technol.*, **34,** 1153 (1962); *ibid.,* **36,** 27 (1964); *ibid.,* **36,** 1261 (1964).

28

*Emulsion Paints**

Emulsion paints are characterized by the fact that the binder is present in a dispersed form in water. In contrast, in a solvent paint the binder is present in solution form. Because of this inherent physical difference, the formulation and handling of emulsion paints differ from conventional solvent systems.

Emulsion paints are composed of a pigment dispersion and a resin dispersion. The pigment dispersion is obtained by milling the pigments into water. The resin dispersion is either a latex formed by emulsion polymerization or a resin put into emulsion form. The pigment dispersion and the resin dispersion are blended to produce an emulsion paint. Surfactants and protective colloids are included to give stability to the product.

Since the binder is present in dispersed form, the external water phase controls the viscosity of the system. The molecular weight of the polymer which is in the internal phase does not affect the viscosity of the system. In a solvent system, as the molecular weight of the binder is increased, the viscosity increases, sometimes beyond practical limits of application. Usually higher molecular weight in the polymer gives better physical performance. Polymers with high molecular weights and improved properties can be readily utilized in emulsion systems.

The unique properties of emulsion paints which have promoted their popularity are ease of application, minimal odor, easy cleanup and outstanding performance in many applications.

The three principal latex systems used today are styrene-butadiene,

Figure 28.1.   Particle size comparison (average microns).

*By C. R. Martens, The Sherwin-Williams Co., Cleveland, Ohio.

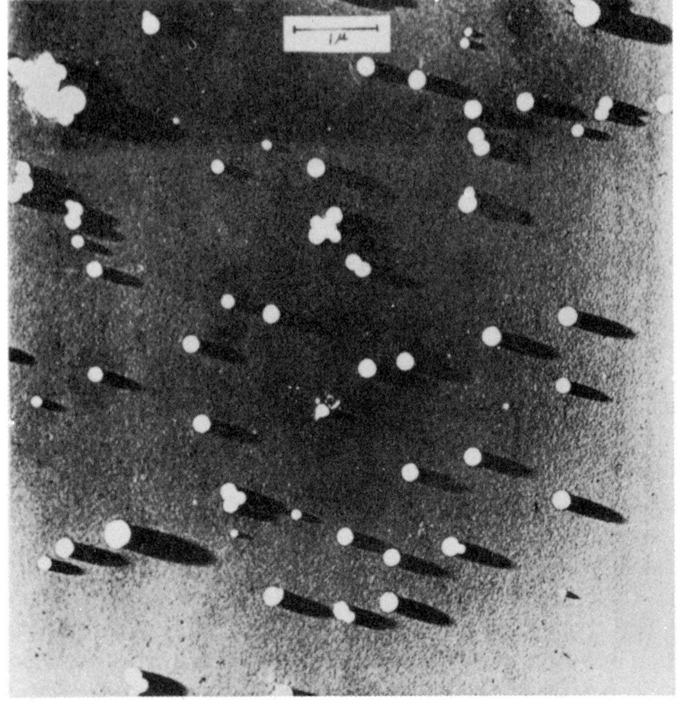

Figure 28.2.   Magnification of latex particles in colloid dispersion. (*Courtesy Dow Chemical Co.*)

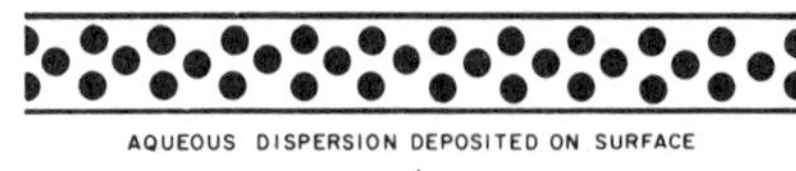

Figure 28.3.   Film formation from an emulsion.

polyvinyl acetate and acrylics. Resins such as alkyds, epoxies, etc., can be produced in emulsion form by post-emulsification. The size of the emulsion particles ranges from 0.1 to 10 μ.

INGREDIENTS OF AN EMULSION PAINT

The ingredients in an emulsion paint are as follows:

	Average (%)
Opaque pigment	20.0
Extender pigment	15.0
Pigment	0.3
Protective colloid	1.2
Latex	40.0
Plasticizer[a]	–
Modifier	–
Preservative	0.5
Antirust agent[a]	0.1
pH buffer	0.1
Fungicide[a]	0.5
Coalescing agent	2.0
Defoamer	0.5
Thickener[a]	0.5
Freeze-thaw stabilizer[a]	2.5
Water	16.8
	100.0

[a]Optional in some formulas.

The pigments are ground with a dispersant and a protective colloid with the minimum amount of water. The dispersant aids in wetting the pigment and keeping particles separated. The protective colloid forms an envelope around the pigment particles. It protects the system from coagulation in the presence of an electrolyte or from excessive shear stress. It protects the system during manufacture and storing. As little as 1% protective colloid will suffice to envelope sticky polymer particles to keep them from coalescing.

The latex is then added to the pigment paste along with other ingredients such as defoamers, preservatives, buffers, fungicides, etc. A defoamer eliminates air bubbles or foam. A preservative protects the paint against microorganisms that may be present in raw materials or equipment. The buffer prevents pH drift. A fungicide protects the final paint film from attack by fungi. A coalescing agent is sometimes used to get good film fusion and in reality is a volatile plasticizer. Thickners are added to adjust viscosity. Emulsion paints are usually not stable to freezing, and materials are added to protect against freeze damage. Anti-rust agents are sometimes added to water-type paints to retard rusting of the container or, in the case of metal coatings, to prevent flash rusting on application.

Surfactants[3]

Surfactants have multiple uses in aqueous coatings: (1) A surfactant is used as a dispersant during the grinding of the pigment by providing better wetting of the pigment particles. (2) Surfactants are required as emulsifiers to produce stable interfaces between oil and other materials and water. (3) They impart certain properties to the aqueous coating such as better wetting, flow, etc.

A surfactant is a compound which affects the surface forces of a liquid or a solid in relation to other liquids, gases or solids. The surfactant generally consists of a molecule in which one end is hydrophilic (water loving) and the other end is lipophilic (oil loving).

The sodium salts of fatty acids (i.e., ordinary soap) would be an example of a surfactant:

$$CH_3{-}CH_2{-}CH_2{-}CH_2{-}CH_2{-}CH_2{-}CH_2{-}CH_2{-}CH_2{-}CH_2{-}CH_2{-}CH_2{-}\overset{\overset{\textstyle O}{\|}}{C}{-}ONa$$

Lipophilic   *Hydrophilic*

When a soap is added to water, it lowers the surface tension (i.e., it reduces the interfacial tension between water and air). The resulting solution foams readily and has other attributes of soap solutions. In addition the soap promotes wetting of solids, mixing of immersible liquids such as oil, etc.

Structure of Surfactants. Surfactants are compounds in which one portion of each molecule is hydrophilic (water loving) and the other por-

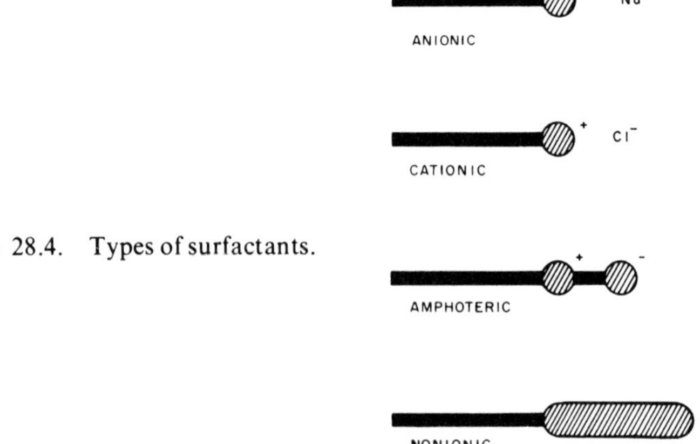

Figure 28.4. Types of surfactants.

tion is lipophilic (oil loving) or hydrophobic (water fearing). The hydrophilic part of the molecule may be a carboxylate, sulfate, sulfonate, alcohol or alcohol-ether. The lipophilic portion of the molecule may be a long hydrocarbon chain in fatty acids, a straight, branched or cyclic hydrocarbon of petroleum, or an aromatic hydrocarbon containing alkyl side chains.

Types of Surfactants. Surfactants may be classified as anionic, cationic, amphoteric and nonionic.

Anionic. The anionics bear a negative charge and migrate toward the anode, or positive pole, while in solution. These are the oldest and best known of surfactants.

$$\left[R-\overset{\overset{\displaystyle O}{\|}}{C}-O^- \right] \quad M^+$$

Soap

$$\left[R-O-\overset{\overset{\displaystyle O}{\|}}{\underset{\underset{\displaystyle O}{\|}}{S}}-O^- \right] \quad M^+$$

Alkyl Sulfate

$$\left[R-\overset{\overset{\displaystyle O}{\|}}{\underset{\underset{\displaystyle O}{\|}}{S}}-O^- \right] \quad M^+$$

Alkyl Sulfonate

Cationic. The cationic group is the opposite of an anionic group. Its surface activity is due to the presence of a long-chain, oil-soluble cation. These are used very little in latex paints since they would neutralize the charges of an anionic surfactant which might be present and cause a break in the emulsion. Cationic surfactants are used to produce stable asphaltic emulsions. Cationic surfactants also possess germicidal anticorrosive, antistatic properties.   The following is the structure of a cationic surfactant.

$$\left[R-\overset{\overset{\displaystyle A_1}{|}}{\underset{\underset{\displaystyle A_3}{|}}{N^+}}-A_2 \right] \quad X^-$$

R represents a hydrophobic group such as a long-chain aliphatic or aromatic group. X represents a negative ion such as Cl, Br, I, or other monovalent ion, and A_1, A_2 and A_3 represent hydrogen, alkyl, aryl or heterocyclic groups.

Amphoteric. Amphoteric or amphobytic surfactants contain both a positive and negative charge. These charges may neutralize each other so that at a given pH the surfactant behaves as if it were nonionic. These surfactants usually exhibit cationic properties in acid solutions and anionic properties in alkaline solutions.

Nonionic. Nonionic surfactants depend chiefly upon hydroxyl groups and ether groups to create hydrophilic action.

$$RC(=O)-O-(CH_2)_n(OH)_n$$

$$RC(=O)-O-(C_2H_4O)_nC_2H_4OH$$

These compounds achieve solubility by being hydrated in water solutions. When the temperature is raised, the forces of hydration are reduced and there is a precipitation of the surfactant lowering its effectiveness. This shows up as a turbidity or cloud. This may lead to erratic results when processing produces a higher temperature in the paint.

Behavior of Emulsions

Lipophilic groups dissolve in the surface of the oil droplets with the hydrophilic groups oriented toward the water phase. There is an interfacial film between the water phase and the oil phase. A strong film prevents coalescence of the dispersed droplets. The charged droplets are stabilized because of repulsion between like charges.

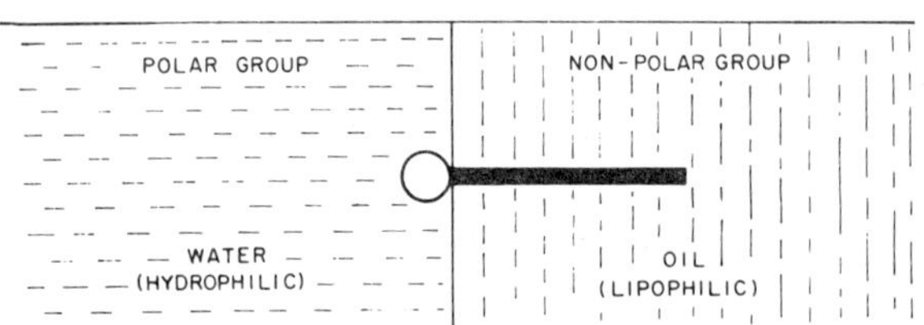

Figure 28.5.   Emulsion interface.

Protective Colloid

Water-soluble resins and gums are used as protective colloids or thickeners in emulsion paints. Their presence in the water phase aids dispersion of pigment, protects the paint against coagulation and settling, and adjusts viscosity.

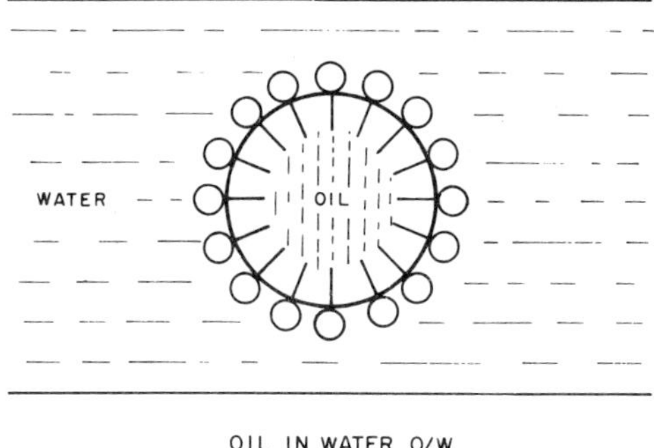

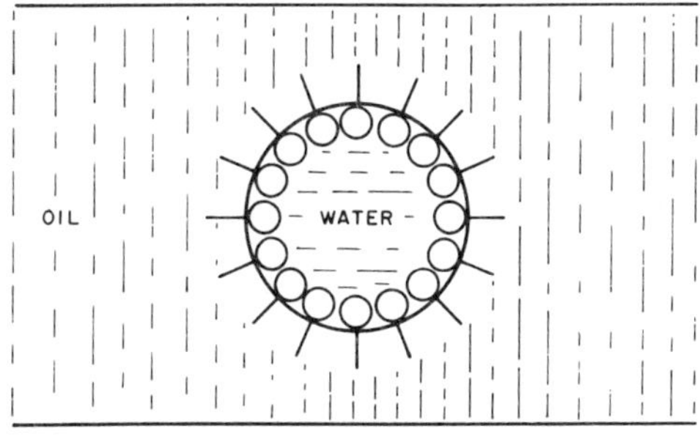

Figure 28.6. Emulsions.

The hydrophilic colloids include starch, sodium alginate, methylcellulose, hydroxyethylcellulose, sodium carboxymethylcellulose, polyvinyl alcohol, ammonium caseinate, sodium polyacrylate. The materials produce high-viscosity solutions at low concentrations. As little as .1 to 1% of protective colloid is usually sufficient.

A portion of the protective colloid incorporated in the grind aids dispersion. By thickening the mix, high shearing action is obtained insuring good mixing. The pigment particles are completely coated with the protective colloid which improves the stability of the paint. A protective colloid is added to an emulsion or latex in manufacture. The protective colloid, because of the increased viscosity it imparts to the water phase, reduces settling tendencies. This improves stability of the system by minimizing the tendency of the dispersed particles to stick together if subjected to stresses by mechanical means, freezing, etc. The water-soluble resin when used as a thickener produces the correct viscosity for application. The type of protective colloid used will affect the flow and leveling of the coating. For many reasons, including economy, it is advantageous

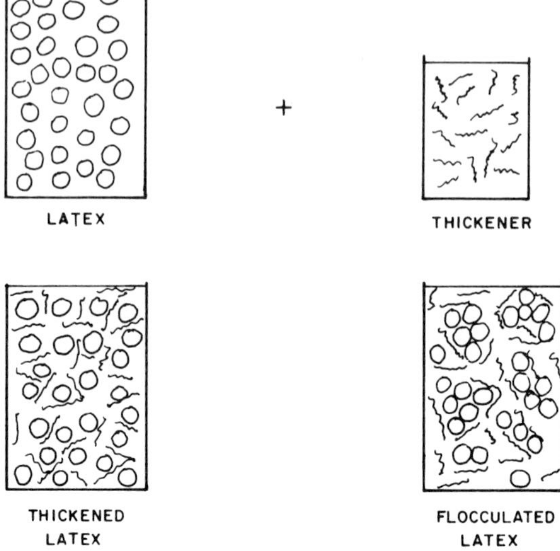

Figure 28.7. Mechanism of thickening.

to keep the amount of protective colloid to a minimum. Protective colloids are relatively high in price compared to the other ingredients used in the paint. Once the film is applied and dried, the protective colloid serves no useful purpose and may actually detract from the performance of the coating. Being water soluble, these materials make the coating more water sensitive, as measured by scrub tests, humidity tests, exposure tests, etc. The protective colloids are usually not compatible with the vehicle or latex, so that when the film dries there is a network of water-sensitive resin throughout the film.

pH

pH is the logarithm of the reciprocal of the hydrogen ion concentration or the hydrogen ion activity. This measure holds for aqueous systems only and has a scale from 0 (strongly acid) to 14 (strongly alkaline), with the neutral point at 7. A unit change in pH corresponds to a tenfold change in acidity.

The pH of a water paint has a profound effect on the physical properties and performance of the system. Buffers are used to maintain the pH level of the paint. Each latex system is stable in specific pH ranges. Surface agents, particularly those of the anionic type, have pH ranges in which they are the most effective. Protective colloids such as carboxymethyl-cellulose, casein, polyacrylates and algenates are soluble only in the alkaline range and become insoluble at low pH. High pH ranges will hydrolyze materials such as proteins. Maintenance of high pH in the package keeps certain bacteria that might cause spoilage under control. The use of permanent alkalies such as sodium hydroxide and other salts to adjust pH should be avoided if possible since water-soluble materials will be left in the film, reducing water resistance. If possible, a volatile alkali such as ammonia should be used for adjusting the pH of the paint.

EMULSION FORMATION

Resin emulsions can be prepared by two different methods: Emulsion polymerization and post-emulsification of the resin. As the term implies, emulsion polymerization is actually two processes carried out in a single operation. Post-emulsification of the resin involves the emulsification of the resin which may be obtained from natural sources or produced from synthetic materials through polymerization by addition or condensation reactions.

Emulsion Polymerization

Most commercial latices are manufactured by a technique known as emulsion polymerization. The monomer is dispersed in water in small droplets and the polymerization takes place within these droplets. Because of good heat transfer through the water phase, the heat of polymerization can be removed easily and polymerization proceeds rapidly. High molecular weight polymers are produced.

Latices in use today are derived from the free radical initiated addition polymerization of unsaturated monomers of the vinyl or maleic type. These are compounds containing the $-C{=}C-$ structures, with various groups such as a chloride, acetate or benzene attached to the carbons. The polymers can be considered as substituted polyethylene; for example, vinyl chloride polymerizes as follows:

$$
\begin{array}{ccccccc}
\text{H} & \text{Cl} & & \text{Cl} & \text{H} & \text{Cl} & \text{H} & \text{Cl} & \text{H} \\
| & | & & | & | & | & | & | & | \\
\text{C} & {=}\text{C} & \rightarrow & -\text{C} & -\text{C} & -\text{C} & -\text{C} & -\text{C} & -\text{C}- \\
| & | & & | & | & | & | & | & | \\
\text{H} & \text{H} & & \text{H} & \text{H} & \text{H} & \text{H} & \text{H} & \text{H} \\
\end{array}
$$

Polymerization. The polymerization process begins with the initiator dissociating into a free radical. The free radical collides with a monomer molecule to form a more complex free radical which continues to combine with additional monomer to form a long polymer chain.

The free radical is formed by the decomposition of an initiator such as benzoyl peroxide, hydrogen peroxide or potassium persulfate.

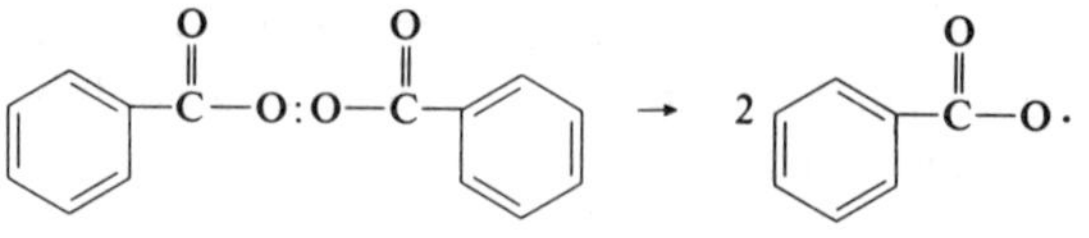

Benzoyl Peroxide

$$S_2O_8 \rightarrow 2SO_4$$
Potassium Persulfate

The second phase of the polymerization is known as propagation. The free radicals react with the monomer to open double bonds.

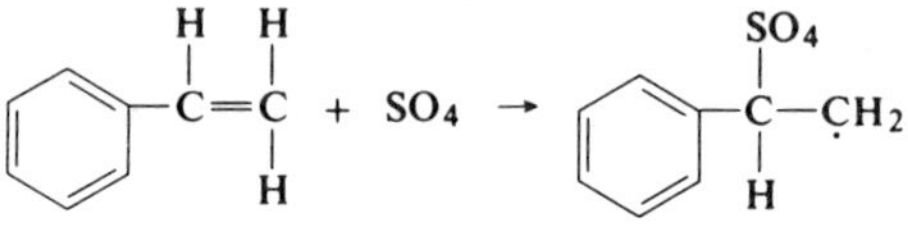

Note that the radical from the initiator actually participates in the reaction. Therefore, one end of a polymer chain is a fragment of the initiator.

The final phase of the polymerization process is termination. This may occur in several ways, such as the combination of two growing chains, combination of a growing chain and a free radical, etc.

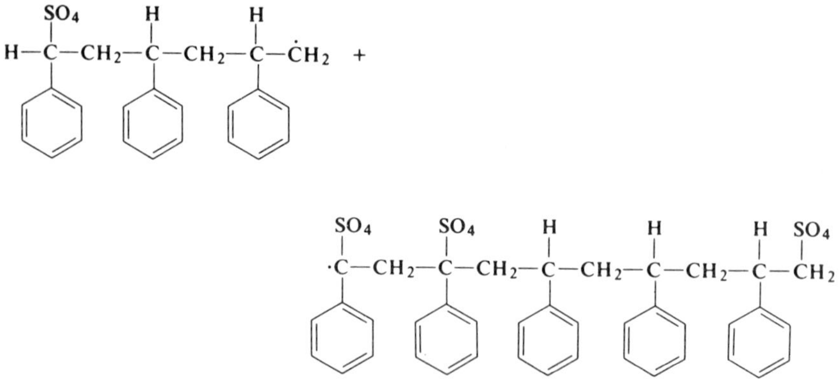

With monomers with a functionality of one (one double bond for molecule), linear polymers are formed. When a molecule with a functionality greater than one is incorporated into the system, branching and cross-linking may occur.

Linear:

$$O-M-M-M-M-----M-M-T$$

Branched:

$$O-M-M-M---------M-M-T$$

$$\begin{array}{c} | \\ M \\ | \\ M \\ | \\ T \end{array}$$

Cross-linked:

$$O-M-M-M-------M-M-T$$

$$\begin{array}{c} | \\ M \\ | \\ M \\ | \\ M \\ O-M-M-M-M-----M-M-T \end{array}$$

where

    O = initiator fragment,
    T = terminator fragment,
    M = monomer.

Cross-linking can be accomplished by the incorporation of monomers with higher functionality, such as divinyl benzene, etc. In a solvent system, cross-linking would produce high viscosities and poor solubility and might even produce gels or insoluble polymers. In a latex which is a dispersed system, the viscosity and solubility characteristics of the individual polymer particles do not affect the viscosity of the latex. In latices a reasonable amount of cross-linking is desirable to toughen and insolubilize the final film.

Many of the commercial latices are produced using more than one monomer to attain the desired properties such as hardness and elongation. For example, vinyl acetate is itself a relatively hard, brittle polymer. It is copolymerized with dibutyl maleate to produce a softer, more elastic polymer. This is known as internal plasticization. The softening point of the polymer in a latex film is important as it controls the fusion or coalescence of the latex particles. For example, a latex that has polymer particles with a softening point above room temperature will, upon evaporation of the water, produce a powdery deposit. If the softening point is close to room temperature, the film may perform satisfactorily in warm weather but fail in cold weather.

Copolymerization. Copolymerization is any polymerization reaction in which two or more different monomers are converted into a homogeneous polymeric product. First the composition of the copolymer depends upon the concentration of the monomer at the locus of reaction. Copolymerization is further complicated by the fact that monomers polymerize at different rates. Therefore, the composition of a copolymer will vary with the degree of conversion. Thus, the polymers formed at the start of the copolymerization may contain a higher proportion of the more reactive monomers than those formed at the end of the polymerization. A latex may contain a wide range of polymer molecules of varying composition and structure. The individual polymerization rates are also affected by temperature.

To prepare copolymers of more uniform composition, various procedures may be used. Sometimes the more reactive monomer is added

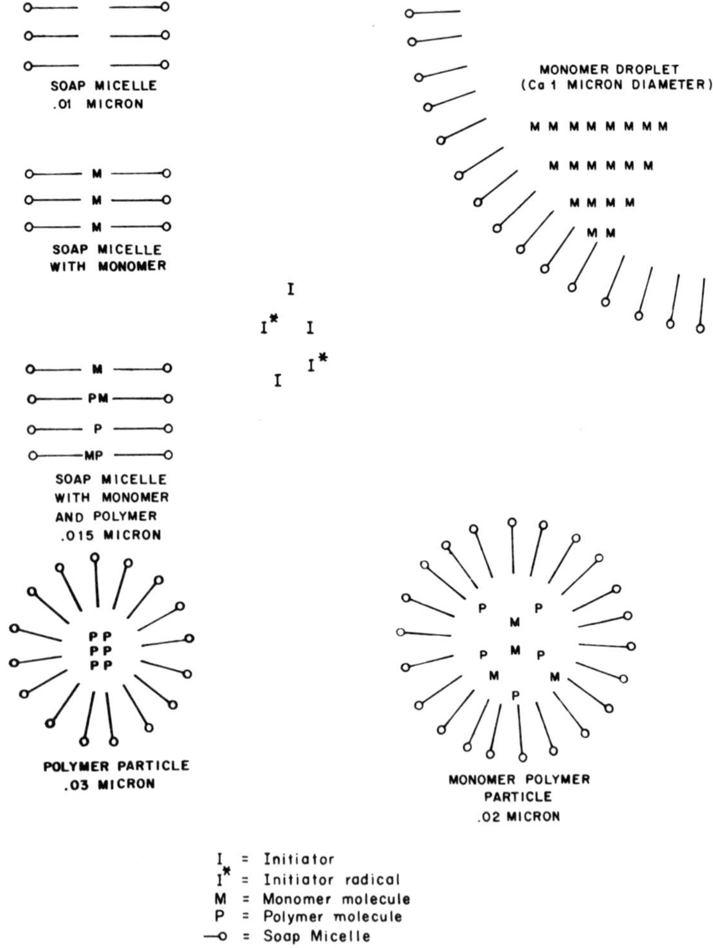

Figure 28.8.   Emulsion polymerization concept.

slowly during the polymerization. Also temperature, pressure and catalyst content may be varied during the polymerization.

One of the advantages of emulsion polymerization is that high molecular weight polymers can be obtained by this method. Increasing the molecular weight of the polymer increases toughness, scrub resistance, solvent resistance, chemical resistance, etc.

Theory. Emulsion polymerization is a complicated process. Harkins' publication[2] did much to explain the theory of emulsion polymerization. Previously, it had been thought that the polymerization took place within the droplet during emulsion polymerization, although this was not substantiated by fact, since the droplet is many hundred times bigger than the final latex polymer particle.

According to the Harkins theory, the monomer is distributed in three different phases in the emulsion systems, i.e. dissolved to a small extent in the water phase, suspended in the form of monomer emulsion droplets, and solubilized in the surfactant micelles.

The process of initiation occurs in the aqueous phase and the monomer radical diffuses into the micelles where polymerization occurs. The monomer emulsion droplets act as a reservoir for monomer. During the polymerization, the monomer emulsion droplets become smaller and the micelles grow larger. It has been proved experimentally that the rate of polymerization is dependent on the number of particles present, which is largely dependent on the concentration of emulsifier.

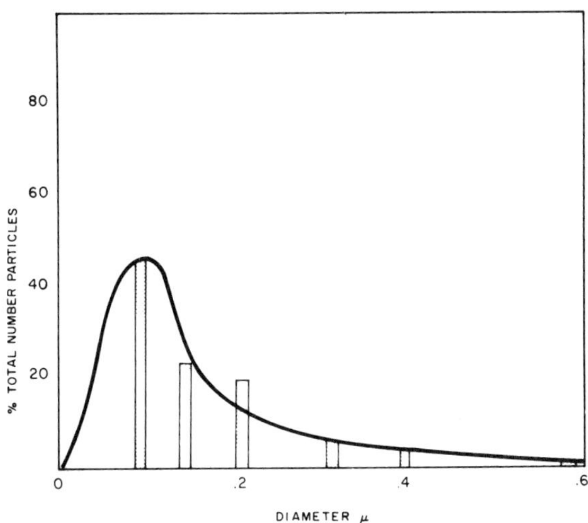

Figure 28.9. Particle size distribution of a typical latex.

The greater the amount of emulsifier present, the greater is the number of micelles present and finally the greater is the number of polymer particles which have a smaller size.

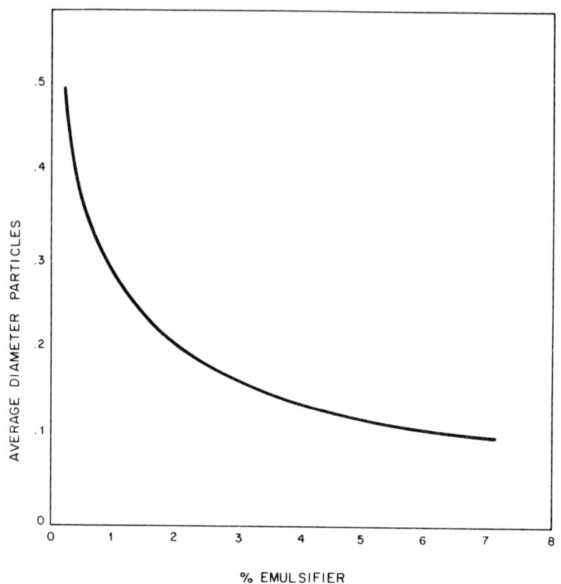

Figure 28.10.   Emulsifier content versus particle size.

Ingredients

The ingredients in an emulsion recipe are as follows:

	Per Cent
Monomers	30–50
Surface-active agents	1–3
Protective colloid	0–3
Initiator	1–3
Modifier	0–1
Buffer	0–1
Water	50–70

Reaction medium—deionized, soft or distilled water is generally used as the continuous phase in an emulsion polymerization. Sometimes small amounts of alcohol or glycols are added as antifreeze materials for low-temperature reactions.

Monomer—one or more polymerizable materials which amount to 45 to 55% of the overall composition of the final latex.

Initiators—these are usually water-soluble peroxides, hydroperoxides and persulfates used to initiate the polymerization reaction.

Emulsifiers—surfactants which may be anionic, cationic or nonionic and are needed to carry out the emulsion polymerization reaction. Their function is to solubilize monomer, suspend monomer droplets and suspend polymer particles.

Modifiers—these materials which may be aldehydes, mercaptans or chlorinated hydrocarbons serve to control the polymerization reaction restricting cross-linking and controlling the molecular weight.

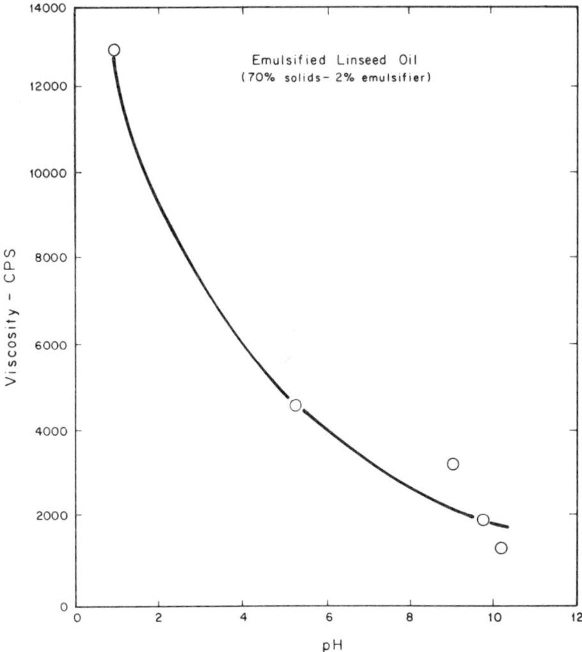

Figure 28.11.   Effect of pH on viscosity.

Protective colloids—water-soluble materials such as polyvinyl alcohol or methylcellulose are incorporated in the batch to stabilize the final latex.

Buffer salts—the pH of an emulsion polymerization batch must be controlled very carefully as the reaction rate, particle size, etc., are effected by pH. These buffers are phosphate, citrate, acetate, carbonate salts, etc.

Post-emulsification

Oils, resin and waxes can be prepared in emulsion form by combining with a surfactant and water under proper mixing procedures.

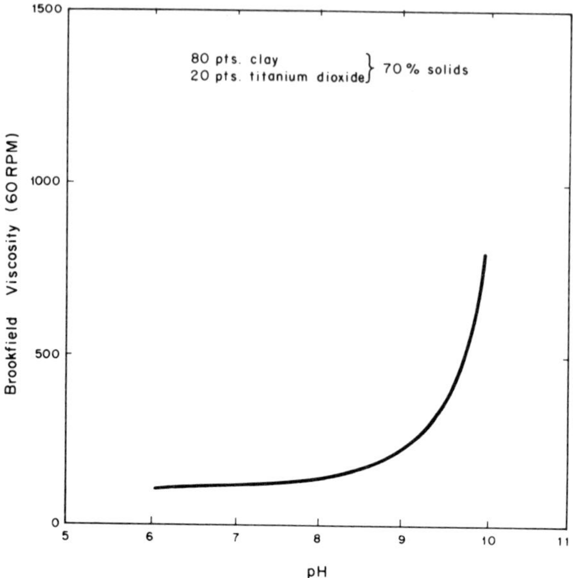

Figure 28.12. Pigment dispersion viscosity versus pH.

For emulsions under normal procedure, the resin, oil or wax must be in a liquid state or converted to a liquid state by the use of heat or the addition of a solvent.

There are several techniques for preparing emulsions in which the resin and water are combined by various mechanical means such as stirrers, mixers, pumps, colloid mills, etc. The techniques are as follows:

Agent in Water. The surfactant is added to the water and then oil is added slowly with agitation. This produces an oil/water emulsion directly.

Agent in Oil. The emulsifying agent is dissolved in the oil phase. This mixture is added directly to water, forming an oil/water emulsion. When water is added to the oil-surfactant mixture, a water/oil emulsion is formed first which inverts to an oil/water emulsion. This method produces fine particle size emulsions. Viscosity is at its maximum value at the inversion point.

Combination. When a combination of surfactants is used, they may be put into both phases. The lypophilic emulsifier is put into the oil phase and the hydrophilic emulsifier is put into the water phase.

In situ. In either oil/water or water/oil emulsions, the fatty acid may be dissolved in the oil and the alkali in the water. The soap then forms at the interface.

Alternate Addition. In this method, the oil and water are added alternately in small portions to the surfactant.

STABILITY OF EMULSIONS

Emulsions must have good stability. The stability depends upon the particle size, the difference in density of the two phases, the viscosity of the continuous phase, the charges on the particles, the nature, effectiveness and amount of emulsifier used, the conditions of storage including high and low temperatures, agitation, vibration and dilution or evaporation.

Since the particles of an emulsion are fully suspended in a liquid, they obey Stokes law unless they are changed. The rising and settling of the particles are termed creaming and sedimentation, respectively.

The phenomena of creaming receives its name from the common example—the separation of cream to the top of the bottle in unhomogenized milk. This is not a break but rather a separation into two emulsions, one of which is richer in the disperse phase. When the richer phase goes to the bottom, the phenomenon is called downward creaming. When creaming or sedimation occurs, coalescence is more likely to occur because of crowding or physical contact of the particles. Particle size data on emulsions usually give the average particle size.

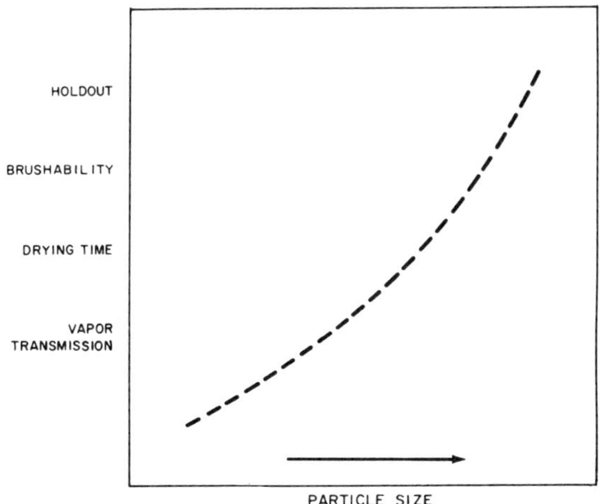

Figure 28.13. Properties improved by increased particle size.

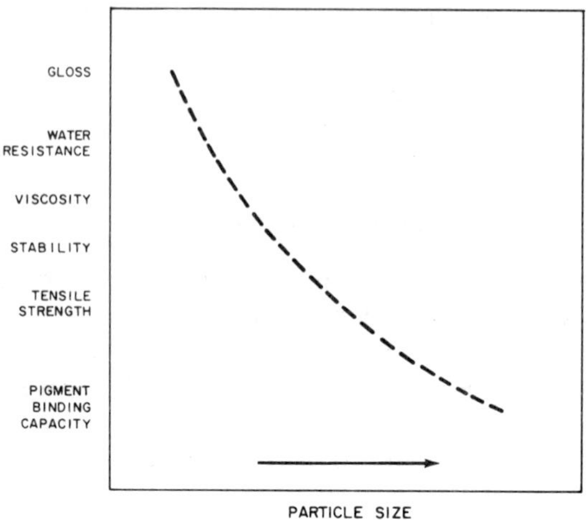

Figure 28.14.   Properties improved by decreased particle size.

Phase Volume

An emulsion consists of many spheres (internal phase) packed into a definite volume of space (external phase). According to solid geometry, an assembly of spheres of equal radius can be placed in a position of packing in two ways. In either case, the spheres occupy 74% of the total volume. Any attempt to exceed this volume will cause breaking or inversion.

REFERENCES

1. Davidson, R. L., and Sittig, M., "Water Soluble Resins," New York, Reinhold Publishing Corp., 1962.
2. Harkins, W. D., *J. Am. Chem. Soc.,* **69,** 1428 (1947).
3. John W. McCutcheon, Inc., "Detergents and Emulsifiers," New York, 1962.
4. Martens, C. R., *Offic. Dig. Federation Soc. Paint Technol.,* **38,** No. 494, 46a (1966).
5. Martens, C. R., "Emulsion and Water Soluble Paints and Coating," New York, Reinhold Publishing Corp., 1964.
6. "Symposium on Surfactants," *Offic. Dig. Federation Paint Varnish Prod. Clubs,* **28,** 417 (1956).

29

*Trade Sales Paints**

Paint products designed for sale to the general public through retail paint outlets (hardware stores, etc.) are known as "trade sales" paints. The manufacturer of a trade sales line should realize that many of his customers are inexperienced in painting practices, and formulate with this fact in mind. The following are some considerations to be weighed in the design of a trade sales line.

(1) Products must be easy to use and should be accompanied by clear directions for use.

(2) Paints must be stable as they may be held in stock for as long as two years.

(3) The products must be safe to use (no toxic materials or low-flash solvents).

(4) The paints should not require "fix-up" by the user.

(5) The products should satisfy current demands, e.g., be available in a number of fashionable colors.

(6) The products must be competitive in price.

(7) Product color as applied should reasonably match color standard card displayed in store.

Trade sales products are also quite often sold to painting contractors, but only if the paints meet the requirements of the professional painter in ease of application, high covering power and competitive price.

A modern trade sales line has a paint to do every job normally needed for the maintenance of the average home. The following outline lists the more important items found in a typical trade sales line.

 (I) Exterior House Paints
  (A) Solvent types for wood (using oils or alkyd resins for binders).
  (B) Water-reducible types (emulsion or latex) for wood or masonry or both (using water emulsions of synthetic resins as binders).

*By H. Roy Hicks, The Sherwin-Williams Co., Gibbsboro, N. J. (Deceased), and D. F. Householder, The Sherwin-Williams Co., Cleveland, Ohio.

   (II) Enamel Undercoaters
  (III) Enamels
      (A) Interior-exterior types (such as porch and floor enamels).
      (B) Interior enamels (sometimes called architectural enamels) supplied in eggshell, semigloss and high-gloss finishes. Architectural enamels are often color matched to the suppliers' flat wall paints.
  (IV) Flat Wall Paints
      (A) Solvent (usually based on alkyd resins).
      (B) Latex (water reducible).
   (V) Varnishes and Stains
      (A) Exterior.
      (B) Interior.

Some trade sales price lists may also include aerosol spray enamels, implement enamels, putty, caulking compounds, metal finishes, roof coatings, swimming pool paints, boat and marine paints, etc. The bulk of this chapter will be devoted to the major usage items, i.e., exterior house paints, enamels and flat wall paints. Enamel undercoaters, varnishes, varnish stains, and stains will be discussed briefly; the miscellaneous low-volume items will be omitted.

Current formulating practices will be reviewed for each item discussed. The reader should be warned that paint formulations are in a continuous state of change. New raw materials, improvements in existing materials, changes in manufacturing techniques, improved painting practices, new substrates to be painted, and cost are just a few of the reasons for formula revision. A short discussion of the major causes of customer complaints will be included under each product type.

EXTERIOR HOUSE PAINTS

This category includes all paints used to protect and decorate exterior surfaces. At one time, all such paints consisted of pigmented oils and varnishes and required turpentine or mineral spirits for thinning. Recently, water-thinnable paints, using latices or water-soluble polymers as binders, have been made available for exterior use. Solvent-type house paints include the following:

  (1) Undercoaters,
  (2) Gloss house paints,
  (3) Flat house paints,
  (4) Shingle and shake paints.

Undercoaters or primers are designed to serve as the first or prime coat on new work, or as the coat under the finish coats on repaint work. An effective undercoater should seal the surface and make it uniform, provide good adhesion for succeeding coats of paint, and offer immediate temporary protection to the surface.

House paint undercoaters are usually based on vehicle systems consisting of blends of heat-bodied and raw linseed oils. The chief reason for combining the two types of oils is to provide control over penetration of the vehicle into the substrate. The use of all raw oil as a vehicle would result in too much penetration, i.e., the oil would tend to be absorbed, leaving the pigment in a relatively dry condition on the surface. A vehicle system of only heat-bodied oil would not penetrate the surface being painted sufficiently to offer good adhesion. For example, heat-bodied oil might not completely bind the powdery chalk which could be encountered in an old paint film on repaint work. Good painting practice would require that chalk on old paint films be removed as much as possible; however, we must be practical enough to admit that for one reason or another, good painting practices are not always followed and that the manufacturer must make his products as "fail-safe" as possible.

Experience has shown that a ratio of about 50% raw oil and 50% bodied oil gives the proper vehicle penetration to an undercoater. Naturally, the 50–50 ratio of bodied to raw oils is not a hard-and-fast rule and may be modified to fit specific job or customer requirements.

The normal PVC (pigment volume concentration) of a trade sales house paint undercoater ranges from 35 to 40%. High levels of zinc oxide are to be avoided in undercoater formulations, as zinc oxide has a tendency to absorb and release moisture, thus causing dimensional changes in the paint film. Oil films containing high zinc levels also become brittle on aging.

The prime pigments used in house paint undercoaters are leaded zinc and rutile titanium dioxide. Since complete hiding is rarely a requirement in undercoaters, the level of the more expensive titanium dioxide is seldom more than $1\frac{1}{2}$ lb/gal. About 50% of the dry content of a house paint undercoater is extender or filler. Factors such as gloss, viscosity, film integrity, shelf stability and stain resistance depend largely on the proper selection of extenders. Those most commonly used in house paint undercoaters are calcium carbonate (whiting), magnesium silicate (talc), diatomaceous silica and mica. Talc and mica reinforce the film and help reduce grain cracking, while diatomaceous silica provides a desirable characteristic known as "tooth." Nearly all modern house paint undercoaters have an eggshell sheen and provide a surface which succeeding coats may grip. Painters refer to this ability to grip or hold the top coat as "tooth."

Recent studies by Van Loo[9] have shown the importance of priming exterior wood as soon as possible. The panels in Figure 29.1 were exposed for varying periods before priming. Those panels which were protected early have held paint over a four-year period much better than the wood which had been weathered before painting.

Attempts to develop water-thinned undercoaters (for use on bare wood) have been relatively unsuccessful to date. Water-soluble pigments found in some species of wood tend to leach through water-base undercoaters and stain the top coat. The lack of penetration of latex vehicles has required modification of latex primer with oils or alkyds in order to properly bind repaint chalk. Such additions have rendered these primers susceptible to attack by mildew, making further treatment necessary to correct the problem. No doubt the appearance of a satisfactory water-based primer for bare wood will occur in the very near future. However, until such primers become available, most manufacturers will continue to recommend the use of oil or alkyd undercoaters for bare wood.

Relatively few problems encountered in house paint failures are traceable to the undercoater. The most common complaint against house paint undercoaters is that they are slow drying. Slow dry may also be associated with ladder tracking, or ladder marks, on the soft undercoater film. Very often these complaints may be due to high humidity and/or low temperature during the drying cycle. Chronic complaints of slow drying should institute a review of the undercoater formula and especially the drier ratios and combination.

Occasional complaints of oil staining may be entered against oil-based undercoaters. Normally, the problem is corrected with an alkyd undercoater.

Solvent-type Gloss House Paints

The diverse uses of solvent house paints have made it impossible for one formulation to serve all requirements. The major formula types of solvent gloss house paints are:

    (1) Self-cleaning white,
    (2) Trim and tinting white,
    (3) One-coat white,
    (4) Fume- and mildew-resisting white,
    (5) Pastel-colored house paints,
    (6) Dark-colored house paints.

Self-cleaning white house paints are the oil-based products generally used for painting wooden siding. This is a bodied–raw oil formulation with leaded zinc extender, and titanium dioxide pigmentation. The pigment portion contains enough anatase titanium dioxide (usually about one-third of the active pigment) to allow a gradual erosion or chalking

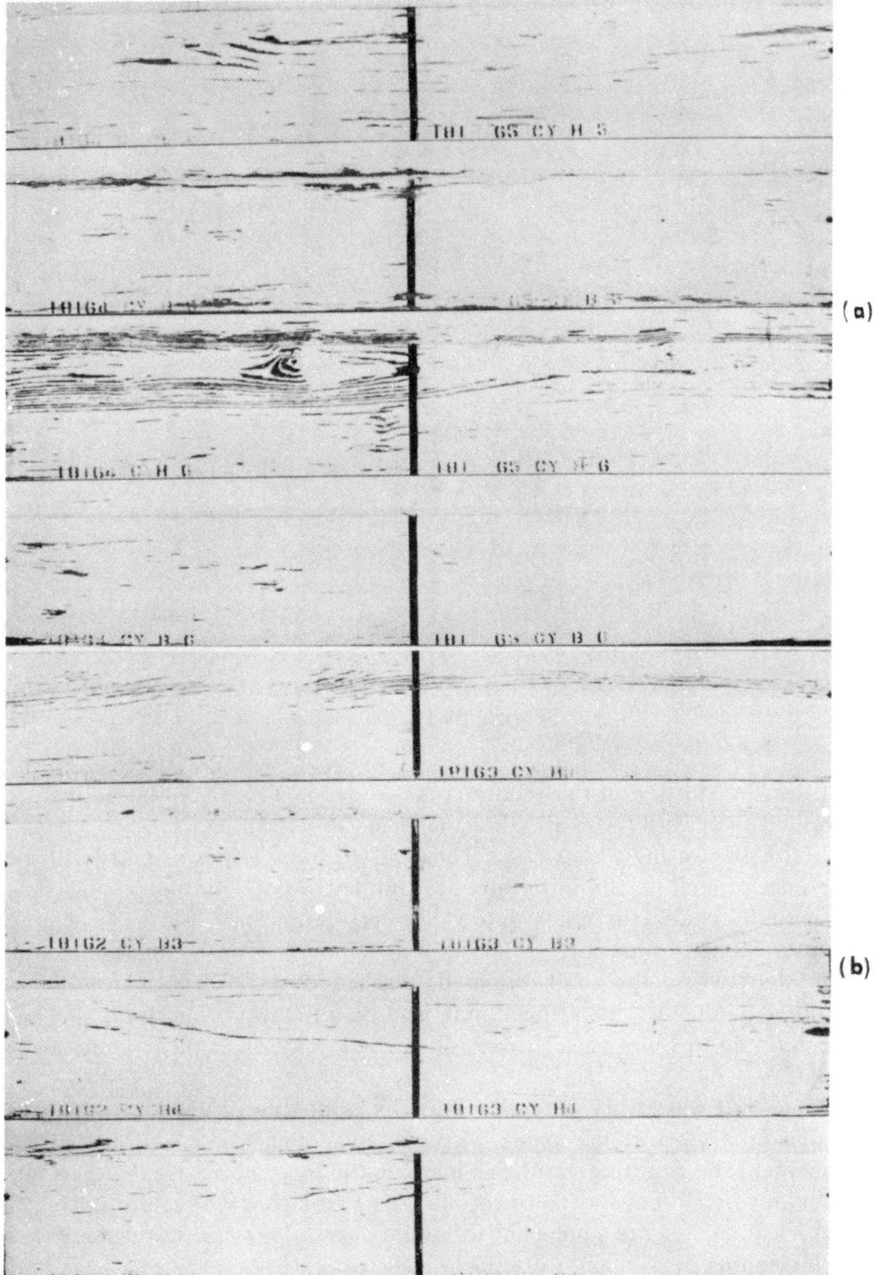

Figure 29.1. After being exposed on paint test fences near Chicago, as indicated, the panels shown above were painted with a first quality oil house paint system and exposed for an additional four years. (a) Boards weathered one month before painting. (b) Boards weathered one week before painting. (c) Boards not weathered before painting. (*Courtesy M. Van Loo, The Sherwin-Williams Co., Chicago, Ill.*)

Figure 29.1. *Continued*

of the film starting in the second year of exposure. Dirt and stains are washed away with the chalk, thus the term "self-cleaning."

Trim and tinting whites differ from self-cleaning whites primarily in the absence of anatase titanium dioxide. While trim and tinting whites may eventually chalk, the chalk rate is far less than that of a self-cleaning white. In many situations, chalking may be an undesirable characteristic, especially where the chalk might be washed down over brick or dark-colored paint. A "chalk face" may also be a liability when the paint has been tinted to some pastel color, since chalk may obscure or change the color.

One-coat whites are generally made by removing or reducing the zinc oxide in a gloss house paint formula and replacing it with titanium dioxide. The resulting paint has increased hiding and is used mostly for repaint work to cover a color in one coat. One-coat whites are normally self-cleaning and are somewhat softer and less mildew resistant than either self-cleaning or trim and tinting whites.

Fume- and mildew-resisting whites are special house paints which resist fume discoloration in industrial areas and mildew of warm humid climates. Lead and mercury are especially sensitive to fume discoloration and are not used in fume- and mildew-resisting whites. These formulations usually contain a high level of zinc oxide (2 lb/gal or more) for mildew protection.

Pastel-colored house paint formulations are very similar to trim and tinting whites. The titanium dioxide level may be somewhat lower than in the white formula, since some hiding is contributed by the colored pigments. The chalk rate of these paints is intentionally low to prevent chalk masking of the color.

Dark-colored house paints are essentially very long oil alkyd enamels. The use of an alkyd vehicle rather than oil greatly improves color and color retention. The PVC of dark-colored house paints is normally from 12 to 15%. Although it is possible to formulate some colors (black for example) at a PVC much lower than 12%, such paints lack the mechanical strength of paints in the 12 to 15% PVC range.

In the discussion of house paint undercoaters, the vehicle was described as a 50–50 blend of raw and bodied linseed oils. House paint finish coats are also based on blends of raw and bodied linseed oils. Bodied oil in a house paint top coat, or finish coat, serves several purposes; e.g., it increases flow, increases gloss and gloss retention, aids pigment wetting, decreases dirt collection, speeds drying, increases exterior durability and gives greater mold resistance.[9] The raw–bodied oil ratio used in house paint top coats is usually about two parts raw oil to one part bodied oil by volume.

Alkyd resins are considerably more resistant on exterior exposure than the various combinations of raw and bodied oils. Most alkyds have a tendency to react with zinc oxide and other heavy metal compounds used as mildew inhibitors. Because of this reactivity, it has been difficult to protect alkyd house paints against attack by mildew. Recently, completely organic mildew inhibitors have been offered to the paint industry. If these materials prove successful, no doubt the use of alkyd house paints will become more widespread.

The extenders most commonly used in house paint top coats are whiting, talc and mica. The proper formulation of exterior paints is a subject still very much under debate. Most formulations are based on the exposure experience of the formulator and his raw material suppliers. Table 29.1 shows the chief differences in the various types of gloss house paints mentioned.

Conventional oil house paints are fairly glossy when first applied and weather to a relatively flat finish, while protected areas retain their gloss. Flat house paints are formulated to a low sheen so that there is relatively little gloss change on weathering and therefore a uniform appearance throughout the life of the paint job. Unfortunately, flat house paints have a tendency to collect dirt more than glossy paints and are more susceptible to attack by mildew (because they offer a larger surface area for

TABLE 29.1.   Comparison of Composition of Gloss House Paints

Product	PVC	Binder	Per Cent Extender	Per Cent Lead Pigment	Per Cent TiO_2	Type TiO_2
Gloss house paint for three-coat system	33	Raw and bodied oil	40	35	25	Anatase and rutile
Trim and tinting white	40	"	45	40	15	Rutile non-chalking
One-coat white	35	"	35	40	25	Anatase and rutile
Fume- and mildew-resisting white	34	"	20	None[a]	20[b]	Rutile
Pastel-colored house paint	37	"	50	40	10	Rutile non-chalking
Dark-colored house paint	12–15	Alkyd resin	30	None	70	Rutile non-chalking
			30	None		

[a] Usually replaced by zinc oxide to the extent of about 40% of the pigmentation.

[b] Often made up of about $\frac{1}{3}$ pure TiO_2 and $\frac{2}{3}$ titanium calcium pigment.

the collection of mildew spores). Unless they are very skillfully formulated, flat house paints may be difficult to apply without lap marks and sheen differences. The low sheen of flat house paints does mask imperfections in the painted surface.

Shingle and shake paints are essentially the same as flat house paints but are usually supplied at lower solids and thinner viscosities. Flatness of this general type of house paints is achieved by shrewd choice of extenders and pigments and formulation at high PVC's (usually above 50%). Vehicles used in the low-sheen house paints (including shingle and shake paints) may be either oil combinations (as in gloss house paints) of alkyd resins. Alkyd resins are used in this type of paint more generally than oils, probably because of the superior color retention of alkyd paints on exterior exposure; however, mildew continues to be a problem.

Most of the field problems encountered with solvent house paints are due in some way to moisture. The tighter construction of modern homes traps moisture, which in many cases can escape only by forcing itself through the walls and paint film on the outside of the house. Showers, automatic washers, driers, humidifiers, home aquariums, cooking and even breathing contribute to the water vapor which must be released from the average home. Faulty flashing, poor caulking and clogged gutters are common construction faults which may contribute to the quantity of water trapped in a building.

Unless it is provided with some other route, moisture vapor will condense somewhere beneath the outside paint film in cool weather. The exact location of the condensation depends on the type and amount of insulation, the heat gradient from inside to outside, etc., or in other words, the location of the dew point within the outside wall of the structure. When outside temperatures go up, the condensed water returns to the vapor state and literally pushes the paint off the outside surface of the building. This action is often seen as blisters in the paint films; each unbroken blister will contain a quantity of water.

Not too many years ago, most homes had open fire-places in nearly every room. While they may not have been the most satisfactory form of heating, they did provide exits for large amounts of water vapor. In modern houses, moisture vapor is supposed to be released primarily through attic vent space. The free open attic vent space should be at least $\frac{1}{150}$ of the ceiling area of the structure. This means that a house of 1100 square feet of ceiling space should have about $7\frac{1}{2}$ square feet of open vent space in the attic.[4] Sometimes wooden siding itself may be vented by forcing wooden or metal wedges between the boards or by drilling holes in the siding.

Many house paint failures are due to faulty application. Paint should not be applied when the ambient temperature is too low or when the humidity is excessively high. A relative humidity of 50% and a temperature of 75 to 80°F are considered good for exterior painting. At times very high temperatures and extremely low humidity may cause difficulties in paint application. Contaminants on the surface to be painted, such as moisture, dirt, grease, etc., will prevent proper adhesion of the coating.

The thickness of the paint film greatly influences its performance. Too thin a film will not protect the surface, while an excessively heavy film may not dry thoroughly. Four to five mils (0.004 to 0.005 inches) is considered optimum film thickness for oil house paints. This is usually achieved by applying one coat of primer and one or two oil top coats. Emulsion house paints almost always require two top coats over a solvent-based primer, and often three coats may be necessary to achieve the desired film thickness.

Studies by Browne[1-3] show that the performance of exterior paints is greatly influenced by the type of wood painted. The different species of wood vary considerably with respect to their suitability for painting. Classification of some of the more commonly used woods as to their suitability for painting (according to the studies by Browne) are shown in Table 29.2. These classifications are based on tests made with two types of paints. Type A fails by checking and crumbling (typical of a white lead and oil paint), and type B fails by cracking and flaking (as would be expected of paints with a high level of zinc oxide).

TABLE 29.2. Classification of Woods as to Paintability

Group I.	Woods giving best performance with both type A and B paints.
	Southern cedar
	Western red cedar
	Southern cypress
	Redwood
Group II.	Woods on which type A paints perform well, but type B does not perform as well as on woods in Group I.
	Northern white pine
	Sugar pine
	Western white pine
Group III.	Woods on which the performance of both type A and B paints is not as good as with Group I.
	White fir
	Eastern hemlock
	Western hemlock
	Ponderosa pine
	Sitka spruce
Group IV.	Woods on which both type A and B paints do not perform as well as on those of Group III.
	Douglas fir
	Southern yellow pine

This does not imply that any and every house paint will give good service on the Group I woods. Poor paints give consistently poor service. Even the best house paints will not always give good service on the woods listed under Group IV. Browne states that Group IV woods require "the most care in the selection of suitable paints."

Emulsion House Paints

Emulsion house paints are based on either styrene-butadiene and polyvinyl acetate copolymers, or acrylic polymer water emulsions. Because it is exceptionally resistant to alkali and is somewhat too brittle for use over wood, styrene-butadiene exterior use is mostly for masonry paints. Acrylic polymers are more resistant to deesterification than the polyvinyl acetate polymers and, for this reason, have been used widely in exterior paints.

Compared to oil house paints, emulsion paints have lower solids, less covering power (requiring at least two coats normally); however, they are more easily applied and the cleanup of brushes and tools is very easy. There is less fire hazard, and emulsion paints can usually be recoated on the same day.

Water-soluble pigments, in some woods (redwood and cedar, for example) leach out and stain emulsion paints. These paints also have a tendency to cause grain cracking on some woods. Grain cracking and water staining may be avoided by using an oil primer or by painting only over previously painted surfaces, but these steps increase the possibility of blistering and peeling because of moisture problems. To prevent rusting, ferrous metals (as found in nail heads, screws, etc.) must be primed with solvent-based materials before receiving emulsion paints.

Early emulsion house paints had poor adhesion to even lightly chalked surfaces. Many manufacturers now incorporate small amounts of oil or resin to improve adhesion over chalk.

The relatively high PVC's found in exterior emulsion paints seemed to indicate fairly porous films. Many manufacturers deduced that the porosity of the emulsion paint films would allow free passage of moisture vapor and thus solve the problem of blisters. Experience has shown that while these films do transmit considerably more water vapor that an oil film (of lower P.V.C.), they have a limit and when this limit is passed, emulsion paints will blister also.

Pigment dispersion in water systems is somewhat more complicated than in solvent systems. As will be discussed later, dispersion of many of the most common pigments is accomplished only through the aid of various wetting agents. If not properly used, dispersing agents may give rise to field complaints of low opacity (pigment flocculation), poor color, bad adhesion, sensitivity to moisture, etc.

Emulsion paints are subject to freezing and if the freeze-thaw cycle is repeated a number of times, the latex vehicle may coalesce or "kick out." Various materials, including glycols, are put in emulsion formulations to give some protection. Even so, care should be exercised to avoid freezing emulsion paints.

UNDERCOATERS FOR ENAMELS

The function of an undercoater for an enamel is to provide a base suitable for recoating with enamels or other interior paints. The modern enamel undercoater closely resembles a quality alkyd flat and usually differs only in the ability to "hold out enamel" and to sand freely. Many alkyd flats sold to the professional trade can be used as enamel under-coaters. Thus, a room may be sprayed with alkyd flat and the wood trim later enameled with no need for special undercoaters.

The normal PVC range for enamel undercoaters is on the order of from 30 to 50%. Flatness is important to prevent creeping and crawling of enamel top coats. Good sanding qualities are usually achieved by in-cluding relatively large amounts of zinc stearate (30 pounds or more per 100 gallons of undercoater).

The vehicle portion of the enamel undercoater is similar to that of a quality eggshell enamel, being primarily medium oil alkyd of the variety used in flat wall paints with some long oil aklyd used at times to improve flexibility.

A few years ago when plaster walls were common in private dwellings, it was necessary to protect the paint from the alkali of the plaster with a wall primer-sealer which was often made from a tung oil-modified phenolic resin. Today, wall primer-sealers are used chiefly to make ceiling and wall surfaces uniform.

The advent of nonpenetrating flat wall paints has reduced the need for wall primers and sealers, and the use of alkyd flats as enamel undercoaters has brought about a definite change in the formulation of first coats for interior surfaces. Considerable effort is being devoted to the development of an emulsion flat which would serve as an enamel undercoater for interior wood, as a primer-sealer for dry walls and as a flat top coat. The fast drying of an emulsion undercoater would allow complete finishing of a room in a single day. The problems to be overcome with interior emulsion undercoaters for wood are similar to those encountered with exterior wood, though usually not so serious. Woods containing water-soluble pigments may tend to stain emulsion undercoaters, and the water in the emulsions may give rise to unsightly "grain raising" in some varieties of wood.

Unlike the exterior situation, where water is a problem for the life of the coating, interior emulsion undercoaters offer a moisture problem only during application and drying. Certain compounds of lead seem to be effective in reducing wood staining, and the search for a completely non-toxic material for this purpose is continuing.

ENAMELS

Interior-exterior Types

One of the chief problems in the formulation of exterior enamels is the control of chalking. Chalking in colored enamels may cause a "chalk face" or white to mask the original color, or when used for trim, a danger exists that colored chalk may stain lighter colors of the same building.

The use of long oil alkyd vehicles in colored enamels for wood is almost universally accepted. Porch and floor enamels require more hardness and are usually shorter in oil length than general-purpose exterior enamels. Some alkyd resins are very slippery when wet with water and if used carelessly in floor enamels may constitute a safety hazard. Modification with oleoresinous vehicles or with pigments to give nonskid characteristics may allow the use of alkyds in floor enamels. Since alkyd resins are liable to attack by alkali in cement and masonry floors, many producers offer a special undercoater to be used on such surfaces before the application of the alkyd floor enamel. Such undercoaters may be unpigmented or lightly pigmented solutions of chlorinated rubber or other resins which resist mild alkali well. Excellent floor enamels are being made on vehicles of chlorinated rubber, solvent-type styrene-butadiene resins, emulsion styrene-butadiene resins, catalyzed epoxy resins, epoxy esters, polyester resins and urethane resins.

One of the most common complaints encountered in the field in the area of colored exterior enamels is the loss of drying speed with age. This is especially chronic in dark colors where the surface area of the pigments used is extremely large (such as in dark greens, blacks, deep reds, etc.). It is generally believed that the driers in the formulation are absorbed on the pigment surface and are therefore not available to the vehicle system.

Interior Enamels

Most of the high-quality interior architectural enamels of today are formulated with color-retentive alkyd resins. In the past, formulations often included rosin or rosin esters to lower cost and increase gloss. The price of phthalic anhydride is now so low that a color-retentive long oil soya alkyd is an exceptionally good buy and often priced below rosin or rosin products. Elimination of rosin in enamel formulations has resulted in a loss of initial gloss, increased gloss retention and vastly improved color retention.

A typical long oil soya alkyd for architectural enamel use is about 60 to 65% oil and about 25% phthalic anhydride. The solvent used is usually mineral spirits.

Some enamel vehicles may contain from 30 to 35% (based on total vehicle solids) of a medium oil length soya alkyd (30% phthalic anhydride). The use of the medium oil length alkyd increases flow in the enamel and reduces brush marks in the dried enamel film. Quite often a semigelled flat wall alkyd is used for the medium oil length vehicle. Poor choice in the selection of the medium oil length vehicle can easily result in poor application characteristics in the final enamel.

Drier composition and concentration depend to a great degree on the enamel vehicle composition. The tendency today is to avoid the use of lead drier in interior enamels and to use calcium and cobalt instead. The dark color of manganese driers and their tendency to stain white and bright colors has reduced their use for interior products. Because of the potential danger to children, the use of toxic materials (such as mercury or pigments containing arsenic) is avoided in interior architectural enamels.

The PVC of a high-gloss enamel may be as high as 25%, depending on pigmentation. It is generally agreed that even though it is possible to obtain opacity in some colors and black with extremely low pigment volumes, best general performance results when enamel PVC is kept above 10 to 12%.

Rutile titanium dioxide is the principal hiding pigment used in most gloss white enamels. Even the highest-quality products seldom contain more than 3 pounds of rutile titanium dioxide per gallon. Low-cost and lower-quality enamels may often be formulated with titanium calcium pigments.

Relatively small proportions of zinc oxide in enamels (0.1 to 0.5 lb/gal) help reduce after-yellowing in white enamels but may cause a surface haze which lowers gloss. The use of zinc may also result in hard, brittle films. The choice of the proper zinc oxide pigment is very important since some grades may react with alkyd vehicles to form seeds.

The use of high-gloss enamels on large surfaces (such as walls and ceilings) may accentuate irregularities in the surface. Because of this problem, semi-gloss and eggshell enamels have largely replaced high-gloss products in popular use. Reduction of gloss is accomplished by increasing the PVC. Normally, that of a semigloss enamel is from 30 to 40%, while an eggshell enamel ranges from 40 to 55%. Considerable variation in PVC is possible in each gloss range depending on choice of extender and pigments. Use of large particle size pigments or extenders with high oil absorption causes rapid gloss reduction at relatively low PVC's. As the gloss of the product is reduced, it is possible through the careful selection of extenders and pigments to get increased hiding power without increasing the hiding pigment. This is due to the opacity of the air-solid interface of the extender and is known as "dry hiding" since it is not apparent until the film dries. This principle is much used in the formulation of low-sheen and flat paints.

Many white enamels contain small amounts of blue, violet, black or other colored pigments which tend to make the enamels appear whiter and greatly increase the hiding of the product. Drier choice and concentration in enamels in all gloss ranges should be carefully controlled. The rate of dry of enamel films may greatly influence gloss and gloss retention.

Discoloration is one of the most common field complaints entered against architectural enamels. Natural gas heat may cause discoloration of products containing lead. Ammonia, moisture and fumes from cooking are a few common household agents which may discolor paints. Cigarette smoke has recently been found to severely yellow most architectural enamels and is probably the most potent yellowing agent in the average home. The author is not aware of any sure method to avoid enamel discoloration by cigarette smoke through formulation.

In recent years, the great challenge to the trade sales formulator seems to have been the successful production of an emulsion enamel. One of the primary problems is that water is such a fast-evaporating solvent that it is difficult to brush a water-based enamel without leaving objectionable brush marks. As Payne[7] mentions, "The production of a high-gloss finish requires that the surface of the coating consist of a thin layer that is almost free of pigment and extremely smooth. Emulsion coatings contain considerable water after coalescence has occured and since water is insoluble in the resin phase it must force its way out through escape channels. These channels contine to shrink after the coating dries and disturb the smoothness of the coating; therefore, it is not likely that high-gloss finishes will be made with emulsions."

Convertible water-soluble polymers may be part of the answer to high-gloss, water-thinned finishes, but the problem of the fast evaporation rate of water must still be resolved. These materials have been on the market for a relatively short time, and at this stage it is impossible to evaluate them completely.

FLAT WALL PAINTS

Alkyd

Alkyds for modern flat wall paints are usually cooked almost to the gell point. It is not unusual to find an alkyd with a viscosity of Z to Z_3 (G.H.) at 30 or 35% solids. They are generally soya alkyds of about 30% phthalic anhydride content with light color (normally 4 to 6) and low acid values (4 to 6). In some cases, the very low acid values encountered in alkyd flat vehicles may cause pigment dispersion problems. This may be noticed when brush and roller are used together and especially when the paint is a medium depth color. Usually, the greater work applied to the paint by brush application will cause this area to be deeper in color when a roller is used. Many commercial dispersion agents are available to correct such dispersion problems; however, some study may be necessary to select the most suitable product for a specific case.

There are two basic reasons for cooking flat wall alkyds to such extremes in viscosity. The increased polymer size obtained in semigelled resins prevent vehicle penetration. Since modern construction involves the use of many porous materials for walls and ceilings (such as paper products, sheetrock, fiberboard, etc.) in combination with nonporous joint cement, plaster, etc., nonpenetrating vehicles contribute greatly to uniformity of color and sheen in alkyd flats over surfaces of varying porosity. The pigment load in alkyd flats must be relatively high with a correspondingly high PVC; thus viscosity may be achieved in the flat paint at fairly low vehicle solid levels.

The PVC of alkyd flats is almost always above the CPVC (critical pigment volume concentration). The voids between pigment particles, not filled with resin solids, provide considerable opportunity for solid-air interfaces and therefore dry hiding.

The use of titanium calcium pigment is more common in flat wall paints than in other trade sales products. The coarse, dry-ground varieties of calcium carbonate contribute greatly to flatness (low sheen) but are difficult to keep in suspension. Talc, diatomaceous silica and colloidal silica are effective flatting agents commonly used in alkyd flats.

The ideal alkyd flat film should have no low-angle sheen and thus would obscure surface irregularities on large planes, such as the unbroken lines of a large ceiling. The mechanics of hiding of titanium dioxide limit the amount and particle size of the extenders and pigments used, if one-coat opacity is to be maintained. As with most paint formulating problems, a compromise usually allows some low-angle sheen, normally about 6 or less on an 85° Photovolt glossmeter, with a small loss of hiding.

Choice of extender pigments may greatly influence such paint characteristics as degree of penetration of the substrate, color uniformity, sheen uniformity, washability and light reflectance. Some tend to orient themselves with brush strokes and cause irregularities in color and sheen. Kaolin and other clays are choice extenders for use in water emulsion paints, since they contribute to hiding power; however, they may settle badly in alkyd paints and are seldom used there.

Alkyd flat wall paints are generally preferred by professional painters for areas of hard wear or for difficult to paint locations. The cost of applying paint today is from four to twenty times the cost of the paint itself; therefore, the ability of a paint to do a one-coat job is of considerable interest to a painting contractor.

Latex

The earliest known attempts to paint were made with simple mixtures of colored pigments and water. These paints could not resist scuffing, rubbing, etc., so later artists mixed their pigments with milk or egg binders. Some paint made in this fashion are still in relatively good condition after 500 years.

The use of emulsion techniques to prepare synthetic vehicles or binders allows easy handling of polymers of very high molecular weight. The viscosity of the emulstion is not dependent on that of the dispersed polymer, but rather on that of the protective colloid used.

Most of the interior latex flats are based on one of three basic emulsified polymer types.

> Styrene-butadiene
> Polyvinyl acetate
> Acrylic polymer

Styrene-butadiene is extremely resistant to alkali and water. The cost is relatively low, but the manufacture of the polymer is complex. An aging styrene-butadiene film tends to continue to cure and thus may become brittle and yellow.

Polyvinyl acetate is sensitive to water and alkali, since it is based on an ester monomer. Cost of this polymer is moderate and manufacture is relatively simple.

While the acrylic emulsion polymers are also based on esters, they are not as sensitive to moisture and alkali as the polyvinyl acetate polymers. The acrylic polymers transmit UV light and are extremely resistant to discoloration.

The formulation of emulsion paints is a science in itself. The presence of water sets up the possibility for all sorts of reactions between the components. Pigment loading of emulsion paints, even with properly selected pigments, is limited by the mechanical stability of the emulsion binder. Generally, it is found that most flats are pigmented in the 30 to 50% P.V.C. range. Because of the relatively low solids in emulsion flats, there is usually some difficulty in obtaining efficiency of TiO_2 pigments at levels above $2\frac{1}{2}$ lb/gal. Aluminum silicate and various clays contribute to hiding and are easily suspended in emulsion paints. Various grades of diatomaceous silica are common as "flatting agents" in emulsion flats. Some grades of talc and calcium carbonate may be used in water-thinned flats. Divalent ions should not be released by pigments used in emulsion paints.

In a solvent paint, air on the surface of the pigment particle is replaced with oil or resin in the pigment dispersion stage of paint manufacture. In emulsion paint production, the air on the pigment surface must be replaced with water or protective colloid dispersed in water. The process is complicated by the fact that most paint pigments are hydrophobic and resist "wetting" with water. The use of wetting agents allows relatively easy dispersion of the most commonly used pigments. Improper use of wetting agents may cause pigment flocculation and loss of opacity, color and hiding; increased sensitivity of the cured film to moisture; impaired adhesion, and many other problems.

The use of soaps, wetting agents and protective colloids in emulsion vehicles makes them foam when agitated. Many commercial "antifoaming" agents are available to reduce foaming in emulsion paints. Small amounts of kerosene or mineral spirits have been used effectively for this purpose. Usually, the commercial antifoaming agents are more efficient and do not create dispersion problems if properly used.

As previously mentioned, care should be taken to protect emulsions and emulsion paints from freezing.

While the basic polymer in an emulsion paint is not usually susceptible to attack, the protective colloid, plasticizer or other paint components may be food for bacteria. Various proprietary compounds furnish protection against bacterial attack.

VARNISH

Exterior

The use of unpainted wood in exterior construction has created a demand for a clear finish which will preserve the natural beauty of the wood. The primary problem is not that of finding or making a varnish capable of withstanding exterior exposure, but rather of preventing degradation of the wood itself by UV light. The best finishes currently available are tung-phenolic varnishes. Their relative success is probably due to their dark color which protects the underlying wood from light. Even under several coats of tung-phenolic varnish, the wood at the varnish interface normally breaks down within two years and the varnish film loses adhesion. Possibly some method will be found to make clear varnish films sufficiently UV absorbent to protect underlying wood. Efforts in this direction include work with low-opacity pigments of various types, as well as the use of the UV absorbers now on the market.

Interior

Alkyd resins are probably the most widely used single product for interior varnish work. Alkyds are easily modified with other vehicles or with extender pigments to give a wide range of film characteristics. Satin-finish varnishes are very popular for use on paneled walls. The low sheen of these products is usually achieved through the use of flatting pigments such as colloidal silica, diatomaceous silica and sometimes stearate soaps.

Shellac has long been the standard for finishing wooden floors, although it is gradually being replaced by alkyds, lacquers and more recently polyurethane resins. Shellac is fast drying and easy to sand, and it does not yellow appreciably on aging. The chief disadvantage in the use of shellac is its poor resistance to water. White shellac is supplied as "3-pound cut" or "4-pound cut." The 3-pound cut is lower in solids and viscosity and is ready for brush application. The 4-pound cut shellac requires about 25% reduction with denatured alcohol before it can be applied by brush.

Shellac gradually loses its ability to dry with age, and for this reason, many manufacturers date their shellac when it is packaged. White shellac more than a year old should be avoided.

Some varnishes, especially shellac, pick up small quantities of iron from processing equipment, containers, etc., and when used on light woods with high tannic acid content (such as oak), an unsightly black stain develops. Often the stain may be removed by applying a wash coat of oxalic acid in either the varnish or solvent.

STAINS

The vehicle composition of a quality exterior stain is similar to that of a good house paint, but with very low solids content. Raw and polymerized linseed oils are used in a ratio of about 3:1. Vehicles for interior stains are usually very low solids alkyd resins. Pigment loads are kept very low in stains to insure a degree of transparency and to avoid undesirable film build. Lightfast pigments, such as iron oxide, titanium dioxide and carbon black, are used to color exterior stains. The wide variety of colors required and the reduced need for lightfastness allow the use of organic dyes and pigments in interior stains. The chief function of a stain is to furnish color, and seldom is opacity a required or even desirable characteristic. The popular use of prefinished wood paneling in recent years has greatly reduced the volume of stains sold through trade sales outlets.

SUMMARY

Most complaints against interior products are due to faulty application, poor surface or surface preparation or unsatisfactory package condition of the product itself. As mentioned before, many trade sales products may remain on a shelf for as long as two years before they are used. Even the most carefully formulated and manufactured product may occasionally undergo storage changes which make its use doubtful, e.g., pigment settling, color change, loss or gain of viscosity, and loss of drying ability.

It is difficult to predict what may happen to a paint over an extended period of time. Many formulators run short-term stability studies on their more promising formulations. In these tests, samples of the paint are stored at elevated temperatures (120 to 130°F) for a period of from one week to several months. The idea behind this sort of test is that chemical reactions in the paint will take place more rapidly at elevated temperatures. Unfortunately, temperature is not the only factor involved in paint stability, and while these accelerated stability tests are certainly useful, they by no means indicate the behavior of a factory batch of paint on extended aging. Sometimes cycles of heat and cold will trigger undesirable reactions in paints where heat alone does not. The effect of alternate freeze-thaw cycling on emulsion paints has been mentioned. In some formulations, the seeding of zinc soaps will take place only if the paint is subjected to alternate cycles of heat and cold. Continued storage at elevated temperatures will not produce seeds even after 30 days.

Much effort has been devoted to finding an accelerated pigment settling test that can be correlated with actual field conditions. The usual approach has been to design some sort of special container, which is filled with paint and centrifuged for a specified time at a specified speed. The paint is then compared to a sample of known performance. Even this type of test does not accurately predict the tendency of a paint toward pigment settling. Many factors are involved other than the density of the pigment combination and the viscosity of the paint vehicle system. One formulator known to the author runs settling tests on his paints by putting unknown and standard samples in the trunk of his car for a few days. Apparently the type of agitation provided by this unique test causes pigment sedimentation in some types of paint.

Color change in the package on prolonged storage is a very difficult problem to resolve and at times may even be hard to detect. Although instrumentation in color work has become very popular in the past few years, the less expensive and more widely used instruments do not give absolute readings and therefore the results may not be reproducible after a few years. Many systems of preserving color standards for practical production use have been suggested and tried. One of the less complicated systems has been reported by Huey, of The Sherwin-Williams Co., who has proposed that actual paint chips of the desired color be stored in a common household food freezer to prevent color drift. Naturally, some precaution must be taken to avoid condensation of moisture on the chips and the chips must not be allowed to thaw until they are needed for color standardization work. Once used, such frozen chips must be discarded.

The maintenance of color standards is becoming increasingly important to the trade sales producer. As time goes by, the public seems to become increasingly color conscious. As many as 2000 separate colors are not unusual in a modern trade sales line. No dealer is eager to carry 2000 colors in stock in latex flat, alkyd flat and enamels of several sheens. By using an intermix system, where a number of basic deep colors (10 to 12) in each type or quality of paint may be mixed with white and/or each other, a large number of colors are made available, and the dealer need stock relatively few items. This system may be augmented by adding small amounts of universal tinting colors to the various mixes to obtain a still larger range.

To correlate such a system in latex flat, alkyd flat, semigloss enamel and perhaps eggshell enamel is a chore of considerable magnitude. Very quickly one learns that dry hiding, vehicle composition, extender pigment content, etc., all have a marked effect on the appearance of a color. Even though an alkyd flat, a latex flat, and a semigloss enamel may all contain exactly 2 pounds of a specified titanium dioxide, the addition, of say, 2 oz/gal of universal tinting color blue to each of the paints may well result in three different shades of blue. The problem does not end here, for even after the formulations have been adjusted to give good visual color matches when tinted, a method of color control must be established. More often than not, good visual color matches between products of widely varying sheen (such as an alkyd flat and a semigloss enamel) will give readings on a colorimeter which may appear to be only vaguely related.   Thus, if color instrumentation is contemplated as a means of control, separate standards should be established for each color in each type of paint.

Often formulators are satisfied because they may have successfully adjusted their two or three basic formulations so that they give excellent visual matches when tinted with blue. Chances are very good that these same three paints will not give acceptable matches when tinted with identical amounts of other colors.

Impossible as it may seem, most reputable paint manufacturers have, in some way, resolved these color problems. If all the problems have not been completely solved, at least we have found enough compromises to allow the sale of color-correlated paint lines with as many as 3000 possible predictable color mixes.

Loss or gain of viscosity of paints is sometimes attributed to "after-wetting" of pigments or the reaction between components of the paint. Certainly, if the problem is due to a chemical reaction, then the accelerated temperature storage test should uncover problems rapidly. Not so easily resolved are the possibilities of improper dispersion of the pigment or of mutual solubility of multiple vehicle systems.

Fortunately, loss of drying of a paint is usually easily corrected by the addition of driers. Exactly what to do with a semi-dry garage door (while receiving advice from the irate homeowner) at minimum expense, trouble and inconvenience to the homeowner can be a perplexing problem.

Although not too many homes are currently built with plaster walls, the painting of plaster is still a problem to the trade sales formulator. At one time, the chief worry in dealing with a plaster surface was that the extreme alkalinity would attack the paint. Alkyds, oils and many other paint vehicles are organic esters and are subject to deesterification by strong alkali. Washing the plaster wall with a solution of zinc sulfate was the standard treatment if the wall was suspected of being highly alkaline.

In many cases, the plaster was so smoothly troweled that the paint film could find no tooth to develop adhesion; as a result, the stresses and strains set up in the drying of the paint literally pulled it off the wall. This was especially true with early emulsion paints which had poor adhesive properties.

Today, it is not uncommon to run into problems caused by the deterioration of old plaster. In these cases the plaster has started to chalk, and the bonds with paint films on it are lost. Usually no problem occurs until a fresh coat of paint is applied and new forces of stress and strain break the weak attraction between the paint and the old chalky plaster.

As with exterior paints, moisture behind an interior paint film will cause trouble. Sometimes the problem may be the leaching of water solubles (as was mentioned in alkyd flat paints), or the paint may be forced off the surface if the hydrostatic pressure is sufficient.

Painting the wide variety of surfaces found in modern construction can offer some challenging problems. Dry wall construction, so widely used today, presents quite a range of porosities. Although many of the paints on the market give satisfactory results on such surfaces, we should not be too surprised when a problem of sheen or color difference does occasionally occur. Usually, this is easily resolved by the use of a wall primer (either alkyd or latex) to render the surface uniform.

In recent years, the use of plastic wall coverings has become widespread. These materials are usually vinyl sheeting and are very nonporous. They are being successfully painted; however, it would be wise to test a prospective paint thoroughly before suggesting it for this use.

The equipment used to manufacture trade sales paints has undergone considerable change since World War II. The trend is to replace time with horsepower. The use of high-speed dispersion equipment in which 2000 gallons of paint may be ground in a matter of 20 minutes or less has brought on some problems. Such high-speed, high-power dispersion develops a great amount of heat, and even though some provision for cooling is provided, localized heating may occur near the center of the tank, or cooled paint may cling to the sides of the tank and form a very efficient insulator. Temperatures as high as 180°F are not unusual in the tanks of high-speed dispersion equipment. At such temperatures, reactive materials tend to act quickly. Great care should be exercised in the selection of materials to be processed in high-speed equipment.

As mentioned earlier, the parade of new raw materials being made available to the trade sales formulator seems endless. New methods of application and improvements of old methods allow and even demand changes in thinking. New building materials at times seem to be eliminating the need for specific trade sales products; yet, a few years later these same materials may create a new market perhaps because the housewife has become tired of the color.

Two-component systems were formerly reserved for use by industrial or industrial maintenance painters because they were "too hard to use." Today, an impressive volume of these two-component products is sold to and successfully used by homeowners.

While we may be blessed with an abundance of new materials and knowledge, there is an equally impressive list of problems to be solved. Southern pine and Douglas fir are still difficult to paint. We have yet to find a completely satisfactory method of dealing with knots in exterior wood. The trade sales market for fire-retardant paints has scarcely been touched. The production of decorative effects through the use of incompatible systems is an open field. Inorganic paint systems are in their infancy. Everyday, the list of unsolved problems seems to grow and with it, the opportunity increases for the alert and well informed.

Acknowledgment.   The author is indebted to Mr. W. W. Cranmer, The Sherwin-Williams Co., Gibbsboro, N. J., for his help in preparing the outline and organizing the material for this chapter.

REFERENCES

1. Browne, F. L., in (Mattiello, J. J., editor) "Protective and Decorative Coatings" Ch. 18, Washington, D. C., U. S. Government Printing Office, 1945.
2. Browne, F. L., "Wood Properties that Affect Paint Performance," Washington, D. C., U. S. Department of Agriculture Forest Products Laboratory, No. 1053, reviewed and reaffirmed June 1958.
3. Browne, F. L., "Miscellaneous Publication No. 629," Washington, D. C., U. S. Department of Agriculture Forest Service, issued July 1947, revised June 1962.
4. Hines, Edward Lumber Co., "A Case for Wood," Chicago, Ill., Edward Hines Lumber Co.
5. Matiello, J. J. (editor), "Protective and Decorative Coatings," Vols. 1–6, New York, John Wiley & Sons, 1941.
6. Parker, D. H., "Principles of Surface Coating Technology," New York, Interscience Publishers, 1965.
7. Payne, H. F., "Organic Coating Technology," Vol. 2, p. 1130, New York, John Wiley & Sons, 1961.
8. Payne, H. F., "Organic Coating Technology," Vols. 1 and 2, New York, John Wiley & Sons, 1954 and 1961.
9. Van Loo, M., "Paints for Today's Architect," Presentation at the Engineering Institute, University of Wisconsin, Madison, Wisconsin, Nov. 30 to Dec. 1, 1961.
10. Von Fischer, W. (editor), "Paint and Varnish Technology," New York, Reinhold Publishing Corp., 1948.

30

*Industrial Finishes**

Industrial finishes, by definition, are paint products which protect and decorate a manufactured article. These coatings can be applied to many different substrates such as metal, wood, paper, plastic and rubber. They may be part of an appliance, automobile, furniture or toy. An industrial finish differs from a decorative finish with respect to its function, since protection may be the dominant requirement and appearance secondary. There are also finishes which are essential to the function of the product, i.e., those which are chemically resistant, light reflecting, photosensitive or insulating in nature.

Industrial finishes, therefore, constitute a complex field because of the continued growth and availability of new pigments, vehicles, solvents and additives which can be used by the formulator. This is a highly competitive field where formulation and manufacturing must combine to meet customer requirements at the lowest possible cost. Recent trends, however, indicate that progressive users of industrial finishes are willing to pay higher prices for increased performance.

Because of the vastness of industry and its multitude of products, complete coverage of the subject is beyond the scope of this chapter. A brief history, a review of some of the available film formers and some typical finishing systems, and a general treatment of the application of these systems will be covered. Some of the problems facing the industrial formulator in this work will also be considered.

HISTORY

The first stages in the development of the modern industrial finish occurred before the end of World War I. At that time paints and varnishes were based mainly on naturally occurring vehicles and a range of pigments and dyestuffs which was quite small by today's standards. It was

*By Floyd C. Bertsch, The Sherwin-Williams Co., Chicago, Ill.

impossible to formulate finishes which could compare to today's products in appearance unless extremely long drying schedules were used and a multiplicity of coats employed. For example, a durable automotive finish consisted of 12 coats or more, each being allowed to dry for 24 hours and each rubbed down to remove blemishes caused by deposition of dust particles during the long drying period.

Other faster-drying finishes in which conditions of use were not stringent were based on linseed oil varnishes, dammar, asphalt and lime-hardened rosin. The standard clear wood finish was a solution of shellac in denatured alcohol which necessitated large numbers of skilled polishers who built up a finish by applying numerous coats.

The disadvantages of these products were manifold. The amount of labor required in their application was large and long drying schedules limited output. On the other hand, the cost of labor was low, Since the tastes of the period were for darker colors, available materials, especially in the pigment area, were acceptable for their purpose. After 1918, the cost of labor rose rapidly and the era of mass production began. Industry began to develop rapidly.

A major step was taken when cellulose nitrate became commercially available, along with suitable solvents, for the manufacture of lacquers. It then became possible to prepare fast-drying finishes which performed comparably to oleoresinous air-drying paints. Finishing schedules were cut from weeks to hours, and consequently many manufactured articles could be offered at reduced cost to the public. The farsighted paint manufacturer soon realized that his future lay largely in technical development. Employment of chemists and technicians in well-equipped laboratories became an accepted feature of the paint industry.

Other resins, such as phenolics, alkyds and urea-formaldehyde, followed. Greater latitude in the choice of solvents was made possible by increased activity in the petroleum industry. Film formers such as silicones, epoxies and vinyl copolymers appeared. The pigment industry was also active. Numerous improved pigments enabled the preparation of white and pastel shades. Finishing processes were revolutionized by the development of improved methods of application. This constant progress has made a wide variety of coatings available to the industrial finisher.

FILM FORMERS

Film formers are the backbone of all industrial finishes. Some of the more important types are discussed in the following sections.

Alkyd Resins

These resins may be modified with a number of oils available on today's market. Some of these are soya, linseed, dehydrated castor and coconut oils. They may be combined with such resins as acrylics, vinyltoluene, silicones, amino resins and phenolics. It is because of this latitude of compatibility that oil-modified alkyd resins are used in the largest volume in industrial finishes. They are of relatively low cost and have a variety of properties. Choice of the type of alkyd resin depends on the end use and the desired final cost.

Acrylic Resins

Acrylic resins have become extremely popular within the last few years as a type of industrial finish film former. They have been used in the plastics industry for many years. Some of the reasons for their use are their toughness, good weathering ability, and resistance to abrasion and chemical attack. They are generally considered to be superior to alkyd resins in the area of gloss retention. For many years, industry relied on amino-modified alkyds, phenolics and epoxy systems for physical and chemical resistance properties. Acrylics of the thermosetting type now meet these needs at competitive costs.

There are basically two types of thermosetting acrylic resin, defined by the points of reaction for cross-linking with resins of the epoxy or amino type. They are either carboxyl or hydroxyl in nature. Both kinds have become very popular in the appliance field because of excellent physical and chemical properties.

Thermoplastic acrylic polymers found large-scale application in the automotive industry beginning in 1957. They were of the lacquer type and exhibited as their outstanding properties nonyellowing, superior gloss retention, corrosion and gasoline resistance.

Amino Resins

These resins are generally employed as cross-linking agents in baking finishes. They are used in proportions up to 30% of the total vehicle binder. They may be used with alkyds, epoxies, thermosetting acrylics, phenolics and other heat-reactive resins. Melamine and urea-formaldehyde are the most popular examples of this type. They are products of the condensation of urea or melamine with formaldehyde.

Epoxy Resins

Epoxy resins are noted for their excellent corrosion resistance. They are among the best film formers for abrasion and chemical resistance. Because of poor exterior properties, primarily due to severe chalking, epoxies are used mainly as primers or undercoaters. They have found acceptance as primers in the appliance field because of their excellent detergent and corrosion resistance at film thicknesses of 0.5 mil dry or less. The epoxy resin is usually cross-linked with a melamine or urea resin upon baking at schedules of 20 to 30 minutes at 350 to 425°F.

Epoxies have also been used to a great extent in air dry maintenance finished where the resin is catalyzed by amines or polyamides. Epoxies will undoubtedly share in the continued growth of chemical and maintenance coatings, and in those products where high performance and durability are required.

Phenolic Resins

Although these were synthetic resins, they are still important because of their excellent chemical resistance. When combined or coreacted with epoxy resins, they form a coating resistant to chemical attack, high temperature and abrasion.

Vinyl Resins

Vinyl resins are generally copolymers of monomers such as vinyl chloride, vinyl acetate and vinylidene chloride. They have become very important since World War II as plastics and adhesives as well as film formers for industrial coatings. As a class, vinyls exhibit toughness, abrasion resistance, color retention and good exterior durability. These properties have enabled them to be used as process equipment finishes, appliance finishes (to a limited extent), collapsable tube coatings and, more recently, coil coatings.

The coil coatings industry, which coats continuous metal strip, has had rapid expansion and along with it has grown the volume of vinyl resin usage. The term vinyl coatings encompasses several basic types including solution vinyls, organosols and plastisols.

Miscellaneous Resins

There are many other classes which are utilized in industrial formulations for one reason or another. Some of them are:

(1) Nitrocellulose—Used as a film hardener in conjunction with other resins and plasticizers.

(2) Polyesters—Good heat stability and resistance to color change on exposure in the saturated type.

(3) Silicones—Also used in heat-resistant finishes; superior to the polyesters in this respect. They have good resistance to weathering.

(4) Urethanes—Only recently introduced but growing rapidly since their basic ingredient toluene diisocyanate continues to decrease in cost. Urethanes have outstanding toughness, abrasion and chemical resistance. Their main drawback, however, is poor exterior properties due to severe chalking.

(5) Chlorinated rubber—Usually used in conjunction with alkyds to form a coating resistant to abrasion, moisture, acid and alkali.

AIR DRY INDUSTRIAL AND AUTOMOTIVE REFINISHING SYSTEMS

A large portion of industrial finishing consists of air dry primer surfacers and top coats for surfaces which are too large to cure the film by baking. This area involves the initial finishing and refinishing of automobiles, trucks, trailers, farm implements and road equipment. Substrates are largely iron and steel where the surface preparation ranges from cleaning and phosphating to no cleaning whatsoever. Finishing systems also vary in quality from primer surfacer plus several coats of enamel to single coats of enamel only.

Air dry primer surfacers are required for proper adhesion of top coats. This is especially important in the auto refinishing field where repairs are made on both small and large areas. Substrates, whether bare steel or old finish, are power sanded to provide a smooth surface offering good adhesion to the new finish.

An example of a fast-drying primer-surfacer with good adhesion plus easy sanding and good holdout properties for the enamel is Formula No. 1.

FORMULA NO. 1.   Automobile Refinishing Primer-Surfacer

Material	*Pounds*
Synthetic iron oxide, black	100
Titanium dioxide	10
Barytes	200
Talc	200
Aluminum silicate	50
Half-second nitrocellulose	100
Modified soya alkyd resin	200
Tricresyl phosphate	30
Butyl acetate	130
Ethyl alcohol	75
Toluene	130
	1225
Weight per gallon	11.9 lb.
Pigment weight	45.6%

Enamel top coats are used largely for refinishing because they have higher solids and afford better gloss and film build per coat than lacquers.

Refinishing enamels do not possess the gloss retention and weather resistance of factory-applied enamels which are cured at approximately 250°F. Ovens necessary to use these factory-type baking enamels are large and generally not available in refinishing shops.

An automobile refinishing enamel is represented by Formula No. 2.

FORMULA NO. 2.   Automobile Refinishing Enamel

Material	Pounds	Supplier
Rutile titanium dioxide	260	du Pont, R-902
Soya lecithin	2	
Modified tall oil benzoate alkyd	615	The Sherwin-Williams Co., XAC16
Lead naphthenate, 6%	24	
Manganese naphthenate, 2%	2	
Cobalt naphthenate, 6%	2	
Methyl ethyl ketoxime, 17%	6	
Guaiacol, 18%	6	
Mineral spirits	75	
Hi flash naphtha	27	
	1019	
Weight per gallon	9.84 lb	

Farm and road equipment finishes are made up largely of a primer-sealer used over steel casting components and an enamel top coat for the unprimed sheet metal and primed casting. The sealer was for many years either a lacquer or short to medium oil length alkyd. Enamels were generally the medium oil alkyd quality which gave reasonable exterior exposure. The manufacturers who required a fast-drying enamel top coat generally used a styrenated or vinyltoluene-type alkyd, sacrificing almost all durability, i.e., less than six months 45° south Florida exposure.

The above materials were either completely air dried because of the ware size or partially assembled, spray applied and force dried at temperatures of 150 to 200°F.

Recent years, however, have seen many changes in this industry. New resins have made possible resistance properties and applications of primer-sealers which excel greatly over their predecessors. Manufacturers have found that new demands for humidity, salt spray, water immersion, oil, gasoline and ethylene glycol resistance can now be met. Enamel holdout with little or no sanding is also possible, giving the complete unit a finish which is highly desirable, protected sufficiently from fuel and lubricating compounds as well as the outdoor elements.

Recent years have also seen the farm and road equipment manufacturer installing efficient cleaning and phosphating equipment along with baking ovens capable of attaining metal temperatures of 200 to 225°F following primer-sealer application. This allows the coatings formulator to choose from many of the urea or melamine cross-linking agents made available by raw materials suppliers. Formula No. 3 shows the type of primer-sealer which has been used successfully by this industry.

**FORMULA NO. 3.   Farm and Road Equipment
Primer-Sealer, Tan**

Material	Pounds
Yellow iron oxide	150
Zinc chromate	70
Zinc oxide	35
Barytes	50
Diatomaceous silica	175
"Bentone" 34[a]	3
Modified soya alkyd resin at 50% NVM	400
Lead naphthenate drier, 24%	15
Cobalt naphthenate drier, 6%	5
Guaiacol, 18%	10
Mineral spirits	175
	1088
Weight per gallon	10.55 lb
Per cent pigment	44.4%

[a]Registered trademark National Lead Co., Chemicals Division.

Present-day primer-sealers are applied by brush, conventional and electrostatic spray, dipping, flow coating and more recently by electrodeposition, sometimes known as electro-coating.

Enamel Topcoats

Much of the equipment manufactured today has only a single heavy coat of air or force dry enamel applied by spray or flow coat. Enamels are formulated to adhere well to cleaned but sometimes unphosphated steel. Soya or linseed alkyds of short to medium oil length are generally used in these enamels because they yield the best exterior durability. These alkyd enamels air dry within 3 to 4 hours and force dry in 20 to 30 minutes at 175°F. Force drying schedules of higher temperatures may also be employed with increased hardness and very little sacrifice of gloss or color. Phthalic and isophthalic anhydride are the principal polycarboxylic acids used in alkyds of this type. Farm machinery, road and earth moving equipment is generally offered in colors which exhibit high visibility from a safety standpoint. Colors are generally in the red, green, blue, yellow or orange family. Red enamels are usually based on molybdate orange for hiding; the shade is controlled with quinacridone or "Thiofast" reds. Lower-cost red enamels having more brightness and fairly good color retention are made by toning molybdate orange with some of the more lightfast BON maroons and manganese-treated reds.

Green enamels are made with chrome greens or blends of iron blue and chrome yellows. Chrome greens are basically physical mixtures of chrome yellow and iron (Milori) blue. Blue enamels are generally made of rutile nonchalking titanium dioxide toned with phthalocyanine or iron blue.

Yellow enamels are best formulated using chrome yellow pigments of the light or lemon to medium shade variety. Chrome yellows are relatively low in cost with good exterior durability, making them desirable in a cost-conscious industry. Orange shades are usually based on molybdate orange pigments. Trim enamels are generally available in shades of white, gray or metallic.

Color and gloss retention have become increasingly important. These properties have been improved greatly by new resin and pigment technology. Many manufacturers who at one time required a minimum 60° gloss of 60 to 80 after six months Florida exposure now require a gloss retention in the area of 80 to 90 after the same period of exposure. Others require the same 60° gloss after 12 months, which formerly was the minimum for six months.

These enamels are generally applied by conventional, airless or electrostatic-airless spray. *Spray application* adapts itself well to this type of manufactured product. Sufficient wet edge must be retained to allow the spray operator to apply a coat at one point and be able to blend in another coat upon returning to the initial sprayed area, without decrease of gloss due to excessive dry spray. This, of course, can be controlled effectively by choice of solvents in the formulation.

Another form of spray application in this and other areas of industrial coatings is the airless-electrostatic method. Recent developments by spray equipment manufacturers now make possible the addition of electrostatically charged particles with airless atomization. This method reduces overspray and consequently reduces paint losses. Unlike many electrostatic applications, solvent polarity does not play an important role in this type of application. One should, however, be aware of solvent evaporation since atomization and good "wrap around" depend largely on the paint particle keeping its "wetness" on its way to the ware.

Many farm and road equipment manufacturers along with other industrial finishes have for many years used the low-cost method of application *by dipping*. The formulation of air and force dry dipping enamels involves some unique problems. Surfaces of dip tanks are completely exposed to the atmosphere causing constant oxidation. This oxidation of the drying oils present in the paint will cause poorly formulated materials to gel or seed.

Properties which must be considered are maximum freedom from silking and flooding, plus good stability and flow. Silking and flooding are caused by the pigment, the vehicle, or both. The particle sizes of the various pigments used in the formulation should be similar. Pigments of large particle size and low absorption, such as inorganic types, tend to wet more easily than the finer organic types. Some vehicles also cause more silking than others due mainly to differences in surface tension. Grinding vehicles should have low surface tension. Additives or wetting agents, which are numerous, are very effective in the control of both flooding and floating. Wetting agents decrease the tendency of the pigment and vehicle to separate, which can cause great problems if allowed to get out of hand. Because a high volume of material is usually involved in the dipping process (up to 7000 gallons), it becomes extremely important that the customer and coatings supplier have an intensive preventive maintenance program. Accelerated stability programs of various types are available which can forewarn of any danger in an approaching "broken" tank.

Flow coating as a means of application has also been accepted readily by industrial finishers. It is especially adapted to large articles such as farm and earth moving equipment, automobile frames and appliance finishing in general where priming of cabinets and many of the interior components is necessary for corrosion resistance. Dipping some of the larger components listed above would require large dip tanks. These components shaped so that spray painting would not be practical. In theory, flow coating may be considered a modified form of dipping. However, rather than being passed through a tank filled with the coating material, the articles to be finished pass along the conveyor into the flow coating chamber. They pass by a number of spray nozzles which are activated by tripping devices. This causes the paint to flow only when the article is in position to receive it. Excess material is allowed to return to the reservoirs, pumped back through filters to the main source of supply and used again.

Some of the same problems arise in flow coating as in dip coating, e.g., aeration and consequent solvent loss. Most flow coaters contain viscosity reading devices which control this important property by the automotic addition of reducing solvents. Vapor tunnels, enclosed chambers through which the coated articles pass immediately after the finish is applied, are common. These tunnels contain solvent vapor which has evaporated from the wet material just applied, or a controlled amount of vapor may be present at the base of the tunnel. This vapor retards evaporation of the solvent in the coating so that "flow-out" occurs and excess material drips off. Vapor tunnels are necessary where lower film thicknesses are desired, e.g., in appliance primers where highly baked coatings such as epoxies and acrylics are used in the prime coat. They are also necessary where line speeds are fast and the article enters the oven soon after leaving the coater.

In the heavy equipment industry, however, vapor tunnels are not generally used since too much flow occurs in areas where good protection is necessary. This lack of film is termed solvent washing. To eliminate some of this problem, formulators have elected to produce coatings, especially enamels, which can be applied without the use of vapor tunnels. Application viscosities are increased from 18 to 24 seconds on a No. 2 Zahn cup at 77°F to 50 to 70 seconds on a No. 4 Ford cup at 77°F. Good flow-out with minimum solvent washing is attained by the use of flow additives rather than vapor tunnels.

A typical equipment enamel is represented by Formula No. 4.

FORMULA NO. 4.   Equipment Enamel, Yellow

Material	*Pounds*
Medium chrome yellow	180
Yellow iron oxide	50
Rutile titanium dioxide	5
Soya lecithin	2
Modified soya benzoate alkyd	636
Methyl ethyl ketoxime, 17%	2
Lead naphthenate, 24%	14
Cobalt naphthenate, 6%	6
Silicone solution, 1%[a]	2
Xylene	130
	1027
Weight per gallon	9.91 lb

[a] Dow Corning DC200,500 centistokes.

APPLIANCE FINISHES

The term appliance covers a broad spectrum of products ranging from small hand units to large commercial freezers. In today's expanding economy, the individual can buy, and demands, an ever-increasing number of appliances. Each year the manufacturer produces and ships an increasing number of units. The appliances which constitute the major portion of the field are air conditioners, dishwashers, freezers, refrigerators, washers and dryers.

There are certain definite requirements for these appliances. Air conditioners, generally the window type, are placed in a partially exterior environment and must, therefore, be resistant to rusting and possess some degree of gloss retention. Freezers and refrigerators are used in an interior environment and must have good resistance to staining from foods that will be stored in them. Dishwashers, washers and dryers must have good resistance to detergents and bleaches. For all appliances, the desired qualities are basically a hard, tough, durable finish with good visual appearance. Prior to the 1950's nondrying oil-modified amine alkyds provided the best-quality finish. The introduction of thermosetting acrylics, however, revolutionized the industry's choice of coatings.

Thermosetting acrylics are generally addition copolymers of styrene or vinyltoluene, esters of acrylic acid, and acrylic or methacrylic acid. In the case of hydroxyl types, however, acrylamide is substituted for most of the acrylic or methacrylic acid, thus giving a potential cross-linking point with an available hydroxyl group. A hydroxyl acrylic, therefore, refers to the potential cross-linking point created by the condensation of the amide group with formaldehyde. Carboxyl acrylic refers to the carboxyl group present in the acrylic acid portion of the molecule.

A brief review of the monomers which go into an acrylic vehicle and their function follows:

(1) Styrene—Advantageous from a cost standpoint. It contributes hardness, stain resistance and general improvement in chemical resistance.

(2) Ethyl acrylate—Contributes light stability and flexibility to the polymer.

(3) Methacrylic acid—An internal catalyst in the hydroxyl types and the point of cross-linking in carboxyl acrylics.

(4) Acrylamide—The point of cross-linking for hydroxyl types and the direct factor in their thermosetting properties.

(5) Cross-linking agents—Either epoxy or amino, or both, for carboxyl and hydroxyl types.

By combining the above monomers and cross-linking agents, one is able to bring about the most popular appliance finish. Variation of the amounts of styrene and ethyl acrylate change the physical properties of the enamel so that one can choose the product suited to his need. High amounts of styrene limit the exterior gloss retention but add to the resistances necessary in home appliances. Decreasing the amount of styrene and increasing the ethyl acrylate improves flexibility and exterior exposure to the point where thermosetting acrylics are now applied to post-formed metal house siding and building materials, a subject which will be covered later in the chapter.

A typical thermosetting acrylic appliance enamel is shown in Formula No. 5.

The basic substrate of the appliance industry is steel, either cold rolled or galvanized. Because outstanding resistance properties are required of appliances, steel is always given a phosphate metal treatment. In addition to the basic phosphate-treated metal, a large boost in resistance over even

FORMULA NO. 5.   Acrylic Appliance Enamel, White

Material	Pounds
Rutile titanium dioxide	300
Thermosetting acrylic resin[a]	500
Bisphenol A-epichlorhydrin epoxy resin[b]	45
Xylene	105
Aromatic naphtha, 100 flash	50
Butyl "Cellosolve"	30
Raybo 3[c]	0.5
	1030.5
Weight per gallon	10.0 lb
Bake schedule	25 min at 350° F

[a]"Acryloid" AT51; registered trademark, Rohm & Haas Co.
[b]"Epon 1001" or equivalent; registered trademark, Shell Chemical Co.
[c]Raybo Chemical Co.

the finest metal treatment can be obtained by *priming the surface* with a thermosetting acrylic or solution epoxy primer. Good metal treatment and primers produce the highly resistant substrate for a quality appliance system. Occasionally however, where a good metal treatment is available and detergent and salt spray resistance are not prime requisites, one can use a one-coat thermosetting acrylic enamel. A solution epoxy primer formulation is given in Formula No. 6.

FORMULA NO. 6.   Appliance Epoxy Primer, Gray

Material	Pounds
Rutile titanium dioxide	95
Lampblack	2
Bisphenol A–epichlorhydrin epoxy resin	210
Urea-formaldehyde resin	130
"Cellosolve" acetate	340
Aromatic naphtha, 100 flash	90
Xylene	40
	907
Weight per gallon	8.90 lb
Bake schedule	20 min at 400° F

Appliance coatings have been applied by several methods. Recent stress on high production rates has brought about the use of automatic systems such as flow coating and electrostatic spray. The majority of the epoxy-type primers are applied by the flow coat method, usually at dry film thicknesses of 0.5 to 1.0 mil. Vapor drain tunnels are generally used in this application since epoxies, and sometimes thermosetting acrylic-type primers, cannot flow out properly with the use of surface-active agents alone. Baking schedules are generally in the area of 20 to 30 minutes at 400 to 450° F.

Enamel top coats, whether they be modified alkyds, thermosetting acrylics or polyesters, are most often applied by automatic electrostatic spray units. The most popular are the rotating bell types, the reciprocating disk and, at times, the reciprocating air and airless gun types. Obvious savings are realized by the manufacturer because overspray is minimized and the overall appearance of the finished appliance is much improved. Rejects are also cut down due to the elimination of human error in manual spraying.

For electrostatic spray, polarity and ohms resistance of the solvent are important. Devices such as ohms resistance meters are available from electrostatic equipment manufacturers who can recommend ranges for best performance of the equipment and type of coating. The resistance of a formulation can generally be controlled by the use of various hydrogen-bonding solvents.

Coil coating is the continuous coating of metal strip, usually applied by an automatic roller coater. This method of application came into its own in the late 1940's, but at that time production lines were generally coating narrow strip for venetian blinds and awnings. Present coil coating lines may be more than 5 feet wide, with line speeds in excess of 150 ft/min. In 1965, 50 million dollars of coil coating materials were applied to pre-coated, post-formed stock.

Some of the basic requirements of a coil coating are: Excellent *adhesion* which is related to *good post-forming ability* such as is the case with metal house siding and awnings. The coating must be *decorative* and possess *long-term durability*.

Substrates are generally one of three types: chromate-treated aluminum, phosphatized steel and treated galvanized steel.

Aluminum has in recent years become the most popular of the precoated substrates. By itself, it does not corrode rapidly and, like galvanized steel, its corrosion product is white in nature. It is therefore not as objectionable as the red rust of steel. Alloys of aluminum are also used to lower cost and add additional structural strength. These tend to corrode more freely than pure aluminum and require coating from a standpoint of protection and decoration as called for by public demand. House siding is now cited and sold with a durability of 20 to 30 years. Coating is a minor component, dollarwise, of industrial finishing, especially when coated by the coil process.

In the coating of *phosphatized steel*, protection through long terms of exposure becomes a main factor. Use is generally confined to roof deck products where support brought about by its rigidity plays an important role. The coating system generally includes a corrosion-resistant primer for best protection.

Galvanized steel has become popular since its corrosion resistance is better than conventional phosphatized steel. However, it also requires treatment for maximum adhesion and protection. It has a definite cost advantage over aluminum and has now become widespread in use. Consistency of substrate from lot to lot has been its greatest problem thus far. A two-coat operation is, therefore, necessary to overcome this weakness. This also overcomes the problem of rapidly degrading adhesion upon exposure. Epoxy-type primers are most typically used. A suggested epoxy primer is given in Formula No. 7.

FORMULA NO. 7.   Roller Coat Epoxy Primer

Material	*Pounds*
Strontium chromate	40
Rutile titanium dioxide	280
Bisphenol A–epichlorhydrin epoxy resin	185
Urea-formaldehyde resin, 50% NVM	90
Aromatic naphtha, 100 flash	230
Butyl "Cellosolve"	220
	1045
Weight per gallon	10.18 lb
Bake schedule	60 sec at 500° F

Three general types of top coat enamels are used in the precoated strip field. The *alkyd-amine* systems offer the lowest cost but are limited by flexibility. This increases the tendency toward corrosion in areas where forming is severe. *Vinyl* finishes offer excellent exterior durability and flexibility but are rather costly due to low application solids which are approximately 25% based on volume solids versus 45% in an alkyd system. Vinyl coatings contain vinyl chloride–vinyl acetate copolymer resins as the binder. The *thermosetting acrylic* finish is the newest entry into this market. It has become one of the most desired coatings because of its excellent durability, approaching the flexibility of vinyls at the cost of the alkyd-amine.

Representative formulations with the qualities described above are given in Formulas No. 8, 9 and 10.

Vinyl dispersion or *organosol*-type coatings have also recently found success in this area. They are suspensions of very finely divided vinyl chloride resins in organic liquids which are not capable of dissolving the resin at room temperature. The organosol formula is similar to the solution-type vinyl in composition except it contains a higher molecular weight vinyl chloride resin. Organosols do not become a continuous film

FORMULA NO. 8.   Coil Coating Alkyd, Gloss White

Material	Pounds
Rutile titanium dioxide	350
Modified dehydrated castor oil alkyd, 55% NVM	520
Isobutylated melamine, 60% NVM	155
Aromatic naphtha, 100 flash	45
Butyl "Cellosolve"	40
Diethylaminoethanol	5
	1115
Weight per gallon	10.80 lb
Bake schedule	90 sec at 420° F

FORMULA NO. 9.   Coil Coating Vinyl, Gloss White

Material	Pounds
Rutile titanium dioxide	200
Vinylite VYHH-"Bakelite"	85
Vinylite VMCH-"Bakelite"	85
Plasticizers	40
Isophorone	300
Aromatic naphtha, 100 flash	275
	985
Weight per gallon	9.55 lb
Bake schedule	90 sec at 400° F

FORMULA NO. 10.   Coil Coating Thermoset Acrylic Enamel, Gloss White

Material	Pounds
Rutile titanium dioxide	300
Thermosetting acrylic resin, 50% NVM	600
Bisphenol A–epichlorhydrin epoxy resin	25
"Cellosolve" acetate	40
Aromatic naphtha, 100 flash	40
Isophorone	55
	1060
Weight per gallon	10.30 lb
Bake schedule	80 sec at 500° F

until they are subject to heat. In most instances, a metal temperature of 360° F must be obtained to effect fusion and film formation. The greatest advantage of the organosol film is its flexibility. When coated over an epoxy primer, the film will withstand a reverse impact sufficient to fracture its substrate, 0.24 aluminum, without film crazing or loss of adhesion.

Another advantage over solution vinyls is the ability to formulate at volume solids of 50 to 55%, thus offering lower unit cost.

Another of the newer coatings presently being offered to the coil coatings field are modified *silicone* resins. These products are formulated either by external addition of silicone to established quality alkyds and acrylics, or by internal polymerization with film formers.

WATER REDUCIBLE COATINGS

These have been of interest to the manufacturer of fabricated ware and the coatings supplier for more than 20 years. During this time, industrial users of protective coatings have desired its greatest advantage, nonflamability. Basic research toward this end has been in process continually, striving toward resistance properties equal to their solvent counterparts.

Several problems have been encountered with these coatings. *Flash rusting* on ferrous substrates occurred almost immediately after application. Here the coating solvent, water, reacted with the substrate to form rust spots. These spots, when examined under the microscope, revealed minute cracking of the film in the spot which was the point for early failure when exposed to outdoor elements. *Adhesion*, at best, was only fair after considerable air drying time, 8 to 24 hours. *Odor* was rather strong, caused by wetting agents necessary to hold the system in solution. *Resistance* properties, such as salt spray, water immersion and humidity were very weak.

Recent polymer chemistry has brought about water reducible systems in the acrylic, alkyd and polyester classifications which do not present the above problems. They have been widely accepted in primer systems, allowing application of conventional solvent-containing enamels after a minimum amount of curing time. Formula No. 11 is a typical water reducible primer.

FORMULA NO. 11.   Water Reducible Baking Primer Red

Material	*Pounds*
Red iron oxide	100
Barytes	180
Talc	100
"Arolon" 304,[a] 43% NVM	500
Manganese drier solution, 6%	2
Water	240
	1122
Weight per gallon	11.2 lb
Bake schedule	30 min at 300°F

[a] Registered trademark, Archer Daniels Midland Co.

Water reducible enamels have been most successful where a baking schedule is employed. Vehicle systems are generally of the water-soluble types which may be cross-linked by melamines or ureas. The field of protective coatings now employs a method of application known as *electrocoating* or electrodeposition, whose potential has become very real in recent years of resin development. Numerous articles have recently appeared on the subject; however, the process itself was granted a patent some 30 years ago for deposition of rubber from latex.

To date, only baking-type systems are marketable. In addition, color and performance latitude is not as great as with conventional paint systems. Even though electrodeposition is still in its infancy, if the breakthroughs of the past several years are an indication of the rate of development of new systems, the time is not too far away when this method of application will enjoy a large part of the industrial finishes market.

WOOD FURNITURE FINISHES

These constitute a highly specialized area of industrial finishing and involve the production of a number of unique materials. In general, furniture woods are classified into one of two classes: the porous or open-grain species, such as oak, mahogany, walnut, elm and chestnut, and the nonporous, such as maple, beech, gum, poplar, magnolia and cotton-wood. Birch and cherry are considered borderline cases. Pores of highly prized furniture woods, when filled, accentuate the configurations of the surface.

Woods for furniture manufacture are brought to the optimum moisture content by kiln drying, the goal being equilibrium with the air under conditions of service. For most areas, the optimum moisture content is about 7%.

Furniture finishes may be one of two types: the transparent or clear finishes characterized by visibility of the wood surface, and opaque finishes or "paint," so-called regardless of the fact that most of the coating materials are lacquer.

The clear finish system usually modifies the color and appearance by *stains*, followed by *sealer, filler* if needed, and finally clear *top coats*. Satisfactory finishes require a number of operations. The following is a typical factory schedule for the finishing of medium- to high-priced mahogany or walnut furniture.

(A) Non-grain-raising stain, sap color. No drying time.

(B) Non-grain-raising stain, body color. One-half hour air dry.

(C) Thin coat of lacquer sealer, sometimes called a wash coat; let dry 30 minutes.

(D) Application of filler by spray or brush. Wipe. Let dry overnight.

(E) Glazing lacquer sealer. Dry 1 hour.
(F) Wiping glaze. Dry overnight.
(G) Sanding lacquer sealer. Dry 1 hour. Dry sand.
(H) Flat or gloss lacquer, two coats. Dry overnight.
(I) Sand with 360 sandpaper and naphtha.
(J) Rub with rubbing compound, apply wax or polish.

These lacquer top coats are regarded as the standard finish for high-grade furniture. The principal disadvantage of lacquer is its low solids and subsequent high unit cost. The use of synthetic baking clears on high-grade furniture is growing rapidly. Some woods and finishes may require a number of materials not mentioned in the above system, such as bleaches and pigment-oil stains.

The following brief review of some of the materials which go into a clear furniture finish system will indicate the uniqueness of this industry.

Stains

Properties most desired in a furniture stain are clear, rich color plus uniform depth, lightfastness, absence of grain raising, nonbleeding and rapid drying. Clarity and richness are brought about by the use of aniline dyes. These may be dissolved in alcohol, aromatic hydrocarbons and water, or in a mixture of alcohols, aromatic hydrocarbons and glycol ethers. The latter combination is known as the N.G.R. or non-grain-raising type. Other stains known as pigment uniforming or wiping stains utilize pigments as the main coloring agent. They are found predominantly in the semi-opaque earth color groups such as raw and burnt siennas and umbers.

Sealers

These are generally applied after the staining operation prior to filling. Filler applied directly over stain clouds or smudges the rich color of the stain. A preventive is a very thin coat of lacquer sealer, which permits complete removal of the filler. The nonvolatile content is formulated at $2\frac{1}{2}$ to 8%. The low solids avoids sealing or bridging the pores which would prevent proper filling. Finish manufacturers are striving to find shorter, more economical ways of arriving at the desired result. A good example is the combination of staining and wash coating in a single operation. Greater skill and care in application, however, are necessary to obtain uniform color.

There is also the sanding-type lacquer sealer which is applied after the coloring and filling process is complete. Basically this is a clear gloss lacquer with the addition of a sanding agent. Sanding agents are metallic soaps, such as zinc stearate. Lacquer sealers are supplied ready to spray at 18 to 21% nonvolatile matter. The properties desired in a sealer of this type are quick drying, easy sanding, good build, toughness and durability. Formula No. 12 is a typical easy-sanding lacquer sealer.

FORMULA NO. 12. Lacquer Sealer

Material	Pounds
$\frac{1}{2}$-sec nitrocellulose (dry)	64
Raw castor oil, 100% NVM	16
Dibutyl phthalate, 100% NVM	16
Maleic resin, 100% NVM	90
Zinc stearate	14
Methyl isobutyl ketone	160
Butyl acetate	80
Ethanol	60
Butanol	60
Toluene	320
Xylol	120
	1000

Fillers

Fillers have a twofold purpose: (a) to produce a smooth surface and (b) to enhance the beauty of the wood by making the pattern stand out more clearly. They are supplied in paste consistency. In production, they are sprayed and allowed to "set" before being wiped across the grain. Unbodied linseed oil and a small amount of short oil varnish frequently make up the vehicle portion. Fillers consist of approximately 75% pigment and 25% vehicle. Pigments generally are made up of 95% extender pigments. The remainder are color pigments, generally siennas, umbers or Van Dyke brown.

Top Coats

The protective film is completed by the application of several coats of clear lacquer which may be applied by spray. One hour of drying time is generally allowed between coats, with rubbing of the final coat after an overnight dry. Clear gloss lacquers are similar in composition to the clear base in sanding sealers at 21 to 27% solids. Flat lacquers are gloss lacquers to which a flatting agent such as colloidal silica is added.

Most desired properties in top coat lacquers are quick drying, good leveling, build, rubbing, and resistance to printing, checking or cracking. Since it is difficult to obtain the optimum in all properties, lacquer formulations vary, depending on the properties the users want stressed.

A method of increasing solids and build is to reduce the viscosity of the lacquer with heat and spraying while it is hot. Little or no change is necessary in the nonvolatile composition of the cold spray lacquer. A change, however, to slower evaporating solvents is necessary. The nonvolatile matter may be increased to 35% with viscosity ranges of 80 to 120 seconds in a No. 4 Ford cup at 77°F. Hot spray temperatures are in the 145 to 165°F range. Formula No. 13 is a hot spray lacquer top coat.

FORMULA NO. 13.   Hot Spray Lacquer Top Coat

Material	Pounds
$\frac{1}{2}$-sec nitrocellulose	120
Raw castor oil	30
Dibutyl phthalate	30
Maleic-modified rosin ester resin	70
Nonoxidizing phthalic alkyd resin	100
Butyl acetate	250
Ethanol	80
Butanol	64
Xylol	256
	1000
Nonvolatile material	35%

One of the most severe problems has been the lack of alcohol and stain resistance in furniture finishes. There has recently been a definite switch from lacquers to thermosetting lacquer using amines. Nitrocellulose, being acidic by nature, cures the system by reacting with the urea resin. This coating has been referred to as a converting lacquer.

Recent emphasis on increased production of plywood wall paneling has led to coatings yielding similar appearance in two coats when compared to the multistep high-cost furniture system described earlier. Use of fast-curing, heat-converting coatings has led to the adoption of this type of coating wherever possible. There is presently some compromise in clarity and leveling to get rapid drying. Constant striving, however, toward improved production rates and continued effort in coatings technology will overcome these difficulties. Efforts are also being directed toward increased use of curtain coaters, roller coaters and high-intensity curing ovens.

CONCLUSION

There are many facets of industrial finishes which cannot be covered within the scope of this chapter. There are those finishes which are called general industrial—can lining coatings, lacquers for paper, flamboyant finishes, hammer finishes, wrinkle finishes, solventless coatings, zinc-rich coatings and many more. It is certain that technological advances in this area have progressed to admirable heights within the last half century. One can be sure that advances in industrial finishes within the next generation will be even more pronounced.

REFERENCES

1. Keenan, H. W., "Paint Technology Manuals," Vol. 3, Ch. 15, New York, Reinhold Publishing Corp., 1962.
2. Parker, D. H., "Principles of Surface Coating Technology," Chs. 42 and 43, New York, Interscience Publishers, 1965.

31

*Maintenance Paints**

Industrial maintenance paints are used to protect equipment and structures from deterioration and corrosion by industrial atmospheres which may involve wet or high-humidity conditions, chemical fumes and spillage, and abrasion. Ability to provide decoration, while important, is secondary to that of protection.

Under the field conditions usually associated with construction activities or plant operations, ideal conditions and labor for application are seldom obtained. Thus it is important in formulating such coatings to provide as much application latitude as possible, consistent with maintaining adequate environmental resistance. This explains also the amount of discussion in this chapter detailing application requirements.

PAINT TYPES AND SELECTION

There are two mechanisms of paint protection: barrier and galvanic (sacrificial).

Barrier Protection

Conventional paints, which are basically organic film-forming resins pigmented if required to the desired color, provide the barrier type of protection. The paint film forms an adherent, resistant, conforming barrier between the environment and the structural material to be protected. The success of such protection requires that the coating be applied and maintained as a continuous film, without skips, holidays, pinholes, abrasions or mechanical injuries which would expose the substrate to the environment.

Sacrificial Protection

The mechanism of sacrificial protection[1] is well known and has proved efficient in the use of galvanized coatings to protect steel. Zinc-rich paints[2]

*By Kenneth Tator, Kenneth Tator Associates, Corapolis, Pa.

similarly provide coatings for steel which are substantially metallic zinc, the dried film containing between 85 and 95% by weight of zinc metal.

When a metal such as zinc is in electrical contact with a metal more noble in the electromotive series, such as steel, a galvanic cell is established and electric current will flow when an electrolyte contacts this by metallic couple. In a zinc-rich paintsteel substrate system the electrolyte is provided by atmospherically contaminated water, rain or dew. Under these conditions, electric current will cathodically protect the substrate from corrosion, the zinc being preferentially corroded and wasted away in the process. It is for this reason that galvanic protection is sometimes called sacrificial protection.

Thus with galvanic protection, it is not as essential as with barrier protection to have pinhole-free coatings, as protection will be provided at any minute areas of exposed substrate by electrical action.

COATING TYPES

An industrial plant is a varied complex of exposures in which almost every type of paint used in domestic and institutional painting has a place. Offices, cafeterias, locker and shower rooms, shops and many warehouses require the normal materials used for similar purposes in institutional painting. Such trade sales products are often entirely satisfactory for those production areas in which the atmospheres are relatively dry and clean. In other production atmospheres, however, specialized coatings are required. The generic types commonly used in industrial maintenance are shown in Table 31.1.[3]

Reactive coatings must all undergo some chemical or physical reaction after application in order to develop their final resistances and protective qualities. For example, oil-base paints would remain a useless liquid film without such reaction with oxygen of the air. Maintenance neoprenes, by reaction, actually vulcanize into a rubber.

Thermoplastic coatings are those which will reversibly soften with increase in temperature and harden with drop in temperature. Also, since these materials undergo no chemical or physical change after application, their resistances are not changed by application. Since most, if not all, of them were originally deposited from solutions of organic solvents, it follows that they can again at any time be dissolved in these same solvents.

Resistances

Thus a number of generalities regarding coating performance may be made. If resistance to organic solvents is required, it would be wise to avoid the plastic types and to select one of the reactive types. The same conclusion applies to heat resistance.

TABLE 31.1. Coating Classification

A — Reactive	*B — Plastic*
Must undergo a chemical reaction after application to develop fullest resistance	*Possesses equal resistance before application* *Reversibly thermoplastic: repeatedly softens in heat, hardens in cold*

(1) Oil Derived
   (a) Straight oil
   (b) Alkyd
   (c) Phenolic-tung
   (d) Epoxy-ester
(2) Sacrificial Metal
   (a) Zinc rich
(3) Silicones
(4) Thermosetting
   (a) Phenolic
   (b) Urethane
   (c) Epoxy
   (d) Furane
(5) Vulcanizable Rubber
   (a) Neoprene
	(1) Wax and Grease
	(2) Bitumen
	   (a) Asphalt
	   (b) Coal tar
	(3) Rubber Base
	   (a) Butadiene-styrene
	   (b) Chlorinated rubber
	(4) Vinyl
	   (a) Chloracetate

All industrial coatings must possess some degree of resistance to acids, as the prevalence of coal-burning equipment has been responsible for predominating acidic atmospheres in industrial areas. In general, it can be stated that the resistance to acids will increase as one proceeds from top to bottom down either the reactive or plastic column in Table 31.1.

All coatings in the plastic group have some degree of resistance to alkalies (caustic), which again increases as one goes down the column. However, it will be noted that some of the best alkali-resistant coatings are in the bottom half of the reactive group; so if one proceeds with the epoxies down that group into and up to the top of the plastic group, an approximate decreasing order of alkali resistance will be followed.

It will be appreciated that no simple table can be an infallible guide to coating selection. Any single generic film-forming type is itself available in different grades and properties; furthermore, it must be formulated with a considerable breadth of choice in respect to pigments, fillers, plasticizers and other modifying agents, each of which can vary in specific resistances and properties. Thus even within a generic type, there will be performance variations, depending upon formulation. For example, epoxy coatings are outstanding for their alkali resistance; however, if an epoxy resin is compounded with alkali-susceptible pigments, or if a caustic-resistant epoxy finish coat is used with a caustic-susceptible primer, failure is bound to result. Thus Table 31.1 should be used only as a weathervane to indicate the probable direction, but within the indicated types, testing should determine the most suitable formulation for use.

DESCRIPTION BY GENERIC TYPES

Oil Derived

In this group are coating materials made with saponifiable drying oils or saponifiable derivatives of drying oils; hence, caustic resistance is poor. The straight oil types, represented principally by house paints, have only small industrial use for the painting of wood in noncritical exposures.

Phenolic-tungs, epoxy esters and particularly the alkyds have wide industrial application for industrial decorative painting. Where somewhat greater resistance to moisture and slightly greater resistance to chemicals than possessed by the alkyds are required, the phenolic-tungs or epoxy esters are used at a premium in material cost. The epoxy ester is favored over the phenolic-tungs where pure whites or light pastel colors are required.

These oil-derived coatings are in general the least expensive, in both material cost and application, of the maintenance coating types and are easy to apply by conventional procedures.

Sacrificial Metal

These zinc-rich coatings[2] are the only current paint types which protect by the galvanic mechanism.

They exist in two general types: *inorganic*, in which the binder or vehicle for the zinc powder is inorganic, such as metallic silicates; *organic*, in which epoxies, chlorinated rubbers, butadiene-styrenes, urethanes or vinyls are the binder.

The inorganic types possess excellent resistance to high temperatures and to organic solvents, and so have additional wide use as coatings for stacks, breechings and manifolds, for the lining of tanks both on ship and land, and for holding and transport of organic solvents and petroleum products.

On the other hand, the organic zinc-rich coatings have a somewhat greater latitude in surface preparation and application, and are used in preference to inorganic types where these cannot be rigidly controlled.

Since protection depends upon the establishment of a galvanic cell, it is essential that there be electrical contact between the base steel, the zinc coating and the environmental electrolyte. Thus there should be no insulating layer of rust or other surface residues on the steel surface before the zinc-rich coating is applied; the quality of surface preparation is even of greater importance with the use of galvanic coatings than with barrier coatings.

The mechanism of galvanic protection requires that the zinc coating be slowly consumed in providing protection. Thus the protective thicknesses will gradually diminish during their service life by natural weathering or galvanic protection; the rate of such dissipation[4] being greater in those environments of greatest corrosive severity. For protection in such environments, it has been found advisable to overcoat the zinc-rich coating with a barrier type of coating to prevent non-useful zinc dissipation.

Such use as a sacrificial base or "primer" for organic systems has been proved to extend the maintenance-free life some four or five times.

Silicones

The principal use of the silicones in industrial maintenance is for high-temperature coatings, as their chemical resistance is generally poor, particularly to alkalies. Either by themselves or as a decorative topcoats for the zinc-rich coatings, they are used for stack coatings and other hot surfaces.

Thermosetting Resins

The straight phenolics and the furanes are not extensively used in maintenance painting, although these have successful application as process equipment linings, usually as baked coatings.

The urethanes and epoxies, while also useful as baked equipment linings, are principally used as cold-set coatings for industrial maintenance. Properties, resistances and uses of these two types of cold-set coatings are substantially parallel; they possess adequate resistance to acid, excellent resistance to alkali and good resistance to organic exposures, combined with resistance to moderate heat.

Neoprene

In spite of the "chlorinated rubber" designation, maintenance neoprene coatings are the only ones in general industrial usage which produce a true rubber film. Thus this material is useful against fumes and occasional spillage of acids and other chemicals in which sheet rubber lining is normally used to hold the chemical. Because of their rubbery nature, they are outstanding for abrasion resistance.

Waxes and Greases

Waxes and greases are used primarily for temporary protection, although they have use for protection of sliding surfaces, such as slip and roller expansion joints. They are easily removed by solvent. Such complete removal is required for painting, as they are incompatible with paint coatings. While considerable protection from humidity may be obtained with ordinary lubricating greases, it is advantageous to formulate incorporating corrosion inhibitors.

Bitumens

The bitumens are basically divided into asphalts and coal tars. This distinction is not always recognized in their industrial use, but it is important.

Asphalt has excellent resistance to sunlight, weathering exposures and direct summer heat, whereas coal tar pitches tend to flow and alligator in such exposures to such an extent that their use is not recommended for direct weathering.

Coal tar pitches, on the other hand, have greater moisture resistance than the asphalts generally formulated for industrial coating; the usual asphalt coatings will tend to lose adhesion to the substrate in continuous water or wet soil immersion.

Thus basically the asphalts should be used in weathering exposures above ground, and the coal tar pitches should be employed in preference to asphalt for continuous wet or damp exposures, water immersion and soil burial.

Rubber Base

The two members of this group, chlorinated rubbers and butadiene-styrene coatings, either are, or could be, derivatives of natural or synthetic rubber. However, the form in which they are used in coatings is no longer a rubber, but a tough, hard resin possessing little of the resilience and elasticity of rubber. The properties and uses of these are substantially similar; they have adequate chemical resistances when properly compounded, and excellent caustic resistance which has led to their wide use as coatings for concrete.

Vinyl Chloracetate

These solution copolymers, when properly formulated, have excellent resistance to both acids and alkalies but poor resistance to organic solvents and to heat. Although when originally introduced they were characterized by low film build and poor adhesion to steel, current formulations are supplied which can provide easily the minimum requirement of 2 mils dried thickness per coat, and system primers are abundantly available which give adhesion under almost all common surface conditions.

PRINCIPLES OF EFFECTIVE MAINTENANCE PAINTING

In severe industrial exposures, the performance of maintenance paints is determined as much by the adequacy of application as by the suitability of the paint materials.[5] An appreciation of the application requirements is necessary not only to attain economical performance from even the best formulated coating, but also to permit manufacturers to formulate coatings of maximum utility and ease of application.

Failure Mechanism

Let us use as an example the most prevalent type of industrial substrate, steel, and explore the failure mechanisms. If there are any breaks or discontinuities in the barrier paint film, protection will be lost at these points, and the environment can enter and attack the substrate. The result of such attack by humidity, moisture or most chemicals in the presence of moisture is rusting. On an average (and this varies considerably with type of exposure), one unit volume of steel in the process of corrosion will yield twenty volumes of rust.

Thus at any discontinuity caused in the initial application, or by abrasion or injury during subsequent service, a considerable expansion and prying action will occur under and against the edges of the paint film immediately contiguous to it. This prying and undercutting action will progressively expose more base steel to the environment, with more generation of prying corrosion product; thus paint failure will spread continuously from areas of initial discontinuity, until the total paint protection is completely lost.

Thus it is evident that maintenance of barrier continuity is essential.

Coating Thicknesses

Since the chance of obtaining continuity increases with increasing thickness of the coating, it is not surprising to find that there is a minimum thickness below which it is difficult to obtain protective continuity with any coating. Pierce[6] (see Figure 31.1) has shown this minimum to be 5 mils. This does not mean that protective continuity cannot be obtained at lesser thicknesses. Industrial structures and equipment have a high normal proportion of surface irregularities, such as projections, prominences, sharp edges and crevices, over which a liquid paint will pull thin by surface tension. Therefore 5 mils is required on the measurable, flat surfaces in order that the lesser thicknesses over these irregularities will still be enough for protection.

Number of Coats

Conceivably it is possible to attain the minimum thickness of 5 mils in a one-coat application. However, no work of man is perfect, and a certain percentage of holidays, skips, pinholing and air entrapment can be expected as normal. Thus it is desirable to apply multiple coatings, so that each succeeding coat will cover and seal any imperfections in the preced-

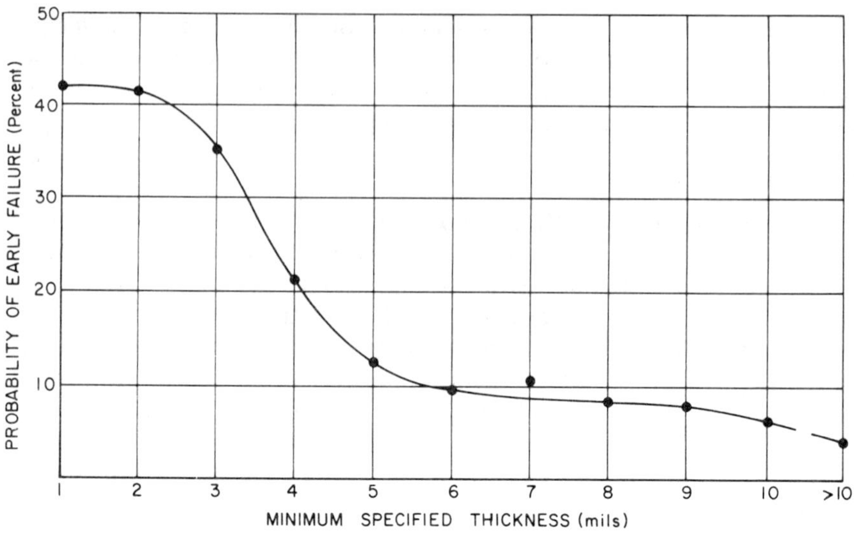

Figure 31.1.   Probability of early paint failure.

ing coats.   It has been proved[7] that a three-coat system provides the highest surety of continuity with minimum application costs.   Thus the now well known rule, "a minimum of 5 mils, in a minimum of three coats."

Preventive Maintenance

Initially applying a paint system to these 5-mil, three-coat specifications will not suffice.   It is necessary that this protective barrier be maintained throughout the entire service life of the application.   Reduction of thickness by chalking, weathering, or service conditions of mechanical injury or abrasion may reduce the barrier thickness below the required 5 mils in local or general areas.

Thus, a regular system of preventive maintenance should be established in which the paint application is inspected periodically during its service life, and any thin or failing areas cleaned and rebuilt to minimum protective thicknesses.

SUBSTRATE MATERIALS

Maintenance painting is predominantly concerned with the protection of steel surfaces, but involved in varying degrees are other types of material: concrete and other masonry, wood and nonferrous metal surfaces. The requirements for protection for other substrates may differ from that for steel.

With steel, the voluminous prying corrosion product is an important factor in the progression of paint deterioration. This factor may be non-existent, or at least of lesser importance, with other substrate materials.

Concrete and Masonry

As moisture and normal weathering do not adversely affect concrete, in these exposures concrete need not be painted except for decorative purposes; for these purposes, the requirements of barrier continuity need not be rigidly adhered to.

Concrete can slowly pass moisture vapor, and in fact has a certain degree of hygroscopicity in itself. Thus, where there may be a moisture differential or gradient through the concrete, blistering may occur in any impervious coating applied to the dryer side. Under such conditions, the moisture-permeable latices of acrylic or vinyl will pass such moisture vapor without blistering. In fact, these latices constitute one of the most useful decorative coatings for concrete and masonry.

Where concrete or other masonry is to be used in an environment which will deteriorate it, a resistant, adherent barrier system, applied and kept free of discontinuities must be used as the protective.[8] Even under such conditions, it is very seldom that the "corrosion product" of concrete is of the voluminous prying type, and gradual spread of deterioration at any coating discontinuity will be by undercutting rather than by prying action.

The surface condition of concrete and other cement products is always alkaline, and thus alkali-resistant paints should always be used.

The normal inhibitive pigments effective for retarding steel corrosion are not effective in preventing deterioration of concrete, and therefore no specially formulated primer need be used. It is advantageous in many cases to thin the first coat of finish to the point where it can be worked or scrubbed into the pores and irregularities of the concrete surface, in order to provide a good base for the subsequent unthinned coats.

Wood

Paint deterioration over wood is normally caused by the fact that wood is basically a dimensionally unstable, hygroscopic material, which breathes, absorbs and releases moisture in response to varying environmental conditions of temperature and humidity, shrinking as moisture is released and expanding as it is absorbed.

As a result of dimensional changes, a tightly adherent nonelastic paint will crack and check, and a paint with indifferent adhesion may flake or peel. Where moisture passage into the wood is possible through unprotected backs and edges of the wood or breaks in the coating, moisture pressures will build up behind the paint film and cause blistering, loss of adhesion, and ultimate cracking and peeling.

As normal weathering has no catastrophic effects upon wood, impounded moisture pressure can be harmlessly relieved by paints which can pass reasonable quantities of moisture vapor. Conventional linseed oil house paints possess this ability to a normally adequate degree; the acrylic and vinyl latices are substantially blister-resistant under almost all conditions.

As woods of ordinary industrial construction are of the "soft" variety and of rough surface texture, it is essential that a primer be first used which will penetrate, fill and adhere to the surface. The linseed oil-rich wood primers serve this purpose well.

For normal atmospheric exposures, lead-free, oil-base or latex house paints may be applied over this primer. For continually wet, humid or chemical exposures which would attack the wood or the fastenings, it is desirable to finish with a coating resistant to the exposure. In many cases, this would include such materials as vinyls and epoxies possessing solvents which would wrinkle and lift the linseed oil wood prime and filler. In such instances, it is desirable to interpose a barrier coat, such as an epoxy ester, between the chemical-resistant finish and the primer.

Nonferrous Metals

Where nonferrous metals are involved, it must be assumed that these metals were selected for their resistance to the exposure. If the selection was proper, we can expect little or no effect from discontinuities in the paint overcoating and may assume the painting is used for decorative purposes only. In such cases, the only problem involved is to insure that the adhesion of the paint to the substrate is adequate.

Many of the corrosion-resistant metals acquire their corrosion resistance by initial corrosion of the metal, from which the corrosion product forms a tight coating over the surface,[1] which prevents further corrosion. Such base materials as stainless steel, nickel, copper and brass, in which the corrosion product formation is relatively insignificant, present no great difficulties in painting, and generally any desired paint system may be applied without special procedures. In the case of painting over aluminum, it generally suffices to use non-lead-containing primers, as lead compounds may be cathodic to, and cause corrosion of, the aluminum.

In the case of zinc, the relatively greater proportion of corrosion product and its alkalinity require special treatment. Best performances over galvanized zinc, and even over zinc-rich coatings, are obtained when a specially formulated primer is used.[9] These are commonly of the zinc dust-zinc oxide, cement-base, two-package wash primer, and certain proprietary one-package wash primer types. Satisfactory experience is also obtained over galvanized surfaces when latex paints are used as the sole overcoating. Difficulties with the zinc dust–zinc oxide and cement-base primers may be encountered when overcoating with chemical-resistant finish coats, such as vinyls, epoxies or urethanes, by solvent attack and lifting of the primer. In such instances, two-package wash primes or a suitable one-package proprietary are satisfactory.

EFFECT OF EXPOSURE

It is conventional in coating technology to consider exposure only to the extent that the coating formulation will be resistant. This is certainly important; no paint system, no matter how well applied, will perform satisfactorily if it is not basically resistant to the exposure. However, there is a very important effect of environment not universally appreciated. It has been pointed out that the amount and type of base corrosion product affects performance, in respect to the rapidity with which deterioration will progress from points of inadvertent barrier discontinuities.

As an illustration, it is known that the same paint system will fail faster in a marine exposure than in a rural one. It may be observed that the type of rusting on unprotected steel in a rural exposure is loose and powdery, whereas that in a marine exposure is a heavy, dense, adherent, striated scale. From the very appearance of the rust, it is obvious that the corrosion product in the rural atmosphere has little strength to pry up paint adjacent to coating discontinuities, and hence progressive deterioration from such discontinuities will be slow. On the other hand, the nature of the corrosion product in a marine exposure possesses this prying ability to a high degree; hence, paint deterioration originating from a discontinuity will be rapid.

In addition, the chemical composition of the corrosion product, and particularly the content of water-soluble contaminates in the corrosion product, may tend to set up osmotic pressure cells at the interface between the coating and the substrate. The rust in a marine exposure certainly will be contaminated with chlorides, whereas the rust in a rural exposure will be relatively free of water-soluble contaminates.

All organic coatings have a real, although low, rate of moisture permeation. All will be wetted for prolonged periods of time with substantially pure water in the form of rain or dew. During these periods of wetness, conditions for osmotic pressure cells are established; paints applied over surfaces containing water-soluble materials will be prone to blistering by osmotic reaction.

There are conditions under which no visible corrosion product will be formed over unprotected steel except possibly as a surface tarnish. Under such conditions, the steel has become passivated and will not corrode. If steel can be maintained in this condition by a constantly favorable environment, no paint protection is required except for decorative purposes. Under such conditions, the paint formulation and application should be sufficient only for resistance to contaminants and environment.

Thus it is good to examine the type of rust formation on unprotected steel when estimating the protective life of any paint system, when deciding on the required application latitude of the paint materials to be used, and when determining the quality of paint application and extent of preventive maintenance required. While there are exceptions, depending principally upon concentrations and temperatures, the following examples may be a guide:

(a) Some environmental conditions result in dense, prying corrosion product with water-soluble contaminants, in which no discontinuities may be tolerated: chloride-containing environments, such as hydrochloric acid fumes, salt processing, marine exposures.

(b) Some exposures cause a light, nonprying type of corrosion product in which some small discontinuities can be tolerated: rural, ammonia, relatively dry chlorine and certain concentrations of sulfuric acid.

(c) Some exposures produce a passive surface in which application is not critical: certain concentrations of sulfuric acid, nitric acid, phosphoric acid and caustic.

The foregoing are some examples. Unfortunately in actual application nothing strictly follows classic rules. Sulfuric acid under certain conditions may cause a passive surface which would require paint for decorative purposes only, and under other conditions would form a light fluffy corrosion product. However, in the latter case the sulfuric acid is an active corrosive and even though the prying action of the corrosion product is negligible, progressive paint failure will occur by solution of the steel under the coating.

Similarly, caustic soda under certain common conditions will present a passive surface requiring no painting for corrosion protection, but under conditions of constant high humidity, active corrosion product of the nonprying type may be formed.

PAINT SYSTEM AND APPLICATION

Industrial painting should always be considered as a system, comprising not only a primer, intermediate, and finish coat, but also including the surface preparation and quality of application. Protection in a rural exposure does not require the quality in any of these respects that would be necessary for the head frame for a Gulf Coast salt mine. Knowledge of these different requirements in application is as important to the paint chemist as to the painter; application latitudes must be built into the paint without sacrifice of resistances to exposures.

Surface Preparation

In all instances it is necessary to remove loose rust and other incompatible residues. In mild exposures and with proper selection of primer, adequate performance usually can be obtained by painting over adherent rust. For severe exposures of prying corrosion types, a surface preparation equivalent to a near-white blast[10] is mandatory.

It has been stated that mill scale makes a good paint base; this may be true within certain narrow conditions of service and environment. As a general rule, painting over mill scale[11] will lead to premature failure caused by moisture and oxygen permeating through the coating and oxidizing the mill scale to the more voluminous common rust, thus destroying adhesion of the paint film to the substrate and setting up an osmotic reaction, causing failure by flaking and blistering.

Priming

Basically a primer is a "glue" to attach the resistant finish coats to the surface. As such, it should possess wettability and penetrability to adhere itself to and into the surface, and compatibility with the finish coats to insure their adhesion to the primer.

Inhibitive pigments are often incorporated in primers to retard steel corrosion and undercutting when exposed in humid or wet environments. These are generally effective only in the presence of such humid exposures and generally have no value, or are even detrimental, in many chemical exposures. Too many instances have been observed where red lead chromate inhibitive primers have been used in caustic exposures with caustic-resistant finish coats. Inevitably the caustic will find its way to the primer through discontinuities, attack the caustic-susceptible pigmentation, and cause failure of the entire system by loosening the primer from the unaffected finish coat, from which it peels in sheets.

As the function of the primer is basically that of an adhesive, and as the best adhesives must possess some degree of polarity, it is apparent that we cannot obtain in the primer the complete chemical inertness required of a true industrial finish. Furthermore, since many primers contain inhibitive pigmentation, which must be reactive to be effective, it is apparent that a well-performing primer can never be considered a resistant, inert coat. Thus the primer must be protected in severe industrial exposures with finish coats which are inert and resistant to the exposure. One coat will not suffice, as inadvertent pinholes, air accumulations, or holidays and skips will leave the less-resistant primer exposed to the environment. A second finish coat must be applied to cover such discontinuities in the first finish coat, thus reestablishing the minimum requirement of a three-coat system.

The thickness required of the primer is only that sufficient to accomplish its purpose of providing mutual adhesion to the substrate and following coats, and providing an adequate supply of any required inhibitor; 1 to $1\frac{1}{2}$ mils is normally adequate.

Intermediate and Finish Coating

In industrial painting, intermediates are usually of the same composition as the finish coats, although preferably differing in color or tint in order to provide contrast between coats. Both then are designed to have optimum inertness and resistance to the intended exposure.

Since the primer is usually applied at about 1-mil dry thickness and there is a minimum requirement for total barrier thickness of 5 mils, it is apparent that the intermediate and finish coats should be adequate in build, and applied in sufficient number of coats to yield in their aggregate more than 4 mils dry thickness. In order to obtain this thickness easily in a minimum number of coats, the intermediate coat formulation is sometimes "bulked" with fillers, increasing its permeability. In such cases, it is important that adequate thicknesses and coats of resistant seal coats be applied over such "bulking" intermediate coat.

REFERENCES

1. International Nickel Co., "Corrosion in Action," 1955.
2. Steel Structures Painting Council Specification SSPC-PS 12-64P, "No. 12 Zinc-Rich Coating System in Preparation."
3. Tator, K., "Organic Coatings Indicator Chart," *Chem. Eng. Prog.* (August, September, October, November 1955).
4. Tator, K., "Corrosion Resistant Zinc Paints," *Prod. Finishing*, 64 (February 1963).
5. Tator, K., "How Protective Coatings Fight Corrosion," *Chem. Eng.*, 143 (December 1952).
6. Pierce, R. R., "Key to Savings in Painting Costs," *Chem. Eng.*, 149 (May 1952).
7. Young and Gerhart, "Film Continuity of Synthetic Resin Coatings," *Ind. Eng. Chem.*, 1277 (November 1937).
8. American Concrete Institute, Committee 515, "Guide for Protection of Concrete Against Chemical Attack by Coatings and Other Corrosion Resistant Materials," February 1966.
9. International Lead Zinc Research Organization, Project ZC-65, "Evaluation of Effect of Zinc Surface Conditions Upon Painting."
10. Steel Structures Painting Council Specification SSPC-SP10-63T, "Near-White Blast Cleaning."
11. "Billion Dollar Paint Thief," *Factory*, 140 (August 1959).

Supplemental Reading

Steel Structures Painting Council, "Good Painting Practice," Vol. I, 1954.

32

*Aluminum Pigments and Paints**

HISTORY

Aluminum and other modern flake metal pigments are descendants of gold and silver leaf powders whose production and use as pigments are described in manuscripts dating back to the early days of the Christian Era. Under the hammer of the skilled gold beater, nuggets or other small pieces of malleable gold and silver were driven out into foil or leaf between sheets of leather. Common gold leaf has a thickness of about 0.1 μ and will transmit green light. This is roughly the thickness of many of the particles in the extra-fine lining grade of aluminum pigment. Disintegration of the metal leaf to pigment was accomplished by rubbing the leaf on a polished slab of hard rock, e.g., porphyry, using a muller of similar rock and diluted honey or the like as a vehicle. As other metals and alloys (e.g., brasses, bronzes and tin) became available, these were converted to pigments in the same way.

At the time aluminum pigments became readily available, they were denoted as aluminum or silver bronzes because of their relationship to the physical character of the gold bronzes which were well known at the time. This is still used to some extent, particularly in art circles.

Early in the nineteenth century the demands for gold bronze pigments for gilding objects of art became so great that the needs could not be met. Gold bronze pigments which are derived from copper-zinc alloys cost about 25 dollars/lb at that time. The great difference in cost between the raw brass at a few cents per pound and the gold bronze pigment attracted the attention of the great British inventor, Henry Bessemer,[6] who may be regarded as the founder of the modern metal flake industry. In about two

*By Alex F. Knoll, Chief Chemist, Alcan Metal Powders Inc., Elizabeth, N. J.

years, Bessemer developed an all-mechanical process for making the pigment and thereby monopolized the field for many years. In time, other mechanical processes developed in Europe. German technology in this field became the most advanced and aggressive and remained so until the beginning of World War I.

About 1890, soon after aluminum became available on a commercial scale and at an attractive cost, the metal appeared as a pigment. The mechanical stamping methods used were attended by the development of considerable dust. While this was no hazard in the production of copper alloy (gold bronze) pigments, the use of dry stamping methods was accompanied by a very great hazard when applied to aluminum. Because of the great chemical activity of aluminum, dust clouds form explosive mixtures with air over a wide range of concentrations.

The aluminum pigment industry was plagued with fires and explosions until the development[14] of the wet-milling process about 1930 by E. J. Hall (not related to Hall the inventor of the basic electrochemical process for producing the pure metal). The Hall Paste Process for producing aluminum pigment was an outstanding advance. The quality of E. J. Hall's work is manifested by the fact that its use is worldwide, and the current equipment and operations differ only in minor ways from those he devised 35 years ago.

E. J. Hall's process used as a raw material the product of an earlier invention of his,[13] a process for safely producing finely divided aluminum powder of high purity in granular form. In this process, molten aluminum metal of high purity is atomized in a nozzle of special design by hot compressed air. The nozzle design is such that the atomized powder stays in the explosive range with air only in relatively minute quantities and only for a fraction of a second before the powder is diluted below the explosive limit by a large volume of cold air. The aluminum grain so produced is relatively coarse and has no pigment value as such. Despite this drastic exposure, it suffers little oxidation, to the extent of only about 0.2% aluminum oxide. It is used in pyrotechnics, rocketry, as fuel as a filler in cold solders, in plastics, as a chemical reagent and as a raw material for conversion to various kinds of aluminum pigments.

Producers of aluminum and other flake metal pigments are part of the metallurgical, rather than the chemical, industry. Present sales are roughly 30 million pounds of aluminum paste and about 5 million pounds of aluminum powder of various kinds.

METHODS OF MANUFACTURE

Ball-milling methods are used in modern production operations. The older and much less efficient stamping methods are still used in Europe in the manufacture of gold bronze. Labor costs are lower in Europe, and gold bronze dust does not form explosive mixtures with air.

The Hall Paste Process has no equal in the production of aluminum pigments. Its safety, versatility and efficiency in producing a large number of types, grades and varieties without by-product have left it without competition.

The raw material used in the paste process is usually aluminum grain. In some instances where the cost warrants it, scrap foil may be used. This is shredded or pelletized for use in the mills.

With the aid of a fatty acid lubricant and mineral spirits or similar high flash point hydrocarbon as a vehicle, the metal particles are hammered by impact of the balls in the mill into thin flaky particles. In the case of the leafing type of pigment, the stearic acid also serves to develop the all-important leafing film on the surface of the flakes and to enhance their luster. Nonleafing types of aluminum pigment may be made by the use of an unsaturated fatty acid or a similar nonacidic polar lubricant. Treatment of the leafing type with an active acid or other polar compound to destroy the leafing film is an alternative method.[2]

After completion of the milling cycle, determined by the degree of fineness required, the contents of the mill are washed out with petroleum solvent of high flash point, screened and filtered; the resultant press cake is then adjusted to the desired nonvolatile content by addition of solvent. The mill product may be converted to a dry aluminum flake powder for those uses which require that the pigment be solvent free.

PROPERTIES AND CHARACTERISTICS OF THE PIGMENTS

Flake metal pigments differ in many ways from inorganic and organic pigments. An understanding of these differences is required for their best use in coatings technology.

Particle Size Characteristics

Aluminum pigments are essentially aluminum metal in the form of thin flakes, irregular in outline and, relative to other pigments, large in two dimensions and quite small in the third. For example, the standard lining grade of aluminum pigment covers a range of from 1 to 100 μ in length and breadth, and averages about $\frac{1}{2}$ μ in thickness. The relatively large size in two dimensions is necessary to obtain a metallic appearance. Despite the small size of the third dimension, the metal is opaque. If the dimensions of the aluminum particles were those of a typical nonmetal pigment, the coating would be dark and would lack the desired metallic appearance. The opacity of extremely thin metals permits good covering or hiding properties in paint films of comparatively low pigmentation.

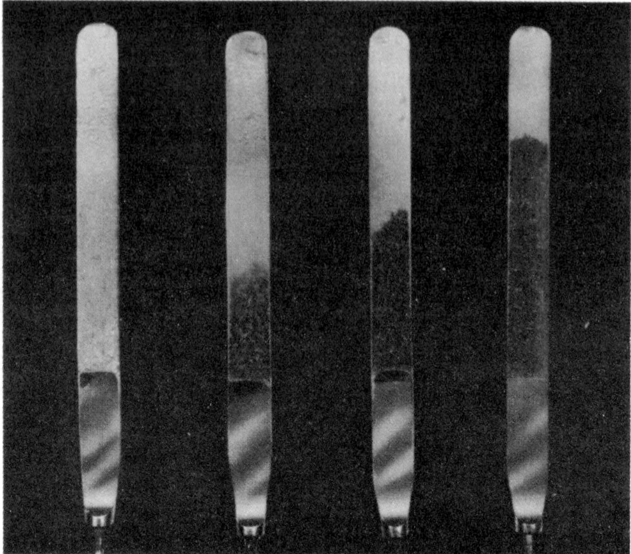

Figure 32.1. Spatula leaf test to evaluate leafing grade of aluminum pigments.

The leaf-like character of the flakes promotes orientation and consolidation of the aluminum particles in the paint film as the paint dries and sets.

A rough estimate of the particle size or grade of aluminum pigments may be obtained with testing sieves. The 100- and 325-mesh sieves are used to define the grades. Typical grade data are given in Table 32.1.

An auxiliary means of defining the grade is the water-covering capacity.[10] The pigment is first washed with solvent to remove the volatile thinner and excess fatty material. A weighed portion of the dry powder thus obtained is then spread on the surface of water in a tray. The powder is contained and compacted between two glass barriers which form part of the setup. By proper technique, a sharply defined area one flake in thickness can be obtained. From the weight of powder used and the measured area, the water-covering capacity of the power can be estimated. Results are usually reported on a dry powder basis.

Light Reflectance

The light reflectance properties of aluminum coatings are useful in evaluating both pigments and coatings. While the human eye is an extremely sensitive instrument for evaluating light reflectance, it does not yield numerical values. The most useful instruments of evaluation are 20 and 60° glossmeters and instruments which include an integrating sphere for measuring total reflectance (combined specular and diffuse reflectance).

A clean, smooth sheet of aluminum will show a total light reflectance of about 85%. Coatings made with aluminum pigments will not reach this level. About 75% is the limit. Only the two flat sides of flake particles reflect. The edges are dark and tend to absorb light, as these have a rough character. As the finer pigments have a greater proportion of edge per average particle, they tend to give darker coatings. This is indicated on the reflectance values given in Table 32.1. However, the finer particles give smoother films and tend to show greater specularity, or metallic gloss. This is true of all coatings in which the pigments are in the leafed condition.

The total reflectance of coatings in which the pigments do not leaf tends to be several per cent lower. In this type of coating, the aluminum particles are immersed in the vehicle film. Any color in the vehicle will detract from the overall reflectance. In addition, since the difference in the refractive index between the pigment and the binder is less than that between the pigment and air, the reflectance will be lower.

The visibility of aluminum paint films is related to the character of the reflectance. Highly specular or chrome-type finishes tend to look dark unless the reflected light is viewed at the proper angle. Such finishes are desirable when one requires a finish simulating bright metal. Best visibility for exterior and interior use is promoted by the use of the matte variety of pigment of larger particle size.

Heat Reflectance

Aluminum metal is a good reflector of heat. Properly prepared aluminum coatings can show an emissivity as low as 0.15. Aluminum sheet will show 0.05 at 100°F.[25] Common aluminum paint films range from 0.25 to 0.70. Other paints range from 0.92 (white) to 0.95 (black). As with light reflectance, the lowest emissivity (highest reflectance) is obtained with pigments of the greatest particle size.

Aluminum pigments and paint films are, for all practical purposes, nonconducting. The oxide film is a good insulator. If the voltage is high enough, arcing will occur and the insulation will break down.

Water Permeability

The high resistance of aluminum pigments and paints to transmission of water vapor is well recognized and often useful. In general, the standard lining grade is preferred.

Density

Aluminum metal has a low density (2.7). Aluminum pigments, paints and paint films are therefore light in weight in comparison with other pigments and their paints. This fact, along with great hiding power and opacity, makes aluminum paint preferable where weight is a consideration.

Chemical Characteristics

Aluminum pigment particles are essentially aluminum metal having the usual activity of this metal. The particles also carry a layer of aluminum oxide and stearic acid or other fatty lubricant as a result of processing conditions.

Aluminum-pigmented paints are useful in painting interiors exposed to acid fumes. Acid conditions are common in paper mills, nitration and pickling plants. Aluminum paints have excellent salt spray resistance, also. Aluminum paints should not be exposed to strongly alkaline conditions. Aluminum paints formulated with a phenolic resin varnish[24a] may be used under conditions of mild alkaline exposure.

Leafing Layer

Where the lubricant is a fatty acid, this seems to be chemically combined with the aluminum oxide layer. In the case of leafing pigments, a considerable body of physical and chemical evidence[16] indicates that the leafing property resides in a firmly adherent, oriented monomolecular layer of aluminum distearate on the surface of each flake pigment particle. In contact with a liquid or vehicle with a surface tension of about 25 dynes or more, leafing or flotation takes place. Contact angle phenomena similar to these control mineral flotation. The physical principles which apply are the same.[22]

Leafing Stability

The leafing layer is quite stable. Aluminum pigments may still leaf perfectly after many years storage. It is important for long leafing life that these pigments be stored in air- and moisture-tight containers. Properly formulated aluminum paints will maintain leafing for months and even years.

The leafing film is, however, subject to impairment or destruction by chemical and physical action. Aluminum pigments can be deleafed by long exposure to air. Oxygen undermines the leafing film with a heavy layer of aluminum oxide and causes it to lose its adhesion to the metal and its proper orientation.

The presence of acids stronger than stearic in the vehicle or solvent leads to replacement of the stearate film by the stronger acid, whose aluminum complex gives no contact angle. The strength of the acid is of greater consequence than its amount. Phthalic, maleic, acetic and other low molecular weight acids are strong deleafers.

Mechanical action of the kind developed in roller mills and long ball mill grinds deleafs aluminum. The leafing grades of aluminum should be dispersed by simple mixing. Short-time, high-speed propeller agitation is satisfactory. Any of these factors should be carefully studied before adoption for routine use.

Moisture in aluminum paints and pigments can cause trouble. Water reacts with aluminum to produce hydrogen. In sealed containers, this can give pressures sufficient to blow the lid or expand the container. Also, it can cause loss of leaf or complete destruction of the leafing film. Since aluminum pigment producers are well aware of this, water in aluminum paste is rarely a problem. Aluminum pigments are specified to contain a maximum of .1% water. The aluminum stearate, which is part of the aluminum pigment surface, adsorbs water and is present in sufficient quantity to fix up to about .2% water at a vapor pressure so low as to render the water substantially inactive. Moisture in aluminum paint vehicles should be controlled to a maximum of .05%. The distillation method for moisture determination[1d] or, alternatively, the Karl Fischer method[1c] may be used as a control.

Lead driers may be used in job-mixed aluminum paints. In ready-mixed aluminum paints, lead driers should be avoided. Lead driers used in the proportion necessary will deleaf aluminum pigments in a few days. An exchange reaction between the stearate radical of the leafing film and the acid radical of the lead drier results in the formation of a nonleafing film on the flake surfaces. Ready-mixed aluminum paints are formulated with cobalt driers in amounts which have substantially no effect on the leafing properties of the pigment.

Vehicles and solvents that have no effect on leafing at room temperatures can deleaf aluminum at higher temperature. Xylene, for example, will deleaf aluminum quickly near its boiling point. The leafing layer is sensitive to solvents over about 120°F.

In ready-mixed paints and in paints used under conditions where loss of leaf occurs, the addition of stearic acid in amounts of 1 or 2 oz/gal may help preserve leaf. Additions in excess of this may lead to soft, cloudy paint films. Proprietary leafing stabilizers are also available.[19c]

Toxicity Hazards

Aluminum pigments are substantially nontoxic. In fact, aluminum dusts have been used in the treatment of silicosis.[11] The inhalation of large amounts of dust should be avoided. Also, the dust creates a serious explosion and fire hazard. Aluminum pigments made with fatty acids meeting Department of Agriculture requirements are available.

ALUMINUM PIGMENT PRODUCTS

Aluminum pigments are sold as paste or powder. The product of the Hall Paste Process is a cake comprising aluminum flake and a volatile petroleum thinner, usually mineral spirits. The cake can be cut to the required nonvolatile content with the same or another kind of thinner. Or it may be dried to a flake powder for applications where a volatile component would give trouble. Cutbacks of the powder with a variety of special thinners, plasticizers, etc., may be obtained from pigment producers, usually on a custom basis, for use in pigmenting plastics, for textile coatings and the like.

TABLE 32.1.   Leafing Aluminum Paste Pigments

ASTM Designation	*Type II Class A*	*Type II Class B*	*Type II Class C*
Common trade designation	Varnish grade	Lining grade	Extra fine lining grade
Minimum nonvolatile content, %	65.0	65.0	65.0
Maximum retained on 325 mesh, %	15.0	1.0	0.1
Maximum easily extracted fatty and oily matter, %	3.0	3.0	3.0
Maximum total impurities other than fatty and oily matter, %	1.0	1.0	1.0
Leafing min., %	55	50	50
Other Typical Properties			
Water covering[10] (dry basis)	8000	15,000	25,000
Total reflectance[5]	73	70	63
Gloss (60°)	85–120	85–180	150–250
Bulking value	1.47	1.47	1.49

Standard Products

Aluminum pigments have become fairly well standardized since the advent of the paste process. There are many detailed specifications published, all probably based on the well-known ASTM D-962-49. Federal, state and local specifications usually conform to this. This specification is undergoing revision which will represent a consolidation of the older specifications. It will include paste and powder pigments of both leafing and nonleafing form.

Table 32.1 gives the typical ASTM classification for the leafing aluminum pastes and shows also the general properties of this type of pigment. Leafing aluminum powder pigments as well as nonleafing pastes and powders will be classified similarly in the revision of ASTM D-962. This will list four types, each with three grade classes.

Special Products

Aluminum pigment producers have a number of varieties of each conventional grade. The variations include pastes of different nonvolatile content, solvent type and surface polish. Such varieties are important particularly in specialty finishes and in the plastics, automotive and industrial finishing fields. Special pigments for exterior paints to minimize highlighting and lap marks are available also. The trade literature of pigment producers should be consulted. Aluminum pigments have advanced far beyond the silver bronzing paint pigment of the early 1900's.

TESTING ALUMINUM PIGMENTS

Methods for testing aluminum pigments have become well standardized under the administration of the ASTM. Designation D-480-59 T ("Tentative Methods of Sampling and Testing Aluminum Powder and Paste") gives details on the necessary equipment and the test procedure. The methods include content analyses for purity and adulterants and working properties. Since the specifications are readily available, it is not necessary to cover these tests here,[1] but a few observations may be helpful.

The spatulas specified in the leafing test are not available currently from any laboratory supply house. They can be cut to size from spatulas appearing in the catalogs of chemical supply houses.

Screen test values are an occasional source of dispute between laboratories. The purchase of a U. S. National Bureau of Standards certified testing sieve (325 mesh) for use as a primary standard is recommended. In addition, samples of the three ASTM classes of aluminum paste, carefully calibrated with the National Bureau of Standards certified sieve should be kept as secondary standards for calibrating and checking the sieves used in routine testing. Factors for correcting the routine test sieves should be limited to a range of 0.85 to 1.15. Any sieves requiring factors beyond this range should not be used for the screen test.

A simple, reliable method for determining the total moisture content of aluminum pigments is based on the well-known distillation method.[1d] This method, however, includes in the final result both the free water and any water combined with the aluminum oxide component of the aluminum pigment. The total content is usually less than 0.10%. A suitable apparatus for carrying out the test is available.[1d] About 300 to 500 grams of aluminum paste or the equivalent of powder is mixed with 500 to 700 ml of anhydrous hydrocarbon thinner, toluene, xylene, mineral spirits, or the like. The thinner can be conveniently dried in the apparatus itself by simply boiling the required quantity in the flask until all the contained water is transferred to the trap where it can be discarded. Alternatively, the Karl Fischer method[7] may be used.

Measurements of total reflectance may be made with any instrument with an integrating sphere component, which uses magnesium oxide as the diffusing surface. The Taylor-Baumgartner reflectometer[5] or the Bausch and Lomb "Spectronic 20" colorimeter are typical examples of such instruments. There are many others. Any glossmeter is suitable for obtaining gloss values.

The water-covering test[10] is relatively simple in principle, but obtaining reliable results requires considerable experience and great care.

ALUMINUM PAINTS

Aluminum paints are no longer limited to the simple pigmented gloss-oil bronzing liquid of the early days of this century. A great variety of aluminum-pigmented maintenance, industrial and trade sales coatings are now available. These are applied by a brush, spray, dip, roller, knife and a number of other methods.

Maintenance Paints

Leafing aluminum pigments are the usual choice of aluminum maintenance paints. This kind of aluminum paint, for maximum performance, depends on the formation of a thin, smooth, oriented and substantially impermeable metal layer at the surface of the coating, as well as layer-like orientation of the flakes within the body of the film. For maximum leafing and its preservation, the binder, volatile solvent and drier which comprise the vehicle must be carefully selected.

Vehicle Composition.   Vehicles may be made with a variety of vacuum-bodied drying oils ("OKO" types), drying oil-resin combinations, or synthetic resin solutions. Raw and boiled linseed oil and the like do not make satisfactory vehicles as they are too fluid and do not possess a high enough yield value or plasticity. They will run and sag. Vacuum-bodied linseed, soya, fish oils, etc., make satisfactory vehicles, but they tend to yield soft films and are slow drying. For areas not subject to pressure or abrasion they do very well. Oleoresinous varnishes made with a variety of synthetic drying oils and synthetic resins (paracoumarone, phenolic and ester gum) are extensively used, as are petroleum and terpene resins. Natural resins are seldom used now.

Vehicles of the cold-cut variety show the best leaf retention and are the usual choice for ready-mixed aluminum paints. These vehicles are low in acid number. Cooked varnishes are preferred because of generally better durability. However, they tend to be higher in acid number and should be pretested for leaf retention before use in ready-mixed paints.

Short and medium oil alkyd and epoxy ester resins and similarly modified polymers do not give good leafing and leveling paints when brushed. They may, however, give satisfactory leafing and uniformity of appearance on spraying. Their acidity is generally of too great activity for use in ready-mixed vehicles. They are used extensively in nonleafing finishes. These and similar types of high polymeric binder seem to give films that set easily. This interferes with leafing despite satisfactory viscosity and surface tension conditions. Nitrocellulose and many other lacquer-type vehicles behave similarly. Long oil alkyd and epoxy ester vehicles can give aluminum paints of good brushing characteristics and may have several weeks of leafing maintenance.

Aluminum vehicles for interior use are of the short oil or the straight resin type for greater hardness and faster drying. Medium oil vehicles are preferred for metal and other nonabsorbing surfaces, and long oil vehicles for similar use on exterior wood surfaces.

Vehicle Acidity. Oil or resin acids, when present in vehicles, tend to deleaf aluminum pigments. As the acid components of the vehicles is almost always more active than stearic acid, the more active acids tend to decompose the stearate leafing layer and render the pigment non-leafing. If the acid value of the solids component of the vehicle is kept below 6, serious loss of leafing will not occur unless the acid is very active—acetic acid, for example. Proprietary additives and procedures have been developed to control the effect of acidity in some types of formulations.[3,12,21]

Driers. Lead, zinc and calcium driers tend to destroy the leafing film. In ready-mixed paints, cobalt is the preferred drier. Small amounts of manganese, iron and zirconium driers may be used. In job-mixed paints, where leaf retention is limited to a matter of a few days at the most, lead driers may be used.

The following illustrates the relative deleafing action of lead and cobalt driers: In a 35-gallon oil length cold-cut varnish showing 80% leafing (15 minutes spatula test), substitution of .5% lead drier (the normal amount) in the same vehicle caused the leafing to fall to zero in one week. The same amount of cobalt drier (about ten times the normal quantity) reduced the leafing only 5% in four months. In a comparison of .05% lead to the same amount of cobalt (normal for the vehicle), the paint containing lead drier lost 10% leafing, while the cobalt lost none in seven months.

Thinners. Mineral spirits and similar petroleum thinners are commonly used in aluminum paints. Solvent naphtha, xylene and similar aromatic solvents, because of their higher surface tension, enhance the bright appearance of the paint. Odorless mineral spirits and other solvents of lower surface tension tend to suppress leafing. The leafing and reflectance of the dry paint film is not affected to the extent expected from the relative brightness of the fluid paint. As the solvent evaporates, the surface tension of the binder component becomes a controlling factor.

Alcohols and other polar solvents are generally not recommended for use in ready-mixed aluminum paints. They tend to introduce moisture, with the usual results, and also tend to react with the leafing film. In job-mixed or nonleafing applications, they may give satisfactory results. All formulations, however, should be pretested for stability to gassing and color change. There is a tendency for the aluminum to darken on continual contact with active solvents.

Moisture. Moisture in aluminum paints may cause gas pressure in sealed containers and loss of leaf in leafing paints. As aluminum pigments are self-scavenging and rarely contain free moisture, the development of gas in aluminum paints is generally the result of moisture in the vehicle or in the container. The moisture control of the vehicle should be kept below .05%. The distillation[1c] or Karl Fischer[1d] methods may be used for the determination of moisture.

Certain dehydrating agents have been recommended as additives to control the minimal amounts of moisture. "Syloid Al" and "Attaclay"[19a] in amounts up to about 2 oz/gal may be used. These additives are usually fine enough to remain unnoticed in the paint film. The use of too much additive may result in poor drying as a result of adsorption of the drier.

Aluminum Pigmentation. In general, about 2 pounds of class A or B, or 1 pound of class C aluminum paste, or the equivalent of dry powder, is used in 1 gallon of aluminum mixing vehicle. The latter is usually formulated to have about 50% nonvolatile content. Generally, the durability of an aluminum paint is improved by a higher pigment solids-to-vehicle solids ratio. A range of from 25 to 45% of aluminum in the dry film covers the range of commercial maintenance paints. Less pigment gives greater permeability, higher gloss, less rub-off; more pigment yields bright low-gloss coatings of superior durability. There is a tendency for this type of coating to collect dirt or to flake in areas where the surface is not exposed to occasional washing by rain. Rub-off is greater and adhesion is poorer with increased pigmentation.

Formulas. Formulas for maintenance aluminum paints are readily available from vehicle component manufacturers and from pigment producers.[19d] A number of federal specifications are useful in this connection.[23,24]

Heat-resistant Aluminum Paints

The heat resistant property of aluminum paint depends on the resistance of the metal and its oxide coating to temperatures up to about 1000°F. These paints are generally of two types.

In the most common type, which can be used only on clean, bare steel, the vehicle component acts merely as a carrier to deposit a thin coating of aluminum flakes on the steel surface. During exposure to heat, the vehicle component of the paint disappears by volatilization or decomposition to non-carbon-depositing substances. The aluminum particles adhere to the base by partly diffusing into and alloying with the iron. Simple solutions of hydrocarbon resins and chlorinated hydrocarbon plasticizers are representative.

The second type is meant for service up to 750°F. Here the binder remains as part of the protecting film. At temperatures up to about 400°F, short and medium oil alkyd or phenolic varnish vehicles are often used. For higher temperatures, modified or straight silicone resins, silicates and titanium-organic components form the binder.

Type I has only limited resistance to exterior exposure. Type II may be used on exterior surfaces if properly formulated.

Formulas for both these types may be obtained from resin producers.[19d,e] A federal specification[24f] gives useful information.

Trade Sales and Aerosol Paints

These paints are formulated with leafing-type pigments. Since leaf retention and stability are of great importance, freedom of the vehicle from moisture and active acidity is a necessity.

Trade sales aluminum paints are usually based on simple resin solutions or cold-cut, short oil varnishes. Polished or luster standard lining pigments are preferred. Pigmentation is at the rate of 1 to $1\frac{1}{2}$ lb/gal of paste on a 50% solids vehicle.

Aerosol paints are based on simple resin solutions. The extra-fine lining-type of paste or powder pigment is preferred. Representative formulas are given in reference.[17]

Colored Aluminum Paints

In recent years, colored aluminum paints have become increasingly popular for both exterior and interior maintenance use. Aluminum pigmentation contributes to the exterior durability of the coatings, lightens the color and adds an attractive satin-like character. In essence, these coatings are made by the addition of about $1\frac{1}{2}$ lb/gal of the standard lining grade of non-leafing aluminum paste to a typical colored exterior varnish, usually a medium oil alkyd. Typical requirements for an untinted non-leafing aluminum paint are given in reference 15. Tinting pigments may be added. Formulas can be obtained from pigment suppliers also.

Roof Coatings

Recently aluminim-pigmented roof coatings have become increasingly popular. They are now available in bituminous and resin-base vehicles, natural aluminum or colored aluminum, fibrated with asbestos or non-fibrated. Bituminous-base coatings usually employ petroleum asphalt cutbacks in mineral spirits. Gilsonite or other compatible resin is often added to raise the softening point.

Straight unfibrated cutbacks at about 50% nonvolatile content pigmented with $2\frac{1}{2}$ to 3 lb/gal of ASTM Type II Class A Aluminum paste yield an attractive aluminum-colored brushing paint. Mineral spirits is used to adjust viscosity. This class of pigment yields the best leafing and brightest appearance.

A typical fibrated aluminum pigmented cutback may contain about 50% nonvolatile content, about half of which is asbestos (7T or 7M grade) plus aluminum. Roughly 20% of the combined pigment and fiber is aluminum.

The fibrated type of coating is generally pigmented with the standard lining grade (ASTM Type II Class B) of aluminum paste. As the fiber interferes with free leafing, pigment strength rather than leafing rate is required. Therefore a finer pigment is used.

Informative specifications covering fibrated coatings will be issued in due course.[24d]

Colored roof coatings are now available. Principles of formulation are given in reference 18.

Aluminum Priming Paints

In systems requiring more than one coat of aluminum paint, past practice has required the use of Prussian Blue or the like to tint the first coat, usually of the leafing type, to obtain color contrast as a guide to complete covering with the next coat.

Recent practice is the use of a nonleafing aluminum paint[15] in the base coat. This may be fortified with lead-silico-chromate or other anticorrosive pigments if desired. Formulas are available.[4, 19b]

Industrial Finishing

Novel effects can be obtained by the use of aluminum pigments in special vehicles. This field is an active one, and the development of new pigments for these critical applications requires the constant attention of aluminum pigment producers.

Hammer Finishes

This type of finish is obtained by dispersing a small amount of non-leafing aluminum pigment in a fast-setting vehicle with the proper solvent balance. Styrenated alkyds are commonly used. The hammer effect is given by small circular patches in which the flakes lie flat and reflect light, surrounded by sharply defined perimeters where the flakes stand on edge. Hammer finishes are applied at film thicknesses of about .7 mil or less. The finest grade of nonleafing aluminum paste must be used to obtain smoothness and high gloss. A variety of colored hammers can be obtained by dispersing colored pigments in the vehicle along with the aluminum. An excellent treatment of hammer finishes may be found in reference 20.

Colored Metallescent Enamels

Metallescent effects can be obtained by the addition of small amounts of nonleafing aluminum paste to highly colored transparent enamels. To obtain the maximum brightness, highlighting, or flop and three-dimensional effect, pigments of the largest particle size consistent with film thickness are required. This type of finish has considerable use in the automotive field. In film thicknesses of about $1\frac{1}{2}$ mils, finely screened varieties of the coarser grades are used. At film thicknesses under 1 mil, the finer grades are employed. The finer pigments tend to darken and reduce the metallic character of the finish.

Flamboyant Finishes

These are two-coat colored systems where the maximum metallic character is obtained. They are popular as novelty finishes on bicycles and similar sport goods. They are obtained by the application of a transparent, colored finish coat over a bright, smooth aluminum base coat. Both leafing and nonleafing aluminum pastes are used, depending on the adhesion requirements of the top coat. When applicable, the luster variety of standard lining aluminum paste is used.

Brilliant or Chrome Finishes

These finishes are based on the polished or luster variety of extra-fine lining aluminum paste dispersed on a rapidly drying vehicle of high leafing potential. As these finishes mar and fingerprint easily, they must be used where exposure to handling is not a factor. In properly formulated finishes of this type, a clear topcoat may be used to prevent marring at the expense of some loss in metallic character.

APPLICATION METHODS

Brushing is the simplest method for small work and intricate jobs. Aluminum paint spreads easily. A minimum of brushing avoids streaking and agglomeration. Brushing is generally not a satisfactory method of applying nonleafing aluminum paints. A streaky uneven surface results.

Air spraying is the most commonly used method for applying leafing and nonleafing aluminum coatings. In general spray tends to yield a brighter, but less specular finish with leafing paints. The pressure and nozzle should be chosen to avoid fogging of the paint.

Airless spraying is becoming increasingly used, particularly on large-scale work to avoid losses and overspraying. Standard and extra-fine lining pigmented paints work well. Coarser pigments often clog the spray gaps, but they may give satisfactory results with nozzle gaps selected for the grade.

Hot spraying of leafing aluminum paints is not practical as the paint rapidly loses leaf. Nonleafing paints should be no problem.

Electrostatic deposition with air atomizing has been used, for some years, in the application of leafing and nonleafing aluminum paints. More recently, electrostatic atomization and deposition has been used more extensively for coating applications. Difficulties have been encountered, however, particularly with nonleafing aluminum pigments. The pigment tends to settle out in the line, and thereby causes short circuits at the high voltages required. Application can succeed in paints pigmented at less than about 4 oz/gal. In view of the highly specialized requirements of this coating method, advice on the subject of formulation and applicability is best obtained from the equipment manufacturer.

Dip and flow coating methods are successfully used in the application of leafing aluminum paints. However, loss of leafing can be a problem. Best results are obtained with large operations where the daily paint replacement is high, upwards of 30% by volume. Air agitation and entrainment should be avoided as this introduces air and moisture which causes loss of leafing. Any agitation should be kept at a minimum. The vehicle should be formulated for maximum leaf preserving properties. The addition of stearic acid at 1 or 2 oz/gal or a proprietary stabilizer may be helpful. These methods of application are not practical with nonleafing coatings. Streaking and other nonuniformities result.

Roller-coating methods are applicable to both leafing and nonleafing coatings. In general, the finest grades of both types are used since roller coating is applied at a film thickness of about $\frac{1}{2}$ mil.

Hand roller coating is practical for applying leafing and nonleafing maintenance paints. These paints are applied at film thicknesses which make the use of the coarser grades of pigment practical.

REFERENCES

1. American Society for Testing and Materials, 1916 Race Street, Philadelphia, Pa. ASTM Designations (a) D-962-49; (b) D-480-59 T; (c) D-1744-64; (d) D-95-62.
2. Babcock, G. M. (to Reynolds Metals Company), U. S. Patent 2,309,377 (Jan. 26, 1943).
3. Babcock, G. M., Rethwisch, F. B., and Woolsey, W. P. (to Reynolds Metals), U. S. Patent 2,904,525 (Sept. 15, 1959).
4. Babcock, G. M., and Rethwisch, F. B. (to Reynolds Metals), U. S. Patent 2,701,772 (Feb. 8, 1955).
5. Baumgartner, G. R., *Gen. Elec. Rev.*, 525 (November 1936).
6. Bessemer, H., "An Autobiography," p. 53, London, 1905.
7. British Standard BS 388: 1964, p. 9, "Specification for Aluminum Pigments," British Standards Institution, 2 Park Street, London, W. I.
8. Brown, M. H. (to Aluminum Company of America), U. S. Patent 2,848,344 (Aug. 19, 1958).
9. Dowell, R. B., *Offic. Dig. Federation Soc. Paint Technol.*, **36** 244 (1964).
10. Edwards, J. D., and Mason, J. B., *Anal. Ed.*, **6**, 159 (1934).
11. Edwards, J. D., and Wray, R. I., "Aluminum Paint and Powder," Third ed., p. 205, New York, Reinhold Publishing Corp., 1955.
12. Fleming, C. S., U. S. Patent 2,312,088 (Feb. 23, 1943).
13. Hall, E. J. (to Metals Disintegrating Company), U. S. Patent 1,659,291 (Feb. 14, 1928).
14. Hall, E. J. (to Metals Disintegrating Company), U. S. Patent 2,002,891 (May 28, 1935).
15. Keane, J. D. (editor), "Steel Structures Painting Manual," Vol. 2, p. 203, Steel Structures Painting Council, Pittsburgh, Pa., 1964; Specification SSPC 101-64 T.
16. Knoll, A. F., *Offic. Dig. Federation Paint Varnish Prod. Clubs*, No. 318, 447 (1951).
17. Knoll, A. F., *Aerosol Age*, **10**, 22 (1965).
18. Luyk, K. E., and Rolles, R., *Offic. Dig. Federation Soc. Paint Technol.*, **37**, 1055 (1965).
19. Manufacturers* of raw materials who are sources of information on vehicles and formulas:
    (a) Drying agents:
        Davison Chemical Company, Baltimore, Md.
        Minerals and Chemicals Philipp Corp., Menlo Park, N. J.
    (b) Priming pigment additives:
        Holland-Succo Color Company, Holland, Mich.
        Mineral Pigments Corp., New Kirk, Md.
        National Lead Co., New York, N. Y.
    (c) Proprietary Additives:
        Aluminum Co. of America, Pittsburgh, Pa.
        Byk-Gulden, Inc., Hicksville, N. Y.
        Raybo Chemical Company, Huntington, W. V.
    (d) Resins and Oils:
        Allied Chemical Corp., Plastics Division, Morristown, N. J.
        Archer Daniels Midland Co., Minneapolis, Minn.
        The Baker Castor Oil Co., Bayonne, N. J.
        Cargill, Inc., Minneapolis, Minn.
        Hercules Powder Co., Wilmington, Del.
        Marbon Chemical Division, Borg-Warner Corp., Washington, W. V.
        Neville Chemical Co., Pittsburgh, Pa.

*Addresses of raw material producers may be found in *Chemical Week*, Buyers Guide Issue, 1966, Part 2, (Oct. 16, 1965).

Pennsylvania Industrial Chemical Corporation, Clairton, Pa.

Reichold Chemicals, Inc., White Plains, N. Y.

Rohm & Haas Co., Philadelphia, Pa.

Velsicol Chemical Corporation, Chicago, Ill.

Washburn-Purex Company, Chicago, Ill.

(e) Silicone Resins

Dow Corning Corporation, Midland, Mich.

Union Carbide Corporation, Silicones Division, Waterford, N. Y.

20. Northwestern Society, *Official Digest Federation Soc. Paint Technol.*, **33**, 1408 (1961).

21. Rolles, R. (to Aluminum Company of America), U. S. Patent 3,085,890 (April 16, 1963).

22. Taggart, A. F., "Handbook of Mineral Dressing," Vol. 1, Sec. 12, p. 37, New York, John Wiley & Sons, 1945.

23. U. S. Department of Interior, Bureau of Reclamation, "Paint Manual," p. 85, 144, U. S. Government Printing Office, Washington, D. C., 1961.

24. U. S. Government Paint Specifications: (a) CGS-52-P 2a; (b) TT-V-81d; (c) TT-V-119; (d) TFC-001079; (e) CE-1409 Paints P2, P3, P4, V-102; (f) MIL-P-20087.

25. Wilkes, G. B., "Heat Insulation," p. 190, New York, John Wiley & Sons, 1950.

33

*Aerosol Coatings**

DEFINITION

Webster defines the noun *aerosol* as "a suspension of fine solid or liquid particles in air or gas, as smoke, fog or mist." With the vastly expanded use of aerosols in the past decade, the word has come to have wider uses. It is used not only as an adjective (as in "aerosol container") but also to identify many products which are dispensed by pressure in a stream, a powder, a heavy spray or a jet squirt.

Today the word aerosol describes not only a wide variety of products and materials, but also industries and even publications. The entire pressurized product industry is referred to as the "aerosol" industry. We have aerosol trade associations, aerosol magazines, aerosol valves, aerosol overcaps—none of which even remotely resemble "smoke, fog, or a mist."

DESCRIPTION

Basically, an aerosol consists of a gastight, valved container filled with a formulation of propellant and active ingredient. When the valve is operated, the gas pressure in the package pushes the propellant and active ingredient through the orifice in the valve. As it leaves the valve orifice, the propellant instantly expands or literally "explodes," rapidly breaking the active ingredient into extremely small particles. Thus fine mists, coarse sprays, foams or puffs are produced—depending on the choice of the container, valve, valve actuator, propellant and active concentrate.

The variables involved determine the end product.

The majority of pressurized packs are two-phase systems. They consist of a single liquid phase and a single gaseous phase. Two-phase systems, when the liquid consists of a homogeneous mixture of product and liquid propellant, account for the greatest portion of pressurized packs manu-

*By Joseph Marchbank, Sprayon Products Company, Cleveland, Ohio.

factured today. When the valve is operated, the vaporization of the liquid propellant causes the product portion of the mixture to be scattered into particles whose size is determined principally by the propellant/product ratio. In "space sprays," such as room deodorants, bactericides and insecticides, the amount of product in the liquid phase varies from 2% to no more than 20% by weight.

By increasing the amount of product beyond 20% to a maximum of 75%, with a corresponding reduction of propellant in the liquid phase, "surface sprays" result. This term is used as the particles can no longer remain airborne because of their larger size. Hair sprays, suntan oils, moth proofers, lubricating oils, paints and varnishes and other coatings are typical examples of "surface sprays."

There are many variations and subjects that could be discussed. Since this chapter concerns paints and coatings, we will ignore the three-phase systems which apply mainly to foams such as shave creams, upholstery cleaners and other aqueous items.

COMPONENTS

A complete aerosol package consists of four basis components:
(1) The container,
(2) The valve and its actuator or nozzle,
(3) The propellant or mixture of propellants,
(4) The base product.

These components will be discussed briefly only as they apply to protective coatings. Other considerations will be ignored to pinpoint the subject.

Containers

Aerosol containers are available in a wide range of sizes to meet the diverse uses and needs of products marketed. The sizes range from 2 to 32 ounce capacity and are available made of tin plate, aluminum, glass and plastic. Most aerosol paints are packaged in metal containers of the two-piece and three-piece type.

The primary function of an aerosol container is to contain the pressure required to dispense the product properly and hold the product without deterioration during the required shelf life. The filled container must comply with I.C.C. Tariff Regulations covering pressurized containers (presently I.C.C. Tariff No. 19), and must be tested as specified for the product.

The three-piece tin plate container is suitable for most coating materials and is in widespread use. It is the least expensive of the containers. The weight of the tin plate ranges from about $\frac{1}{4}$ to 1.25 pounds; however, in most cases $\frac{1}{4}$- and $\frac{1}{2}$-pound tin plate are used. Where a solvent-based product is used, the tin plate generally affords sufficient protection. However, should corrosive solvents or moisture be present, then the use of a protective coating should be investigated. It should be kept in mind that the use of a container with an internal coating does not eliminate the possibility of corrosion. It does substantially reduce the rate of corrosion and will extend the shelf life of the container. Unfortunately, the organic coating may contain pinholes or may become scratched or damaged during the fabrication process, leaving exposed metallic surfaces. Generally speaking, the tin plate container will suffice for protective coatings.

Valve and Nozzle

The valve plays a vital role in pressure container effectiveness. It must be leakproof, provide the proper spray or stream, offer the desired spray or discharge rate, and perform satisfactorily throughout the life of the product. The valve must be selected with great care since it, along with the actuator, is responsible for delivery of the product in its desired form. The valve consists of a mounting cup, stem, housing, gasketing and spring. The valve mounting cup consists generally of 1-pound electrolytic tin plate with a flowed-in gasket. The stem and housing components are generally made of nylon or "Delrin" and may contain various orifices to deliver the container contents at the desired rate. Many variations are available.

Certain valves contain a valve gasket which usually is made from buna N or neoprene. These compounds have been found to be most resistant to solvents found in aerosol paints. The spring generally used in aerosol valves is made of stainless steel.

Each valve is fitted with a dip tube which extends to the bottom of the can. Polyethylene and polypropylene have been used successfully for this application. The diameter of this tubing is important although most aerosol paints utilize tubing having an inside diameter of 0.120 to 0.122 inch. Smaller-sized tubes are often referred to as capillary tubing, but they find little use with aerosol paints.

As indicated previously, the actuator is responsible, along with the valve, for delivering the product in its desired form. There are many actuators available, and many of them have been used for aerosol paint products. The actuators are made of polyethylene, polypropylene or nylon, and may contain one or more internal and external orifices. An actuator for a paint aerosol generally will have one orifice and may vary in size from .016 to .020 inch. The possibilities and variations are much too numerous to detail.

In every instance, the propellant must be chosen carefully to obtain the desired product and package effectiveness. Two classes of propellants are available—liquefied propellants and nonliquefied or compressed propellants.

Propellants

The nonliquefied or compressed gas propellants include nitrous oxide, carbon dioxide and nitrogen. At usable temperatures and pressures, these three materials are gases and exist in a container only as a gas. They have relatively low solubilities and, as a result, act as a piston to force the product through the valve opening. The pressure of these aerosols tends to drop rapidly during use. Since their solubility is limited, they are not present in the product in high enough quantities to offer atomization when expelled through the valve. Although they offer good cost saving possibilities or potential, they have not proved suitable for dispensing protective coatings.

The liquefied propellants are used practically all the time in coating aerosols. The large number available can be grouped into four categories: fluorocarbons, chlorinated hydrocarbons, straight-chain aliphatic hydrocarbons and mixtures.

The aerosol industry was founded on the fluorocarbons, and although there are a large number available, propellant 12 and 11 are the two used in paint aerosols. The chief advantages of fluorocarbon propellants are low toxicity and nonflammability.

Certain chlorinated hydrocarbons often are considered propellant components, although they function more as solvents or to lower the pressure of the high-pressure propellant portion. The chief materials in this class are methylene chloride and 1,1,1-trichloroethane.

Straight-chain hydrocarbons employed as paint aerosol propellants include propane and isobutane. Often the propellant consists of mixtures of these components. One of the chief advantages of straight-chain hydrocarbon propellants is their low cost. They are, of course, flammable. In order to utilize the advantages of the different propellant components, much use is made of various mixtures such as nonflammable mixtures of fluorocarbons and hydrocarbons. In addition, there are other materials used with the fluorocarbons which produce blends that offer solubility and cost advantages. Dimethyl ether and vinyl chloride blended with propellant 12 have proved to be useful tools to the aerosol formulator.

With the preceeding brief discussion of the components of a paint aerosol, we can proceed to discuss the formulation of a paint to make up the active ingredient. This discussion involves the coating material, solvents and propellant. These components cannot be separated when the liquid contents of an aerosol are considered.

During the early days of the aerosol industry, paint aerosols were formulated by simply taking existing nonaerosol paints and diluting them with propellant. Little, if any, attention was given to solubility, settling or valve clogging. As a result, many of these products failed due to one or more of the above mentioned characteristics. Precipitation of the resins due to the nonpolar character of the propellant resulted in inferior products which in many instances were nonfunctional. Settling of paint pigment and other solids occurred which would not redisperse upon shaking or which, if redispersed, would form relatively large particles or seeds. Fortunately, most of these problems have been overcome, as evidenced by the rapidly increasing sale of aerosol paint products and other coatings.

The paint formulator must think of an aerosol coating as a special formulation as he does when he works on a dip coating material or one for electrostatic spray. The special requirements dictate a modified approach to the proper formula. He must think of the complete formulation including the propellant. The propellant is basically a very fast-evaporating solvent with very poor solubility properties. The kauri-butanol value of most propellants is less than 25.

PAINT FORMULATION

The formulation of an aerosol paint must take the following factors into consideration:
   (1) Solubility,
   (2) Evaporation rate of solvents,
   (3) Settling or suspension,
   (4) Float and flocculation of pigments,
   (5) Grind,
   (6) Viscosity.

These properties refer to any paint coating whether it be a nitrocellulose or acrylic lacquer, a long oil alkyd or a vinyl-toluene-modified alkyd.

The solubility of the base polymer must be considered to allow for the poor solution characteristics of the propellant which must be thought of as a diluent. This is often accomplished by the use of aromatic hydrocarbons in place of aliphatic or by utilizing polar solvents. Solvents such as acetone and methyl ethyl ketone have worked successfully. Adjustments must be made in the total solvent blend to maintain solvency of the film-forming polymer in the overall solvent-propellant mixture. The evaporation rates of solvents in the aerosol must also be adjusted to permit good breakup or atomization, satisfactory flow-out, and to avoid running and sagging. Normally the solvents required need to be much faster in evaporation rate than those for normal spray use.

One of the greatest faults of paints in aerosols is hard settling of the pigments. The pigment settles so firmly that on agitation it is redispersed as hard particles or seeds. Suspension agents must be incorporated to permit only soft moist settling that redisperses to a smooth, homogeneous mixture. Bentones, "M-P-A" and many other agents have proved adequate. The formulator must judge each system of suspension agents in the particular vehicle involved. It is prudent to avoid suspension agents that tend to flocculate as they will cause color float and even loss of gloss. The aerosol will show the same color and gloss variations as a dipping material. A good test is to reduce and spray the material under consideration, then make a flow out of the remaining reduced material. If a color or gloss variation is apparent, the aerosol spray out will closely resemble the flowout. Suspension is a property that cannot be over emphasized.

An aerosol paint should have a North Standard grind of 7 to offer the best gloss. An aerosol is applied in multiple thin coats so that a grind of 6 may show up as being grainy due to the particles protruding through the surface. The valves will handle grinds of 3 to 4 without clogging so these precautions regarding grind have nothing to do with the operation of the aerosol. They are strictly concerned with appearance. A normal spray application lays down a wet film in excess of 2 mils which allows the liquid to float and literally drown the coarse particles. An aerosol spray application is much thinner and is dry enough that the particles protrude through the surface, resulting in a grainy appearance. For best appearance, the grind should be maintained at or close to a North Standard value of 7.

The viscosity control will vary with the polymer as in normal spray application. Some coatings atomize well at high viscosities while others resist breakup and require further reduction to obtain the desired atomization. This principle applies to aerosols as well as to conventional use.

With rare exceptions, almost any type of coating can be packed in an aerosol can. This statement must be qualified since the product must be inherently stable. The aerosol will not add any property that the base material does not have. It is a means of delivering a protective coating in a convenient manner.

34

*Paint and Varnish Removers**

As paint formulations improve, it becomes more of a problem to remove coatings in order to apply a new coat of paint. Special stripping compositions have been developed to keep up with advances in modern paint technology.

Coatings which are particularly difficult to remove are catalyzed epoxies, polyurethanes and thermosetting acrylics. The older the coating, the more difficult it is to strip.

There are two broad types of paint removers: the solvent type and the chemical type which is usually alkaline in nature.

Paint Removal

Paint and varnish removers which are usually applied by brush but can be applied by spray or other methods must contain a thickener so that the paint and varnish remover will stay on a vertical surface, such as a wall, long enough to soften the paint. For coatings that are difficult to remove, the painted part is immersed in a tank of the stripper.

The problem of removing a coating is often very complex. Some of the factors which influence the problem of a stripping operation are:

(1) Type of film former,
(2) Thickness of coating,
(3) Type of surface applied,
(4) Primer used,
(5) Type of pigment,
(6) Curing time, temperature and method,
(7) Age of coating.

Formulation

The correct paint and varnish remover for a given application is the one which does its task satisfactorily at the lowest cost. An inexpensive ma-

*By C. R. Martens, The Sherwin-Williams Co., Cleveland, Ohio.

terial which requires the use of heat, is slow acting, or takes several applications may cost more than a higher-priced remover.

When choosing a paint remover, the nature of both the coatings and the substrates must be taken into consideration. Frequently a single stripper must work on a variety of materials, whose composition is often unknown. In this case, methylene chloride strippers are generally used. However, when a paint remover is needed for a specific application such as stripping of rejects of a coating of known composition, then several considerations should be made.

Most common substrates are unaffected by solvent-based strippers. Some exceptions are plastics and certain aluminum alloys which are corroded slightly by chlorinated hydrocarbons or their decomposition products. Some of the modifiers incorporated into organic solvents may have undesirable effects. For example, amines may discolor wood or react with copper or cadmium surfaces, and acids may corrode ferrous and magnesium alloys. Aqueous strippers should not be used on wood since water raises the grain. Caustic rapidly attacks aluminum and zinc alloys.

Solvent Paint and Varnish Removers

The most popular type of paint and varnish remover is the solvent type. These removers contain a primary solvent plus a co-solvent, activator, thickener, evaporation retarder, corrosion inhibitor and emulsifier (for water rinsable type).

A number of factors must be considered in formulating a solvent-type remover. All of the desirable features may not be realized in one product, but the goals are rapidity of action, correct viscosity, nontoxicity, nonflammability, low odor, clean rinsability, noncorrosiveness and package stability.

Solvents. Since the active ingredient is the solvent, most attention is paid to its selection. By far the most popular paint remover solvent is methylene chloride. It is practically nonflammable, thus minimizing fire hazards, one of the greatest dangers in paint removal. Methylene chloride is one of the least toxic of the chlorinated hydrocarbons. It is also one of the most efficient paint removing solvents.

In general, the efficiency of a chlorinated solvent decreases as the chlorination or the chain length of the organic radical increases.

One method of rating the efficiency of a solvent for paint removal is the time required to wrinkle a standard (oleoresinous) paint film. As methylene chloride is the most effective, it is given the top rating as shown below.

Methylene chloride	100
Chloroform	69
Ethylene dichloride	45
Trichlorethylene	36

Monochlorobenzene	36
Carbon tetrachloride	24
o Chlorobenzene	22
Propylene dichloride	15
Trichlorobenzene	10

Other solvents which will soften paint films are (in approximate decreasing order of efficiency): ketones, esters, aromatic hydrocarbons, alcohols and aliphatic hydrocarbons. Many less common solvents are useful for specific applications, but they are too costly for widespread use. Some of these are: 2-nitropropane, dimethylformamide, dimethyl sulfoxide, tetrahydrofuran and 1,1,2-trimethoxyethane.

Co-solvent. Combined with methylene chloride, methanol is one of the less expensive and most frequently used co-solvents. The general rule to follow in adding co-solvents to methylene chloride is that polar solvents reduce stripping time, nonpolar solvents do not. Small concentrations of polar solvents are more effective than large concentrations except when used on air-dried phenolic coatings and shellac.

Activators. Activators enhance penetration of the solvent into the protective coating. With a methylene chloride remover, the addition of enough water to saturate the methylene chloride will reduce the time as much as 90% for some formulations. A small amount of acetic acid (0.2%) will speed up removal further.

Amines and other alkaline ingredients are often used as activators. Some of these are ammonia, monoethylamine, morpholine, etc. Amines, however, sometimes react with chlorinated solvents and stain wood.

Thickeners. The thickener must meet four major qualifications: (1) It must impart high viscosity at low solids concentration. (2) It must be compatible with the blended composition. (3) It should maintain uniform viscosity during storage. (4) It should form a soft, nonadhering film upon drying. The thickened system should be thixotropic for easy application and nonsag properties on vertical surfaces.

A number of substances, organic and inorganic, are used as thickeners. The most important are organics such as methylcellulose, ethylcellulose, cellulose acetate and nitrocellulose. Also used as thickeners are materials such as bentonite, metallic soaps, starch, zein, casein, polyacrylate esters, etc. If surface-active agents are used, methylcellulose is practical in a flush-off formula. The viscosity of the thickener is influenced not only by the solvent but also by the co-solvent. The thickener in flush-off formulations should be water soluble or at least water dispersible in the presence of emulsifying agents.

Penetrants. Wetting agents, primarily amines, not only promote wetting power, but also augment the penetration of the coating and make rinsing easier. Other agents used are petroleum sulfonates and ethylene glycol monobutyl ether.

Evaporation Retarders.   Evaporation retarders in paint and varnish removers are waxes of petroleum origin. Upon application the wax crystallizes out to form a film on the surface, retarding the evaporation of the solvent. The concentration of wax should be kept at a minimum, as all wax must be removed from the surface before repainting since wax will soften, decrease adhesion and possibly hinder the drying of the new paint.

A typical formulation is shown in Table 34.1.

TABLE 34.1.   Paint and Varnish
Removers (Wash-off Type)

Methylene chloride	71	gal
Toluene	3	gal
Methanol	12	gal
Di-triisopropanolamine	9.5	gal
Water	1.5	gal
Methylcellulose (4000 cps)	13	lb
"Areskap"[a]	33	lb
Potassium oleate	22	lb
Paraffin (122–124° F, mp)	16.5	lb

[a] Registered trademark, Monsanto Co.

Chemical Removers

A boiling solution of caustic soda, at a concentration of a few pounds per gallon, is an effective and inexpensive paint remover, and is often used for general stripping purposes. Additives such as sequestering agents, surfactants and activators are added to the caustic to increase stripping rates. Gluconic acid and alkali metal gluconates are good sequestering agents in highly alkaline solutions. Sequestering agents aid in the removal of paints containing oxide pigments and keep metallic ions from precipitating in alkaline solution. The function of the surfactant in these removers is to act as a wetting agent and to emulsify and remove the partially decomposed paint film from the surface. Surfactants that are stable in hot alkalies are sodium resinate, fatty acid soaps, sodium lignosulfonate and petroleum sulfonates. Activators such as phenols and their sodium salts are added to speed removal.

Table 34.2 shows a formula for a dry alkaline paint stripping compound.

Nonchlorinated Solvent Paint Removers

Because of their low cost, strippers based on solvents other than chlorinated hydrocarbons are still used for removing oleoresinous finishes and less resistant finishes. They are of the lacquer solvent type containing aromatic solvents such as benzene and toluene with co-solvents such as

TABLE 34.2. Dry Alkaline Paint Stripper

	Weight (lb)
Sodium hydroxide	85
Sodium lignosulfonate	6
Sodium gluconate	5
Cresylic acid	3
"Nacconol"[a]	1
	100

[a] Registered trademark, Allied Chemical Corp., National Aniline Division.

methanol, and sometimes ketone solvents. Activators, thickeners and waxes of the same kind used with methylene chloride can be incorporated.

Phenols and chloroacetic acids are highly active chemicals which are useful in specific situations. Specific materials which are commercially available at low cost are cresylic acid (mixed cresols) and a crude mixture of mono-, di- and trichloroacetic acids.

Removal of Epoxies. For the removal of epoxy coatings, acidic-type materials are particularly effective. One of the most effective strippers for this purpose is a solution of concentrated nitric acid in dimethyl sulfoxide used in the temperature range of 120 to 130°C.

Mechanism of Paint Removal

In order to understand why a particular paint remover is more effective at stripping certain types of coatings than others, it is helpful to consider how a paint remover works. Simple dissolution of the film is almost never observed except when shellac or lacquers are removed with certain solvents. Usually the coating is swelled, softened and raised so that it can be quickly removed by scraping or flushing with water. In solvent-type removers, the film is swelled by the solvent. In many types of coatings (e.g., oils or alkyds), materials such as alkali saponify or break down the ester linkage, thereby aiding removal. The remover attacks the weakest part of the film and increases penetration.

Methods of Paint Removal. There are four basic methods used in stripping paints. A hot flow-on method is most economical for removing multiple layers of paint from large areas. The remover, usually an alkaline type, is heated to temperatures of 180 to 212°F and applied through a spray head which is a perforated pipe or rake. The runoff is collected and recirculated. A second method is a steam gun, in which an alkaline-type remover is applied at 5 to 10 pounds of pressure.

Tank stripping is applicable where small work must be stripped. This is an excellent method for reclaiming parts in volume. Parts remain immersed until the paint is loosened. Agitation will speed paint removal. For tank immersion stripping, a welded steel tank with an overflow dam to carry away loosened paint is recommended. Where stripping is done with hot solutions, a closed steam coil or immersion-type gas burner is required. Once paint has been loosened, a high-pressure rinse effects complete removal.

Manual brushing is used when dealing with large vertical or inverted surfaces where it is not desirable to set up a hot flow on a steam gun stripping operation. The paint is removed manually by brushing on a viscous solvent-type stripper. After the paint is softened, it is loosened by steam or water pressure.

REFERENCES

1. Durney, L. J., *Prod. Finishing*, **24**, No. 6, 39–44 (March 1960).
2. McLaughlin, P. H., *Paint, Oil, Chem. Rev.*, 8–10 (Jan. 8, 1959).
3. Spring, S., *Metal Finishing*, **57**, No. 4, 63 (1959).
4. U.S. Federal Specification TT-R-230a, Remover, Paint (Alkali Type for Hot Application), August 1962.
5. U.S. Federal Specification TT-R-243, Remover, Paint (Alkali–Organic Solvent Type), April 1963.
6. U.S. Federal Specification TT-R-248, Remover, Paint and Lacquer (Solvent Type), March 1965.
7. U.S. Federal Specification TT-R-251g, Remover Paint (Organic Solvent Type), November 1965.

35

*Surface Preparation**

Surface preparation is fundamental to good painting practice. If it is done correctly, proper adhesion is obtained and the coating system gives its maximum performance. If it is neglected, or if the method of preparation is poorly selected, immediate or premature failure of the coating may result. Surface preparation may also be required for acceptable appearance of the coated object.

If corrosion products or other surface contamination is not removed, failure of the coating may result through one or more of the following mechanisms: (1) Simple mechanical interference with proper adhesive contact between coating and substrate will result in coating failure, if the contaminant loses adhesion to the substrate. (2) Absorption of moisture or corrosive chemicals by osmosis or by simple wicking action is followed by substrate corrosion and/or film degradation. Paint failure results in either event. (3) Foreign matter, substrate flaws or breaks in the coating may provide conditions for electrochemical corrosion of ferrous surfaces. In the presence of moisture and oxygen, such spots may act as cathodic half-cells; anodic sites develop on the substrate which then dissolves and forms oxidation (corrosion) products.[43] Maximum coating performance requires elimination of corrosion products or foreign matter and the greatest practical surface uniformity.

The selection of a surface preparation method is dictated by the type of substrate-coating combination to be used. As a rule, the modern, high-performance coatings are characterized by poor wettability and high cohesive strength, and tend to pull away from a poorly prepared surface. Vinyls, epoxies and zinc-rich coatings are in this category. Oil paints, on the other hand, are noted for their ability to tolerate poorly prepared surfaces. Substrates such as metal, masonry and plastics differ from each other in surface preparation and coating requirements. The type of object

*By Kenneth C. Waldo, The Sherwin-Williams Co., Cleveland, Ohio.

to be painted is important; structural steel, for example, would tolerate harsher preparation methods than would thin steel sheet.

Surface preparation may include one or more of the following steps shown in Table 35.1.

TABLE 35.1.   Degrees of Surface Preparation

(1) Removal of loosely adhering foreign matter, corrosion products or old paint, usually by hand methods.

(2) Removal of tightly adhering foreign matter which may later loosen and destroy adhesion, or which may serve as corrosion sites. Power equipment or chemical treatment is usually required.

(3) Creation of a surface roughness (etching, profile, anchor pattern) to insure mechanical adhesion.

(4) Modification of the surface by chemical treatment, to make it more compatible with the coating.

SURFACE PREPARATION METHODS

Hand Cleaning

The usual purpose of hand cleaning is to remove loosely adhering dirt, grease, oil, tar, rust, old paint, chemical fallout or other foreign matter with a minimum of time and expense. It might be used in one of these cases: nominal clean-off of large areas; preparing areas too small to justify more elaborate methods; preparing areas whose thickness, quality or closeness to an unlike surface precludes the use of harsh cleaning methods; to avoid contamination of nearby equipment by abrasives; where chemical fallout or work schedules limit the time available for surface preparation; where the limited life or value of an object makes more thorough preparation uneconomical.

The coating to be applied must have good wetting properties and at least some toleration for surface contamination. Hand cleaning is most acceptable for paints of the long oil alkyd, phenolic or oil type, although it is often used with others. Loose material may be defined in terms of the time necessary to clean a standardized area of the proposed work. One specification defines loose mill scale, loose rust or loose paint as that removed by a 2-minute wire brushing of a test area of 2 square feet.[31]

Tools for handwork include bristle brushes, brooms, chisels, scrapers, sandpaper, emery paper, steel wool, etc.

Hand washing with water or scrubbing under the stream from a low-pressure water hose is a suitable means for removing loosely adhering foreign matter. Cleaning agents may be added to the water, but washing must be followed by thorough rinsing to insure complete removal of any water-soluble residues. Low-pressure steam or medium-pressure (200 to 1000 psi) water cleaning equipment, which may include provision for application of chemical cleaning solutions or paint strippers, may also be used, and many designs are available.[15,21,37]

Solvent wiping is a common method of removing light oil and grease films. It is adequate in many cases, but the rags and solvent must be changed frequently or it may result in a more thorough contamination of the entire surface. High flash point solvents such as mineral spirits or high flash naphtha are preferred, to minimize the fire hazard, and suitable protective equipment and ventilation should be provided.

Flame cleaning by itself is not sufficient where coatings requiring excellent surface preparation are to be used, but it is an effective method for removing mill scale or thick films of old paint, oil or grease, before final surface preparation.[36]

Model specifications for the various methods are available from the Steel Structures Painting Council[50,51] and the National Association of Corrosion Engineers.[31]

Power Cleaning

Power cleaning may be used for one or more of the following purposes: (1) efficient removal of loosely adhering foreign matter where only nominal surface preparation is needed, (2) removal of tightly adhering scale or other foreign matter which may later lose adhesion or promote under-film corrosion and thus lead to coating failure, and (3) roughening of the surface to improve adhesion.

Power tools of many types are available and include various wire brushes, disks, hammers and scalers. A rotary scraper, with steel blades rotating at high speed, is sometimes useful for removing rust, old paint and mill scale.[41] A needle gun, which conforms to the surface, may prove useful for cleaning irregular surfaces such as bolt heads, weld seams and corners.[12] Power tool cleaning alone is not considered satisfactory for badly pitted surfaces, or where an anchor pattern is needed.[36] Care should be taken that rotary tools such as wire brushes do not have a polishing effect; they may sometimes soften old coatings or other foreign matter and leave fused particles firmly embedded in pits in the metal. This fused-in foreign matter may not readily be noticed and will accelerate corrosion.[45]

Water blasting involves the use of a water stream, at 3000 to 7000 psi, supplied by a high-pressure pump. The water stream will not remove tightly adhering paint or mill scale, but may be useful for flushing off loose paint, loose rust or dirt.[4,26] Such equipment would be useful in areas where dust from other types of power cleaning operations could not be tolerated or for preliminary cleaning. This equipment is not suitable for the application of detergent solutions or paint strippers because of its high fluid consumption.

Abrasive Blasting

Directing abrasive particles against a surface at high velocity is one of the most effective means of surface preparation. Many studies have demonstrated that application to a blasted surface greatly prolongs the life of maintenance coatings. Although the increase in service varies, a good maintenance paint system applied to blasted steel should last many times longer than when applied to a hand-cleaned, weathered surface.[11,16,50]

Abrasive Classification. There are several ways of classifying abrasives. One of the most common classifications is in terms of source.[13]

(1) Natural abrasives: This category includes inorganic abrasives such as sand and flint. Sand is the abrasive most often used for surface preparation for painting and will be discussed in detail below. Organic or agricultural abrasives such as ground walnut shells or corn cobs are also included. These are used mostly for the cleaning and polishing of delicate surfaces. They will not remove mill scale or tightly adhering corrosion products from steel.

(2) Manufactured abrasives: These include ore reduction by-products such as mineral slag (a copper or lead slag by-product) or utility slag (from certain coke burning furnaces). These are characterized as fast-cutting, medium-durability abrasives. The particles are often extremely sharp and needlelike, and may become embedded in the steel substrate and have an adverse effect on coating performance. Variations in raw materials and furnace operations cause wide variations in particle characteristics.[45] Nonmetallic abrasives such as silicon carbide and aluminum oxide are fast cutting, durable, uniform, and supplied in many types and particle sizes. They are suitable for an operation where they can be recovered and reused.[46] Glass beads are finding increasing use as abrasives.[27] Metallic abrasives in shot or grit form are made from steel, iron and other metals. They are initially expensive but are reusable and very efficient.[46]

Sands. Sands[24] differ a great deal in their effectiveness. The reason for this is better understood if the nature and origin of sand is considered. Sand usually may be classified as quartz or chert (flint or flint-like silica), with a silica analysis of 90 to 100%. Both are very resistant to weathering and survive the breakdown of associated rock formations. As rock weathers, they are washed downstream and eventually settle out due to changes in river velocity and accumulate in beds or "terraces." At this point, the sand has undergone the processes of formation, weathering (including both physical and chemical processes) and passage in a stream.

Its resistance to shattering on impact (i.e., its toughness) is related to stresses set up during formation, weathering and stream travel. Its particle size and shape are related to the distance it has traveled in the stream and the accompanying wearing process, which makes the particles round and smooth. These three factors, toughness, particle size and particle shape, determine its effectiveness as an abrasive.

The absolute performance of a given sand cannot be predicted from the history of the sand bed because two conflicting forces are at work: (1) The amount of soft material eliminated increases as the distance the sand travels increases. (2) However, as the sand travels, the particles become more rounded, and their cutting effect is diminished. Hence, the only sure means of predicting performance is to run a practical test on different lots of sand and select one on the basis of relative performance.

A sample size of 400 pounds is suggested, with tests run on a large, fairly flat surface, such as the wall of a large tank, taking care that nozzle air pressure and other variables are kept constant.[24] The use of specially selected sand is recommended; low-cost, unclassified river sand is often used, but may contain soft or nonangular particles which produce much dust but little cutting action, and mud or clay which is embedded in the surface and causes loss of primer adhesion and corrosion.[42,46]

Selection of Abrasives.[13,24,32,50] Factors to be considered include the type of metal, quality (rough steel, fine parts, etc.), operating conditions (exterior, cabinet, etc.), abrasive breakdown rate, suitability for recovery, initial surface condition and desired final condition, and depth of anchor pattern needed for the proposed surface-coating combination. The variables in abrasives include:

(1) Hardness—hard particles have a fast, deep cutting action.

(2) Size—larger particles cut deeper, but a finer particle abrasive is usually faster. Typical particle size recommendations are 16 to 40 mesh or 30 to 50 mesh.

(3) Uniformity—important for either a uniform deep etch or a light polishing action.

(4) Shape—angular particles have a cutting action, whereas round particles clean mainly by impact and have more of a polishing effect.

(5) Specific gravity—denser particles cut deeper at equal velocity.

(6) Chemical composition—must not contaminate surface so as to reduce coating performance.

(7) Color—may affect visual control of blasting progress or impart color to surface.

(8) Availability and cost—including long-term cost, if abrasive is recoverable.

Blasting Equipment.[11,13,36,46,50] Blasting equipment operates on one of three principles; pressure, suction or centrifugal force. In pressure equipment, which is the most common for field use, the abrasive is carried in a high-pressure stream of fluid, normally air, although water alone could be used. In the suction method, the abrasive is picked up by the vacuum created as the fluid passes through a jet. In the centrifugal force method, the abrasive is hurled from the blades of a rotating paddle wheel. Pressure equipment is generally preferred for field use because it provides a combination of mobility and high production rate.

Pressure equipment may be either gravity feed, in which the abrasive drops from a hopper into the fluid stream, or pressure feed, in which the abrasive is fed into the stream under pressure. Most field equipment is of the gravity feed type. Suction feed equipment is mostly used for light-duty work with fine abrasives where cleaning, polishing or a light etch is needed, or for small areas such as welds, since it gives a relatively low impact and low production rate compared to pressure blasting.[11,13]

Wet blasting is used to reduce dust, or to avoid excessive abrasion of the surface. A disadvantage is that a rust inhibitor must be used to avoid flash rusting of steel.[11,36,50]

Centrifugal (wheel blast) equipment may afford low-cost, effective blast cleaning, but it is best suited for permanent installation in a shop where the work can be brought to it in a system designed for conveyor line use.[50]

Design of blasting equipment is of considerable importance. The cost of safe, efficient equipment, in place of unsafe, inefficient homemade rigs, is more than offset by the savings in labor and downtime costs. Materials of construction, nozzle type, air line sizes and air pressure, as well as selection of the correct abrasive, are all important to a successful operation.[13]

Sandblasting specifications are important, to prevent misunderstanding of terms such as "good commercial blast," etc. An agreement on definitions with the sandblasting contractor promotes achievement of the coating program's objectives.[14,40] An inspector should insure that dust and spent abrasive has been removed from the less accessible portions of the work, where it might be trapped under the paint and cause film failure. If the abrasive is being recycled, it must be periodically checked for degradation and contamination, and correct adjustment of other operating variables must be verified.[40] The anchor pattern should be uniform and of the proper depth. For example, a 2-mil profile is recommended for many maintenance coatings, to be followed by 6 mils (dry film) of paint. If the profile is too shallow, cleaning may be incomplete or adhesion inadequate; if it is too deep, peaks may protrude above the coating and serve as corrosion sites.[11] Care must be taken that thin steel is not sandblasted too deeply or it will buckle. A helpful set of pictorial standards for blast cleaning and other surface preparation methods has been agreed upon by the Swedish Standards Association (SIS 05 59 00-1962), American Society for Testing and Materials (D-2200-63T) and the Steel Structures Painting Council (SSPC-Vis 1-63T).

Chemical Cleaning

Chemical cleaning procedures may be used to replace other methods, as when a drastic treatment is necessary, and acid pickling is used as an alternative to sandblasting, or where milder cleaning procedures are needed to avoid surface damage that might result from mechanical cleaning methods. Chemical cleaning methods are frequently used in conjunction with other methods of surface preparation, when more than one type of soil is present and mechanical methods alone are inadequate. Chemical cleaning methods might be used in preparation of any type of surface and with any type of coating, since the term is taken here to include a wide variety of treatments, which are chemical in nature, or in which chemicals are employed as solvents.

Chemical cleaning procedures may lead to damage of some metals and alloys. Examples are the pitting and hydrogen embrittlement often associated with acid pickling or the reactivity of aluminum with improperly formulated alkaline cleaners. Before recommending use of strong acids or alkalies for cleaning metal, it is necessary to insure that the conditions under which such damage may occur are understood and avoided.[47]

Often placed in the "chemical cleaning" category are acid pickling, alkali cleaning, acid cleaning, proprietary paint strippers, emulsion cleaning and solvent cleaning.

In pickling,[31,44,50] an acid solution is used to dissolve or loosen rust, mill scale, etc. Use of the relatively strong, hot, pickling solutions is usually inconvenient in the field, leading to erratic results because of the difficulty of maintaining good control of time, temperature and rinsing procedures. Under shop conditions where these variables may be carefully controlled, uniform surface preparation is obtained and results are in some cases comparable to those obtained by sandblasting, with the advantage that the pickling solution can penetrate to inaccessible spots which could not be reached by an abrasive blast.

Sulfuric, hydrochloric or phosphoric acid solutions are most commonly used, but nitric, hydrofluoric or organic acids are sometimes specified.[47]

Oil and grease interfere with the pickling action and must be removed beforehand, usually by solvent or alkali cleaning.

Effectiveness is primarily governed by time, temperature and acid concentration. Temperature, acid type and acid concentration are usually constant in a shop process, and soaking time is varied depending on the completeness of preparation needed.

The pickled surface is contaminated with iron salts and acidic residues which will cause paint failure if allowed to remain. Sometimes an alkaline rinse is used to neutralize the surface, but this itself may leave residues which cause failure, and a rinse in warm water is more common. A phosphate or chromate treatment may follow to further prevent rusting and form an improved base for the shop primer. In place of phosphating, a passivation treatment such as sodium nitrite may be used. The pickling solution may be treated with inhibitors to prevent pitting of the cleaned metal.[5] Selection of acid and inhibitor depends on the type of metal to be cleaned and on the scale or other deposits on it.[5,47] Foaming agents may be added to the pickling solution to reduce evaporation losses and minimize objectionable acid fumes.

In a typical process, covered in the Steel Structures Painting Council's SP-8,[51] scale and oxide removal is accomplished by use of an inhibited 5% (by volume) sulfuric acid solution, at a temperature of 170°F. After descaling, the steel is given a 2-minute rinse in 170°F water, then an inhibitive treatment in a weak dichromate-phosphoric acid bath at 190°F. The hot parts dry quickly and can be shop primed while still warm.

Methods less often encountered in connection with painting include use of concentrated aqueous alkaline solutions, molten caustic and electrolytic alkaline or acid baths.[47,50,51]

Alkaline cleaners are characterized as economical and easy to use, but they require thorough rinsing and may detract from corrosion resistance. They are available in a variety of established and proprietary formulations; common ingredients are soda ash, caustic soda, silicates, phosphates and surfactants. Emulsifiable petroleum solvents may be added. They clean by dissolving water-soluble soils, by emulsifying or saponifying fats and oils, by wetting hydrophobic soils, etc.[50]

Conditioners may be added to improve receptivity to phosphating. Where phosphating is to follow, cleaner selection is critical, as excess alkalinity may lead to formation of a coarse-grained, unusable phosphate deposit.[30]

Thorough rinsing is necessary and, on steel, a chromic or phosphoric acid rinse is suggested to prevent flash rusting and insure the elimination of the last traces of alkalinity.[50]

Alkaline cleaners are most effectively used in agitated soaking tanks, electrolytic cleaning tanks, or with pressure or steam spray equipment. Temperatures of 160 to 200°F are common. Because of the corrosive nature of the solutions and the need to use them hot for best results, hand application is used only as a last resort.

Proprietary paint strippers may be similar in formulation to the alkaline cleaners mentioned above, or they may be acidic; they can also be water or solvent based. In addition to the active chemicals, special-purpose additives such as evaporation retarders and pitting inhibitors may be included. Application may be by flow-on or immersion or in high-pressure steam or water equipment.[8,15,18,21,22,37,48]

Emulsion cleaners consist of a solvent such as kerosene and an emulsifier dispersed in water. Effectiveness may sometimes be increased by addition of a mild alkali. They tend to leave a slight oil film on the surface, even after thorough rinsing. This must be removed by washing if the paint will not tolerate it.

Solvent cleaning[47,50] is used as an alternative to the water-based cleaning systems mentioned above, when those systems are too slow or ineffective, or would damage the surface or leave a residue which interferes with subsequent painting. Solvent cleaning is effective when the soil is primarily oil, grease, fat or wax which cannot be readily removed by a water-based system. It may be used for preliminary cleaning, e.g., before sandblasting or acid pickling, or as the final step before phosphating or painting. Methods of use include hand wiping, immersion, spray and vapor degreasing.

Immersion cleaning is done in dip tanks, which may be operated hot or cold and may be open or part of a closed system. The cleaning tank should be followed by one or more rinse tanks to reduce the contamination from dirty solvent. Immersion cleaning is aided by agitation; in ultrasonic immersion cleaning, for example, the extreme turbulence caused by cavitation of the solvent promotes efficient soil removal.[48]

A variation on the immersion process is the diphase system, in which a water-immiscible solvent such as trichloroethylene is covered by a layer of water. The organic solvent removes the oil-type soil and water-soluble residues are picked up by the water layer. A water spray follows to finish cleaning water soluble residues from the work. The diphase system's main use is in paint stripping and other preliminary cleaning.[19]

Spray application may be practical in an operation where the solvent can be recovered and reclaimed.

Vapor degreasing is now a widely used method of solvent cleaning. The solvent is vaporized in a heated sump; as parts are suspended in the condensation zone, clean vapor condenses on the cold metal and carries away soil as it runs off (Figure 35.1).

Desirable characteristics of the solvent include high vapor density, stability, low specific heat, low boiling point, low toxicity and nonflammability. Certain chlorinated hydrocarbons represent a good compromise of these properties; trichloroethylene is by far the most used. Perchloroethylene finds much use and methylene chloride is used to some extent.

Many equipment designs are in use, and include open or closed tanks, and setups for continuous or intermittent operation. Figure 35.2 shows a three-stage vapor degreasing unit. Work is first immersed in boiling solvent on the left, then in clean, warm solvent (center), and finally, it is

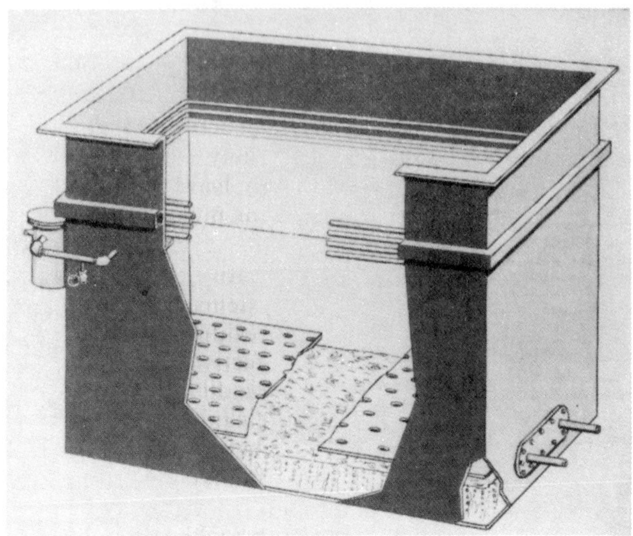

Figure 35.1. Straight vapor degreaser. (*Courtesy E. I. duPont de Nemours & Co., Electrochemicals Dept.*)

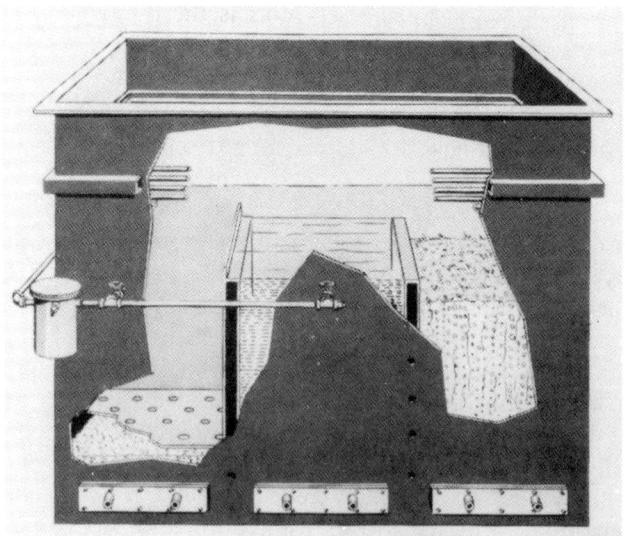

Figure 35.2. Three-stage degreaser. (*Courtesy E. I. du Pont de Nemours & Co., Electrochemicals Dept.*)

vapor-rinsed in the right-hand compartment. Equipment design for solvent cleaning should insure that solvent loss to the surroundings is held to an acceptable level for reasons of safety and economy. Conditions for formation of hazardous or damaging by-products must be understood and avoided. For example, heat, light or contact with some metals may decompose trichloroethylene to form hydrogen chloride and other harmful by-products. Stabilizers minimize this effect in commercial grades. Trichloroethylene may form explosive by-products with strong alkalies such as sodium hydroxide.[19]

METAL PRETREATMENTS

Metal Conditioning

The terms conversion coating, metal conditioning and metal pretreating are used somewhat loosely and interchangeably to describe the various phosphoric acid, chromic acid and proprietary treatments which are used to inhibit corrosion and improve paint adhesion. In addition, there are some film-forming materials which are classed as pretreatments.

Conversion Coatings. A conversion coating is defined as a uniform crystalline or amorphous deposit formed on a properly prepared surface by a chemical reaction with the base metal. This class includes many known or proprietary chromate and phosphate treatments.

Phosphating (phosphatizing)[6,7,39] is the most common of these treatments. The mechanism is believed to involve dissolving of the substrate metal by the acidic solution, neutralization at the solution-substrate interface and precipitation of a coating of secondary and tertiary phosphates. The treatment is intended to: (1) improve paint adhesion—the conditioned surface is often more compatible with the paint and its increased roughness helps adhesion; (2) improve corrosion resistance through improved adhesion and the rust-inhibitive nature of the deposit (it is especially effective in resisting undercutting from salt corrosion or detergents); (3) insure proper surface preparation (cleaning, oxide removal, etc.). A tightly adhering conversion coating is a practical guarantee that previous surface preparation has been adequate, since the conversion coating will not form correctly on an improperly prepared surface. The process may also be used to improve the lubricant retention properties of metals.

A conversion coating is mandatory for some coating requirements and is recommended for improved performance in many cases. The process has extremely wide application, ranging from treatment of reconditioned steel drums and structural steel to fine parts and coil coating stock.

A conversion coating will improve performance of most paints, and its use is desirable with many types of vinyl, epoxy and plastisol coatings.

Metals frequently phosphated include steel, iron, zinc, galvanized steel and aluminum.

Chromate treatments are also widely used. As in phosphating, a complex inorganic salt is deposited in a thin, tightly adhering layer on the metal. Aluminum, zinc, galvanized steel, cadmium and magnesium are often chromated.

Copper, brass and stainless steel may receive chromate or other specialized treatments. Terneplate is sometimes given an oxalate treatment.

A great many proprietary modifications of the basic phosphate and chromate conversion processes have been developed for specific metals and alloys and application methods. There are many experienced suppliers, including Amchem Products, Inc.,[2] International Rustproof Co.,[22] and Parker Rust Proof Co.[38] The Dow Chemical Co. is a major source of information on treatments for magnesium.[17]

Types of Phosphate Coatings.[6]   Four main types are in current use:

(1) Iron phosphate is a very thin amorphous, iron oxide–iron phosphate film which is iridescent brown or blue in appearance. It provides minimal corrosion resistance but good flexibility and adhesion at low cost and is used where economy is more important than corrosion resistance. Deposits of up to 50 mg/sq ft are common, and suppliers' literature shows suggested ranges of 25 to 120 mg/sq ft.

(2) Crystalline zinc phosphate is most commonly used as a paint base. The finish is usually a dull, light gray. It tends to provide poor flexibility and adhesion, but corrosion resistance is good and it is used where this is more important than adhesion. Deposit weight: 200 mg/sq ft (range 150 to 2000).

(3) Microcrystalline zinc phosphate has a darker, shinier appearance than crystalline zinc phosphate and superior adhesion and corrosion resistance. It is equal or superior to iron phosphate in flexibility and adhesion when used as a prepainting treatment. Deposit weights: spray 150 to 300 mg/sq ft; dip 250 to 600 mg/sq ft.

(4) Manganese phosphate gives a coarse, black deposit, which improves oil absorption and is used mainly to prevent metal-to-metal galling and scuffing in break-in service of moving parts. Deposit weight: up to 5000 mg/sq ft. Both iron and zinc phosphate treatments reduce the effect of variations in metal quality, but superior results are generally achieved in the zinc phosphate.

Phosphating is accomplished with dilute aqueous solutions composed of: (1) a primary phosphate of an alkali metal, zinc, or manganese; (2) an accelerator, which reduces processing time[20] (a treating time of 1 to 5 minutes is common for commercial formulations; a number of inorganic oxidizing agents, such as chlorates, and various organic compounds are used in the proprietary formulations); (3) phosphoric acid.

Chromate rinses are often applied after phosphating to improve corrosion resistance.[6,30] Excessive coating weight has an adverse effect on gloss, adhesion and flexibility of the paint film.[7,20]

Application. The most common methods in production line work are immersion (using a series of tanks) and spray (using a series of booths). Some phosphating solutions are formulated for brush or roller application. The steps in a typical process are shown in Table 35.2.

TABLE 35.2

(1) Cleaning (mechanical, chemical, etc.)	(5) Rinse
(2) Alkaline pretreatment cleaner (proprietary)	(6) Chromic acid wash
(3) Rinse	(7) Drying
(4) Phosphating	(8) Paint application

Proper phosphate deposition requires close control of the variables, which include:

(1) Cleaning method: The phosphating solution must be able to reach and attack the base metal without interference from cleaning residues; hence, careful selection of cleaning agents is required.[6,30,39]

(2) Application method: This affects the rate of deposition and quality of the phosphate structure.

(3) Formulation: Acid concentration, phosphate concentration, use of accelerators and other additives all influence quality and amount of the deposit.

(4) Time: This affects the total amount of deposit.

(5) Solution temperature: This has a critical effect on reaction rate and quality.

Considerations in the selection of a conversion process are: (1) type of metal and its condition; (2) final condition desired, based on the coating to be used and on the end use of the painted article; (3) process control limitations, including allowable or practical surface preparation, cleaning facilities, available equipment and desired production rate.

Other Chromate and Phosphate Treatments. The use of dilute chromic and phosphoric acid rinses as corrosion inhibitors has been mentioned in connection with pickling and alkaline cleaning.

There are other treatments which lie on the borderline between these inhibitive rinses, chemical-solvent cleaning and conversion coating. They may involve either phosphate or chromate treatment. They provide a degreasing, cleaning and etching action, and leave a slight phosphate or chromate deposit on the surface. They are used when only a nominal improvement in corrosion resistance and adhesion is needed..

Cold phosphate treatment is in this class. It involves the use of a mixture of phosphoric acid, a water-soluble grease solvent (butyl "Cellosolve," for example), a wetting agent and water. These mixtures may be used to remove rust and grease films, fingerprints, etc. They deposit a very thin phosphate coating which often improves adhesion and temporarily prevents rerusting between surface preparation and priming steps. Typical formulations are given by Spruance,[50] in SSPC-PT 2-64,[51] and in MIL-C-10578B, "Compound, Metal Conditioner and Rust Remover (phosphoric acid type)."

Basic zinc chromate–vinyl butyral wash primer consists of a vinyl butyral resin solution pigmented with basic zinc chromate, which is mixed just before use with phosphoric acid solution. It is often used in place of a conversion coating, when application of the later is not practical. It undergoes a reaction with the surface, similar to that of a conversion coating, and the film forming vinyl resin offers additional protection. The resulting film is moisture-sensitive and is not suitable as a "shop" or field primer, as it must be coated within a few hours for best results. Excellent results have been obtained through correct use of this type of formulation, which is often referred to as the "WP-1" or "Formula 117, Bureau of Ships" wash primer. Details of its use and formulation are covered in SSPC-PT 3-64,[51] and in MIL-P-15328B (supersedes MIL-C-15328B).

Other Pretreatments. A wetting oil treatment is sometimes used on rusted steel when all of the rust cannot be removed. It is not suitable for use with all types of coatings or under all service conditions, and detailed specifications for its use, such as SSPC-PT 1-64,[51] should be consulted. The wetting oils are essentially raw or boiled linseed oil, with or without driers, which are believed to penetrate through the rust and form a bond with the surface, permitting adhesion of paints which would not adhere directly to a rusty surface.

SURFACES

The type of surface affects the selection of a surface preparation method. Considerations include:

(1) Type of object: Those which are easily damaged or which need a smooth finish must be treated differently from those which can tolerate loss of substrate or a rough finish.

(2) Initial condition: A clean, active surface requires relatively little treatment, whereas one with a high degree of contamination or inertness needs more intensive or more specialized preparation.

(3) Proposed coating system: They have widely varying surface preparation requirements (anchor pattern, special pretreatments, etc.)

(4) End use: In general, the more abuse the coating must endure, the better must the surface preparation be, to obtain adequate adhesion. Final appearance must also be considered.

METALLIC SURFACES

Cast iron and mild steel are relatively simple to paint and the surface preparation method depends on such factors as initial surface condition and coating type. The more rigorous methods such as abrasive blasting, flame cleaning, power cleaning and pickling are likely to be used on structural steel. Parts to be finished on a production line will often receive coatings which require special preparation methods, so that chemical cleaning and application of conversion coatings are more likely to be encountered. Alloys often present special coating problems; rigorous surface preparation may not be permissible and the alloy may present special adhesion problems. For example, a wash primer of the WP-1 (MIL-P-15328B, 5 May 1961) type is often recommended for stainless steel.[51] Specialized conversion coatings may be used.[47]

Zinc-coated Steel

Galvanized steel often presents an adhesion problem, and many paints tend to flake from it on aging. This is true not only of oil-based paints which are believed to lose adhesion because of the formation of zinc soaps at the paint-metal interface, but also of many paints which are not considered saponifiable.[1] However, it is usually possible to obtain satisfactory adhesion. In the past, many treatments have been recommended which are now generally recognized as being of doubtful value or even harmful. Scouring or abrading is likely to cut through the zinc coating and is self-defeating. Washing with hydrochloric acid, vinegar or copper sulfate solutions was formerly often recommended, but those recommendations have now been largely abandoned. Solvent wiping with mineral spirits or naphtha is usually recommended to remove any oil, grease or ordinary dirt. Fingerprints may be removed with a 50/50 mixture of alcohol and mineral spirits or naphtha or other effective solvent blend.

Weathering is often recommended as the best surface preparation for galvanized steel; during weathering, the zinc reacts with the atmosphere to form a protective film which minimizes further reactivity between the zinc and the coating. Current thinking favors instead the use of various phosphating pretreatments or a wash primer.[51]

Galvanized steel is often given a stabilizing treatment at the mill to prevent "wet storage" or "white rust" staining during storage and shipment. The treatment interferes with phosphate deposition, and phosphating should be confined to untreated galvanized steel.[9] There is also some question of the advisability of using "cold phosphate" treatments such as MIL-C-10578B (SSPC-PT 2-64)[51] or WP-1 (MIL-P-15328B) wash primers (SSPC-PT 3)[51] on treated galvanized steel. However, specially treated grades of galvanized steel are available which may show improved compatibility with wash primers. Thorough advanced testing of any proposed surface preparation–pretreatment–coatings system is needed. The "Steel Structures Painting Manuals"[50,51] contain up-to-date detailed discussions of recommendations for painting galvanized steel, and a report issued by the American Iron and Steel Institute, "Paintability of Galvanized Steel"[9] represents an exhaustive study of the problems of painting exterior galvanized steel. The comments on painting difficulty with hot dipped galvanized steel apply in varying degrees to other forms of zinc-coated steel. Galvanneal (hot dip, heat treated to produce a zinc-iron alloy surface) may be untreated or may have a mill-applied phosphate or other surface treatment.

Electrogalvanized steel (generally a much thinner zinc coating than found on hot dip) may be untreated or may have a mill-applied phosphate coating or other treatment. Anticipated changes in technology make it desirable to consult steel suppliers for advice on current treatments, and to test the specific type of zinc-coated steel before making recommendations. Spray metallized zinc-coated steel and zinc-rich primers require zinc-resistant paints, suitable barrier coats or weathering.

Aluminum

Drastic methods should be avoided as it is generally desired to erode as little of the surface as possible. Care must be taken in treating aluminum with caustic or acid, but it may be cleaned with properly selected and used acid and inhibited caustic solutions. Solvents, steam, inhibited caustic, acid pickling and nonmetallic abrasive cleaning are among the methods used.[47,51] Many proprietary cleaners are available for use with specific phosphating and chromating pretreatments.[2,18,22,38,48]

Copper, Brass, Bronze and Other Metals

In general, copper, brass and bronze do not present any particular surface preparation or painting problems. Depending on the condition of the surface, solvent cleaning, sanding, wire brushing, pickling or sandblasting might be recommended.[51]

Lead requires only solvent cleaning and light wire brushing or sanding.[51] Terne plate (tin-lead alloy on steel), if new, should be solvent cleaned to remove any oil film or, if weathered, wire brushed to remove rust spots. Wash primers are sometimes recommended.[51]

Tin and tin plate should be solvent cleaned to remove any oil film. Some paints do not adhere well to tin plate and a wash primer may be needed.[51]

Magnesium may be cleaned by solvents, inhibited alkaline or acid cleaners, sanding or nonmetallic abrasive blast cleaning.[51] A special wash primer or other pretreatment is required and would influence the choice of cleaning method.[17] Many alloys have highly specialized coating requirements, and surface preparation must be selected to fit the individual case.

NONMETALLIC SURFACES

Cellulosic

Natural Wood.   *New*: Fill holes with putty or plastic wood patch. Seal knots and pitch streaks with shellac or knot sealer. Sand as needed. Brush off dirt and dust. *Previously painted*: Remove loose paint with wire brushes, scrapers, etc. Blow torches and flame scrapers are sometimes used for removing paint from wood in low fire hazard areas. Dirt, chemical contamination and paint chalk must be removed by washing with water, or water and a cleaning aid such as trisodium phosphate followed by rinsing. Mold (mildew)* must be removed or it will usually survive painting and appear on the new finish. Table 35.3 shows one solution suggested for mold removal.[34] This solution is quite strong, and the literature of many paint companies suggests milder solutions, such as one-third cup of detergent powder and one-half cup of hypochlorite per gallon, or two-thirds cup of trisodium phosphate per gallon.

TABLE 35.3.   Mold Control Solution

$\frac{2}{3}$ cup trisodium phosphate
$\frac{1}{3}$ cup detergent powder
1 quart 5% sodium hypochlorite
3 quarts warm water
1 gallon (approximately)
Scrub on full strength. Rinse thoroughly.

Plywood.   Unfinished exterior or interior grades of plywood are handled much as new wood, above. Sand with the grain. Some plywood has an "overlay" or resin-impregnated fiber sheet bonded to its surface. Common grades are "medium-density," "high-density" and "acrylic" overlay. Other types of factory-applied finishes such as lacquer, wax or

---

*Often called "mildew," the black growth found on paint films and structural surfaces is usually a mold.

pigmented coats may be encountered. Any of these may interfere with adhesion and require special treatment. The supplier or the American Plywood Association[3] should be consulted for current recommendations. When instructions are not available, a starting point is cleaning with steel wool and solvents such as turpentine or proprietary solvent blends recommended for biting into old paints, followed by tests of the proposed coating system.

Other Cellulosic Materials. Cellulosic material such as pressed chip board, composition board, hardboard ("Masonite,"* etc.) may require little or no special preparation or may present an adhesion problem because of mill-applied treatments. Solvent wiping and sanding might be beneficial, but the manufacturer should be consulted for current preparative methods.

Masonry and Concrete

General considerations in painting these surfaces are that they must be clean, dry and preferably present a nonreactive surface. Construction defects and weathering damage must be repaired to insure that moisture from behind or within the construction cannot seep through and destroy paint adhesion.[23] Mildew must be removed by washing with sodium hypochlorite, or other strong cleaning solution.

Stone: Remove soil and loose material by washing, brushing or blasting. *Brick*: Remove grease, dust, efflorescence or other foreign matter. It may be washed with solvent, household detergent solutions, or with one of the many proprietary masonry cleaners. These may be acidic or alkaline in nature and contain wetting and sequestering agents to fit a particular situation. The efflorescence often found on brick is harmful to paint and must be removed by washing or mechanical cleaning.

Cementitious Surfaces—Concrete, Cement and Mortar. Portland cement is obtained from the pulverized clinkers of hydraulic calcium silicates with lesser amounts of calcium sulfate, aluminates, magnesium oxide, etc.[28,52]

Portland cement concrete is generally understood to be a mixture of cement, coarse aggregate, fine aggregate, water and air voids. It may contain additives such as accelerators or retarders. The liquid portion of the wet cement paste has an alkalinity of 0.5 to 1.0N, but as the cement cures this is dissipated through curing reactions and reaction with the air (carbonate formation).[28] A period of two to six months is usually required for the alkalinity to drop to a negligible level. Three months weathering of new concrete, mortar or stucco before painting is suggested by a number of paint suppliers. When these surfaces must be painted before

---

*Registered trademark Masonite Corporation, Chicago, Illinois.

weathering action can eliminate the excess alkalinity, a wash with a phosphoric acid–zinc chloride solution is suggested.[25,52] This treatment may not be desirable with some paints, and each should be evaluated, especially in the case of latex paint. A zinc sulfate wash (3 lb/gal) is sometimes suggested where oil-based paints are to be used.

Weathering is also often necessary to remove form-saving oils from new concrete. Many of these are of a nondrying nature and would interfere with paint drying and adhesion. At least the outer one-sixteenth inch should be dry before painting; a moisture meter should be used in case of doubt.[33] Concrete surfaces need to be roughened sufficiently; they should be free of dust and any thin surface layer of powdery cement (*laitance*), to permit adhesion; any holes in the surface must be opened up so that they can be sealed properly.[29]

Surface Preparation Methods:[33,29]

(1) Preliminary cleaning: *Oily* or *greasy material* must be removed before other surface preparation is attempted, as it will hinder the action of chemical agents such as acid etching solutions or may be driven further into the surface by physical preparation methods such as sandblasting, power grinding or brushing. This soil may be removed by alkaline detergent, steam and detergent or solvent cleaning. *Acid contamination* may be removed by an alkaline cleaner followed by a rinse, and *alkali contamination* by steam and/or detergent cleaning.

(2) Sandblasting: Wet or dry sandblasting is an efficient method for either light or thorough preparation (see section on poured concrete, p. 638).

(3) Hand or power tools: *Wire brushes* are suitable for fast removal of loose material but are not effective in opening air pockets. *Impact tools* will open up the surface, but are slow, and used mostly for touch-up work. *Power grinders* may be as effective as sandblasting in opening up voids but are slow and best suited for touch-up work.

(4) Acid etching: Loose concrete and oily contamination must first be removed. A typical etching solution is made from 1 part (by volume) of muriatic acid (30% strength) and 2 parts (by volume) of water. This is applied at 50 to 75 sq ft/gal, held 2 to 3 minutes and thoroughly rinsed off.

(5) Sacking: A sack coating may be used in some cases. This is a mixture of cement, sand and mortar which is hand rubbed into the surface of green concrete. It seals voids and bubbles and presents a rough surface which can be coated easily. If it is used on aged, hardened concrete, poor adhesion will result. A sack coating will provide good adhesion for thin, flexible nonshrinking coatings but should not be used with thick coatings which have a tendency to shrink, as the sack coating may be pulled away from the base concrete.

Types of Concrete:[29,33]

(1) Poured concrete and precast slabs are characterized by the presence of air bubbles, which might be as large as 2 inches in diameter at, or just below, the surface. The opening to the surface is usually small and impossible to seal permanently because the air will eventually force its way out, no matter how many coats of paint are applied.

For heavy-duty service, involving continuous immersion, frequent spillage or severe abrasion, the air pockets must be well opened and filled, to provide a smooth, pinhole-free coating. *Sandblasting* is the best method of opening up the air pockets. *Sacking* might be suitable, as noted above, where a thin coating may be used. *Hand* or *power tool methods* are usually inadequate or too slow. *Acid etching* does not open air pockets sufficiently for correct application of heavy-duty coatings.

For moderate service, when poured concrete will be subjected only to occasional spillage, light abrasion or chemical fumes, *hand* or *power tool preparation* is adequate. It removes loose or powdery material but does not do a thorough job of opening voids. Dust trapped in the voids is best removed by air blast or vacuum. A *sack finish* could be used as an alternative on green concrete. Acid etching will open up the voids sufficiently on a horizontal surface, but it is not practical for a vertical surface because of fast runoff and the need for several applications. *Light sandblasting* will open up the surface enough to insure adhesion under moderate service conditions. For *architectural service* where weather is the primary source of attack and only nominal resistance to abrasion, chemicals or immersion is required, sealing of all voids is not essential. *Sacking, hand tool* or *power tool cleaning* is usually sufficient to provide a clean surface, rough enough to insure adhesion of architectural coatings.

(2) Concrete block has a rough surface with large open voids, but it does not have the subsurface air pockets common to poured concrete. Removal of superficial dust and dirt is usually all that is required, and hand cleaning is generally sufficient for heavy-duty, medium-duty or architectural service. Water washing or chemical cleaning may be required if the block is contaminated with free alkali, grease or dirt. A block filler should be used for efficient paint utilization and best appearance.

(3) Sprayed concrete such as "Gunite"* forms a hard, dense surface with few, if any air pockets. It usually requires only hand cleaning, but if very smooth, may need acid etching or sandblasting for proper adhesion of some coatings.

(4) Concrete floors are usually troweled to give a smooth, air pocket-free surface. If not troweled, they would be treated as poured concrete.

---

*Registered trademark Gunite Concrete & Construction Co., Kansas City, Mo.

Steel-troweled floors are usually too smooth for proper adhesion and must be roughened. *Acid etching* and *sandblasting* are the most effective methods for roughening a very smooth surface and for removal of the loose powdery cement layer invariably present on new concrete floors. *Mechanical cleaning* may be sufficient for light-duty service on rough, clean and dust-free floors.

Mortar[23] for masonry is usually a combination of cement, sand and hydrated lime or other ingredients for correct working and performance properties. Surface preparation and painting problems with it are similar to those with cement and concrete.

Stucco (cement plaster)[23] is made with portland cement, mortar sand, water, and sometimes pigments, clays, etc. The need for weathering to reduce alkalinity is the same as for concrete. Aged stucco may require washing or mechanical cleaning to remove dirt or loose material.

Asbestos-cement Board.[25,35] Asbestos-cement board may be one of two types; (1) air cured, which is strongly alkaline in nature and requires alkali-resistant paints; (2) steam autoclave cured, which has only a low residual alkalinity and on which alkali-sensitive paints such as alkyds may be used. Surface preparation is usually confined to removal of loose dust by hand cleaning and solvent or water wash as needed to remove oil or other soil.

Plaster. New plaster should be allowed to dry thoroughly, perhaps about one month, before painting. Loose dust should be brushed off. Soil should be removed from aged plaster by washing with detergent and water, followed by thorough rinsing and drying. The paint supplier may recommend removal of calcimine or glue sizing by scrubbing with hot water. If coatings with high cohesive strength, such as vinyls or catalyzed epoxies, are to be applied to an aged, very smooth plaster, it may be necessary to roughen it for best adhesion.

Glass, Porcelain, Glazed Tile

Properly selected clear varnishes and paints will adhere well enough to these surfaces for many uses, e.g., blocking out windows, decorative purposes or color coding. Where other approaches have failed, etching the surface with abrasives or hydrofluoric acid–wood flour paste may prove beneficial.

Plastics[10]

Surface preparation recommendations may change quickly in the rapidly advancing plastics industry. For that reason, where performance is critical, or where paints, inks or adhesives must be applied to plastics on a commercial scale, it is adviseable to determine the exact type of plastic and to consult directly with its supplier for current recommendations.

It is possible to make a few generalizations; the plastics which a coatings formulator is most likely to encounter may be roughly separated into three practical groups: (1) thermoplastics, excluding polyolefins; (2) polyolefins; (3) thermosetting resins.

Thermoplastics.   Examples are vinyls, thermoplastic acrylics ("Plexiglas,"* "Lucite"**) polystyrene, cellulose acetate, cellulose acetate butyrate (CAB), acrylonitrile-butadiene-styrene (ABS) and acetal plastics. Forms include pliable film, rigid and semirigid packaging materials, rigid parts and structural units.   Surface preparation normally includes solvent wiping.   Proprietary treatments or primers are sometimes needed.   Light sanding or abrasive cleaning might be used in some cases.   As a rule they are solvent sensitive and a carefully balanced wiping solvent is needed which will clean and bite into the plastic enough to promote adhesion but which will not cause crazing (formation of fine surface cracks).   Overtreatment with solvent may produce a crazing tendency which is visible only after the coated article is in use.   Mold release agents and migrating plasticizer may prevent initial adhesion or cause loss of adhesion on aging. Solvents such as mineral spirits and V.M. & P. naphtha may be tried at the start, and the detailed recommendations of the plastics suppliers should be consulted.   Adhesion should be confirmed by an aging test of at least 72 hours, preferably one week.   Accelerated aging in an oven may be desirable where extended or severe service is anticipated.

Polyolefin Plastics.   These are discussed separately from other thermoplastics because they are distinctly different from the solvent-sensitive plastics discussed above.   Polyethylene and polypropylene are the ones most likely to be encountered at present in coating applications.   This includes film, containers and an increasing variety of other manufactured articles.   These plastics are so smooth and resistant to most paint solvents that the surface must have an oxidizing treatment before paint or ink will adhere to it.   This is done by flame, chemical or electronic treatment.

Flame treatment is the most popular at present for economic reasons. The plastic is briefly exposed to the blue portion of a gas flame.   Simple equipment may be used for testing or for small production jobs.   Articles may be passed through the flames of burners in a circular arrangement or the flame from a laboratory burner may be used.

In a typical commercial flame treating process, polyethylene film is passed over a water-cooled drum while the hot gasses from a burner flame impinge on the outer surface of the film.[49]   In electronic treatment, the film is subjected to the discharge from a high-voltage electrode.

In either flame or electronic treatment, it is necessary to apply the coat-

---

*Registered trademark Rohm & Haas Co., Philadelphia, Pa.
**Registered trademark E. I. du Pont de Nemours & Co., Wilmington, Del.

ing or adhesive as soon after the treatment as possible, as the plastic will in time regain its inert character. Overtreatment must be avoided or the plastic will undergo excessive degradation or distortion.

Chemical treatment is used much less than flame or electronic treatment at present, but many proprietary chemical treatments have been developed in recent years, and it is to be expected that these will occupy a more prominent place in the future.

Thermosetting Plastics. Members of several chemical classes are grouped here, as they share the same surface preparation recommendations. Included are unsaturated polyesters, alkyds, phenolics, alkyd ureas, alkyd melamines, thermosetting acrylics, epoxies and polyurethanes. They are encountered in the form of structural panels, packaging materials, parts, toys and coatings.

Solvent wiping to remove mar-proofing agents may be sufficient. However, many present adhesion problems which require specially formulated finishes. Wiping solvents often suitable include toluol, mineral spirits and V.M. & P. naphtha. Selection must be tailored to fit the solvent resistance of the plastic and its pigmentation. Finishes may require heat curing or baking to obtain adhesion.

COATINGS

Ideally, superior methods (such as sandblasting and surface conditioners) should be used to obtain maximum performance from any coating. However, slow drying, good wetting coatings may be used successfully on hand-cleaned surfaces which have had a minimum of preparation. Certain oleoresinous and alkyd maintenance paints would be of this type. Air dry epoxies, polyurethane coatings, the short oil alkyds and baked systems require better surface preparation. This may include sandblasting, chemical cleaning or the use of a surface conditioner. As a rule, catalyzed epoxies and vinyls require still better surface preparation, and sandblasting or use of surface conditioners may be desirable or even mandatory. Zinc-rich coatings, whether organic or inorganic, generally must be applied to a sandblasted surface. High-build mastics, whether applied to masonry or metal, usually require a well-roughened surface and sandblasting is normally necessary.

REFERENCES

1. Ambler, C. W., *J. Paint Technol.*, **38**(497), 343 (1966).
2. Amchem Products, Inc., Ambler, Pa.
3. American Plywood Association, Tacoma, Wash. 98401.
4. Anon., *Mater. Protect.*, **1**(6), 70 (1962).
5. Anon., *Mater. Protect.*, **5**(1), 8 (1966).

6. Barrett, L. D., *Mater. Protect.*, **3**(7), 20 (1964).
7. Barrett, L. D., private communication (July 1, 1966).
8. Beck Equipment & Chemical Co., Cleveland, Ohio, 44111.
9. Bigos, J., Greene, H. H., and Hoover, G. R., "Paintability of Galvanized Steel," American Iron & Steel Institute, 150 East 42nd St., N. Y., reprinted from "Finishes for Metals," No. 1005, Building Research Institute, Inc., 1963.
10. Billmeyer, F. W., Jr., "Textbook of Polymer Science," New York, John Wiley & Sons, 1962.
11. Bley, R. E., *Offic. Dig. Federation Paint Varnish Prod. Clubs*, **31**(417) 1257 (1959).
12. C. J., Breitenstein Co., Buffalo, N.Y. 14221 ("von Arx Air Gun").
13. Clementina, Ltd., 2277 Jerrold Ave., San Francisco, Calif. (Keefe, K. F., "Blast-Off," Form 3064, 1962).
14. Cody, L. W., *Offic. Dig. Federation Soc. Paint Technol.*, **33**(433), 199 (1961).
15. DeVilbiss Co., Toledo, Ohio.
16. Doolittle, A. K., *Mater. Protect.*, **2**(11), 32 (1963).
17. Dow Metal Products Co., Div. Dow Chemical Co., Midland, Mich.
18. DuBois Chemicals Div., W. R. Grace & Co., Cincinnati, Ohio.
19. E. I. Du Pont de Nemours & Co., Inc., Electrochemicals Dept., Wilmington, Del.
20. Freeman, D. B., *Prod. Finishing (London)*, **16**(12), 66 (1963).
21. Gray Co., Inc., 60 Eleventh Ave., N.E., Minneapolis, Minn.
22. International Rustproof Co., Div. Lubrizol Corp., Cleveland, Ohio 44132.
23. Kuenning, W. H., *Offic. Dig. Federation Soc. Paint Technol.*, **35**(461), 584 (1963).
24. Lambert, W. R. *Mater. Protect.*, **4**(5), 11 (1965).
25. Lauren, S., *Offic. Dig. Federation Soc. Paint Technol.*, **35**(462), 654 (1963).
26. E. L. Lester & Co., Houston, Texas 77021.
27. Microbeads Div., Cataphote Corp., Jackson, Miss.
28. Mielenz, R. C., *Offic. Dig. Federation Soc. Paint Technol.*, **35**(461), 563 (1963).
29. Montle, J. F., and Tarlas, H. D., *Mater. Protect.* **2**(7), 36 (1963).
30. Moore, C., *Am. Paint J.*, **47**(51), 22 (June 17, 1963).
31. National Association of Corrosion Engineers, 2400 West Loop South, Houston, Tex., 77027 "Guide to the Preparation of Contracts & Specification for the Application of Protective Coatings," Task Group T-6J-1, Tune, N., Ch., 1962.
32. N.A.C.E., Task Group T-6G-1, Bennett, P. J., Ch., *Mater. Protect.*, **3**(7), 76 (1964).
33. N.A.C.E.; Task Group T-6G-3, Rudolph, H. T., Ch., *Mater. Protect.*, **5**(1), 84 (1966).
34. National Paint, Varnish & Lacquer Assn., Inc., Washington, D.C. 20005, Circular 786, "Mildew," February 1960.
35. New York Society for Paint Technology, Technical Subcommittee No. 64, Lauren, S., Ch., *Offic. Dig. Federation Soc. Paint Technol.* **33**(442), 1346 (1961).
36. Oakes, E. W., *Mater. Protect.*, **2**(33), (1963).
37. Oakite Products, Inc., New York, N.Y. 10006.
38. Parker Rust Proof Div., Hooker Chemical Corp., Detroit, Mich. 48211.
39. Rohrer, J. B., *Paint Varnish Prod.* **54**(4), 68 (1964).
40. Rohwedder, J. L., *Mater. Protect.*, **4**(7), 11 (1965).
41. Roto-Scraper Co., Pasedena, Calif. 91101.
42. Schantz, R., private communication (July 29, 1966).
43. Shanks, F. W., *J. Paint Technol.*, **38**(497), 351 (1966).
44. Shores, J. G., *Mater. Protect.*, **5**(2), 71 (1966).
45. Simpson, C. M., private communication (Aug. 20, 1966).
46. Skaggs, T. D., *Mater. Protect.*, **2**(10), 71 (1963).
47. Spring, S., "Preparation of Metals for Painting," New York, Reinhold Publishing Corp., 1965.

48. Turco Products, Inc., Div. Purex Corp., Ltd. Wilmington, Calif.
49. U.S. Industrial Chemicals, Co., Div. National Distillers & Chemicals Corp., N.Y., "Printing on Polyethylene" PTD 13-660.
50. "Steel Structures Painting Manual," Vol. 1, "Good Painting Practice" (Bigos, J., editor) Steel Structures Painting Council, Pittsburgh, Pa., 1954.
51. Steel Structures Painting Manual, Vol. 2, "Systems & Specifications" (Keane, J. D., editor), Second ed., S.S.P.C., 1964.
52. United States Dept. of the Interior, Bureau of Reclamation, "Concrete Manual," Seventh ed., Denver, Colo., 1963.

36

Application

PART A. BRUSH, ROLLER, DIP AND FLOW COAT*

The user of paint and related coatings is ordinarily concerned with processes for applying a uniform film, usually no more than a few thousandths of an inch thick. Many processes are used, but in principle these reduce to just two: (1) Direct contact—the coating fluid is brought into initial, intimate contact with the substrate. Included are brushing, roller coating, flow coating, and dipping, and all of the variations on these methods. (2) Atomizing—the coating material is discharged toward the substrate and atomized to a greater or lesser extent before it strikes the surface. The atomization enables application of a uniform film. Conventional, airless and electrostatic spray are of this type.

Spray application and electrodeposition, which is a specialized variation of dip tank coating, are dealt with in a separate section; the other methods are covered here.

Brushes

The basic components of a paint brush are:

(1) Bristles—natural bristle or synthetic filament.

(2) Setting—formerly vulcanized rubber, now usually epoxy resin, in which the butt ends of the bristles are set.

(3) Divider (or plug)—a wooden wedge placed into the setting, to provide proper paint reservoir space and to help shape the brush contour. Oversize dividers are sometimes used to give an appearance of fullness in a cheap brush with insufficient bristles for its size.

(4) Ferrule—a metal band in which the setting is anchored, usually nickeled tin plate.

(5) Handle—usually wood. It may be straight or rounded ("beavertail" style) for easier handling. Handles are available in a variety of lengths for different jobs.

*By Kenneth C. Waldo, The Sherwin-Williams Co., Cleveland, Ohio.

The Natural Bristle Brush.   The best natural bristle, such as Chinese hog bristle, has a "flagged" tip; i.e., the end of each bristle splits into two or more parts. The flag promotes paint retention, allowing a heavier loading of paint on the brush than would be possible if the bristles ended in a smooth tip. The flag also increases the number and fineness of bristle contacts with the surface, leaving finer brush marks that flow together evenly. In addition, hog bristle is tapered from butt to tip, so that it has uniform resiliency throughout its length.

Chinese hog bristle was the accepted standard of quality before World War II. When it became unavailable, brushmakers turned to Korea, Formosa and India, as well as to European sources, for natural hog bristle. Excellent natural bristle brushes are still in common use, but there are also many poor-quality hog bristle brushes on the market.

Synthetic Bristle.   Nylon is the generally accepted synthetic bristle today. The original nylon bristles were not tapered and were blunt tipped. As a result, paint holding and spreading properties were poor, compared to hog bristle brushes. Tapered nylon filament is now available, and since the development of machinery to split or flag the tips, it has been possible to formulate nylon brushes which are competitive with the best hog bristle brushes in paint holding capacity and smoothness of application. Nylon has the added advantage that it absorbs only about 7% water and is entirely suitable for use with latex paint. Hog bristle may absorb as much as 65% of its weight in water, and becomes soft and floppy in latex paint.

Bristle Formulation.   Too stiff a brush will leave excessive brush marks. One that is too soft is hard to control and requires greater exertion to use. A blend of stiff and soft bristles of different lengths is needed to give the proper degree of resiliency and wearing quality according to the length of the brush. A short-bristled brush needs a higher percentage of softer bristles. A long-bristled brush needs enough stiff bristles to hold its shape and minimize exertion on the part of the painter.

A brush must contain the proper weight of bristle for good performance. Since the bristle is the most expensive component of a brush, a cheap brush usually will be found to be deficient in bristle content and, therefore, in paint holding and application properties.

Hand Rolling[2,5]

The paint roller and tray have replaced the brush in many cases of hand application. The principal reasons are:

(1) The method is easier to learn. Variation in results obtained by different users is minimized.

(2) The application is faster.

(3) The equipment cost is lower than that of the better-quality brushes needed for the same production rate.

Mechanism. When a paint-saturated roller moves across a surface, thousands of fibers continuously expand and contract, abrading and wetting the surface and metering out paint at a rate controlled by the roller fabric design.

Components. There is some variation in quality and design features of the tray, roller frame, cover holder (cage or cylinder type), handle, and extension handles, but the roller cover is the most complex part.

Covers. The original roller covers were made from lamb's wool pelts (lamb's wool shearlings). Many other materials have been tried; some of the more common ones are shown in Table 36A.1.

TABLE 36A.1. Roller Cover Materials

(1) "Dynel"[a] (nylon reinforced)	Excellent for general use (latex or hydrocarbon-thinned paints). Unsuitable for use in ketone-thinned paints.
(2) Mohair	Excellent solvent resistance. Limited to short nap.
(3) "Dacron"[b]	Soft nap minimizes bubbling for smooth surfaces.
(4) Lamb's wool (pelt)	High paint capacity. Solvent resistant. Fibers mat in latex paint.
(5) Rayon	Solvent resistant. Softens in latex paint. Mostly for low-cost "throwaways."
(6) Carpet	Solvent resistant. For stippling and for viscous mastics and adhesives.
(7) Frieze	Texture paints.
(8) Miscellaneous	Polyurethane foam.

[a] Registered trademark, Union Carbide Corporation.
[b] Registered trademark, E.I. du Pont de Nemours & Co.

Lamb's wool softens and mats in latex paint and, as a result, has largely been replaced by synthetic fiber covers for this application. An almost unlimited variety of versatile or highly specialized covers may be made from the synthetic fibers now available, and it is to be expected that these will increasingly replace the natural cover materials.

(A) Core Types: The following is a list of the kinds of core materials used in roller covers and the advantages of these materials.

(1) Plastic—good for latex paints, but poor solvent resistance.

(2) Phenolic-fiberboard—moderate cost, good solvent resistance.

(3) Plated steel wire—very durable, but suitable with a limited number of fabrics.

(B) Cover Sizes: The standard core diameter is $1\frac{1}{2}$ inches, but $2\frac{1}{4}$-inch covers are now popular with the professional painter, as they hold over 50% more paint. Lengths of 7 and 9 inches are most common. Other

TABLE 36A.2. Cover Recommendations

| | Type of Surface | | |
Coating Type	Smooth	Semi-smooth	Rough
Alkyd flat	1,4	1,4	1,4
Alkyd enamels (gloss, semigloss)	2,4	4	4
Enamel undercoater	2	1	
Epoxy	2,4	4	4
Interior latex paint	1	1	1
Stipple	5		
Texture	6		
Varnish	2		

| | Typical Nap Lengths (in.) | | |
| | Surface Quality | | |
Fabric	Smooth	Semi-smooth	Rough
(1) "Dynel" (with nylon)	3/8	3/4	1-1/4
(2) Mohair	3/16		
(3) "Dacron"	5/16		
(4) Lamb's wool (pelt)	1/4	1/2–3/4	1–1-1/2
(5) Carpet	1/4		
(6) Frieze	1/8		

sizes in use include 4, 14 and 18 inches. Low-cost "throwaway" 3-inch trim rollers are useful.

(C) Nap Length: The desirable nap length increases with the surface roughness. The generally accepted classification according to nap length is:

(1) Short nap for smooth surfaces: $\frac{3}{8}$ inch or less. Use on dry wall, smooth metal, smooth plaster, sanded woodwork.

(2) Medium nap for moderately rough surfaces: $\frac{3}{8}$ to $\frac{3}{4}$ inch. Use on sand-finished plaster, dry wall, light stucco, sandblasted metal.

(3) Long nap for rough surfaces: $\frac{3}{4}$ to $1\frac{1}{2}$ inch. Use on heavy stucco, rough masonry, concrete block, cinder block, brick, wire fences.

(D) Formulation Principles: With the synthetic fibers it is possible to formulate covers which look superficially alike but differ drastically in performance and cost. Terms used in defining cover characteristics include:

(1) Paint discharge capacity—net delivered quantity of paint.

(2) Paint discharge rate—weight of paint per square foot or mils wet film per square foot.

(3) Leveling effect—influence of fabric formulation on ultimate film smoothness.

(4) Conformability to surface—ability of fabric to adjust to surface irregularities and to insure uniform coverage of nonuniform surfaces.

Each of these factors is governed by the co-relation of a specific paint-fabric combination. Contributions of the fabric to these characteristics are:

(1) Fiber cross section—irregular or flat fibers hold more paint than regular, round fibers. Denier (weight in grams per 9000 meters of fiber); as this value increases for a given fiber, so do fiber cross section and stiffness, which affect leveling and conformability. The denier may not give a valid comparison of different fibers, since they vary in specific gravity, and hence in weight per unit of cross section.

(2) Degree of crimp—affects paint discharge rate.

(3) Fabric density—quantity of fiber per unit area. Affects paint holding capacity. Leveling improves as density increases, but paint holding capacity falls off above a certain density, so a compromise is necessary.

(4) Fabric finish—heat setting or other physical treatments may be used to alter crimp or other fiber qualities.

New covers should be broken in by soaking in solvent and squeezing out the excess or by rolling out the first load of paint on newspaper. This wets the fibers and eliminates lint or other dirt. Used covers also give better results if this practice is followed as they generally contain some paint residue, no matter how well they seem to have been cleaned, and will give more consistent results after the first load or two of paint is worked through the cover fabric. Load the roller just short of dripping, apply in a wide pattern, then roll out to a uniform appearance. In many covers the nap tends to lay in one direction, and a rougher paint film is laid down when rolling against the nap; hence, finish off with the stroke that gives the smoothest finish. On large jobs, or with fast-drying paint, rinse the roller in solvent occasionally to prevent accumulation of thickened paint, or work with two rollers, letting one soak in solvent while the other is being used. Clean the cover well after use, or discard it.

Special Equipment. Specialized equipment and accessories of interest include:

(1) Extension poles for hard-to-reach places such as ceilings, high walls, tanks and ship hulls.

(2) Oversize painting trays for large rollers. (A wheelbarrow can be converted into a paint tray for large rollers.)

(3) Paint mitt—a glove with a cover fabric surface for piping and other irregular surfaces.

(4) Divided rollers which conform to pipe contours.

(5) Rollers hooked up to a paint reservoir, so that the painter need not stop to reload the roller.

Specifications.   There is little standardization in the roller cover industry, because of the availability of a large number of fibers for blending
cover fabrics.   A federal specification MIL-R-17987B defines a 100%
"Dynel" cover.

Machine Roller Coating

This method is widely used for painting large, flat sheets of metal,
paper, plywood, wallboard, etc., and aluminum or steel coil stock.   The
material is fed into the roller-coating unit by hand or conveyor.   The ap-
plicator roller may be clad in rubber, gelatin, or composition, and either
is in direct contact with the paint supply or picks it up from a transfer roll.
The setup may be for coating one or both sides of the stock.   Viscosity,
line speed and roller tolerances must be closely controlled for successful
application, but once this is done, economical, fast coating is possible.
Most of the coil stock used by the coil coatings industry is painted by this
method.

Dipping

In the simplest form of this process, an article is hand dipped in an open
tank of coating material, withdrawn slowly enough so that the excess coating runs off, and hung up to drain and dry.   From that beginning, the
process may become as sophisticated as needed.

In high-production industrial painting, the work may be carried on a
conveyor, immersed into and withdrawn from the tank, and carried
through a drying oven, so that it is handled only at the beginning and end
of the conveyor line.   Dipping is especially advantageous for irregularly
shaped articles which would be difficult to coat completely by other
methods.

Since a coating in an open tank is subject to evaporation loss and
possibly air oxidation, the opening should be as small as the size of the
articles to be dipped will permit.   There should be provision for covering
the tank when it is not in use.   If the movement of parts through the tank
is insufficient to keep the paint stirred, an agitator or recirculating pump
must be provided.   Agitation should be such as to minimize foaming.
Optimum withdrawal rate depends on the size of the article to be coated
and on coating viscosity and solvent evaporation rate.

Information on tank temperature, viscosity and solids content is fundamental to good dip tank practice.   Viscosity control is important to maintain proper draining time; hence, temperature should be controlled by the
use of insulation and jackets or coils for heating and cooling.   Since the
coating will tend to thicken through solvent evaporation or air oxidation,
periodic adjustment with fresh coating material or solvent is necessary.   If

viscosity adjustments have been made solely by adding thinner, it is possible for the solids content to drop below the level needed for proper gloss, hiding and film thickness. Therefore, solids content should be monitored by periodic checks on the specific gravity.

Where a tank is operated on a long-term basis with repeated additions of new lots of paint, samples should be taken periodically for accelerated stability tests in order to anticipate any drift in stability or other quality factors.

If it is necessary to shut down a tank for any length of time, it should be covered. If the coating is subject to skinning when not in use, the surface should be covered with a thin layer of solvent or antiskinning agent.

Flow Coating

The coating material is poured over the substrate, and the excess is allowed to drain off. In practice, the coating material is discharged onto the work from a series of jets. The excess is permitted to drain off or is removed by a squeegee, etc., and recycled. This is generally a conveyorized process. It is useful for flat work and also for items with irregularities which would make uniform film thickness difficult to achieve by spraying (louvers, for example). The setup should include a chamber rich in solvent vapors, which promotes good drainage.

Removal of Excess Paint. *Centrifugal Finishing.* Small parts are flow coated or dipped, and the excess coating is removed by rotating them at high speed in a metal basket.

Scraper (Squeegee) Process. After application by dip, flow coater, roller coater, etc., excess paint is removed by a scraper blade. This is called the gasket process when the scraper takes the form of a collar or gasket around cylindrical objects such as pipes or rods.

Tumbling

This method is used for small objects which cannot be efficiently spray or dip coated. The objects, with a small amount of paint, are rotated in a drum or box-shaped container until uniformly coated, then dumped onto screens to drain and dry, or are dried while tumbling. The coating must be specially formulated for flow, tack and drying properties for use in this process.

Silk Screen Process

In silk screen processing, a coating is applied to a substrate by forcing the fluid through a fabric or metal screen with a squeegee. If portions of the screen have been blocked out by a stencil film, then only the desired outline is transferred to the substrate.

Construction of the screen is fast and economical, compared to setting up other methods of printing colors and complex designs.

In a simple form, a fabric screen is stretched tight and fastened to a wood frame. A design is cut into a film of paper or plastic which is then adhered to the screen. The screen is placed on the substrate and a thick paint, especially formulated for the proper flow characteristics, is forced through the screen with a squeegee. The result is a paint deposit which reproduces the unfilled portion of the screen. The screened stock is then air or force dried. Multiple-color copy may be produced by applying paint from a series of screens with overlapping stencils. The process may be manual or automatic, employing a machine to screen the ink. Runs vary from a few to several thousand copies. The process first gained popularity for making short runs where lithographic techniques were impractical. Originally it was used on flat stock, but it has since been adapted to printing on curved surfaces such as steel drums and plastic bottles.

The process is now used for printing on coated or uncoated wood, metal, paper, textiles and plastics. A few examples are bottles (plastic, glass), flat signs (metal, wood, paper, plastic), food wraps, radio and appliance dials, T-shirts, illuminated plastic signs and printed circuits. Many of these substrates have special ink formulation and surface preparation requirements.

Flock coatings are produced by screening an adhesive on the substrate and loading it with short textile fibers which have been agitated to give them a static charge. The charged fibers are oriented to produce a velvet- or suede-like finish.

The principal screen fabrics are:

Silk	Wire
Nylon	Cotton

Table 36A.3 shows some representative specifications. Durability is an important factor in fabric selection, both for ability to withstand use and for recoverability of the screen when the stencil is no longer needed.

Stencil Types:

(1) Lacquer soluble—stencil film is softened and adhered to the screen with lacquer solvents.

(2) Water soluble—same as above, but water replaces lacquer solvent.

Methods of Stencil Preparation:

Hand Cutting. The design is cut with a knife and the stencil is then adhered to the screen.

TABLE 36A.3.   Screen Fabrics[a]

Fabric Type	Fabric No. (mesh count/in.)	Mesh Opening (in.)	Open Area (%)	Characteristics and Uses
Nylon	70–465	.0098–.0010	51–21	Monofilament. Harder to get stencil adhesion. Tends to stretch. Recoverable, durable, moderate cost.
Silk	74–200	.0094–.0025	47–23	Multifilament. Easy stencil adhesion. Less expensive than nylon but less durable.
Wire	70–508	.0106–.0010	55–25	Strong, pliable, outstanding durability. Higher cost for ultra-critical reproduction.
Polyester (multifilament)	74–196	.0092–.0028	43–32	Low cost. Suitable for textile screening.
Cotton	—	—	—	Low cost. Suitable for textile screening.

[a] Ranges shown are representative of those noted in trade literature.[4, 6] Actual variation in sizes available is much larger.

Photographic:

(A) Photographic Film: A specialized film is used, which can be adhered to the screen after exposure and development.   Two types of film are in use: (1) Nonsensitized—not light sensitive as purchased.   The type requires sensitizing in a potassium dichromate or proprietary sensitizing bath before use.   (2) Presensitized—light sensitive.   This requires dark storage and has limited storage life, as with other ready-to-use photographic film.

A photographic or hand-drawn positive is placed on a glass plate and the film is placed over it, with the film's clear plastic backing next to the positive.   The assembly is securely fastened in a mounting frame, then exposed through the glass.

After exposure, the film is placed in a "neutralizer" bath, which neutralizes the sensitizer and accomplishes some hardening.   This is followed by a water wash to remove the water-soluble, unexposed portion of the gelatin.

The still-soft film is carefully adhered with water, gelatin side down, to the inside of the screen, and the plastic backing is peeled off.   Once dried, the developed gelatin is firmly fixed in the screen and is ready for use.

(B) Photosensitive Liquid Emulsion: A photosensitive liquid emulsion is applied to a clean, dry screen by squeegee, etc.   At least two coats are applied; one on the front (topside), and one on the back (underside). A second coat applied to the back permits heavier ink deposits for relief effects.

There are two types of emulsions: (1) *Nonsensitized*—not sensitive to light; indefinite shelf life.   A potassium dichromate sensitizer must be added to it just before use.   (2) *Presensitized*—contains sensitizer as supplied. It has the advantage that the operator does not have to stock the sensitizer and add it himself, but the emulsion is light sensitive and has a limited shelf life even when stored in the dark.

After drying, the emulsion-coated screen is placed in contact with a photographic or hand-made positive, held firmly in place by a vacuum clamp or other mounting. The screen is then exposed through the positive to a light source of predetermined intensity for a specified time.

After exposure, the screen is washed with water. The unexposed emulsion washes out of the screen, whereas the exposed portion remains; no further developing or hardening operation is needed and the emulsion film will harden on drying.

Wood Graining

Highly authentic wood grain patterns are reproduced by a photographic process and applied in a production finishing operation to plywood, particle board, composition board, etc., and metal. The coated stock is used for furniture, television and radio cabinets, and paneling. A typical process includes:

(1) Filler—applied by reverse roller coater and cured.

(2) Base coat—applied by roller coater.

(3) Wood grain highlights—applied by gravure printing (using color-keyed printing presses, one, two, or more colors may follow each other, depending on the degree of authenticity desired).

(4) Clear top coat—applied using roller or curtain coater.

(5) Final curing.

The entire process takes only a few minutes including cure time. Curing is accomplished with infrared or other heat sources. Partial cure may be used between the intermediate steps to enable application of the next coat.

Fluidized Bed Process

A bed of powdered, fusible plastic is kept in a state of suspension and agitation by an upward-rising stream of air or inert gas. When an object preheated above the melting point of the powder is immersed in the bed for a few seconds, the particles fuse to the object to form a continuous film. The object must contain sufficient heat to produce complete fusion of the resin particles. Like conventional dipping processes, it is useful for coating irregularly shaped objects.

The process is primarily used for metallic objects but might be adapted to some nonmetallic objects which could withstand the necessary heating. Uniform coverage and film thickness are controlled by immersion rate, immersion time and temperature of the object. The need for surface cleanliness and proper surface preparation is the same as for conventional dipping or other coating operations.

In its simplest form, the fluidized bed coater is an open-topped tank divided horizontally by a porous plate, on which the resin particles rest. Clean, dry air or inert gas is forced into the lower part of the tank and diffuses upward, keeping the resin particles in a state of suspension. Objects are preheated by an oven or other heat source. Temperatures of 300 to 800°F are common. Vibrators and air flow control equipment may be added.

Melting point, flow properties, particle size and particle shape are a few of the resin characteristics which must be controlled.

The process is especially useful for applying films of 5 to 50 mils of resins which would be difficult to apply from solution, such as fluorocarbons and polyolefins, but it is not limited to these.[3]

Other methods[3] of applying powdered coating particles include:

(1) Powder spray coating—coating particles are sprayed onto a preheated object.

(2) Flame spray—a localized portion of the substrate is preheated with a gas flame; coating particles passing through the flame are partially melted and fuse on striking the preheated surface.

Decalcomania

Historically, the term refers to a process for transferring prints from paper to another substrate. The term now generally refers to any process for applying previously prepared copy to a substrate.

Decals for Opaque Surfaces. These are used for furniture and wall decoration, etc. The copy is screened or printed over an adhesive-covered backing paper. This is usually followed by a clear overprint varnish. In use, the decal is moistened and the flexible film bearing the copy is slid off, carrying with it enough adhesive to insure its sticking to the surface.

Decals for Transparent Surfaces. These are used for windows and glass doors, windshield stickers, etc. A common type is designed so that the copy face will be applied directly to, and read through, the glass; hence a layer of adhesive is applied directly over the final copy.

Pressure-sensitive Decals. Adhesive-backed paper or plastic signs, from which the protective paper backing is peeled before use, are also called decals. In addition to their commonplace use for bumper stickers, they are increasingly popular for preparing lettering and decorative striping for commercial use (trucks, railroad rolling stock, etc.) and are replacing conventional stenciling in many cases, because of the increased ease and speed of application.

Drying

Where air drying is not satisfactory because of production schedules or because the coating requires a bake, common alternatives include steam, gas or electric-heated ovens and banks of infrared lamps. Factors that must be considered in designing a force-dry or baking installation include: (1) removal of combustion fumes, which might cause dulling or wrinkling; (2) removal of solvent vapors which, if allowed to accumulate, may affect drying, gloss or color; (3) insuring that solvent fumes do not constitute a health or fire hazard; (4) uniform surface temperature on all parts of the coated object, so that a uniform cure is obtained; (5) freedom from drafts which cause nonuniform surface temperature; (6) close temperature control, once the system is set up.

The installation may be conveyorized so that objects coming from a coating station continue nonstop through the drying section, or racks of freshly coated objects may be moved to a closed oven for a specified time.

Radiation Curing[1]

Ionizing radiation creates free radicals and ions which initiate polymerization, cross-linking or other reactions and may be used to cure properly formulated coatings and plastics.

In a typical process, commercial alternating current is transformed to about 300,000 volts direct current; electrons are generated in a high-vacuum accelerator tube and, as a thin, horizontal beam, pass out through a thin window made of titanium or aluminum.

Rapid and complete curing of a system such as unsaturated polyester/styrene is possible, and with less heat buildup than would be experienced in a catalyst-cured system. Because of the low heat buildup, the process is suitable for curing 100% solids coatings on wood, wood products, metals, plastics, rubber and fabrics. Line speeds of up to 150 ft/min seem practical at present. The coating is applied by conventional means, such as direct and reverse roller coaters.

Because of its newness, technical problems, and high installation cost, the process now has only limited use. However, it seems certain that it will be increasingly used for high production volume, continuous curing of coatings and plastics.

REFERENCES

1. Anon., *Chem. Eng. News*, **45**(8), 48 (1967).
2. EZ Painter Corporation, Milwaukee, Wis. 53207, TECK, 11-2-65.

3. Kutik, L., *J. Paint Technol*, **38**, 493 (1966).
4. Kressilk Products, Inc., 73 Murray St., New York, 10007.
5. Painting & Decorating Contractors of America, Inc., Chicago, Ill., "Painting & Decorating Craftman's Manual and Textbook," Fourth ed., 1965.
6. Tobler, Ernst & Trabler, Inc., 71 Murray St., New York, 10007.

PART B. SPRAY PAINTING*

The title "Spray Painting" is really not inclusive enough to describe all the materials and processes of atomizing liquids for protective, decorative and technical purposes. By this is meant that materials other than paints are sprayed, and special types of equipment are available to spray-apply coatings and chemical formulations.

Continuing developments in spray equipment parallel the paint and chemical industry's requirements for proper application of today's modern paints, coatings and resins. The growth of these two industries in many instances is associated with specialization resulting in precise coatings or paints for particular results. Spray equipment meets application needs with various systems.

The five systems that are extensively used for paint and coating application, both in product finishing and maintenance painting, are:
(1) Conventional air spray,
(2) Airless spray,
(3) Hot spray,
(4) Dual component (catalyst spray),
(5) Electrostatic spray.

Conventional Air Spray

This is the oldest of all systems. It is the most universal and is used more than all the others.

The principle of atomizing with compressed air permits conventional equipment to be practical in outfit sizes ranging from one using only a $\frac{1}{3}$-hp compressor and low-pressure, low-volume gun to one requiring a multihorsepower compressor supplying a high-volume gun.

Most any liquid and even semiliquids can be sprayed with a conventional system. Abrasive paints and coatings with aggregates are handled and sprayed with no difficulty.

The simplest conventional system is the suction type. This setup consists of a gun with a suction air cap whose design produces a vacuum action around the center hole. The attached cup containing the paint has a vent hole in the lid. When air rushes through the cap center hole with

*By Irvin B. Thomas, The De Vilbiss Company, Toledo, Ohio.

the fluid tip centered there, the vacuum developed siphons the paint from the cup.

A suction feed hookup is limited to spraying low-viscosity paints and the volume of paint sprayed is small. This means that suction feed is used most by auto, furniture and appliance refinishers and craftsmen using materials adaptable to this setup.

For high-speed spraying, a pressure feed lineup of equipment is used. The paint is forced to the gun through hose from a pressure tank or pump. Gun-attached pressure cups can also be used. The name of the hookup describes the operation—paint is supplied by pressure. In the case of the pressure tank and pressure cup, air pressure is regulated into these containers to push the paint up and out the fluid tube to the gun.

In contrast to limited suction feed, pressure feed is unlimited. Extremely heavy paints and coatings can be pumped to the gun. Medium- or light-viscosity materials can be supplied from a pressure tank or pumped in volumes to meet the needs of even the highest production.

Great progress has been made in pump design and performance for supplying paints in large amounts for maintenance painting and product finishing.

Airless Spray

This system of spraying has made a big impact on the industry. While conventional spraying is still most universal, more painting contractors each year are switching to airless for large commercial work. The airless system is a simplified hookup available in air-operated pump style or electric oil motor type requiring no compressed air.

The air-operated high-ratio pump requires a compressor. The electric airless model is of a design requiring no air compressors for pumping or spraying. In either type, the requirement is to pump paint to high pressure generally beyond 1000 psi and commonly to 3000 psi.

It is evident that considerable size is necessary in the equipment itself to develop this pressure and to pump a volume of paint equal to production requirements. Size alone and high capacity somewhat preclude airless for auto, furniture and similar refinishing. At this time, conventional spraying is used for these smaller jobs.

Atomization. It is necessary to digress from our discussion of the other spray painting systems—hot spray, dual component and electrostatic—in favor of an explanation of the atomization processes in conventional and airless systems. Both of these systems are better understood and used to their fullest advantage when correctly evaluated as to how paints and coatings are atomized.

Atomization is the breaking up of a stream of paint by some source of energy into small spheres. In conventional spray, the atomization is done by air pressure. Compressed air is a source of energy. As compressed air is directed through the air cap, it is aimed at the material stream to form the droplets into a pattern which now is directed onto the surface while the gun is moved.

Not all paints atomize alike. The viscosity of the material is one factor that determines the atomizing pressure required. Heavy and viscous paints need high pressure, while low-viscosity paints will break up with low air pressure.

The other factor determining the air pressure necessary for uniform atomization is the fluid flow from the gun. For example, to atomize a 32-oz/min flow of enamel may require 70 pounds of air pressure; to atomize 16 oz/min of the same enamel only 45 pounds of pressure would be needed. Variations in air pressure affect the degree of atomization, and spray operators must exercise good judgment in selecting pressures.

The energy in airless atomization is hydraulic pressure. When paint is pumped to high pressure and then directed through the small orifice of the cap on the gun, it comes out of that orifice at extremely high velocity. The sudden release of the high velocity causes the paint stream to split apart into atomized particles. The design and angle of the orifice are determinants of pattern size.

It could be said that atomization in the conventional system is done with outside forces, i.e., by compressed air jets, whereas the atomization in airless spraying is caused by the release of the internal velocity of the paint stream.

What really happens in airless spraying is that the material tears itself apart when released through the orifice. Consequently, the paints and coatings used in airless have to be shearable. Most paints will atomize well with airless. Viscosity is a factor in sprayability to the extent that heavy paints or very viscous materials are difficult to atomize uniformly, whereas lower-viscosity paints break up very well. Examples of easy-to-atomize formulations are alkyd flat and latex wall paints. Difficult-to-atomize materials are adhesives, chlorinated rubber and similar cohesive types. Enamels and lacquers atomize and spray very well with the airless process.

Earlier, reference was made to the pattern of the atomized paint. The patterns formed by the air cap on a conventional gun and the spray cap on an airless gun are similar. Figure 36B.1 shows typical patterns. The dimensions "length" and "width" are measured in inches. The size pattern produced by a cap is a matter of cap design. On a conventional cap, the jets of air from the horns define the pattern; on an airless cap, the angle machined at the orifice determines the pattern dimensions.

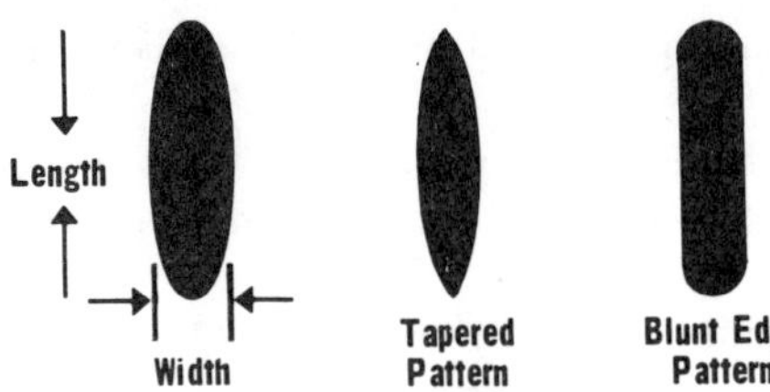

Figure 36B.1. Spray pattern shapes and dimension identification.

Since there are varied requirements to suit production needs, different caps are fitted on guns for wide or narrow patterns or for atomizing a high- or low-volume fluid flow. There is really no one universal cap. Consequently, the selection of the right cap for job conditions is important.

Hot Spray

This is the system of atomization of paint at elevated temperature. Note the use of the words "atomization of paint." The description is purposely made this way to clear up the common misconception that hot paint is applied onto a surface. The basis for hot spray is the result of paint being influenced by temperature changes and the way viscosity determines atomization. While paint is very receptive to temperature changes and readily accepts heat, it is not a good retainer of heat.

At the instant of atomization with a paint temperature of 140°F, Figure 36B.2 shows how heat is lost as the pattern continues toward the surface with the paint arriving in a chilled-down condition. The primary reason for the heat dissipation is the refrigerating effect of solvent loss during atomization. This occurs because the paint within the heater is generally raised to a temperature beyond the boiling point of the fast-evaporating solvent. Immediately upon release, this hot solvent expands and vaporizes

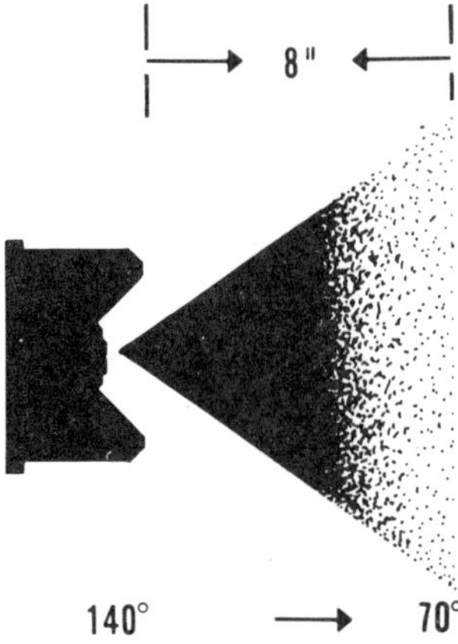

Figure 36B.2. Heat loss in spray pattern in hot spray process.

out of the pattern—this is a cooling action. The atomizing air also has a cooling effect.

Hot spray produces several very beneficial advantages, among which are viscosity control, lower spraying pressures and heavier film build when desired.

Using heat for viscosity control eliminates the variables of viscosity due to spray area temperature changes and the possibility of incorrect viscosity due to human error in mixing.

There is naturally a limit to the mechanical capabilities of paint heaters, and their potential is approximately a 100° rise. This means if the temperature of the unheated paint is 50°F, the heater will raise it to 150°F.

It is not necessary to heat paint excessively to obtain viscosity benefits. Figure 36B.3 illustrates the relationship between viscosity drop and heat

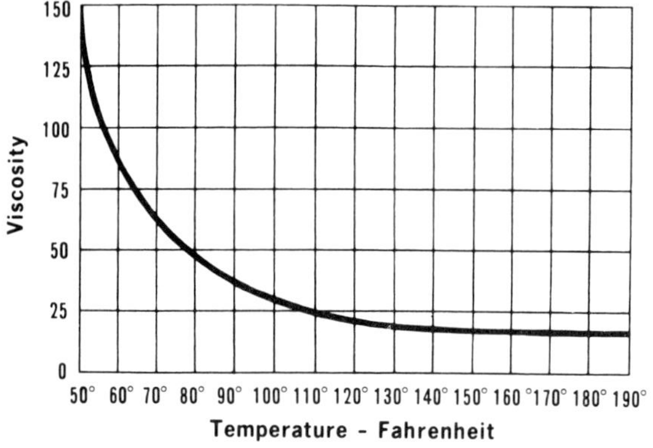

Figure 36B.3. Typical viscosity curve showing viscosity range when paint is heated.

applied. It is important to note that at about 130°F the lowest viscosity is attained and additional heat does not have any significant effect. In practice, most materials would be heated to a temperature range to overcome heat loss in the hose and to assure adequate temperature of paint at the gun.

Hot spray is an ideal system for lacquers, enamels, resins and solvent-type paints, with a limitation to materials that are in the paint classification. Coatings of a mastic or similar extremely heavy viscosity are rarely hot sprayed.

Because atomizing air pressure in conventional spray and hydraulic pressure in airless spray are selected and adjusted to produce uniform atomization based on viscosity, it is evident that a lower-viscosity material can be atomized at lower pressures. Consequently, a heated and less viscous paint can be broken up at lower pressure. Lower pressures make for economy because a lower level of overspray is generated. Any reduction in overspray and bounce-off is advantageous because of less contamination to surroundings, better atmosphere for the operator, and fewer demands on the spray booth.

Dual Component

The term catalyst spray is often used to identify this system. However, a distinction is necessary to differentiate the spraying of catalyzed paints from true two-component or dual-component formulations.

There are many two-package paints or coatings that are mixed in a common container just prior to use. The mixture is designed with long enough pot life to allow for spraying with regular spray equipment. This is not the type of coating that requires equipment to keep the components separate.

Those formulations that have zero or extremely short pot life must be sprayed with equipment that will keep the resin and catalyst separate until the instant of application. This is dual-component equipment.

Both conventional and airless systems are available for spraying dual components. Since the heart of any system is the spray gun, this item is specially designed and manufactured with separate fluid passageways. Some models keep the components apart inside the gun so that mixing takes place in the atomized pattern outside the gun. In another type, the materials join in a mixing chamber in the gun and are emitted together from the gun. This type requires the attachment of a purge or cleaning line into the mixing chamber so that when spraying is stopped this gun area can immediately be flushed clean.

Use of dual-component spray equipment demands accurate controls, such as calibrated fluid ratios and more precise spraying techniques than used in ordinary spraying. Noncorrosive metals and durable plastics are used in the tanks, pumps, regulators, hoses and guns.

Electrostatic Spray

The most popular area for electrostatic spraying is in product finishing. While portable power sources and hand-operated guns are available, the electrostatic system is sometimes impractical for maintenance or refinish painting because of surface, kind of paint or production speed.

The surface to be painted must be conductive. This means that if it is not conductive, it must be pretreated with a coating or solution that will attract the electrostatic charge. It is evident that many products from our factories lend themselves to electrostatic finishing because they are metal and conductive. In contrast, the various surfaces painted for maintenance or decorative reasons in too many cases are not conductive.

The types of paints used in product finishing are very economically and successfully applied by the electrostatic system.

Multiple gun setups are arranged on a production line to assure high-speed finishing. A single electrostatic spray gun is slow compared to a single conventional or airless setup; therefore, the latter systems are best for meeting the production speeds of maintenance painting.

Spraying Today's Paints

In today's technology, there are two major users of spray: product finishers and refinishers. This distinction is made for reasons of selling and distribution of the equipment itself and because of the types of paints and coatings applied.

These classifications influence the development, engineering and style of equipment. Product finishing spray equipment becomes part of a finishing department in a plant and is installed like other tools in a somewhat permanent location. Spray equipment for refinishing is more portable.

The spraying capacity of product finishing equipment is generally higher than refinishing setups because in factories the items to be sprayed are brought into the spray booth continuously at regular intervals either on powered conveyors or in batches. It is common on conveyorized spray lines to have a nozzle combination on a gun atomizing a quart of paint per minute distributed uniformly over a 14-inch-long pattern. Both manual and automatic guns are capable of sustained high production.

The product finishing paints used today on metal, plastic and wood surfaces are beautifully sprayed, meaning that the combination of paint chemistry and spray equipment style assures the user of preselected good results.

In the early days of spraying, different guns and nozzle combinations were needed to apply lacquer or enamel. Today because of similarity in viscosities in lacquer, enamel, epoxy and other product finishes in like formulations, a standardized gun and nozzle combination will spray all of them.

In contrast, spraying of paints in the maintenance painting profession requires attention in a larger degree to selection of guns and nozzle combinations. This is necessary because of the wide range of viscosities of maintenance paints and coatings from lacquers to mastics. Too, while high production is important in maintenance painting, the fluid flow rate from a gun is adjusted to somewhere between 1 and $1\frac{1}{2}$ pints/min for most jobs. Both conventional and airless systems are capable of spraying greater volumes than these, but the lower rates are mentioned in the interest of being practical.

Another division in the refinishing field is concerned with spraying automobiles, trucks, boats, business machines, canteen dispensers and similar renovated and repaired items. In this category are washing machines, dryers, freezers and separate electric and gas motors which are repainted after repair. These service operations represent a large share of the refinish market.

The spraying of automobiles and trucks is done by independent body shops and car dealers in standard design but specially engineered spray booths. The spraying of vehicle lacquers and enamels in these production shops is accomplished mostly with a suction feed gun. Pressure tank setups are used particularly in fleet repainting shops because all units in the fleet are painted the same color. Of course, this volume method of handling paint has advantages over the suction gun method which lends itself best to the rapid and frequent color changes that are a necessary part of job shop spraying.

Figure 36B.4 shows a typical auto refinishing spray booth with scientifically placed lights and an air replacement unit on the right at ceiling

Figure 36B.4. Auto refinishing spray booth with air replacement unit on right and exhaust fan on left.

line and an exhaust outlet on the left. Clean, heat-tempered air in adequate volume is drawn across the car during the spraying operation.

Appliance refinishing is done in floor-type booths. The spraying is again suction feed for job work or pressure feed for volume repainting.

Automation and Spraying

The general subject of automation must be considered from the viewpoint of product handling method, spray machines and spray equipment.

Automation starts with the handling method, and therefore, the selection of the spray machine is influenced by the plant's handling process.

The following is a list of types of handling methods and items to be sprayed:

Chain-on-edge machines — Parts that can be sprayed as they rotate but must be dried or baked on the machine can be handled on this type of device.

Example: Automobile wheels, electric motors, shells, and containers such as aerosol cans and 5-gallon pails.

Continuous roll — Coil steel, paper, fabric and other materials are sometimes coated as they are rewound.

Example: Decorating fabric or paper, bonding foam to fabric.

Fabricating machine — Occasionally, it is desirable to spray on the part as it leaves the fabricating machine.

Example: Coating or marking pipe, lubrication of a billet as it enters an extruding die, adding moisture to paper.

Lay down conveyor — Articles can be easily conveyed horizontally by laying them on a cross-flight, cable or pin-type conveyor.

Example: Coating plywood, porcelain enameling stove parts, adhesive on panels and honeycomb for laminating.

Overhead conveyor — This is the most popular because it makes vertical turns as well as horizontal and, therefore, can be used to convey parts through many operations. The parts may be nonrotating, indexed or rotated.

Example: Sheet metal parts such as washers, dryers, refrigerators, automobile fender and hoods are conveyed through the washers, metal preparation, paint application, bake (ovens can be overhead) and on through fabrication.

Pallet-type conveyor — Large and/or heavy articles are often conveyed on platforms riding on chains or wheels. They can be at floor level or elevated.

Example: Automobile bodies, furniture, case goods, caskets, heavy machinery.

Rotary table machines — Parts that must be rotated but can be removed wet are sprayed on a rotating table with workholder on spindles. The parts can be loaded and unloaded continuously for higher production.

Example: Decorating glassware, bonding cement on clutch disks.

The following is a listing of standard spray machines in relation to handling methods for various products and items:

Following gun mechanism — Following guns are sometimes used to coat small parts, to coat the interior of cavities not requiring an extension and to provide a longer spray duration.

Example: Decorating glassware, coating inside of aerosol cans.

Rotary arm transverse — This device is used to coat flatwork at the higher conveyor speeds up to approximately 450 sq ft/min. The coating is not as uniform as that obtained with the transverse machine.

Example: Plywood, leather, sheet steel.

Short stroke reciprocator

This device is used to provide a short reciprocating motion for electrostatic guns.

Example: Any object on an overhead conveyor that can be sprayed electrostatically.

Stationary gun or guns

Occasionally, a stationary gun or gang spray will apply a satisfactory coating to a product. It is the least expensive of the various machines. There is a minimum speed which will permit the required pattern length without flooding. However, there is no maximum speed since additional guns in series will provide adequate coverage. There is always the possibility of streaks whenever it is necessary to blend two or more patterns together to cover wide surfaces.

Stationary guns are often used on rotating parts such as on the single spindle, rotary table and chain-on-edge machines.

Example: Priming of lumber, coating pipe, lubricating dies, spraying identification marks.

Transverse machines

The reciprocating guns of the horizontal and vertical transverse machines are used to produce a uniform coating on flatwork. However, a slight contour and some edges can be coated with the guns at compound angles. These machines will spray at the rate of about 120 sq ft/ for a first-class finish and about 240 sq ft/min for finishes such as ceiling tile and wallboard.

Example: Automobile sides and tops, stove parts, shingles (metal and wood), adhesive on metal, wood and fabric.

Vertical moving mechanism

This device is used to move extensions in and out of a part, or a gun or guns over the outside of a part. It is necessary to use this in conjunction with a following device or to have the part stationary at the time of spraying. It is generally used in the chain-on-edge machines. However, a slowly moving vertical transverse machine may be used for parts on an overhead conveyor.

Example: Shells, 5-gallon pails or other containers requiring extensions.

The relationship of spray equipment to automation is summed up as follows:

Airless

Although airless lacks the flexibility to easily accommodate the set conditions of an automatic process, it does have advantages, particularly where there are limited exhaust facilities.

Conventional air spray

Conventional air spray is particularly well suited to automation because of its flexibility of adjustment and ability to handle nearly all materials. It can invariably be adjusted to satisfy the conditions predetermined by an automatic process.

<table>
<tr><td></td><td>Additional film build, higher-quality finishes and greater economy can be had by the addition of a paint heater.</td></tr>
<tr><td>Electrostatic</td><td>Electrostatic is ideally suited to automation since it recognizes the area rather than the particular shape of a part. Therefore, different parts with like areas can be intermixed on the line with the same equipment settings. It has the added advantage of being more economical than the other methods.</td></tr>
</table>

The Professional Sprayman

The operation of spray equipment in many operations today requires trained personnel. Not only do the factors of economical use of paints and perfection of finishes dictate the need for know-how, but the high degree of mechanization of spray equipment itself demands technical aptitude.

PART C. ELECTROSTATIC SPRAYING AND ELECTRODEPOSITION*

The use of electric forces to aid in the deposition of various types of materials onto surfaces as decorative and protective coatings first came into practical use in the late 1930's. Since then, as might be expected, the original methods have been improved and the field of application broadened. As a result of this effort, electrostatic forces are no longer considered simply an adjunct to the method of applying liquid finishes but now, in their own right, occupy a recognized position in the finishing field. Newer methods, such as the electrodeposition of resins from liquid baths, have been developed and show much promise. Electric forces are also being used to charge and disperse powdered resin materials in air which are then attracted to work as finishes.

Electrostatic Spraying

Behind the operation of all electrostatic spraying methods is the fundamental law of electrostatics which states that bodies carrying an excess of like electrical charges (negative-negative or positive-positive) will repel each other, whereas bodies charged oppositely (negative-positive) will attract. Applied in its simplest form to a spraying operation, this can be interpreted to mean that a small isolated particle of paint, if charged negatively, will be attracted to and deposited upon an adjacent positively charged object. To establish an operable painting system, therefore, it is only necessary to produce paint particles, charge them electrically and bring them into a position near a surface bearing the opposite charge. The electrical forces can then effectively act to deposit the particles on the surface. The various available electrostatic methods differ from one

*By Emery P. Miller, Ransburg Electro-Coating Corp., Indianapolis, Ind.

another only in the manner in which these required steps are accomplished.

Processes and Equipment.   In the original electrostatic process which was made practicable around 1940, electrostatic fields were used to charge and deposit paint particles formed with standard air gun atomizers.   This process is still in use today both in its original and modified forms.   These forms can generally be classified as No. 1 process units.

Another process was subsequently developed in which the electrostatic fields actually play an important part in the atomization of the material in addition to bringing about the charging and depositing of the formed particles.   These are referred to as No. 2 process units.

Whereas both types of processes were originally introduced for only automatic finishing systems, hand spray guns using both methods are now available.

No. 1 Type Automatic Units.   In the original No. 1 type unit, as shown in Figure 36C.1, the articles to be coated are supported on a grounded

Figure 36C.1. Material introduced into the space between electrodes and articles on the conveyor by air spray guns is electrostatically charged and deposited on the articles.

conveyor which carries them into a spraying enclosure. In the enclosure, a suitable electrode system is positioned along both sides of the path of the articles. Each electrode of the system is made of a series of fine wires and extends parallel to the path of the article for several feet. The electrodes, which are about 12 inches away from the article surface, are connected to each other and to a high-voltage supply which maintains them at a high negative charge (100,000 volts) with respect to the article. The voltage on the electrode and its shape create in the space between the article and the electrode an abundance of negatively charged air particles which are urged toward the articles and away from the electrodes. An ordinary air atomizer is arranged to spray finely divided paint particles into this space parallel to the conveyor. By collision with the air particles causes the paint particles to become negatively charged. This is known as charging by ion bombardment. Once the paint particles are charged, they too will be urged away from the electrode toward the article. Their path, as determined by the force of the atomizing air, is thereby modified, and much of the spray which would normally miss the article is now collected on it.

This attraction to the article results in a marked increase in the efficiency of this type of operation as compared with unaided air spray.

Any material lost with this process results from the fact that the electric field forces cannot overcome the initial momentum imparted to the particle by the atomizing air. These particles are then blown through the field into the exhaust. Therefore, an atomizing air velocity as low as possible is desirable. On some articles where the paint losses were as high as 70% with air spray, the use of this method of electrostatic spraying reduced losses to 35%. The electrostatic wrap of overspray onto surfaces not exposed to the direct spray of the guns permits the automatic spraying of items like bicycle frames which cannot be automatically coated by other methods without considerable complications. These advantages have led to rather widespread adoption of this method.

The original method of electrostatic spraying was characterized by the fact that the paint particles were produced by an independent method, charged by ion bombardment, and then deposited by the action of the electrostatic field; it is thus typical of a group of processes having these specific features. Whereas the guns were originally positioned to spray the material parallel to the article path so that the field forces could act on the particles for a long time, in a second modification of this method the guns are positioned so as to spray through the electrode perpendicularly toward the articles on the conveyor. In this form, whose layout is shown in Figure 36C.2, the paint is also atomized by air and receives its charge by bombardment with air ions from the electrode. Because the paint particles as they are projected toward the work are in the field for only a

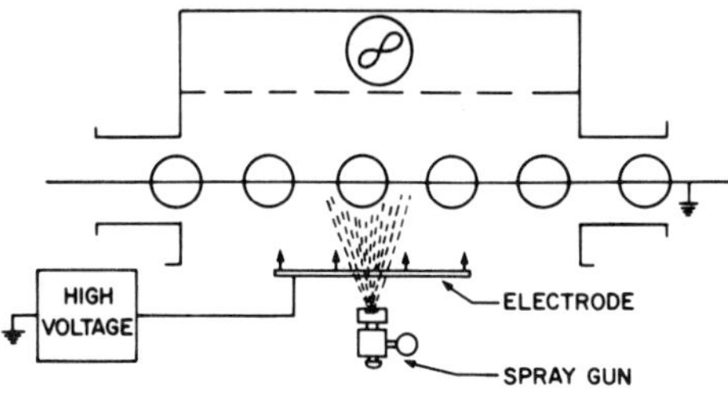

Figure 36C.2. Material sprayed perpendicular to the conveyor through an electrode by an air spray gun is charged and attracted to the articles.

comparatively short time, they generally receive less charge and are more apt to escape the deposition action of the electric field.

In another modification, the electrode is supported from the gun as shown in Figure 36C.3 in the form of a circle of sharp points surrounding the gun head. Here again the field is established and the paint is charged by raising the gun, and hence the points, to high voltage with respect to the article.

In this form, the spray gun is held at high voltage. It cannot be adjusted during operation. Paint supplied to the gun from a pressure pot must not be too conducting if the pot is grounded. Leakage through the paint line from gun to ground would short the electric field.

In still another modification, the electrode on the front of the guns as shown in Figure 36C.3 is replaced by a sharpened point which extends from the fluid tip. The entire gun is raised to high voltage with respect to the article. Air ions produced from the sharpened point bombard and charge the paint particles as they are produced by the air stream.

These four modifications of No. 1 type units of electrostatic equipment have had reasonably wide application. Their use will result in paint savings over nonelectrostatic methods. They are automated processes in which the parts are carried to the painting process, painted, and then carried on to subsequent operations. Because of the automatic character, the units can be installed so as to be unavailable to plant personnel. The elements at high voltage can be adequately shielded from contact. Where the spray gun is charged, it cannot be touched while in operation. When

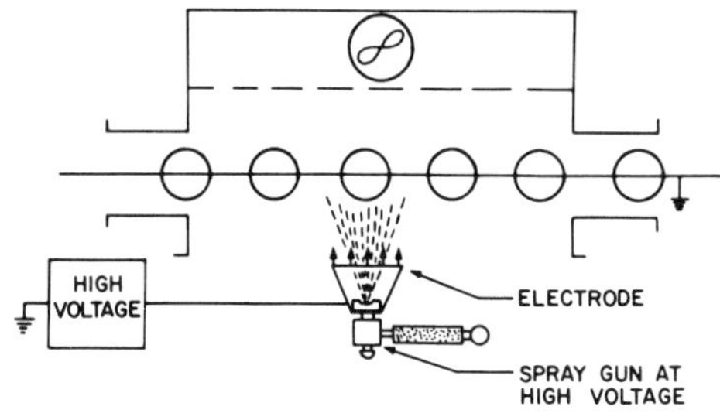

Figure 36C.3. Material is sprayed from a gun equipped with a pointed electrode. The gun and electrode are charged to high voltage to charge the spray and deposit it on the articles.

the gun is charged, consideration must also be given to the electrical character of the coating material. A conducting material being supplied from a grounded paint tank will short-circuit the electric supply to ground by way of the paint line. Under these conditions it is necessary to insulate the paint pot from ground and guard it from being contacted if satisfactory operation is to be obtained. As long as this possible shorting out of the voltage is taken into account, the paint can have any electrical character since the particles are formed by air forces, and charging of the particle occurs by ion bombardment of the isolated preformed particle.

Whereas the original of these four methods is the most efficient, it does lack one principle advantage of the other modifications. Because the gun is not aimed directly at the parts, it cannot be triggered in conformity with empty spaces on the conveyor. This has not proved to be a major handicap.

With all of these methods there is some overspray loss into the surrounding atmosphere. They must be used in standard spraying enclosures operated at reduced air velocity. If the production volume is large, water-curtained booths may be required.

No. 2 Type Automatic Units. With any of the air atomizing, bombardment charging methods, a sizable portion of the paint is still lost from the process. The electric forces cannot overcome the mechanical direction given to some of the paint particles by the atomizing air, so these will escape precipitation on the article. To avoid this loss, a fully electric method now referred to as Ransburg No. 2 process was developed. Three modifications of parts of this equipment—the blade, the bell and the disk—are available at present for automatic finishing. These types, because they do not have the disturbing effects of the air on the deposition process, made it possible for the first time to obtain application efficiencies in the neighborhood of 100%.

(A) Blade-type Atomizer: In the blade modification of the No. 2 process, the atomizer shown in Figure 36C.4 takes the form of a wedge having a plane surface with a smooth sharpened forward edge. The blade is best suited to the coating of extended horizontal surfaces. The edge of the blade is horizontal and coextensive with the surface to be coated. The plane behind the edge is inclined upward from the edge, away from the surface to be coated. The atomizer is mounted on insulators so it can be connected to a high-voltage supply and thus held at high potential with respect to the surface. Coating material to be applied is allowed to flow onto the plane and across it by gravity to the sharp forward edge where it appears as a uniform film. When the electric field is applied to the atomizer, the liquid film on the edge is formed into a series of

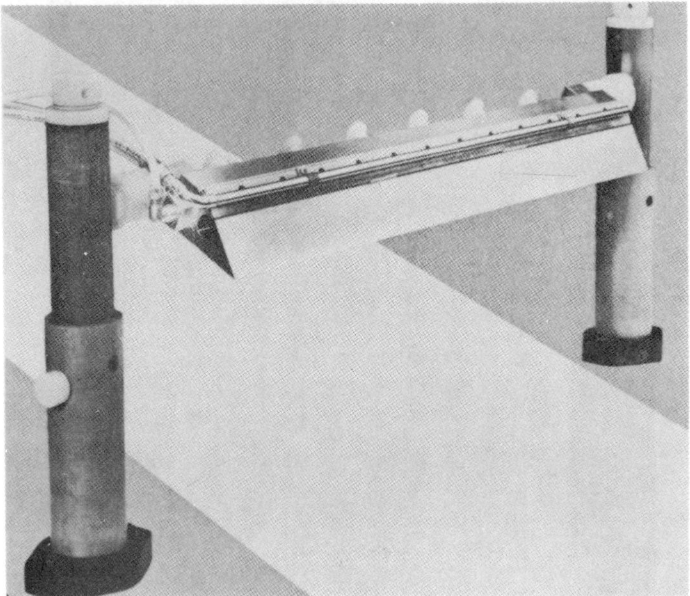

Figure 36C.4.   No. 2 process blade-type atomizer.

pointed projections equally spaced along the entire edge.   From each of these projections small particles of paint are drawn in rapid succession. Because the electric field is very high at the edge, each leaving particle carries with it a maximum charge.   The particle is urged away from the atomizer and toward the article with maximum electric force.   The particles follow along the lines of electric attraction extending from the blade edge to the surface.   All the paint leaving the blade lands on the surface as useful coating.   The only added force present in this process is that of gravity which is responsible for forming on the atomizer the film of paint which is presented to the electric field.   As opposed to the situation which exists in the air-electrostatic processes, it will be noted that in this process the field is responsible for forming the particles from the liquid, charging the particles and depositing the particles on the object.

(B) Bell-type Atomizer: In the bell modification of the No. 2 process as shown in Figure 36C.5, the atomizer is in the form of a bell or funnel. It is mounted so that it can be rotated about its central axis.   The forward sharpened edge of the bell is positioned opposite the objects to be coated and spaced from them at a distance of about 1 foot.   The bell and its rotator are supported on a suitable insulator so they can be connected to a high-voltage supply which charges them negatively to a voltage of

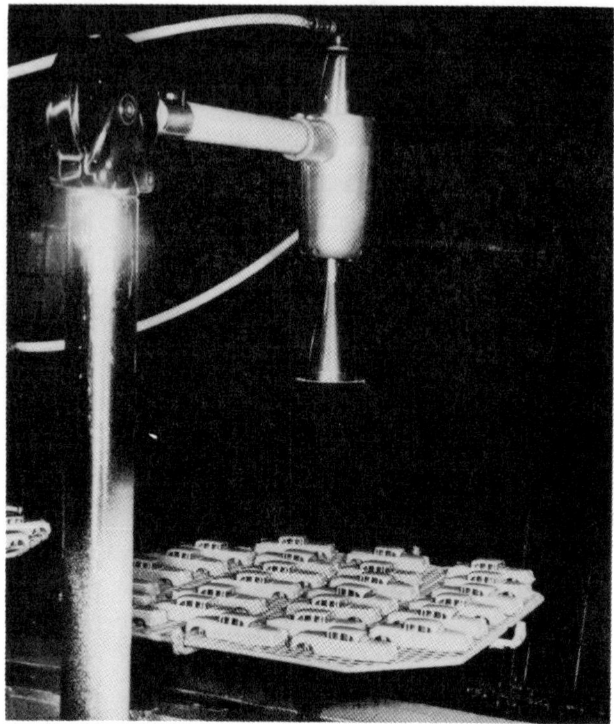

Figure 36C.5.   A single No. 2 bell-type atomizer spraying downward.

about 100,000 volts with respect to the grounded work pieces. Coating material to be applied is introduced onto the inner surface of the bell. Rotation of the bell spreads this material into a thin film whose leading edge is formed at the forward edge of the bell. Again in this modification the electric field at the bell edge forms the liquid into a series of projections from which charged paint particles are drawn. They travel to the work under the action of the forces of the field with practically 100% efficiency. The path followed by the particles will be the resultant path dictated by the forces of the field and the rotational forces of the bell. A hollow conical spray pattern results.

For the coating of small individual objects this offers no difficulty since the electric wrap of the spray onto the article will produce adequately uniform coatings. However, such a hollow cone pattern will not produce a uniform film of paint on a flat surface. It is common practice where large surfaces are being coated to combine the emission from several bells suitably disposed at one side of the article path. Such an arrangement is

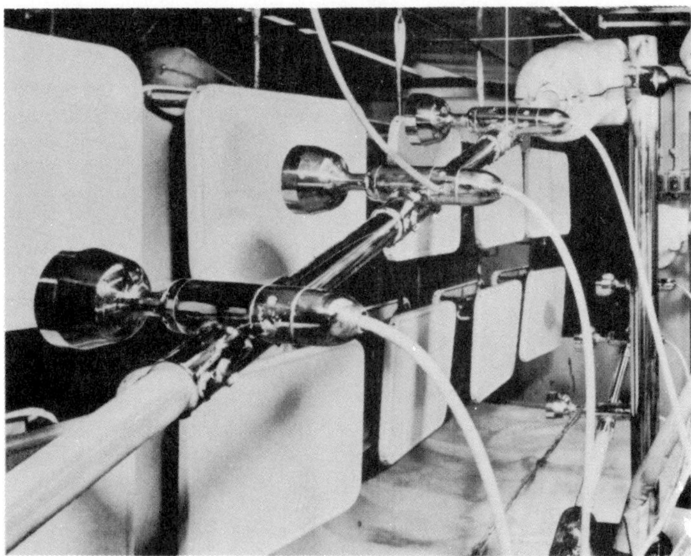

Figure 36C.6.   A triple head No. 2 bell-type unit installed to coat table tops.

seen in Figure 36C.6. The wrap of material onto surfaces not directly exposed to the spray can be readily seen in this photo. The 2-inch-deep edges of these table tops are completely covered. This is a characteristic feature of electrostatic application.

Uniformity of pattern from bell-type atomizers can also be obtained, as seen in Figure 36C.7, by mounting the bells on a reciprocator arranged to move them up and down over the work surface as it moves by.

The bell, unlike the blade, can be used to paint in all directions—upward, sideways and downward—with equal facility. The rotation of the head must be fast enough to counteract the forces of gravity and to form the liquid into a thin, uniform film from which the particles are atomized by the electric field. Without the field, the coating will leave the bell edge radially in comparatively large drops and will not arrive on the article.

(C) Disk-type Atomizer: Operating on the same electrical principles as the blade and the bell, the disk-type atomizer, as seen in Figure 36C.8, consists of a sharp-edged disk mounted on a suitable rotator. The entire atomizer is mounted on an insulating support and raised to a high negative potential with respect to the parts to be coated. The conveyor carries the parts about a loop of almost 360° which is concentric to the disk. The paint to be applied is introduced onto the surface of the disk. Rotation forms it into a thin film flowing across the disk surface to its exposed edge.

Figure 36C.7. No. 2 bell atomizers are moved to obtain uniform film on refrigerator shells.

Figure 36C.8. A stationary No. 2 process disk coats door hardware.

At the edge, the electric forces form the liquid, just as in the case of the blade, into the liquid projections spaced about the disk circumference. The charged, atomized paint particles are drawn from these projections by the field. The particles are highly charged. They are attracted to the work. Any one part will receive coating during its entire passage about the loop. Many passes of the spray pattern over the work give excellent uniformity.

If the disk is uncharged, the coating material will be thrown from its edge in a rather sharp band of relatively large particles. Some of these, by pure mechanical direction, will be thrown onto the article. When the disk is charged, however, all the particles leave the edge of the disk bearing the same electrical charge. They will be smaller and will repel each other. A spray fan 6 to 8 inches high is thus formed at the surface of the work. Objects of this total height can, therefore, be coated without moving the disk. For slightly taller objects, the disk can be tilted so the pattern will sweep over the object as it moves about the loop. For still more extensive surfaces, the disk and its drive are mounted, as seen in Figure 36C.9, on a reciprocator arranged to move it up and down as the objects go about the loop. In this way, complete uniformity over the sides of refrigerator cabinets can be produced without difficulty.

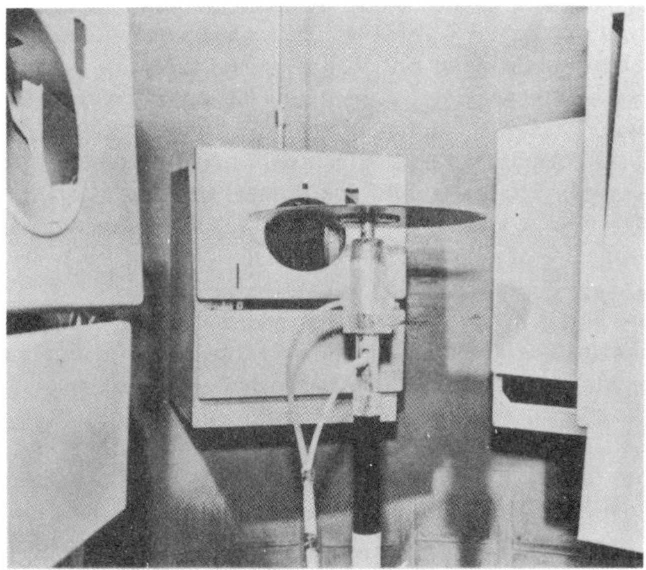

Figure 36C.9.   A No. 2 process disk atomizer is reciprocated at the center of a loop to coat dryer jackets.

If the objects are cylindrical or lend themselves to rotation, a friction bar rotator about the loop engages with rotating members on the hooks to rotate the parts as they are carried about the loop. For multi-sided objects such as washer jackets, the parts can be indexed to expose the several sides for equal intervals to the spray from the disk.

(D) No. 2 Process Characteristics: The electric field acts in the disk, bell and blade atomizers to produce the small atomized particles from the liquid film. The electrical properties of the material will have an influence, therefore, on the operation of these units. The material should not be too electrically conductive or too nonconductive. Test devices are made which read the resistivity of the material. The range of resistivity which can be expected to give satisfactory operation is indicated on the meter face. Almost all types of material in use today can be adjusted by alteration of solvent balance so that they can be applied by one or the other of the No. 2 methods. Water-base materials can be applied with the disk but not with the bell or blade. Materials which have a high metallic content will tend to short-circuit the electrical supply to ground. In these cases, it may be necessary to insulate the paint supply so that it too can be at high voltage. If this is required, the supply should be enclosed to guard it against contact by plant personnel.

The viscosities of materials sprayed by these processes must be lower than those used with normal air spraying techniques. A viscosity of 20 to 25 seconds in a Zahn No. 2 cup will be usable.

Since the blade, disk and bell all operate on the same basic principle, their respective efficiencies are essentially the same. When suitably arranged with respect to the parts on the conveyor so as to meet their individual physical demands, they can be used to coat all types of items with an efficiency close to 100%. Bicycle frames, washing machines, clothes dryers, refrigerators, automotive parts, building siding, shelving and stove parts are but a few of the items being finished by one or the other of these devices.

With the No. 2 process units the overspray is practically negligible. Spraying enclosures are required, however, because solvent vapors released at atomization must still be exhausted. Exhaust velocities, however, can be much lower than needed for air methods.

The disk, bell and blade as well as the air-electrostatic methods described thus far are all automatic arrangements. They are designed and installed to do a specific finishing operation on a specific part or group of parts. In this sense, they are inflexible. Where complete coverage on all areas cannot be obtained with the automatic setup, the added efficiency of the process makes it economical to precede or follow it with a manual supplementary operation.

Electrostatic Hand Guns.   In gaining the increased efficiency possible with electrostatic methods, some of the flexibility of operation is sacrificed.   The atomizers are not able to be handled since they are at high voltage during operation.   They cannot be moved readily so as to coat various types of objects or to coat various faces of rather complicated parts.   It is desirable to have a device that has all the flexibility of the hand air gun and yet has the application efficiency of the electrostatic bell and disk unit.

(A) No. 2 Process Hand Gun: Such a device was developed in the form of the No. 2 process hand gun shown in Figure 36C.10.   This device

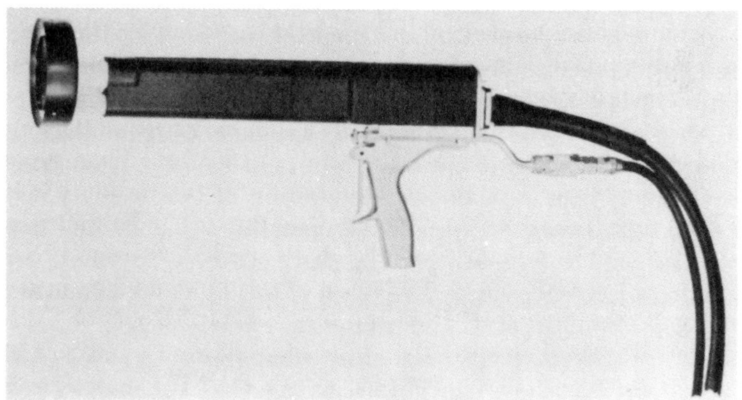

Figure 36C.10.   No. 2 process hand gun.

consists of a plastic central body with a handle at one end and a plastic rotating bell at the other.   The bell is rotated by a motor in the plastic body.   It is surface coated with an electrically resistive coating.   A small brush contact slides on the outer bell surface.   A high-voltage cable entering the handle of the gun connects the voltage supply with the brush. Paint from a pressure pot is brought to the head through the gun handle and a suitable trigger valve.

When the bell is held near a grounded surface to be coated and the voltage is turned on, a strong electric field is established between the bell edge and the surface. Coating material fed to the bell through the valve in the handle is distributed on the inner surface of the bell by the rotation. The field, as in the automatic bell unit, causes this material as it appears at the forward sharpened edge of the bell to be drawn into points and to be atomized into charged particles which are then deposited on the surface.

Specific electrical features are built into the gun so the operator can handle the unit without danger of shock. While holding the gun by the grounded handle with one hand the operator can touch the bell with the other hand without experiencing any shock. The bell can also be touched to the surface being coated without any electrical discharge which could start a fire. Any hand electrostatic device that is to be successfully used must possess these features.

This gun has all the advantages of the No. 2 system and all its limitations. The coating material cannot be too conducting. If it is, the unit will be electrically shorted to ground, the bell edge will no longer be charged and the unit will cease to operate. This gun is ideally suited to the coating of open, tubular objects. Its high efficiency allows it to be used in areas where only evaporated solvent removal is provided. It can be used to finish hospital beds, office furniture and similar items without removing them from their use area.

(B) Air-Electrostatic Hand Guns: By a suitable design of the unit, it is possible to combine the electrostatic features of the No. 1 ion bombardment-type equipment with the air atomization of the ordinary air hand gun. Such design must be safe for the operator to handle and dare not cause a discharge to ground when the sharp needle electrode is brought close to the object being sprayed. Typical of this type of equipment is the Ransburg electrostatic air (R-E-A) gun shown in Figure 36C.11.

This device is an appropriately constructed air spray gun and has all the parts of such devices—fluid nozzle, air cap, etc. The major portion of the gun at its forward end is made of electrically nonconducting material. A metal handle at the rear of the gun is grounded. An operator handling the unit is, therefore, always at ground potential.

The ion creating electrode is a short, small-diameter, pointed metallic element extending through the fluid orifice. This electrode is charged to a high negative charge (approximately 60,000 volts) with respect to the articles being sprayed. Since there are no other metal parts of the gun adjacent to the electrode, the maximum electrical field will exist at the electrode tip. A large number of negatively charged air ions will be produced near the electrode end. The electrode tip is located so that the air atomizes the paint issuing from the fluid orifice at about the same location. The paint particles are created in and pass through this intense region of ionization. A high charge is thus given to the air atomized particles by ion bombardment. Such a device will have increased efficiency over the ordinary air gun although it will not be as efficient as the No. 2 type device. The mechanical blast of the atomizing air still blows some of the material away from the control of the electrical attraction. This same blast, although it is the cause of the gun's decrease in efficiency,

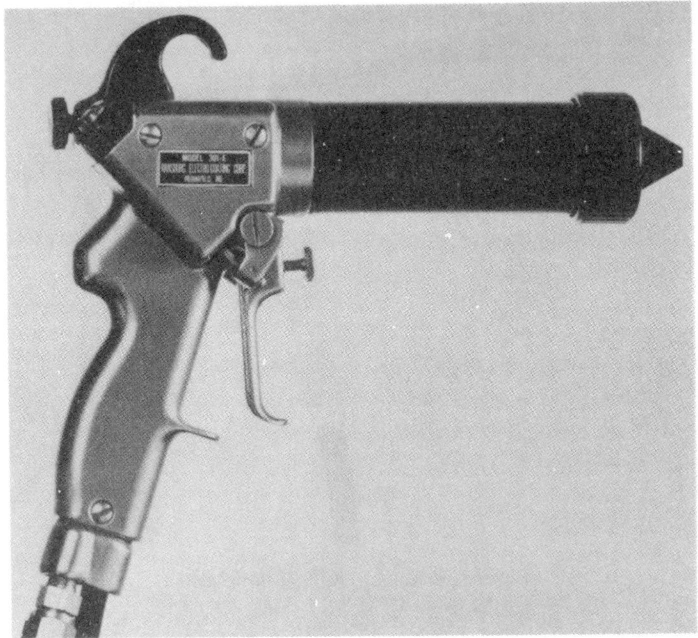

Figure 36C.11.   R-E-A hand gun.

is advantageous in some cases where penetration into recesses or concealed areas is needed. Efficiency is sacrificed in these cases to obtain increased penetration.

(C) Hydraulic Electrostatic Hand Guns: It is also possible to combine No. 1 type ion bombardment electrostatic charging with guns that produce the particles by "pressure" atomization. In these guns, the forward end of the gun is again plastic and the electrode system is a small, pointed element positioned to one side of the orifice from which the material issues as a spray. Typical of such guns is the Ransburg electrostatic hydraulic (R-E-H) gun shown in Figure 36C.12. The plastic body of the gun is supported on a grounded metallic handle. At the forward end of the plastic body, the fluid orifice and charging needle are located. The paint, under a pressure of about 1000 psi, is introduced at the rear of the plastic body. An internal valve at the front end of the gun and operated by the trigger releases this material through the very small orifice. When released, the material breaks into a very fine spray just ahead of the front of the gun in the vicinity of the electrode end.

The charging electrode is connected to a high-voltage source by a cable entering through the handle of the gun. Since the electrode is a thin sharp

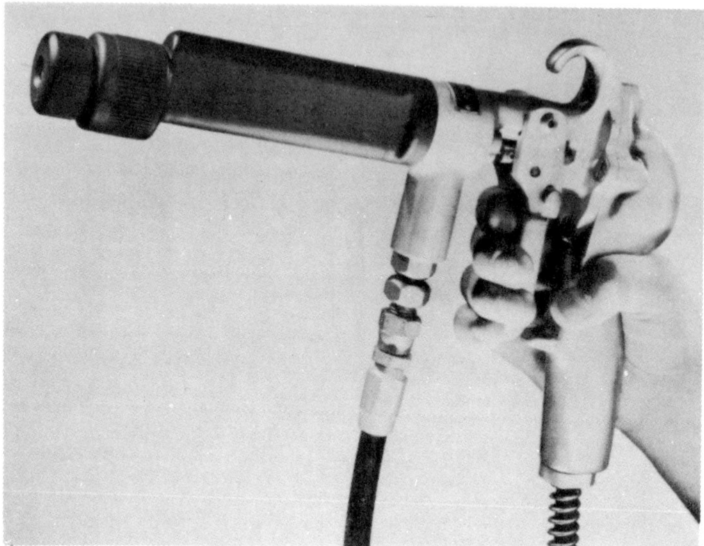

Figure 36C.12.   R-E-H hand gun.

element, an abundance of air ions will be produced about its forward end. These ions will bombard and charge the paint particles as they are formed. They will be attracted to the work. Since the electric forces are again competing with the mechanical forces of particle formation for control of the particle path, some of the particles will escape deposition. This type of gun, therefore, although more efficient than the standard pressure atomizer, is not as efficient as the No. 2 type unit. It has the advantage of being able to handle large quantities of material.

With all the hand guns, safety is of primary importance. Since they must be handled during operation, the operator must be free to touch the charged elements without getting into difficulty. It is also necessary that the gun not cause any electrical discharge when the charging electrode is brought near the object being coated. All the hand guns described have these characteristics built into them.

Since the charging electrode and the materials are in contact at the front of the gun, the use of conducting materials leads to complications. The conductivity of the paint column will short the electrode to ground and thus destroy its charging effect. If the paint supply is insulated from ground, this shorting can be avoided. When this is done, however, the paint supply is at high potential and this introduces the possibility of electrical discharges which are not considered safe for hand-operated devices. The use of conducting materials with hand devices, therefore, is not recommended.

Guns of the R-E-A and R-E-H types have been modified for use as automatic spray devices in order to take advantage of their safety characteristics. These devices are mounted alongside the conveyor carrying the parts to be coated. One or more guns can be used for each operation. They can be stationary or mounted on suitable reciprocators as desired. The swinging of the parts on the conveyor, which shortens the electrode-object distance and causes sparking in other automatic systems, causes no difficulty with these guns.

Electrostatic Application of Powders. Finely ground particles of many types of materials such as epoxies, polyvinyl chloride, polyethylene, nylon, polyesters and many others, both thermoplastic and thermosetting, are now available. These can be applied to surfaces and fused by heat to form a plastic coating on the surface. This method of coating has an obvious advantage over application of coatings from the liquid phase. No solvents are required so their cost is saved. Heavier films can readily be formed because runs and sags seldom occur. Materials such as nylon and polypropylene which cannot easily be put into solution can conveniently be applied by this method.

The same electrostatic forces described above for the deposition of liquid particles can also be used to deposit powdered materials. It is only necessary to electrically charge the powder particles and bring them adjacent to the object or surface to be coated, the electrical attraction will then cause the particles to collect on the surface. If the surface is preheated, the particles will fuse on it as they arrive. Film build will continue as long as adequate temperature exists. The surface can be post-heated to fuse the film into a smooth surface.

If the electrical properties of the powdered material are correct, the powder can be applied to a cold surface. Such powder, when charged, will be attracted to the cold surface and will adhere, if it retains some of its charge. Subsequently arriving particles will be either collected or repelled, depending upon the magnitude of the retained charge. A limited thickness of powder will thus be formed. It will adhere on the surface until the part can be carried to an oven where heat is applied to fuse the powder.

(A) Electrostatic Powder Equipment: Guns to apply powders utilizing electrostatic forces have been developed. The Ransburg electrostatic powder (R-E-P) gun shown in Figure 36C.13 is typical of this type of device. This gun has a grounded metallic handle to which a plastic forward body is attached. The plastic body consists of two tubular members. At the forward end of this tubular body, a plastic rotatable diffuser is attached. The diffuser is rotated by an air motor in the handle. Powdered material, fluidized in an air fluid bed, is carried to the gun by an air pump and a suitable tube extending between the bed and the gun. The powder

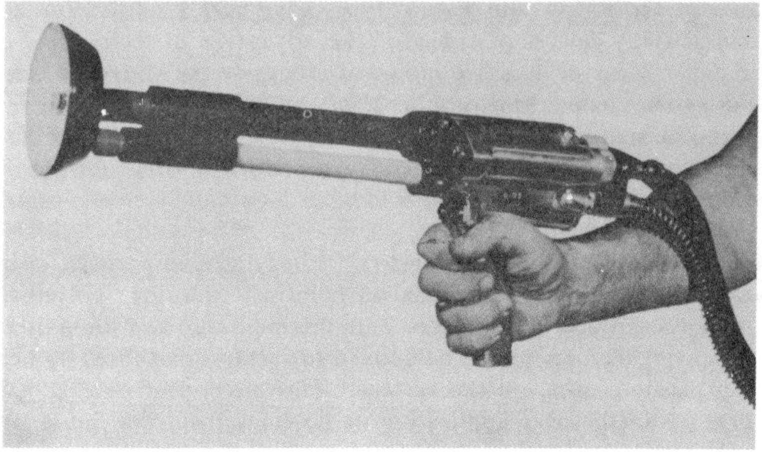

Figure 36C.13.   R-E-P gun.

enters the gun at the handle and moves through the upper body tube to the center of the diffuser.

High voltage from a voltage supply comes to the gun handle through a cable. This cable extends through the second tubular body member and contacts the resistive coating on the outside of the diffuser through a brush contact.

A third tube connected to the rear of the handle carries air to the gun motor to produce rotation of the diffuser.

When the gun is positioned with its diffuser near a surface to be coated, pulling the trigger turns on the high voltage, charges the diffuser, causes the diffuser to rotate and delivers powder to the inside of the diffuser. The powder moves across the diffuser surface to its conducting edge. At this point, it will be charged. It will then be attracted to the oppositely charged surface. This action can be seen in Figure 36C.14.

(B) Powder Gun Use: Such hand guns are finding many applications in industry. They are being used to coat appliance parts against chemical corrosion, to insulate electrical components and to dust materials to prevent self-adhesion. An ever-widening interest indicates more extended future application.

An automatic version of this powder gun has been developed. A group of such guns supported in fixed position is being used to coat various objects automatically. Pipe destined to be installed below ground is being coated with an epoxy protective film using such an arrangement. The pipe, up to 12 inches in diameter, is cleaned and then heated to about 500°F. It is carried lengthwise through the coater at speeds up to 80

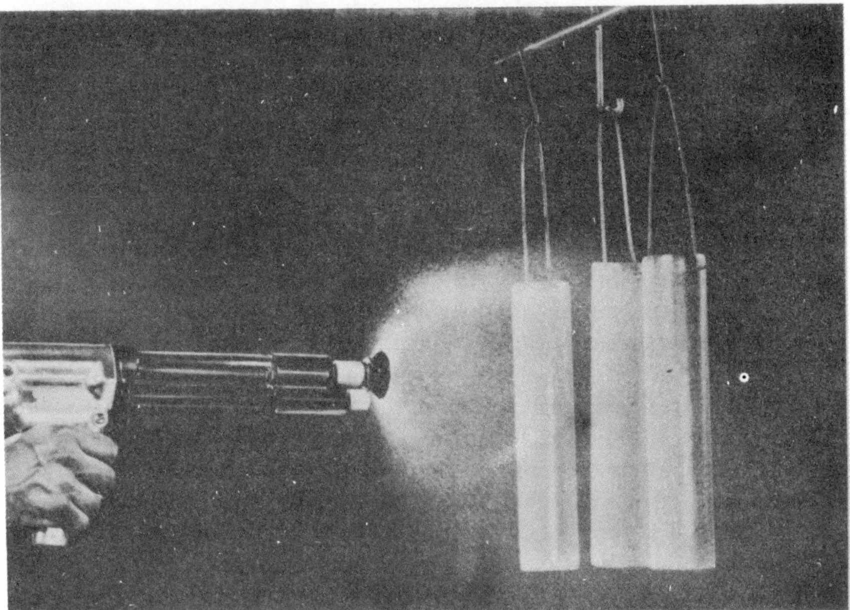

Figure 36C.14.   R-E-P gun being used to apply powder to cylindrical objects.

ft/min, depending on its diameter. In the coater, epoxy powder, introduced about the pipe by electrostatic guns, is electrostatically attracted to the pipe surface. It fuses to a 13-mil film and is then quenched in a water shower. Complete protection automatically applied in a continuous manner is the merit of this process.

Other applications of this comparatively new technique of coating will undoubtedly develop as new powders are produced and prolonged usage gives industry a broader knowledge and understanding of the method.

Electrodeposition

In addition to being used to charge and precipitate liquid paint particles onto surfaces, electric forces are now being used to deposit or "plate" resin coatings onto article surfaces from liquid baths. This process has been variously referred to as electro-deposition, electropainting or paint plating. The last term is perhaps more descriptive since this process is in many outward respects similar to metallic electroplating.

Process. Electrodeposition is essentially a dipping process which has been supplemented by electric forces. The coating material is a low-viscosity, water-base material containing, among other lesser components, resins, electrolytic stabilizers and pigments in rather low percentages and in suitable form. This material is placed in a tank whose cross section is adequate to allow clearance between all parts of the tank and the article to be coated. It is made long enough to provide sufficient immersion time to accumulate the desired coating.

The part to be coated is hung from a conveyor by a support which is electrically insulated from the conveyor. The part is carried to the tank and immersed in the bath. Once in the bath, the part is made the positive electrode (anode) in a dc circuit. The negative electrode (cathode) is either the tank itself or another suitable electrode immersed in the bath. The current is applied between these two electrodes for a period of several minutes. The article is then removed from the bath and allowed to drain. On the surface of the part, deposited by the passage of the current, there will be a tacky resin coating plus some residual loose bath material. This loose material is removed by rinsing the part with a deionized water shower. The part is then placed in an oven where the deposited resin layer is fused to a final coating.   Following the arrangement shown schematically in Figure 36C.15, this process can be carried out in a completely automatic fashion.

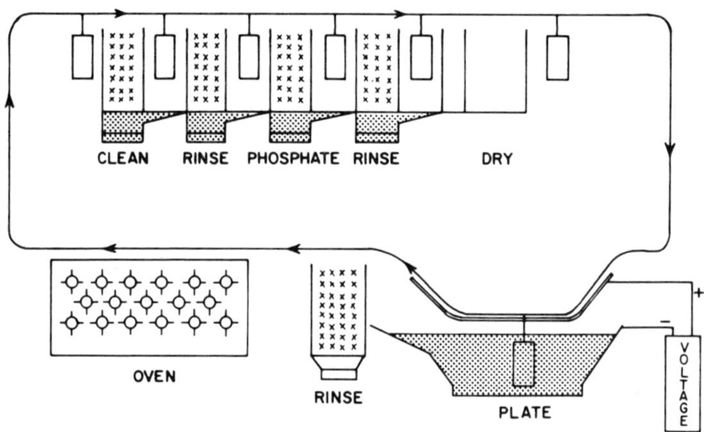

Figure 36C.15.   Schematic of the arrangement of electrodeposition system.

Characteristics of the Method.   This method of finishing has numerous advantages not obtained with other methods. Complete overall coverage of exposed surfaces with a uniform film is obtained. Because the deposited film is insulating, the film builds gradually to about 1 mil and then deposition essentially ceases on that area. Uncoated areas will continue to accumulate material. Because of this, inside or shielded areas will also be coated to a greater or lesser degree depending upon the "throwing power" of the bath. It is not unusual for metal between welded overlapping seams to be coated, if the liquid can penetrate into the area. In the coating of auto body shells and similar parts having deep enclosed areas, auxiliary electrodes are inserted into cavity areas of doors, trunks, etc., to obtain coatings in these areas.

`Since the materials are essentially water based, no fire hazard exists. The materials are supplied at high solids and reduced with deionized water. Minimum storage is therefore required, and the cost of solvents is greatly reduced.

Operation of the process is extremely simple and easily controlled. It can be completely automatic.

The process is very efficient. Practically all solids in the bath are deposited as useful material. Only the small portion of loose material withdrawn with the part is lost from the process.

These advantages over other methods make this process extremely attractive for large-scale painting operations.

Materials. The materials which can be used with this process are somewhat limited due to the requirements placed upon bath characteristics by the process. In addition to meeting all the physical requirements imposed on the finished film by the use to which it is to be put, the materials must be capable of being formulated into a suitable bath which has long-term stability. The total requirements are therefore numerous and serve to limit the applicability of the process at this stage. Nevertheless, materials have been developed that are operable, and further development by coatings manufacturers will broaden the nature and number of materials available.

Today the materials being applied are principally one-coat baking materials. Red oxide primers based on water-soluble epoxy esters, alkyd-modified phenolics and oil-based resins have been used. Similar grey and black primers have also been used. These have found major use as protective undercoats in the automotive field. They are also being used as one-coat finishes on machine parts, castings and automotive wheels. Several white and pastel one-coat materials have also been developed and have received limited use.

Tank Design. Since the heart of the system is the material tank, careful thought must be given to its design. A schematic of a typical tank layout is shown in Figure 36C.16. The overall size of the tank must be such as to allow the largest parts to be completely immersed in the bath with a clearance space to the tank or electrode of about 6 to 10 inches. The length of the tank will be determined by the type of article, the film thickness required, the type of paint and the operational voltage. These factors

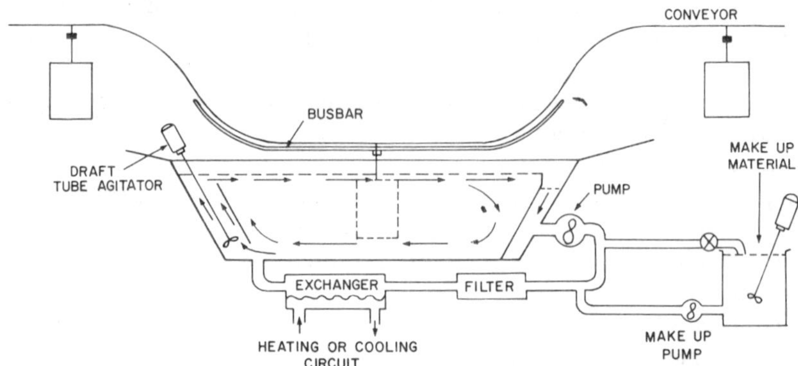

Figure 36C.16.   Electrodeposition tank and associated equipment.

are usually determined by the supplier of the material, and careful consultation between the material supplier and equipment supplier is needed for correct design.   The immersion time is usually between 1 and 3 minutes.

The bath is a low solids material having about 7 to 10% solids.   It must be circulated constantly to avoid settling.   Circulation also provides a homogeneity in the tank during deposition.   Bath bearing adequate solids must replace depleted bath adjacent to the article surface.   Circulation also helps remove gases released at electrodes during operation.   An average circulation rate of six volume changes per hour is normal.   This is usually obtained by a combination of side-mounted draft tube mixers and an auxiliary circulation system.

The auxiliary circulation system is necessary so that a portion of the bath can be passed through filters to remove foreign or undispersed material and through heat exchangers to permit temperature control.   The additive solids material which serves to maintain the bath against depletion during operation can also be added to this circulation system remote from the main tank.   Adequate time for dispersion of this material into the bath before it reaches the main tank is thus provided.

Voltage Supply.   The voltage supply used with presently available materials should be capable of delivering 40 to 250 volts dc.   The current required is a function of the area of the part being coated.   As the film accumulates, its resistance builds and the current at fixed voltage decreases.   At initial immersion of the part, the current required may be as much as 20 amperes/sq ft.   This will fall to approximately 3 amperes/sq ft when a film of 1 mil has been deposited.   The deposition obeys Faraday's Law, and for an average material, 60 coulombs of electricity will be required to deposit 1 gram of solid material.   This is about 180 coulombs/

sq ft of surface plated to a film of 1 mil. The supply should be designed to provide reasonable reserve above normal calculated load. A unit to coat small parts such as switch boxes need only supply 100 amperes, while an auto body line requires thousands of amperes.

The current is transferred to the article from an electrical bus bar by means of a drag pickup system associated with the support hook and arranged to contact the bus. The length of the bus along the conveyor will determine when the part is connected into the plating circuit.

Tank Control.   The complicated nature of the phenomenon in the tank requires that the material in the tank be rather carefully controlled if consistent results are to be obtained over extended operation. This control, although required, is not complicated to achieve.

The temperature of the bath must be maintained at the temperature specified by the supplier. Normally this requires keeping the bath in the neighborhood of 75 to 85°F. Since electrical power is constantly introduced into the bath by the plating current, means must be provided for removing this heat. A heat exchanger in the paint line, through which chilled water is circulated, is desirable. A means of switching this to heated water should also be available if the operation area is apt to fall below operating temperatures during downtime.

Solids in the bath are constantly being removed by the plating process. These must be replaced by the periodic addition to the bath of additive materials. These are usually high solids materials which are reduced to near tank solids content in an auxiliary mixing tank before being added to the main bath. Observation and measurement of the rate of electrical power consumption will serve to establish the required rate of solids addition. This can be done automatically if desired. Paint producing companies test their materials extensively before offering them as usable materials. Tank stability of the materials is a most important characteristic. It is usual that a material will only be considered satisfactory if it remains in usable condition after undergoing 18 or 20 solid turnovers. A turnover is the complete usage and replacement of the bath solids. This process of solids replacement in a good bath material should, therefore, offer no difficulty if done in accordance with manufacturers' recommendations.

As the solids are removed by operation of the system, other electrical changes take place in the tank. The pH of the tank gradually increases due to amine release at the cathode. This changes the electrical character of the solution and hence the nature of the operation. The pH of the bath should be controlled. This can be done by formulating the additive materials as high solids, low amine material to compensate for this amine increase. Another technique is to use an auxiliary cathode in a tank separated from the main tank by a semipermeable membrane designed to selectively isolate the formed amines so they can be removed from the cathode tank.

Metal Preparation.   Much of the success of the operation of an electro-deposition unit depends upon the preparation of the metal in advance of the plating process. The appearance of the final film, its adherence to the base material and its durability are all factors which are directly related to the quality of metal preparation.

The parts should be free of scale, rust and welding flux. Since the plating process depends upon the electrical resistivity of the surface of the part to be plated, local variations in the character of the surface will be reflected in variations in final film character.

Grease removal is a necessity. Contaminating materials carried into the bath will alter its properties even though the part may plate satisfactorily with some oil or grease on its surface.

Parts can be phosphated to improve the quality of the finished product. It must be carried out in such a way as to produce uniform, complete, overall treatment. Most satisfactory for use with electrodeposition on steel parts is a zinc phosphate coating of about 150 mg/sq ft.

On nonferrous materials, electroplating is equally dependent on metal preparation. Zinc can be cleaned and phosphated in essentially the same manner as steel. Aluminum is normally covered with an oxide layer that is detrimental to the adhesion of any subsequent coating. It can be elec-trocoated, however, without any processing other than cleaning, although higher voltages than normal may have to be used.

Other metals can be handled with varying degrees of difficulty if the paint formulation is specifically designed for the proposed use.

Rinsing.   After any metal preparation, a final rinse in deionized water is needed to be certain that chromic acid residue is not left on the part to mar the finish or be carried into the bath to alter its electrical properties. Parts should be either completely wet or completely dry before entering the bath. A splotchy part will have a splotchy appearance. A wet surface area will cause initial local variation in bath concentration, and this will plate differently from an adjacent dry area.

After plating, the loose material on the plated surface must be removed by a water rinse to that the film, on entering the bake oven, will be free of surface contamination. This can be a city water rinse followed by a short deionized water rinse to conserve water and improve the finish.

For best appearance in the final film, the parts should have all excess liquid removed from the surface with an air blast before entering the oven.

Safety.   Since currents of high value are available at voltages up to 300 volts, all normal precautions for such electrical systems should be taken to protect plant workers from contacting the conducting elements of the system. The unit is completely automatic so the tank can be surrounded with an enclosure having doors which are interlocked to the voltage supply. Signs and quick-acting overload devices should be incorporated in the unit.

Some gases are developed as the process operates. Oxygen and hydrogen in limited quantities can be generated at the electrodes. Amine fumes can come from the bath materials as they are handled and agitated. Reasonable ventilation should be installed in the area so that these gases do not accumulate to the detriment of plant personnel required to be in the area.

Since the materials are principally nonflammable, the ordinary fire hazards are not present with this type of operation. It should be noted, however, that the process is electrical and should, therefore, be isolated from other painting areas in which hazardous operations are being performed.

Summary. Despite the apparent complication of the process over ordinary dipping operations, electrodeposition offers so many advantages that it undoubtedly will be widely accepted. In Europe, at the present time, many units have been installed. Some have been installed in this country. These units are being used to finish automobile bodies, wheels, gasoline tanks, horns, blowers, heater parts and rocker panels. In the appliance field, small panel sections, gear housings and compressor cases are being finished. A unit still in the initial stages has been installed for the application of white materials to aluminum extrusions. In the electrical field, units have been installed to coat fixture boxes, meter bases and similar conduit hardware.

These applications are serving to pave the way for broader use. In anticipation of their increasing market, numerous paint companies have installed experimental units and are investigating numerous materials. A number of companies have invested considerable effort and capital to be in a position to handle the intelligent installation of the equipment. Everything points to widespread activity in this area.

37

Production

PART A. RESIN AND VEHICLE MANUFACTURE*,**

A paint manufacturer makes many of the resins and vehicles used in protective coatings. In most cases these are produced in solution or emulsion form. They are easy to handle in this form and are incorporated into paints and varnishes in this state. Solid resins are much more difficult to handle and must be put into solution before use.

Figure 37A.1.   Portable alkyd reactor.

*By C. R. Martens, The Sherwin-Williams Co., Cleveland, Ohio.
**The author wishes to thank the Brighton Corporation for help in preparing this material. Photographs and drawings courtesy Brighton Corporation.

There are many different operations carried out in the resin reactor. These may be the bodying of oil, the production of oleoresinous varnishes, driers, alkyds or other condensation resins, acrylic or other addition resins and synthetic latices.

These operations require heating, agitation, removal of reaction by-products and cooling. Many different reactions take place in the kettle some of which are:

(1) Polymerization   (bodying of oil)
(2) Isomerization   (thermolizing oils)
(3) Dispersion of resin in oil   (oleoresinous)
(4) Cracking of resins for solubility   (congo)
(5) Distillation of components   (oils)
(6) Esterification   (alkyds, etc.)
(7) Polymerization of monomers   (acrylics—solution and emulsion)
(8) Chemical reaction   (driers)

All of the above operations are batch processes. Along with the reactor, auxiliary equipment such as pumps, filters and storage tanks are required for handling resins and vehicles in a modern paint plant. The actual reaction conditions required to produce vehicles are described in Chapters 3 to 15 of this book.

Resin Reactors

Resin reactors vary in size from 500 to 10,000 gallons. These are well designed to give exact process control. A typical unit is shown in Figure 37A.2.

A resin reactor must in most cases be designed to handle a large number of vehicles, operate over a wide range of temperatures and be easy to clean.[9]

Stainless steel is generally used for the reactor because it is durable and produces light-colored resins. Other metals have certain disadvantages: Copper discolors the resin and loses strength at high temperatures; aluminum is attacked by alkali cleaning compounds and loses strength at high temperature, and "Monel" metal discolors the resin and will crack under certain conditions.

For the general-purpose resin reactor, type 316 stainless steel is preferred. Type 304 stainless steel is satisfactory where high acid conditions are not prevalent for a prolonged length of time. Surfaces should be polished for ease of cleaning.

The reactor is similar to batch reactor equipment used in the chemical industry. Kettles are fitted with an agitator, a manhole, sightglass, lines to charge liquid reactants, condensers, temperature measuring devices,

Figure 37A.2.   Resin reactor.

sampling devices, a discharge line and a source of heat. The manhold or charging port is used for charging solid raw materials and for access to the kettle for cleaning and repairs.

Good agitation is required so as to attain intimate contact of the reactants, thus speeding the reaction. Good agitation also improves heat transfer. Water is formed during the production of condensation polymers and this must be removed or the reaction is retarded.

Agitators.   The usual procedure is to use a turbine-type agitator operating at proper speed to give the correct degree of agitation. Too rapid agitation will cause discoloration and too slow agitation will not give the proper heat transfer at the interface and heat-up time will be increased.

Figure 37A.3 shows five different types of agitators. The first three (a), (b) and (c) are used for batch viscosities up to 20,000 cps. Type (d) is for heavier bodied resins. Types (a) and (b) are used in jacketed or indirect

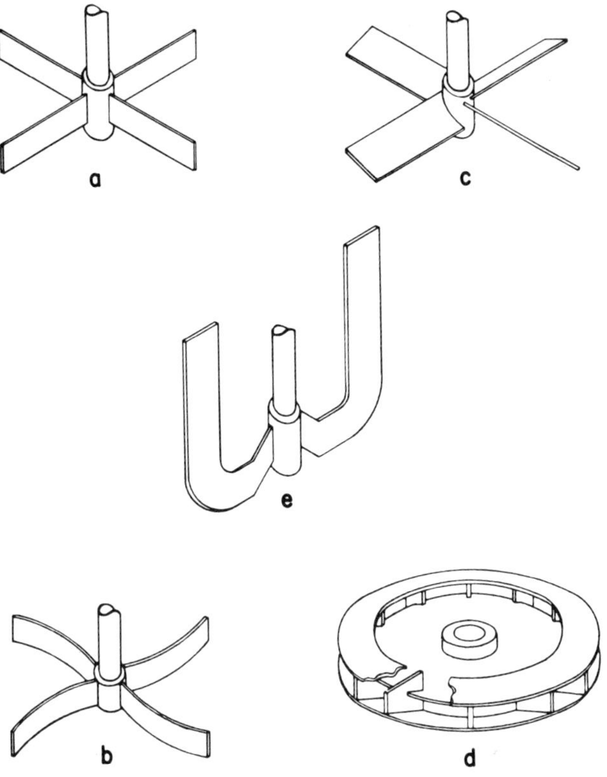

Figure 37A.3.   Agitators.

heated reactors. Type (a) develops greater pumping action but consumes more power and produces higher shear. Type (b) has less shear and is used primarily for emulsions. Type (c) is used on direct fired kettles because it produces a downward thrust giving good velocity flow across the bottom of the kettle which is subjected to the greatest amount of heat and the highest temperature. This type of agitator keeps solid materials in suspension and prevents settling at the bottom of the kettle. Type (d) is the shrouded-type turbine which is used in applications similar to type (a). Theoretically, it produces more flow, but it is also more difficult to clean.

Turbines generally have four or six blades and diameters between 40 and 50% of the reactor diameter. They are vertically mounted in the center of the reactor and require vertical baffle plates attached to the side wall to prevent swirling.

A well-designed mixer has the agitator shaft suspended by ball or roller bearings in the drive, uses a water-cooled packing box or mechanical seal and requires no foot bearing inside the reactor.

The agitator drive should be variable speed, the high speed for maximum heat transfer in heating and cooling, the low speed for small batches and minimum agitation when emptying.

Auxiliaries. The reactors are fitted with various types of condensers, depending on the resin requirements, and with fume disposal equipment.

The condensers are usually water-cooled shell and tube types, either vertically or horizontally inclined, with vapors condensing inside the tubes. Their purpose is to recover liquids which the process heat has vaporized and displaced from the batch.

In some types of reactions, the condensate is either withdrawn from or returned to the reactor in its entirety. In some cases, a two-component immiscible system is obtained from an azeotrope as in the solvent method of processing alkyds. The two materials are separated in a decanter; the water separates to the bottom and is removed. The solvent, which might be xylene, is returned to the batch.

Where fume disposal equipment is required for a reactor, a water washing system is generally used to remove unreacted vaporized solids and

Figure 37A.4.   Fume disposal equipment.

liquids from the venting gases. Water is sprayed into the vapor chamber removing fumes and solid particles (see Figure 37A.4).

Heating Methods

Heat application can be classified as direct if the energy is transferred directly to the reactor from the source which may be electricity or a combustible fuel, and indirect if there is an intermediate such as a fluid or vapor.

The thermal efficiencies of the various system are as follows:

Electricity	75–90%
"Dowtherm"	55–65%
Gas	50–65%
Oil	35–50%

The cost of electrical heating is high except in a few areas where power costs are low enough to be competitive with gas and oil. Electrical heating is clean and can be easily controlled. There are several methods by which electricity can be utilized, i.e., immersion, induction and on external surface.

Immersion heating elements present a potential hazard from burnout due to corrosion of the thin protecting metal sheath, discoloration of the resin from local overheating and failure of the elements as a result of resin buildup. Induction heating[1] in its simplest form is wall heating in which the reactor vessel is the core and secondary winding of the transformer. The primary winding of a transformer is a coil which is fitted around the outside of the reactor. The reactor must be made of ferromagnetic metal for induction heating, and since this is unsuitable for contact with resin, the vessel must be made of stainless clad steel.

The third form of electrical heating is with external surface heating elements which are clamped to the outside of the reactor; the heat is transferred by conduction.

Direct firing is the least expensive method of heating and has the fastest rate of heat up.[3] Direct firing has the disadvantage of a tendency toward local overheating which chars material on the bottom of the kettle and affects the color of the resin.

Oil and natural gas are the most economical fuels for this type of heating. The heating should be "radiant" to eliminate flame impingement and hot spots. One arrangement is the "black body" style in which the bottom of the kettle and the furnace is coated with carbon or another black flame-resistant compound.

Indirect Heating.[4,5] The most widely used indirect heating system is based on a material called "Dowtherm"* which is a eutectic mixture of

*Registered trademark, The Dow Chemical Co.

diphenyl (26.5%) and diphenyl oxide (73.5%).[2] The boiling point of "Dowtherm" is 496°F at atmospheric pressure and 670°F at 80 psi. "Dowtherm" is used in both liquid and vapor form; it is preferred to steam because only moderate pressures are required to reach high temperatures. "Dowtherm" is also used as a coolant in certain steps of the resin reaction (see Figure 37A.5).

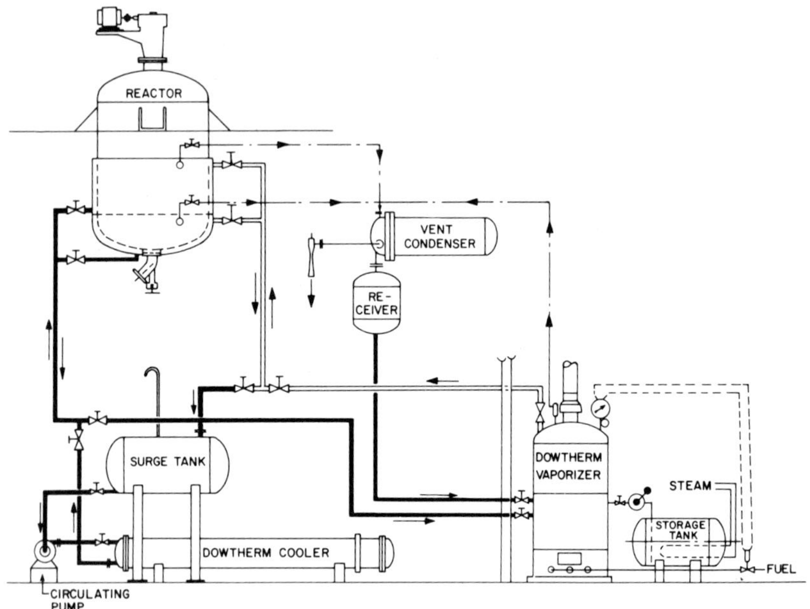

Figure 37A.5. "Dowtherm" system.

Controls

An automatic control system is a part of any large modern resin processing system. Careful control of temperature is required. In practically all installations where each reactor has its own heat supply, a sample control system consisting of a measuring element, a controller and a final control element is adequate to keep the controlled variable near the set point.

Two types of resin reactors are generally available. Those heated by steam which have maximum operating temperatures up to 350°F are classified as low-temperature reactors (see Figure 37A.6). These are used for the manufacture of phenolics, ureas, melamines, acrylic resins, and acrylic and polyvinyl acetate latices.

High-temperature reactors heated to temperatures of 670°F can be used

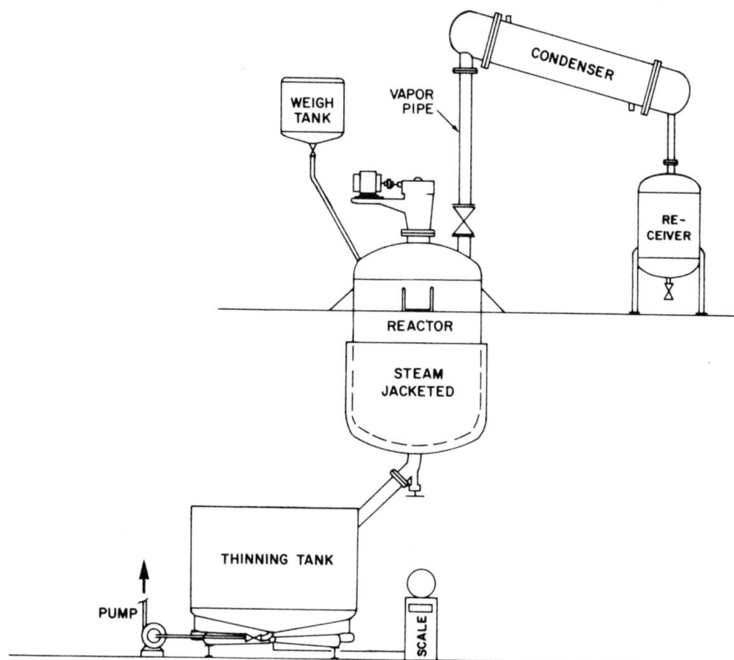

Figure 37A.6.   Low-temperature resin production system.

for producing the above resins in addition to alkyds, polyesters and ester gum, and for the heat bodying of drying oils (see Figure 37A.7).

Other Equipment

Thinning Tanks.   After the resin has been produced it is partially cooled and dropped into solvent. The tank used for this purpose is water cooled, fitted with a condenser and an agitator and mounted on scales so that it is possible to add accurate amounts of solvents. Since most alkyds are thinned to 50% solids, the thinning tank is usually at least twice the capacity of the reactor.

Pumps.   The paint and varnish producer uses mainly centrifugal, rotary and reciprocating pumps.[10]

A centrifugal pump in its simplest form involves an impeller rotating within a casing. This method of pumping is quite satisfactory for paints, varnishes and solvents, and since the impeller revolves very rapidly, a motor can usually be coupled directly to the shaft of the pump. A centrifugal pump is not suitable for handling paste products and is not normally self-priming.

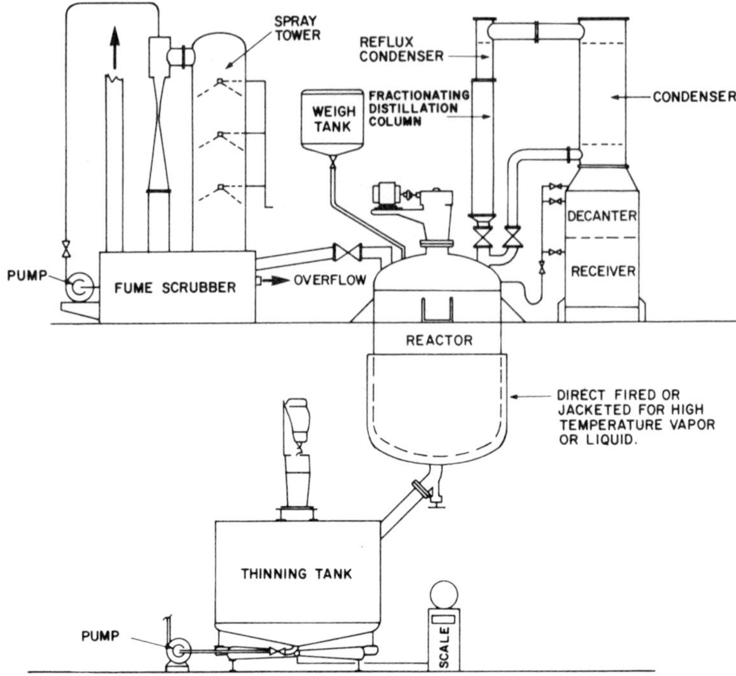

Figure 37A.7.   Resin production system.

Rotary pumps can handle liquids of various viscosities without pulsations and deliver a positive quantity of liquid under varying conditions. In a rotary pump casing, two or more impellers are used in the form of gears in which they rotate with very small clearances. Pigments will cause wear of the pump, but replacement costs are low.

Reciprocating pumps operate by the motion of a piston through a cylinder. The high maintenance costs and pulsating discharge have largely eliminated this pump in favor of the rotary type.

Filters.[6-8]   Insoluble foreign matter is removed from resin solutions and varnishes with filter presses or centrifuges. Varnishes and resins are filtered hot using a filter aid.

REFERENCES

1. Anon., *Paint Manufacturing*, **29**, 18 (1959).
2. Anon., *Paint Varnish Prod.*, **49**, No. 8, 53 (1959).
3. Anon., *Paint Varnish Prod.*, **49**, No. 12, 53 (1959).
4. Anon., *Paint Varnish Prod.*, **51**, No. 7, 73 (1961).
5. Lerstad, W., *Paint Varnish Prod.*, **51**, No. 9, 69 (1961).
6. Arnstein, L. R., *Paint Varnish Prod.*, **54**, No. 9, 35 (1964).

7. Golden, R. E., *Offic. Dig. Federation Paint Varnish Prod. Clubs*, **30**, 518 (1958).
8. Puglisi, S. J., *Offic. Dig. Federation Paint Varnish Prod. Clubs*, **30**, 490 (1958).
9. Rubin, I. E., *Paint Varnish Prod.*, **52**, No. 1, 67 (1962).
10. Scala, A., *Paint Varnish Prod.*, **49**, No. 11, 77 (1959).
11. Helgeson, R. L., *Offic. Dig. Federation Paint Varnish Prod. Clubs*, **30**, 509 (1958).
12. Jackson, T. M., Jr., *Offic. Dig. Federation Paint Varnish Prod. Clubs*, **30**, 501 (1958).
13. Mills, W. G. B., *Paint Manufacturing*, **33**, 29, 149, 401 (1963).

PART B.   PAINT MANUFACTURING*

Paint manufacturing, which was originally a simple mixing operation supervised by practical men has become a specialized branch of the chemical industry usually supervised by chemical engineers employing all the latest techniques of process control and industrial engineering to produce uniform products at low manufacturing costs.

Some manufacturers specialize in a standard line called trade sales products which are sold to the consumer through hardware and paint stores. Other manufacturers are specialists in custom-made products for different industries. Paint manufacturing plants vary in design and equipment based on market and product specialization although there are wide areas of common raw materials and common processing equipment used in the industry.

Raw materials, other than pigments, originally used in the paint industry generally had an agricultural origin and came from many different countries. Varnishes used as paint vehicles were made from China wood oil that came from China, perilla oil from Manchuria, and linseed oil from Argentina, Canada and the United States. Fossil resins like Kauri, Congo, East India, Manila and Damar came from Africa and the Far East in addition to rosin from the United States. Solvents were turpentine, mineral spirits and coal tar solvents. White pigments were white lead, zinc oxide and lithopone. Colored pigments were iron oxide and earth pigments and a limited number of organic colors.

The trend toward the present use of chemical raw materials and the domination of the industry by chemists and chemical engineers started in the 1920's with the advent of nitrocellulose lacquer. Present chemical raw materials are more numerous and more uniform than before, finished product quality has been improved, and customer requirements are more demanding—all of which has upgraded the processing and quality control requirements of manufacturing. Despite escalating costs of labor, salaries and manufacturing expense, the industry has been able to keep manufacturing costs per gallon relatively constant by improving labor productivity and increasing batch size.

*By Gordon Mutersbaugh, The Glidden Co., Cleveland, Ohio.

Plant Location

Paint plants were originally located in a few large cities like Chicago, Philadelphia and Cleveland where the formulating and processing knowledge of the industry was centralized, just as the rubber industry was located in Akron. Today, new plants are usually located outside large cities where land is less expensive and city fire and operating restrictions are less burdensome. Area supply of skilled workers is less important than formerly since workers can be trained and are more mobile.

Paint manufacturing is a truly competitive industry. There are over 1500 paint companies in the United States which cover a broad range of sizes and products. There are six or eight large national companies which operate plants in different areas of the country and set the pattern for the industry as far as research and advertising are concerned. Their combined sales amount to only 10 to 15% of the industry total. Freight rates and customer service generally limit the area that can be served by any one plant although product and market specialization even of smaller companies permit competition within larger areas.

Plant Design

Paint manufacturing is still largely a batch operation due to the large number of raw materials and finished products manufactured, many of

Figure 37B.1.   Paint plant (Glidden Co., Reading, Pa.)

which must be custom formulated and processed although every effort has been made to increase batch size to reduce costs. Early plants were usually four-story buildings (see Figure 37B.1) in which the manufacturing process was designed to flow by gravity from the top floor where vehicles were stored in tanks or drums, and pigments or other dry materials were stored in bags. The mixing of pigments and vehicles took place on the third floor; the grinding or dispersing process occurred on the second floor; the manufacture of the finished product, warehousing and shipping took place on the first floor.

Since World War II, new paint plant design has been radically changed. The development of material handling equipment, particularly in relation to raw materials and warehousing of finished products, and high labor costs requiring maximum mechanization have changed plant design. Most new plants are laid out to handle raw materials and finished stock in one-story buildings at truck height levels using fork trucks, conveyors and storage racks to reduce material handling costs.

Processing is done in a two story building or in a plant laid out on a hillside using three levels (see Figure 37B.2). The top level is for storage of raw materials, the intermediate level for processing, and the lower level for filling and warehousing. By this method, the process flow is conducted

Figure 37B.2. Paint plant (Glidden Co., Atlanta, Ga.)

through the plant by gravity without the investment and operating cost of elevators.

New plants are designed with explosion-proof motors and electrical equipment to take care of the increased fire and explosion hazards of many of the new raw materials used by the industry.

New plants are generally designed for two-shift operation to reduce investment and manufacturing cost and to improve service.

Paint Manufacturing Equipment

Before 1930, the paint industry mixed pigments in vehicles in 50 to 75 gallon heavy-duty paste mixers which were usually arranged in tandem on the floor above the stone mills or roller mills that were used for grinding the paint in paste form. The paste was transferred through chutes to the floor below for grinding. In addition to these paste mixers which were more generally used, certain products were mixed in edge runners which were large stone or iron wheels running on edge in a large pan. Smaller batches were often mixed in pans called pony mixers (i.e., a heavy-duty mixer with vertical steel fingers).

Stone mills were an adaption of the grist mill principle where a round, horizontal stone usually about 30 inches in diameter pressed on a similar revolving stone. Both stones were cut with grooves or furrows with a carborundum wheel by a man known as a mill dresser. The paste to be ground was introduced through the eye or center of the top stone and traveled to the periphery after being rubbed between the stones. The fineness of grind was controlled by pressure applied to the stones. Stone mills produced good-quality paint, but the output of each mill was low and labor requirement per gallon was high since it required an operator for each eight to ten stone mills.

Roller mills were also largely used by the industry for grinding paint. They were usually three-roll, water-cooled, horizontal mills. The early rolls were made of granite and were replaced with steel rolls made in the Krupp works in Germany. Each roll initially had about three-quarters of an inch of chill which was gradually reduced as the rolls had to be ground and trued up with use. High-speed roller mills were developed that had six times the output of the original stone mills. The operator developed skill in adjusting the rolls to produce a product of required fineness. Roller mills had high maintenance cost and high labor operating cost although they produced a good-quality product.

During the era from, roughly, 1930 until World War II the pebble mill and steel ball mill was the workhorse of the industry (see Figure 37B.3). The pebble mill was used by the ceramics industry before being adopted by the paint industry. It is a revolving cylinder of different sizes lined

Figure 37B.3. Ball mills, Glidden Co.

with buhrstone or a ceramic lining and filled about half full of selected hard pebbles or specially prepared ceramic stones. The speed of the mill varies with size and is regulated so that the stones will cascade inside the mill. Control of the viscosity of the material to be ground and other operating factors are important for getting the best results in the shortest grinding time.

The steel ball mill is similar to the pebble mill except that the grinding media are steel balls. The inside of the mill is made of tough chrome manganese steel and sizes over 90 gallons are water jacketed to take away heat. Steel ball mills grind in about half the time required by pebble mills. Investment, power and maintenance costs are higher. They are generally used in the industry for grinding blacks, oxide primers and dark colors. There are problems of color degradation and iron contamination which must be considered where steel ball mills are used.

Since World War II, the pigment suppliers have developed pigments that are easy to disperse. The paint industry is therefore rapidly going to high-speed dispersion mixers (Figure 37B.4) and sand mills (Figure 37B.5), which permit batch manufacture in a matter of minutes compared with the slower pebble and steel ball mills where the batch was usually in the mill 24 hours or longer. With proper selection of pigments and

Figure 37B.4. High-speed dispersion mixers.

dispersion technique, a high percentage of products can be made in high-speed dispersion mixers which require relatively small amounts of building space and which permit operating flexibility and improved service to customers. The mill has special appeal to the manufacturer who wants to increase output without adding to present building size. The mill is a high-speed vertical shaft with a horizontal blade that revolves in the product to be dispersed.

The sand mill is a cylinder of various sizes filled with fine sand, glass beads or small pellets. The sand is agitated by a vertical motor-driven shaft with metal disks spaced at intervals. The batch is usually pumped through the mill from the bottom at a speed which will produce the required fineness of dispersion. The batch must be premixed. While both the high-speed mill and the sand mill are dispersion mills depending on the selection of pigments that are easy to disperse, the sand mill is capable of developing higher gloss and a higher degree of dispersibility of some pigments and is generally used for making automobile and other similar high-grade enamels. Mills which depend on high shear, like the dough mixers, or mills depending on the vibrating principle have had limited use for specialized products.

After the pigment has been dispersed, the batch of the product as sold is manufactured in mixers of various sizes. Batches of 200 gallons or less

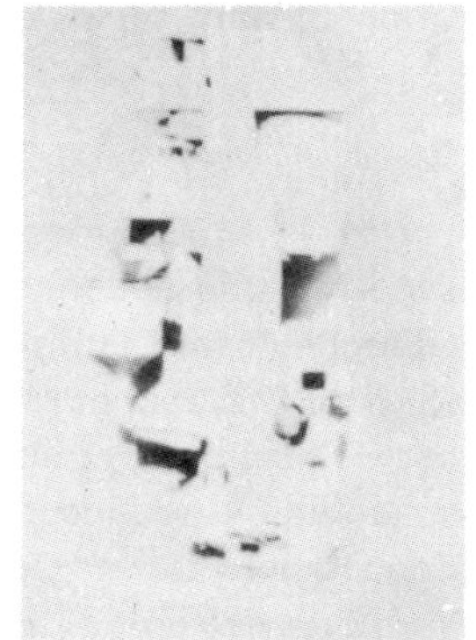

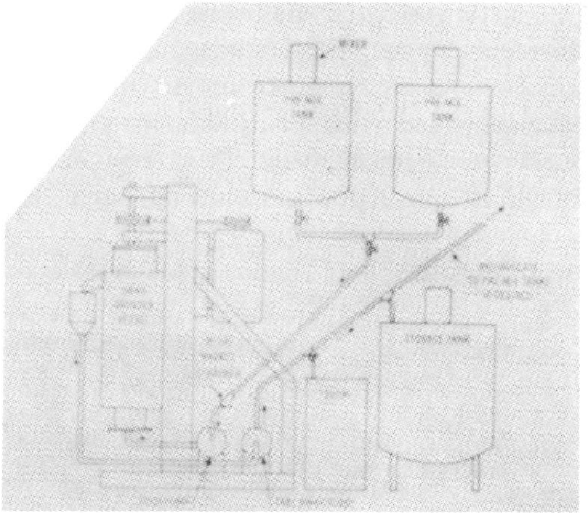

Figure 37B.5.   Sand grinding.

are usually made in portable kettles in which the final ingredients are added and the color is shaded to standard. A sample of the batch is taken to the quality control laboratory, and batch adjustments are made. These kettles are mixed by moving them under vertical agitators during the thin-down and shading operation. Larger batches are usually made in station-ary mixers. Much progress has been made in reducing manufacturing cost by making the largest batch that sales requirements and inventory limita-tions permit. Most manufacturing costs are on a per batch basis; thus the larger the batch, the lower is the manufacturing cost per gallon.

Some mixers are round while others are square with rounded corners. Agitators are of the paddle type which give best results for certain prod-ucts that require minimum air entrapment. Some products can be made with high-speed agitation which usually mixes thoroughly in a shorter time.

Paint Filling

Originally many paints were sold in paste form and were mixed by the user with sizable quantities of oils, liquids or thinners. These paste products were usually filled directly from the mill. Since World War II, most products have been sold ready to use, or they require only a small amount of thinning to adjust to optimum application consistency. This relatively thin viscosity has helped to reduce manufacturing cost and makes it possible to use more efficient filtering methods and labor-saving automatic filling machines (see Figure 37B.6). Automatic filling machines are desirable not only to reduce labor cost but to insure accuracy of measurement. After the batch has been approved for quality, it is usually strained before packaging. The fineness of straining depends on the type of product. Metal cloth strainers of varying mesh sizes are generally used. A machine which vibrates a metal or nylon screen is often used. Some products are filled through cloth bags or forced under pressure through felt cylinders for products which require the ultimate in cleanliness.

Long runs of products that are put into 1-gallon (or smaller) containers

Figure 37B.6.   Automatic-filling machines.

are usually filled automatically in a line that includes automatic labeling, bailing, corrugated box casing and gluing. Many products are filled in 5-gallon metal pails or 55-gallon drums. These are usually filled by weight. The drums and pails have special organic coatings to minimize metal contamination.

Material Handling Equipment

Most of the unskilled manual labor has been eliminated in the paint industry since World War II by installing material handling equipment. Each job is studied, often by an industrial engineer, and equipment is installed which will complete the operation with the least amount of labor.

Fork trucks, both electric and gas operated, have taken over most of the former work of handling raw materials and finished products.

In some cases, conveyors have been installed to move products from one operation to another.

Pumps, pipelines and storage tanks have eliminated much of the manual labor and shrinkage losses of handling liquids and solvents in drums, although technical complexity still requires handling drums of liquids and solvents used in smaller quantities.

Most pigments can be purchased in bags on pallets so it is possible to handle a pallet load at a time with a fork truck rather than handling one bag at a time by hand. Many raw materials are available in pellet form, to avoid excessive dust, or in larger-size fiber drums. There has been much talk of bulk handling of pigments but little actual progress. Only a few raw materials are used in large enough quantities, and in those cases there are usually technical disadvantages to handling the material in this manner; also, the cost of bringing the pigment from the point of storage to the point of use generally does not justify the investment.

Further mechanization and use of material handling equipment is limited at each plant by the inability to restrict the number of raw materials and standardize the number of finished products manufactured and stocked.

Physical Distribution

Much progress has been made in reducing manufacturing and filling costs, but the cost of operations covering the time the product is filled until it is delivered to the customer, which includes receiving, warehousing, inventory control, warehousing and shipping, has been increasing.

Investment in inventory of raw and finished products often exceeds investment in the plant, and good inventory control is essential for manufacture, scheduling, service to customers and product turnover. Computers are making progress in this area which has been a clerical operation in the past. The labor cost of warehousing is being reduced by the use of racks, lift trucks for picking orders and other labor-saving devices. Trucking costs have been increasing, and savings can be made by consolidating and scheduling shipments.

The industry has seasonal sales patterns. Production for stock in low sales months helps to maintain constant employment and maximum employee productivity with minimum investment in plant and equipment. This requires annual planning, scheduling of equipment and control of inventory.

Personnel

Supervisory personnel originally came from the promotion of outstanding workers. Workers were hired for their attitude and physical ability regardless of education. Today, the heavy work is done by machines. Worker requirements have changed. Workers should be high school graduates who make the minimum number of mistakes. Top supervision should be by chemical engineers who have a technical knowledge of the products manufactured and an engineering knowledge of equipment and industrial engineering. The future of the industry depends on its ability to attract competent chemists and chemical engineers. Workers are generally unionized and receive more in fringe benefits than they did in wages 20 years ago. In spite of the rapid increase in hourly rates, labor costs represent less than 10% of the cost of the product.

Quality Control

Quality control was originally simple and inexpensive. The shader approved the color of his own batches, and drying, viscosity, gloss and other tests were held within practical tolerances. Today, expensive and time-consuming technical tests have been set up to insure maximum uniformity. Color is controlled by computers, and every effort is made to test the product to duplicate the application experience of the customer. Despite this enlarged effort, the increasing technical complexity of the product and the method of customer application leaves much to be desired in anticipating customer dissatisfaction by increasing time and money spent for quality control.

Every effort is being made to buy more uniform raw materials and to process them under better understood and better controlled processing conditions to insure greater product uniformity, rather than increasing the number and complexity of quality control tests which will reject more batches.

Maintenance, Safety and Good Housekeeping

The increase in mechanization requires employment of more mechanics to service and repair equipment. Manufacturing within closer tolerances requires that scales and meters must be accurate and all equipment must be well adjusted and in a good state of repair.

All equipment must be kept clean to avoid problems of contamination which are much more serious because of the incompatibility of many products and the use of the same equipment for making many types of products.

Good housekeeping, plant and equipment maintenance, and cleanliness are essential not only for appearance sake but, with so many different raw materials and so many different finished products, also for low-cost, high-quality products. They must become a part of the day-to-day routine enforced by supervision and accepted by workers.

Some of the new raw materials have resulted in new health and fire hazards that are getting increasing management attention, along with normal safety precautions. Modern plants are designed with explosion-proof electrical equipment and sprinklers. All equipment is grounded since static electrical fires have been common in the past. Experience has shown that if a fire does start, automatic sprinklers are the best way to put it out before it becomes serious. Adequate ventilation and elimination of toxic and potentially explosive solvent fumes are also essential.

A well-run paint plant must have a safety committee that periodically inspects the plant, looking for all unsafe equipment and worker practices that must be corrected. This requires constant attention as the problem is never permanently solved. All dirty rags must be stored under water to avoid spontaneous combustion.

Each department must have fire drills so that each man knows what to do in case of a fire.

Workers should be properly protected with goggles, safety shoes and safe clothing when handling caustic or other dangerous chemicals.

Workers should not be sent into tanks until they have been cleaned, gas masks have been supplied and more than one extra man is present in case a worker is overcome.

Workers should be examined periodically by a doctor. While some of the old health hazards of the industry are known, many of the chemicals are so new that health hazards are not always understood.

38

*Paint Testing**

DEFINITION

Basically, testing is an application of the scientific method to a determination of the appropriate properties or qualities of a given substance. By expressing the parameters found in mathematical terms, we obtain a precise value for these characteristics. It is then possible to make critical comparisons with similar systems and judge their worth in the light of their proposed end use. Thus, whether we are assaying an ore or evaluating a complex paint system, essentially we are attempting to discover the intrinsic worth of a given material or collection of materials usually for a specific purpose.

IMPORTANCE OF TESTING

The necessity and consequent importance of materials testing, whether the material be a cement, ceramic or an organic coating, should be self-evident. Actually without comprehensive testing and the interpretation of the results obtained, progress in the application of newly developed raw materials and unique application of existing products would be seriously hampered. Indeed even the quality of existing materials would be in doubt. The results of development become meaningful only through an accompanying evaluation; without it, research is sterile. In a practical sense, this is the only method by which we can obtain a reasonable indication of the performance to be expected from our product or positively ascertain its composition. Unfortunately, such testing is too often considered a monotonous, routine operation which is but a prelude to the more glamorous phases of research, development and application.

CLASSIFICATION

Paint testing is sometimes divided into two classifications. The first category, akin to the quantitative analysis of chemistry, has to do with a

**By K. Schreiber, Pittsburgh Plate Glass Co., Cleveland, Ohio.*

determination of compositional constants. This includes such qualities as solids content (by weight or volume), viscosity, color, gloss, hardness, etc.

The second category, similar to qualitative analysis, is concerned with the performance characteristics of the coating. Prominent among such qualities are environmental behavior, chemical resistance, etc. The tendency is to assign those properties which we can most positively determine and which can easily be duplicated (e.g., viscosity) into the first category. Those which require more elaborate testing and a good bit of interpretative judgement are put into the second category. Obviously this classification is arbitrary and merely of academic interest. As more sophisticated test methods and instruments become available, the transition from the second to the first class is accelerated.

Another classification sometimes employed divides the field into raw materials testing and the evaluation of the finish itself. The latter can be further subdivided into laboratory and environmental testing.

HISTORY OF TESTING

Historically there has been a tremendous increase in the entire field of testing in the last twenty years. This has come about for several reasons. First of all, the resin and pigment chemists have advanced the state of the art so tremendously that the sheer number of possible raw materials and their combinations has multiplied many times over. There is no doubt that the unprecedented amount of applied research in the polymer field has helped the paint industry. Urethanes, polyester resins and vinyls are but a few examples of fields which have grown tremendously. It is only through complete and exacting analysis of the properties of these new materials that we can logically arrive at a judgment as to their value and application in the coatings field.

Secondly, our general technology outside of the paint field has advanced so rapidly that many new possibilities for the application of coatings have arisen. It has been truly stated that our scientific information, and the products which flow from it, are growing at an exponential rate. The use of fiber-glass reinforced plastics, as one example, has grown tremendously in the last few years. To coat such a material requires a highly specialized type of finish possessing unusually good adhesion, impact resistance, humidity resistance, etc. Similarily, the increased use of aluminum and zinc die castings has imposed new demands for coatings which are adaptable to these substrates and yet are economically feasible. Again these requirements can only be met through an exhaustive testing program involving new and improved compositions.

Application methods have advanced and broadened considerably, thereby opening up new fields for applying recently developed or existing coatings. For example, such processes as electrophoretic deposition, vacuum metalizing, fluid bed application, flame spraying, etc., have widened the possibilities for the coatings industry and challenged its resources for the development of more rigorous and comprehensive evaluations as well as new test methods.

Thirdly, as we have advanced technologically, there has been an accompanying increase in the demands made upon coatings. In short, the performance requirements have been raised and the trend continues in this direction. Part of this has undoubtedly resulted from a realization that in today's market the cost of labor has multiplied many times over that of materials. Thus, in the maintenance field, it is now accepted that it is cheaper to employ an expensive silicone-modified alkyd, for example, than to utilize a conventional unmodified alkyd finish and have to refinish every other year. Economically, this opens the door to materials of proven durability, chemical resistance, etc., which have previously had limited application because of their high cost.

The highly competitive automobile industry has acted as a propulsive force in new and improved coatings and their evaluation. Where automobile manufacturers were once content to simply cover engine parts with an inexpensive oleoresinous black finish, today they are demanding 100-hour salt spray resistance (along with other qualities) as a minimum.

Manufacturers of office machines costing thousands of dollars are not satisfied with previous standards of abrasion and impact resistance. Their specifications requirements have risen considerably—and rightfully so.

Even the advertising field has capitalized on the results of the testing laboratory. Thus Armco Steel has recently featured a guarantee on their factory-enameled wall panels for fifteen years against "blistering, peeling, cracking, flaking, checking, and chipping, and ten years against excessive color change." Stran Steel warrants the finish on one of its lines "against chipping, cracking, blistering or peeling and against chalking in excess of ASTM D-659-44 number eight rating for a period of five years."

Color has been equated with style or fashion. As a result, the demand for new and improved pigments of proven durability has grown tremendously. This, in turn, has expanded the demands made upon pigment testing. The quality of present-day colors is a tribute to the analyst's labors in exposing and successfully evaluating thousands of new pigments.

ELEMENTS OF TESTING

The complete testing procedure is composed of four integral elements. In the order of their application, these are as follows:

(1) The discernment of the qualities or properties of the system, or material, which are most relevant to its proposed end use.

(2) The selection of the appropriate test methods for these qualities.

(3) The actual test procedure. Details of the tests themselves will be found in ASTM and Government Specifications and "The Physical and Chemical Examination of Paints, Varnishes, Lacquers and Colors" by Henry A. Gardner.

(4) Interpretation of the results obtained and a critical comparison with similar test results obtained from related materials so that conclusions can be drawn.

Unless these elements are recognized and intelligently coupled, our determinations may be considered incomplete or misleading.

Discernment of Appropriate Qualities

Indispensable to the process of evaluation is a prior judgement as to which qualities or properties are of prime importance in the service life of the system or material being tested. Needless to say, this selection can only be made correctly when a complete and accurate knowledge of these conditions is known. All too often, this information is relayed to the laboratory through one or more intermediaries and is understandably unreliable. In other instances, the service life details are unknown. Obviously under these circumstances, a wide discrepancy can exist between test conclusions and actual performance.

The vagaries of consumer application are well known—and unpredictable. Who could predict that a finish designed for resistance to the corrosive atmosphere of a chemical plant would be applied to steel piling in marine service? Such misapplications cannot possibly be forseen in the testing laboratory. Frequently the ultimate consumer will stress properties which to him seem of overriding importance, but which really may not be so essential to the ultimate life or integrity of the finish. The automotive industry's emphasis on gloss retention would be an example of this. At times, the industrial finisher, or the ultimate consumer, will insist on the paint manufacturer's meeting certain requirements in his paint specification, but will still hold him responsible for a breakdown in the field. Actually, the failure may be due to a factor or combination of factors which are completely outside the specification! This merely indicates that where the human element enters the finishing picture, it is difficult to completely envisage what conditions will actually be encountered. Nevertheless, a judgment from available information must be made; its validity determines the success of failure of the testing steps to follow.

Selection of Test

Following the assessment, a choice must be made as to which of the particular tests available will evaluate these selected qualities with the greatest degree of accuracy. Here compromise is necessary since any performance test method is, at best, an approximation of actual service conditions. This follows from the fact that we have isolated a single quality, evaluated it and drawn conclusions from it alone. It necessarily neglects other minor, related factors which, in reality, would have a bearing upon the results obtained. Sometimes these cumulative factors are sufficient to invalidate the results obtained with the single test. For years, a manufacturer of automobile steering wheels relied on Taber Abraser values and a "flannel rub test" to ascertain quality, but found so little correlation that the test was abandoned. Factors like resistance to perspiration, oil and grease and aging of the finish were of such consequence that he had to resort to testing the wheels in actual use on taxicabs.

Testing the resistance of a marine finish, e.g., by subjecting it to the conventional salt spray or saltwater immersion test, would probably be as close an approximation as could be achieved practically—particularly when compared with similar finishes tested under identical conditions. We must recognize, however, that results are *relative*, not absolute. This comparative value of much of our testing is noted time and again by the ASTM.

In testing traffic paints, for example, the statement is made: "This method of test is intended for testing the *relative* values of service of traffic or pavement-marking paints" (ASTM D-713-46).

In the marine finish mentioned earlier, no definite conclusion could be reached that since the finish had withstood 500 hours, it would withstand a minimum of five years in service—or even one year! Such associated variables as exposure to ultraviolet light, fungicidal growth, degree of salinity, water temperature and other minor factors must also be considered.

Nevertheless, such compartmentalized or "fragmented" testing is necessary. Short of actually running exposures for protracted periods of time, i.e., literally duplicating exposure conditions, such a procedure is valuable. A good bit of the progress made in the coatings industry has necessarily been based on such tests and corroborated in field service. Furthermore, it must be recognized that our industry is not unique in employing this approach. Such fields as the plastics and rubber industry face identical problems and solve them by similar, somewhat arbitrary premises and conclusions.

A good bit of the problem of "fragmentation" can be overcome by selecting a number of qualities for evaluation, rather than a single principal one, and rating them as to their relative importance. The overall conclusion as to the relative worth of a particular finish can then be established by means of this weighted average. The value of this more comprehensive approach is recognized and employed in an increasing number of industrial specifications since it comes closest to giving a true answer to industry's question: Which coating is best for my use?

To give an answer within a reasonable period of time, resistance tests must frequently increase the severity of the attacking medium. Understandably, the conditions of service cannot be reproduced exactly. Even in time-consuming panel weathering, compromises must be made since one is usually geographically limited in the number of locations available. Furthermore, even the same location has greatly varying amounts of insolation from year to year.

Many attempts have been made to shorten this test period by the use of the weatherometer or to obtain an indication of expected service through other means. A recent study entitled "Possibility of Predicting Exterior Durability by Stress/Strain Measurements"[9] attempts to solve this problem in a unique fashion. The Cleveland Condensing Type Humidity Cabinet employing the cyclic method appears to be a reasonable method for predicting the resistance of painted metal surface to degradation induced by or dependent upon condensation in outdoor exposures."[2]

The testing laboratory frequently has to employ testing procedures which are unrealistic but nevertheless dictated by a specification. These can be costly. Thus a power lawn mower manufacturer who stipulated that his air-dried finish had to be gasoline resistant within 2 hours ended up with a needlessly expensive finish. Perhaps the best technique for countering such specifications is to develop two materials, of which one meets the demands and the other is realistic. A comparison of costs points out the extra expense involved with the former coating.

Customers will at times devise tests which they believe meet the unique conditions of use of their product. These can occasionally be valuable. Some of the automobile manufacturers, for example, have found that shooting a BB pellet at a finished panel duplicates the effect of gravel hitting the side of a rocker panel.

Too often, however, such tests are amateurish attempts to duplicate what the customer considers to be his unusual service conditions. This "paper clip gadgetry" should be discouraged. Such esoteric procedures usually are not reproducible, have no meaning to anyone other than the party involved and fail to meet the scientific standards toward which the coatings industry is striving. Some years ago the author encountered such a "test" for paint coverage which had been set up by the testing laboratory of a large appliance manufacturer. The test results were used in a very complex mathematical formula which had been developed. While imposing in appearance, it was found that in practice the results were inconsistent and not reproducible. A simple Pfund Cryptometer test would have been more practical.

In this connection, the advantages of ASTM methods are apparent.

Test procedures which have been devised by this organization represent a prodigious amount of effort and the best thought of thousands of highly qualified individuals. Furthermore, these tests are constantly being reviewed, revised and expanded. Also, results can be duplicated by other laboratories and command the respect of all well-informed technical men.

Conduct of Test

Needless to say, the test procedure itself should be carried out with extreme care, accurate calibration of instruments and control of variables insofar as possible. With critical determinations, tests should be run in duplicate or triplicate. It has been noted that even in a seemingly infallible machine operation, such as a spectrophotometer, operator error is substantial. If the nature of the test is such that a subjective judgement is necessary, the use of multiple operators is advisable. In running a pencil hardness test, for example, it was found that one of the biggest variations was the "personal touch" from one technician to another.[4]

Since a paint film is influenced by such a wide variety of conditions, i.e., age, humidity, temperature, etc., it is apparent that particular attention must be given to controlling and standardizing there factors as much as possible. The substrate is a neglected influence in this respect. As has been pointed out,[3] ferrous surfaces are completely nonuniform. This condition drastically affects qualities of paint adhesion, humidity and salt spray resistance. It is possible to overcome this deficiency to some extent by employing the same lot of steel or using a treated metal. For customer problems and development, the use of this metal is preferable.

Working with a natural product such as wood, its variations and defects affect the results to an even greater extent. A cold-check test run on one type of coated wood, or wood with varying moisture content, will give results which differ from those with another type of wood. This variance is noted in ASTM D-1211-57-T which states: "Except when elaborate precautions are taken to control the moisture content of the wood before and after finishing and during the test, failures will often occur as a result of dimensional changes in the wood due to moisture change rather than temperature change."

It is characteristic of the paint testing field to frequently make determinations in an area where human judgement, and therefore error, must be called upon to an unusual degree. Determining the degree of dry is such an operation. Years ago, the Federation of Societies for Paint Technology sent identical samples of an air-drying finish to the testing laboratories of half a dozen prominent paint manufacturers. They requested dust-free, set-to-touch and print-free times. The results reported evidenced a disconcerting variation. Unfortunately there is not even universal agreement on the meaning of the terms.

Of all the common paint tests, accurate determination of drying time is one of the most difficult. This is due not only to the subjective nature of tactile perception, but also to the effect of a number of important variables such as humidity, room temperature, age of sample, film thickness, air circulation and nature of the substrate. In this case, as well as others of a similar nature, instrumental methods which yield a definite value of time or dimension should be employed wherever possible. Thus in the instance cited, Gardner Laboratories has a Circular Drying Time Recorder which operates on the principle of dragging a small, weighted metal sphere over the wet surface of the paint at a regulated, timed rate. The evident change in pattern is a measure of the dry. Their Drygraph, Traffic Paint Drying Time Wheel (ASTM D-711-55) and the Sanderson Drying Time Meter are other applicable instruments. While these and other similar instruments do not duplicate sensory reaction, they do at least give a definite numerical time unit which is of value when compared with other results obtained in similar fashion.

Interpretation of Results

The best chosen set of tests and the most precise results are vitiated if conclusions are incompletely and incorrectly drawn. One of the most common errors in this area is the failure to recognize the inherent limitations of a particular test method as compared with service conditions. This is particularly true when attempting to equate the results of laboratory equipment testing with exposure to the vicissitudes of a natural environment.

No analyst with any degree of experience would be prepared to translate Florida exposure tests into durability in other parts of the country, or hot detergent tests into the expected life of a washing machine finish. However sales and advertising pressures may be sufficiently strong that such conclusions are expected in unequivocal terms. The best and most logical procedure here is an interpretation based on comparative results with a product which has a well-substantiated service history. The conclusions which follow may be hedged—but they will be accurate.

Because of the "fragmentary" nature of a particular evaluation, generalized conclusions simply cannot be drawn. Even where a particular test is seemingly duplicated in service (e.g., an impact test), we must be aware that what passes in the laboratory may fail in the customer's hands. The best that can be said is that under particular laboratory conditions of age, film thickness, temperature, etc., finish A withstands 45 inch-pounds of inverse impact with a particular instrument while finish B only takes 35 pounds impact under these conditions. The presumption is that the former finish will perform better in service. Actually this may not be true. In practice, however, we must fall back on the testing apparatus at hand and the results so obtained and make our presumptions as positive statements. This procedure of course is followed routinely. In the analysis of a problem, however, the presumptive character of the conclusions must be considered. Sometimes assumptions which seem perfectly logical are not confirmed in practice. For example, it was recently found that certain coatings actually show an increase in adhesion, or rather cohesion, after water immersion.[10]

SAMPLING

The technique of accurate sampling is an important consideration in any test program. Acquiring a representative portion of the batch or lot is fundamental to the entire testing operation. Many possible mistakes and incorrect assumptions can be made. For example, a sample of a particular lot of a pigment is sent directly from the pigment supplier to the paint manufacturer. It may be representative of the lot as manufactured, but does it necessarily correspond to that same lot after transit and storage?

The recommended procedures of analytical chemistry can be used to advantage with dry solids, i.e., the use several samples followed by mixing by the quartering method. Contamination of solvents by water, rust, oil, etc., when transported in multipurpose tank wagons is not unknown. Sampling solely from the top of such a conveyance is not accurate. Similarly, extracting one lump of hard resin from a barrel disregards the classification that may have taken place and the possibility that the dust at the bottom may contain dirt and other contaminants.

Paint consumers are notoriously inaccurate in their sampling procedures. Attempting to check viscosities of one lot of paint against another when the sample has been taken from the last few gallons of a 55-gallon drum is just one hurdle which the analyst must face and take into consideration. It would be easy to say that such obvious violations of all rules of sampling should be rejected. In practice, the analyst does the best he can, explains the situation fully, and hopes for an improvement in the future!

KEEPING RECORDS

An effort should be made to keep accurate and complete records of test conditions and results. Control laboratory records are frequently the only resource when a complaint in the field is received. An experienced paint man realizes that valuable information can be garnered from an analysis of past records and results. Duplication of effort over a period of time is minimized particularly if indexing and cross-indexing are employed. In keeping records, such indefinite terms as "very good," "good," "fair," etc., should be avoided and a numerical rating employed. ASTM ratings for blistering and chalking should be used where applicable. Where series of similar tests are being repeated over a period of time, a form listing all pertinent properties to be rated saves time and assures completeness.

ESTABLISHMENT OF STANDARD VARIANCES

In the testing of either raw materials or a finished product, the establishment of clear-cut limits, or standard variances, on the properties evaluated should be considered mandatory. The alternative is a haphazard, nebulous implied set of limitations which are interpreted differently at various times by different individuals.

The criteria for these limits should be based not on penalizing a supplier for small lapses in his standards, but rather on a realistic evaluation of how far a given property can be "off standard" before the quality of the end product is perceptibly affected. This same philosophy holds true for shipment to customers. Such a fundamental property as solids content is important but even here it is easy to be too arbitrary. If a 60% solids is specified and one lot is determined to be 58.5%, but the next one is 61.59%, is there enough difference here to warrant returning the low solids lot? While such a procedure may establish a reputation for accurate determinations and insistence on quality, it also creates delays, ill will and added expense.

A corollary to setting up a set of test standards and their limits is to have both your suppliers and your customers informed (and agreed) as to what these conditions are. Recently a large paint manufacturer followed his usual practice of running Ford No. 4 cup viscosities within specified limits at 80°F. It was not until the customer had rejected his material several times for inadequate viscosity that he ascertained that the customer's laboratory was conducting his viscosity readings at "room temperature." Needless to say, this was a "summer problem."

COLOR TESTING

The field of color testing offers an opportunity for setting limits which can benefit all concerned. For years, shading and color matching have been the domain of individual, almost personal, judgment. Certainly there is a great future for color standards and limitations based on comparision of spectrophotometric curves or preferable tri-stimulus values, and an excellent presentation of this subject has been given.[7] The transition to this type of specification, however, is not going to be rapid. The cost involved in purchasing a spectrophotometer and training an operator to use it with skill and imagination is an obstacle to many small paint manufacturers and consumers. The tendency of paint consumers using this method for the first time is to set tolerances which are unrealistically close. This can be overcome by making the change to this instrument a cooperative venture between the producer and consumer. It has been pointed out that "the precision of spectrophotometric colorimetry varies with the care taken in the calibration and operation of the spectrophotometer."[1]

INCORRECT STANDARDS AND TEST PROCEDURES IN INDUSTRY

In meeting nongovernment specifications, it must be recognized that in many cases, as previously mentioned, the standards which must be met and the methods of evaluation are devised by individuals outside the paint industry itself. Lacking the necessary background, they rely upon standard ASTM procedures at best and upon proprietary improvisations at worst. Usually the paint chemist must follow these procedures regardless of their general quality or appropriateness. Hastily drawn specifications which call for standards of hardness, for example, where abrasion resistance is really required are misleading and should be called to the attention of the parties involved.

The coatings industry has an obligation to educate other industries along these lines, although admittedly this is not an easy task. As indicated, the ASTM standards should represent a goal toward which we can strive, keeping in mind that the misinterpretation of these tests is as damaging as employing the wrong test. If making progress along these lines is not forthcoming, we will find ourselves at a disadvantage in competing with other types of coatings, such as plating and plastics, where more exacting and appropriate test procedures have been devised and applied.

One of the greatest stumbling blocks in a customer's setting up test methods is the profusion of tests available for a given property. Possibly this is an indication of an inquisitive, healthy industry. However, it has created confusion since it is difficult for the nontechnical man to discriminate under these circumstances. It would seem that a critical evaluation of present tests with definite recommendations as to which particular test is best would be worthwhile.

We in the industry are not blameless. Such a simple operation as determining viscosity has been hampered by the use of temperatures ranging from 72 to 80°F in our own laboratories. The impression sometimes given to paint consumers is that a given manufacturer uses a particular method to "prove" the superiority of his product over his competitor's. The situation is not improved by paint consumers or producers who set up paint specifications built around their own product using specialized tests of dubious value.

Realistically we must recognize that present-day paint testing as practiced in all but the best equipped and largest paint manufacturer's and consumer's laboratories leaves something to be desired. As the industry has progressed from an art to a science, some of its test methods have not caught up with it! Specifications or test procedures drawn around such terms as "general appearance," "satisfactory dry," etc., are not appropriate for today's modern finishes and sophisticated testing instruments.

ECONOMIC LIMITATIONS

Fundamentally the cause of the technological lag in our field in the past has been due, at least partially, to economic factors. Unlike the chemical industry as a whole, the coatings industry has been primarily a low-profit field which has simply not been able to afford the necessarily expensive testing laboratories, equipment and qualified personnel to staff them. Consequently, the technical literature on testing has been limited. Another factor has been the reluctance of many paint companies to divulge their findings, even in the field of testing, to others (see reference 5).

Fortunately, this situation is gradually improving. There is a trend away from the multitude of small paint companies which has been so characteristic of our field for years. Consolidation and the economic realities of a more specialized field have taken their toll. The remaining companies in the field are recognizing the necessity of complete testing and evaluation. More important, there is growing technical communication in the field due to the efforts of such societies as the Federation of Societies for Paint Technology, the American Chemical Society and the American Society for Testing Materials.

INHERENT TECHNICAL LIMITATIONS

A factor which has limited the effectiveness of testing, or at least complicated it, concerns what we may designate as the nondeterminant nature of the coatings we are dealing with. For the most part, their performance and qualities are not immutable or absolute. Their characteristics are directly related to a number of highly variable, interdependent conditions. This is probably more true for coatings than for most other engineering materials.

Basically a composite system is involved—the coating and its substrate. As is well known, such a fundamental property as adhesion may vary appreciably with the same coating when applied over different types of metallic surfaces; therefore, it is difficult to assign an absolute "adhesion factor" to a given resin. In a practical sense, in this area we are also dependent upon such variables as cleanliness, roughness and degree of oxidation of the substrate. Film thickness is another variable which influences such related qualities as impact resistance, extrudability and bendability.

Even when these factors are controlled and different lots of the same paint formula are involved, the degree of dispersion influences the results. In a classical study of the relationship between surface area and particle size of pigments, it was shown that properties such as tinting strength, consistency, jetness of mass tone and bronzing are influenced by these dimensional changes.[6]

More important is the fact that thermosetting resins exhibit varying degrees of oxidation and polymerization (which influence physical properties) depending upon the variables of time and temperature of cure. If hardness in such a system, for example, is dependent upon such widely varying conditions, it is apparent that we can only establish a value for a particular system under a definite set of rigidly controlled conditions. Other well-known factors which influence performance are humidity, age of coating, exposure condition, etc.

This enormous array of variables, all of which must be controlled if a testing program is to be completely accurate, obviously imposes severe demands on time, attention and equipment. In practice, this handicap has been overcome to a *limited* extent by evaluating various films at the same time, using identical testing equipment and controlling such apparent factors as film thickness, etc., as best we can. In general, the relative results obtained are reproducible. We must not lose sight, however, of their relative, parochial nature and consequent limitations.

STATISTICAL ANALYSIS

Statistical methods can be of considerable help in the field of testing. They lend themselves particularly well to the analysis and interpretation of a large number of tests of a qualitative or semiquantitative nature. Their use in the field of environmental testing and corrosion control is unusually appropriate and the determination of paint drying time also benefits from this approach.[8] Through the use of such methods, it is possible to set up standard deviation data and separate constant experimental error from the effect of variations in the material. As the field of testing grows, its use as a time saver and interpretive tool becomes more important. The ASTM Manual on Quality Control of Materials (Publ. 15-C, January 1951) and subsequent publications are useful in this area.

REFERENCES

1. Billmeyer, F. W., Jr., *J. Paint Technol.*, **38**, 726 (1966).
2. Cleveland Society for Paint Technology, *Offic. Dig. Federation Soc. Paint Technol.*, **37**, 1392 (1965).
3. Grossman, G. W., Jr., *Paint Ind. Mag.*, **76**, No. 9, 7 (1961).
4. Houston Society for Paint Technology, *Paint Technol.*, **38**, 691 (1966).
5. Kiefer, D. M., *Chem. Eng. News*, **43**, No. 5, 86–96 (1965); No. 6, 80–92 (1965).
6. Kienle, R. H., *Offic. Dig. Federation Paint Varnish Prod. Clubs*, No. 300, 11 (1950).
7. MacAdams, D. L., *Offic. Dig. Federation Soc. Paint Technol.*, **37**, 1487 (1965).
8. Prane, J. W., *Paint Ind. Mag.*, **76**, No. 8, 5 (1961).
9. Schurr, G. G., Hay, T. K., and Van Loo, M., *J. Paint Technol.*, **38**, 591 (1966).
10. Walker, P., *Offic. Dig. Federation Soc. Paint Technol.*, **37**, 1561 (1965).

39

*Fire Protection, Safety and Health**

In the coatings industry, because of the flammable materials handled, fire and explosions are a constant menace. Present also are the dangers of moving machinery such as mixers, mills, conveyors, etc. The handling of raw materials, manufacturing and handling of finished goods can also be hazardous.

It is desirable that raw materials storage, manufacturing and finished stock storage be separated from each other.

HOUSEKEEPING

Good housekeeping must be the part of any efficient operation or safety program.

The daily removal of waste and combustible material is recommended. If not immediately removed, such materials should be kept in fire-resistant enclosures. All cleaning rags and filter clothes should be disposed of in metal cans filled with water. No combustible material should be near heat sources such as furnaces, steam pipes or radiators. Containers should be clearly marked identifying the contents. All tanks should be covered. Aisles should be kept clear so that traffic can move smoothly without accidents. Spills must be cleaned up immediately to prevent falls.

Building must be equipped with the proper fire control devices such as sprinklers and fire extinguishers.

FIRES AND EXPLOSIONS

Paint plants contain large amounts of organic solvents, potential fuel for fires. All that is needed is a spark or contact with high temperatures to

*By C. R. Martens, The Sherwin-Williams Co., Cleveland, Ohio.

ignite these combustibles. These may come from the following sources:

(1) Static electricity—from moving vehicles or materials. Proper grounding will prevent this.

(2) Static spray—can be caused by a vertical drop of solvent of 18 inches or more. The conductivity of the solvent affects the static build up. Divert the flow of solvents to the side of the equipment. Containers used for transferring solvents should be grounded.

(3) Struck spark—use nonferrous tools to eliminate.

(4) Electrical—use explosion-proof motors and switches—permanent electrical connections.

(5) Chemicals—reactive combinations and spontaneous combustion produce temperatures in excess of ignition points.

Static Electricity

Static electricity is one of the most prevalent hazards in the paint industry for starting fires. The basic method for lessening the hazard of static electricity is to provide a means by which the electrical charges are led harmlessly away as rapidly as they are produced. Provisions for a continuous conductive path from the point of accumulation to an earth ground will permit the charge to be drained off. The resistance to the earth should not be more than 3 ohms. The best ground is a metallic water pipeline.

Grounding and bonding of all operations and buildings is necessary. Precautions are taken from the receiving point of solvents; tracks and cars are bonded. Buildings, equipment and pails used for transferring solvents are grounded.

All equipment, solvent lines, and drive shafts using belts should also be positively grounded to a water line or a driven electrical ground rod. The belt should be made of conductive rubber and the pulley of steel.

Flammable solvent fill pipes should go to the bottom of tanks, drums etc., to minimize the generation of static electricity in the free fall of flammable solvents. Buildup occurs when these solvents break up into small droplets or spray. As this occurs, static is built up and a spark can be created. This is known as "static sprays." This usually occurs during the initial filling of a vessel when the vapors are in the explosive range. This results in a flash fire or an explosion.

The conductivity of wash solvents can be increased by the addition of small amounts of polar solvents. This reduces the static electricity hazard. Wash solvents should have a high flash point and contain a small amount of a polar solvent such as ethanol to reduce resistance down to 0.4 megohm or lower.

Struck Sparks

To eliminate struck sparks, nonferrous tools should be used. Lids on mixtures should be of a nonferrous metal such as aluminum. Gear-driven agitators should have bronze pinion gears. Platforms around mixtures should be of aluminum plate or some other nonsparking surface. Concrete floors should be made nonsparking and conductive.

Electrical Equipment

Arcing of electrical equipment is another source of ignition that has caused fires. Fires can be caused by short circuits, arcing in switches and in electrical motors in flammable vapor-air atmospheres. All electrical equipment and wiring should be explosion-proof per Class I, Group D—National Electric Code.

Explosive Range

The explosive range includes all concentrations of a mixture of flammable vapor or gas in air (usually expressed in per cent by volume) in which a flash will occur or a flame will travel if the mixture is ignited. The lowest percentage at which this occurs is the lower explosive limit, and the highest percentage the upper explosive limit. If such a mixture is confined and ignited, an explosion results.

	Flash Point, Closed Cup (°F)	Explosive Limits Volume	
		Lower	Upper
Acetone	0	2.1	13.0
Amyl acetate	76	1.1	7.5
Benzene	12	1.4	7.1
Butyl acetate	72	1.4	7.6
Castor oil	445	—	—
Dibutyl phthalate	315	—	—
Ethyl acetate	24	2.2	11.0
Ethyl alcohol	55	3.5	19.0
Glycerine	320	—	—
Linseed oil	435	—	—
Mineral spirits	100	—	—
V.M. & P. naphtha	20–45	0.9	6.0
Toluene	40	1.3	7.0
Xylene	63	1.0	6.0

Autoignition Temperatures

The autoignition temperature of a substance is the lowest temperature required to initiate combustion in the absence of a spark or flame. The autoignition temperature has no relation to flash point as shown on the next page.

	Flash Point (°F)	*Autoignition Point (°F)*
Mineral spirits	104	473
Toluene	40	1026
Xylene	63	900
Ethyl alcohol	55	750

High temperatures may be encountered in varnish cooking operations as well as localized high temperatures in high-speed dispersion operations.

Nitrocellulose

Nitrocellulose is one of the most flammable of materials used in the paint industry. When wet with water it will not burn. Nitrocellulose becomes flammable when the water is extracted and replaced with alcohol or when dry. Nitrocellulose in a dry state burns very rapidly. When dry, it can be ignited by friction or impact. Drums of nitrocellulose should be handled with care and stored in isolated areas.

HEALTH

The manufacture of paints, varnishes and lacquers is not considered a dangerous occupation, but health hazards do exist in any plant where liquid and dry materials are handled.

The physical nature of certain pigments and inerts will cause irritation to the respiratory system. Many dusts are potential explosion hazards. Hoods over machinery or respiratory masks should be provided when such conditions exist.

Most solvents are toxic to some extent in the vapor-air mixtures that are inhaled. Therefore, it is essential that adequate ventilation be provided. Acceptable concentrations of various solvents are as follows:

Safe Concentration of Chemicals
(8 hours of exposure per 24 hours)

	Parts of Vapor per Million Parts of Air
Benzene	100
Toluene	200
Xylene	200
Methyl alcohol	200
Butyl alcohol	200
Amyl alcohol	56
Acetone	200
Ethyl acetate	140
Butyl acetate	400
Amyl acetate	400

Another aspect of the health problem is dermatitis. Some of the solvents, resins, pigments and other materials may cause dermatitis. Protective clothing and equipment should be worn by the worker.

Hazardous Materials

Particularly hazardous materials handled in paint manufacturing operations are given below. Doubtful mixtures should never be handled in a confined space such as a ball mill.

Aluminum Powder. Mixtures of aluminum powder with oxide pigments such as Venetian red are hazardous. Mixtures of aluminum powder with technical-grade oleic acid have been known to detonate. Fires have been caused by processing aluminum powder mixtures with polar solvents on roller mills. Rusty surfaces coated with aluminum paints will spark when struck with a nonferrous tool.

Amines. Basic tertiary amines catalyze the decomposition of molten maleic anhydride to such an extent that may result in an explosion. The addition of amines to cotton waste containing epoxy resins may result in spontaneous combustion. Nitrocellulose will ignite on contact with certain amines.

Oxides. Mixtures of propylene oxide and carbon black have caused ball mill explosions. Mixtures of propylene oxide and caustic on mixing will flash. Explosions have occurred when ethylene oxide was passed into an amine solution.

Chlorinated Materials. Trichloroethylene when mixed with caustic soda decomposes into chloroacetylene gases which will ignite or explode on contact. Chlorinated rubber and zinc oxides react exothermically at temperatures of 400°F.

Peroxides. Organic peroxides are extremely flammable. Peroxides should not be mixed with driers, strong acids, strong bases, amines or reducing agents.

Drying Oils. The spontaneous ignition of drying oils may occur in contact with cellulosic materials. Also, pigments such as chrome green, chrome yellow, iron oxides, lamp black, Prussian blue, red lead, toluidine red, and para red when wet with a small amount of drying oils may cause fires by spontaneous combustion.

Railroad Shipping Labels for Hazardous Materials. Red label—inflammable liquids that give off ignitable vapors at 80°F or lower (open cup).

Yellow label—inflammable solids other than those classified as explosives which cause fires by friction, moisture or chemical changes.

White label—corrosive liquids, strong mineral acids or other corrosive liquids that are liable to cause fires when mixed with chemicals or organic material.

Red or green label—includes all flammable or non-inflammable gases under pressures exceeding 25 psi at 70° F.

REFERENCES

1. Sax, N. I., "Dangerous Properties of Industrial Materials," New York, Reinhold Publishing Corp., 1963.
2. *Fire Protection Publication*, National Fire Protection Association, 60 Batterymarch Street, Boston, Mass.

Glossary

ABRASION RESISTANCE: The resistance of a surface to being worn away by rubbing or friction. Abrasion resistance is not necessarily related to hardness but more related to toughness.

ACID NUMBER: The number of milligrams of potassium hydroxide required to neutralize the free fatty acid in 1 gram of fat, oil, wax or resin.

ACRYLIC: A family of synthetic resins made by polymerizing esters of acrylic acids.

ADHESION: The ability of one substance to stick to another. There are two types of adhesion: mechanical, which depends on the penetration of the surface, and molecular or polar adhesion, in which adhesion to a smooth surface is obtained because of polar groups such as carboxyl groups.

ALIPHATIC: The name applied to petroleum products which are straight-chain hydrocarbons derived from a paraffin-base crude oil.

ALKYD: A synthetic resin which is the condensation product of a polybasic acid such as phthalic, a polyhydric alcohol such as glycerin and an oil fatty acid.

ALLIGATORING: A form of paint failure in which cracks form on the surface layer only. It is caused by the application of thick films where the underlying surface remains relatively soft.

AMPHOTERIC: Possessing both basic and acidic properties.

ANIONIC SURFACTANT: One which has a negative charge and migrates toward the anode or positive pole while in solution.

ANTIFOULING: An antifouling paint is one which contains toxic or poisonous substances to prevent growth of barnacles on the hull of ships or other objects submerged in water.

ANTIOXIDANT: A material which when added to a varnish or an oil retards or prevents oxidation and drying.

ANTISKINNING AGENT: A type of antioxidant, usually volatile, which when added to a varnish or an oil will inhibit the surface skinning in the container.

ASPHALT: A complex mixture of solid and semisolid hydrocarbons. First applied to a natural mixture, or bitumen, of petroleum origin. Now includes asphalt-like residues, or pitches, from petroleum coal tar or lignite.

BALL MILL: A revolving cylindrical metal mill which uses steel balls, stones or other media to grind or disperse the pigment in the vehicle.

BINDER: The nonvolatile portion of a coating vehicle which is the film-forming ingredient used to bind the pigment particles together.

BLEEDING: A condition which exists when the color of a dye, stain or pigment passes through the top coat producing a stain. It occurs when the pigment is somewhat soluble in the vehicle of the top coat.

BLISTERING: A paint film failure usually caused by application of paint on a surface containing an excessive amount of moisture or other volatile material.

BLOOM: A bluish fluorescent coat which forms on the surface of some films.

BLUSHING: A term applied to lacquers when they become partially opaque, cloudy or translucent upon application or drying. Fast-evaporating solvents may cool the film enough to cause water condensation, precipitating solid materials.

BODY: A practical term used to give a qualitative picture of consistency.

BREAK: The break in an oil is the flocculent material or foots which separate on long standing or upon heating. Varnishes are said to break when a portion of the resin separates out.

BULKING VALUE: The bulking value of a pigment is usually expressed in gallons per 100 pounds of pigment.

CATALYST: A substance which accelerates the speed of a chemical reaction.

CATIONIC SURFACTANT: One which bears a positive charge and migrates toward the cathode or negative pole while in solution. Seldom used in latex paints.

CHALKING: The presence of a loose powder on the surface of a paint after exposure to the elements.

CHECKING: A phenomenon manifested by slight breaks in the surface of the paint film.

CHROMA: Color intensity or purity of tone, being the degree of freedom from gray.

COALESCENCE: The fusing together of a latex film upon evaporation of water.

COHESION: The attractive force between like molecules. It is the force that holds the molecules of a substance together.

COLLOID: A material composed of ultramicroscopic particles of a solid, liquid or gas dispersed in a different medium which can be a solid, liquid or gas.

COLOR: A generic term referring inclusively to all colors of the spectrum, and white and black. Color is described by three properties: hue, lightness and saturation.
  (1) Hue (color, character, dominant wavelength), blue, green, red, etc.
  (2) Lightness (brightness, reflectance, value) position on the gray scale between pure black and pure white.
  (3) Saturation (purity, grayness, cleanliness, muddiness, chroma), purity or intensity of color.

COLOR RETENTION: Color stability after exposure to the elements.

COMPATIBILITY: The ability of two or more materials to mix with each other without separation.

CONDENSATION RESIN: A generic term applied to resins formed by splitting off water or other substances during resin formation, e.g., alkyds, phenolics, urea and melamine resins, etc.

CONSISTENCY: Relative stiffness, body of a composition in bulk.

COPOLYMERS: Mixed polymers or heteropolymers. Products of the polymerization of two or more substances at the same time.

CORROSION: Detrimental change in a material usually a metal under conditions of exposure.

COVERAGE: Hiding power of a paint usually expressed in square feet per gallon.

CRAWLING: The tendency of a liquid to draw up because of high surface tension.

CRAZING: Fine lines or minute surface cracks occuring on painted surfaces due to unequal contraction in drying or cooling.

CROCKING: Removal of color on abrasion or rubbing.

DEFOAMER: Products used for controlling undesirable foams.

DILATANCY: The properties of some materials in which the resistance to flow increases with agitation.

DILUENT: A liquid which is blended with an active solvent to reduce cost.

DISPERSION: Any heterogeneous system of solids, gases or liquids.

DRIER: Any catalytic material which when added to a drying oil accelerates drying or hardening of the film.

EFFLORESCENCE: A phenomenon whereby a whitish crust of fine crystals forms on a painted surface. These are usually sodium salts which diffuse through the paint film from the substrate.

EMULSIFIER: A material which when added to a mixture of dissimilar materials such as oil and water will produce a stable homogeneous emulsion.

EMULSION: A suspension of fine particles or globules of a liquid within a liquid.

EMULSION POLYMERIZATION: The process of polymerization taking place in the presence of water to form a latex.

ESTER GUM: The glycerol ester of rosin acids.

EXOTHERMIC: Evolution of heat energy.

EXTENDER: A pigment which contributes very little hiding to the system, but does reinforce the film and alter the gloss.

FALSE BODY: Thixotropic flow property of a suspension or dispersion. A composition which thins down on stirring is said to exhibit false body.

FIRE POINT: The temperature at which a material takes fire without flame from an external source.

FLASH POINT: Lowest temperature at which a substance in an open vessel gives off enough vapors to produce a flash of fire when a flame is passed near the surface.

FLOATING: Separation or layering of pigments either in the liquid state or in coatings.

FLOCCULATION: Formation of clusters of particles separated by relatively weak mechanical forces.

FOOTS: Sludge deposited during standing or aging of vegetable oils.

GAS CHECKING: The fine checking, wrinkling or frosting of a tung oil or varnish film under certain drying conditions. It is believed to be caused by rapid absorption of oxygen on the surface or impurities in the atmosphere.

GLOSS: The shine, sheen or luster of the surface of a coating. *Specular gloss*: the ratio of reflected to incident light at specified angles of incidence. Most common are angles of 20, 60 and 85°. See also SHEEN.

GRAIN RAISING: Roughness of wood caused by the application of stains or other materials. Fibers of the wood raise due to swelling by water or certain solvents.

HAZE: The dullness of a surface removable by polishing. It usually results from faulty solvent balance or incompatibility of ingredients.

HIDING POWER: The ability of a paint to obscure the background over which it is applied.

HOMOPOLYMER: A polymer produced from a single monomer species.

HUE: (Color, character, dominant wavelength) Blue, green, red, etc.

HYDROLYSIS: The interreaction of water with a material resulting in decomposition.

HYDROPHILIC: Attracted by water or water loving.

HYDROPHOBIC: Repelled by water or water hating.

HYGROSCOPIC: Absorbs and retains moisture from atmosphere or other sources.

IMMISCIBLE: Not miscible. Any liquid which will not mix with another liquid, in which case it form two separate layers or exhibits cloudiness or turbidity.

INCOMPATIBLE: This term is applied to liquid and solid systems to indicate that one material cannot be mixed with another specified material without separation or impairment of properties.

INERT: The term applied to various extended pigments such as asbestine, barytes, silica, calcium sulfate, mica, talc, etc. In general, they have poor hiding power but they are inert from a chemical and physical standpoint. While they contribute some desirable properties to a paint, they are primarily used to lower the cost.

INFRARED: Rays which are longer than visible red light rays and are felt as heat. Infrared radiant energy is used to cure coatings.

INHIBITOR: Any substance which slows or prevents chemical reaction or corrosion.

INTENSITY: The intensity of a color is its purity or saturation. For example, an intense red is one which is a very strong, pure red color.

INTUMESCENT: To swell, bubble up or foam as the result of heat. Intumescent coatings expand, thereby insulating the surface and slowing burning.

IODINE NUMBER: Oils, waxes and resins have the property of absorbing iodine because of their unsaturation. The amount of iodine absorbed measures the degree of unsaturation.

LACQUER: A term which usually indicates that the material dries by evaporation and forms a film from the nonvolatile constituents.

LATENT HEAT: The heat absorbed by a substance to effect a change in state without change in temperature. For example the latent heat of fusion of ice water is 80 cal/g.

LATEX: A generic term describing stable dispersions of resin particles in a water system.

LEAFING: The ability of an aluminum or gold bronze to exhibit a brilliant or silvery appearance. The flat pigment particles align themselves more or less parallel with the coated surface.

LIFTING: The softening and penetration of a film by solvents of another film resulting in raising and wrinkling.

LIGHTNESS: (Brightness, reflectance, value) Position on the gray scale between pure black and pure white.

LIPOPHILIC: Oil loving.

LIVERING: An increase in the consistency of a paint resulting in a rubbery or coagulated mass.

LUSTER: The gloss of a finish.

MALEIC RESINS: A class of resins obtained from the condensation of maleic anhydride with rosin, terpenes, etc.

MASS TONE: The color produced by a single color dispersed full strength in a suitable vehicle.

MELAMINE RESINS: Synthetic resins which are the condensation products of formaldehyde and melamine.

MELTING POINT: The transition point between the solid and the liquid state. Expressed as temperature at which this change occurs.

METHACRYLATE RESINS: A class of resins produced by the polymerization of methacrylic esters.

MICRON: One-thousandth of a millimeter.

MILDEW: Organic surfaces exposed to high temperature-humidity atmospheres are attacked by fungus growth. This is a dark discoloration, usually a mold type of fungus but more commonly called "mildew."

MINERAL SPIRITS: A petroleum fraction with boiling range between 300 and 400°F.

MOLECULE: The smallest unit of a substance.

MONOMER: An individual molecule capable of being polymerized to a polymer.

MOTTLING: A film defect associated with spraying. Appears as circular imperfections.

NAPHTHAS: Hydrocarbons of the petroleum type containing substantial portions of paraffins and naphthalenes.

NITROCELLULOSE: A substance produced by the treatment of cotton or wood fibers with nitric acid. Used in lacquers.

NONPOLAR SOLVENTS: The aromatic and petroleum hydrocarbon groups characterized by low dielectric constants are referred to as nonpolar solvents.

NONVOLATILE MATTER: The portion of a material which does not evaporate at ordinary temperatures.

OIL: A general term for a water-insoluble viscous liquid.

OIL ABSORPTION: The quantity of oil required to wet a definite weight of pigment to form a stiff paste.

OLEORESINOUS: Indicating a material which has been made by the combination of an oil and a resin.

OPACITY: Hiding power or the degree of obliteration.

OPAQUE: Impervious to light or not translucent.

ORANGE PEEL: A pebbled film surface similar to the skin of an orange in appearance. It is caused by too rapid drying before leveling takes place.

ORGANIC: A compound containing carbon and hydrogen and in some cases other elements such as oxygen, nitrogen, sulfur, phosphorus, the halogens, etc.

PASTEL: A tint. A color to which white has been added.

PEBBLE MILLS: Mills usually lined with porcelain or buhrstone in which flint pebbles or porcelain balls are used as the grinding medium.

pH VALUE: The numerical expression used to describe the hydrogen ion concentration. The pH value denotes the degree of acidity or alkalinity.

PHENOLIC RESINS: A class of resins produced as the condensation product of phenol or substituted phenol and formaldehyde or derivatives.

PIGMENT: A finely divided, insoluble substance which imparts color to the material to which it is added.

PIGMENT VOLUME CONCENTRATION: The ratio of the total volume of pigment to the total volume of nonvolatile matter in a paint.

PINHOLING: The appearance of fine, pimply elevations or tiny holes on a coating.

PLASTICIZERS: Materials which are added to resins to soften and improve flexibility.

POISE: A fundamental and absolute unit of viscosity measurement. A substance is said to have a viscosity of 1 poise when a force of 1 dyne is required to move a surface of 1 sq cm at a speed of 1 cm/sec relative to another plane surface separated from it by a layer 1 cm thick.

POLAR SOLVENTS: Solvents such as alcohols, ketones, etc., which contain oxygen, etc. These have high dielectric constants.

POLYMER: A large molecule formed when many molecules are linked together by polymerization.

POLYSTYRENE RESINS: Synthetic resins formed by polymerization of styrene.

PRIMARY COLORS: In theory, those colors from which all other colors and white may be made. The primary colors in visible light are red, green and blue. The so-called pigment primaries, each absorbing a light primary, would then be blue green or cyan (minus red), majenta (minus green), and yellow (minus blue). Because of deficiencies in the available cyan and majenta colorants, confusion developed, so that red, yellow and blue are now often referred to as the pigment primaries.

REDUCER: A volatile compound which is employed to bring coatings to the proper consistency. Also called thinner.

REFRACTIVE INDEX: A measure of the deviation from normal that a beam of light undergoes upon passing through a given substance.

RELATIVE HUMIDITY: A method for expressing the amount of moisture in the air. It is expressed in per cent of saturation and is an indication of the proportional amount of moisture that could be absorbed or held at that temperature.

RESIN: An organic polymer in the form of a crystalline or amorphous solid, or viscous liquid, of either natural or synthetic origin.

RHEOLOGY: The science of the flow of viscous fluids.

ROLLER COATING: The method of applying coatings by means of rubber or steel rolls.

ROLLER MILL: A paint mill consisting of hardened steel rolls which revolve in opposite directions. It disperses dry pigments into a vehicle.

ROSIN: The resinous product obtained from various pine trees which contains primarily abietic acid.

SAGGING: The tendency of a wet paint film to flow downward and become thicker on vertical surfaces.

SATURATION: Purity or intensity of color. Degree of freedom from grayness.

SEEDS: Undesirable particles which develop in a liquid coating by partial gelation of the vehicle or by agglomeration of pigment particles.

SHADE: The difference in appearance between colors of similar hue.

SHEEN: A specular reflectance taken at a low angle, usually 85°.

SILKING: Parallel hairlike striations showing in coatings.

SKINNING: The formation of a solid surface layer on a liquid varnish or paint when exposed to air.

SOLUTION: A homogeneous liquid mixture, the proportion of whose constituents may be varied within certain limits. A solid is said to be in solution when the molecules of the liquid have exceeded the attraction of those of the solid.

SOLVENT: The term applied to products which dissolve the film-forming constituents.

SOLVENT POWER: The strength of dissolving power of a solvent.

SPAR VARNISH: A very durable waterproof varnish for exterior use.

SPECIFIC GRAVITY: The ratio of the weight of an equal volume of a substance to the weight of an equal volume of water at stated temperature.

SPEWING: Migration of components to the surface of a coating usually because of incompatibility.

SPREADING RATE: The area of a surface over which a unit volume of paint will spread; usually expressed in square feet per gallon.

STRENGTH: The strength of a pigment is its opacity or tinting power.

SURFACE TENSION: The tension exhibited by the free surface of liquids measured in dynes per centimeter.

SYNTHETIC RESIN: Complex, substantially amorphous organic semisolid or solid materials built up by chemical reaction of simple molecules.

TACK: Pull resistance exerted by a material adhering. In liquids, tack is a function of viscosity.

TALL OIL: A by-product material from the manufacture of wood pulp. It is a mixture of fatty acids and rosin.

THERMOPLASTIC: The term applied to resins which soften and flow when heated.

THERMOSETTING: The term applied to resins which become hard after heating and cannot be resoftened.

THIXOTROPY: The property possessed by certain materials which thin out when subjected to stress and again thicken on standing.

TINT: A color produced by mixing a colored material, dye or pigment with white pigment or paint.

TINTING STRENGTH: The coloring power of a given quantity of pigment or paint.

TONE: A modification of a full-strength color (mass tone) secured by blending with other colors.

TOUGHNESS: The ability of a material to take bending, impact, etc., without cracking.

ULTRAVIOLET: The light rays which are outside of the visible spectrum at its violet end.

UREA-FORMALDEHYDE RESINS: Synthetic resins obtained by the reaction between urea and formaldehyde in the presence of acid or alkaline catalysts.

VAPOR PRESSURE: The outward pressure of a mass of vapor at a given temperature. It is the index of the volatility of a liquid.

VEHICLE: The liquid portion of any paint, enamel or lacquer.

VINYL: A general term applied to a class of resins such as polyvinyl chloride, acetate, butyral, etc.

VISCOMETER: Any instrument that measures viscosity or the internal friction of fluidity of a liquid.

VISCOSITY: A property of fluids which can be described as the resistance to flow.

WEATHEROMETER: A laboratory device used to estimate the life of coatings by subjection to ultraviolet light, water, etc.

WRINKLE FINISH: A varnish or enamel film which exhibits a novelty effect similar to fine wrinkles.

Abrasive cleaning, 622

Abrasives, 623

Accroides resins, 266

Acid number, 446

Acids, 37–40

Acrylic resins
adhesion, 117
application, 122
compatible with other resins, 118–119
cross-linking, 124–126
glass transition temperature, 116
hardness, 114–115
industrial, 556
properties, 113–120
solvents for, 325
stability, 113
thermosetting, 124–125
types, 112–113

Acrylics
appliance enamel, 565
coil coating, 568
concrete sealer, 130
house paint, 131

Additives, silicones, 243–245

Adhesion
acrylics, 117
vinyls, 154

Adhesives, vinyl, 178

Aerosol, 9–10
components, 608
containers, 608
definition of, 607
formulation, 611–612
propellants, 610–611
valves and nozzles, 609–610

Agitators, 692–694

Air pollution, 11

Airless spray, 657

Alcohol solvents, 312

Alcoholysis, 43

Alkyd
calculations, 43
coil coating, 568
combination with other resins, 33
definition, 33
esterification, 34
flat wall paints, 545–546
industrial, 555
long oil, 47
manufacture, 44
medium oil, 46–47
preparation, 41–42
production, U.S.A., 7
raw materials, 34–40
short oil, 44–46
silicone, 254–255
structure, 42
types, 43

Aluminum, 634

Aluminum paints
colored, 601–602
hammer finishes, 603
heat resistance, 237–238, 601
history, 589–590
maintenance, 598–599

Aluminum pigments
heat reflectance, 593
leafing, 594
leafing stability, 594–595
light reflectance, 592–593
particle size, 591–592

Aluminum soaps, 410

Amino resins, industrial, 556

Aminoplast resins
alkyds, 51
chemistry, 50
other resins, 58
uses, 49

Amphoteric surfactants, 516–518

Anhydride, 37–40

Aniline point, solvents, 290

Anionic surfactants, 516–517
Antimony oxide, 356
Appliance finishes, 563–566
  acrylic, 565
  silicone, 242
Application, 13–14, 586–588, 604
Aqueous baking enamel, 64
Aqueous systems, preservation, 423–427
Arylide maroon, 378
Asphalts, 271–272
Atomization, 657–659
Autoignition temperatures, 726–727
Automotive
  enamels, 559
  refinishing, 558
  surfacer, 61, 558

Barrier protection, 575
Benzidine
  orange, 369, 380
  yellow, 367, 380
Bitumens, 580
Black-bone, 379
  carbon, 379
  lamp, 379
Blown oils, 29
Blue toners, 372
BON (β-oxynaphthalic) reds, 375, 384
Bone black, 379
Brass, 634–635
Bronze, 634–635
Brush application, 644–645

Cadmiums
  orange, 368–369
  red, 377, 383
  yellow, 366–380
Calcium driers, 398
Carbazole dioxazine violet, 373
Carbon black, 379, 491
Cationic surfactants, 516–517
Cellulose
  carboxymethyl, 145
  ethyl hydroxyethyl, 145
  hydroxyethyl, 145
  nitrate, 136–137
  structure, 132
Cellulose acetate, 141
  butyrate, 142
    plasticizers, 276
Cellulosic derivatives, 132–134
Chemical cleaning, 625–629

Chemical resistant coatings, 196
Chlorinated rubber
  milling, 193
  other resins, 195
  plasticization, 189
  structure, 188
Chroma, 461
Chromate treatments, 631–632
Chromaticity coordinates, 463
Chromes
  greens, 370, 381
  orange, 368–381
  yellow, 364
Chromium oxide, 370, 381
  hydrated, 370, 381
Coalescence, film, 19–21
Coatings
  black heat-resistant, 241
  coil, 8
  conversion, 629–631
  future, 8
  heavy duty, 92
  missile, 241
  properties, 7
  sacrificial metal, 578–579
  thickeners, 581
  types, 577
  wood, 8
Cobalt driers, 397
Coconut oil, 26
Coil coating, 8, 566
  acrylic, 568
  alkyd, 568
  silicone, 256
  vinyl, 568
Color control—instruments, 472–475
  formulation, 485–492
  measuring instruments, 409
  order systems, 467
  pigments, 444
  reflectance, 1
  testing, 719–720
  tolerance, 471
Compatability—plasticizers, 278–279
Computer
  color matching, 492
  formulation, 504
Concrete, 583, 636–639
  sealer—acrylic, 130
Congo resins, 259–261
Containers, aerosol, 608

Copolymerization, 121, 523
Copper, 634–635
    phthalocyanine blue, 372, 382
    phthalocyanine green, 371, 381
Cost, paint, 496
Cottonseed oil, 26
Coumarone-indene resins, 108
Critical pigment volume concentration,
        498–499
Cross-linking, acrylics, 124–126

Dammar resin, 264
Decals, 654-655
Dehydrated castor oil, 26
Dinitranaline orange, 369, 380
Dipping, 649–650
Dispersion, high speed, 704
Dowtherm, 695–696
Driers, 394–404
    acids, 396–397
    epoxy esters for, 98
    formulation, 501
    metals, 397–399
    oils for, 401
    recommendations, 401–402
    selection, 399
    silicones, for, 232, 234
    testing, 400
    water-based paints, 400
Drying, 655
    mechanism, 394
    phenomena, 30

East Indies resins, 266
Electrodeposition, 9, 14, 683–689
Electrostatic spray, 661–662, 666–683
Elemi resins, 265
Eleostearic acid, 25
Emulsification, post-, 527–528
Emulsion
    polymerization, 521–527
    polymerization—vinyl, 150–151
    polyvinyl acetate, 184–185
Emulsion house paints, 540–541
Emulsion paints
    acrylics, 126–128
    ingredients, 515
Emulsion stability, 529
Enamels, 62, 542–545
    appliance, 62
    aqueous baking, 64
    epoxy phenolic baking, 100
    epoxy polyamide brushing, 93
    gloss white architectural, 411
    industrial, 560–563
    silicone exterior, 254
    thermosetting appliance, 130
Epoxies
    amine coating, 98
    baking finishes, 98
    esters, 95–96
    formulation, 96–97
    government specifications, 95
    manufacturing, 97
    phenolic coatings, 99–100
    powder coatings, 102–103
Epoxy primers, 565
    roller coat, 567
    thermoplastic coatings, 104
    thermosetting acrylics, 102
    zinc rich, 95
Epoxy resins
    amine cured, 86
    coal tar coatings, 93–94
    curing, 85–87
    curing agent selection, 88–89
    industrial, 556
    manufacture, 83
    markets, 82
    physical characteristics, 84
    properites, 84
    selection, 89
    solvents for, 325
    structure, 83
Equipment
    blasting, 623–625
    pigment dispersion, 509–510
Etherification, 53
Ethylcellulose, 143
    plasticizers for, 277
Evaporation, 299–306
    cooling, 304
    film thickeners, 306
    films, 15
    humidity, 305
    rate, 303
    temperature, 303
Explosive range, 726
Exposure, 585
Extended pigments, 364
Extraction of oils, 26

Farm equipment, primer sealer, 560
Fatty acids, 24–25, 37–41
Fillers, furniture, 572
Films
  coalescence, 19–21
  cure, 17
  evaporation, 15
  fixation, 14–15
  formation, 514
  formers, 555
  fusion, 19
  rheological, 16–17
Finish coating, 588
Fire-retardant coatings, 9
Fire-retardant paints, 202
Fires and explosions, 724–725
Fish oil, 26
Flamboyant finishes, 603
Flat wall paints, 545–546
  alkyds, 545–546
  latex, 546–548
Flatting, 416–417
Flatting agents, 412–417
Flow coating, 562, 650
Flow modifiers, 407–412
Fluidized bed, 653–654
  coatings, 9, 14
Formulation
  aerosols, 611–612
  computer, 504
  example, 501–504
  mill base, 506–508
  paint and varnish removers, 613–614
  principles, 497
Functionality, 209
Fungicides, 438–440
Furniture
  stains, 571
  top coats, 572
  wood fillers, 572
  wood sealers, 571–572
Fusion, film, 19–21

Gas chromatography, 449
Glass transition temperature, 116
Gloss, effect on color, 482–484
Government specifications—epoxy, 95
Green toners, 371
Grinding—sand, 705

Hammer finishes, 603

Hansa yellow, 367, 380
Hazardous materials, 728–729
Health, 727
Heat-bodying linseed oils, 29
Heat reflectance, aluminum pigments, 593
Heat-resistant aluminum paints, 601
Heat stabilizers, 172
  vinyl, 161
Heavy-duty coatings, 92
Hexamethylol melamine, 51
Hiding power, 329–330
High-speed dispersion, 704
Hot spray, 659–661
House paint, acrylic emulsion, 131
Housekeeping, 724
Hue, 461
Hydrated chromium oxide, 370, 381
Hydrocarbon resin, description, 107
Hydrophilic, 518

Indanthrone blue, 372, 382
Indices of refraction, paint materials, 348
Initiators, 121
Inks, polyvinyl acetate, 185
Instruments, color measuring, 469–471
Iodine number, 446
Iron
  blues, 371, 382
  driers, 398
  red, 376
Isocyanates, 206
  blocked, 206

Kauri-butanol value, 290
Kauri resin, 261–262
Ketone solvents, 311

Lacquers
  acrylic, 123
  acrylic aerosol, 129
  acrylic spray, 129
  cellulosic, 133
  soft nonferrous metals, 129
  steel coating, 130
  white aluminum, 129
  wood, acrylic butyrate, 130
Lakes, 363–364
Lampblack, 379
Latex
  flat wall paints, 546–548
  particles, 514
  particle size, 526

Lead driers, 397–398
Lead phosphate, dibasic, 354
Leaded zinc oxide, 355–356
Leafing, aluminum pigments, 594
Licanic acid, 25
Light reflectance, aluminum pigments, 592–593
Light sources for color matching, 484–485
Light stabilizers, 161
Linoleic acid, 25
Linolenic acid, 25
Linseed oil, 25
    heat-bodying, 29
Lipophilic, 518
Lithols, 374, 384
    rubine, 375, 384
Lithopone, 357–358

Maintenance
    coatings, two package, 90
    paints, aluminum, 598–599
    preventive, 582
    primer, 91
Manganese driers, 398
Manila resin, 262–265
Manufacture, 496
    aluminum pigments, 590–591
    paint, 699–709
    pigments, 339
    resin, 690–691
Marine paints, chlorinated rubber, 199–201
Masonry, 583, 636–637
    paints, 201–202
McAdams tolerances, 475–479
Melamine resins, 60
Mercadium
    orange, 369
    red, 377
Metal preparation, 688
Metallic soaps, 390–394
Metallic surfaces, 633–635
Metals, nonferrous, 584
Methylcellulose, 144
Micelle, 524
Microorganisms, 422–423
Mill base formulation, 506–508
Mineral violet, 373, 383
Moisture testing, 454
Molecular weight control, 121
Molybdate orange, 368, 381
Monomers, reactivity, 120–121

Munsell system, 467–468

Natural resins, 258
Neoprene, 579
Nickel azo yellow, 366
Nickel titanate, 365
Nitrocellulose lacquers, viscosity, 318
Nitrocellulose plasticizers, 277
Nitrocellulose solvents, 321–324
Nonferrous metals, 584
Nonionic surfactants, 516, 518
Nonreactive white pigments, 356–362

Ochers, 378
Oils, 37–41, 445–449
    absorption, pigments, 443
    blown, 29
    copolymer, 30
    drying, 41–42
    epoxidized, 30
    extraction, 26–28
    fatty acid composition, 27
    maleic treated, 29
    processed, 28–29
    refining, 28
    testing, 31, 445–451
    treated, 28–29
Oiticica oil, 26
Oleic acid, 24
Opacity, 349
Organosols, 165–170
Ostwald system, 468

Paint and varnish removers, formulation, 613–614
Paints
    classification, 4
    filling, 706–707
    film deterioration, 431
    industry, growth, 2
    manufacturing, 699–709
    performance, 496
    removal, 613, 617–618
    sales—dollars, 2
    sales—gallons, 2
Paper coatings, vinyl, 186
Para reds, 374
Particles
    latex, 514
    size, 513, 525
        emulsions, 529–530

Performance, paint, 496
Peroxide value, 447
Petroleum resins, 108
pH, 520, 527, 528
Phase volume, 530
Phenolic
  resins, 100%, 80
  industrial, 557
  production, 68
Phenolic resins, types, 71
Phenols, 429
Phenylmercury salts, 428
Phosphate treatments, 629–632
Pigment dispersion
  definition, 505
  equipment, 506
Pigment volume concentration, 498, 500
  vinyl resins, 162
Pigments
  binder demand, 337
  chlorinated rubber, 192
  classification, 327
  color, 444
  extender, 364
  formulation, 501
  function, 328
  hiding power of white pigments, 350
  inorganic, 363
  manufacture, 339
  nitrocellulose, 139–140
  oil absorption, 337, 443
  from ore, 339
  organic, 363
  properties, 343
  reactivity, 336
  silicones, 234
  supplemental, 331–345
  testing, 443–435
  tinting concentration, 498, 500
  tinting properties, 386
  tinting strength, 349, 444
  titanium, 358–362
  types, 328–329
Pitches, 271–272
Plasters, 639–641
Plasticization, theory, 274
Plasticizers
  acrylics, 118
  chlorinated rubber, 189–192
  compatibility, 274, 278–279
  efficiency, 275

ethylcellulose, 277
nitrocellulose, 139, 277
permanence, 275
stability, 275
testing, 451–452
vinyl, 160
Plastisols, 171
Plywood, 635–636
Polymer testing, 451–452
Polymerization
  acrylics, 120–122
  emulsion, 521–527
  vinyl bulk, 148
  vinyl emulsion, 150
  vinyl solution, 149
  vinyl suspension, 149
Polyols, 34–37
  urethanes, 207
Polyurethanes
  chemistry, 207
  prepolymers, 211
Polyvinyl acetate, 183–184
  acetate resins, 175
  formal resins, 176
Powder coatings, 102
Premixers, 311
Preservation, aqueous systems, 423–427
Preservatives, 427–430
  phenols, 429
  phenyl mercury salts, 428
  sulfur compounds, 429
  testing, 427
Preventive maintenance, 582
Primers
  automotive, 412
  chlorinated rubber, 200
  epoxy, 565
  heat-resistant metal, 241
  silicone, 240
  wash, 182
Priming, 587
Processed oils, 28–29
Propellants, aerosol, 610–611
Protection
  barrier, 575
  sacrificial, 575–576
Protective colloids, 518–520

Quinacridones
  reds, 376
  violet, 373, 383

Radiation curing, 655
Raw materials, 496
Reactive white pigments, 350–356
Reactivity, monomers, 120–121
Reactors, resin, 691–692
Red
    cadmium, 377, 383
    iron oxide, 376, 383, 490
    lead, 377
    mercadium, 377
    quinacridone, 376
    thioindigo, 378
    toluidine, 374
Refining oils, 28
Reflectance, 461
    colors, 1
    values, 1–10
Removal, paint, 613
Resin
    manufacture, 690–691
    reactors, 691–692
    testing, 451–452
Resinated toners, 363
Rheology, coatings, 16–17
Roller application, 645–649
Roof coatings; aluminum, 602
Rosin, 267–269

Sacrificial metal coatings, 578–579
Sacrificial protection, 575–576
Safflower oil, 25
Sampling, 718
Sand blasting, 622–625
Sand grinding, 705
Sanding aids, 417–420
Saponification number, 446
Saturation, 461
Sealers, wood furniture, 571–572
Seeding, 404
Siennas, 378
Silicones, 579
    chemistry, 224
    manufacture, 225
Silk screen process, 650–653
Skinning, 404
Solubility
    acrylics, 117
    vinyl resins, 154–159
Solubility parameter, solvents, 286–289
    resins, 289
Solvency, 285–293

Solventless coatings, 94
Solvents
    acrylics, 325
    alcohol, 312
    chlorinated rubber, 194
    classification, 281–285
    epoxy, 325
    ester, 308
    evaporation rate, 316
    flash point, 319
    formulation, 501
    glycol ether, 311
    hydrocarbon, 283–285
    ketone, 311
    lacquer, 310
    nitrocellulose, 138–139, 321–324
    odor, 320
    oxygenated, 309
    paint and varnish removers, 614
    resin solubility, 313–314
    selection, 89
    strength, 317
    testing, 452–453
    vinyl, 324
    viscosity, 293–296
    water sensitivity, 320
Soya oil, 26
Spectrophotometers, 469
    curves, 458–460
Splash zone coatings, 103
Spray
    airless, 657
    electrostatic, 661–662, 666–683
    hot, 659–661
    painting, 656–666
Stains, 549
    furniture, 571
Static electricity, 725
Stearic acid, 24
Strontium yellow, 365
Styrene-butadiene resins, 110
Sulfur compounds, 429
Surface preparation, 587
Surfactants, 516
    amphoteric, 516, 518
    anionic, 516–517
    nonionic, 516, 518
Suspending agents, 405–406
Swimming pool paints, 201

Tall oil fatty acids, 26

Tank configuration, 510–511
Terpene resins, 109
Testing, history, 711–712
  moisture, 454
  plasticizers, 451
  resins, 451
  solvents, 452–454
Thioindigo red, 378
Thixotropy rating, 17
Titanium pigments, 358–362
Toluidine reds, 374, 458
Toxicity hazards, 595
Traffic paint, chlorinated rubber, 203
Triglycerides, 22, 24, 448
Tristimulus colorimeters, 470–471
Tristimulus values, 463
Tumbling, 650
Tung oil, 26

Ultramarine blue, 372, 382
Umber, 378
Undercoaters, enamels, 541–542
Urea, 51
  resins, 60
Urethane coatings, two package, 215
Urethanes, drying oil modified, 210

Varnishes, 548
Vat dyes
  oranges, 369
  yellows, 367
Vehicle formulation, 501
Vinyl chloride
  acetate solution resin, 153
  chloroacetate, 580
  coil coating, 568
  solution resin, 151
Vinyl resins
  adhesion, 154
  dispersion, 165
  heat and light stabilizers, 161
  industrial, 557
  modifiers, 156
  plasticizers, 160, 276
  solubility, 154–159

solution viscosity, 319
solvents, 324
viscosity, 158
Violet—mineral, 373, 383
  organic toner, 373
  quinacridone, 373, 383
Visible spectrum, 456–459
Volatility, 297

Water-based paints
  driers, 400
  production, 10
Water passage, 435
Water permeability, aluminum pigments, 593–594
Water reducible coatings, industrial, 569–570
Water repellants, silicone masonry, 245
White enamel, 459
  architectural gloss, 411
White lead
  basic carbonate, 351
  basic silicate, 353–354
  basic sulfate, 352–353
White pigments
  nonreactive, 356–362
  reactive, 350–356
Wood, 583–635
  coatings, 8
  finish
    catalyzed, 63
    vinyl, 179
  furniture finishes, 510–513
  graining, 653

Yellow iron oxide, 366, 380, 489
Yield value, 17

Zinc-coated steel, 633–634
  driers, 398–399
  leaded, 355–356
  oxide, 354–355
  sulfide, 357
  yellow, 364–380
Zirconium driers, 399